THE ROUTLEDGE HANDBOOK OF GEOSPATIAL TECHNOLOGIES AND SOCIETY

The Routledge Handbook of Geospatial Technologies and Society provides a relevant and comprehensive reference point for research and practice in this dynamic field. It offers detailed explanations of geospatial technologies and provides critical reviews and appraisals of their application in society within international and multi-disciplinary contexts as agents of change.

The ability of geospatial data to transform knowledge in contemporary and future societies forms an important theme running throughout the entire volume. Contributors reflect on the changing role of geospatial technologies in society and highlight new applications that represent transformative directions in society and point towards new horizons. Furthermore, they encourage dialogue across disciplines to bring new theoretical perspectives on geospatial technologies, from neurology to heritage studies.

The international contributions from leading scholars and influential practitioners that constitute the *Handbook* provide a wealth of critical examples of these technologies as agents of change in societies around the globe. The book will appeal to advanced undergraduates and practitioners interested or engaged in their application worldwide.

Alexander J. Kent is Honorary Reader in Cartography and Geographic Information Science at Canterbury Christ Church University, UK, and leads the Coastal Connections project for World Monuments Fund and English Heritage.

Doug Specht is a Chartered Geographer and a Reader in the School of Media and Communication at the University of Westminster, UK.

THE ROUTLEDGE HANDBOOK OF GEOSPATIAL TECHNOLOGIES AND SOCIETY

Edited by
Alexander J. Kent and Doug Specht

LONDON AND NEW YORK

Designed cover image: Buildings in Amsterdam coloured according to year of construction from the Dutch cadastral register (January 2015), for the project Smart CitySDK. Map made with TileMill by Bert Spaan, Waag Society.

First published 2024
by Routledge
4 Park Square, Milton Park, Abingdon, Oxon OX14 4RN

and by Routledge
605 Third Avenue, New York, NY 10158

Routledge is an imprint of the Taylor & Francis Group, an informa business

© 2024 selection and editorial matter, Alexander J. Kent and Doug Specht; individual chapters, the contributors

The right of Alexander J. Kent and Doug Specht to be identified as the authors of the editorial material, and of the authors for their individual chapters, has been asserted in accordance with sections 77 and 78 of the Copyright, Designs and Patents Act 1988.

All rights reserved. No part of this book may be reprinted or reproduced or utilized in any form or by any electronic, mechanical, or other means, now known or hereafter invented, including photocopying and recording, or in any information storage or retrieval system, without permission in writing from the publishers.

Trademark notice: Product or corporate names may be trademarks or registered trademarks, and are used only for identification and explanation without intent to infringe.

British Library Cataloguing-in-Publication Data
A catalogue record for this book is available from the British Library

Library of Congress Cataloging-in-Publication Data
Names: Kent, Alexander, 1977– editor. | Specht, Doug, editor.
Title: The Routledge handbook of geospatial technologies and society / edited by Alexander J. Kent and Doug Specht.
Other titles: Handbook of geospatial technologies and society
Description: Abingdon, Oxon; New York, NY: Routledge, 2023. | Includes bibliographical references and index.
Identifiers: LCCN 2022038857 (print) | LCCN 2022038858 (ebook) | ISBN 9780367428877 (hardback) | ISBN 9781032431284 (paperback) | ISBN 9780367855765 (ebook)
Subjects: LCSH: Geographic information systems—Social aspects. | Social sciences—Methodology.
Classification: LCC G70.212 .R68 2023 (print) | LCC G70.212 (ebook) | DDC 910.285—dc23/eng20230110
LC record available at https://lccn.loc.gov/2022038857
LC ebook record available at https://lccn.loc.gov/2022038858

ISBN: 9780367428877 (hbk)
ISBN: 9781032431284 (pbk)
ISBN: 9780367855765 (ebk)

DOI: 10.4324/9780367855765

Typeset in Bembo
by codeMantra

Dedicated to Gordon Petrie (1930–2021)
Emeritus Professor of Topographic Science and
Honorary Research Fellow,
University of Glasgow
Via, Veritas, Vita

CONTENTS

List of figures   *xii*
List of tables   *xxiii*
List of contributors   *xxiv*
Acknowledgments   *xxxvii*

    Introduction   1
    Alexander J. Kent and Doug Specht

PART I
Origins and perspectives of geospatial technologies   5

1 Latitude, longitude, and geospatial technologies to 1884   7
    Matthew H. Edney

2 The photo-mechanical era of cartography: a recollection   23
    William Cartwright

3 The roots of GIS   42
    Michael F. Goodchild

4 Positivism, power, and critical GIS   50
    Wen Lin

5 Geospatial standards: an example from agriculture   60
    Didier G. Leibovici, Roberto Santos, Gobe Hobona, Suchith Anand,
    Kiringai Kamau, Karel Charvat, Ben Schaap and Mike Jackson

6 Technology, aesthetics, and affordances   76
   Philip J. Nicholson

7 Race and mapping   86
   Catalina Garzón-Galvis and Beth Rose Middleton Manning

8 Feminist geography and geospatial technologies   98
   Meghan Kelly

9 Mapping the subaltern   109
   Penelope Anthias

10 Geospatial technologies and rural and indigenous spatial knowledges   122
   María Belén Noroña

11 Social constructivism and geospatial technologies: neogeography, big
   data, and deep maps   133
   Barney Warf

PART II
Understanding geospatial technologies   145

12 Mobile mapping   147
   Gordon Petrie

13 Airborne and ground-based laser scanning   171
   Mathias Lemmens

14 Drones and unmanned aerial vehicles (UAVs)   188
   Faine Greenwood

15 Airborne photogrammetric mapping   202
   Gordon Petrie

16 Digital elevation models (DEMs)   226
   Oluibukun Gbenga Ajayi

17 Extended reality (XR)   236
   Łukasz Halik and Alexander J. Kent

18 Free and Open Source Software for geospatial applications (FOSS4G)   246
   Rafael Moreno-Sanchez and Maria Antonia Brovelli

19	APIs, coding and language for geospatial technologies *Oliver O'Brien*	256
20	Spatial analysis and modelling *Timofey Samsonov*	263
21	The geovisualization of big data *Nick Bearman*	291
22	Machine learning and geospatial technologies *Izabela Karsznia*	299
23	Artificial intelligence for geospatial applications *Vit Vozenilek*	313

PART III
Applications of geospatial technologies — 329

24	Location matters: trends in location-based services *Georg Gartner*	331
25	Mapping buildings and cities *Tomasz Templin*	343
26	Underground mapping *Aurel Saracin*	358
27	Geospatial technology and food security: forging a four-dimensional partnership *Hillary J. Shaw*	384
28	The past, present and future of technologies for improved water management *Leonardo Alfonso*	394
29	Ocean mapping: taxonomies of the fluid geospatial *Rupert Allan*	403
30	Geospatial technologies in transport: shaping and recording everyday lived experiences *Nigel Waters*	424

31	Geospatial technologies in electrical systems *Ivan Bobashev*	437
32	Geospatial technologies and public health *Fikriyah Winata, Sara McLafferty, Aída Guhlincozzi and Yiheng Zhou*	450
33	Applications of GIScience to disease mapping: a COVID-19 case study *Leah Rosenkrantz and Nadine Schuurman*	463
34	Geosurveillance and society *Rob Kitchin*	476
35	Geospatial technology and journalism in a post-truth world *Amy Schmitz Weiss*	486
36	Advancing sustainability research through geospatial technology and social media *Yaella Depietri, Johannes Langemeyer, Derek Van Berkel and Andrea Ghermandi*	494
37	Crisis and hazard mapping *Amelia Hunt*	505
38	Humanitarian relief and geospatial technologies *John C. Kostelnick*	520
39	Geospatial technology and the Sustainable Development Goals (SDGs) *Doug Specht*	532
40	Maps of time *Menno-Jan Kraak*	541
41	Geospatial technologies in archaeology *Alexander J. Kent and Doug Specht*	554
42	Mapping planetary bodies *Trent Michael Hare*	562

PART IV
New ontologies and strategies for geospatial technologies — 577

43	Toward the democratization of geospatial data: evaluating data decisioning practices *Victoria Fast, Nikki Rogers and Ryan Burns*	579

44	Developing geospatial strategies *Mark Iliffe*	589
45	Map thinking across the life sciences *Rasmus Grønfeldt Winther*	600
46	Spatial anthropology: understanding deep mapping as a form of visual ethnography *Les Roberts*	613
47	The quantum turn for geospatial technologies and society *Daniel Sui*	623
48	The Locus Charter: towards ethical principles and practice for location data services *Denise McKenzie and Ben Hawes*	632

Index   *643*

FIGURES

1.1	Images of early navigational instruments, 1805	9
1.2	The geometry of the mariner's astrolabe in determining the Sun's altitude	10
1.3	Use of the cross staff to determine the Sun's altitude	11
1.4	Use of the back staff	12
1.5	Detail of the transversal scale on the sight arc of a back staff, c. 1775	13
1.6	An English octant by John Bleuler, 1775–1790	14
1.7	Diagram of the working of an octant	15
1.8	A sextant by Riggs & Brother, Philadelphia, 1820–1840	17
1.9	Chronometer no. 2620 by John Bliss, New York, c. 1880	18
1.10	Diagrams from "The Time Ball, Royal Observatory, Greenwich", 1844	19
1.11	Detail of record of a track on James Horsburgh, *The Chart of the East Coast of China*	21
2.1	Staedtler Mars Lumograph drafting pencils	26
2.2	Sharpening paper pad on paddle stick	26
2.3	Triangular engineering scale	27
2.4	Ten-point divider	27
2.5	Rolling parallel rule	28
2.6	(1940). Map -334 P series – City of Hobart	28
2.7	Klimsch camera lenses and bellows, with overhead steel rails	29
2.8	Klimsch camera copyboard	30
2.9	A typical drawing room, Lands Department, Brisbane, 1953	31
2.10	Graphic lettering pen set	31
2.11	Rotring isograph pen	32
2.12	A draughtsman with the 1st Topographical Survey Troop, at the 1st Australian Task Force (1ATF) Base, puts the finishing touches to a map	33
2.13	French curve drafting set	33
2.14	Wrico lettering guides	34
2.15	Pen, lettering guide, contour pen and ink topographic map, reproduced using the diazo process	34
2.16a	Letraset 'rub-on' type	35
2.16b	Letraset 'rub-on' type	35

Figures

2.17	Transfer lettering	36
2.18	Hotham and Falls Creek Alpine areas	36
2.19	Scribing using stable-base plastic 'scribecoat'	37
2.20	Colour separation scribing of the intermediate contours for a USGS topographic map using a freehand scriber	38
2.21	Application of transparent adhesive-backed type to the culture lettering separate of the Miller Peak, Arizona 7.5-minute topographic map	38
2.22	Using an airbrush for manual hillshading, c. 1952	39
2.23	Photo-lithographers conducting quality checks on a processed printing plate	40
5.1	APSIM WPS FME combined scientist and non-scientist client implemented using FME workspace	66
5.2	A BPMN diagram describing the workflow for the APSIM WPS	67
5.3	Workflow design of a generic model to be used in food security and sustainability	68
5.4	Snapshot of the FOODIE VGI profile	71
5.5	Revised FOODIE VGI profile of SDI4apps based on principles of Citizen Sensors	72
8.1	Google Trends results for the search term 'feminist GIS'	99
9.1	Map showing traplines from the Inuit Land Use and Occupancy Project	111
9.2	Outline map of black and indigenous communal land claims along Nicaragua's Caribbean coast	112
9.3	Major stationary sources of air pollution and minority population in the Bronx	117
9.4	Israeli 1958 overprint (purple) on a British 1946 map, marking the destruction of Palestinian villages	118
10.1	Catequilla hill near the city of Quito during June's solstice (21st June 2003)	124
12.1	The principal hardware components of a vehicle-mounted mobile mapping system	148
12.2	Diagram showing a system comprising imaging cameras, a GPS receiver and a measuring odometer (with its optical encoder)	149
12.3	The successive models of the Ladybug multiple camera units that are in widespread use on mobile mapping systems	150
12.4	(a) A ring of eight cameras for use in a multi-camera system; (b) the integrated R5 nine-camera system that is mounted and operated on Google's Street View mapping cars; and (c) the R7 camera unit from Google with its 15 cameras providing panoramic coverage from each successive exposure position of the mapping platform	150
12.5	(a) Diagram showing the construction of a SICK 2D laser scanner; (b) a SICK LMS291 laser scanner; and (c) this SICK LMS291 scanner has been placed on its side to measure a vertical profile within its 2D plane to the side of the mapping vehicle	151
12.6	(a) The RIEGL VQ-450 2D scanner; (b) the Lynx 2D scanner from Teledyne Optech; and (c) the Z+F Profiler 9012 scanner instrument	152
12.7	(a) A twin-headed VMX-1HA 2D scanner unit with a pair of 'full circle' scanners from RIEGL; (b) a Teledyne Optech Lynx SG scanner unit equipped with twin 'full circle' scanners	153
12.8	(a) Diagram showing the main external features of the Velodyne HDL-64E spinning laser scanner; (b) showing the comparative sizes of the HDL-64E, HDL-32E and VLP=16 multi-laser spinning lasers	153

12.9	(a) The principle of location by the simultaneous measurement of ranges (R1–R4) from satellites at known positions in space to the GPS receiver mounted on the mobile mapping platform on the Earth; (b) showing how the range circles from four satellites intersect to define a unique position on the Earth; and (c) in fact, the minimum requirement to fix the location (position) of the platform is three ranges	154
12.10	Illustrating the concept of differential GPS	155
12.11	An early example of the Litton LN200 IMU	156
12.12	(a) The main components of an Applanix POS/LV GPS/IMU sub-system used in a mobile mapping system; (b) a block diagram showing the connections between the various components of the Applanix POS/LV sub-system	156
12.13	(a) An early Street Mapper mobile mapping system from IGI; (b) a more modern Pegasus Two system from Leica Geosystems	157
12.14	Examples of compact mobile mapping systems	157
12.15	Examples of portable backpack mobile mapping systems	158
12.16	(a) A GeoSLAM Zeb hand-portable laser scanner for close-range mapping measurements; (b) a GeoSLAM Zeb Discovery backpack mapping unit combining a laser scanner; a cartographic grade GPS for accurate positional location; and a four camera Pulsar unit from NC Tech providing 360-degree imaging and scanner coverage	158
12.17	(a) One of the original camper vans operated as a mobile mapping vehicle by TeleAtlas; (b) a modern fleet of TomTom mobile mapping vehicles based in Poland	160
12.18	(a) TomTom maps are heavily used by drivers in vehicles fitted with satnav devices (left) and cell phones (right); (b) a TomTom map of Manhattan, New York being used for traffic planning	160
12.19	(a) An early mapping car being used for map revision by Navteq (b) a later Here mapping vehicle complete with a multi-camera unit, a laser scanner and a GPS receiver	161
12.20	(a) An Ovi Maps product being used for navigation in Paris; (b) a Navteq map of Karachi, Pakistan	162
12.21	Google Street View coverage	163
12.22	(a) A current model of the Google Street View mapping vehicle; (b) a pedal powered tricycle equipped with the same suite of sensors as a Google mapping car; and (c) two different models of the Google Trekker backpack	164
12.23	(a) An Apple Maps display with road navigation directions on an Apple iPhone; (b) an Apple Maps mapping vehicle equipped with a multiple camera unit and three Velodyne laser scanners	165
12.24	(a) Data depicting the highway infrastructure at a busy road junction in Atlanta, Georgia, USA	166
12.24	(b) Detail of the road markings, crossings, poles and overhead wires that has been extracted from these data	166
12.25	Siteco Road-Scanner mobile mapping system is equipped with additional high-definition cameras	166
12.26	(a) A railroad speeder (track maintenance car) equipped with a Teledyne Optech Lynx mapping system; (b) a Rail Mapper vehicle equipped with a	

Figures

	Z+F Profiler 9012 2D laser scanner, a multi-camera unit and a GPS receiver; (c) a Z+F Profiler 6007 scanner system mounted on the front of a train	167
12.27	(a) A survey launch equipped with a Topcon IP-S3 mobile mapping system; (b) a model showing the land and underwater features of a harbour area	167
12.28	(a) The building facades that are acquired through mobile mapping need the roof patterns acquired by airborne imaging in order to create a 3D model of this block of properties; (b) A small part of the 3D City Model of the Historic Peninsula of Istanbul. The data have been collected at LoD 2	168
13.1	Whether an incident EM beam is (partly) reflected, absorbed or transmitted depends on its wavelength	172
13.2	Part of the highly detailed digital terrain model (DTM) of the Netherlands available as open data	173
13.3	During an airborne Lidar survey GNSS and IMU continuously measure the six parameters of the exterior orientation	173
13.4	Diagram representing the nucleus of a point cloud	174
13.5	Application of a geodesic morphological ground filtering method on the National Elevation Model of an urban landscape type in the Netherlands; the left part depicts the DSM and the right part the DTM after ground filtering	175
13.6	During propagating through the atmosphere at approximately the speed of light, the laser beam diverges and its energy is weakened by scattering and absorption by air molecules	176
13.7	Relationship between height above ellipsoid (h) and height above geoid (H), with geoid undulation N = h-H	177
13.8	Various scanning mechanisms used in airborne Lidar systems (top) and the resulting scan patterns on the ground	178
13.9	Principles for capturing of multiple returns and full-waveform digitization (FWD) of a tree	179
13.10	Miniaturization of laser scanners allows UAS Lidar	179
13.11	Three commercially available terrestrial laser scanners, mounted on tripods	180
13.12	Intensity image of a TLS scan of Delft's historic centre where passing pedestrians and cyclists cause spikes and occlusion	181
13.13	Relationship between TLS-centred polar coordinates and Cartesian coordinates	183
13.14	Basic system components of a mobile mapping system	184
13.15	NavVis 3D Mapping Trolley (left); Leica Pegasus backpack (centre); and Handheld GeoSLAM ZEB-REVO mobile 3D scanner	184
13.16	Comparison on data collection characteristics of airborne laser scanning (ALS) and mobile laser scanning (MLS)	185
14.1	The descent of drones	191
15.1	(a) A Wild RC30 aerial film camera with its tracking and navigation sight at front; (b) the projection of the ground through the perspective centre (the camera lens) results in a frame image being exposed in the focal plane of the aerial camera; and (c) a block of aerial photography comprising several strips of overlapping photographs and showing both the forward (60%) and lateral (20%) overlaps	204
15.2	(a) The aerial photographs that overlap in the flight direction provide stereo-coverage of the ground and allow the formation of a stereo-model of the terrain; (b) a Kern PG-2 analogue stereo-plotting instrument; and (c) a Zeiss Planicomp analytical plotter	205

Figures

15.3	(a) A Leica DMC III large-format aerial digital camera mounted on a PAV gyro-controlled mount; (b) a Vexcel UltraCam Eagle Mk. 3 large-format camera showing the multiple lenses of the design; and (c) a diagram showing how the large-format panchromatic frame image is built up from the 15 smaller medium-format images	207
15.4	(a) The arrangement of the linear arrays in the focal plane of the ADS pushbroom scanner; (b) a perspective view of the imaging geometry of the ADS pushbroom scanner as it generates its continuous strip coverage of the terrain; and (c) a Leica ADS80 laser scanner with its control electronics cabinet and display monitor	208
15.5	(a) An IGI DigiCam medium-format camera installed in the grey Wild gyro-stabilized mount; (b) the latest Phase One iXM-RS150 medium-format camera; and (c) a Teledyne Optech CM-11K medium-format camera producing 86 megapixel images	209
15.6	(a) A GeoTechnologies' SF-DMC small-format digital camera; (b) a Canon EOS small-format camera equipped with a Canon GP-E2 GPS receiver for the measurement of positioning data; and (c) a Canon EOS 5DS camera being inserted into an FAA-approved pod to be mounted on the side or the undercarriage of a Cessna light aircraft	210
15.7	Diagrams showing the ground coverage of frame photography acquired by (a) twin (fan); (b) triple (fan); (c) quadruple (block); and (d) pentagonal (Maltese Cross) multi-camera systems	211
15.8	(a) An IGI Dual-DigiCAM system with the twin medium-format cameras tilted; (b) the DLR-3k three-camera system makes use of a fan of three Canon EOS small-format digital cameras; and (c) an IGI Quattro-DigiCAM system	212
15.9	The layout of various small-format multi-camera systems developed by Track'Air and designed to acquire systematic oblique digital aerial photography	213
15.10	Multi-spectral camera systems all using four small-format digital cameras with parallel optical axes to produce separate red, blue, green and near infra-red images	213
15.11	(a) Diagram showing the overall principle of airborne laser; (b) the relationship between the main hardware components of an airborne laser scanner system	214
15.12	Ground measuring patterns: (a) the saw-tooth scan pattern; (b) the sinusoidal pattern; (c) the raster scan pattern; and (d) the elliptical pattern	215
15.13	Powerful high-performance airborne topographic laser scanners capable of conducting mapping operations from high altitudes	216
15.14	(a) Teledyne Optech's compact Orion low-altitude laser scanner and CS medium-format camera; (b) Leica CityMapper integrated system; and (c) RIEGL VQ-1560i-II integrated system	217
15.15	(a) The overall concept and design of an airborne bathymetric laser scanner; (b) the operating principle of the Hawkeye airborne bathymetric scanning system	219
15.16	Powerful bathymetric laser scanners for use over deeper water	220
15.17	Bathymetric laser scanners for operation over shallower water	220

15.18	(a) Block of stereo-models linked together by aerial triangulation; (b) compilation of the vector line data required for map production; and (c) diagram of a stereo-viewing system to form the visual stereo-model in a DPW that can be viewed and measured in 3D by the photogrammetric operator	221
15.19	(a) The orthophotograph is formed by the orthogonal projection of the 3D stereo-image data contained in the stereo-model on to the datum plane; (b) an alternative method to generate an orthophotograph is to project a single photograph on to the DTM; and (c) diagram showing how the orthophoto acts as the base layer for all the other thematic map layers other data themes within the geospatial database	222
16.1	Flowchart summarizing the step-by-step procedure of extracting ALOS Phased Array band Synthetic Aperture Radar (ALOS PALSAR) DEM, SRTM DEM, and TanDEM-X	230
16.2	The step-by-step process of generating DEM from drone-acquired nadir images using Agisoft photoscan	232
17.1	Simplified representation of a 'virtuality continuum'	237
17.2	An example of an augmented reality (AR) user interface on a smartphone and its method of user interaction	238
17.3	An example of a virtual reality (VR) user interface via an HMD (head-mounted display) and its method of user interaction	238
17.4	An example of the difference between augmented reality (AR) and mixed reality (MR)	239
17.5	Level of detail (LOD) for buildings according to the OGC CityGML 2.0 standard	240
17.6	The various types of hardware, movement, and visualization involved in creating a sense of illusion with XR technologies	241
17.7	The different AR display technologies, using video or optical methods	242
20.1	Basic spatial data models for objects (a), fields (b), networks (c) and global distributions on a sphere (d)	265
20.2	Spatial relations (examples)	269
20.3	Spatial operations (examples)	271
20.4	Non-spatial operations (examples)	272
20.5	Grid operations	272
20.6	Data model transformations (examples)	273
20.7	Application of spatial analysis and modelling to visualization and mapping	275
20.8	Application of spatial analysis and modelling to object recognition and data enrichment	276
20.9	Application of spatial analysis and modelling to study spatial distribution	278
20.10	Application of spatial analysis and modelling to derive polygons	279
20.11	Application of spatial analysis and modelling to quantify dominance	280
20.12	Application of spatial analysis and modelling to explore dynamics and evolution	281
20.13	Application of spatial analysis and modelling to analyse movement and trajectories	284
20.14	Application of spatial analysis and modelling to study spatial heterogeneity	285
21.1	Cartograms are a really useful way of showing population related data	294
21.2	Example of a map (a) and a space-time cube (b) showing the GPS trajectory of a Galapagos Albatross	294

Figures

22.1	District groups considered in the research outlined in white	305
22.2	Maps of Tarnowski, Dębicki and Brzeski districts: (a) source data from GGD; (b) atlas map used as evaluation reference; (c) RF; (d) DL; (e) DT_GA; and (f) DT	307
22.3	Maps of Chojnicki and Bytowski districts: (a) source data from GGD; (b) atlas map used as evaluation reference; (c) RF; (d) DL; (e) DT_GA; and (f) DT	308
22.4	Decision tree generated for all districts with the use of the DT_GA model	308
22.5	Decision tree generated for all districts with the use of the DT model	309
23.1	AI systems can analyse, recognize, make decisions and solve problems without human intervention	314
23.2	Machine learning and deep learning related to Weak and Strong AI	316
23.3	Principle of neural networks	317
23.4	Differences between machine and deep learning	320
23.5	Esri applies AI in image classification to conduct traffic and pedestrian movement planning in Cobb County, Georgia	321
23.6	A proposed framework for the integration of AI and big data in smart cities to ensure livability	322
23.7	AI SIS Technology System according to Gunafu	323
23.8	The role of AI in achieving the Sustainable Development Goals	325
24.1	Schematic illustration of location based services as interface	333
24.2	Research Challenges of LBS	336
24.3	Accessing traces of historical information about European culture of archive collections through LBS	337
24.4	Semantic LBS making use of an 'annotated' world	338
25.1	Figures illustrate common problems for mapping buildings from the ground (left – street-level photo) and air (right – aerial photo with building footprints)	344
25.2	The five LoDs of CityGML. The geometric details and semantic information changes with consecutive levels. The LoD1+ proposed by some authors was additionally presented in the figure	346
25.3	Comparison of virtual city models at various level of details	348
25.4	The mapping techniques used at the specific LoDs of CityGML	349
25.5	Data sources and platforms used for creating LoD1 building model	350
25.6	Data sources and platforms used for creating LoD1+ building model	350
25.7	Data sources and platforms used for creating LoD2 building model	352
25.8	Data sources and platforms used for creating LoD3 building model	353
25.9	Visualization of typical point cloud acquired using the terrestrial and aerial (helicopter) mobile laser platform	354
25.10	Methodology for preparing LoD4 building model based on different measurement platform and sensors	355
26.1	Types of underground utilities	359
26.2	Cluster of cables and underground pipes	359
26.3	Underground marking of the route of a pipe with metallic tape	359
26.4	Electronic markers for underground utilities	360
26.5	Principle of wave propagation in the soil	361
26.6	GPR systems with a single antenna	362
26.7	GPR systems with antenna arrays	362

26.8	Locating a pipe perpendicular to the investigation route	363
26.9	Locating a pipeline along the investigation route	363
26.10	Underground investigation tapes	364
26.11	Field computer screen of a GPR system with two antennas	364
26.12	Pipelines visualization into a tomography (in red)	365
26.13	Establishing the investigation routes according to the surface elements	366
26.14	Ways of marking on the ground the detected positions of the underground utilities	366
26.15	Direct georeferencing of GPR records using GPS	366
26.16	Integrated Mobile Mapping + GPR solutions	367
26.17	System calibration and parameter setting for GPR investigation	367
26.18	Marking on the screen the types of underground networks according to the standards	368
26.19	Screen windows in Object Mapper software	369
26.20	Screen windows from the ViewPoint application	369
26.21	User interface of the EKKO Project software	370
26.22	User interface of the GPR-SLICE software	371
26.23	View sections with GPR-SLICE software	371
26.24	3D view with GPR-SLICE software	372
26.25	3D Visualization menu from GPR-SLICE software	372
26.26	3D view with Radar Studio – Post Processing GPR Software	374
26.27	Information viewing windows with GeoPointer X Software	374
26.28	Instant 3D viewing with IQMaps application	375
26.29	DX Office Vision software user interface	375
26.30	Post-processing window with DX Office Vision software	376
26.31	Information structure in 'Utility Network'	376
26.32	Graphical view in ArcGIS 'Utility Network'	377
26.33	Geospatial data management with Bentley Systems	378
26.34	Example map with all underground utilities	379
26.35	AR view with 'vGIS Utilities' application	380
26.36	Viewing underground utilities on the tablet with 'AugView'	381
26.37	Trimble 'SiteVision'	381
27.1	Spurious correlation, persistent over time, between variables 1 and 2	389
27.2	Use of social transects to investigate food insecurity in Birmingham, UK	390
28.1	Infographics of the information flow in a Hydroinformatics system for the case of flood risk management	395
28.2	System to record pipe noises to be used for machine learning, to recognize anomalies in household water pipes.	398
28.3	Methodology to generate data of critical scenarios to feed the AI platform	399
28.4	Model outputs for different times of the day, showing instant pressures for a critical leak event at node wNode_355 (arrow)	400
29.1	Mercator's projection of the world	404
29.2	Gnomonic projection	404
29.3	Principles of echo sounding	406
29.4	'Upwelling' or 'sandbank'? At the time of photographic capture, this uncharted island in the territorial waters of Cuba did not appear on any publicly available chart	408

29.5a	Bearing ranges in St Simon's Sound. St Simon's Sound 'Roadway'	408
29.5b	Bearing ranges in St Simon's Sound. St Simon's Sound 'Roadway' night-time navigation vectors, with ranges (aligned lights), channel marker colours and flash intervals	409
29.6a	Passage from Cancun to Key West, annotated with narrative. Vector Chart	410
29.6b	Passage from Cancun to Key West, annotated with narrative. Raster Chart	410
29.6c	Grib File: predictive meteorological modelling: geo-referenced weather forecasting 'Grib' files	411
29.7	Sextant handling	412
29.8	Sextant ordnances. Celestial positioning terms	412
29.9	To arrive at a Destination seen ahead on a bearing of 10 degrees, (approximately northwards), a ship might have to take a heading of 320 degrees, because of the set of a current from 270 degrees (from the West), which will cause drift of the ship eastwards towards an Obstacle found at a bearing of 50 degrees	415
29.10	AIS in La Manche/The English Channel. AIS – the relational map of hazards (or help). La Manche/English Channel	417
29.11	IALA Maritime Buoyage System: Regions A and B. Rules for Mariners	418
31.1	Satellite photograph of Earth at night from outer space	438
31.2	Simplified schematic of how modern electrical grids bring electricity from power plants to customers	439
31.3	United States national average wind map	443
31.4	United States national global horizontal solar irradiance map	444
31.5	Visualization of electrical distribution	445
31.6	Example 150-metre service area of each transformer along the distribution network, extending 25 metres away from the line in either direction	446
31.7	Flowchart of GIS workflow, including suitability analysis, network analysis, and interpolation	447
32.1	Hotspots of infant low birthweight incidence in Brooklyn, NY based on 1-kilometre and 2-kilometre (radius) search windows	451
32.2	Grocery store availability across different racial/ethnic groups in Chicago, Illinois	452
32.3	Grocery stores within ten-minute walking areas in census tracts with different racial/ethnic groups	453
32.4	County-level distribution of COVID-19 case per 100,000 population in the United States	455
32.5	Distribution of current asthma prevalence (%) among adults aged >18 in census tracts	457
32.6	Monthly average of Daily Max-8 Hours Ozone concentration (ppm) in Chicago	458
32.7	Map of density of public libraries and population by county	460
33.1	John Hopkins animated global map of daily confirmed cases of COVID-19	465
33.2	Confirmed cases of COVID-19 per 100,000 people in Ottawa by Ward	465
37.1	Smawfield, M. (2015) Nepal Earthquake 2015: Humanitarian Information Review and Analysis (HIRA), Nepal	508
37.2	Qatar Center for Artificial Intelligence & Standby Task Force (2015) MicroMap of Nepal Earthquake – April 2015, Kathmandu	508
37.3	Médecins Sans Frontières (2014) Guéckédou – Base Map, Guinea	510

37.4	Allan, R., Gayton, I., Monk, E.J.M. and Yee, K-P. (2019) "Surveyed Villages Within the Catchment Area of Nixon Memorial Methodist Hospital Based in Segbwema, Sierra Leone"	511
37.5	MapAction and UNOCHA (2011) LIBYAN ARAB JAMAHIRIYA - Who, What, Where by Cluster (14th March 2011)	512
37.6	Reine Hanna Medair (2014) Informal refugee settlements, West and Central Bekaa, Lebanon (August 2014)	513
37.7	HOT, MapAction, REACH & UNHCR (2018) Facilities Map – Rhino Camp Settlement – Zone 7 – Arua District – Uganda	514
38.1	A Web map displaying real-time status of confirmed cases of the novel coronavirus (COVID-19) around the world as of 20th March 2020	522
38.2	Earthquake Web map, centred on the Pacific tectonic plate, by the US Geological Survey (USGS)	523
38.3	Landscan population dataset for estimating populations at risk during humanitarian crises	525
38.4	Satellite images of San Juan, Puerto Rico before (left) and after (right) Hurricane Maria in September 2017	526
38.5	Humanitarian OpenStreetMap Tasking Manager for a crowdsourced project to map roads, buildings, and waterways in response to the Colombian refugee crisis	529
39.1	Sustainable Development Goals measurable through geographic information	534
40.1	Temporal perspective: (a) Ortelius' map of Iceland from 1587; (b) OpenStreetMap's Iceland (2021)	542
40.2	Old and modern classics: (a) Minard's 'Carte Figurative des pertes successives en hommes de l'Armée Française dans la campagne de Russie 1812–1813'; (b) real time flight information with situation at 03-02-2021 UTC 08:00	543
40.3	Nature of time: (a) linear – changing municipal boundaries Steenwijkerland; (b) continuous – wind patterns above the Faroe Islands	544
40.4	Change: (a) existential change – the appearance or disappearance of phenomena; (b) attribute change – qualitative or quantitative change; (c) locational change – move or shrink/expand of phenomena	545
40.5	Display of change: tunnel built on the Faroe Islands 1960–2020: (a) single map – I. no temporal hierarchy/II. temporal hierarchy; (b) series of maps – I. Period 1960–1975, II. Period 1975–1990, III. Period 1990–2005, IV. Period 2005–2020	546
40.6	Map types to represent change: (a) flow map of Faroe's migration to the island of Streymoy in 2019; (b) choropleth map of Faroe's relative population change between 1985 and 2020 (inset slope chart with absolute change); (c) diagram maps of Faroe's relative population change between 1985 and 2020; (d) Space-Time-Cube of Minard's 'Carte Figurative' (compare with Figure 40.2a); (e) cartogram of travel times from Steenwijk to other railway station in the province of Overijssel	547
40.7	Comparing situations before and after: (a) Faroe's tunnels built in the past and those planned for the future; (b) the situation in the European airspace on 7th March 2020 and on 7th April 2020 after the COVID outbreak	549
40.8	Animation of tunnel building in the Faroe Islands	550

40.9	Temporal exploration: comparing the age of tunnels and their length in an interactive linked view environment highlighting he oldest and newest tunnel: I. Tunnels by length; II. Tunnel map; III. Scatterplot length versus year; IV. Tunnels by age	551
40.10	Temporal story map: building of the Eysturoyar tunnel	552
41.1	A typical GIS interface for an archaeological project, with layers of data arranged in a hierarchical structure (left) and visualized on screen as a map (right)	555
41.2	Aerial photograph of the pyramids at Giza taken by Eduard Spelterini in 1904 from a hot air balloon about 600 metres above ground	556
41.3	Lidar image of the grand plaza of the Maya city of Tikal in present-day Guatemala	558
41.4	A 3D visualization and digital model of the south shelter at Catalhöyük, Turkey	558
42.1	Lunar topographic maps of the Mare Imbrium region extracted from the Lunar Map Series, 1976	563
42.2	Example image mosaic of Raditladi crater	564
42.3	Spice framework	565
42.4	The potato-like shape of the asteroid Eros	567
42.5	Extracted portion of the original map (Hare *et al.*, 2015) centered on Mare Nectaris	568
42.6	Extracted portion of the original global geologic map of Mars (Tanaka *et al.*, 2014) showing Valles Marineris (canyon system more than 4,000 km long)	570
42.7	Showing JPL's MarsTrek (above) and ASU's JMARS (below) as initial consumers and GIS interfaces for the MarsGIS planetary spatial data infrastructure (PSDI) services	573
43.1	Data decisioning practices that enable to restrict (left to right) data dissemination	580
44.1	Community mapping and identification of areas prone to flooding in Dar es Salaam	593
44.2	The integrated geospatial information framework (as referenced by the United Kingdom)	594
45.1	Galápagos map embedding the first three steps of the five-step theory or model of allopatric speciation	602
45.2	The Src enzyme regulates and triggers many biochemical signalling pathways involved in cell division, survival, motility, and adhesion	605
45.3	This is figure 3 of Bertin's composite map of 'Interdepartmental Migrations in France' (1954), which grants a synoptic view of the migration landscape in mid-1950s metropolitan France	607
45.4	Two basic experimental strategies for producing gene expression maps of human brains	608
46.1	Map showing locations showing in the amateur film Old St John Market and Town Scenes	618
46.2	Screenshot of Google Earth map of Liverpool film locations featuring embedded video of the amateur film Old St John Market and Town Scenes	619
46.3	Screenshot of Google Earth map showing the location of the Dee Estuary	620

TABLES

8.1	D'Ignazio and Klein's (2020a) feminist principles for data feminism	100
13.1	Cross-matrix confronting a small fraction of the points in the filtered point cloud with ground truth	176
19.1	Summary of different kinds of API frameworks	259
22.1	ML models performance for all districts	307
26.1	Network type and frequency	360
26.2	Type of GPR antenna and frequency	361
28.1	Sample of the seven first records out of 6.2 million for the considered city	399
30.1	Mobility Report for Calgary, Canada and the United States 17th November to 29th December 2020	433
32.1	Examples of the use of geospatial tools in environmental health assessment	458

CONTRIBUTORS

Editors

Alexander J. Kent is Honorary Reader in Cartography and Geographic Information Science at Canterbury Christ Church University and has been actively engaged in the geospatial community for over 20 years, as President of the British Cartographic Society, Editor-in-Chief of *The Cartographic Journal*, Founding Chair of the Commission on Topographic Mapping and of the World Cartographic Forum (both International Cartographic Association). His publications include *The Routledge Handbook of Mapping and Cartography* and *The Red Atlas: How the Soviet Union Secretly Mapped the World*. Alex has served as an advisor to the UK and Canadian National Commissions of UNESCO and is currently leading a project with World Monuments Fund and English Heritage to develop strategies for managing coastal heritage sites.

Doug Specht is a Chartered Geographer and a Reader in the School of Media and Communication at the University of Westminster. His research examines how knowledge is constructed and codified through digital and cartographic artefacts, focusing on development issues in Latin America and Sub-Saharan Africa. He also writes on pedagogy and education, and holds Advanced Teacher Status.

Contributors

Oluibukun Gbenga Ajayi is a Senior Lecturer in the Geoinformation Technology section of the Department of Land and Spatial Sciences at the Namibia University of Science and Technology (NUST), Windhoek, Namibia. He holds a bachelor's and master's degree with First Class Honours and Distinction respectively, and also a PhD in Surveying and Geoinformatics. He is the recipient of the Len Curtis Award 2018 presented by Taylor and Francis and the Remote Sensing and Photogrammetry Society (RSPSoc), United Kingdom, for authoring the most outstanding technical paper published in *International Journal of Remote Sensing*, which was adjudged the best scientific paper in the open literature of remote sensing during the year 2017. He specializes in geospatial modelling and remote sensing applications.

Contributors

Leonardo Alfonso is a Civil Engineer of the National University of Colombia, currently holding the position of Associate Professor at IHE Delft. He has a combined consultancy and research experience in water modelling. He holds an MSc degree from UNESCO-IHE and a PhD degree from Delft University of Technology and UNESCO-IHE. Since November 2018, he is coordinating the Hydroinformatics Specialisation, which is part of the master's degree Water Science and Engineering and Erasmus Mundus Flood Risk Management. His research interests include technologies for data collection, hydrometric networks design, model-based optimization and decision making under uncertainty using Value of Information concepts.

Rupert Allan is an ocean skipper, maritime conservationist, humanitarian lead and spatial designer. His work is characterized by people-centred creativity. As a boat builder and master of the British-flagged ocean-going ketch Sandpiper, his interest in off-grid marine engineering technologies and OpenSource navigation has taken him 10,000 sea miles of maritime passage-making in small boats. Rupert leverages this technical expertise in alternative technology and geospatial data ethics for humanitarian operations, working with MSF, UN, and Humanitarian OpenStreetMap Team (HOT). He associates his research with the Royal Geographic Society, the Sustainable Earth Institute (Plymouth University) and the Manson Unit (MSF UK).

Suchith Anand is Chief Scientist at the Global Open Data for Agriculture and Nutrition. He is an internationally recognized expert in geospatial science, providing guidance and advice to governments and international organizations on open education, open data and open science policies. He has authored a wide range of publications; from journal papers, scientific reports and book chapters, to keynote presentations and international strategy documents. He has mentored over 1,000 emerging leaders in sustainable development programs and initiatives worldwide through GeoForAll.

Penelope Anthias is Assistant Professor in Human Geography at Durham University. She also holds a research position at the Department of Food and Resource Economics, University of Copenhagen. Penelope completed her PhD in Geography at the University of Cambridge in 2014, followed by a Ciriacy-Wantrup Postdoctoral Fellowship at the University of California, Berkeley. Penelope has spent several years conducting ethnographic research on indigenous land claims in the Bolivian Chaco region. Her first book, Limits to Decolonization: Indigeneity Territory and Hydrocarbon Politics in the Bolivian Chaco, was published by Cornell University Press in 2018.

Nick Bearman completed his PhD in GIS at the University of East Anglia in 2011 and now works in the public and private sectors providing GIS training and consultancy. He delivers bespoke training to clients including University of Liverpool, ONS, UCL, University of Southampton and American University of Beirut, Lebanon and is Cartographic Editor for the journal Cartography and Geographic Information Science. Previously Nick has held post-doc and teaching positions at the Universities of Exeter, Liverpool and UCL. Nick is a Fellow of the Royal Geographical Society, a Chartered Geographer (GIS) and a Fellow of the Higher Education Authority.

Derek Van Berkel is Assistant Professor of Data Science, Geovisualization and Design. His research examines the human dimensions of land-cover/land-use change and ecosystem services at diverse scales. It aims to use spatial analysis and geovisualizations of social and environmental data and spatial thinking to develop solutions for today's most pressing environmental challenges. Within this growing body of interdisciplinary work, he leverages social theory, big data, machine learning, spatial-temporal computer modeling (e.g., agent-based and cellular automata) and spatial statistics. This work addresses critical domains including (i) landscape management (e.g., weighing tradeoffs between ecosystem services); (ii) conservation and recreation planning (e.g., identifying special and valued places); and (iii) development of sustainable and resilient communities (e.g., health and well-being of communities).

Ivan Bobashev is an energy systems researcher and GIS designer with a background in applied economics and development. By answering the fundamental question 'where?', Ivan is advancing various energy reliability and renewable energy projects around the world. Ivan works with a variety of institutions dedicated to systemic impacts in the world of energy from UN agencies, renewable energy developers, outdoor apparel industry leaders, academia, small-medium sized enterprises, and international NGOs. Ivan holds a Master of Development Practice (MDP) from the University of California, Berkeley and a Bachelor in Economics from the University of North Carolina, Chapel Hill.

Maria Antonia Brovelli is Professor of GIS and The Copernicus Green Revolution for Sustainable Development at Politecnico di Milano (PoliMI) and a member of the School of Doctoral Studies in Data Science at Roma La Sapienza University. From 1997 to 2011, Maria was Head of the Geomatics Laboratory of PoliMI (Campus Como) and from 2011 to 2016, the Vice-Rector of PoliMI for the Como Campus. Currently, she is the coordinator of the Copernicus Academy Network for PoliMI and Head of the GEOLab. She is also Vice President of the ISPRS Technical Commission on Spatial Information Science, a former member of ESA ACEO (Advisory Committee of Earth Observation); Co-Chair of the United Nations Open GIS Initiative; Chair of the UN-GGIM (Global Geospatial Information Management) Academic Network; and a curator of the geospatial series 'AI for Good' organized in partnership with 40 UN Sister Agencies.

Ryan Burns is Assistant Professor in the Department of Geography, University of Calgary, and a Visiting Scholar at University of California, Berkeley. His research at the intersection of GIScience and critical human geography is primarily concerned with the social, political, and institutional inequalities of digital technologies, particularly within urban settings.

William Cartwright was Professor of Cartography at RMIT University. He retired from the University in 2020, after 40 years in academia. He is principal of Cartwright|Geo|Consulting. Before becoming an academic he spent 11 years in the mapping industry. He holds a Doctor of Philosophy and a Doctor of Education and six other university qualifications – in the fields of cartography, applied science, education, media studies, information and communication technology and graphic design. He was President of the International Cartographic Association (2007–2011) and Chair of the Joint Board of Geospatial Information Societies (now UN-GGIM: Geospatial Societies) (2011–2014). In 2013 he was made a Member of the Order of Australia (AM) for service to cartography and geospatial science as an academic, researcher and educator.

Karel Charvat graduated in theoretical cybernetics. He is a co-chair of OGC Agriculture DWG, member of International Society for Precision Agriculture, Research Data Alinace,

Club of Ossiach, CAGI, and CSITA. From 2005 to 2007 he was President of European Federation for Information Technology in Agriculture Food and Environment (EFITA). He was an organizer of many hackathons, and founded the concept of INSPIRE hackathons. He works on implementation on national INSPIRE Geoportal. Now he is also active in Plan4all association. He has extensive experience in ICT for environment, transport, agriculture and precision farming. He is also one of the promotors of Open and Big Data in Agriculture in Europe and has participated in the following projects: Wirelessinfo, Premathmod, EMIRES, REGEO, RuralWins, Armonia, aBard, EPRI Start, Ami@netfood, AMI-4For, Voice, Naturnet Redime, Mobildat, SpravaDat, Navlog, c@r, Humboldt, WINSOC, Plan4all, Habitats, Plan4business, SmartOpenData, FOODIE, SDI4Apps, AgriXchange, FOODIE, SDI4Apps, OTN, DataBio, SIEUSOIL, STARGATE, Polirural, AfraCloud, SKIN, ENABLING, Liverur, and Innovar.

Yaella Depietri is Research Fellow at the Technion – Israel Institute of Technology and at the University of Haifa, both in Haifa, Israel. She earned her PhD in Environmental Sciences at the Institute of Science and Technology (ICTA), Autonomous University of Barcelona (UAB) (Spain) in 2015 jointly with the United Nations University, Institute for Environment and Human Security (UNU-EHS), Bonn (Germany). She then was a Post-Doctoral Fellow at The New School in New York City (USA) and, later, at the Technion. Her research focus is on urban areas, ecosystem services, social media, disaster risk reduction, and wildfires.

Matthew H. Edney has degrees in geography and cartography from University College London and the University of Wisconsin-Madison. After teaching at SUNY Binghamton, he moved to the University of Southern Maine in 1995; in 2007 he was appointed Osher Professor in the History of Cartography. He has also directed the History of Cartography Project, in Madison, since 2005. He is the author of many works on mapping and map history, most recently *Cartography: The Ideal and Its History* and *Cartography in the European Enlightenment*, edited with Mary Pedley, Vol. 4 of *The History of Cartography* (both University of Chicago Press, 2019).

Victoria Fast is an Associate Professor in the Department of Geography at the University of Calgary. Their field of research is broadly urban GIS, an interdisciplinary mix of geographic information science, human geography, and urban studies. Drawing on 'spatialized' expertise in participatory and volunteered GIS, open data, mapping, analysis, and modelling, Dr. Fast's research is driven by the goal of ensuring all individuals – and especially the disabled community –have just and sustainable access to the built and digital environment unrestricted by social, economic and cultural forces, such as income, gender, race, and disability.

Georg Gartner is a Full Professor for Cartography at the Vienna University of Technology and head of the Research Group Cartography. He was Dean for Academic Affairs for Geodesy and Geoinformation at Vienna University of Technology. He is responsible organizer of the International Symposia on Location Based Services and Editor of the Book Series *Lecture Notes on Geoinformation and Cartography* by Springer and Editor of the *Journal of LBS* by Taylor & Francis. He served as President of the International Cartographic Association, as Vice-President of the Austrian Society of Geodesy and Geoinformation and is a member of the Academic Network of United Nations Global Geospatial Information Management.

Contributors

Catalina Garzón-Galvis (Andean South American) is Principal Practitioner at Community Research and Education in Action for True Empowerment (CREATE) in unceded occupied Chochenyo Ohlone lands. Catalina's work includes coordinating community planning, participatory action research, and participatory curriculum development partnerships with community-based organizations and coalitions for environmental health and justice. She received her BA in Environmental Sciences and her master's in City and Regional Planning from UC Berkeley. Her doctoral research focuses on power and privilege dynamics in participatory action research partnerships for environmental justice in the San Francisco Bay Area.

Andrea Ghermandi is an Associate Professor at the Department of Natural Resources and Environmental Management, and the Director of the Natural Resources and Environmental Research Center at the University of Haifa, Israel. He is an alumnus of the Global Young Academy and (co-authored >50 publications in peer-reviewed journals). He received a PhD in Analysis and Governance of Sustainable Development from the University of Venice in 2008. An environmental engineer by training, his research spans over a range of fields including ecosystem services valuation and mapping, and the use of research synthesis techniques and crowdsourced digital data in environmental studies.

Michael F. Goodchild is Emeritus Professor of Geography at the University of California, Santa Barbara. He received his BA degree from Cambridge University in Physics in 1965 and his PhD in geography from McMaster University in 1969. He was elected member of the National Academy of Sciences in 2002, and Foreign Member of the Royal Society and Corresponding Fellow of the British Academy in 2010; and in 2007 he received the Prix Vautrin Lud. He serves on the editorial boards of ten other journals and book series, and has published over 15 books and 550 articles.

Faine Greenwood is an expert on civilian drone technology and its uses in humanitarian aid, research, and disaster response. Greenwood has conducted research on civilian drones at the Harvard School of Public Health, New America, the World Bank, and the Massachusetts Department of Transportation, among other organizations.

Aída Guhlincozzi is an Assistant Professor in the Departments of Geography and Women's and Gender Studies at the University of Missouri. Her research includes health care accessibility for vulnerable populations, including linguistically-appropriate physicians for the Latine community, and disability health geographies, including those related to autism. Her research uses a mixed methods approach of interviews, surveys, and GIS analysis. She is also an advocate for diversity, equity, inclusion, and justice in Geography and the Geosciences at-large, and holds a camp for middle schoolers of colour to engage with Geography and GIS at Mizzou.

Łukasz Halik is a Polish cartographer, geographer and academic. He is a Lecturer in Cartography and Geomatics at Adam Mickiewicz University in Poznań (Poland) and Vice-Chair of the Commission on Topographic Mapping International Cartographic Association. He was a member of the working group in the Ministry of Digital Affairs (Poland) devoted to distributer ledgers and blockchain technology, and his research focuses on Extended Reality (XR).

Trent Michael Hare is a Cartographer with the Astrogeology Science Center of the US Geological Survey in Flagstaff, Arizona, where he has worked since 1989. He has been technical lead for project development and management of cartographic and science research tasks for numerous NASA science and exploration programs. His work includes scientific and cartographic development of efficient and effective Geographic Information System (GIS) analyses, dataset interoperability, creation of geospatial tools, cartographic representations and metadata. Trent is an active member of NASA review and advisory panels for space science, space exploration, and planetary cartography.

Ben Hawes is Engagement Director of the Benchmark Initiative on ethical use of location data. Technology policy consultant, and adviser to the Connected Places Catapult and Digital Leaders Cities Group. Previously a technology policy official in the UK Government, working on smart cities, Internet of things, government relationships with major tech companies, the 2017 Hall-Pesenti Review of Artificial Intelligence, the 2011 Hargreaves Review of Intellectual Property, the 2010 revision of the Government's Guidance to Ofgem, and the Digital TV Switchover Plan.

Gobe Hobona is the OGC's Director of Product Management, Standards. In this role he provides oversight of OGC Application Programming Interface (API) evolution and harmonization activities. He is also accountable for the Compliance Program and Knowledge Management. His previous work at Envitia, Newcastle University and the University of Nottingham covered topics such as spatial data infrastructure, metadata harmonization, geographic information retrieval, the Semantic Web, autonomous systems, and Earth observation. He holds a PhD in Geomatics from Newcastle University. He is a professional member of both the Royal Institution of Chartered Surveyors (RICS) and the Association for Computing Machinery (ACM).

Amelia Hunt is an independent researcher who has worked with humanitarian and health organizations around the world, including Humanitarian OpenStreetMap Team in Tanzania, UNOCHA, the British Red Cross, and London School of Hygiene and Tropical Medicine. She has previously written on crowdsourced mapping in crisis zones, focusing on collaboration, organization, and impact. Amelia currently works in Australia for WentWest and as an independent Humanitarian Communications Specialist.

Mark Iliffe is a Geographer working in the United Nations Statistics Division, primarily supporting the United Nations Committee of Experts on Global Geospatial Information Management (UN-GGIM). As part of a small Secretariat team, he supports and facilitates dialogue and work between Member States in developing global geospatial frameworks, standards and norms that can be implemented at the national level. Before joining the United Nations, he was a co-founder of the 'N-LAB' – a centre of excellence at the University of Nottingham in International Analytics and consulted for the World Bank in East Africa, where he led the Ramani Huria mapping initiative, among other geospatial activities across the region. He holds a BSc in Computer Science, an MSc in Geospatial Engineering, and a PhD in the Digital Economy from the University of Nottingham. In his spare time, he is an avid motorcyclist and whiskey enthusiast.

Mike Jackson is Emeritus Professor of Geospatial Science at the University of Nottingham Geospatial Institute and a consultant in the area of geospatial intelligence. Previously, he was Founder and Director of the Centre for Geospatial Science, University of Nottingham; Director of the Space Department at QinetiQ; and CEO of Laser-Scan Holdings PLC. Mike is a non-executive Director of the Open Geospatial Consortium (OGC) and past President of the Association of Geographic Information laboratories for Europe (AGILE). He has published over 100 papers and book chapters in the areas of geographic information systems, geo-intelligence and remote sensing. He has a PhD from Manchester University and holds fellowships of the Royal Institute of Chartered Surveyors and of the Royal Geographical Society.

Kiringai Kamau is the lead advisor to Kenya's Agriculture Transformation Office (ATO) in data sourcing and agribusiness; supports governments, universities, research organizations, private sector, farmers and development partners. He lectures agricultural economics at the university of Nairobi, is a social entrepreneur, and promotes capacity building in ICT and Agriculture. He infuses agripreneurship in value chain aligned ICTs integration using the GODAN Programme for Capacity Development in Africa (P4CDA). He is an open data champion in promoting student led, agricultural sector transformation through farmer based, farmer-owned agribusiness hubs that provide future focused dynamics for intra-country rooted economic transformation.

Izabela Karsznia received her PhD at the Faculty of Geography and Regional Studies, University of Warsaw in 2010. Since 2010 she has been working as an Assistant Professor at the Department of Geoinformatics, Cartography and Remote Sensing, Faculty of Geography and Regional Studies at the University of Warsaw. She actively cooperates with national and international mapping agencies on projects concerning cartographic generalization and spatial data infrastructure development. Between 2014 and 2015 she was working as a Postdoctoral Fellow at the GIS Unit, Department of Geography, University of Zurich. Her research interests concern Geographic Information Systems, historical GIS, automation of cartographic generalization, and spatial data infrastructure development.

Meghan Kelly is an Assistant Professor at the Department of Geography, University of Durham. She received her PhD at the Department of Geography, University of Wisconsin–Madison in 2020 before completing a postdoc in the Critical Geospatial Research Lab at Dartmouth College. She works at the intersection of map design, feminist theory, and digital storytelling. In addition to her scholarly work, Meghan is a practising cartographer and more information can be found on her website: http://meghankelly-cartography.github.io/.

Rob Kitchin is a Professor in Maynooth University Social Sciences Institute and Department of Geography. He was a European Research Council Advanced Investigator on the Programmable City project (2013–2018) and a Principal Investigator on the Building City Dashboards project (2016–2020). He is the (co)author or (co)editor of 34 academic books and (co)author of over 200 articles and book chapters. He has been an editor of *Dialogues in Human Geography*, *Progress in Human Geography* and *Social and Cultural Geography*, and was the co-Editor-in-Chief of the *International Encyclopaedia of Human Geography*.

John C. Kostelnick is a Professor of Geography in the Department of Geography, Geology & the Environment, and also serves as Director of the Institute for Geospatial Analysis and Mapping (GEOMAP) at Illinois State University, Normal, Illinois, USA. He holds a PhD in Geography from the University of Kansas. His primary research interests include crisis and hazard/risk mapping, GIS integration into science and society, map design, and cultural mapping. His work in humanitarian mapping includes design of the global standard for cartographic representation of landmines, minefields, and mine actions on maps.

Menno-Jan Kraak is Professor of Geovisual Analytics and Cartography at the University of Twente/ITC. Currently he is ITC's vice-dean of capacity development. He was President of the International Cartographic Association (ICA) for the period 2015–2019. He has written more than 200 publications, among them the books *Cartography: Visualization of Geospatial Data* (fourth edition 2020, CRC Press) and *Mapping Time* (2014, Esri Press) and *Mapping for a Sustainable World* (2020, United Nations).

Johannes Langemeyer is an Established Researcher and Principal Investigator at the Institute of Science and Technology, Spain. He is a geographer by training (Humboldt Universität zu Berlin) and holds a PhD in Sustainability Science from the Stockholm University (Stockholm Resilience Centre) and in Environmental Science and Technology from the Universitat Autónoma de Barcelona (UAB). His interdisciplinary research addresses ecosystem services, justice, and resilience primarily in urban environments. He further develops integrated assessment approaches in the context of green infrastructure planning and nature-based solutions.

Didier G. Leibovici's expertise is in geospatial data analytics and after 15 years' research in leading UK universities (Oxford, Leeds, Nottingham, Sheffield), five years at IRD (France), two years at Sanofi-Recherche (France), four years at INSERM (France) working within interdisciplinary and international context for European research programmes with UK, France, LMIC (in Africa and South-Asia), he is setting up GeotRYcs, a geo-spatial-temporal data scientist consulting service. Didier has a PhD in Biostatistics and a master's degree in computing-science; his scientific production in data analysis and geospatial science are on spatiotemporal data modelling and analysis within different contexts, such as epidemiology, agriculture and agro-ecological monitoring, dynamics in population studies, location-based citizen crowdsourcing of environmental information within interdisciplinary projects. Didier's interests are in challenging the potential of interoperability developments to manage cross-domains scientific models involving geospatial data from heterogeneous sources, their metadata information to qualify the scientific workflow and also part of the data analysis.

Mathias Lemmens has over 35 years of research and teaching experience in geodata acquisition technologies and geodata quality. His popular book *Geoinformation – Technologies, Applications and The Environment*, published by Springer in 2011, provides a unique and in-depth survey of the geomatics field. As Editor-in-Chief of GIM International since 1997, he writes on numerous state-of-the art Lidar, photogrammetric and GNSS technologies. He is author of over 500 publications, including scientific and professional papers, books, book chapters and lecture notes. In November 2019 – on the occasion of his retirement as Director MSc Geomatics for the Built Environment at Delft University of Technology, the Netherlands – he published the book: *Points on the Landscape – Twenty Years of Geomatics Developments in a Globalizing Society*. He is also author of the monograph *Introduction to Pointcloudmetry – Point Clouds from Laser Scanning and Photogrammetry*, published by Whittles Publishing in 2023.

Wen Lin is a Senior Lecturer in Human Geography in the School of Geography, Politics and Sociology at Newcastle University. Her research interests include critical GIS, public participation GIS, and urban geography. Her main research centres on examining the intersection between the development and usage of geospatial technologies and the sociopolitical conditions in which these practices are situated. In particular, she has worked on three related themes: investigating the sociopolitical implications of recent mapping practices combined with Web 2.0 technologies; examining GIS-related practices in urban planning agencies; and examining public participation GIS practices in urban governance.

Beth Rose Middleton Manning (Afro-Caribbean, Eastern European), is Professor of Native American Studies at UC Davis in unceded occupied Patwin lands. Her research centres on Native environmental policy and Native activism for site protection using conservation tools. Her broader research interests include rural environmental justice, climate change, and Afro-Indigenous intersections in the Americas. Beth Rose received her BA in Nature and Culture from UC Davis, and her PhD in Environmental Science, Policy, and Management from UC Berkeley. She has published two books with University of Arizona Press, *Trust in the Land: New Directions in Tribal Conservation* (2011) and *Upstream: Trust Lands and Power on the Feather River* (2018).

Denise McKenzie is a strategic advisor, partnership builder, and presenter with over 20 years of experience with the global geospatial community. She works internationally to evangelize the benefits, value, and application of location data across government, private sector, and academia and her experience covers a broad range of domains including smart cities & IoT, agriculture, defence, sustainability, insurance and development. This diversity ensures that she works where geospatial meets mainstream technology. Currently, she is a co-Director of the Benchmark Initiative operating through Ordnance Survey's Geovation accelerator, exploring the ethical use of location data, and is in the process of completing a master's in Sustainability with the University of Southampton. In the broader geospatial community, she is the Chair of the board of directors for the Association for Geographic Information (AGI) in the UK, is a member of the Global Advisory Board for the Location Based Marketing Association and PLACE, and a steering committee member for Women in Geospatial+. Previously she has worked with the Victorian State Government in Australia on geospatial innovation projects and as the Head of Outreach and Communication at the Open Geospatial Consortium where she led work such as the UN-GGIM Geospatial Standards Guides and strategic partnership engagement.

Sara McLafferty is Professor Emerita of Geography and Geographic Information Science at the University of Illinois at Urbana-Champaign. Her research investigates place-based inequalities in health and access to healthcare and employment opportunities for women, immigrants, and racial/ethnic minorities in the United States. She has also written about the use of GIS and spatial analysis methods in exploring inequalities in health and access to health care, and she co-authored *GIS and Public Health* with Ellen Cromley.

Rafael Moreno-Sanchez is Professor in the Department of Geography and Environmental Sciences at the University of Colorado Denver. He received a bachelor's degree in forestry in 1982 in Mexico and a PhD in natural resources management in 1992 in the United States. He worked for the National Institute for Forest Research (INIFAP) in Mexico until 1996. Since then, he has been teaching and doing research at the university level in the US in the areas of Natural Resources Sustainable Management and Geographic Information Science and Technology. He combines these two areas to support decision-making and planning in the areas of land use, natural resources management, and sustainable development in different locations around the world.

Contributors

Philip J. Nicholson is a Senior Researcher at the Association for Decentralised Energy. He is a human geographer and artist. His academic work attends to the aesthetics of geospatial media via creative practice, play, experimentation, and narrative and received his PhD in Human Geography from the University of Glasgow in 2018 for a thesis that looked at GIS from an arts and humanities perspective. Since then, he has worked on developing Creative Geovisualization as a method of working collaboratively and creatively to map social, economic and environmental datasets in postcolonial, post-independence, and UK policy contexts.

María Belén Noroña is a scholar and activist committed to bridging scholarship with the realities of diverse populations, particularly indigenous people and marginalized women. In the last ten years, she has conducted participatory and collaborative research with grassroots organizations on topics related to ethnic territoriality, indigenous ontology and epistemology, and oil-related violence. Through her involvement with non-profits and rural grassroots organizations in countries like Ecuador, she has led research relevant to both scholarship and the concerns of marginal communities. Currently, she is an Assistant Professor of Geography at Pennsylvania State University.

Oliver O'Brien is a Senior Research Associate at the UCL Department of Geography's Geospatial Analytics and Computing group. His principal project role within the group is as Centre Technical Manager for the ESRC Consumer Data Research Centre, a multi-university collaboration which obtains, integrates and derives unique insights from retail business datasets. His research specialisms include urban mobility (including bikeshare and escootershare), digital cartography and data visualization. His Web mapping platform of choice is OpenLayers and he has long been a data contributor to the OpenStreetMap project.

Gordon Petrie (1930–2020) was Emeritus Professor of Topographic Science at the University of Glasgow. He joined the University in 1958 after a short career as a professional land surveyor with the Directorate of Overseas Surveys and taught and researched in photogrammetry, remote sensing and surveying until his retirement in 1995. He continued to publish, for example on the history of mapping and on photogrammetric and laser instrumentation, and received the Fairchild Photogrammetric Award from the American Society for Photogrammetry and Remote Sensing, the President's Medal from the Photogrammetric Society and the Bartholomew Globe from the Royal Scottish Geographical Society. This book is dedicated to his memory in recognition of Professor Petrie's contribution to advancing education and research across the geospatial sciences.

Les Roberts is Reader in Cultural and Media Studies at the University of Liverpool. The core focus of his work is situated within the interdisciplinary fields of spatial anthropology and spatial humanities. He is author/editor of a number of books that engage with these research areas, including *Deep Mapping* (2016), *Film, Mobility and Urban Space* (2012), *Liminal Landscapes* (2012), and *Mapping Cultures* (2012). His latest monograph, *Spatial Anthropology: Excursions in Liminal Space*, was published by Rowman and Littlefield in 2018 (paperback in 2020). More information at www.liminoids.com.

Nikki Rogers holds a master's degree in Geographic Information Systems and is passionate about equity in data access within and across regions, municipalities, and communities. Her prior research has also focused on how GIS has and can contribute to understanding factors influencing pedestrian collisions. She believes that spatial analysis of pedestrian collisions can help lead the way for changes in the built environment that can increase commuter safety for frequently marginalized groups.

Leah Rosenkrantz has a PhD from the Department of Geography at Simon Fraser University. Her research involves exploring issues of global public health through both a spatial and 'platial' lens. Leah has worked on projects related to trauma registries in low- and middle-income countries, antimicrobial resistance, and most recently, COVID-19.

Timofey Samsonov is a leading research scientist at Lomonosov Moscow State University (MSU), Faculty of Geography, Moscow, Russia. He obtained his PhD in Cartography from MSU in 2010 on the topic of multiscale hypsometric mapping. His research is primarily focused on the new methods for automated generalization and visualization of spatial data. He collaborates frequently with specialists in climatology, hydrology and oceanology to build cartographic and computational tools for spatial analysis of hydrometeorological data. Timofey is also active in the Commission on Mountain Cartography and the Commission on Generalization and Multiple Representations within the International Cartographic Association (ICA).

Roberto Santos is a team leader at the Digital Research Service, University of Nottingham. Roberto is responsible for the operational management of a team of research software engineers and is currently involved in several projects across the university. Roberto holds a PhD in Geography, and his expertise includes an overlap of Software Engineering, Geographic Information Science, Agriculture and Meteorology.

Aurel Saracin is an Associate Professor at the Faculty of Geodesy within the Technical University of Civil Engineering Bucharest. He obtained a PhD scientific title in civil engineering with the thesis *Topographical Technologies Used for Tracing, Conducting and Performing of Underground Buildings* in 2000. Aurel participates in research to investigate the characteristics and routes of utility networks using classical procedures and GPR techniques. He introduced and promoted the current techniques of investigation for mapping underground urban networks among students in the faculty during the course 'Special surveying', and master's students in the course 'GeoRadar for the built environment'.

Ben Schaap is a Soil Scientist and a Data Science Alliance Manager for the Wageningen Data Competence Centre. He develops and gives guidance to strategic implementation to advance data applications within the context of food systems. Ben is co-leading the GO FAIR Food Systems Implementation Network and co-developed the Farm Data Train, compatible with the FAIR Data Train concept. In cooperation with the Dutch Ministry of Agriculture Nature and Food Safety he has contributed to GODAN Secretariat and was acting as the Research lead. He has a background in agronomy, and he has worked climate change adaptation related research projects.

Nadine Schuurman is a GIScience researcher research focuses on spatial epidemiology: understanding the spatial distribution of health events and health services in support of improved population health and health care provision. Since 2002, her research has focused on the application of GIS to bettering our understanding of health conditions and health services resource allocation. She is also deeply interested in global health. In each case, her goal is to generate policy-relevant evidence to assist health policy makers and administrators understand and rationalize choices in support of improved prevention and treatment of disease, and better access to care.

Hillary J. Shaw is Visiting Fellow in the Centre for Urban Research on Austerity at De Montfort University, Leicester. He has authored many scholarly articles and books. These include *The Consuming Geographies of Food: Diet, Food Deserts and Obesity* (Routledge, 2014), which articulates how sustainable political and economic structures for feeding the future global population of 10 billion are achievable. A recent monograph, *Corporate Social Responsibility, Social Justice and the Global Food Supply Chain* (Routledge, 2019), explores CSR in relation to the impact of food policy on society and the environment, focusing on food retailing and its global supply chains.

Daniel Sui is Professor of Geography and Public & International Affairs at Virginia Tech. He also serves as Vice President for Research and Innovation. His main research interests are theoretical foundations of geographic information science, the future of work, and geography of innovation. Sui is a 2009 Guggenheim fellow and recipient of distinguished scholar award by both AAG (2014) and CPGIS (2019).

Tomasz Templin is the Assistant Professor at University of Warmia and Mazury in Olsztyn, Poland. He has 20 years of professional experience in diversified service industries (surveying, geoinformatics, satellite geodesy, water sciences, transportation, environmental protection). His research focuses on the distributed and mobile geographic information systems, data modelling, spatio-temporal analysis, Virtual and Augmented Reality, and 3D scanning. In several years, he was involved in a number of R&D projects funded by private companies, public institutions, national and international agencies.

Vit Vozenilek is a Full Professor of Geoinformatics and Cartography at Palacky University Olomouc, Czech Republic. His research and education involve geosciences and spatial technologies, mainly spatial modelling and geovisualization. He has participated in many national and international projects, published more than 300 scientific articles, 27 books, and nine atlases. He is a head of the Department of Geoinformatics at Palacky University Olomouc, Vice-President of International Cartographic Association, and a member of editorial boards of numerous international scientific journals.

Barney Warf is Professor of Geography at the University of Kansas. His research and teaching interests lie within the broad domain of human geography. Much of his research concerns telecommunications, particularly the geographies of the internet, including the digital divide, e-government, and Internet censorship. He is also interested in political geography, including elections, voting technologies, and the US electoral college. He has also studied a range of topics such as international producer services, fibre optics, offshore banking, military spending, religious diversity, and corruption. He has authored or edited numerous books, three encyclopaedias, book chapters, and refereed journal articles.

Nigel Waters is Professor Emeritus of Geography at the University of Calgary, where he taught from 1975 to 2007 and was the Founding Director of the Masters in GIS Program and the Transportation Theme School. During 2007–2014 he was a Professor in the Department of Geography and Geoinformation Science and the Director of the Center of Excellence for

Geographic Information Science at George Mason University (GMU). He is currently an Affiliate Faculty Member in the College of Science (GMU) and an Adjunct Professor of Civil Engineering (Transportation) in the Schulich School of Engineering, University of Calgary.

Amy Schmitz Weiss is a Professor in the School of Journalism & Media Studies at San Diego State University. She received her PhD from the University of Texas at Austin in 2008. She teaches journalism courses in basic writing and editing, multimedia, Web design, data journalism, mobile journalism, sensor journalism, media entrepreneurship and spatial journalism. Her research interests include spatial journalism, online journalism, media sociology, news production, multimedia journalism and international communication. Her research has been published in several peer-reviewed journals, as book chapters and in a book she has co-edited.

Fikriyah Winata is an Assistant Professor in the Department of Geosciences at Mississippi State University and gained her PhD from the University of Illinois at Urbana-Champaign. She uses GIS, spatial analysis, and statistical methods to examine disparities in access to healthy food in Chicago, Illinois, and the availability of primary healthcare services in Indonesia. Prior to attending graduate school, she was a GIS instructor and GIS specialist at Esri Indonesia with multiple GIS certifications. In her dissertation research, she draws on the concepts of therapeutic landscapes and networks in health geography to understand the health and wellbeing of women who work in domestic employment.

Rasmus Grønfeldt Winther is Professor of Humanities at the University of California, Santa Cruz and Affiliate Professor of Transformative Science at the GeoGenetics Section of Globe Institute at University of Copenhagen. He works in the philosophy of science and philosophy of biology, and has interests in epistemology and political philosophy, cartography and GIS, and science in general. Recent publications include 'A Beginner's Guide to the New Population Genomics of *Homo sapiens*: Origins, Race, and Medicine' in *The Harvard Review of Philosophy*; 'Mapping the Deep Blue Oceans' in *The Philosophy of GIS* (Springer); *When Maps Become the World* (University of Chicago Press, 2020); and *Our Genes: A Philosophical Perspective on Human Evolutionary Genomics* (Cambridge University Press, 2022).

Yiheng Zhou is a master's student in Health Informatics at Yale University. Her current research includes developing advanced algorithms to derive healthcare insights and facilitate the adoption of data-driven approaches, focusing on biomedical images and electronic health records. She holds a BS in Geography and Geographic Information Science from the University of Illinois at Urbana-Champaign, where she applied GIS and spatial analysis to wildlife conservation, disease surveillance, and disaster management.

ACKNOWLEDGMENTS

Many people have contributed towards the publication of *The Routledge Handbook of Geospatial Technologies and Society* and the whole venture would not have been possible without their support. We owe each of our authors our thanks for their willingness to share their expertise, their diligence in attending to our requests, and their patience in bearing with us as we managed this substantial new volume through the challenges brought by the onset of a global pandemic. In particular, we are especially grateful to Stuart Granshaw and to Stewart Walker for their assistance in finalizing the chapters by Gordon Petrie that he submitted before his passing. We thank the reviewers of this volume for all their suggestions that have enhanced its scope and sharpened its focus, and Andrew Mould, Egle Zigaite, Claire Maloney, and all the team at Routledge, whose professionalism has ensured that this project has remained an exciting and rewarding experience. Finally, and most importantly, in bringing this project to fruition, we owe a huge debt of gratitude to our friends and families. They have inspired, encouraged, and put up with us for more than we could ever have wished as we devoted our time to editing this *Handbook*.

INTRODUCTION

Alexander J. Kent and Doug Specht

This *Handbook* presents a wide-ranging introduction to geospatial technologies and offers a critical evaluation of their impact on society within a variety of international and multi-disciplinary contexts. As a companion volume to *The Routledge Handbook of Mapping and Cartography*, it aims to provide a relevant and comprehensive point of reference for research and practice in this dynamic field. It covers key principles, techniques and applications and explores how geospatial technologies have shaped – and continue to shape – the world around us.

Geospatial technologies are concerned primarily with location. They involve the collection, distribution, storage, analysis, processing and/or the presentation of spatial data that relates to the Earth, and are essential for all aspects of mapping, from land surveying to remote sensing. Geospatial technologies incorporate geographic information systems (GIS), global-navigation satellite systems (GNSS), terrestrial laser scanning (TLS), virtual reality (VR) and augmented reality (AR), photogrammetry, geographical artificial intelligence (GeoAI) and many other tools, systems and processes. Since everything – object, thought, decision or action – occurs somewhere, these technologies are essential for supporting all functions of society, from its infrastructure to its immunization. Increasingly, geospatial technologies are employed in other fields to provide additional layers of understanding or analysis; for example, by mapping the networks and distribution of people and their comments using social media. As disciplinary boundaries become ever more fluid, the *Handbook* encapsulates the immense scope for multi-disciplinary approaches to offer new applications of geospatial technologies.

The origins of these technologies lie in the earliest human attempts to orientate ourselves and to understand the environment. Although our tools have continually developed to advance these goals, their use by governments (especially for the purposes of taxation and defence) has seen their most transformative impact on societies. In particular, the growth of European colonial possessions and the requirements of rapid urbanization in the eighteenth and nineteenth centuries refashioned landscapes, which demanded greater precision and industry in the production of maps and surveys. The major conflicts in the first half of the twentieth century inspired the combination of photography with mapmaking, while the early years of the Cold War saw the deployment of aerial and satellite reconnaissance, further enhancing the remote acquisition of geospatial information and intelligence.

Today, we have come to rely on geospatial technologies more than ever in order to meet the challenges of managing our planet more sustainably while facing the global threat of climate change. From sensor-driven mapping that allows the monitoring of remote environments to the use of AI to analyse traffic flows, these technologies have become embedded in the lives of much of the world's population. We are as reliant on them for maintaining our food, water and energy security, as we are for navigating our way home.

One of the key advantages that geospatial technologies offer their users is the ability to gain location-based knowledge from a geographically remote perspective. The evolution of tools for advancing the understanding of real or imagined environments has sought to deliver experiences that meet the changing needs of society and for a diverse range of applications, including search-and-rescue, urban planning and gaming. Consequently, this volume draws attention to the growth of important new digitally enabled technologies, including 3D mapping, free and open-source software (FOSS), unmanned aerial vehicles (UAVs), machine learning and social media. These are each set within the wider theoretical, practical and social contexts of their use. The successful utilization of geospatial data relies upon effective methods of geovisualization to communicate the right information to the user at the right time and enable them to make the right decisions, whether the goal is to save lives or to explore a fantasy landscape. However sophisticated geospatial technologies become, there will always be a requirement for their output to be presented clearly and effectively through the application of good design.

As geospatial technologies are agents of change, the *Handbook* recognizes the common threads of power, control, participation and representation in various applications around the globe. Geospatial data can transform knowledge in contemporary and future societies and the book examines the rise of participatory geospatial technologies that enable citizens to be the creators and managers of their own data. More broadly, it explores the dialectic relationship with society and geospatial technologies, encompassing important questions surrounding ethics and the future of the geospatial sector at large.

The global impact of COVID-19 has highlighted the relevance of this dialectic further, especially in the effective capture and communication of information about the geographical dimension of the pandemic. Although a number of chapters in this volume discuss how the pandemic has influenced specific applications of geospatial data, every aspect of our dealing with geospatial technology has, in turn, been altered by the pandemic. Existing tools have been put to new uses, such as contact tracing or remotely monitoring the health of travellers, and, likewise, societies have changed their relationship with these tools. This has led to a greater awareness of technologies working in the background of our everyday lives, but, also to a palpable tension between a deep sense of gratitude for how these technologies have limited the spread of the pandemic and an unease at their level of intrusion and surveillance. If the twenty-first century has woven geospatial technologies into the fabric of society more tightly, the pandemic has highlighted for many just how close this interweaving has become. It has never been more crucial to understand not only how these technologies work, but how they interact with our daily lives and experiences.

Organized into four inter-related sections, the *Handbook* comprises 48 chapters from leading authors drawn from around the globe, including eminent scholars, cutting-edge researchers, influential practitioners and industry experts. In turn, these sections aim to provide a survey of historical and theoretical approaches to geospatial technologies, how they work, their real-world applications and some fresh perspectives on their future.

The first section, *Origins and perspectives of geospatial technologies*, offers a multi-disciplinary foundation for understanding society's changing approaches to spatial problems. Its 11 chapters offer critical perspectives on the motivations behind the development of geospatial technologies and examines the interests they serve. The section discusses the rise of the importance of geospatial knowledge and its relationship with science, its use (and abuse) by different groups and the role of key technologies in shaping the way the world is modelled today. It accomplishes this through a diversity of themes, including aesthetics, race, positivism, feminism and social constructivism, and provides the theoretical basis of the volume.

The next section, *Understanding geospatial technologies*, introduces and demystifies the key systems, techniques and devices in use today. It offers technical depth in its explanation of how these technologies work, from data capture through drones and laser scanners to data processing, analysis and visualization using artificial intelligence and machine learning. Its wide-ranging survey of geospatial technologies allows an appreciation of the scope of methods and tools they encompass and an understanding of the principles behind their implementation. Overcoming the challenges of gaining meaningful knowledge from ever-increasing amounts of geospatial data, which often requires creative solutions, forms a common theme throughout this section.

The third section of the *Handbook* focuses on applications. Its authors examine how geospatial technologies are used to shape the ongoing transformation of space in our everyday lived experiences and beyond. Its 18 chapters explore the mapping and modelling of a diversity of phenomena, including cities, social media, hazards, time, oceans, underground spaces and planets. The section also examines the specific application of geospatial technologies to secure the supply of essential commodities and their infrastructure, such as food, water, transport and power, as well as examining how these technologies are contributing towards public health, humanitarian operations, journalism and the UN Sustainable Development Goals (SDGs). Chapters addressing important current topics, such as mapping COVID-19 and geosurveillance, provide new insights into the developing role of geospatial technologies in society.

In the fourth and final section, *New ontologies and strategies for geospatial technology*, the volume explores how ways of constructing and geospatial knowledge are being challenged. It discusses the associated complexities of breaking away from traditional methods and models of understanding and organizing space to create alternative views of the world and their uses and users. The section incorporates chapters that focus on the emergence of new ontologies through neurology, spatial anthropology and quantum science, as well as how advances in the democratization of geospatial data and strategies for their implementation are shaping the future of the geospatial industry. The section concludes with a critical discussion of ethical principles and practice in geospatial technologies and their embodiment through the Locus Charter.

Together, these sections provide a comprehensive reference point for understanding geospatial technologies as well as a substantial resource for the critical reflection of their application. As new developments continue to emerge, the *Handbook* will provide a sure foundation for informing their design, development and use.

PART I

Origins and perspectives of geospatial technologies

1
LATITUDE, LONGITUDE, AND GEOSPATIAL TECHNOLOGIES TO 1884

Matthew H. Edney

Before the twentieth century, location was a *local* and *relative* quality. To *locate* was to place and arrange, or to find something relative to known spatial landmarks; *locations* of things were specified by reference to other things. Early geospatial technologies, in their broadest sense, were accordingly devices to keep track of things. They were devices like the itineraries that helped ensure that one stayed on course along an intended route, or the logbooks that allowed one to reconstruct the route one had taken and to determine one's position via 'deductive reckoning' (or 'dead reckoning'). Even the coordinate system of latitudes and longitudes was first applied to the Earth by ancient astronomers/astrologers to define terrestrial locations with respect to the planets and stars. Places on the Earth were situated by calculating differences in latitude and longitude from itinerary distances and the Earth's size, perhaps adjusted by celestial observations of latitude; alternatively, itineraries could be graphically plotted on maps and the latitudes and longitudes of particular places interpolated. Many repeated calculations and interpolations generated geographical tables, like those in Claudius Ptolemy's *Geography* (c. 150 CE) and in medieval Islamic *zījes* (compendious astronomical tables). Early modern geographers continued the same basic practices.

This chapter traces the development of geospatial technologies – instruments, tables, and larger systems – used to determine latitude and longitude directly and further considers the application of those technologies to mapping and mobility. It does not consider the more complex astronomical practices and larger instrumentation developed for more place-bound determination of position. It ends with the final stabilization of the global framework in 1884 with the international acceptance (mostly) of a specific zero meridian. Further reference should be made to the key histories of navigation, longitude determination, instrumentation, and hydrography (Taylor, 1971; Bennett, 1987; Andrewes, 1996; Chapuis, 1999; Mörzer Bruyns, 2009; Dunn and Higgitt, 2016; Withers, 2017; Schotte, 2019) and also *The History of Cartography* (esp. Edney and Pedley, 2019; Kain, forthcoming) on which this chapter draws extensively. Images of instruments are also readily available online, especially through the website of the National Maritime Museum, Greenwich (http://www.rmg.co.uk/collections), and other specialized collections (see figure captions).

Location at sea I: latitude

Before the nineteenth century, navigation at sea was overwhelmingly experiential; mariners navigated through their learned awareness of the nature of the sea and the winds, with the added use among mariners in the shallow waters off northern Europe of the lead-line to examine the character of the sea floor. Oceanic mariners used the heavens as well. They associated key coastal features with the altitude – the angle of elevation above the horizon – of Polaris (the North Star). Importantly, they made no explicit connection between stellar altitudes, a celestial measurement, and the concept of terrestrial latitude. This is evident in the *kamal* used by Arab sailors in the Indian Ocean. The *kamal* is a small, thin, rectangular panel of wood with a long string running through its centre; the string bears small knots at equal intervals. Holding one end of the string in their teeth, and the other outstretched, the mariner used their free hand to move the panel until the lower side touches the horizon and the upper side the target star; the number of knots in the string defines the altitude. The small size of the panel meant that the instrument could be used only for the relatively low altitude of Polaris within the tropics.

The Portuguese government began in the 1420s to send out expeditions to determine a sea route around Africa to Asia. Such a concerted effort required the standardization of mariners' experiences. Greatly simplified versions of astronomical instruments were introduced to measure rather than estimate stellar altitude. These instruments were suspended in use so that gravity defined the horizontal (0° altitude). The marine versions were heavier, to hang more steadily in the wind on a heaving ship, and they reduced the many complex astronomical markings to just a simple scale of degrees around their limb (the outer portion of the circle). The 'mariner's astrolabe' (Figure 1.1, top left) comprised a graduated circle with a rotating alidade, or sighting vane. Holding it by the ring, so that the 90° gradation coincided with the zenith (the direction of the vertical), the navigator moved the alidade until they could sight onto the star, and then measured the altitude directly off the circle; when sighting the Sun, the observer turned the alidade until the shadow or image of the Sun cast by the upper sight coincided with the lower (Figure 1.2). The 'mariner's quadrant' (Figure 1.1, centre right) was a quarter circle, held vertically with the centre of arc at the zenith; the mariner turned the instrument until they could sight the star through the two fixed sights on one edge, reading off the altitude at the intersection of the graduated arc with a plumb line hanging from the centre of arc. Both instruments generally measured to 1° of arc, so they made appreciable the difference, up to 3°, between Polaris and the actual celestial North Pole. The common practice was therefore to observe Polaris when it was at the same altitude as the North Pole, as indicated by the horizontal alignment of the two guard stars in *Ursa minor*.

Stellar altitudes were actively coupled to terrestrial latitude only in about 1480. As the Portuguese approached the equator and Polaris became hard to sight as it dropped low to the horizon, they drew on cosmography to define a new technique to determine latitude by reference to the Sun. The 'rule of the Sun' was recorded within the *regimento do astrolobio e do quadrante*, which circulated in manuscript and, from 1509, in print. The mariner observed the Sun's altitude at its noon zenith and read the Sun's daily declination (solar latitude with respect to the equator) from a table. From these, the latitude is readily calculated: if the Sun is to the south of the mariner, latitude is 90° less solar altitude plus declination; if the Sun is to the north of the mariner, then the latitude is solar altitude less 90° plus declination.

The stellar and solar techniques for latitude determination continued in use into the twentieth century, with a long series of instrumental and observational refinements.

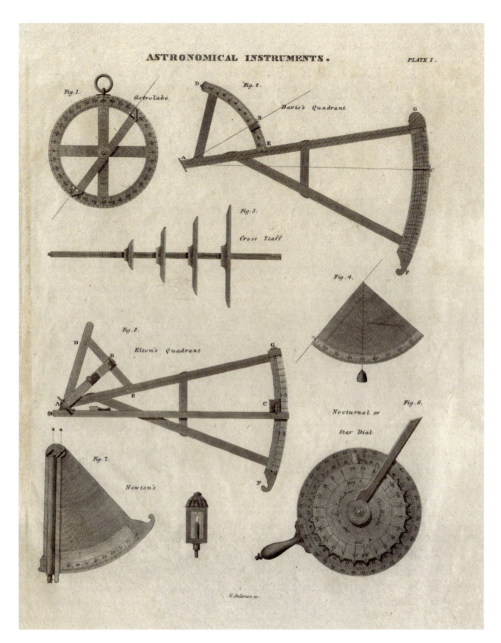

Figure 1.1 Images of early navigational instruments, from Vol.1 of Abraham Rees, *Cyclopaedia, or Universal Dictionary of Arts, Sciences, and Literature* (Philadelphia, 1805). Top: mariner's astrolabe and back staff (Davis quadrant). Centre: cross staff with four cross pieces; Elton's quadrant; and mariner's quadrant. Bottom: Isaac Newton's design for a double-reflecting quadrant and a nocturnal. Courtesy of Adler Planetarium, Chicago (P-27k)

Figure 1.2 The geometry of the mariner's astrolabe in determining the Sun's altitude. Vignette from Willem Jansz. Blaeu, *Het Licht der Zee-vaert* (Amsterdam, 1608), 22; hand-coloured woodcut. Courtesy of the Osher Map Library and Smith Center for Cartographic Education, University of Southern Maine (Osher Collection); Available at: https://oshermaps.org/map/7356.0021

Instrumentally, mariners adopted a series of new angular measuring instruments: the cross staff (or Jacob's staff) in the sixteenth century, the back staff (or Davis quadrant) in the seventeenth, and the octant in the eighteenth.

The wooden cross staff (Figure 1.3) works on the same principle as the *kamal*. The mariner moved the cross piece along the staff until one end touched the horizon (defining 0° altitude) and the other the sun or star; the angle was then read off from the staff's gradation of angular degrees. The gradation of the staff was determined by basic trigonometry:

> the angle subtended at the eye = 2 × cotangent (half the length of the cross piece / distance of the cross piece from the eye).

Up to four cross pieces could be used, of differing lengths, with the longer pieces used in ever higher latitudes, each with its own gradation of angles (Figure 1.1, centre left).

Staring at the Sun is not, of course, good for the eyes. An Englishman, John Davis, published the design of a new instrument in 1595: the back staff (Figure 1.1, top right). Because it measured altitudes up to 90° of arc, it was often called a quadrant. The mariner stands with their back to the Sun (Figure 1.4), holding the handle at E; they move the 'sight vane' (F) on

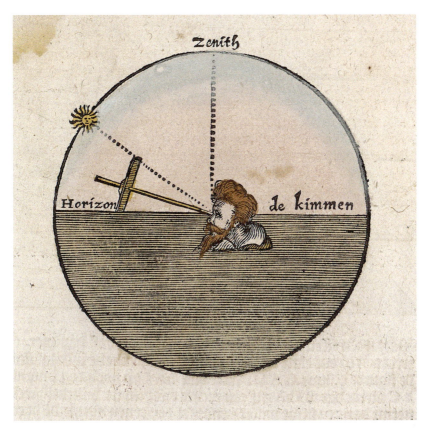

Figure 1.3 Use of the cross staff to determine the sun's altitude. Vignette from Willem Jansz. Blaeu, Het Licht der Zee-vaert (Amsterdam, 1608), 23; hand-coloured woodcut. Courtesy of the Osher Map Library and Smith Center for Cartographic Education, University of Southern Maine (Osher Collection); Available at: https://oshermaps.org/map/7356.0021

the 'sight arc' (DE), graduated to 30°, and the 'shadow vane' (G) on the 'shadow arc' (BC), graduated to 60°, until they see the horizon through the horizontal slit in the fixed horizon vane (A) and simultaneously the shadow cast by the shadow vane (G) falls on the horizon vane. While the shadow arc was commonly graduated to 1°, the larger sight arc was more finely graduated and perhaps also bore a transversal scale for greater precision (Figure. 1.5). Added together, the values read off each arc give the Sun's altitude. Made from hardwoods, back staffs were cheap and used through the 1780s. However, its size meant the back staff was cumbersome, especially in the wind, and its use of the Sun's shadow meant that it was useless for observing stars.

Designs began to be advanced in the later seventeenth century for more compact instruments that could observe stellar as well as solar altitudes (Figure 1.1, bottom left). The use of mirrors to reflect the target compressed the 90° arc from the horizon to the zenith to an arc of just 45° (one eighth of the circle). Properly called octants, they are also known as '(double-reflecting) quadrants'. John Hadley's 1731 design proved effective and was rapidly adopted (Figure 1.6). His octant featured two sets of sights and mirrors to be used when the target (Sun or star) is before or behind the observer (Figure 1.7). In both cases, the mariner lifts the sight vane to their eye, so that they can look through the half-silvered horizon glass, bringing

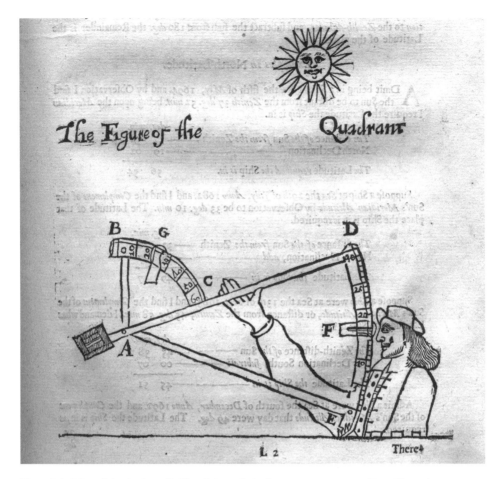

Figure 1.4 Use of the back staff. The sight and shadow arcs were generally calibrated to measure slightly more than the ideal 30° and 60° indicated here. From Samuel Sturmy, *The Mariners Magazine* (2nd ed.) (London, 1679), 75. Courtesy of the Charles Babbage Institute, University of Minnesota

the horizontal divide in the mirroring just to the horizon; they then move the index, which rotates the fully mirrored index glass at the instrument's centre of arc until the image of the Sun or star is visible in the mirrored upper half of the horizon glass; when the lower limb of the Sun or the star appears to touch the horizon, the index is clamped and the mariner reads the altitude off the scale. The instrument is gently rocked to 'skim' the Sun or star over the horizon to ensure that the observation is made when the instrument is indeed vertical. A retractable shaded glass can be inserted between the horizon and index glasses when viewing the Sun. The geometry of back observation compressed the altitudes still further, allowing the 45° arc to measure 180°, albeit with necessarily reduced precision. Octants were originally equipped with a transversal scale, which permitted reading the instrument to one minute of arc (fore observation). Later improvements included the addition of a vernier scale, to permit reading to seconds of arc, screw threads for fine motion of the index, and artificial horizons (enclosed glass troughs filled with mercury or spirit levels) to be used in place of the horizon glass should the horizon be obscured. Telescopes were rarely added.

Figure 1.5 Detail of the transversal scale on the sight arc of a back staff perhaps made by Benjamin King in Salem, Massachusetts, c. 1775. The main scale along the arc's limb is graduated to 5′ of arc. The transversals – the diagonal lines – permitted readings to ½′. In this image, the index reads between the second and third gradations between 17° and 18°, so slightly more than 17°10′. The increment is read by following the index to its intersection with a diagonal line and one of the 11 arcs concentric to the main scale; the index and transversal lines intersect at the seventh concentric arc, so the increment is 7 × ½′ for a total value of 17°13½′. Courtesy of the Collection of Scientific Instruments, Harvard University (Inventory 5259)

The increasing precision of these instruments required a number of observational refinements. Beginning in the sixteenth century, observations of Polaris were refined by the use of a nocturnal, an instrument that measured the angular difference between Polaris and the true celestial pole (Figure 1.1, bottom right). The provision after 1690 of tables of stellar declinations in official ephemerides (see below) permitted observers on land to observe stars for latitude in the same manner as mariners used the Sun; they also permitted the observation of the Sun or stars to determine local time. As watches became common in the later 1700s, mariners could observe the Sun's altitude before and after noon and the time in between, rather than trying to catch it *exactly* at its zenith, although the necessary calculations were complex. There was also an extension of this 'double altitude and elapsed time' technique to observing the altitudes of known stars on either side of the meridian. Despite the precision of instruments, by 1800, skilled navigators were able to determine their latitude to 1′.

Location on land: longitude

The equivalency of differences in longitude and local time was known to the ancients, who timed lunar eclipses to determine longitudinal differences; the same technique was advocated in sixteenth-century Spain, but lunar eclipses that could be seen simultaneously in Spain and the New World were very rare. When, in 1610, Galileo Galilei first turned a telescope onto the planets and found that Jupiter had moons (satellites), he quickly realized that their eclipses by that planet's body could be used as a celestial clock. Giovanni Domenico Cassini's 1668 precise table (ephemeris) for the motions of Jupiter's brightest moon, Io, led

Figure 1.6 An English octant by John Bleuler, 1775–1790. The fore sight vane is on the right-hand side of the instrument; on the left-hand side, from bottom to top, are the back sight vane and back horizon glass, the fore horizon glass, the shade, and the index glass at the centre of arc. The scale is graduated to 20′ of arc, the vernier reading to 1′. Mahogany frame, brass index arm and fittings, and ivory scale and vernier. 44 × 36.9 × 7.3 cm; radius 39.7 cm. Courtesy of the Collection of Scientific Instruments, Harvard University (Inventory 5304)

to his recruitment to establish the Paris Observatory with the specific goal of implementing this technique to permit the field determination of longitude. Cassini's refined tables of the 'eclipses of Jupiter's satellites' began to be published in 1690 in the Paris Observatory's annual ephemeris, the *Connoissance des temps*.

The field observer required three main pieces of equipment. First, an astronomical quadrant to observe the Sun or stars and determine local time; second, a good pendulum clock to hold that time; and, third, a telescope to observe Io in its passage about Jupiter. Given the rather primitive optics of the period, the telescope would have to be in the order of 4.5–5.5 metres in length. The observer recorded their local time for the moment Io passed (immersed) behind Jupiter; the tables gave the local time in Paris predicted for the same

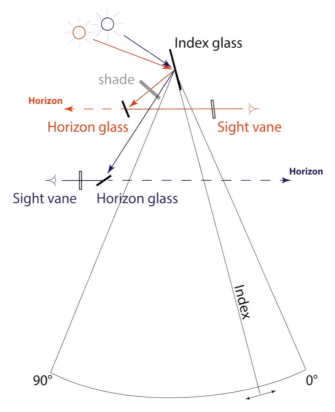

Figure 1.7 Diagram of the working of an octant. Red: fore observation, with the sun or star before the observer. Blue: back observation, with the sun or star behind the observer. The horizon glasses are fixed; the observer moves the index arm to make the index glass rotate and so bring the image of the sun or star down to the horizon glass. The movable shade is used only in observing the sun. See also Mörzer Bruyns (2009: 26–27)

eclipse; comparison of the two local times gave the difference in longitude from the Paris Observatory to the observer in the field. Observation required a great deal of time: the pendulum clock had to be properly stabilized and calibrated, and then many observations would have to be taken of scarce eclipses over several months to give a mean result that would, it was hoped, correct for the indeterminate errors known to exist within the tables.

Even if eclipses of Jupiter's satellites occurred with sufficient frequency to be an aid to navigation, the necessary instruments were quite unsuitable for shipboard use: the ship's motion would upset the clock's pendulum and would make it impossible to hold such a small target as Io in the telescope's sights. Nonetheless, the technique was used by observers on land to determine increasing numbers of precisely and reliably determined longitudes through the eighteenth century: from 90 positions in 1699, 220 in 1753, 390 in 1777, to 870 in 1785 (Chapuis, 1999: 26–27).

Location at sea II: longitude

The early modern determination of longitude at sea is fraught with myth and misunderstanding. From the practising mariner's point of view, knowledge of longitude was *not* needed to avoid disasters like the famous loss of Sir Cloudesley Shovell's flotilla in 1707. Moreover, it was no help to be able to fix position at sea if coastlines were not accurately situated within the same reference system. The development of systems to determine longitude at sea was pushed by landbound natural philosophers, by those involved in imperial marine expansion who needed greater certainty in relocating newfound lands, and by mariners following long-distance routes that required significant course corrections out of sight of land. Two techniques were eventually developed, one astronomical, the other horological.

Lunar distances

The astronomical solution – the method of 'lunar distances' – uses the Moon as a celestial clock: the mariner measures the distance (angular separation) between the Moon and a star at a local time, which is compared, via tables, to the local time at Greenwich for the same lunar distance. To perfect the technology so that it could be used in navigation required a huge investment of resources by the British state. The famous Longitude Prize, offered in 1714, represented only a small fraction of the overall expenditure. The Greenwich Observatory was founded in 1675 specifically to define the locations of the fixed stars with great precision, to provide a reference by which to track the Moon's motion; the result was John Flamsteed's momentous star catalogue, posthumously published in 1725. The parameters for the Moon's motion had to be precisely determined, which was accomplished in 1755 by Tobias Mayer, professor of astronomy at the University of Göttingen, which George II had founded in 1737; the Greenwich Observatory then took on the task of continually updating those parameters to take into account the Moon's constantly shifting orbit under the combined gravitational influence of the Earth, Sun, and other planets. The French astronomer Nicolas-Louis de La Caille refined the computational procedures in 1761, permitting the Greenwich Observatory to begin publishing all the necessary tables and methodological guides in 1767, as the *Nautical Almanac*. In addition, the entire technological assembly had to be successfully tested at sea, which was accomplished by Nevil Maskelyne (1763–1764) and then by James Cook on his first Pacific voyage (1768–1771).

The final component of the technological assemblage was an adequate instrument. Mayer had proposed the use of a repeating circle, in which a number of measurements are taken, each on a different portion of a fully graduated circle. Jean-Charles Borda implemented the design c. 1775; it proved versatile for many uses but added to the complexity of timing the observation of lunar distances. The English initially preferred the octant. Designed to be held vertically to measure solar or stellar altitudes, the octant could also be held awkwardly at an angle: the mariner sighted the limb of the Moon through the horizon glass and adjusted the index so that the reflection of the selected star touched the Moon. But it would at times be necessary to measure a lunar distance of more than 90°, and back observation was too imprecise.

The solution, introduced by John Bird in the 1740s, was to extend the instrument's arc to 60°, so as to measure up to 120° with fore observation; the result, the sextant, was more compact and easier to handle than the octant (Figure 1.8). As the instrument was widely adopted later in the eighteenth century, the sextant differed further from the octant. The sextant was generally cast in one piece from brass, to be sturdy and to not deform with heat and humidity; it had no sight vane and horizon glass for back observation; its limb was generally engraved by high-precision dividing engines (introduced by Jesse Ramsden in 1766) with a vernier; and low-power, achromatic telescopes were standard, as were artificial horizons. Octants remained in widespread use for basic altitude observations, for latitude, but the more refined sextant was used for lunar distances.

The method of measuring lunar distances was complex. Two observers had to measure, pretty much at the same time, the altitudes of the Moon and the chosen star, which also

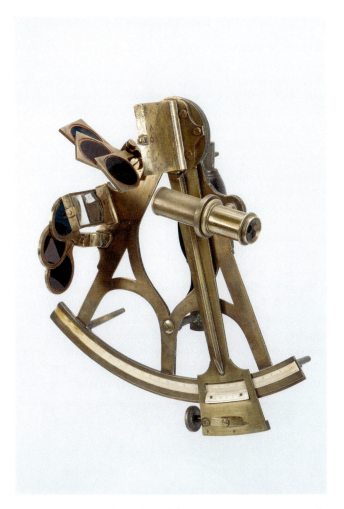

Figure 1.8 A sextant by Riggs & Brother, Philadelphia, 1820–1840. Cast in a single piece from brass, with rare ivory scale reading to 20′ and vernier reading to 30″. 12 × 23.4 × 24.3 cm. Courtesy of the Collection of Scientific Instruments, Harvard University (Inventory 5245)

gave local time, as well as the distance between them; corrections then had to be made for refraction and parallax and the observations reduced to the centre of the Moon; the necessary computations were laborious, initially taking a skilled mathematician up to four hours to complete. Practical implementation of lunar distances required: new computational systems so that the Greenwich Observatory could generate the necessary predictive tables of the Moon's motion; pre-printed forms and refined table designs to reduce computation to about 30 minutes; and the active education of mariners in the technology. Although not in widespread use until after 1800, lunar distances continued to be taken well into the twentieth century. It was also adopted for aerial navigation on slower, propeller-driven planes.

Other nations were not restricted to the use of the *Nautical Almanac* and therefore to determining longitude with respect to Greenwich. Maritime nations such as France, Spain, and the Netherlands published ephemerides with the British tables adjusted to their own, newly created, national meridians (Paris, Cadiz, Tenerife, respectively).

Chronometers

The alternate solution to determining longitude at sea was to use a chronometer to hold the local time of departure ports, to which the mariner could compare local time as determined from solar or stellar observations. A small correction for the port's longitude would give the mariner's longitudinal difference from a standard meridian, permitting the chronometers' readings to be related to charts and for mutual checks with lunar distances.

Pendulum clocks – the only devices in the early eighteenth century that could reliably hold time for long periods – were unusable aboard ship. The clocksmith John Harrison, having developed an accurate, spring-driven casement, thought to use this mechanism as the basis for a seagoing clock. Funded in part by the Board of Longitude, he completed three complex chronometers (H.1, H.2, and H.3) between 1735 and 1759; he also made a large watch (H.4) to be used in 'carrying' the local time from the deck, where the astronomical observations were made, to the cabin, where the large H.3 was housed. In a 1760–1761 voyage to Jamaica, Harrison's son found that H.3 did not hold time well enough, but the smaller H.4 did. Eventually, Harrison shared enough information about his construction practices to permit other clock makers to reproduce his work and he was finally awarded the Longitude Prize (split with Mayer for his lunar tables). Yet Harrison's casement was not amenable to mass production. By 1790, John Arnold's simpler design enabled him to ramp up production from two or three instruments annually to hundreds, to meet the East India Company's demand for chronometers to manage deep-water course corrections (Figure 1.9).

The adoption of the chronometer as a standard element of marine navigation was further dependent on a larger astronomical-technological assemblage. Specifically, chronometers had to be regularly tested to determine the specific rates at which they lost time – to adjust

Figure 1.9 Chronometer no. 2620 by John Bliss, New York, c. 1880. This chronometer would run for 56 hours without winding. The main, outer dial shows hours (I…XII) with divisions to each minute. The lower dial shows seconds (0…60); the small upper dial indicates how many hours, in eight hour increments, the chronometer will run. The instrument is mounted on gimbals in a mahogany box, with an inner glass lid to permit the instrument to be read while still protecting it. Brass, 15 cm diameter, 7 cm depth, in box 18.5 cm square. Courtesy of the Collection of Scientific Instruments, Harvard University (Inventory DW0246)

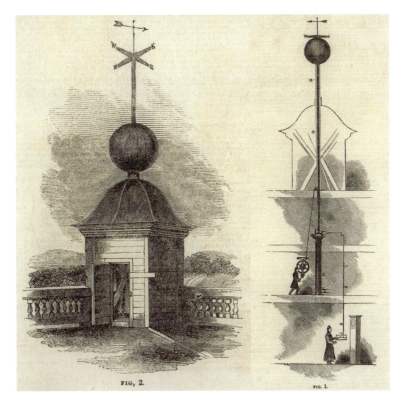

Figure 1.10   Diagrams from "The Time Ball, Royal Observatory, Greenwich" *Illustrated London News* 5 No.132 (9th November 1844): 304. The ball would be raised at 12:55pm, reaching the top at 12:58pm; it would be released at exactly 1:00pm

their use at sea – and to set them to the correct time. Initially, local time was defined by mariners from their own astronomical observations. However, beginning in 1829 at Portsmouth, England, naval depots and ports maintained prominent 'time balls' that were dropped at precise times (generally noon, but 1pm in British territories) as determined by careful onshore astronomical observations; ships in harbour could check and, if necessary, reset their chronometers (Figure 1.10). Time balls thus required the maintenance of an observatory with a regular observation schedule to continually determine time and also to refine the port's longitude; the observatories could also house chronometers onshore while they were regulated. The global proliferation of time balls after 1850 is a mark of the increasingly widespread use of chronometers in marine navigation. (Eventually, time balls and other signals were more widely adopted to allow people to set their own clocks and watches, until displaced by telegraphic and later radio time signals.)

Location on sea *and* land

The ability to determine longitude at sea stimulated new practices and technological systems in the nineteenth century. Practically, Europe's imperial agents adapted the new marine technologies to land exploration. Some used chronometers, once they were sturdy enough to withstand the rigours of land travel. The majority used sextants equipped with artificial horizons to take lunar distances and solar and stellar altitudes. What had once been distinct arenas of mapping and voyaging – the lands versus the seas – were blended together into the singular practice of traversing the same abstracted space (Kennedy, 2013: 6–20). The sextant acquired a profusion of alternative forms to meet the needs of a wide range of users and environments.

Technologically, the new methods were part and parcel of substantial reforms in marine and terrestrial mapping. The use of eclipses of Jupiter's satellites had by 1800 given greater precision to the world's coastlines in smaller scale marine mapping. Murdoch Mackenzie had in the 1740s pioneered the use of onshore triangulations to map local coastlines and to define the locations of landmarks from which the inshore hydrographer could, by resection, precisely locate soundings and other observations. The rapid expansion after 1790 of systematic, triangulation-based territorial surveys permitted the detailed mapping of entire coastlines and inshore hydrography, an effort that required substantial state investment.

These innovations encouraged the complete reformulation of marine mapping. The plane charts that mariners had used since the fifteenth century, criss-crossed by rhumb lines, were replaced by largely empty charts constructed on Mercator's projection with accurately situated coastal outlines. Gerhard Mercator had famously created that projection, which he had used for his 1569 world map, to aid navigators, but its actual use at sea was problematic: to follow a straight rhumb requires constant course corrections based on precise determination of location, which was of course impossible before lunar distances and chronometers. In the 1840s it became common practice for marine navigators to keep close track of their routes and to plot them on the new charts (Figure 1.11). They could do so, without wrecking, because both ships and coastlines were now situated within the same spatial framework.

Well, almost. The absolute determination of longitude required the measurement of the difference between local time and one of several standard times maintained by different nations. The complexity of the situation was evident in the USA, where the federal government established a twofold system in 1850: federally issued astronomical tables and marine charts would continue to use the Greenwich meridian for standard time and therefore zero longitude, but federal terrestrial mapping would henceforth use the meridian of the Naval Observatory in Washington, DC. In 1879, two thirds of the ships carrying on the world's trade used charts and ephemerides based on Greenwich, but this meant that the other third determined time and longitude from another ten meridians (Paris, Cadiz, Naples, Oslo, Ferro, Pulkova, Stockholm, Lisbon, Copenhagen, and Rio de Janeiro) (Withers, 2017: 165). The potential for navigational confusion was great. Moreover, the global spread of the telegraph – the first transatlantic telegraph cable was laid in 1854–1858 – required a universal time to properly integrate signals. These issues, together with the growing adherence to international standards of metrology in science and trade, led to international congresses of geographers in Venice (1881) and geodesists in Rome (1883), where the issues were debated and national rivalries exposed, but both conferences concluded that the existing technological infrastructure meant that it would be simplest, and cheapest, to adopt Greenwich time as the global standard. Eventually, representatives of the world's leading powers met in Washington in 1884, empowered to make a final decision: Greenwich was selected for universal time. But the Greenwich meridian was still not accepted for zero longitude for official mapping by the French (until 1911) and the USA (until 1912), requiring the maintenance of separate ephemerides.

Conclusion

Only after 1900 did *locate* acquire the sense of specifying position within an absolute coordinate system without reference to other spatial features and *location* acquire the further,

Latitude, longitude, and geospatial technologies to 1884

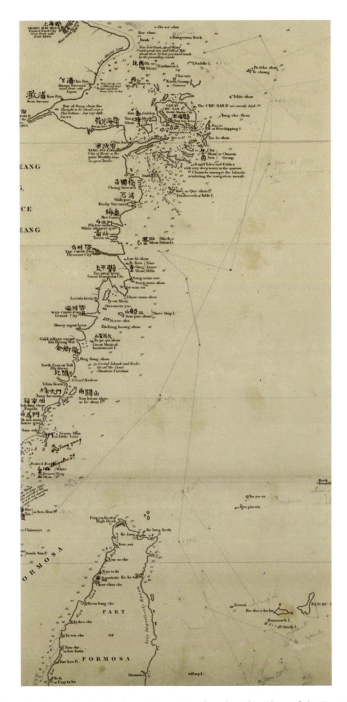

Figure 1.11 Detail of record of a track on James Horsburgh, *The Chart of the East Coast of China* (London: J. & C. Walker for the East India Company, 1835). The daily positions and the tracks in between show the passage of an unknown vessel from east of Taiwan in November 1852, north towards the Yangtze estuary and then south again in late January 1853; the navigator also recorded further information about hazards. Courtesy of the Geography and Map Division, Library of Congress (G7821.C6 1835.J2)

technical meaning of a set of geographical coordinates. Yes, longitude continues to be defined with respect to a conventional reference, but the conventional nature of the Greenwich meridian was hidden under a host of standardized practices, including the radio transmission of time signals after 1905. Even so, those practices remained restricted to specialists; the vast majority of the populations of the industrialized world continued to situate themselves within the world through maps and charts but also still by the time-honoured practice of referring to landmarks.

References

Andrewes, W.J.H. (Ed.) (1996) *The Quest for Longitude: The Proceedings of the Longitude Symposium, Harvard University, Cambridge, Massachusetts, November 4–6, 1993* Cambridge, MA: Harvard University Press.

Bennett, J.A. (1987) *The Divided Circle: A History of Instruments for Astronomy, Navigation, and Surveying* Oxford: Phaidon.

Chapuis, O. (1999) *A la mer comme au ciel: Beautemps-Beaupré et la naissance de l'hydrographie moderne, 1700–1850. L'émergence de la précision en navigation et dans la cartographie marine* Paris: Presses de l'Université de Paris–Sorbonne.

Dunn, R. and Higgitt, R. (Eds) (2016) *Navigational Enterprises in Europe and its Empires, 1730–1850* Basingstoke: Palgrave Macmillan.

Edney, M.H. and Pedley, M.S. (Eds) (2019) *Cartography in the European Enlightenment* (*The History of Cartography: Volume 4*) Chicago: University of Chicago Press.

Kain, R.J.P. (Ed.) (forthcoming) *Cartography in the Nineteenth Century* (*The History of Cartography: Volume 5*) Chicago: University of Chicago Press.

Kennedy, D. (2013) *The Last Blank Spaces: Exploring Africa and Australia* Cambridge, MA: Harvard University Press.

Mörzer Bruyns, W.F.J. (2009) *Sextants at Greenwich: A Catalogue of the Mariner's Quadrants, Mariner's Astrolabes, Cross-staffs, Backstaffs, Octants, Sextants, Quintants, Reflecting Circles, and Artificial Horizons in the National Maritime Museum, Greenwich* Oxford: Oxford University Press.

Schotte, M.E. (2019) *Sailing School: Navigating Science and Skill, 1550–1800* Baltimore, MD: Johns Hopkins University Press.

Taylor, E.G.R. (1971) *The Haven-Finding Art: A History of Navigation from Odysseus to Captain Cook* (2nd ed.) London: Hollis and Carter.

Withers, C.W.J. (2017) *Zero Degrees: Geographies of the Prime Meridian* Cambridge, MA: Harvard University Press.

2
THE PHOTO-MECHANICAL ERA OF CARTOGRAPHY
A recollection

William Cartwright

My career in cartography spans from the era of drawing with pen and ink (using lettering guides on wax-impregnated linen or stable based plastic) to the Web (using computers and contemporary communication systems to produce, publish, and deliver mapping products). To establish a reasonable point to start and finish, I chose to base this on a cartography/mapping sciences journal published in Australia, my home country, and to look at the period between when I began working as a cartographer to the period when the manual production of cartography was overtaken by computers, which was originally called computer-assisted cartography. The journal I used for this guidance is *Cartography* (ISSN 0069-0805), the journal of the Australian Institute of Cartographers (which later became the Mapping Sciences Institute, Australia). This journal was first published in 1954 and ran until 2003, after which *Cartography* and *The Australian Surveyor* merged to become the *Journal of Spatial Sciences* (ISSN 14498596) (2004–current; back issues of *Cartography* are available at http://www.tandfonline.com/loi/tjss19#.VEh6KFcixws).

I joined the workforce in 1967, which is a convenient general starting point to begin this treatise. But, where to end? Issues of *Cartography* included papers on manual map production up until 1978. So, the decision was made to address the period 1968–1978. I will call this 'a nostalgic journey through a decade of pre-computer map production – with an Antipodean 'flavour'. What is provided below is a list of topics of papers published in *Cartography* that focused on manual map production. For completeness, the list includes papers beginning in the first publication year, 1954, to give some 'foundation' information on Australian map production in the previous decade. But, again, this chapter focusses on the decade beginning in 1968.

Whilst not coinciding with my 'begin' date in the industry, it is worth noting those map production-biased papers that were published in the journal prior to 1968, as this provides some foundation for the period when cartography was being formalized in Australia. The Australian Institute of Cartographers was founded in 1923 to 'improve the knowledge and standards of cartography' (Encyclopaedia of Australian Science, 2006), but it was not until 1952 that this organization became a national organization, whereby all state organizations combined to form this national institute. Two years later, in 1954, its first journal was published.

DOI: 10.4324/9780367855765-4

Looking through the issues of the journal, it became evident that papers on the application of computers to map production began to appear around the beginning of the 1970s. So I chose to examine the period between 1954 (the first issue of the journal) until 1978, when papers about map production devoid of computers stop. It is worth noting that as early as Volume 3 (1959–1960), papers were being published about the use of computers in mapping, although these covered the computational aspects of procedures like photogrammetric Block Adjustment. The more general 'where-to?' and papers focusing on the automation of map production started to emerge from around 1969 (Volume 7: 1969–1972, Volume 8: 1973–1974, Volume 9: 1975–1976, and Volume 10: 1977–1978), 'overlapping' with the papers concerned with manual methods of map production.

Cartography (Journal of the Australian Institute of Cartographers) map-production focused papers, 1954–1978

Volume 2: 1957–1958
 Cartography and the development of photolithography
 Xerography and electrostatic printing
 Mechanical composition (pre-printed lettering)

Volume 3: 1959–1960
 Strip mask process
 Xerography and electrophotography – 1959
 Present techniques and modifications for electronic computing (Block Adjustment)

Volume 4: 1961–1962
 Polyester drawing media (Permatrace, Cronaflex, and so on)
 The preparation of a computer programme for internal block adjustments
 Practical production of plastic maps, reliefs, and models in Czechoslovakia
 The drawing and production of maps
 Cartographic scribing
 An outline of map reproduction
 Dyeline or Diazo processes

Volume 5: 1963–1964
 An outline of map reproduction
 Xerography and electrophotography
 Conventional photography
 The preparation of publications for lithographic printing

Volume 6: 1966–1968
 Sizing scales for process cameras
 Copying machines and projectors combined to produce a photographic technique
 Automation and map production in Australia

Volume 7: 1969–1972
 Problems in automated cartography
 Automated cartography in the division of national mapping
 Trends in colour-cartography

Volume 8: 1973–1974
Problems in the construction of small-scale relief models as teaching aids
An automated system for thematic mapping

Volume 9: 1975–1976
Computer-aided map compilation
Improved techniques for the construction of relief models
Future mapmaking methods

Volume 10: 1977–1978
Evaluation of an experimental bathymetric map produced from Landsat data
The teaching of reprographics at undergraduate level in cartography at RMIT (Royal Melbourne Institute of Technology)
An automated standard mapping system
Half-tone photography for thematic maps
Cartographic data banks

The content of the following sections in this chapter were 'guided' by the papers about manual map production that were issued in *Cartography* between 1968 and 1978, being, generally, the years that I worked with those methods of production. These are listed below and are rearranged to accord to the 'flow' of manual cartographic production:

- Preparation of publications for printing and printing maps;
- Manual plotting and scaling and process camera scaling;
- Drawing the map and drawing media;
- Mechanical composition: strip masks, pre-printed lettering, and scribing; and
- Map reproduction and reprographics – processes leading to printing maps.

Plotting and scaling manual and using the process camera

The starting point for manual map production was the compilation sheet. A graticule (lines of latitude and longitude) would be drawn in pencil and then perhaps in coloured ink, and the compiled, measured, or interpreted information (from aerial photographs) would then be plotted. In order to plot the information, a sharp pencil (perhaps a 5H or 6H) (Figure 2.1) was sharpened to a pinpoint with a sandpaper block (Figure 2.2), and an appropriate scale was used (Figure 2.3), sometimes with mechanical instrumentation such as the device illustrated in Figure 2.4, for subdividing distances into equal parts.

If a pre-printed map was to be used later to ink-in updates or corrections (as the printed paper map was also the database for graphically illustrating these changes), a 360° pre-printed protractor was sometimes included in the print. A rolling parallel ruler (Figure 2.5) would be employed to lay out the appropriate bearing to plot, and the parallel ruler was then rolled over the map document to the appropriate location for drawing that bearing onto the map. An example of this type of map is given in Figure 2.6, showing part of the City of Hobart, Australia, drawn in 1940.

During the map compilation process there frequently existed the need to capture information from existing maps, which had been produced at various scales. To convert the original map scales to that needed for the compilation scale (usually converting from a larger original map scale to a smaller compilation map scale, for accuracy) a process camera was employed. These cameras were large and required two rooms of sufficient size to accommodate them – one

Figure 2.1 Staedtler Mars Lumograph drafting pencils
Source: CIA. https://www.flickr.com/photos/ciagov/30579640680/in/album-72157674852500522/.

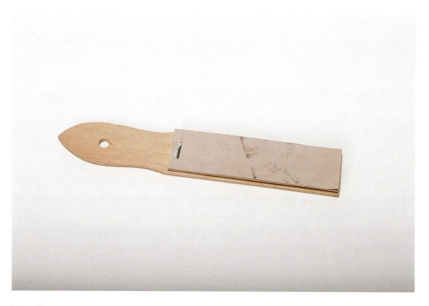

Figure 2.2 Sharpening paper pad on paddle stick
Source: CIA. https://www.flickr.com/photos/ciagov/30844683936/in/album-72157674852500522/.

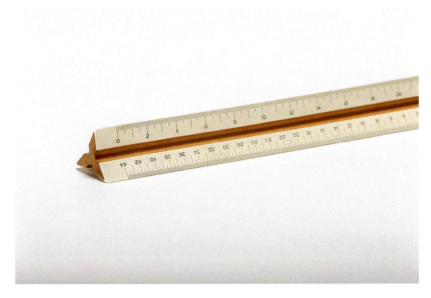

Figure 2.3 Triangular engineering scale
Source: CIA. https://www.flickr.com/photos/ciagov/30245415043/in/album-72157674852500522/.

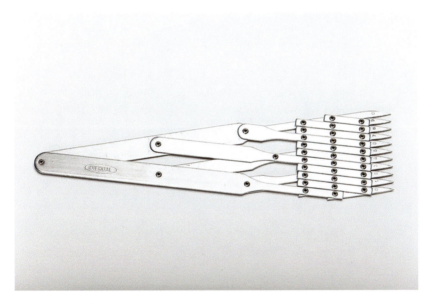

Figure 2.4 Ten-point divider
Source: CIA. https://www.flickr.com/photos/ciagov/30245415273/in/album-72157674852500522/.

Figure 2.5 Rolling parallel rule
Source: CIA. https://www.flickr.com/photos/ciagov/30249068114/in/album-72157674852500522/.

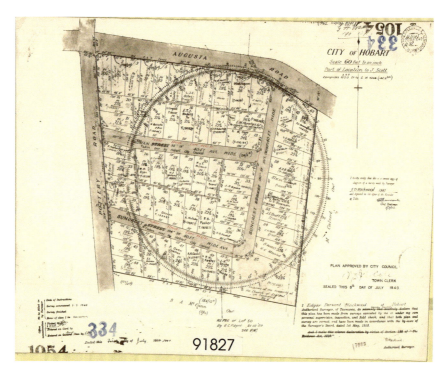

Figure 2.6 (1940). Map –334 P series – City of Hobart, Augusta Rd, Pottery Rd, Truman St, Suncrest Ave, various landholders, surveyor ED Blackwood, draughtsman Lance Duncombe. Libraries Tasmania. 1940–1952

Source: Libraries Tasmania Online collection. https://stors.tas.gov.au/AF718-1-39.

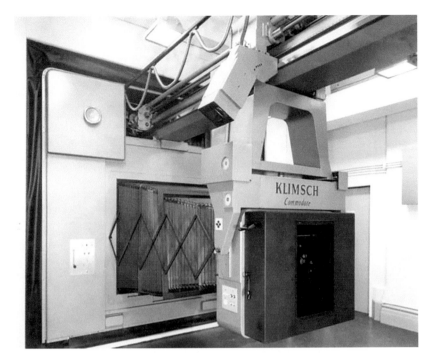

Figure 2.7 Klimsch Camera lenses and bellows, with overhead steel rails. Reproduced with permission. Artefact Gallery, Museum of Lands, Mapping and Surveying, Department of Resources, Queensland, Australia

Source: https://www.qld.gov.au/recreation/arts/heritage/museum-of-lands/artefact-gallery/photographic#gallery-0-11.

room for the so-called 'copyboard', where the original was mounted (sometimes in a vacuum frame), illuminated, and photographed. The other room was a darkroom that housed another vacuum frame that would take the stable-based photographic film. Through the use of multiple lenses and moveable copyboard frames, the image would be scaled, a photograph exposed, and the photographic material processed. This form of 'photo-mechanical' processing contributed greatly to the map compilation process, and later in the map artwork associated with the output of colour separation film. Figures 2.7 and 2.8 show one such camera – a Klimsch used by the Department of Lands, Mapping and Surveying, in Queensland, Australia.

Drawing the map

The ultimate role of manual map production was to produce artwork – drafting film, mylar film, scribe coat, peel coat, and lettering sheets that would be used for making printing plates. Once produced, the artwork would be processed using various photo-mechanical procedures – to combine lettering, line, and areal components into one composite photographic image (and one composite photographic image per printing press impression colour). Drawing and photographic emulsions were produced on stable-base drawing and photographic media. The manuscript map image was drawn or scribed, areas requiring depiction were peeled, and lettering drawn using lettering guides or stuck-up using stable-base polyester film. Photo-mechanical processes allowed for original artwork to be created on another

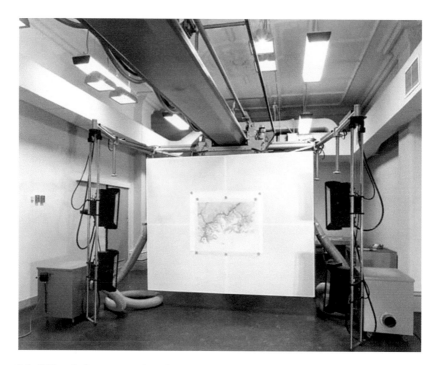

Figure 2.8 Klimsch Camera copyboard. Reproduced with permission. Artefact Gallery, Museum of Lands, Mapping and Surveying, Department of Resources, Queensland, Australia

Source: https://www.qld.gov.au/recreation/arts/heritage/museum-of-lands/artefact-gallery/photographic#gallery-0-11.

medium and then the images later transferred to printing plates. Later, colour separation using filters saw original colour 'separated' onto different film negatives and then transferred to individual printing plates for later re-composition as full-colour reproductions of the original.

In the photo-mechanical era of cartographic production map artwork was drawn or scribed onto stable colour separation bases by using either the 'positive' or 'negative' production process. The 'positive' process was conducted by drawing lines, lettering, tones, and so on using ink, stick-up letters, and the like. The 'negative' process removes a pre-prepared emulsion using scribing tools, and then mates these pieces of artwork with negatives for lettering and tones that have been produced photographically. Work was undertaken in the cartographic drawing room, and this usually consisted of formally arranged desks with drawing boards atop. A typical drawing room is shown in Figure 2.9.

Positive map artwork

The 'positive' process draws lines, lettering, tones, and such using ink, stencils, stick-up letters, and so on. Maps were drawn with pen and ink onto Mylar film that had its surface specifically treated to receive drawing ink. This was initially done using pen (with nibs) and ink. Graphic pen sets were used for line drawing and arcs and circles were drawn with similar equipment (Figure 2.10). During my time working in the manual map production era, the introduction of drawing pens from companies like Rotring (Figure 2.11) was seen

The photo-mechanical era of cartography: a recollection

Figure 2.9 Cartographic Branch, Survey Office, Lands Department, Lands Department, Brisbane, 1953 by Queensland State Archives is marked with CC PDM 1.0
Source: https://www.flickr.com/photos/60455048@N02/36039183801.

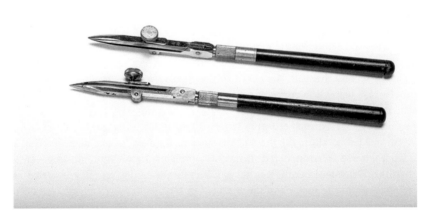

Figure 2.10 Graphic lettering pen set
Source: CIA. https://www.flickr.com/photos/ciagov/30764634372/in/album-72157674852500522/.

Figure 2.11   Rotring isograph pen. Image by PigPog licensed under CC BY-NC-SA 2.0
Source: https://www.flickr.com/photos/55672723@N00/17580748.

as revolutionary. These drawing tools allowed for map drawings and corrections to be made quickly and efficiently – both in the drawing office and in the field. Similar drawing pens were sold by Keuffel and Esser. Figure 2.12 shows a cartographer from the Australian Army's first Topographical Survey Troop, at the first Australian Task Force (1ATF) Base working on a topographic map in Nui Dat, South Vietnam.

Drawing straight lines was done with a straight edge, but drawing arcs required the cartographer to use a curve template selected from a set of French Curves. These wooden boxes of plastic curves were an essential cartographic tool. Drawing arcs in ink actually required the use of two French Curves – one selected with the correct arc radius to guide the drawing tool, and a second, beneath the curve in use, that raised the drawing curve so that ink would not 'sneak' below the curve during the drawing process, spoiling the drawing. A set of French curves is shown in Figure 2.13.

Rotring pens were also used to apply lettering, using lettering guides. A set of lettering guides were needed to allow for the various sizes to be used, as per map drawing specifications. Each letter was 'constructed' from the templates. Lettering templates were sold by companies like Wrico and Leroy (Keuffel & Esser) (Soper, n.d.). Samples of Wrico lettering guides can be seen in the image in Figure 2.14.

A typical monochrome topographic map drawn using this method is shown in Figure 2.15. It was drawn using pens, lettering guides, and contour pens and reproduced using the diazo process, and was produced for the State Electricity Commission of Victoria by the Department of Crown Lands and Survey, Victoria, Australia, in 1954.

Producing lettering via lettering guides and drawing pens was later superseded with the introduction of transfer lettering. Here, individual letters were applied to the drawing media – 'transferring' them – by rubbing over the letters using a glass or plastic burnishing tool. (From my experience, I considered the glass burnishing tool superior to the plastic

The photo-mechanical era of cartography: a recollection

Figure 2.12 Nui Dat, South Vietnam. 1969-09. Corporal Bob Kay of Burwood, Victoria, a draughtsman with the 1st Topographical Survey Troop, at the 1st Australian Task Force (1ATF) Base, puts the finishing touches to a map

Source: Australian War Memorial Photograph Collection. Image Number BEL/69/0601/VN https://www.awm.gov.au/collection/C320232.

Figure 2.13 French curve drafting set
Source: CIA. https://www.flickr.com/photos/ciagov/30880916205/in/album-72157674852500522/.

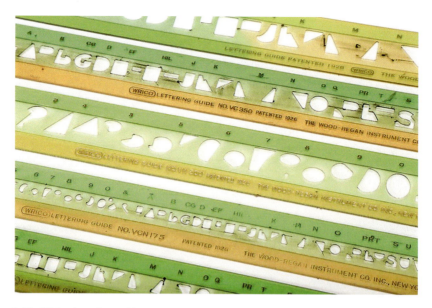

Figure 2.14 Wrico lettering guides
Source: CIA. https://www.flickr.com/photos/ciagov/30880917045/in/album-72157674852500522/.

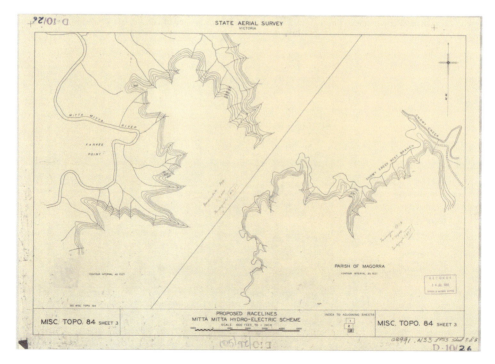

Figure 2.15 Pen, lettering guide, contour pen and ink topographic map, reproduced using the diazo process

Source: State Electricity Commission of Victoria. Department of Crown Lands and Survey. 1954, Mitta Mitta Hydro-Electric Scheme [cartographic material]: proposed racelines/compiled by Department of Lands and Survey [...] for the State Electricity Commission. https://nla.gov.au/nla.obj-2807719132.

The photo-mechanical era of cartography: a recollection

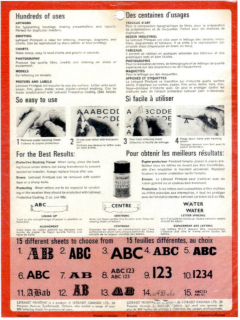

Figure 2.16a Letraset 'rub-on' type. Image 'letraset-front' by bunnyhero, licensed under CC BY-SA 2.0

Source: https://www.flickr.com/photos/40643628@N00/3949786592.

Figure 2.16b Letraset 'rub-on' type. Image 'letraset-back' by bunnyhero, licensed under CC BY-SA 2.0

Source: https://www.flickr.com/photos/bunnyhero/3949007447/.

ones; see instructions provided with the early Letraset 'PrintPaks', shown in Figure 2.16a,b.) Perhaps the most popular brand was Letraset, which supplied not only letters, but also stipples, symbols, and architectural drawing images (Figure 2.17).

'Negative' map artwork

The negative process removes a pre-prepared emulsion using scribing tools, and then metes these pieces of artwork with negatives for lettering and tones that have been produced photographically. Cartographers would produce 'sets' of colour separation artwork for later combination into composite photographic stable-base images for later printing plate production. This involved the creation of many drawings. If, for example, a topographic map, like that shown in Figure 2.18, was being prepared, for its reproduction using flat-colour printing, numerous drawings – linework, areal artwork, lettering sheets, and hillshading – would be completed. If this was an eight-colour print process, using flat-colour printing, the total number of artwork items would in the order of 40 pieces of artwork (8 × 5 colour separations, assuming five-colour impressions).

Linework was done by removing an upper emulsion from the stable-base drawing material using scribing tools. Depending upon specifications, different jewelled-head scribing tips were used to remove this emulsion. This required a skilled technique, as the removal of too little or too much of the upper layer resulted in an inferior line image. The cartographer would work on a light table, with a compilation sheet placed below the scribing material,

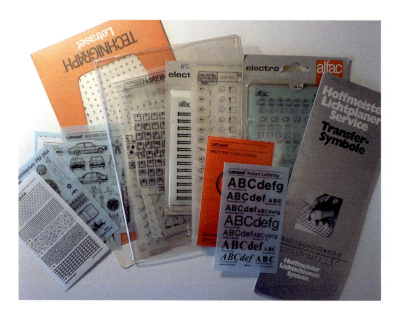

Figure 2.17 Transfer lettering. Image 'letraset' by Enjoy Surveillance is licensed under CC BY-NC-SA 2.0
Source: https://www.flickr.com/photos/65582772@N00/275563957.

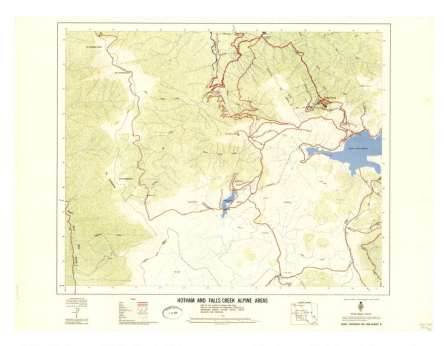

Figure 2.18 Hotham and Falls Creek Alpine areas: part of the shires of Bright and Omeo [...] prepared and distributed by the Department of Crown Lands and Survey, Victoria; compiled in 1967 from State Aerial Survey photography and survey information. Melbourne: A.C. Brooks, Govt. Printer, 1967. Pen, lettering guide, contour pen and ink topographic map. Reproduced using flat-colour printing

The photo-mechanical era of cartography: a recollection

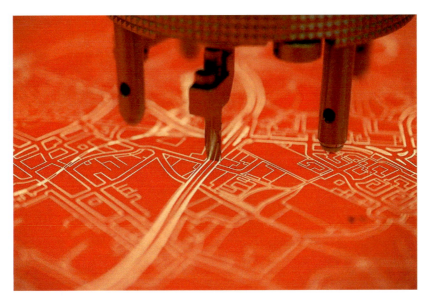

Figure 2.19 Scribing using stable-base plastic 'scribecoat'. Image 'cartographic engraving' by sparkling.spots, licensed under CC BY-NC-SA 2.0
Source: https://www.flickr.com/photos/85861828@N02/854311520.

and the image was basically traced from the image underneath. This was an exacting process that, when completed expertly, resulted in sharp linear definitions. Lines were scribed for communications networks, drainage, contours, and so on. Figure 2.19 illustrates the scribing technique. Once the scribing was completed, it sometimes required finishing or 'touch-ups' using a freehand technique (Figure 2.20).

The lettering colour separation drawing was made by applying transparent adhesive-backed type to clear film overlays. The lettering was pre-composed on lettering machines and the transparent film produced via photo-mechanical methods. Placing the lettering was an arduous and precise task, as lettering, when naming natural features like watercourses, required each letter to be applied individually, so as to achieve artwork outputs that accorded to type placement specifications. The image in Figure 2.21 shows lettering being placed during the production of the Miller Peak, AZ 7.5-minute topographic map at the US Geological Survey.

Topographic maps could also use hillshading to enhance the depiction of terrain. Artist/cartographers used airbrushes to emboss stable-base drawing media, like Mylar, in order to give the impression of shadows that would be created by the terrain being depicted by the topographic map. The picture in Figure 2.22 shows the use of an airbrush for manual hillshading at the US Geological Survey.

Once all colour separation artwork was complete, the compositive negatives or positives were produced. The next stage was photo-mechanical processing, which led to the production of printing plates.

Photo-mechanical production

Photo-mechanical production allowed for the original artwork to be done on another medium and then the images transferred onto photographic film, which were used to expose

Figure 2.20 Colour separation scribing of the intermediate contours for a USGS topographic map using a freehand scriber
Source: US Geological Survey. Public domain.

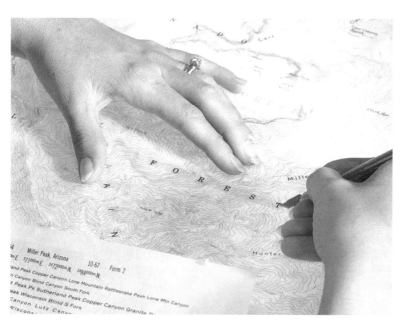

Figure 2.21 Application of transparent adhesive-backed type to the culture lettering separate of the Miller Peak, Arizona 7.5-minute topographic map
Source: US Geological Survey. Public domain. https://www.usgs.gov/media/images/usgs-cartographer-work-5.

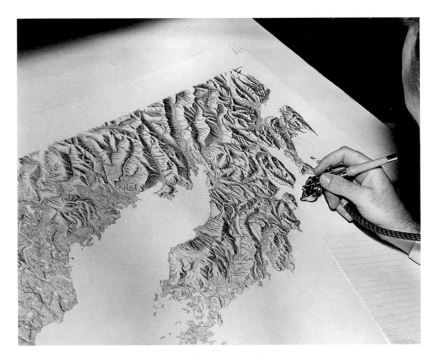

Figure 2.22 Using an airbrush for manual hillshading, c. 1952
Source: US Geological Survey. Public domain. https://twitter.com/USGS/status/709408221807546368/photo/1.

and process printing plates (Cook, 2002). The combination of separate films containing lines, text, and tones combined all drawing map elements into a one negative or positive that would be used to make printing plates. As stated previously, these were drawn as separations, and then combined at this intermediate stage between artwork production and printing. Halftone photography processes, using filter sets, were used to transfer the continuous imagery of airbrush-produced elevation hillshading to halftone dots that could be printed. It is interesting to note that in many cartographic courses, reprographics was included in the programme, as it was an integral part of the map production process during this era (Cartwright and Williams, 1985; Cartwright, 1987; Cartwright and Fraser, 1988).

Prior to the manufacture of printing plates, and then committing to the printing process, a proof was made that allowed for the map to be checked for completeness and correctness. Proofing systems could work with positive or negative composite photographic materials. My particular experience was the use of the Cromalin (a trademark of du Pont) proofing process (Slater, 1975), whereby several sheets of clear photosensitive plastic were exposed, one for each print colour. Once exposed, the Cromalin substrate would remain 'sticky' in the image areas, which were then dusted with colour pigments and could be mixed according to standard printing colour systems, like the Pantone system. When complete, all images were laminated together, providing the facsimile map. The proof was used to check registration, colour, type, errors, and omissions, and whether the map was produced accorded to specifications.

The composite colour separations (usually negatives) were used to expose printing plates, which were then processed and readied for printing. The image in Figure 2.23 shows photolithographers from the Department of Lands, Mapping and Surveying, in Queensland,

Figure 2.23 Photo-lithographers from the Department of Lands, Mapping and Surveying, in Queensland, Australia, conducting quality checks on a processed printing plate

Australia, conducting quality checks on a processed printing plate. The artwork was complete, proofs checked, revisions made, and printing plates processed. Cartographers then handed over the map production process to the printers.

Conclusion

This 'journey' through the manual map production process has hopefully 'painted' a picture of the methodologies, tools, and materials used by cartographers from 1968 to 1978. During this period, the cartographer needed to master the mathematics of mapping, plotting techniques, drawing skills, and photo-mechanical processes. In this way, accurate and 'printable' map artwork could be generated through drawing, assembled via photo-mechanical processes, and accurate materials provided for the printing of maps.

Postscript

I have endeavoured to illustrate this chapter with photographs and images that portray the tools and materials employed. Some of the images were sourced from the archival collection of the Cartography Section of the US Office of Strategic Services (OSS), which produced maps and topographic models for use in military campaigns during the Second World War. Although dating from before the period examined here, these images are nevertheless indicative of the processes, tools and materials used.

References

Cartwright, W.E. (1987) "Paper Maps to Temporal Map Images: User Acceptance and User Access" *Proceedings of the Information Futures: Tomorrow TODAY Conference* Melbourne: Victorian Association for Library Automation, pp.68–87.

Cartwright, W.E. and Fraser, D.D. (1988) "From Manual Map Production to Computer Assisted Cartography" *Proceedings of EDTECH 88: Designing and Learning in Industry and Education* Canberra: Australian Society for Educational Technology, pp.150–159.

Cartwright, W.E. and Williams, B.J. (1985) "Cartography at Royal Melbourne Institute of Technology" *Cartography* 14 (2) pp.104–106.

Cook, K.S. (2002) "The Historical Role of Photomechanical Techniques in Map Production" *Cartography and Geographic Information Science* 29 (3) pp.137–154.

Encyclopedia of Australian Science (2006) "Australian Institute of Cartographers (1923–?)" Available at: https://www.eoas.info/biogs/A001873b.htm (Accessed: 20th March 2021).

Slater, H.L. (1975) "Cromalin, a New Art Medium" *Leonardo* 8 (4) pp.313–316.

Soper, J. (n.d.) "Keuffel & Esser Company" Available at: https://www.mccoys-kecatalogs.com/KELeroy/LeRoy_Production/LeRoy_Soper.pdf (Accessed 29th March 2021).

3
THE ROOTS OF GIS

Michael F. Goodchild

GIS has its roots in the 1960s and 1970s, in a number of projects that applied computers to the handling of geographic data. Some projects used the term geographic information system, while other researchers saw that what they were doing could be part of a larger development and joined the growing GIS community. Exactly who was most responsible for GIS, and who has the best claim to be its parent, is very much a matter of personal opinion. In what follows I tell the story of GIS and its roots from my own perspective, knowing that what I write may not be quite what others might remember (for extensive surveys of the history of GIS, see Foresman, 1998). One could argue that the story is of merely historical interest, and that GIS today has moved far beyond its roots, as is abundantly clear from the chapters of this book. Yet the roots of GIS continue to be important, and in this chapter I show how GIS today is still influenced in critically important ways by the decisions and assumptions that were made at its birth (and see Goodchild, 2018).

Early beginnings: maps and computers

In the mid-1960s, computers were already being used for research and teaching in geography, as I discovered when I took a senior undergraduate course in urban geography at McMaster University, taught by Gerard Rushton. They were being used for conventional statistical analysis of data, and the multivariate methods of factor analysis, multiple regression, and discriminant analysis were becoming popular. Aggregate data from the census, tabulated by spatial units such as the county or census tract, were available and widely used, especially in decomposing the spatial structure of cities. Factorial ecology, for example, as advanced by Brian Berry and others (Berry and Horton, 1970), used computers to extract principal axes from tables of census data, thereby identifying some of the basic underlying dimensions of spatial variability in cities. But while computers were certainly part of the burgeoning quantitative revolution, the emphasis was more on processing tables of numbers than on analyzing the contents of maps.

There were good reasons for this. Building digital representations of maps was a complex and largely manual process that still limited the growth of GIS well into the 1980s. Moreover, only one output device was available at that time, the line printer, which created arrays of symbols on large sheets of paper. Shades of grey could be produced by overprinting symbols, but the "map" was limited by the size of symbols to a spatial resolution of 1/10 by either 1/6 or 1/8 of an inch. The SYMAP package that appeared around 1967 (Chrisman, 2006) used the rows and columns of the line printer as its coordinate system, and users were supplied with 'SYMAP rulers', which could be used to measure locations in divisions of 1/10 inches (for x) and 1/6 inches (for y). The plotter, or computer-driven pen, became available in the early 1970s, and by 1975 it was possible to create maps on the screen of a Tektronix 'storage-tube' terminal, a device analogous to the 'Etch A Sketch' that built a glowing image of lines under computer control until a command to erase was received.

The Canada Geographic Information System (CGIS) (Tomlinson and Toomey, 1999), which I first encountered in 1972, became for me and many others the first truly compelling example of a computer application driven by the contents of maps. In the 1950s a collaboration between the Canadian federal and provincial governments had resulted in a massive mapping project known as the Canada Land Inventory. Its declared purpose was to assess the potential of Canada's land resource, for agriculture, forestry, wildlife, recreation, and urban development. A massive field campaign in much of southern Canada resulted in tens of thousands of maps at 1:50,000, each covering an area subdivided into irregularly shaped zones according to each zone's present use or its future potential for each type of land use. Some maps contained as many as 2,000 zones.

The stakeholders in the project had been promised assessments of quantity of land (i.e., the area currently devoted to each type of land use), and the area that was believed to have potential for each type. Two methods had traditionally been available for measuring area from maps: overlaying transparent sheets of dots and counting the dots within each zone; and using a mechanical device known as a planimeter. Both are tedious and time-consuming, and the results are subject to substantial error. Roger Tomlinson, a geographer working in the Canadian remote-sensing industry and completing a PhD part-time at University College, London, persuaded the Canadian government to contract with IBM to build a system that would digitize the maps and compute areas from the coordinates of zone boundaries. Moreover, the computer could superimpose the contents of different maps of the same area, allowing answers to questions such as 'How much land is used for x on Map 1 but has good potential for y as shown on Map 2?'. Answering that type of question by hand would have been even more tedious, time-consuming, and error-prone, but Tomlinson and IBM could see how it would be possible in principle, given digital representations of each map.

An IBM 360/65 was acquired and dedicated to the project. Maps were scanned using a custom-built map scanner, first transcribing them by hand onto transparent scribe-coat so that only the boundaries were visible to the scanner, and then converting the scanner output to polyline vectors. Because sequential magnetic tape was the only available device for bulk storage at the time, an elegant solution was found to representing the vectorized boundary network. Each segment of boundary between two network junctions (in topological terms an edge or 'arc' between two vertices) was stored in sequence on tape, headed by pointers to the identities of the zone (polygon) on the left and the one on the right. This 'topological' approach later became the 'coverage' structure of early ARC/INFO. It elegantly avoids the 'double-digitizing' problem of a polygon-by-polygon structure, allows the areas of all polygons to be measured in a single pass of the data, and leads to simpler algorithms for polygon overlay. With fine-resolution scanning it was possible through this process to compute areas to far greater accuracy than would have been possible manually.

CGIS was designed for a simple purpose, the calculation and tabulation of map areas. There was no provision for map output, as no suitable output devices existed at the time. But its vision of a system for acquiring and processing large amounts of data from maps provided perhaps the clearest vision from the mid-1960s of what GIS might eventually become; and more fundamentally, why one might want to put a map into a computer.

Finding common ground

My personal experience with CGIS began in 1972, when my first PhD student, J.H. Ross, was hired at Environment Canada and introduced to the project. At that time, it was in serious trouble: the key algorithms (the vectorization of raster scans, calculation of polygon area, and polygon overlay) were proving highly problematic, and the project was far behind in delivering its products to its stakeholders. We suggested that by rasterizing the data using a common grid we could produce the estimates of areas of overlap that the project had promised. The accuracy of the estimates would depend on the grid size, just as dot density affects the accuracy of the old dot-counting method. Moreover, we would have the beginnings of a raster GIS, and could imagine the wide range of functions that a GIS might support, going well beyond the area-measurement vision of CGIS into many techniques of spatial analysis. I started writing code under contract in the summer of 1972, using some sample datasets from Environment Canada.

Meanwhile Roger Tomlinson was also beginning to see additional potential for GIS, and to promote an expanded vision. He organized conferences in 1970 and 1972 (Tomlinson, 1971, 1972) under the auspices of the International Geographical Union's Commission on Geographical Data Sensing and Processing, bringing together people from around the world who were exploring other motivations for putting maps into computers. There was the Harvard Laboratory for Computer Graphics, the developer of SYMAP (Chrisman, 2006). There were landscape architects such as Jack Dangermond who had been inspired by Ian McHarg's concepts of planning (McHarg, 1969) as a process of overlaying layers of data prepared by different disciplines: geology, soil, natural vegetation, surficial and groundwater hydrology, human population, and so on. There was Duane Marble who was using computers and map data to analyse transportation options in Chicago; there was the UK's Experimental Cartography Unit, busy exploring the use of computers in mapmaking; and work was under way in Australia by Bruce Cook and others based largely in remote sensing. Tomlinson's vision was that all of these applications could be supported by a single, monolithic computer application that would recognize and process all of these data types, producing both statistical and map output, and allowing its users to handle spatial data in a formally structured way.

In my view it is this vision, rather than CGIS or any single application, that truly drove the initial development of GIS, and it suggests that Tomlinson's status as the 'father of GIS' should be credited to more than CGIS alone. All these applications and more could have been successful on their own, with custom software to handle specific use cases. But the history of GIS from 1970 to the early 2000s was driven instead by Tomlinson's single-minded vision of a *geographic* information system that would handle and advance all aspects and applications of geographic research, from the vectors of CGIS to the rasters of remote sensing.

Assumptions and workarounds

To achieve the goals of CGIS, and of GIS more generally, in the extremely limited technological environment of the 1960s and 1970s required some very significant insights and shortcuts. One, the topological data structure, has already been mentioned as an outstanding achievement of CGIS and later of early ARC/INFO. Another was the map scanner, which allowed for bulk digitizing, albeit only after the maps had been carefully redrafted by scribing. But there were many more (Goodchild, 2018).

Accuracy and uncertainty

The manuscript maps of CGIS were all of the type that we would now identify as area-class maps, dividing the mapped area into a set of irregularly shaped, non-overlapping, and space-exhausting areas. Such maps are representations of a variable conceptualized as a nominal field, in which every point **x** in space is assigned one of a discrete number of classes $c(\mathbf{x})$. The lines delimiting the areas have no width; in CGIS they were captured as strings of pixels in the scanned images, but then reduced to zero width as sequences of connected vectors, or *polylines*. In reality, however, variables such as potential for use in agriculture do not change suddenly; for example, maps of vegetation tend instead to show gradual transitions or *ecotones*. Only in the case of land use can a case be made that the lines are real and the areas they contain are truly homogeneous, since many of the lines on these maps follow property boundaries and the areas they delimit are managed as units. But even then, there are limits to the accuracy with which property boundaries can be measured, mapped, and scanned. In short, the builders of CGIS assumed that the input maps were the truth and made no allowances for inaccuracies or uncertainties.

This assumption led rapidly to problems in the implementation of CGIS. There are many reasons why a line on the Earth's surface might appear in more than one layer of the database. The coastline or the banks of a river or lake, for example, would appear in all layers, while a property boundary might appear in maps of current and historic land use, and also in maps of potential for forestry or agriculture. But the representations of such common lines will not be the same, since both independently inherit the uncertainties of the mapping process and the scanning. When pairs of such maps are overlaid, the result is a host of very small 'sliver' polygons that can eventually swamp the system (Longley *et al.*, 2015, Table 14.1). Many approaches have been adopted since CGIS to address the problem, typically by replacing slivers by single lines that average the two versions, but at the time sliver polygons were a largely unanticipated but major issue for CGIS.

The problem is almost entirely an issue for vector databases; it will have minimal effect for a raster database as long as the cell size is much larger than the positional uncertainty in boundaries. One corollary of the problem is that it is virtually impossible to specify the spatial resolution of a vector database, since it is a function not of the representation or even of the process of digitizing the data, but of its initial conceptualization as an area-class map (Goodchild, 1994). A cynic is said to have once described area-class maps as showing 'lines that do not exist surrounding areas that have less in common than one might think'. From that rather over-stated perspective, it is important to note that even today it is common practice to create area-class maps of soils, land use, or land cover, and to treat them as the truth when used in GIS.

By the late 1980s, accuracy had become a significant topic of research in GIScience (Goodchild and Gopal, 1989). An important transition occurred in the early 1990s when it became clear that accuracy (or error) was only part of the story. The area-class maps of CGIS were essentially non-replicable, since two independent experts would not produce the same maps – the results would differ not only in the positions of boundaries, but even in the number of areas and the configuration of the boundary network. In part this derives from the definitions of classes, which are likely to contain vagueness and to use terms such as 'mostly' or 'generally'. The response of the research community was to replace accuracy with uncertainty, and to venture into fields such as fuzzy sets (Petry *et al.*, 2005).

While we know much more today about uncertainty in GIS data, the research community continues to be frustrated by the apparent lack of interest in uncertainty among GIS users and developers. The assumption made in the 1960s, that maps represent the truth, survives in many if not most GIS applications. Yet 'the map is not the territory', as Korszybski (1933) pointed out (Janelle and Goodchild 2018); maps almost always limit the degree of detail which they capture from the world, leaving it to the GIS user to assess the impacts of those omissions on the results of an application.

Flattening the Earth

Maps are made by transferring features on the surface of the Earth to paper, creating documents that are easily stored, copied, or shipped. In positioning features on maps, it is necessary to adopt one of a number of nonlinear transformations known as projections (Bugayevskiy and Snyder, 1995), from a location on the curved surface of the Earth expressed in latitude and longitude to a position on paper expressed in x and y coordinates. Over very small areas the effects of the transformation can be ignored, but when a map covers a substantial proportion of the Earth's surface, they become critically important, both to the process of mapping and to the results of any analysis.

The maps that formed the input to CGIS were of small areas, so the flattening involved could be mostly ignored, and measurements made on the maps could be treated as if they were measurements on the Earth's surface. Even here, however, issues arose. CGIS used the UTM projection, which divides the Earth into zones that are six degrees of longitude wide. At the boundaries of zones there will necessarily be discontinuity. This occurs, for example, on the 114th Meridian between UTM Zones 11 and 12, which happens to run through the city of Calgary, Alberta.

Rasters initially became popular in the 1970s following the development of remote sensing; several raster-based GIS were developed at that time, and eventually fully integrated into GIS. To define a raster, it is necessary to lay a grid on a flat surface, and it is impossible to lay a raster on the curved surface of the Earth. It follows that rasters also involve flattening, and that the cells of a raster cannot be of exactly equal area. The spatial resolution of a raster must therefore vary. For example, if a raster is laid on a Plate Carrée projection (equal divisions of longitude in the columns, equal divisions of latitude in the rows), the cells will be approximately square at the Equator, but away from the Equator they will shrink in an east–west direction, eventually to zero at the Poles.

Today we routinely use GIS at scales from the local to the global, and we almost always flatten the Earth to do so, preserving the legacy of a decision made in the 1960s. If instead we had adopted the globe as the basis for GIS, rather than the map (Goodchild, 2018), we could have created a technology that works equally well locally and globally, avoiding the need for the user to address the issues that result from flattening. We could have used latitude and longitude consistently throughout the technology, and even exploited discrete global grid systems (DGGS) (Sahr *et al.*, 2003), which are hierarchical structures that link different spatial resolutions, forming cells at each level that are approximately equal in area and geometry. But in the 1960s the trigonometric functions necessary to handle latitude and longitude would have challenged the computational power of computers, and globes had little practical significance beyond the decorative: they were difficult to construct, difficult to store and ship, and limited to very coarse spatial resolution. Today, with almost unlimited computing power available, the decision of the 1960s continues to impose limitations on what GIS can achieve.

Evolving visions of GIS

Many developments have occurred within the broad field of GIS since the early days of the 1960s and 1970s. One, the merger of raster-based and vector-based GIS into a single, integrated technology has already been mentioned. In what follows I discuss several more that have helped GIS achieve the success and popularity that it now enjoys. Each is in effect a rewriting of the original vision of GIS, taking us far beyond what was envisioned in the 1960s and into the full range of geospatial technologies that are the subject of this book.

Data models

CGIS was built to accommodate one type of data and one data model: the area-class map and the layer of non-overlapping, space-exhausting polygons. Layers of data could be combined to produce statistics, and by the mid-1970s it had become possible to develop models of suitability in the style of McHarg's landscape architecture (McHarg, 1969). Some of the earliest commercial applications of GIS were in natural resources management, especially forestry, but it quickly became apparent that there was an important type of data that could not be readily accommodated: the road network. Modifications were made to the basic data model to allow road segments to be treated as edges or arcs, and to drop the requirement that edges end in junctions, to deal with dead ends. Another set of modifications was required to allow bridges and tunnels, where edges cross without intersecting. By 1990, however, after many such modifications, it was clear that the basic topological or coverage model of vector GIS would have to be replaced. As with any significant advance in commercial software, however, the older model was retained alongside the new one in order to support existing users; the coverage model survives in Esri products to this day.

The object-oriented models that appeared in the 1990s treated the individual point, line, or area feature as fundamental, and focused on representing relationships such as intersection or shared boundaries where appropriate (Zeiler and Murphy, 2010). It became possible to represent variation in the third spatial dimension, and standards were created for representing the structures of entire cities using CityGML or the construction-industry standard BIM (Kolbe and Donaubauer, 2021). By the turn of the century, it had become possible to represent virtually any form of geographic variation, or any geographic phenomenon, in a GIS database. Yet while GIS had become enormously successful across a wide range of human activities, it was clear that many of the more human aspects of the world had been overlooked.

Exploiting the Internet

With the growth of the Internet in the 1990s it became possible to think of a GIS not as an isolated database and processor but as part of a vast network of connected databases and processors. Projects such as the Alexandria Digital Library (Smith *et al.*, 1996) and the Geospatial One-Stop (Goodchild *et al.*, 2007) made it possible to share geographic data in digital form, to support search for geographic data, and to think of GIS as a means of communicating knowledge about the geographic world. The need for digitizing tables and scanners withered away, as users were able to obtain more and more data in digital form. At the same time, it became essential to provide descriptions of data, so that users could assess their fitness for use; in 1992 the first standards for geospatial metadata appeared, and programs were developed to encourage the creation of metadata.

The Internet also made it possible to rely on others to provide processing capabilities, by building client-server architectures. Instead of owning their own GIS, users were able to rely on GIS functions provided by servers. While many of these services were primarily designed for professionals, it was clear that the general public could benefit from GIS services that addressed everyday needs: searching for businesses, or wayfinding. By the late 1990s, it was possible to use Internet services to find hotels in specified areas, or within specified distances from features such as airports. GIS-based services such as Google Maps were introduced in 2004 and were quickly followed by the launch of the iPhone in 2007. Today, many GIS-related aspects of day-to-day existence are supported and facilitated by smartphones. After 2010, many of these services had migrated to cloud technology, which could be readily and transparently accessed from devices that ranged from the desktop and laptop to the smartphone and tablet.

In 2006, Andrew Turner popularized the term *neogeography* (Turner, 2006) to describe the effects of these developments on the relationship between professional experts and the general public. In essence, his thesis was that in the area of GIS services, no effective distinction existed; that the average consumer had become both a user and a provider of GIS data and services. A host of Internet-based programs allowed consumers to contribute to the error-checking and maintenance of large geospatial databases, and projects such as OpenStreetMap (openstreetmap.org) were almost entirely staffed by largely untrained volunteers.

This consumerization of GIS has also had a profound impact on the role of placenames. While coordinates are fundamental to GIS, they play almost no role in everyday life and the minds of people. Almost no one knows the latitude and longitude of their home, but everyone knows their street address. Placenames played no role in CGIS, and even in the early 1990s they were omitted from the set of *framework* datasets of the National Spatial Data Infrastructure. Yet by the early 2000s and with the rise of Internet-based geospatial services, it had become essential to provide interfaces to placenames and street addresses. The *point-of-interest* databases that now provide this interface in mapping and wayfinding services evolved from earlier gazetteers (lists of placenames that originally formed the indexes of atlases) and directories of business services. Today, it has led to a renewed interest in the *platial* (based on placenames) as a counterbalance to the traditional *spatial* (based on coordinates) focus of GIS.

Conclusion

My objective in this chapter has been to offer a personal account of the origins of GIS, and its evolution into a central part of the geospatial technologies that are the focus of this book. It was born in response to a set of well-defined needs that were only loosely connected with the broader society of the 1960s and 1970s. Its emergence as a field was very much the work of one individual, Roger Tomlinson, and his vision of a single computer application that could integrate a number of what (at the time) were largely independent projects, all concerned in one way or another with putting the information from maps into computers.

This is a book about geospatial technology *and society*, yet it has only been in the past three decades that the societal aspects have come to the fore. One stimulus was the critique of GIS that emerged in the late 1980s and early 1990s (for an overview see Pickles, 1995): GIS was expensive and thus exacerbated inequality between the GIS haves and the GIS have-nots; GIS could be used as a tool of surveillance, threatening individual privacy; GIS oversimplified the world, dividing it with infinitely thin lines; and GIS was a mechanical exercise in button-pushing that encouraged an automaton-like behaviour in its users. Another stimulus was the growth of GIS-related consumer services, and the emergence of neogeography.

Today it would be virtually impossible to ignore the societal context of GIS, or the impacts that GIS has had on society.

References

Berry, B.J.L. and Horton, F.E. (1970) *Geographic Perspectives on Urban Systems* Englewood Cliffs, NJ: Prentice Hall.

Bugayevskiy, L.M. and Snyder, J.B. (1995) *Map Projections: A Reference Manual* London: Taylor and Francis.

Chrisman, N. (2006) *Charting the Unknown: How Computer Mapping at Harvard Became GIS* Redlands, CA: Esri Press.

Foresman, T.W. (1998) *The History of Geographic Information Systems: Perspectives from the Pioneers* Englewood Cliffs, NJ: Prentice Hall.

Goodchild, M.F. (1994) "Integrating GIS and Remote Sensing for Vegetation Analysis and Modeling: Methodological Issues" *Journal of Vegetation Science* 5 pp.615–626.

Goodchild, M.F. (2018) "Reimagining the History of GIS" *Annals of GIS* 24 (1) pp.1–8 DOI: 10.1080/19475683.2018.1424737.

Goodchild, M.F., Fu, P. and Rich, P. (2007) "Sharing Geographic Information: An Assessment of the Geospatial One-Stop" *Annals of the Association of American Geographers* 97 (2) pp.249–265.

Goodchild, M.F. and Gopal, S. (Eds) (1989) *Accuracy of Spatial Databases* Basingstoke: Taylor and Francis.

Janelle, D.G. and Goodchild, M.F. (2018) "Territory, Geographic Information, and the Map" In Wuppuluri, S. and Doria, F.A. (Eds) *The Map and the Territory: Exploring the Foundations of Science, Thought and Reality* New York: Springer, pp.609–628.

Kolbe, T.H. and Donaubauer A. (2021) "Semantic 3D City Modeling and BIM" In Shi., W., Goodchild, M.F., Batty, M., Kwan, M.P. and Zhang, A. (Eds) *Urban Informatics* Beijing: Springer, pp.609–636.

Korzybski, A. (1933) *Science and Sanity: An Introduction to Non-Aristotelian Systems and General Semantics* Lakeville, CT: International Non-Aristotelian Publishing Co.

Longley P.A., Goodchild, M.F., Maguire, D.J. and Rhind, D.W. (2015) *Geographic Information Science and Systems* Hoboken, NJ: Wiley.

McHarg, I.L. (1969) *Design with Nature* New York: American Museum of Natural History.

Petry, F., Robinson, V.B. and Cobb, M.A. (Eds) (2005) *Fuzzy Modeling with Spatial Information for Geographic Problems* New York: Springer.

Pickles J. (1995) *Ground Truth: The Social Implications of Geographic Information Systems* New York: Guilford Press.

Sahr, K., White, D. and Kimerling, A.J. (2003) "Geodesic Discrete Global Grid Systems" *Cartography and Geographic Information Science* 30 (2) pp.121–134 DOI: 10.1559/152304003100011090.

Smith, T.R. (1996) "A Digital Library for Geographically Referenced Materials" *Computer* 29 (5) pp. 54–60.Tomlinson, R.F. (1971) *Environmental Information Systems: Proceedings of the UNESCO/IGU First Symposium on Geographical Information Systems, Ottawa, September 1970* Ottawa: International Geographical Union, Commission on Geographical Data Sensing and Processing.

Tomlinson R.F. (1972) *Geographical Data Handling* Ottawa: International Geographical Union, Commission on Geographical Data Sensing and Processing.

Tomlinson R.F. and Toomey, M.A.G. (1999) "GIS and LIS in Canada" In McGrath, G. and Sebert, L. (Eds) *Mapping a Northern Land: The Survey of Canada 1947–1994* Montreal: McGill–Queen's University Press, pp.467–468.

Turner, A. (2006) *Introduction to Neogeography* Sebastopol, CA: O'Reilly.

Zeiler, M. and Murphy, J. (2005) *Modeling Our World: The Esri Guide to Geodatabase Concepts* (2nd ed.) Redlands, CA: Esri Press.

4
POSITIVISM, POWER, AND CRITICAL GIS

Wen Lin

Emergence of critical GIS

The earliest significant root of GIS has been traced to the 1960s, and the first GIS has been widely attributed to the development of a land information system in Canada spearheaded by Roger Tomlinson (Goodchild, 1995). GIS has experienced rapid growth and widespread adoption. Critical GIS is a younger field, which emerged from the critiques of GIS by human geographers in the early 1990s. There are four sections in this chapter. 'Emergence of Critical GIS' provides a brief introduction to critical GIS, whereas sections 'Debates on Positivism and GIS' and 'Power and Critical GIS' elaborate on questions on positivism and power that have often arisen, before the last section on more recent developments in critical GIS.

Schuurman (2000) provides an insightful discussion of the debates that led to the emergence of Critical GIS, which can be described as three waves (see also Schuurman, 2009). The first wave was between 1990 and 1994, and was characterized by polemical debates, often in the form of commentaries between critics of GIS and its defenders (e.g., Taylor, 1990; Goodchild, 1991; Openshaw, 1991; Pickles, 1991; Lake, 1993; Sui, 1994). Critics voiced concerns of epistemological and ontological implications of GIS technologies and analysis as well as how GIS might reinforce existing inequalities (Sheppard, 2005; Schuurman, 2009). For example, Taylor (1990) viewed that GIS was concerned with geographical facts rather than knowledge, and that the emergence of GIS marked a return of empiricism. In response, Goodchild (1991) suggested that GIS and cartography involved fuzziness and generalizations and were not merely a process of data handling. Goodchild (1991) also argued that while some GIS research could be positivist, GIS was more than just a repository of facts and could provoke profound geographical thoughts. In these early critiques, raised by commentators outside the field of GIS, GIS was often equated with positivism (Schuurman, 2000; Sheppard, 2005). Although these exchanges were polemical, the need for more attention to examine changes in GIS technology in relation to the broader social contexts was recognized (Pickles, 1995a).

The second wave was from 1995 to 1998. While many discussions in this period echoed the concerns raised in the first wave, arguing that GIS contained positivist epistemology and prioritized certain forms of knowledge production such as knowledge produced through Cartesian space and Boolean logics (Aitken and Michel, 1995; Pickles, 1995b; Rundstrom, 1995; Sheppard, 1995), this wave saw greater cooperation between social theorists and GIS specialists. This is reflected in the text of *Ground Truth: The Social Implications of Geographic Information Systems* edited by Pickles (1995a). The volume has contributions from both sides analysing the emergence of GIS and associated implications in the contexts of disciplinary practice, the arena of production, marketing strategies and discourses, war practice, and governance (Pickles, 1995b). Contributions addressed new opportunities for positive social action brought by the powerful data handling and mapping capabilities of GIS, but also raised concerns of the dangers of unmediated technical practices. For example, Goodchild (1995: 34) argued that GIS, rather than representing a resurgence of positivism, occupied 'elements of the entire spectrum, from the positivist end to the other end'. He illustrated this further through providing three different perspectives regarding GIS: as a technology, a research field, and a community, all of which contained diverse groups of practices. In conclusion, Goodchild acknowledged the importance of researching social implications of GIS as parts of a GIS research agenda.

In the following chapter, Taylor and Johnston (1995) discussed the rise of GIS as an outgrowth of the quantitative revolution. They pointed out that the quantitative revolution itself was not a unitary monolith within which there were two tensions related to the emergence of GIS. The first tension was between deductive and inductive 'science', for which the use of GIS tended to align more with the former realm thus reflecting empiricism. The second tension was between the so-called 'pure' and 'applied' geography. Taylor and Johnston (1995) argued that the growing emphasis on applications since the 1970s played a key role in the success of GIS, which was practised by a different group from the quantitative revolution: applied quantitative geographers. The authors continued to discuss the possible negative impacts of GIS such as its 'data-led and market-based orientation to its use' (p.60). However, they pointed out that their goal was to argue for the proper use of GIS, noting that while some critiques on quantitative methods were valid, quantitative methods were not necessarily philosophically flawed: 'What they are used for and how to make best use of them within geography depends on the attitudes and mindset of their users and what they want to do with them' (*ibid.*: 62). In these debates, while concerns remained regarding links between GIS and positivism as well as the limits of GIS representation of the world and broader socio-political implications on surveillance, marketing and marginalization of disadvantaged groups, there were more constructive dialogues between the GIS critics and GIS proponents during this period (Schuurman, 2000).

The subsequent third wave was characterized by 'a greater commitment to the technology' from critiques of GIS and a more nuanced analysis of power (Schuurman, 2000: 569), underpinned by dynamic cross-fertilization spawned in the Friday Harbor meeting held in 1993 attended by GIS specialists and human geographers (Sheppard, 2005; Schuurman, 2009). Yet, concerns on epistemological integrity persisted. For example, Wright *et al.* (1997) examined the tension between GIS as a tool and GIS as a science and identified and evaluated three positions shown in these debates: GIS as a tool, toolmaking, and science. They commented that GIS represented a continuum of these positions, and that '[w]hether GIS is a geographical science in and of itself depends on both the rigour with which the tool is employed and the scope of the tool's functionality given the nature of the substantive problem' (Wright *et al.*, 1997: 358). Adopting a broader view of science that it is not confined to a particular epistemology, the authors proposed to view GIS as 'a new kind of science, one that emphasizes visual expression, collaboration, exploration, and intuition, and the uniqueness of place over more traditional concerns for mathematical rigour, hypothesis testing, and generality' (*ibid.*: 358–359). With this broad definition of science, the authors indicated that 'doing GIS' was not necessarily associated with positivism and that it could engage with social theory (Schuurman, 2000). Yet, while Pickles (1997) welcomed Wright *et al.*'s (1997) discussion, he pointed out that these efforts from the field of GIS 'to ask fundamental questions about its own practice and intellectual and practical commitment' were long overdue (p.367). Nonetheless, Pickles (1997) embraced Wright *et al.*'s (1997) call for more work on theory, which was seen as opening 'a space for a more palatable and effective genre of sociotheoretical intervention' (Schuurman, 2000: 584).

These debates of GIS have resulted in greater attention from GIS scholars to the social and political impacts of GIS and how society, in turn, has shaped the development of GIS. These debates in the 1990s catalysed the emergence of the 'GIS and society' research agenda, which was later known as critical GIS (Chrisman, 2005; Sheppard, 2005; Elwood, 2006; O'Sullivan, 2006). More nuanced analyses emerged on themes and issues raised in these earlier debates, such as social constructions of GIS development (e.g., Harvey and Chrisman, 1998) and multiple spatial ontologies concerning database models (Schuurman, 2006). Meanwhile, efforts in research and practice on public participation GIS (PPGIS) proliferated (e.g., Ghose, 2001; Craig *et al.*, 2002; Elwood, 2006; Sieber, 2006).

For Sheppard (2005: 13), 'critical GIS research should be relentlessly reflexive'. Wilson (2017) suggests that critical GIS investigates the intersection of critical social theory and geographic information science (see also Schuurman, 2006; Elwood, 2014), while Pavlovskaya (2018: 42) suggests that there are three ways in which critical GIS can question the status quo: 'challenging the status quo of technology, challenging the status quo of social power, and creating spaces of possibility'.

Debates on positivism and GIS

As noted in the section 'Emergence of critical GIS', one main critique raised by social theorists in the debates on GIS in the 1990s is the epistemological underpinnings of GIS. For example, Lake (1993) argues that GIS is positivist, characterized by universal law-making and an objective and detached way of knowing. Such a critique has stimulated responses and reflections on the epistemological range of GIS (Schuurman, 2000; Kwan, 2002; Sheppard, 2005; Elwood, 2014).

There are two perspectives regarding addressing critiques on GIS-related research as underpinned by positivism. One perspective is to argue that quantitative geography and quantitative analysis in GIS are not inherently positivist (Schuurman, 2001; Sheppard, 2001, 2005). For example, Sheppard (2001) contests the tendency to equate quantitative geography with positivism. He shows that practices of quantitative geography are much more complex and diverse, such as mathematical practices involving fuzzy sets, fractals, and complex theories that are not positivist and that 'statistical analysis also has turned away from the deductive and general to the inductive and computational' (Sheppard, 2001: 544).

Wyly (2009) also argues against the perceived inherent link between positivism and quantitative methodology and proposes the notion of 'strategic positivism'. He acknowledges that quantification has often been used by those in power and echoes Sheppard (2001) regarding the past mistakes made by quantifiers who did not pay attention "to the finer points of distinction between empiricism, positivism, logical positivism, and critical rationalism" (Sheppard, 2001: 538). Meanwhile, Wyly (2009: 316) emphasizes that quantification can be harnessed for progressive purposes and describes recent efforts by 'a small cohort of critical geographers' to build 'a comprehensive infrastructure for a powerful movement that emphasizes analytical rigour, scholarly accountability, and progressive, strategic relevance' as constituting 'strategic positivism'. Strategic positivism 'recognizes the dangers of universalizing, decontextualized epistemological truth claims of the sort advocated by hard-core positivists in the mid-twentieth century' and also 'avoids the oppositional universality of antifoundational thought' (*ibid.*: 316). In other words, while it is acknowledged there is not a single standard of 'abstraction, observation, causal analysis, and generalization' to understand the social world, Wyly (2009: 316) argues that 'the standards of knowledge production need not be in flux every day for every question about every aspect of the social world'.

Shelton's (2018) article provides a useful example of Wyly's 'strategic positivism'. Drawing upon critical GIS and relational sociospatial theory, Shelton (2018) provides a critique of conventional concentrated poverty research, which tends to treat this issue as a separate problem from other concerns such as gentrification or income inequality. Through a case study of Lexington, Kentucky, Shelton (2018) uses a series of graphs and maps based on census data and property ownership records to illustrate spatial patterns of poverty and affluence and links between the two processes of concentrated poverty and concentrated affluence. He argues that this work on the one hand demonstrates the potential for mapping as a way to visualize spatial inequalities following Wyly's (2009) 'strategic positivism', and on the other hand shows the potential 'for visualizing alternative conceptualizations of the spatiality of injustice itself, a way of "theorizing with GIS"' (quoting Pavlovskaya, 2006).

Specifically, Shelton (2018) reconstructs the measurement of concentrated poverty and concentrated affluence. He expands the measurement of 'racially/ethnically concentrated areas of poverty' to not only consider household income in relation to the poverty line, but also relative to the broader urban context (the citywide median household income). Similar adjustments have been made in three other measures such as the 'racially/ethnically concentrated area of affluence' indicator to compare the tract relative to the corresponding citywide level. These measures are shown in a bar chart, first in terms of the number of areas of each measure to depict changes over time from 1970 to 2014. Six maps covering the same six time periods (1970, 1980, 1990, 2000, 2010, 2014) are subsequently presented to show the spatial variations of poverty and affluence. These figures show that concentrated affluence is a more widespread problem than concentrated poverty, although both have been on the rise recently. Two more maps are presented, one on visualizing residential properties in tracts of poverty (satisfying either the racially/ethnically concentrated poverty or relative poverty measurements) owned in tracts of affluence (satisfying either the racially/ethnically concentrated affluence or relative poverty measurements), and the other map on all other locally owned properties owned outside of tracks measured in either concentrated affluence or relative affluence. The analysis and figures show how these two processes are interconnected and co-produced.

A second perspective is that GIS can incorporate other forms of data and knowledge and thus demonstrates a broad epistemological range, as exemplified in the burgeoning work on participatory GIS, feminist GIS, and qualitative GIS. A number of scholars have argued for a notion of epistemological flexibility, for which GIS research may be informed by 'realist, pragmatist, positivist, and other paradigms' (Elwood, 2010: 97). The next section expands on some of these further to address the questions on power in critical GIS research.

Power and critical GIS

Critiques on GIS have raised important 'questions about access, representation, expertise and power' (Elwood, 2006: 694). Rather than rejecting GIS based on these concerns, there have been studies seeking to obtain a better understanding of how GIS technologies and applications are socially constructed and how these processes have produced space, knowledge, and power

(Elwood, 2006). In particular, tremendous efforts have been made in the area of participatory GIS to address concerns of unequal access to GIS and spatial data and the appropriateness of GIS usage (Sheppard, 2005; Dunn, 2007). These efforts seek to include local knowledge in GIS analysis and presentation for a more democratized decision-making process and community empowerment. Meanwhile, it has been well documented that participatory GIS is context-dependent and often has contractionary outcomes regarding empowering marginalized groups (e.g., Harris and Weiner, 1998). Elwood (2002) provides a multi-dimensional analysis of empowerment at multiple scales of interaction regarding community use of GIS. There are three forms of empowerment – distributive empowerment, procedural, and capacity-building – for which there might be variable outcomes concerning community GIS. As such, it has been argued that the question is less about whether GIS is empowering or disempowering, but more about researching how GIS might foster empowerment and disempowerment and who might be impacted within specific contexts (Elwood, 2002).

Research on participation and representation in participatory GIS has called for continuous efforts to refine and extend the conceptualization of 'power' and 'the political' such as drawing insights from feminist geographies and from urban and political geography (Elwood, 2006: 703). For example, the use of GIS-based maps in participatory GIS may not be merely about spatial analysis techniques, but also can serve as a form of communicative media or counter-mapping practices. As such, the notion of visualization is expanded that enables a fuller examination of ways in which knowledge and power are produced and negotiated (Elwood, 2006).

Another area of concern on GIS revolves around its representational capabilities, as GIS tends to project a 'God's eye view' and prioritizes representing a Cartesian space (e.g., Rundstrom, 1995). There have been fruitful discussions in developing qualitative GIS (e.g., Cope and Elwood, 2009), with efforts to incorporate qualitative forms of data and knowledge in GIS. These efforts can be linked to participatory GIS and feminist GIS (Elwood, 2006), the former of which was addressed in the above brief discussion while the latter is elaborated upon below.

Feminist GIS research was also spawned by the critiques of GIS in the early 1990s (e.g., Kwan, 2002; McLafferty, 2002). A feminist approach is underpinned by the recognition of reflexivity, the importance of giving voice to the research subjects, positionality of the researcher, and the importance of diversity and difference (McLafferty, 2002: 265). For example, informed by critical GIS and feminist political and economic geographies, Whitesell and Faria (2019) map the Ugandan wedding industry at three levels: the body, city, and global. They use Uganda's bridal fashion industry as a case study of globalization and intend to understand the power and/in the economy through mapping the linkage between 'global, power-laden networks of business with intimate, embodied, and everyday practices of self-presentation' (*ibid.*: 2). Drawing upon survey data, archival media coverage on fashion, and semi-structured interviews with bridal entrepreneurs, three maps were produced, termed as 'global intimate maps', as 'playful and provocative experimentations with geovisualization methods' (Whitesell and Faria, 2019: 6).

The map at the global scale is a choropleth map showing where the supplies were imported for the traders. In addition to colour coding the areas based on the quantity of supplies, the authors include quotes from students' testimonies and newspaper headlines in the form of flows between the origins of the supplies and Uganda to highlight underrepresented global linkages, as a way of feminist intervention in the map. The second map at the city level adopts a similar strategy to visualize how transnational flows and macroscale shifts have been produced at Uganda's city spaces. Six sites playing a role in the bridal industry are featured on this map, each with an image displayed to depict the scene and a participant's quote beneath each image. The last map zooms in further on a key site of production and consumption for the bridal fashion business and attempts to capture 'the simultaneously neoliberal/feminized and public/intimate spaces operating on and through this street' (p.11) by displaying quotes on an imagery of a particular shopping mall from this site. These quotes are telling regarding both the 'oppressive power relations embedded within the bridal and fashion industry, as well as positive accounts of Ugandan vendors and consumers who strategically navigate and profit from it' (p.12). Providing these maps as 'creative, playful, radical, and instructive expressions' (p.13), the authors argue that these maps help to foreground embodied experiences situated in the bridal markets of Kampala, Uganda, which are yet linked to global economic flows that are influenced by various neoliberal reforms.

Research into questions about expertise and power also involves studies on how GIS is socially constructed at the technological, organizational, and institutional levels (e.g., Martin, 2000; Chrisman, 2005; Harvey and Tulloch, 2006; Lin, 2013). For example, Lin (2013) examines GIS constructions in relation to China's urban governance through an in-depth case study of Shenzhen. Informed by work from GIS implementation, critical GIS, and governmentality studies, the case study analyses how in Shenzhen's urban planning agencies, GIS has been developed from a practice involving goals for digitizing internal organizational workflow into an expanding practice to 'geo-code' the urban landscape. This 'geo-coding' practice works to regulate the urban environment and 'govern at a distance', a particular form of geographic rationality (Rose-Redwood, 2012). The earlier phase of GIS development revolved more around workflow automation while later developments attempted to build the so-called 'urban grid information system' to divide the city into a set of grids, each with a unique code. The system consisted of staff working on each grid to collect and report data to the city management monitor centre, which would provide feedback to the city management committee with connections to other city departments for further actions if needed. The system was praised for being able to streamline or bypass the existing administrative structure that was often compartmentalized when it came to dealing with issues related to the urban environment. These technical developments were intertwined with the transformation of China's urban governance, moving towards a form of governing rationality that 'views the urban environment as a whole and leaves more room for urban governance at lower levels, in stark contrast to the strict top-down hierarchies of pre-reform governance' (Lin, 2013: 917). This analysis shows that such GIS developments have been government-centric rather than citizen-centric, although they were not a monolithic process without contestations including those from the general public.

Critical GIS in the Web 2.0 age: new opportunities and challenges?

From a stormy start, critical GIS continues to grow and evolve (e.g., Elwood, 2009; Schuurman, 2015; Thatcher et al., 2016; Wilson, 2017), including responses to new questions and challenges raised by new geospatial technologies. In particular, the explosive growth of user-generated geographic information facilitated by mobile devices and Web 2.0 technologies since the mid-2000s has drawn significant scholarly attention. Various terms have been used to describe spatial data produced and practices associated with these technologies, such as volunteered geographic information (VGI) (Goodchild, 2007), neogeography (Turner, 2006), maps 2.0 (Crampton, 2009), the geoweb (Scharl and Tochtermann, 2007), and 'smart' technology (Poorthuis and Zook, 2020).

Many have seen new and exciting opportunities brought by these geospatial technologies, including new forms of data production, collection, and dissemination, complementing (or even replacing) conventional spatial data construction. Meanwhile, concerns have also been raised regarding challenges such as issues of data quality, privacy intrusion, and unintended consequences. Scholars have called for more research viewing GIS as more than traditional desk-based mapping technologies to include these online digital technologies. For example, Elwood (2008) suggests that conceptualizations and findings from critical GIS, participatory GIS, and feminist GIS provide useful foundations to advance and inform VGI research, but she also notes that VGI presents new challenges and needs. One example is that regarding the concern of access to spatial data, the production and use of VGI constitute new types of technologies, data, and expertise compared to conventional GIS.

A growing body of work has sought to examine impacts of these new geospatial technologies and data (e.g., Sui *et al.*, 2013). Two broad trajectories can be identified. One is to engage with the technological system such as reworking or expanding the technical capabilities of GIS as well as exploring possibilities provided by new geospatial technologies to lower barriers for access to spatial data and democratize decision-making processes (e.g., Gahegan, 2018; Poorthuis and Zook, 2020). While a host of research has been devoted to examining issues of data quality and spatial infrastructure development, there is a call for more developments of technologies beyond off-the-shelf software and platforms that are 'our own' (Poorthuis and Zook, 2020: 356). A second trajectory is to engage with social theories, including the call for more attention to feminist theory, queer theory, and black geographies (Elwood and Leszczynski, 2018; Gieseking, 2018; Elwood, 2020) and more work on political economy of technological development and use (e.g., Lin, 2018; Thatcher and Imaoka, 2018). For example, Gieseking (2018) calls for queering the geographic operating system as critical GIS intervention, such as building counter-data from PPGIS and PGIS practices and engaging with (re)designing GIS that are affordable and accessible to all (Gieseking, 2018: 65).

There are vibrant areas of increasing interest in using GIS and digital spatial technologies in interdisciplinary investigations, such as digital humanities (e.g., Cooper and Gregory, 2011), digital geographies (e.g., Ash *et al.*, 2018), and critical quantification (e.g., Bergmann, 2013). For example, Cooper and Gregory (2011) develop a literary GIS to explore the spatial relationships between two textual accounts of tours of the Lake District in England by two poets, respectively. They employ 'four main stages' of mapping to illustrate the cartographical readings of the geo-specific texts from these textual accounts: 'base maps' on each writer's movement, 'analytical maps' through density smoothing techniques to visualize geo-specific information such as place name references, 'exploratory maps' to illustrate more subjective and abstract elements such as mapping the writer's emotional responses to different locations, and 'interactive maps' using Google Earth to allow interactions between the map reader and the visualized movements of both writers. Cooper and Gregory (2011) note that while not without limitations, this 'Mapping the Lakes' provides possibilities to open up space for future literary cartographies.

To conclude this relatively short overview of critical GIS, as Sheppard (2005) suggests that critical GIS means being relentlessly reflexive, it is perhaps worth reminding ourselves as critical GIS researchers to 'be careful, be modest, and be critical' (Wyly, 2009: 317) and recognize the importance of 'modest theorizing' (Elwood, 2020: 222).

References

Aitken, S. and Michel, S. (1995) "Who Contrives the 'Real' in GIS? Geographic Information, Planning and Critical Theory" *Cartography and Geographic Information Systems* 22 (1) pp.17–29 DOI: 10.1559/152304095782540519.

Ash, J., Kitchin, R. and Leszczynski, A. (2018) "Digital Turn, Digital Geographies?" *Progress in Human Geography* 42 (1) pp.25–43 DOI: 10.1177/0309132516664800.

Bergmann, L. (2013) "Bound by Chains of Carbon: Ecological – Economic Geographies of Globalization" *Annals of the Association of American Geographers* 103 (6) pp.1348–1370 DOI: 10.1080/00045608.2013.779547.

Chrisman, N. (2005) "Full Circle: More than Just Social Implications of GIS" *Cartographica* 40 pp.23–35 DOI: 10.3138/8U64-K7M1-5XW3-267.

Cooper, D. and Gregory, I. (2011) "Mapping the English Lake District: A Literary GIS" *Transactions of the Institute of British Geographers* 36 (1) pp.89–108 DOI:10.1111/j.1475-5661.2010.00405.x.

Cope, M. and Elwood, S. (Eds) (2009) *Qualitative GIS: A Mixed Methods Approach*. Thousand Oaks, CA: Sage Publications.

Craig, W.J., Harris, T.M. and Weiner, D. (Eds) (2002) *Community Participation and Geographical Information Systems* London: Taylor and Francis.

Crampton, J.W. (2009) "Cartography: Maps 2.0" *Progress in Human Geography* 33(1) pp.91–100 DOI: 10.1177/0309132508094074.

Dunn, C. (2007) "Participatory GIS – A People's GIS?" *Progress in Human Geography* 31 (5) pp.616–637 DOI: 10.1177/0309132507081493.

Elwood, S. (2020) "Digital Geographies, Feminist Relationality, Black and Queer Code Studies: Thriving Otherwise" *Progress in Human Geography* 45 (2) pp.209–228 DOI: 10.1177/0309132519899733.

Elwood, S. (2014) "Straddling the Fence: Critical GIS and the Geoweb" *Progress in Human Geography* pp.1–5 DOI: 10.1177/0309132514543616.

Elwood, S. (2010) "Geographic Information Science: Emerging Research on the Societal Implications of the Geospatial Web" *Progress in Human Geography* 34 (3) pp.349–357 DOI: 10.1177/0309132509340711.

Elwood, S. (2009) "Geographic Information Science: new geovisualization technologies – emerging questions and linkages with GIScience research" *Progress in Human Geography* 33 (2) pp.256–263 DOI: 10.1177/0309132508094076.

Elwood, S. (2008) "Volunteered geographic information: future research directions motivated by critical, participatory, and feminist GIS" *GeoJournal* 72 (3) pp.173–183 DOI: 10.1177/0309132509340711.

Elwood, S. (2006) "Critical Issues in Participatory GIS: Deconstructions, Reconstructions, and New Research Directions" *Transactions in GIS* 10 (5) pp.693–708 DOI: 10.1111/j.1467-9671.2006.01023.x.

Elwood, S. (2002) "GIS Use in Community Planning: A Multidimensional Analysis of Empowerment" *Environment and Planning A: Economy and Space* 34 (5) pp.905–922 DOI: 10.1068/a34117.

Elwood, S. and Leszczynski, A. (2018) "Feminist Digital Geographies" *Gender, Place and Culture* 25 (5) pp.629–644 DOI: 10.1080/0966369X.2018.1465396.

Gahegan, M. (2018) "Our GIS is Too Small" *The Canadian Geographer* 62 (1) pp.15–26 DOI: 10.1111/cag.12434.

Ghose, R. (2001) "Use of Information Technology for Community Empowerment: Transforming Geographic Information System into Community Information Systems" *Transactions in GIS* 5 (2) pp.141–163 DOI: 10.1111/1467-9671.00073.

Gieseking, J. (2018) "Operating Anew: Queering GIS with Good Enough Software" *The Canadian Geographer* 62 (1) pp.55–66 DOI: 10.1111/cag.12397.

Goodchild, M.F. (2007) "Citizens as Sensors: The World of Volunteered Geography" *GeoJournal* 69 pp.211–221 DOI: 10.1007/s10708-007-9111-y.

Goodchild, M.F. (1995) "GIS and Geographic Research" In Pickles, J. (Ed.) *Ground Truth: The Social Implications of Geographic Information Systems* New York: Guildford Press, pp.31–50.

Goodchild, M.F. (1991) "'Just the Facts'" *Political Geography Quarterly* 10 (4) pp.335–337 DOI: 10.1016/0260-9827(91)90001-B.

Harris, T. and Weiner, D. (1998) "Empowerment, Marginalization, and 'Community-integrated' GIS" *Cartography and Geographic Information Systems* 25 (2) pp.67–76 DOI: 10.1559/152304098782594580.

Harvey, F. and Chrisman, N. (1998) "Boundary Objects and the Social Construction of GIS Technology" *Environment and Planning A: Economy and Space* 30 (9) pp.1683–1694 DOI: 10.1068/a301683.

Harvey, F. and Tulloch, D. (2006) "Local-government Data Sharing: Evaluating the Foundations of Spatial Data Infrastructures" *International Journal of Geographical Information Science* 20 pp.743–68 DOI: 10.1080/13658810600661607.

Kwan, M. (2002) "Feminist Visualizations: Re-envisioning GIS as a Method in Feminist Geographic Research" *Annals of the Association of American Geographers* 92 pp.645–660 DOI: 10.1111/1467-8306.00309.

Lake, R. (1993) "Planning and Applied Geography: Positivism, Ethics, and Geographic Information Systems" *Progress in Human Geography* 17 (3) pp.404–413 DOI: 10.1177/030913259301700309.

Lin, W. (2018) "Volunteered Geographic Information Constructions in a Contested Terrain: A Case of OpenStreetMap in China" *Geoforum* 89 pp.73–82 DOI: 10.1016/j.geoforum.2018.01.005.

Lin, W. (2013) "Digitizing the Dragon Head, Geo-Coding the Urban Landscape: GIS and the Transformation of China's Urban Governance" *Urban Geography* 34 (7) pp.901–922 DOI: 10.1080/02723638.2013.812389.

Martin, E.W. (2000) "Actor-networks and Implementation: Examples from Conservation GIS in Ecuador" *International Journal of Geographical Information Science* 14 (8) pp.715–738 DOI: 10.1080/136588100750022750.

McLafferty, S. (2002) "Mapping Women's Worlds: Knowledge, Power, and the Bounds of GIS" *Gender, Place and Culture* 9 pp.263–269 DOI: 10.1080/0966369022000003879.

Openshaw, S. (1991) "A View on the GIS Crisis in Geography, or, Using GIS to Put Humpty-Dumpty Back Together Again" *Environment and Planning A: Economy and Space* 23 (5) pp.621–628 DOI: 10.1068/a230621.

O'Sullivan, D. (2006) "Geographical Information Science: Critical GIS" *Progress in Human Geography* 30 (6) pp.783–791 DOI: 10.1177/0309132506071528.

Pavlovskaya, M. (2018) "Critical GIS as a Tool for Social Transformation" *The Canadian Geographer* 62 (1) pp.40–54 DOI: 10.1111/cag.12438.

Pavlovskaya, M. (2006) "Theorizing with GIS: A Tool for Critical Geographies?" *Environment and Planning A: Economy and Space* 38 (11) pp.2003–2020 DOI:10.1068/a37326.

Pickles, J. (1997) "Tool or Science? GIS, Technoscience, and the Theoretical Turn" *Annals of the Association of American Geographers* 87 (2) pp.363–372 DOI:10.1111/0004-5608.00058.

Pickles, J. (Ed.) (1995a) *Ground Truth: The Social Implications of Geographic Information Systems* New York: Guilford Press.

Pickles, J. (1995b) "Representations in an Electronic Age: Geography, GIS, and Democracy" In Pickles, J. (Ed.) *Ground Truth: The Social Implications of Geographic Information Systems* New York: Guilford Press, pp.1–30.

Pickles, J. (1991) "Geography, GIS, and the Surveillant Society" In Frazier, J.W., Epstein, B.J., Schoolmaster, F.A. and Moon, H. (Eds) *Papers and Proceedings of Applied Geography Conferences* 14 pp.80–91.

Poorthuis, A. and Zook, M. (2020) "Being Smarter about Space: Drawing Lessons from Spatial Science" *Annals of the American Association of Geographers* 110 (2) pp.349–359 DOI: 10.1080/24694452.2019.1674630.

Rose-Redwood, R.S. (2012) "With Numbers in Place: Security, Territory, and the Production of Calculable Space" *Annals of the Association of American Geographers* 102 (2) pp.295–319 DOI: 10.1080/00045608.2011.620503.

Rundstrom, R.A. (1995) "GIS, Indigenous Peoples, and Epistemological Diversity" *Cartography and Geographic Information Systems* 22(1) pp.45–47 DOI: 10.1559/152304095782540564.

Scharl, A. and Tochtermann, K. (2007) *The Geospatial Web: How Geobrowsers, Social Software and the Web 2.0 are Shaping the Network Society* New York: Springer.

Schuurman, N. (2015) "What is alt.gis?" *The Canadian Geographer* 59 (1) pp.1–2 DOI: 10.1111/cag.12163.

Schuurman, N. (2009) "Critical GIS" In Kitchin, R. and Thrift, N. (Eds) *International Encyclopedia of Human Geography* Oxford: Elsevier, pp.363–368.

Schuurman, N. (2006) "Formalization Matters: Critical GIS and Ontology Research" *Annals of the Association of American Geographers* 96 (4) pp.726–739 DOI: 10.1111/j.1467–8306.2006.00513.x.

Schuurman, N. (2001) "Critical GIS: Theorizing an Emerging Science" *Cartographica* 36 (4) pp.1–108.

Schuurman, N. (2000) "Trouble in the Heartland: GIS and its Critics in the 1990s" *Progress in Human Geography* 24 (4) pp.569–590 DOI: 10.1191/030913200100189111.

Shelton, T. (2018) "Rethinking the RECAP: Mapping the Relational Geographies of Concentrated Poverty and Affluence in Lexington, Kentucky" *Urban Geography* 39 (7) pp.1070–1091 DOI:10.1080/02723638.2018.1433927.

Sheppard, E. (2005) "Knowledge Production through Critical GIS: Genealogy and Prospects" *Cartographica* 40 (4) pp.5–21 DOI: 10.3138/GH27-1847-QP71-7TP7.

Sheppard, E. (2001) "Quantitative Geography: Representations, Practices, and Possibilities" *Environment and Planning D: Society and Space* 19 pp.535–554 DOI: 10.1068/d307.

Sheppard, E. (1995) "GIS and Society: Towards a Research Agenda" *Cartography and Geographic Information Systems* 22 (1) pp.5–16 DOI: 10.1559/152304095782540555.

Sieber, R. (2006) "Public Participation Geographic Information Systems: A Literature Review and Framework" *Annals of the Association of American Geographers* 96 pp.491–507 DOI:10.1111/j.1467-8306.2006.00702.x.

Sui, D. (1994) "GIS and Urban Studies: Positivism, Post-positivism, and Beyond" *Urban Geography* 14 pp.258–78 DOI: 10.2747/0272-3638.15.3.258.

Sui, D., Elwood, S. and Goodchild, M. (Eds) (2013) *Crowdsourcing Geographic Knowledge: Volunteered Geographic Information (VGI) in Theory and Practice* London: Springer.

Taylor, P.J. (1990) "Editorial Comment: GKS" *Political Geography Quarterly* 9 (3) pp.211–212 DOI: 10.1016/0260-9827(90)90023-4.

Taylor, P.J. and Johnston, R. (1995) "GIS and Geography" In Pickles, J. (Ed.) *Ground Truth: The Social Implications of Geographic Information Systems* New York: Guilford Press, pp.51–67.

Thatcher, J.E. and Imaoka, L.B. (2018) "The Poverty of GIS Theory: Continuing the Debates Around the Political Economy of GIS Systems" *The Canadian Geographer* 62 (1) pp.27–34 DOI: 10.1111/cag.12437.

Thatcher, J., Bergmann, L., Ricker, B., Rose-Redwood, R., O'Sullivan, D., Barnes, T.J., Barnesmoore, L.R., Imaoka, L.B., Burns, R., Cinnamon, J., Dalton, C.M., Davis, C., Dunn, S., Harvey, F., Jung, J., Kersten, E., Knigge, L., Lally, N., Lin, W., Mahmoudi, D., Martin, M., Payne, W., Sheikh, A., Shelton, T., Sheppard, E., Strother, C.W., Tarr, A., Wilson, M.W. and Young, J.C. (2016) "Revisiting Critical GIS" *Environment and Planning A: Economy and Space* 48 (5) pp.815–824 DOI: 10.1177/0308518X15622208.

Turner, A. (2006) *An Introduction to Neogeography* Sebastopol, CA: O'Reilly.

Whitesell, D.K. and Faria, C.V. (2019) "Gowns, Globalization, and 'Global Intimate Mapping': Geovisualizing Uganda's Wedding Industry" *Environment and Planning C: Politics and Space* DOI: 10.1177/2399654418821133.

Wilson, M.W. (2017) *New Lines: Critical GIS and the Trouble of the Map* Minneapolis: University of Minnesota Press.

Wright, D.J., Goodchild, M.F. and Proctor, J.D. (1997) "GIS: Tool or Science? Demystifying the Persistent Ambiguity of GIS as 'Tool' Versus 'Science'" *Annals of the Association of American Geographers* 87 (2) pp.346–362 DOI: 10.1111/0004-5608.87205.

Wyly, E. (2009) "Strategic Positivism" *The Professional Geographer* 61 (3) pp.310–322 DOI: 10.1080/00330120902931952.

5
GEOSPATIAL STANDARDS
An example from agriculture

Didier G. Leibovici, Roberto Santos, Gobe Hobona, Suchith Anand, Kiringai Kamau, Karel Charvat, Ben Schaap and Mike Jackson

As a preamble, it is worth remembering that geospatial standards relate to: (a) datasets that have geographical location of some sort which can be often represented as a cartographical map or at least being delineated or attached to a geographical area; (b) a set of rules derived from computing science modelling and domain specific science characteristics; and, eventually, (c) processing capabilities allowing to transform and then display the results as a summary, a map, and so on. The definition of a geospatial standard for some data or a service delivery allows data exchange from machine to machine into what is called data interoperability, i.e., understanding the data to be able to operate on it. Different levels of interoperability are often distinguished: syntactic, semantic, structural interoperability. Syntactic interoperability (e.g., format, encoding, protocol and standard) ensures communication and information exchange; structural interoperability (e.g., resolution, spatial accuracy, temporal accuracy, other structural quality, orthorectification) ensures the expected intrinsic characteristic of the data; and semantic interoperability (e.g., classification, measurement attributes, quality of the data, ontology of the data, reference of the information model) ensures an understanding of the meaning of the data and its properties, i.e., enabling users to grasp the knowledge behind the information shared. Thus, different types or levels of standardization may be considered within a particular geospatial standard focusing more often on the syntactic and structural interoperability, with the semantic interoperability adoption coming often from combining multiple viewpoints within a best practice recognized standard and a data specification. Therefore, the level of adoption of standards can vary from one domain to another due to different practices within communities and the range of stakeholders involved.

This chapter is not an attempt to review the evolution of geospatial data standardization in the last 25 years, facilitated since the creation of the Open Geospatial Consortium (OGC) in 1994 and of the ISO/TC211 Technical Committee on geographical information of the International Organization for Standardization (ISO), which was established the same year. It aims to highlight the impact geospatial standards have had so far, as much on their difficult task of harmonizing data exchange between sometimes highly developed disciplinary communities, as on the innovation they enabled. The chapter examines their remits in the current world scenario where the combination of increasing access to high-speed Internet and pervasive computing generates vast amounts of structured and unstructured geographically associated data. These advances are boosting the development of standards, but also expectations in relation to data harmonization as well as best practices for a large range of requirements from applications using geolocated information and geospatial data gathered from various sources.

Three main challenges have always been the drivers for innovation in OGC standards development, conditioning their adoption: (1) cross-disciplinary integration; (2) compatibility between national, regional and international e-infrastructures; and (3) efforts required to make possible data conflation from heterogeneous sources including crowdsourcing. Because of the technological advances listed above, these challenges are pressing more and more as the possibilities appear endless. They exist within lots of thematic domains working with legacy geographic information, surveyed data, remote sensing imagery and population data, along with real-time data coming from a range of sensors, Internet of Things (IoT), big data derived from social media, including crowdsourcing, citizen science and volunteered geographic information. From its multiple facets, the agriculture domain is particularly sensitive to the advances in all these respective aspects, from the above challenges to the use and re-use of heterogeneous data within e-infrastructures that enable data transformation and computational models within a range of applications.

The discussion here reflects activities in the last two years within the Agriculture Working Group of OGC, where different domain experts discussed the needs for and from agriculture standards. These standardization efforts are particularly viewed from current and previous EU-funded projects such as agriXchange, FOODIE, DataBio, Demeter and SIEUSOIL. A few case studies involving agriculture illustrate the many issues that have been raised, resolved or still to be resolved or proposing best practices to make the most of their successful adoption.

Besides traditional bodies such as the Food and Agriculture Organization (FAO), the Consultative Group on International Agricultural Research (CGIAR), the European Federation for Information Technologies in Agriculture (EFITA) and the International Society for Precision Agriculture (ICPA), there is also the Global Open Data for Agriculture and Nutrition (GODAN). This is concerned with decision making that is linked to food security and agriculture sustainability from using agriculture information. There are also initiatives from communities of researchers and industrial bodies such as Research Data Alliance (RDA) working groups and the OGC Agriculture Domain Working Group, which undertake initiatives and goals toward the adoption of data and processing standards frameworks for the benefit of all. Other key players include the World Health Organization (WHO), the World Bank and the United Nations committee of experts on Global Geospatial Information Management (UN-GGIM), who are also shaping initiatives for better adoption of developed standards. The Infrastructure of Spatial Information in Europe (INSPIRE), the Group on Earth Observation System of Systems (GEOSS) and the African GEOSS (AfriGEOSS) are examples of concrete outcomes of these efforts.

After the brief historical notes and generic aspects of geospatial standards, the chapter illustrates the process of designing, developing and implementing a standard. Then, it shows how the adoption of geospatial standards plays a role in cross-disciplinary innovation. A section is dedicated to the issues arising from different national policies and international initiatives for the adoption of standards. New paradigms coming from citizens' involvement are analysed in relation to the additional disruptive data flow. Finally, the chapter explores what the future brings as new research challenges in light of the recent developments in the agriculture and related domains.

Historical background

Questions about the standardization of geospatial information started with the growing success of Geographic Information Systems (GIS) in the 1980s within vendor-locked standards and software. With the advent of Open Source and Open Software, particularly the Geographic Resources Analysis Support System (GRASS), the Open GIS project to develop the vision of data interoperability became the Open GIS Consortium in 1994 and is now the Open Geospatial Consortium. This openness was primordial for the large adoption that we see today and is still the key motivation for its spread (Coetzee et al., 2020).

Many of the standards that are implemented across the geospatial industry have been developed and published by the OGC. The OGC is an international consortium of more than 520 businesses, government agencies, research organizations and universities, driven to make geospatial (location) information and services FAIR (Findable, Accessible, Interoperable, and Reusable). OGC runs a member-driven consensus process that creates publicly available, royalty-free, open geospatial standards.

An OGC standard is a document, established by consensus and approved by the OGC membership, that provides rules and guidelines that aim to optimize the degree of interoperability in a given context. The rules and guidelines are specified to take into consideration community and member requirements, as well as market and technology trends. OGC standards endeavour to improve geospatial interoperability across data, processes, workflows, visualization, products, representation and other aspects of geospatial information management.

The OGC is organized into Standards Working Groups (SWGs) and Domain Working Groups (DWGs). SWGs focus on the development of specifications, whereas DWGs provide a platform for discussing topics of relevance to specific communities of interest. The Agriculture DWG is an example of a DWG within the OGC. The Agriculture DWG was formed with a focus on technology and associated policy issues and other technological interests that relate to agriculture as well as the means by which those issues can be appropriately factored into the OGC standards development process. The goal of the group is to establish a link between experts in the domains of agriculture and technology.

The OGC maintains a close formal relationship with ISO/TC211. This relationship has allowed multiple OGC standards to be adopted as ISO standards. For example, the OGC Geography Markup Language (GML) standard was adopted by ISO as ISO 19136:2007 *Geographic information – Geography Markup Language (GML)* – and the OGC Web Feature Service (WFS) standard was adopted by ISO as ISO 19142:2010 *Geographic information – Web Feature Service*. The OGC also maintains a close formal relationship with the World Wide Web Consortium (W3C). This relationship has led to the joint development of multiple specifications, such as the Time Ontology in OWL, Semantic Sensor Network Ontology and Spatial Data on the Web Best Practices specifications (OGC and W3C, 2017a, 2017b). These specifications are maintained by the Spatial Data on the Web Interest Group – a joint working group of the OGC and W3C.

Designing, developing and implementing standards

Geospatial data are applied in multiple domains to improve the understanding of location characteristics and their impact on various aspects of society. Beyond geospatial technologies to capture data, GIS makes it possible to gather data and relate information, within a planned analysis, on real-world phenomena via their location on Earth. From the data capture to the decision making arising after analysis within a GIS, standards are there to seamlessly and efficiently work through a workflow representing the planned analysis. Within a particular domain, if the FAIR principle (described above) is widely adopted, the latter workflow is also effective in cross-domain workflows. A non-exhaustive list of domains that make use of geospatial technologies, data and standards includes: agriculture, aviation, built environment, citizen science, data science, defence, Earth systems science, emergency services, energy and utilities, geology and geophysical sciences, hydrology, insurance and financial services, land administration, law enforcement, marine, meteorology, simulation and gaming and transportation. This section focuses on soil data, used in a range of applications involving different disciplines and communities in agriculture, as an example of the path of development of a standard from a geospatial data specification.

The endeavour of data interoperability behind development of standards can be very challenging, depending on the domain. Structural and syntactic interoperability consensus may depend on the success of existing data exchange format and standards within a discipline, a community, or directly from the data collection methods (e.g., data capturing devices' capabilities). However, overcoming the semantic hurdle in order to reach domain interoperability is a more difficult task. In other words, legacy data formats and best practices within a domain often incline toward a specific standard framework, but different countries or slightly different communities utilizing the same thematic data may use competitive semantic or ontological frameworks that need converging in order to reach interoperability. Note that this convergence for data interoperability only means to reach compatible standards (i.e., interoperability), not necessarily unifying them. Developments of the European INSPIRE thematic data specifications were possible thanks to the INSPIRE framework and methodology (e.g., working groups, experts, and so on), but with variable progress depending on the thematic data. Reaching data harmonization and FAIRness is conditional to the rate of adoption of the standards and data specifications, and gives a measure of success.

Soil data exchange is a good example of this sometimes long and multifaceted process. In 2010, a markup language for soil data called SoilML was proposed (Montanarella *et al.*, 2010). Pourabdollah *et al.* (2012) integrated the various soil databases approaches of Soil and Terrain (SOTER) Databases, the Food and Agricultural Organization (FAO) classification of 1988 and World Reference Base of Soil Resources (WRB) into the SOTER Markup Language (SoTerML) allowing a flexible characterization and proposed some mapping between the different classifications. SoTerML was designed to be compatible with the Geography Markup Language (GML) standard, an OGC standard built on the Geoscience Markup Language (GeoSciML) for geological data. The INSPIRE soil data specification was established in 2013 and in the same year the Global Soil Map Markup Language (GSMML) was developed (Wilson *et al.*, 2013). Also, in 2013, the ISO published the ISO 28258:2013 standard for the digital exchange of soil-related data. ISO 28258:2013 aims to facilitate the exchange of soil-related data via digital systems between entities such as individuals and organizations. ISO 28258:2013 also uses the GML standard to represent features and to explicitly geo-reference soil data, thus facilitating the use of soil data within GIS. In 2016, the OGC Soil Data Interoperability Experiment was conducted to illustrate the range of different data modelling solutions that are in use, including the SoTerML and the INSPIRE approach (Ritchie, 2016). This OGC Soil Data Interoperability Experiment evaluated existing models and proposed a common core model, including a schema, based on GML, which was tested through the deployment of OGC Web services and demonstration clients.

Hoffman *et al.* (2019) provide a survey of the various standardization activities relating to soil and agriculture. Their study highlights that even with the existence of standards to support soil science and agriculture, there continue to be differences between models adopted by various communities of interest (Batjes *et al.*, 2018; Kempen *et al.*, 2019). To improve interoperability across the different communities, there is a need for joint activities between:

- OGC Agriculture Domain Working Group;
- FAO (Global Soil partnership, i.e., pillar 4 with GLOSIS as the e-infrastructure (see http://www.fao.org/global-soil-partnership/pillars-action/4-information-data/glosis/en/);

- FAO pillar 5 for the harmonization aspects including the soil profile data exchange standard);
- ISO; and
- International Union of Soil Sciences Working Group – Soil Information Standards (IUSS WG-SIS).

Note here the large adoption of XML format (evidenced by use in GeoSciML, INSPIRE and SoilML) as a means of serializing data that are structured according to the data models defined by the data specifications. This is the case for soil data and for a range of environmental data used in agriculture. However, other serialization syntactic formats may also be used to distribute geospatial information over the Internet (e.g., RDF, JSON). Even if in theory one can switch between serialization formats, different modelling communities prefer one or another, which generally facilitates their adoption and evolution.

Emergence of cross-disciplinary enablers

In subdomains of agriculture such as agricultural economics, arable farming, cattle farming, erosion control, mixed farming, soil fertility management, and soil improvement, the geospatial component plays a key role in associating attribute information to a location on the Earth. Such attribute information may include, for example, genetic data, genomics and soil data. Whilst the attribute information may have been collected from a field or farm, in other cases it may be the result of scientific modelling and a simulation, such as in the case of plant growth modelling. Developing data services and e-infrastructure enabling each partner of a multidisciplinary project to visualize data sources from the others is often a driver for data integration generating motivations within teams (Helbig et al., 2017). As a result, strategies and analytics for the visual exploration of complex environmental data are the companions of cross-disciplinary innovation. The geospatial domain structuring of the required attribute information, along with choice of appropriate spatio-temporal scales, allows partners to envisage unforeseen coherence between such multidisciplinary information. These roles also facilitate learning and teaching in agriculture and related environmental studies (Guan et al., 2012). Standards developed and published by the OGC that have been demonstrated to be effective enablers of cross-disciplinary agricultural information exchange include:

- OGC Sensor Observation Service (SOS): This standard specifies a Web service interface for requesting, filtering and retrieving observations and sensor system information. This is the intermediary between a client and an observation repository or near real-time sensor channel.
- OGC SensorThings API (STA): This standard provides an open, RESTful, geospatial-enabled, JSON-based and unified way to interconnect and task IoT devices, data and applications over the Web.
- OGC Sensor Model Language (SensorML): This standard defines models and schema for describing sensor systems and processes and provides information needed for discovery of sensors, location of sensor observations, processing of low-level sensor observations and listing of task-able properties.
- OGC Observations & Measurements (O&M): This standard defines conceptual models for encoding observations and measurements from a sensor, both archived and real-time. Adopted by the International Organization for Standardization (ISO) as ISO 19156:2011.

- OGC Sensor Planning Service (SPS): This standard specifies a web service interface for requesting user-driven acquisitions and observations. This standard allows for tasking sensors.
- OGC Geography Markup Language (GML): The standard specifies an Extensible Markup Language (XML) grammar for expressing geographical features. GML serves as a modelling language for geographic systems as well as an open interchange format for geographic transactions on the Internet.
- OGC Web Map Service (WMS): This standard specifies an interface that provides a simple HTTP interface for requesting geo-registered map images from one or more distributed geospatial databases.
- OGC Web Coverage Service (WCS): This standard specifies an interface that offers multidimensional coverage data for access over the Internet.
- OGC Web Feature Service (WFS): This standard specifies an interface that offers direct fine-grained access to geographic information at the feature and feature property level, through discovery operations, query operations, locking operations, transaction operations and operations to manage stored, parameterized query expressions.
- OGC Web Processing Service (WPS): This standard specifies an interface that provides rules for standardizing how inputs and outputs (requests and responses) for geospatial processing, analysis and computational services should be represented.
- OGC Catalogue Services for Web (CSW): This standard specifies an interface that supports the ability to publish and search collections of descriptive information (metadata) for data, services and related information objects.
- OGC WaterML: This standard enables the representation of water observations data, with the intent of allowing the exchange of such datasets across information systems.

To explore and demonstrate how OGC standards can be applied to various domains, the OGC executes testbed projects involving several different governmental, private and academic organizations. Phase 13 of the OGC Testbed series (Testbed-13) included an experiment to test whether some of the standards listed above could enable an application to offer agricultural yield prediction within a Spatial Data Infrastructure (SDI). The experiment and its outcomes are presented by Schumann (2018).

Agricultural yield prediction relies on a variety of factors, such as crop type, disease, relief, soil and weather. Expert knowledge is required to enable the configuration of the variables and sub-variables involved in crop-yield prediction. To enable the integration of an agricultural yield prediction capability within an SDI, participants of Testbed-13 created a wrapper around the Agricultural Production Systems Simulator (APSIM) product using an implementation of the OGC WPS standard. This standard is particularly suited to interfaces for prediction models as it allows for specification of any number of input and output parameters for a process. The participants also created a workflow, using FME software, for executing the WPS and visualizing the results (Figure 5.1).

Amongst the scenarios explored by the testbed was an agricultural scientist client scenario that demonstrated how geospatial standards could enable climate and agricultural science by facilitating the ingest of geospatial data into crop forecasting models, potentially including temporal and spatial subsets of climate model variables relevant to rain and soil conditions. The focus was on providing access to data from the MERRA-2 mission of the National Aeronautics and Space Administration (NASA), exposed through implementation of the OGC WCS 2.0 standard. Moreover, the client application within the scenario also supported the integration of on-demand crop yield prediction outputs from a WPS implementation that

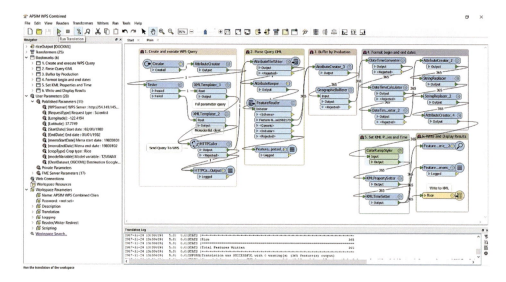

Figure 5.1 APSIM WPS FME combined scientist and non-scientist client implemented using FME workspace

Source: OGC.

had been enabled to provide predictions through running the APSIM crop model. FME was configured to invoke the WPS from the perspective of a scientist user or optionally from the perspective of a non-scientist user. The scientist option employed the full parameter query, whereas the non-scientist mode employed the abbreviated, non-expert query. In both cases, when the client received the WPS response, it parsed the GML contained in the response.

Testbed-13 also examined how a common infrastructure that supports both scientist and non-scientist users could be provided. Using a combination of the Business Process Modelling and Notation (BPMN) and the WPS standards, the testbed participants were able to implement the workflow shown in Figure 5.2.

The workflow enacts a simplistic process for the scientist user that translates the input variables into an APSIM execution document, executes the process and then returns the result. A file containing meteorological information is specified by a scientist within the scientist process. In contrast, for the non-scientist user, the process executes using a few parameters and makes decisions on the type of template APSIM file to use from the crop keyword (a restricted WPS string input). The most suitable file is identified by the workflow engine within the non-scientist process and then augmented with MERRA-2 data for a provided date range.

The architecture for this common infrastructure implemented in Testbed-13 included WCS and WPS implementations. The architecture implemented in the testbed could be extended to include implementations of WFS, SOS and STA to allow for ingesting meteorological information from near-real time observations and forecasts. Further, the architecture could be extended to support geospatial formats such as the OGC GeoPackage standard, which provides a standard way for encoding geospatial concepts in an SQLite database. Both GML and GeoPackage standards support fundamental geospatial concepts such as Coordinate Reference Systems and Simple Feature geometries. These geospatial concepts make it possible to geographically integrate cross-disciplinary data from climate science, agriculture and other fields.

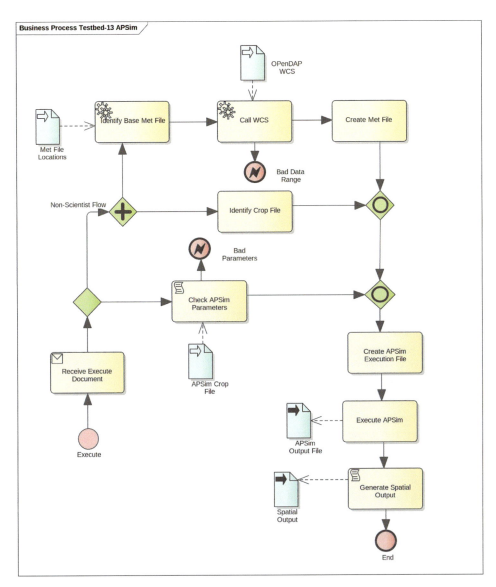

Figure 5.2 A BPMN diagram describing the workflow for the APSIM WPS
Source: OGC.

Note, in Figure 5.1 the FME environment plays the role of the BPMN workflow client and engine environment seen in Figure 5.2. Both make use of existing OGC standards to enable encapsulating within a WPS call of a combination of macro services such as APSIM. Within the GRASP project (Leibovici et al., 2017a, 2017b) a BPMN workflow client was also designed to create the scientific modelling, combining lower-level transformations and computations themselves encapsulated in a WPS (Figure 5.3). In this type of diagram both data and processing are seen as OGC Web services. This enabled the combination of specific sub-models closer to designing bespoke agricultural scientific modelling as well as allowing the use of published services of established models.

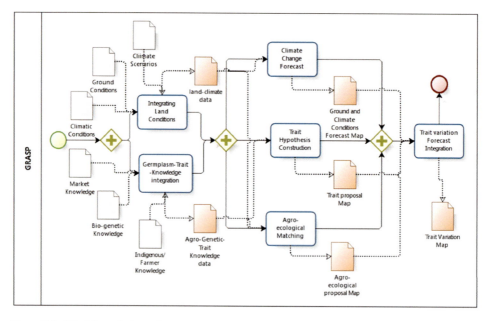

Figure 5.3 Workflow design of a generic model to be used in food security and sustainability: the Genetic Agro-ecological Sustainability Proposal model (BPMN diagram)
Source: Leibovici et al. (2017a, 2017b).

Changing granularity, each submodel can be represented as a BPMN workflow diagram combining even lower-level geospatial computational data transformations encapsulated again in WPS making use of appropriate APIs. In order to facilitate the combination of such lower-level processing facilities, metadata derived from using these APIs brings identification and quality within a standardization framework (see below under 'Future research for standard adoption').

The use and adoption of WPS and BPMN standards in scientific modelling are still in their infancy as the WPS 'libraries' are not yet well-developed towards providing generic processing available for re-use. In terms of processing environment, the complexity (e.g., in crop modelling) has certainly been an important slowing factor as it makes the development of new systems much dependent on existing ones, therefore re-using previous computing environment, i.e., often as monoblocs of written code programs somewhat like APSIM. As multidisciplinary modelling in the environmental domain develops and more and more interoperable geospatial standards become available, communities dealing with complexity such as in crop modelling making use of climate forecasting data and weather information (Jones et al., 2019), genetic data and geo-phenotypic information (Leibovici et al., 2017a, 2017b), smart farming (Colezea et al., 2018)), farming practices and agro-ecological data (Bordogna et al., 2016; McMenzie et al., 2019), and others, all stand to benefit.

The complexity in such geoprocessing modelling justifies the focus on the role of workflow environments such as using the BPMN standard in the development of geoprocessing facilities within the virtual research e-infrastructure such as proposed by the European Open Science Cloud (EOSC), the Open Science Framework (OSF) and services infrastructures such as the GEOSS (Group on Earth Observation System of Systems) (Barker et al., 2019; Enescu et al., 2019).

Connection with national standards

The presence and use of cloud computing resulted and generated an incredible amount of structured and unstructured geographical data, which now represents a benefit or challenge depending on how it is harnessed. Such data call for the use and re-use of standards to maximize its potential for innovation as evidence for decision making at the national, regional and continental levels. Understanding the evolution of standards is then critical, particularly so for the productive sector of agriculture.

ISO/TC 211 publishes a family of international standards that promote an understanding and usage of geographic information to increase the availability, access, integration and sharing of geographic information. The interoperability of geospatially enabled computer systems allows a unified approach to address global ecological and humanitarian problems through geospatial e-infrastructures by then contributing to sustainable development and delivering country commitments to the Sustainable Development Goals (SDGs) identified by the United Nations.

The coordinated development implied here requires an understanding of organizations that use geopolitical decision-making at national, regional and continental levels. This helps support the unhindered flow of digital spatial information that relies on open systems for data delivery within rapidly evolving sectors. Once the organizational framework is in place, a cooperative research and development paradigm allows to evolve the ways of sharing sectoral information promoting administrative and technical support within technologically aligned paradigms. This supports the transfer of the benefits accruing to the private sector for the facilitation of adoption and attendant private sector investment. However, realizing this is not possible if governments do not lead the consumption of GIS solutions. Demonstrations of innovation uptake are at their best when they are led by governments, particularly in the areas of real-time communication deriving from data infrastructure. This then promotes the creation of market demand that the private sector only fortifies and takes the lead when the commercial benefits are established using government demand of services and associated standards.

It is also important to acknowledge that there is a need for expanding capacity development (especially in lower-income developing countries). This will help in integrating OGC/ISO standards with regional and national standards. For example, the INSPIRE directive identifies a number of themes that form a framework for Spatial Information across Europe. One of those themes is 'agricultural and aquaculture facilities'. To enable member nations to conform to the INSPIRE directive, the European Commission led initiatives to develop Technical Guidelines for Data Specifications based on each of the themes (the INSPIRE Data Specification on Agricultural and Aquaculture Facilities – Technical Guidelines are available at https://inspire.ec.europa.eu/Themes/137/2892).

Citizen science: a new paradigm

Over the last few years, citizen science has advanced across many disciplines. In the geospatial domain, the rise of volunteered geographic information (VGI), citizen science (CS) and crowdsourcing (including social media), either providing information on existing geospatial data or being part of a data collection from using a geolocated device, has seen rapid rise in interest. CS is the concept of opening part of scientific work to the general public. There exist a number of terms that are used as synonyms or subcategories like crowd science, crowdsourced science, or in relation to spatial information there is VGI, citizens sense or citizens as sensors. Some of the most widely used applications in this domain include OpenStreetMap (https://www.openstreetmap.org) and Geo Wiki (https://www.geo-wiki.org/). Generally, CS is closely linked to the concept of Responsible Research and Innovation (RRI), but it has a number of similarities with other initiatives or Technik's like Living Labs, Science Shops, Hackathons, Datathons and IdeaThons.

To better understand the full potential of CS-captured data for agriculture, it is necessary to compare different techniques and possibilities of CS and to analyse how CS could be easily harmonized and integrated with other spatial information. The Farm Oriented Open Data in Europe (FOODIE) is a recent open data initiative based in Europe with ambitions to help agriculture across the globe. It aims to increase efficiency and open new opportunities for all involved in planning, growing and delivering food to the marketplace. The FOODIE project will enable farmers to provide their own data by easy-to-use crowdsourcing tools and applications that encapsulate the complexity of the underlying services technology. Mobile or Web applications offer Web-forms collecting specific information from farms with a wizard application allowing a farmer to configure it after a few 'clicks'. Then, the sensor system within a farm can send periodic observations to the FOODIE platform. Challenges and lessons learned from projects like FOODIE further understanding of why CS projects use existing standards or do not, and whether the need to continuously evolve creates barriers for the adoption of standards.

The discussion paper 'StandardizedStandardized Information Models to Optimize Exchange, Reusability and Comparability of Citizen Science Data' (OGC 16–129 2016) provides some answers, as well as proposes a flexible framework: SWE4CS. It makes use of OGC O&M and SWE Common standards suite. The paper also makes use of RDF and SKOS to leverage the Linked Data capabilities of the Semantic Web. The European project COBWEB was one of the first projects to apply these standards for CS as well as focusing on integrating the data quality issues (Higgins *et al.*, 2016; Leibovici *et al.*, 2017a, 2017b). CS is a key enabler for widening participation (especially youth in low-income countries) in agriculture and improving the livelihoods of smallholder farmers. For example, in the European project LandSense, contributing data from farmers are combined with Copernicus remote sensing Sentinel-2 data as a service delivered on the farmer's mobile phone to inform them about geolocated crop health (CropSupport app), with the fusion of standardized information that enables this to happen.

The FOODIE consortium has found a consensus on the content and structure of the FOODIE Data Model. The FOODIE model is a complex model including Core model, Transport Data Model, FOODIE sensor observations profile and FOODIE VGI profile. Within FOODIE data modelling activities, a VGI profile will be developed, enabling farmers and other volunteers to add agri-based content on the top of the base map provided by the FOODIE platform. An example of such an approach is currently used in the United States for the drought monitoring program CLIMAS. The primary assumption is that a farmer or volunteer is not an expert, but has skills for monitoring. For that reason, registration of the volunteers on the FOODIE platform is useful, since reporting options will be then limited by the VGI profile that suits the skills of a volunteer. In that regard, the VGI profile for meteorological monitoring is foreseen as the initial profile.

VGI has often been used for reporting of negative features. Examples in the agriculture domain may be found in the reporting of drought, mildew, wrong application of a pesticide beyond the field, and so on. The FOODIE VGI profile was developed in mid-2016 within this consensus and integrated into the Sensor Observation profile as an additional extension (Figure 5.4), taking advantage of the tested structure and respecting the conception of universal observation with result value of any data type (Řezník *et al.*, 2016).

Geospatial standards: an example from agriculture

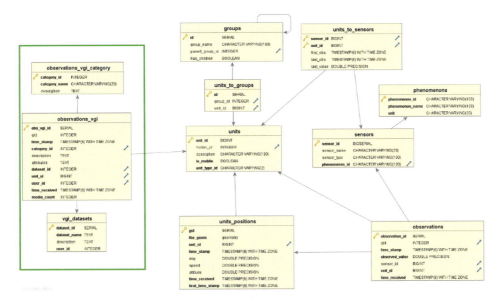

Figure 5.4 Snapshot of the FOODIE VGI profile
Source: Řezník *et al.* (2016).

There are many types of sensor data available on the Web with a variety of formats, encoding and standards. Wide-ranging services and solutions for publishing sensor data from different types of sensors are usually very complicated for non-expert users. The SDI4 Apps address this problem by designing tools and processes for effortless and straightforward utilization of existing sensor services available on the Web. This work has been done in the scope of the Open Sensors Network pilot as part of the SDI4 Apps project (Kepka and Ježek, 2012). The aim of Open Sensors Network is to create an environment where different groups of volunteers (e.g., farmers) can integrate low-cost sensors (meteorological data, air quality, and so on) into local and regional Web sensor networks. Re-using existing sensors operated by third-party providers is currently problematic. The problems range from initial search of services through metadata description of sensors to filtering of found sensor candidates. Therefore, a new module for VGI was designed and implemented. This VGI module contains new services for receiving and publishing data over the Web. The SensLog data model was extended with new tables with emphasis on variability of VGI. A VGI observation is characterized by several mandatory attributes and can be enriched with additional attributes (see Figure 5.5).

Crowdsourced, CS and VGI data, perhaps more than other data due to their recent integration into other SDIs, are likely to expose evolving solutions for data specification. Before convergence of these solutions is established, the cross-transformation from one data model to another would ensure as much working interoperability as possible while promoting its adoption. As this was already recognized in Charvat *et al.* (2014), integration of this type of data with existing SDIs (INSPIRE, GEOSS) can bring benefit to the development of both data models. On the other hand, it required action on both sides, i.e., by official SDIs to introduce lighter API-based interfaces and by CS to put effort into standardization to support easier integration.

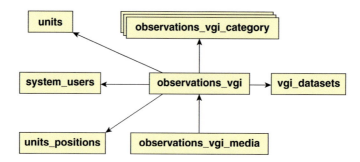

Figure 5.5 Revised FOODIE VGI profile of SDI4apps based on principles of Citizen Sensors
Source: Kepka and Ježek (2012).

Future research for standard adoption

Standards for automatic re-use of data by humans and machines is critical for efficient use of advancement in data science, such as with machine learning and artificial intelligence. Standards do not only propel the automatic re-use of data with interoperability for these emerging data science applications, but will also provide much better provenance information on how these systems are dealing with personal data (GDPR) and use licensed input data (CC-by, proprietary, and so on). The domain-specific communities in the Research Data Alliance are working on specification of domain standards.

Artificial intelligence, and more specifically machine learning, are current areas of interest for the geospatial community. Machine learning models are being developed and applied to geospatial data such as remote sensing imagery to improve feature extraction in support of land cover classification. An example of a project that examined the application of machine learning in geospatial solutions is the OGC Testbed-15 project, which used environmental data from Canada to explore scenarios such as:

- Forest change prediction model (Petawawa Super Site Research)
- Forest supply management decision-maker machine learning models (New Brunswick)
- River differentiation machine learning models (Quebec Lake)
- Linked data harvest models (Richelieu River Hydro)
- Arctic Web services discovery machine learning models

The outputs of the OGC Testbed-15 project are expected to inform future standardization activities. Given the importance of environmental management to agricultural operations, future research should look to apply emerging standards and best practices that integrate machine learning into geospatial solutions. These machine learning applications are often integrated within domain-specific models, and the scientific workflow representing the whole model can be seen as the metadata of the entire geoprocessing, playing a role in the Open Science approach. Open Science relevant to the various scientific communities involved in CODATA (the Committee on Data of the International Science Council or ISC), which exists to promote global collaboration to advance Open Science and to improve the availability and usability of data for all areas of research.

The virtual research environment and initiatives of open science, such as the European Open Science Cloud (EOSC), OSF (Open Science Framework) as well as the GEOSS (Group on Earth Observation System of Systems) in which standards and FAIR principles are paramount, have educational content as well as propose a scientific dissemination and reproducible framework where the workflow is encoded (e.g., using the BPMN standard) and gives a visual semantic as well as executable capacity (via a workflow engine). Within such systems, the workflow itself plays a role as metadata that will need to be standardized, also containing the metadata of each part of the workflow (data or processing). These metadata are driving the analytics (e.g., derived from machine learning) that can be associated with the scientific workflow, extending it towards decision-making, including the attached uncertainty assessment. Eventually, the metadata composition in the workflow contains the mechanisms to run the workflow itself as well as these analytics (Rosser et al., 2018).

The recent development of the OGC API standards is also modernizing the way through which geospatial data is accessed over the Web. An API is an interface that when backed by implementing code satisfies an interoperability need, as documented in a Requirements Specification. Web APIs use an architectural style that is founded on the technologies of the Web (W3C), which has led OGC to develop a series of new specifications, the OGC APIs. The approach taken in the OGC APIs is consistent with the W3C Data on the Web Best Practices. For example, the OGC API – Features standard, approved in October 2019, allows for the dissemination, query and retrieval of geospatial feature data. Documentation about OGC APIs is provided through OpenAPI documents (or its commercial equivalent, Swagger). In the foreseeable future, agricultural solutions may offer implementations of OGC API – Features that disseminate data encoded according to SoilML. A data infrastructure for farming, for instance, could also deploy implementations of the OGC SensorThings API standard to serve observations collected by sensors on the farm. The OGC SensorThings API standard allows for sensors in the Internet of Things to be accessed in order to retrieve observations. For requesting entities, the OGC SensorThings API follows the OData specification of the Organization for the Advancement of Structured Information Standards (OASIS). Future research should explore how such an API-driven architecture could improve the effectiveness of farming operations, agriculture modelling and decision-making.

Conclusion

Currently the standards community in the geospatial domain is experiencing a strong incentive to capitalize on the long history of standards through the advancement of open science and more specifically the adoption of FAIR as the guiding principles for better reuse of (research) data by humans and machines. It is critical that the geospatial community recognizes its existing role as an area where standards are re-used and engages with other scientific communities to re-use the existing standards in, for example, the health sector or in the logistics sector.

Research data scientists in CODATA, GO FAIR, RDA and others are proposing a globally generic standard on how to deal with FAIR Digital Objects in order to establish FAIR compliant e-infrastructures (e.g., Wittenburg et al., 2019). Initiatives such as EOSC and CODATA should be encouraged because they go in the direction of increasing the adoption of the long list of existing geospatial standards and into design of future standards that are able to answer new challenges.

References

Barker, M., Olabarriaga, S.D., Wilkins-Diehr, N., Gesing, S., Katz, D.S., Shahand, S., Henwood, S., Glatard, T., Jeffery, K., Corrie, B. and Treloar, A. (2019) "The Global Impact of Science Gateways, Virtual Research Environments and Virtual Laboratories" *Future Generation Computer Systems* 95 pp.240–248 DOI:10.1016/j.future.2018.12.026.

Batjes, N.H. and van den Bosch, H. (2018) "How Well is the WDC-Soils Serving its User Groups?" *SciDataCon-IDW 2018* Gaborone, Botswana DOI:10.17027/6F38-HN42.

Bordogna, G., Kliment, T., Frigerio, L., Brivio, P. A., Crema, A., Stroppiana, D., Boschetti, M. and Sterlacchini, S. (2016) "A Spatial Data Infrastructure Integrating Multisource Heterogeneous Geospatial Data and Time Series: A Study Case in Agriculture" *ISPRS International Journal of Geo-Information* 5 (5) 73 DOI:10.3390/ijgi5050073.

Charvat, K., Esbri, M.A., Mayer, W., Campos, A., Palma, R. and Krivanek, Z. (2014) "FOODIE – Open Data for Agriculture" *Proceedings of the IST-Africa Conference* Pointe aux Piments, Mauritius, pp.1–9 DOI: 10.1109/ISTAFRICA.2014.6880647.

Coetzee, S., Ivánová, I., Mitasova, H. and Brovelli, M. A. (2020) "Open Geospatial Software and Data: A Review of the Current State and A Perspective into the Future" *ISPRS International Journal of Geo-Information* 9 (2) 90 DOI:10.3390/ijgi9020090.

Colezea, M., Musat, G., Pop, F., Negru, C., Dumitrascu, A. and Mocanu, M. (2018) "CLUeFARM: Integrated Web-Service Platform for Smart Farms" *Computers and Electronics in Agriculture* 154 pp. 134–154 DOI:10.1016/j.compag.2018.08.015.

Enescu, I.I., Plattner, G.K., Bont, L., Fraefel, M., Meile, R., Kramer, T., Espona-Pernas, L., Haas-Artho, D., Hägeli, M. and Steffen, K. (2019) "Open Science, Knowledge Sharing and Reproducibility as Drivers for the Adoption of FOSS4G in Environmental Research" *International Archives of the Photogrammetry, Remote Sensing and Spatial Information Sciences* XLII-4/W14 pp.107–110 DOI:10.5194/isprs-archives-XLII-4-W14-107-2019.

Guan, W.W., Bol, P.K., Lewis, B.G., Bertrand, M., Berman, M.L. and Blossom, J.C. (2012) "WorldMap – A Geospatial Framework for Collaborative Research" *Annals of GIS* 18 pp.121–134 DOI:10.1080/19475683.2012.668559.

Helbig, C., Dransch, D., Böttinger, M., Devey, C., Haas, A., Hlawitschka, M., Kuenzer, C., Rink, K., Schäfer-Neth, C., Scheuermann, G., Kwasnitschka, T. and Unger, A. (2017) "Challenges and Strategies for the Visual Exploration of Complex Environmental Data" *International Journal of Digital Earth* 10 pp.1070–1076 DOI:10.1080/17538947.2017.1327618.

Higgins, C.I., Williams, J., Leibovici, D.G., Simonis, I., Davis, M.J., Muldoon, C. and O'Grady, M. (2016) "Citizen OBservatory WEB (COBWEB): A Generic Infrastructure Platform to Facilitate the Collection of Citizen Science data for Environmental Monitoring" *IJSDIR* 11 pp.20–48.

Hoffmann, C., Schulz, S., Eberhardt, E., Grosse, M., Stein, S., Specka, X., Svoboda, N. and Heinrich, U. (2019) "Data Standards for Soil- and Agricultural Research" *BonaRes Data Centre* DOI:10.20387/BonaRes-ARM4-66M2.

ISO: ISO 19136:2007 "Geographic Information – Geography Markup Language (GML)" Available at: https://www.iso.org/standard/32554.html.

ISO: ISO 19142:2010 "Geographic Information – Web Feature Service" Available at: https://www.iso.org/standard/42136.html.

ISO 19156:2011 "Geographic Information – Observations and Measurements" Available at: https://www.iso.org/standard/32574.html.

ISO: ISO 28258:2013 "Soil Quality – Digital Exchange of Soil-related Data" Available at: https://www.iso.org/standard/44595.html.

Jones, A., Edwards, B., Reusch, K. and Fraccaro, P (2019) "Geospatial Big Data Analytics for Agriculture Using IBM-PAIRS" *Geophysical Research Abstracts* 21 Munich: European Geosciences Union.

Kempen, B., Yigini, Y., Viatkin, K., de Jesus, J.M., de Sousa, L.M., van den Bosch, R. and Vargas, R. (2019) "The Global Soil Information System (GloSIS) – Concept and Design" *Geophysical Research Abstracts* 21 Munich: European Geosciences Union.

Kepka, M. and Ježek, J. (2012) "Server-side Solution for Sensor Data" In Mildorf, T. and Charvat, K. (Eds) *ICT for Agriculture, Rural Development and Environment Where We Are? Where Will We Go?* Prague: Czech Centre for Science and Society, pp.264–274.

Leibovici, D.G., Anand, S., Santos, R., Mayes, S., Ray, R., Al-Azri, M., Baten, A., King, G., Karunaratne, A., Azam-Ali, S. and Jackson, M. (2017a) "Geospatial Binding for Transdisciplinary Research in Crop Science: The GRASPgfs initiative" *Open Geospatial Data, Software and Standards* 2 (20) pp.1–11 DOI:10.1186/s40965-017-0034-3.

Leibovici, D.G., Williams, J., Rosser, J.F., Hodges, C., Chapman, C., Higgins, C. and Jackson, M. (2017b) "Earth Observation for Citizen Science Validation, or, Citizen Science for Earth Observation Validation? The Role of Quality Assurance of Volunteered Observations" *Data* 2 (4) 35 DOI:10.3390/data2040035.

Montanarella, L., Wilson, P., Cox, S., Mcbratney, A., Ahamed, S., McMillan, B., Jacquier, D. and Fortner, J. (2010) "Developing SoilML as a Global Standard for the Collation and Transfer of Soil Data and Information" *Geophysical Research Abstracts* 12 Munich: European Geosciences Union.

OASIS (2014) "OData Version 4.0 Part 1: Protocol" Available at: http://docs.oasis-open.org/odata/odata/v4.0/os/part1-protocol/odata-v4.0-os-part1-protocol.html.

OGC 16–129 (2016) "Standardized Information Models to Optimize Exchange, Reusability and Comparability of Citizen Science Data" *OGC Discussion Paper* Available at: http://www.opengis.net/doc/DP/16-129.

OGC & W3C (2017a) "Semantic Sensor Network Ontology" Available at: https://www.w3.org/TR/vocab-ssn/.

OGC & W3C (2017b) "Spatial Data on the Web Best Practices" Available at: https://www.w3.org/TR/sdw-bp/.

OGC & W3C (2022) "Time Ontology in OWL" Available at: https://www.w3.org/TR/owl-time/.

Pourabdollah, A., Leibovici, D.G., Simms, D.M., Tempel, P., Hallett, S.H. and Jackson, M.J. (2012) "Towards a Standard for Soil and Terrain Data Exchange: SoTerML" *Computers & Geosciences* 45 pp.270–283 DOI: 10.1016/j.cageo.2011.11.026.

Řezník, T., Lukas, V., Charvat, K., Charvat Jr, K., Horáková, Š. and Kepka, M. (2016) "Foodie Data Models for Precision Agriculture" *13th International Conference on Precision Agriculture* St. Louis, MO, 31th July–4th August, 2016.

Ritchie, A. (2016) "OGC Soil Data Interoperability Experiment" *Open Geospatial Consortium*, No. OGC 16-088r1" Available at: https://portal.opengeospatial.org/files/?artifact_id=69896.

Rosser, J., Jackson, M.J. and Leibovici, D.G. (2018) "Full Meta Objects for Flexible Geoprocessing Workflows" *Transactions in GIS* 22 (5) pp.1221–1237 DOI:10.1111/tgis.12460.

Schumann, G. (2018) "OGC Testbed-13: NA001 Climate Data Accessibility for Adaptation Planning" No. OGC 17-022 *Open Geospatial Consortium* Available at: http://docs.opengeospatial.org/per/17-022.pdf.

W3C (2017) "Data on the Web Best Practices" Available at: https://www.w3.org/TR/dwbp.

Wilson, P., Simons, B. and Ritchie, A. (2013) "Opportunities for Information Model Driven Exchange and On-line Delivery of GlobalSoilMap Data and Related Products" *Proceedings of the 1st GlobalSoilMap Conference: Basis of the Global Spatial Soil Information System* DOI:10.1201/b16500-85.

Wittenburg, P., Strawn, G., Mons, B, Boninho, L. and Schultes, E. (2019) "Digital Objects as Drivers towards Convergence in Data Infrastructures" *EUDAT b2share records* DOI:10.23728/b2share.b605d85809ca45679b110719b6c6cb11.

6
TECHNOLOGY, AESTHETICS, AND AFFORDANCES

Philip J. Nicholson

In March 2016, I attended a panel session entitled *Algorithmic Governance* (Crampton et al., 2016) at the American Association of Geographers (AAG) Annual Meeting in San Francisco. In this session, panelists discussed how everyday life is affected by the ubiquity of digital technologies that act according to a programmed repertoire of computerized functions; these functions, whilst 'hidden' from view, nevertheless allow for all manner of activities to take place. One question that emerged time and again from these discussions was: how do we as researchers come to understand how these technologies work, in terms of producing the spaces that we inhabit, and shaping our everyday aesthetic but also affective experiences, whether that be a matter of behaviour or felt emotion? An audience member suggested that one avenue to explore was to look at the actual code that underpinned these technologies to understand how the algorithms were constructed, and the opportunities and constraints that emerged from this. A panel member, Louise Amoore, challenged this method of enquiry, responding that we, as researchers, needed to resist the notion that by 'opening the black box' we could get at some underlying truth concerning the nature of digital technology: an argument continued in her book *Cloud Ethics: Algorithms and the Attributes of Ourselves and Others* (2020). In short, we cannot assume that by gaining access to the original intention of the coder we might be able to presuppose its effect on the world; this is because it is not the code in isolation that will help us understand its agency, but its interfacing with the world.

What struck me about this exchange was the refusal to regard code as foundational to the nature of digital technologies. Although I can appreciate the interrogative search for a key object of analysis that eschews the outward affordances of digital devices so as to 'unlock' explanations as to what a technology can and cannot do, and upon which a research design can be built, I can also understand the frustrations that ensue from engaging with a 'black box' that we cannot peer into because most are protected, proprietary mechanisms unavailable to us, or because we do not have the disciplinary expertise to map out the contents. However, I appreciate that such concerns can reinforce the idea that such a 'black box' is the epicentre that defines all other objects as belonging (or not) to digital technology, as well as from which all causality and affect emanates. There is an idealism to the belief that if we can understand the supposed core (the code) of these digitally mediated agents, then we might be able to extrapolate and understand any possible consequence.

To expand on this issue further, the risk in looking to code as *the* crucial object of analysis pertaining to such technologies is that it gives far too much credit to the programmer, implying that the author of these coded responses, or algorithmically controlled behaviours has, in some way, pre-empted all the ways in which they may affect the world. Such a position does not take into consideration how such 'algorithmically governed' spaces may give rise to unexpected or unintended behaviours. For instance, in the same AAG session Agnieszka Leszczynski raised the case of *Girls Around Me,* a 'hook-up' app that scrapes data from social media platforms to visualize nearby and potentially available women (Leszczynski and Elwood, 2015; Leszczynski 2016). This is just one example of where the affordances of digital technologies – in this case social media platforms like *Facebook* and *Foursquare* – have been exploited in ways not necessarily intended by the creators. What is more, the existence of these affordances suggests that the affective qualities produced by these digitally mediated actors are dependent upon their interactions. Therefore, how digital technologies produce affect can neither be traced to one distinct entity, nor determined by gaining access to some privileged space. If we take this line of critique further, what it undercuts is the notion that to research technology is to search for clear lines of cause and effect and, from their exposition, to draw up a summative explanation of what that technology indeed is.

Technology

GIS technology – computerized machines collecting and analysing data

Traditionally, GIS has been explained as a confluence of spatial science, the evolution of computer technology and the proliferation of personal computers, and a societal need to capture, store and analyse various types of geographical data (Goodchild, 1992). GIS has been cast as a 'computer system for input, storage, analysis, and output of spatially referenced data, (Goodchild, 1988: 560). Today, it is common to speak about GIS in terms of using certain software applications, such as *ArcMap* or *QGIS*, to work with geographical datasets. Working in this way involves importing datasets collected by the user from the field or often downloaded from online archives and repositories, such as local and national government (e.g., population densities, flood maps, fossil-fuel drilling platforms, and so on). In these applications the user can view, manipulate and analyse data to produce outputs, often – but not exclusively – in the form of maps. Data manipulation and analysis involves using built-in tools and downloadable plugins to run processes applying statistics, georeferencing, buffering, overlaying, line of sight analysis, route mapping and so on.

In this kind of encounter with GIS it is common to take GIS 'technology' as meaning the computer hardware and software that enables the doing of GIS. In the most immediate sense, GIS 'technology' is the 'technical stack' (Ash *et al.*, 2018: 37),[1] or the computer that the user sits in front of, the software that the user interacts with, and the peripheral devices that facilitate this interaction.

Under the hood

However, whilst this 'technical stack' does facilitate the doing of GIS, the site for explicit user-manipulation of datasets is determined by a specific ontology embedded in the software. This is the way in which GIS or spatial databases render, contain and store data or knowledge of reality. GIS models reality in a very particular way that presents certain constraints as well as possibilities as to how objects are represented, manipulated and tested.

Arguably, the genealogy of this ontology can be traced back to the ideas of British-American human geographer Brian Berry, specifically his geographic matrix (1964). Berry's matrix defined a version of reality by interpreting observations of the 'real world' through the lens of logical positivism. The assumption was that these observations allow one to know certain things about the real world, and by plotting observations on a matrix the spatial scientist can establish a geographical fact, i.e., something that is known about geographical reality. Modern GIS software inherits this ontology, and it is perhaps most apparent in its use of vector and raster representations of space and their associated attribute tables. What has been a concern for critical geographers has been the ways in which GIS software, configured in this way, with supposed origins in spatial science and the authority to represent the world, privilege scientifically controlled observations as the only appropriate method of knowledge gathering.

Assembling GIS

Taking a second glance, we can see that this work is reliant upon further technologies. Beyond this relatively insular sphere of the solo user and their standalone PC there is a broader horizon still of digital technologies and collections of people, management structures, professional protocols and legal frameworks, disciplinary expertise and professional networks, and so on. These are vital to, but not immediately apparent in, the everyday workings of GIS. They fade from view because they are not involved in day-to-day user activities unless a 'fault' occurs, which is when this extended world of things reveals itself.

On the one hand, we have an ontology that refers to how scientific observations through fieldwork, or lab work, or analysis, become translated into objects, or numbers, or representations, within a GIS. On the other hand is a more expansive notion of an ontology of GIS that geographers have turned to theories of assemblage to account for. What is meant by assemblage here is, 'a way of thinking about phenomena as productivist or practise-based' (Anderson and McFarlane, 2011: 126), and as a rejection of the structural ontologies mentioned above. This means thinking of GIS as a collection of hardware, software, people, emotions, skills, legal and commercial interests and practices within the world, as well as a way of world-making.

Following this idea of GIS as assemblage, some geographers have paid attention to the physical manifestations of data infrastructures, such as intercontinental fibre-optic cables, data centres, Internet exchanges, broadening the scope for considering how GIS is constituted outwith the confines of specific software applications (Ash et al., 2018). Others have focused on the relationships between the body, computer interfaces, screens and the spatialities therein, paying attention to how screens manifest digital environments via various affective qualities, leading to 'spatial understandings, embodied knowledge, political awareness and social relationships of users' (Ash et al., 2018: 33). Some scholars (Kwan 2007; Aitken and Kwan 2010; Nicholson 2018) have begun to think about the workings or the *doing* of GIS in terms of the emotional dispositions, frustrations, anxieties and affective atmospheres of GIS as producing a material and embodied residue that must be taken into consideration when we consider what GIS is.

Critical geographers have noted the commercialization of GIS. But they have also explicitly sought to move away from conceptualizing and talking about GIS in reference to a set of desktop applications (e.g., created by Esri) and instead prefer to apprehend GIS as practice (Preston and Wilson, 2014). Here, doing GIS is more about approaching a research project with a particular mindset, or set of diverse research skills and interests, than the deployment of a set of technical or computer-based skills. This attention to GIS as practice has informed the Participatory GIS (PGIS) project, for example, to import more inclusive, socially responsible, and representative approaches into the doing of GIS research in human geography (Dunn, 2007).

Similarly, studies of the 'Geoweb' focus on the ways in which such technologies can facilitate more democratic and participatory practices and can mobilize political agency allowing non-expert citizens to gather, contribute and structure their own spatial data in open databases and repositories, and to deploy GIS for various cultural as well as political and economic ends. For Wilson (2011), users engaged in work with geospatial technologies develop a literacy that becomes part of their own cyborgian (Haraway, 1991) formation of *their* geocoding body. Their geocoder perspective becomes digitally constituted, as 'their vision is linked to code' (Wilson, 2011: 363). This broader ontology of GIS allows us to not just question what GIS can do beyond making maps and visualizing information, but also what the implications of using GIS are.

Ash *et al.* (2018) propose that geographers consider this 'assemblage of the digital' as a multiple 'digital *despositif*', or 'ensemble consisting of discourses, institutions, architectural forms, regulatory decisions, laws, administrative measures, scientific statements, [and] philosophical, moral and philanthropic propositions' (Gordon, quoted in Ash *et al.*, 2018: 13). Relevant research questions would then address: the exclusions inherent in applications of such technology – who and where is captured?; the exercise of power relating to who has access to digital tools and technologies; unpacking the power relations of the 'interlocking technical stack'; resisting the perpetuation of masculinist, Western-centric and colonial perspectives in economies and cultures of big data, algorithmic governance, digital technologies and so on; and challenging claims to citizen-centric and democratized data regimes in a world where corporate and political powers are increasingly taking ownership of such data.

In recognizing this dynamic ontology, of course, what becomes apparent is that the objects one is dealing with do not remain stable. In the case of GIS, one has the ability to more precisely map terrain and communities; however, this mapping or rendering of communities and terrain may lead to land management decisions that permanently change environments, or make people more vulnerable to, for example, state and/or corporate surveillance (Pickles, 1995).

Aesthetics

Aesthetics as the way in which the material qualities of the world become known

Feagin (1999: 12) defines aesthetics as 'the branch of philosophy that examines the nature of art and the character of our experience of art and of the natural environment' and describes Baumgarten's definition of aesthetics as 'for the study of sensory experience coupled with feeling, which he argued provided a different type of knowledge from the distinct, abstract ideas studied by "logic"' (*ibid.*). Here, aesthetics refers to a taking-in of the world as it presents itself to us – through images, sounds, smells and so on – interpreted via individual perspectives that are trained, cultural, social, and deeply personal. To find a register for discussing aesthetics that is productive for GIS scholars today, Hawkins (2019: 162) chooses to eschew notions of aesthetics as they pertain to artistic renderings in exchange for the term 'aesthesis', to refer to 'the senses as, related to attention, to how we know and attend to the world'.

What does GIS look like?

In terms of aesthetics, when understood through the lens of practice, GIS is for the most part (but not exclusively) a visual experience. GIS interfaces employ a particular way of ordering graphic or visual information to communicate to the user, and so GIS becomes a practice of experiencing the world by 'seeing' the world through digital abstraction. One way in which information is ordered is by a grid. Grids are used in various ways, not least in the form of the graticule (to order geo-located objects on the map viewer), but also in the form of tabulated attributes (where these objects are defined in a table), and more broadly in the graphical user interface, where windows, side panels and drop-down menus all conform to a convention of visualizing information according to an ordered arrangement of straight lines and spaces.

For Schuurman (2002), decisions made by the GIS user – the way in which they individually conceptualize space, and the way in which GIS methodology and practice evolve – are directly influenced by the architecture of the graphical display:

> [These are] driven by the available options on the menu bar of the application. Thus, while the capacity for visual/intuitive interpretation may be expanded by the nature of GIS output, innovative treatment of the data may be restricted by the range of choice associated with the pull-down menus
>
> (Schuurman, 2002: 259)

Software design, and the possibilities presented in the form of graphics and text, influence how data is read, manipulated and understood, and often shape the direction of travel for the user. Tools, processes and functions that are presented as near-to-hand seem sensible in their arrangement and position on the screen but can also lead to pre-determined outcomes. For instance, upon launch, Esri's *ArcMap*[2] application presents the user with three main windows, which are, from left to right, 'Layers', 'Data Frame' and 'Arc Toolbox'. Each of these windows renders the user's apprehension of the data according to their own utility function – opening up some possibilities but closing down others – having consequences for how the material qualities of the data are understood. To evoke an adage, 'If your only tool is a hammer, then every problem looks like a nail'. For instance, the 'Layer' window renders data in discrete slices that can be overlaid on top of one another, hidden and revealed, promoted and relegated in the field of view in the 'Data Viewer'. The 'Data Viewer' renders vector graphics and raster images in two dimensions (often in the form of a map) as if viewed from above. Here the user can zoom, pan, select and draw via haptic interactions with the computer mouse. Some scholars have noted how such map-like representations can position a 'view from nowhere' evoking a totalizing objectivity. For Harris (2006: 119), satellite imagery in particular offers 'a totalizing, objectifying transcendent gaze, and allows one to transcend the subjective'.

In the 'Arc Toolbox' the user can choose from a list of preprogramed processes that can manipulate data according to its coded attributes. For instance, these processes can join one dataset to another, reconfigure geometry, perform statistical analysis, and so on. Data apprehended via the 'Arc Toolbox' becomes mutable according to how it is coded in numerical, statistical, and relational terms.

More than the eye

Aesthetic experience is not entirely determined by the ways in which the world presents itself to us. That is, individuals bring with them their own (often visual) literacies and biases that have been forged through experience and training. The way that we see, interpret and attach meaning to images, for example, can rely upon disciplinary, social, cultural and also deeply personal constructions. Consider the following extract from an interview with a remote sensing specialist about his day-to-day practice doing GIS. This individual explained that, to him, the images that he analyses are just grids of numbers. From his perspective, his job is just to make sense of that data according to his trained, disciplinary perspective. That is, the way in which he has learned through formal education and as part of his job to see and interpret images by,

> taking measurements through the electromagnetic spectrum of what's happening on the ground [...]. It doesn't matter what the image is. What I'm looking at is its colour, shape, features, trying to get aligned with the control points [...]. It doesn't matter that it's spatial!
>
> *(Remote Sensing Consultant, 20th August 2014)*

Individuals learn to read imagery in particular ways and come to construct a way of seeing aligned with the disciplinary concerns of – in this case – remote sensing, that are then brought to bear on the image. Moreover, because GIS practice is often interdisciplinary, mobilizing different disciplinary perspectives (e.g., scientific fields), the aesthetic experience and the ways in which meaning is attached to images are not consistent.

Beyond the grid

The aesthetics of grids, lines, windows, toolboxes, layers, complicated interfaces, and so on, persist for the most part in commercially available software applications. However, there are efforts amongst critical geographers and software developers – underpinned by an interest in the doing as well as critical appreciation of aesthetics – to import ideas from the arts and humanities (Sui, 2004).

Story mapping, for instance, has become a popular way to visualize spatial information. Caquard (2011: 136) sets the 'story map' apart from the 'grid map': where the grid map makes the landscape 'dream proof' and the story map encourages 'wonder'. Story maps deploy narrative as a powerful aesthetic device for spatializing personal and embodied experience to create emotionally charged maps. Others (Nicholson et al., 2019) have noted how story maps can also deal with some of the uncertainty in interdisciplinary work. Furthermore, Knowles et al. (2015), paying attention to the limitations of GIS methodologies for humanistic research, aims to take account of the subjectively perceived and relational understandings of space, developing an 'Inductive Visualization' that is contextualized as an alternative to GIS that can be done without access to computers. The claim here is that collaborating via analogue mediums (such as pen and paper) can inspire more creative articulations of space than GIS software often allows.

Moreover, there is a distinct push by the GIS community, as well as a growing cohort of artists working with GIS, to 'hack' the aesthetics to understand GIS beyond a technologically mediated form of praxis. Geographer Mei-Po Kwan (2007), for instance, has written about her own creative endeavours with GIS, applying artistic effects to GIS maps, as she claims, to 'explore the aesthetic potential of GIS by experimenting with various artistic styles and techniques' (Kwan, 2007: 28). This kind of experimental (Last, 2012) approach to research is emblematic of how some geographers have come to adopt artistic modes of representation, such as using creative media. For Hawkins (2011: 473), creative modes of thinking and doing can be useful in reinforcing relational, post-humanistic, assemblage understandings of space by querying a 'Cartesian subjectivity', by which it is meant that artistic practice is often well suited to destabilizing structural ontologies.

Unpacking the aesthetics of GIS enables an understanding of how conventional GIS interfaces can configure and sometimes predetermine the experience of users and the outputs they create. What is more, the import of creative modes of producing and communicating geographic knowledge can broaden the scope of these aesthetics and enable the creation of different geographical imaginaries. But how does this change the experience of the GIS user or practitioner in terms of how their practice is configured? How does the way in which users and practitioners apprehend or experience GIS steer the user in one direction or another? That is, how does a reconfiguration of the methods of communication or how information is presented to us determine what is made possible, or 'afford' certain outcomes?

Affordances

Affordances as design

The term affordance was coined by the psychologist James Gibson (1977) to account for how animals read and act upon cues to take advantage of their environment. For example, for a human, an object can be perceived as a seat, or affords sitting, or demonstrates its 'sit-on-able-ness', when it has a flat surface that is knee height above the ground. According to Gibson (1977: 57), affordances are always read from the perspective of the individual. Gibson explains that one can be equipped to take advantage of the affordances of environments either 'by nature or by learning'. However, it is the work of designer Don Norman, notably his book *The Design of Everyday Things* (1990), that has been most influential on how user experience and interface designers think about affordances (see Gaver, 1991). Norman believed that products can be designed to take advantage of people's inherent ability to read affordances in that they provide signs as to their operation:

> When affordances are taken advantage of, the user knows what to do just by looking: no picture, label, or instruction is required. Complex things may require explanation, but simple things should not. When simple things need pictures, labels, or instructions, the design has failed.
>
> *(Norman, 1990:9)*

In the most rudimentary sense, the affordances of modern GIS software applications give clues to the user by means of graphical signifiers such as drop shadows and rollover effects on buttons and menu items to hint to their 'click-on-able-ness'. Similarly, the user might come to understand that they can zoom in and out in a map viewer using the scroll wheel on their mouse because of a perceived correspondence between the backwards and forwards motion of their finger and the movement of the representation of terrain on the screen. Furthermore, such interfaces also draw upon a user's familiarity with graphical user interfaces more generally so that they know, for example, that clicking on a field heading in a table sorts the data alphabetically and numerically.

Opening up possibilities

However, the way in which geographers discuss affordances are less about how software enables data to be manipulated within a closed environment, and more to do with how GIS as a knowledge-making assemblage (technology and practice) facilitates material affects in the world beyond. Here, affordances are thought of in terms of what the technology or practice allows to happen. Affordances encompass what is made possible, both intended and unintended. These may be desirable or undesirable effects, sometimes revealing biases in their design. Geographers have used the term affordances to account for what is made possible by particular methodological approaches. For example, Preston and Wilson (2014) define five affordances of qualitative GIS research practice. These are: (1) 'breadth' in the variability of data types that can be incorporated; (2) the 'depth' of understanding afforded to research participants' individual experiences; (3) 'iterativeness and flexibility' in the ability to reassess and reframe objectives of the research as new insights become apparent; (4) the potential for 'visualizations' as with conventional GIS but also allowing a de-centring of the 'God's eye view' and creating new forms of dialogue; and (5) reorienting the research practice as 'process' rather than towards an end product.

Going one step further, Leszczynski and Elwood (2015) discuss affordances as the potential ways in which a technology can be used apart from (and including) the 'design features' of the technology. They give an example of the way in which the aforementioned *Girls Around Me* app was designed to scrape location data from check-ins on *Foursquare* for female-sounding names. The app renders that information to a smartphone screen, so that the user will be able to locate venues with a high concentration of women. Such knowledge, they argue, can afford a sense of security in women users as to the use of space. This app exemplifies for them how the 'affordances' in the design of the *Foursquare* app and other 'open' social networking apps and in how they manage the 'data flow' can be used for means not intended by the creators. What is afforded here is not just simply a malfunction but has material consequences exploiting and making certain bodies vulnerable. This particular affordance allowed for, as Leszczynski (2016) terms it, 'the rendering of vulnerable bodies' through 'affordances' of a masculinist heteronormativity, exemplifying the way in which different groups or individuals have conflicting understandings of what is afforded.

Conclusion

At this point, I would like to return to the panel session at the AAG that I discussed in the introduction to this chapter. Following Louise Amoore's refutation of the legitimacy of the 'black box' as a site at which (if we are allowed access) we might understand the nature of digital technologies, her interlocutor in the audience offered an alternative position from which to understand digital technologies, that is, as 'a complex system of emergent properties'. This alternative rendering of digital technologies, such as GIS, allows us to take into consideration their agency – for instance, their ability to evoke certain affects – but also to figure their affects as being co-constituted in relation to other objects, materials, bodies, spaces, and so on.

GIS has traditionally been understood as a collection of computer technologies for the capture, storage and analysis of geographic data, underpinned by a specific ontological understanding of reality. Geographers have come to understand GIS beyond the confines of hardware and software, and instead as constituted as part of a broader assemblage as practice. This chapter, following a reading of Baumgarten's definition of aesthetics, has described how GIS configures practices of experiencing the world by 'seeing' the world through digital abstraction according to grids, straight lines and disciplined visual literacies. However, some geographers and artists have looked to rethink these aesthetic conventions with the import of ideas from the arts and humanities, notions of experimentation, appropriation, and using narrative to structure spatial information. Furthermore, theories of affordances have been discussed in terms of how they can elucidate how individuals read and act upon clues in their environment, how user experience and interface designers understand the term as relating to how users apprehend the utility of objects, and how geographers have used the term to discuss what is made possible – both intended and unintended.

This chapter has outlined a reconfiguration in the way that GIS is interrogated in respect to technologies, aesthetics and affordances. Here, technology is apprehended in terms of what it can do, rather than what it is made of; various aesthetics are framed in terms of how they are read and made sense of, instead of the ways in which they are rendered onscreen; and affordances as to what these aesthetics allow to happen, rather than how they are designed.

Notes

1 The technical stack, sometimes referred to as the 'solutions stack', is the interlocking set of, 'platform, operating system, code, data, interface' (Ash *et al.*, 2018: 37) that makes the work possible.
2 *ArcMap* is an 'all-in-one' desktop GIS application. The reader might be familiar with applications like Microsoft *Word* or Adobe *Photoshop*. These applications are geared towards creating particular outputs, equipping the user with sets of toolbars, preprogramed formatting options, and the ability to directly manipulate digital media. If *Word* is for word processing and *Photoshop* for image manipulation, then *ArcMap* is for the processing of spatialized data.

References

Aitken, S.C. and Kwan, M.-P. (2010) "GIS as Qualitative Research: Knowledge, Participatory Politics and Cartographies of Affect" In DeLyser, D., Herbert, S., Aitken, S., Crang, M. and MacDowell, L. (Eds) *The SAGE Handbook of Qualitative Geography* London: SAGE, pp.287–304.
Amoore, L. (2020) *Cloud Ethics: Algorithms and the Attributes of Ourselves and Others* Durham, NC: Duke University Press.
Anderson, B. and McFarlane, C. (2011) "Assemblage and Geography" *Area* 43 (2) pp.124–127 DOI: 10.1111/j.1475-4762.2011.01004.x.
Ash, J., Kitchin, R. and Leszczynski, A. (2018) "Digital Turn, Digital Geography?" *Progress in Human Geography* 42 (1) pp.25–43 DOI: 10.1177/0309132516664800.
Berry, B. (1964) "Approaches to Regional Analysis: A Synthesis" *Annals of the Association of American Geographers* 14 (1) pp.1–7 DOI: 10.1111/j.1467-8306.1964.tb00469.x.
Caquard, S. (2011) "Cartography I: Mapping Narrative Cartography" *Progress in Human Geography* 37 (1) pp.35–144 DOI: 10.1177/0309132511423796.
Crampton, J.W. and Miller, A.(2016) "Algorithmic Governance" in *American Association of Geographers Annual Meeting*, San Francisco, CA.
Dunn, C.E. (2007) "Participatory GIS – A People's GIS?" *Progress in Human Geography* 31 (5) pp. 616–637 DOI: 10.1177/0309132507081493.
Feagin, S.L. (1999) "Aesthetics" In Audi, R. (Ed.) *The Cambridge Dictionary of Philosophy* (2nd.ed.) Cambridge: Cambridge University Press, pp.11–13.
Gaver, W.W. (1991) "Technology Affordances" *Proceedings of CHI'91* New Orleans, pp.79–84 DOI: 10.1145/108844.108856.
Gibson, J.J. (1977) "The Theory of Affordances" In Shaw, R. and Bransford, J. (Eds) *Perceiving, Acting and Knowing: Toward an Ecological Psychology* Hillsdale, NJ: Erlbaum, pp.67–82.
Goodchild, M.F. (1988) "Geographic Information Systems" *Progress in Human Geography* 12 (4) pp. 560–566 DOI: 10.1177/030913258801200407.
Goodchild, M.F. (1992) "Geographical Information Science" *International Journal of Geographical Information Systems* 6 (1) pp.31–45 DOI: 10.1080/02693799208901893.
Haraway, D. (1991) *Simians, Cyborgs and Women: The Reinvention of Nature* London: Free Association Books.
Harris, C. (2006) "The Omniscient Eye: Satellite Imagery, 'Battlespace Awareness', and the Structures of the Imperial Gaze" *Surveillance and Society* 4 (1–2) pp.101–122 DOI: 10.24908/ss.v4i1/2.3457.

Hawkins, H. (2011) "Dialogues and Doings: Sketching the Relationships Between Geography and Art" *Geography Compass* 5 (7) pp.464–478 DOI: 10.1111/j.1749–8198.2011.00429.x.

Hawkins, H. (2019) "Towards a GeoHumanities GIS?" *Transactions in GIS* 23 (1) pp.161–163 DOI: 10.1111/tgis.12499.

Knowles, A. K., Westerveld, L. and Strom, L. (2015) "Inductive Visualization A Humanistic Alternative to GIS" *GeoHumanities* 1 (2) pp.233–265 DOI: 10.1080/2373566X.2015.1108831.

Kwan, M. (2007) "Affecting Geospatial Technologies: Toward a Feminist Politics of Emotion" *Professional Geographer* 59 (1) pp.22–34 DOI: 10.1111/j.1467–9272.2007.00588.x.

Last, A. (2012) "Experimental geographies" *Geography Compass* 12 (12) pp.706–724 DOI: 10.1111/gec3.12011.

Leszczynski, A. (2016) "Algorithmic Governance" *AAG Annual Meeting* 29th March to 2nd April, San Francisco, CA.

Leszczynski, A. and Elwood, S. (2015) "Feminist geographies of new spatial media" *The Canadian Geographer* 59 (1) pp.12–28 DOI: 10.1111/cag.12093.

Nicholson, P.J. (2018) *Snap, Pan, Zoom, Click Grab, and the Embodied Archive of Geographic Information Systems* (PhD thesis) University of Glasgow.

Nicholson, P.J., Dixon, D., Pullanikkatil, D., Moyo, B., Long, H. and Barrett, B. (2019) "Malawi Stories: Mapping an Art-science Collaborative Process" *Journal of Maps* 15 (3) pp.39–47 DOI: 10.1080/17445647.2019.1582440. Norman, D.A. (1990) *The Design of Everyday Things* New York: Doubleday.

Pickles, J. (1995) *Ground Truth: The Social Implications of Geographic Information Systems* New York: Guilford Press.

Preston, B. and Wilson, M. (2014) "Practicing GIS as Mixed Method: Affordances and Limitations in an Urban Gardening Study" *Annals of the Association of American Geographers* 104 (3) pp.510–529 DOI: 10.1080/00045608.2014.892325.

Schuurman, N. (2002) "Women and Technology in Geography: A Cyborg Manifesto for GIS" *The Canadian Geographer/Le Géographe canadien* 156 (3) pp.258–265 DOI: 10.1111/j.1541–0064.2002.tb00748.x.

Sui, D. (2004) "GIS, Cartography, and the 'Third Culture': Geographic Imaginations in the Computer Age" *The Professional Geographer* 56 (1) pp.62–72 DOI: 10.1111/j.0033-0124.2004.05601008.x.

Wilson, M.W. (2011) "'Training the Eye': Formation of the Geocoding Subject" *Social and Cultural Geography* 12 (4) pp.357–376 DOI: 10.1080/14649365.2010.521856.

7
RACE AND MAPPING

Catalina Garzón-Galvis and Beth Rose Middleton Manning

This chapter speaks to a growing body of academic and political work that addresses race and cartography, advocating for a shift in understandings of space as entwined with histories of exclusion and attempted erasure. The land acknowledgements in the authors' biographies invoke non-settler conceptions of space, while recognizing that the institutions named are within Indigenous homelands and are part of a heritage of invasion (see Blenkinsop and Fettes, 2020; Lee *et al.*, 2020). Elevating Black, Indigenous and People of Colour leadership, this analysis of race and mapping asserts the need for recognition, reparations and rematriation. The term 'reconciliation' is not used, since some harms wrought by racism, settler-colonialism, attendant land uses and desecration cannot be reconciled (see Tuck and Yang, 2012). Indigenous and decolonial scholars, critical geographers and cartographers of race and ethnicity invite reflection on what changes are possible, even while recognizing the context of irreparable harms, power imbalances and incommensurability with a settler-colonial state.

Mapping, racism and settler-colonialism

Mignolo (1992: 59) posits that early European maps of the Americas are 'cognitive and culture-relative artifacts used not only for wayfinding but also for the colonization of space and of the mind'. Maps created by explorers like Columbus rendered known that which was previously unknown to them by extending a geographic consciousness necessary for European colonial expansion. The hierarchical ordering of space is evident in the privileged orientation that Europe is given relative to the Americas, mapping conventions derived from Christian ideology (T-O maps modeled after the cross), and depictions of wild creatures on the periphery reinforcing the othering of non-European places and peoples (see Livingstone, 2010). Mignolo (1992: 59) contends that 'economic expansion, technology and power, rather than truth, characterized European cartography early on, and national cartography of the Americas at a later date'.

Indigenous peoples resisted the imposition of foreign spatial epistemologies attempting to displace them. Mundy's *The Mapping of New Spain* (1996) examines maps created by Nahua cartographers as part of early colonial land surveys, whose emphasis on Indigenous social relations and toponyms rendered them incomprehensible to Spanish administrators. Nahua maps conveyed a storied landscape instead of a *terra nullius,* unsettling a colonial imaginary of the Americas as empty lands devoid of civilization (see Bauer, 2016). These maps, made by 'community leaders who were incisively aware of the political and legal ramifications and uses of land titles, territorial maps, and boundary narratives' (Craib, 2000: 26), revealed resistance to the colonization of space. Sociospatial relations conveyed in Nahua maps were suppressed by Spanish settlers via land grant maps that privileged Western demarcations of private property.

Dividing space in this way had the dual impacts of subsuming Indigenous perspectives on space and imposing confining racialized systems of sociospatial organization. Though socially constructed, racial hierarchies have had profound material effects on people of colour by engendering spatial stratification predicated on dehumanization via simultaneous treatment as property and exclusion from property rights (Harris, 1993). Structural racism disadvantaged Black and Indigenous peoples by expropriating land and exploiting labour through laws and policies designed to increase white settler property. While those categorized as Black were subjected to the 'one drop rule' to maximize labour exploitation, fraction-based blood quantum and termination policies were applied to Indigenous peoples to accelerate alienation from rights to land and self-determination (Day, 2015).

Racial classifications were foundational to the dispossession and genocide of Indigenous peoples through cartographic practices of erasure. Settler-colonial states engaged in cartographic dispossession by replacing Indigenous place names with European language toponyms and excluding culturally significant places like sacred sites (Craib, 2017). Bauer (2016) calls out settler attempts at 'firsting', or asserting themselves as discoverers with the rights to name Indigenous places that already have names. Smith (1999: 53) underscores the complicity of Western mapping practices with genocide by depicting Indigenous homelands on settler-colonial frontiers as a *tabula rasa*: 'for Indigenous Australians to be in an "empty space" was to "not exist"'.

Settler-colonial geographies deployed environmentally deterministic and culturally reductionist narratives to justify ongoing imperialist expansion based on racialized logics of exclusion (Peet, 1985; Winlow, 2019). Western mapmaking applied a Kantian nature/culture binary and Linnaean systems of biological classification to propagate an ideology of white supremacy, 'literally and metaphorically mapp[ing] those boundaries [...] that separate and exclude the world of privilege from the world of the "other" along racial lines' (Kobayashi and Peake, 1994: 226). Winlow (2006, 2009) examined how Taylor's zones and strata theory and Ripley's *Races of Europe* rationalized racialized spatial orders via the use of shading and geometric divisions in cartographic depictions of migration, physical differentiation and the distribution of social ills, conventions that continue to influence data visualizations created with geospatial technologies today.

Mapping, dislocation and racialized exclusion

Hunt and Stevenson interrogate cartography as an instrument of the settler-colonial state since 'the map is a form of knowledge that has the power to dispossess' by "produc[ing] space [...] that is seemingly abstracted from the lived experiences of those who actually occupy it' (2017: 4–5). This abstracted representation of spatial knowledge marked a departure from embodied and experiential ways of knowing. Instead, mapping propertied interests produced spatial knowledge that entrenched state power by excluding and dislocating Indigenous and communities of colour.

Redlining practices in the United States demonstrate how mapping ingrained racialized value-hierarchies into the commodification of property to perpetuate residential segregation and racial inequality (Massey and Denton, 1993). Within the 1930s redlining maps produced by the Home Owners Loan Corporation (HOLC), neighbourhoods were assigned an 'A, B, C, or D' grade and a colour designation; the colour red denoted neighbourhoods assigned the lowest grade of 'D'. Accompanying area descriptions defined a neighbourhood's 'desirability' according to perceived racial composition and physical conditions. These designations were then applied by government agencies and investors to apportion higher value properties to whites.

The systematic exclusion of people of colour from economically valorized neighbourhoods consolidated an inequitable distribution of intergenerational wealth along racial lines (Oliver and Shapiro, 1999) and disproportionately concentrated undesirable land uses, environmental hazards, and health disparities in communities of colour (Bullard, 1994). Landscapes and peoples in what environmental justice scholars refer to as 'sacrifice zones' were rendered economically and socially dispensable through discriminatory housing and investment practices (Lerner, 2012). Historically, redlined neighbourhoods still contend with adverse impacts associated with higher pollution exposures such as elevated rates of chronic health conditions like asthma and cancer, lack of access to health insurance, and reduced life expectancy (Nardone *et al.*, 2020).

Racialized disadvantage was also manufactured via *tabula rasa* planning and policies that enabled infrastructural expansion of the settler-colonial state through subsequent waves of dispossession and displacement. Deemed 'Negro Removal' by civil rights activists, 1960s urban renewal policies facilitated white flight to affluent suburbs at the expense of historically redlined Black neighbourhoods targeted for freeway expansion (Aaronson *et al.*, 2017). Indigenous reservations and homelands were likewise targeted for the expansion of hydroelectric dams and the construction of the Keystone XL pipeline (Lawson, 2009; Estes, 2019; Gilio-Whitaker, 2019). Disenfranchisement, dispossession and dislocation produced profound emotional pain, traumatic stress and intergenerational trauma in Black and Indigenous communities (Cajete, 2000; Fullilove, 2001) – what decolonial scholars refer to as the 'colonial wound' from repeated subjection to racialized dehumanization (Fanon, 1961; Mignolo, 2005).

Mapping resistant ways of being and knowing

Centring the agency of peoples experiencing racism and colonization is integral to understanding how marginalized spatial knowledges can empower to transform oppressive conditions (Fanon, 1959; Smith, 1999; Collins, 2002). Reflecting on Indigenous sense and depth of place, Johnson (2012: 831) argues that 'what is missing from our place worlds is a connection to the significant cultural histories and moralities […] stored within our storied landscapes'. Allen *et al.* (2019: 1010) posit that place is better suited than landscape as an analytic for Black geographies, since 'placemaking builds and positions communities within society even as these communities build and position place with specific values toward a sociopolitical re-ordering of society and space'. Discussing how Black girls navigate practices of confinement, Butler (2018: 31) asserts that 'conceptions of Blackness are tied to reclaiming a sense of belonging, weaving one's self into genealogies of resilience, and conjuring new imaginings of existing'.

Racial formations are rooted in the specific historical trajectories of places as well as resistance to the appropriation of spatial knowledges to exploit land and labour (Gilmore, 2002; Omi and Winant, 2014). According to Johnson (2012: 833), 'when we are engaged with place, we are carrying out an act of remembrance, a retelling of the stories written there, while also continually rewriting these stories'. Dispossession, migration and cultural alienation due to forced assimilation complicate a confounding of place with location and identity in that 'locating indigeneity in relation to a specific place overlooks indigenous peoples' contemporary and historic mobility' (Pulido, 2018: 313). Mignolo's (2012) concept of 'border thinking', as generated from geopolitical and ontological boundaries like the colour line, directs us towards lived experiences of embodied resistance as place-based knowledge. McKittrick (2006) and Goeman (2013) challenge us to centre the experiences of Black and Indigenous women engaging in mapping practices as a means of engendering unbounded possibilities for being in the world.

Counter-mapping as power-building

Counter-mapping engages communities historically exploited by or excluded from Western cartographic representations and practices in deploying mapping tools to challenge the dominant narratives, policies, and decisions that these inform. Peluso (1995: 386–387) posits that countermaps 'greatly increase the power of people living in a mapped area to control representations of themselves and their claims to resources'. Counter-mapping contests the attempted historical erasure of relationships to land and rights to place by centring experiential ways of knowing and questioning settler-colonial boundaries used to consolidate power and privilege. Countermaps challenge master narratives of race and power relations by 'tell[ing] the history of Western research through the eyes of the colonized' (Smith, 1999: 2).

Counter-mapping can enlist Western geospatial technologies (GT) in utilizing government data to dispute the conflation of historically devalued places and peoples. The 'Mapping Inequality' and 'Renewing Inequality' projects digitized US redlining maps and generated data visualizations of displacement patterns due to urban renewal policies (Cebul, 2020; Nelson et al., 2020). The National Community Reinvestment Coalition (2020) leveraged these projects to analyse racial disparities in adverse health outcomes like COVID-19 rates and to advocate for prioritizing historically redlined neighbourhoods for housing investments and pandemic response resources.

Counter-mapping can apply GT to integrate community-generated data in seeking accountability for settler-colonial violence. Lucchesi (2019) mapped the locations of cases of missing and murdered Indigenous women and girls (MMIWG) in North America and conducted pathway analyses to reconstruct their life courses. The Sovereign Bodies Institute (2020) deployed this MMIWG database to advance justice for impacted Indigenous women and girls, underscoring how gender-based and state-sanctioned violence concentrated MMIWG cases along the Keystone XL pipeline. Lucchesi (2018: 24) aims 'to effectively communicate the varied scale of the loss and violence suffered, and […] through the mapmaking process itself, to find healing'

Counter-mapping also deploys GT to elevate resistance to ongoing racial violence and dislocation. The Anti-Eviction Mapping Project combined digital storytelling, crowdsourcing and eviction data to create story maps with Black youth and families experiencing gentrification (McElroy, 2018). Activists crowdsourced the locations of police killings, Black Lives Matter (BLM) protests and contested confederate statues since the murder of George Floyd by Minneapolis police (Romo and MacLachlan, 2020). Journalists mapped statue removals of Spanish colonizers by Indigenous rights activists working in solidarity with BLM activists to connect contemporary police brutality with the genocide of California Indians (Ortiz et al., 2020). The Yellowhead Institute (2019) mapped mining claims in First Nations territories to advance reparations as part of a larger #LandBack movement to rematriate

Indigenous lands. These are but a few examples of counter-mapping as digital organizing to support contemporary racial justice and Indigenous rights movements.

Indigenous decolonial mapping

Rose-Redwood *et al.* (2020: 152) define decolonial mapping as 'spatial practices and cartographic techniques that center on Indigenous relationships and responsibilities to land, including but not limited to spatial narratives, place ontologies, more-than-human relations, navigational guidance, and territorial demarcations'. Beyond spatial practices, Indigenous cartography may be expressed in three or more dimensions, grounded in ceremony, song, dance, story and embodied practice (Pualani Louis, 2017). Pearce and Louis (2008: 110) call for GT applications that are transformative rather than translational or assimilationist in relation to Indigenous spatial knowledge by 'finding ways to shape [Western] structures in order to convey the structures of Indigenous cartographies'.

Political maps of California and associated state histories foreground settler geographies and narratives. Indigenous geographies, including toponyms, narratives, histories and relationships between places remain, but are not indicated on settler maps. Bauer (2016) foregrounds Indigenous narratives of place, drawing on oral histories from Wailaki, Concow, and Paiute people recorded as part of a 1930s WPA project. The rich narratives they share draw a new map of the state now known as California, one that reflects the processes of creating different peoples in their respective homelands, and their interactions with other beings to form the landscapes, mountains and rivers traversed today. Bauer's narratives unsettle settler histories, reflecting on settler narratives as one very recent and limited story of California. Further, they reflect on the primacy and continuance of Indigenous narratives, which shaped the landscapes settlers came into contact with and guided their very actions on the land.

After years of attempting to draw a map of Native homelands in California that would not ignore any single nation's understanding of its own boundaries and shared areas with other nations, the Native American Heritage Commission released its Digital Atlas of California Native Americans in 2020. The atlas offers boundaries as blurred areas of overlap between Native nations, indicating understandings of shared use areas, rather than the strict political or linguistic boundaries imagined by non-Native anthropologists. By rejecting long-accepted anthropological maps in favor of listening to Native nations' own understandings of their histories and homelands, the NAHC atlas foregrounds Indigenous epistemologies of place. As such, it recognizes diverse nations' rights to be involved in decision-making processes about the management of shared homelands.

Participatory mapping

Mapping as an empowering or decolonial practice requires a radical shift in methodologies used to generate and represent spatial knowledges (see Asselin and Basile, 2018). Louis (2007) outlines four principles for Indigenous methodologies to guide geographic research: (1) relational accountability, or responsibility to honour relationships within the research process and with the land; (2) respectful (re)presentation, or a commitment to deep listening, cultural humility and respect for decisions made by Indigenous communities about what and how to research; (3) reciprocal appropriation, which requires that research benefit both Indigenous communities and the researcher; and (4) rights and regulation, or research processes guided by Indigenous protocols that recognize their intellectual property rights.

Participatory mapping approaches operationalize principles of accountability by emphasizing critical reflection on power and privilege dynamics in knowledge co-creation (Ayala *et al.*, 2005; Fox *et al.*, 2006; Harris, 2016; Muhammad *et al.*, 2017). Cultivating critical cartographic literacy in Indigenous communities entails activating an 'understanding of the epistemological divide between Western and Indigenous cartographic systems' (Johnson *et al.*, 2005: 90). Incorporating iterative cycles of awareness, reflection and action into the aims, design and implementation of mapping projects can support mutual co-learning between community cartographers and outside researchers (Garzón-Galvis *et al.*, 2019). Partnership agreements can delineate protocols for shared decision-making, structures of collaboration and community benefits including the equitable distribution of funding and ownership of co-created knowledge (Minkler, 2004).

Participatory approaches can incorporate body mapping, mental mapping and life mapping techniques to deepen awareness of shared lived experiences (De Jager *et al.*, 2016; Campos-Delgado, 2018). Story mapping and power mapping can generate a shared analysis of the root causes of these experiences based on how community members themselves interpret them (Littman *et al.*, 2021). Introducing maps and information created with GT can foment critical dialogue by ground-truthing outsider representations based on lived experiences. This resulting layering of outsider and community knowledge can inform collective action planning to advance positive change by enlisting outside researchers as allies (Moore and Garzón, 2010).

Mapping visions for community revitalization in West Port Arthur, Texas

Community assets and hazards mapping has been used for decades by Environmental Justice (EJ) organizations to document place-based knowledge and ground-truth geospatial data on environmental and health conditions collected by government agencies (Sadd *et al.*, 2014; Huang and London, 2016; Corburn, 2017). Garzón-Galvis and Moore developed a popular education approach to community assets and hazards mapping that combines storytelling prompts with hands-on activities using poster-sized printouts of aerial base maps overlaid with transparency paper to inform visioning and action planning (Moore and Garzón, 2010). Prior to each workshop, they worked with community partners and participants to tailor facilitators' agendas and materials, like defining community boundaries and map legends.

In 2011 Hilton and Marie Kelley of Community In-Power and Development Association (CIDA) invited Garzón-Galvis and Moore to facilitate a mapping workshop in West Port Arthur, Texas, a historically redlined Black neighbourhood on the Gulf Coast contending with multiple health risks from refinery and petrochemical facilities. Participants were each asked to share a community treasure that they value and an environmental problem in their community. After discussing the relationships between assets and hazards, participants identified opportunity sites where land use changes that meet community needs could be located. For example, a vacant lot initially identified as a problem due to illegal dumping became viewed as a potential location for locally owned business incubators. Transparency overlays containing stickers representing treasures, problems and opportunity sites were then juxtaposed to inform action planning. Participants presented their maps to community leaders at the Texas Environmental Justice Encuentro and to regulatory agencies and decision-makers to catalyse environmental remediation and economic reinvestment that would benefit West Port Arthur residents.

Visualizing alternative land futures: towards rematriation of Indigenous lands

The Indigenous women-led Sogorea Te Land Trust (STLT) in Oakland, California is changing the narrative about the heavily urbanized East San Francisco Bay Area. STLT visual renderings, oral histories and multifaceted activism for site protection reassert the East Bay as a storied Indigenous homeland with sacred sites to be respected and protected. Before forming STLT, co-founders Corrina Gould and Johnella LaRose were active in cultural protection around San Francisco Bay, with a focus on protecting ancient shellmounds and other culturally important places threatened or desecrated by development. This work continues through both Indian People Organizing for Change (IPOC), Confederated Villages of Lisjan and STLT. Hence, STLT is committed to rematriating land to Indigenous women-led restoration and stewardship to address settler-colonial alienation and dispossession.

In collaboration with artists, IPOC created artistic renderings of the Bay Area that visualize alternative futures for the land. A 2017 image of the West Berkeley Shellmound – now covered by a parking lot and proposed for a high-rise, mixed use development – offers a vision of a park covered with Native plants, bisected by a restored creekbed, that includes an educational and community facility honouring the history and reverence of the site. The visualization of urban spaces as Indigenous spaces, in both Oakland with STLT and in Los Angeles with the Mapping Indigenous LA Project (see Senier, 2018), provides a decolonial perspective on urbanized landscapes, revealing their layered identities as diverse Indigenous homelands in need of protection and stewardship.

In 2019, STLT erected a traditional arbor, in East Oakland, on land pending transfer to STLT from the non-profit Planting Justice. According to STLT co-founder Gould, the construction of the arbor is especially significant because it is the first time in 250 years that a ceremonial structure has been in Lisjan homelands in the East Bay. This first property transferring to STLT is a sovereign space, 'where Lisjan people can practice their spirituality without outside interference – a real part of the definition of rematriation'. As an urban Indigenous women-led land trust on Lisjan Territory, the STLT works with the Tribe to revitalize culture, language and ceremony, and simultaneously works to ensure that other urban Natives have access to ceremonial places and land to grow their traditional medicines and foods.

This work offers a grounded possibility to apply the sentiment expressed by California Governor Newsom in Executive Order N-15–19, which recognizes the genocidal impacts of the development of the state of California and commits to investment in truth and healing processes with Indigenous peoples. With EO N-15–19, there is an opportunity to examine detailed histories of the state's most iconic cities, landmarks and natural features, and to address the violence and erasure inherent in their naming and development. EO N-15–19 holds the possibility of Indigenous land rematriation/repatriation, and the work of Native land trusts like STLT show concrete processes of how and where to bring that to fruition.

Mapping allotments in Mountain Maidu homelands

In 2019, the Maidu Summit Consortium celebrated the long-awaited transfer of Tásmam Kóyom or Humbug Valley to Mountain Maidu people. Tásmam Kóyom is one of several mountain valleys threatened by flooding for hydroelectric development. Maidu countermaps foregrounding both cultural understandings of space and more political aspects of land rights asserted the importance of the site and contributed to advocating for its protection and, ultimately, repatriation.

Cultural maps, such as the Salt Song Map (Klasky, 2009) or the Honey Lake Maidu map of Worldmaker's Trail (Morales, 2009), are often multi-dimensional, containing histories, relationships and responsibilities expressed in land formations, songs and practices. These maps counter the silences of colonial maps by affirming place as homeland, foregrounding relationships between humans and the surrounding landscape that extend far before settler invasion. More explicitly political Indigenous counter-maps emphasize the assertion of Indigenous property rights.

Maidu elder Lorena Gorbet developed a detailed, hand-drawn map of parcels under a northeastern Sierra reservoir, Lake Almanor, at the headwaters of the North Fork Feather River stairway of power and the California State Water Project (Middleton, 2010; Middleton Manning, 2018). Beginning with Gorbet's map and conversations with community members, Middleton Manning contributed to researching Indian allotment files and tracing their cancelation and sale by the Bureau of Indian Affairs, in collusion with hydropower and timber companies operating in the region. The resulting GIS layer of allotment parcels may be used on tribal and other maps to assert Maidu histories of government-recognized land rights in Plumas and Lassen counties. While a political counter-map of allotment parcels is limited to areas that the government once recognized as Maidu land, it can still be used to advocate for co-management, representation and land rights.

The broadest scope to assert Indigenous rights may be in the combination of cultural and political counter-maps. However, concerns over site protection, protection of traditional knowledge and attention to the uses that maps may be put to once created limit the type of knowledge community members may want to place on a map. These decisions about what to share spatially, how and with whom must be made within the community.

Mapping sacred sites in Winnemem Wintu homelands

The Winnemem Wintu Tribe was not duly consulted or compensated when the 1941 Central Valley Project Indian Lands Acquisition Act authorized seizing their allotment lands for Shasta Dam construction along the McCloud River to supply water to California cities and large-scale agriculture (Farnham, 2007). The dam resulted in the submersion of 90% of their ancestral homelands and ensuing dispossession from sacred sites, salmon runs and traditional lifeways (Dadigan, 2014). McTavish (2010) documented how maps depicting tribal territorial boundaries, allotment lands, eminent domain powers and withdrawal of government recognition from *rancherias* (California Indian reservations) under termination policies eroded Winnemem Wintu claims to land and cultural resources.

In 2011, Winnemem Wintu Chief Caleen Sisk and her son Michael (Pomtahatot) Preston invited Garzón-Galvis and Moore to build on a tribal oral history project by training youth to use GPS (global positioning system) units to map the locations of sacred sites at risk of being submerged due to proposed expansion of the Shasta Dam (DataCenter, 2015). Participants created a data security protocol to protect tribal ownership of the geospatial information gathered when sharing recommendations for sacred site protection with state water and energy infrastructure decision-makers and non-tribal members (SLFP, 2011). The Tribe leveraged this mapping in advocating against proposed dam expansion that would destroy at least 40 sacred sites and in developing a plan to restore traditional Chinook salmon runs (WWT, 2017). The Winnemem Wintu have also held ceremonies at endangered sacred sites and organized a 300-mile prayer run from the Bay Area to the McCloud River headwaters to raise awareness about threats to the water, salmon and traditional lifeways (WW, 2019).

Towards an anti-racist decolonial praxis

Forging an anti-racist cartography necessitates transforming toxic environments within academia and research institutions for people of colour engaging in activist cartography and mapping for racial justice, for which their white colleagues are all too often rewarded while they contend with denials of tenure and career advancement (Mahtani, 2014). Strengthening educational pipelines and institutional supports for students and faculty of colour is imperative to addressing the discipline's settler-colonial legacy by problematizing the binary of privileged (predominantly white) cartographers surveying racialized mapped 'subjects'. Peake and Kobayashi (2002) lay out an agenda for an anti-racist geography that includes aligning research, teaching and institutional practices with racial justice movements to centre the meaningful participation of both directly impacted communities and scholars of colour.

Lest participation alone become construed as a panacea, an anti-racist cartography should also heed cautionary tales from participatory mapping practitioners. Fox *et al.* (2006: 105) call these recombinant forms of hegemonic power and privilege the ironic effects of spatial information technologies: 'the more we map, the more likely it is that we will have no choice but to map'. The desecration of sacred sites whose locations have been digitized underscores the dangers of continued appropriation of spatial knowledges to perpetrate settler-colonial violence. Navigating acts of cartographic refusal in beleaguered communities (see Sylvestre *et al.*, 2018) entails an ethics guided by the political principle of 'nothing about us, without us' that respects the right to not be mapped and rooted in the rightful return of land and resources: 'Until stolen land is relinquished, critical consciousness does not translate into action that disrupts settler colonialism' (Tuck and Yang, 2012: 19). An anti-racist decolonial cartography as grounded theory expressed spatially moves from mapping to collective action that builds power to dismantle the institutions underlying both racism and settler-colonialism.

Acknowledgments

The authors would like to thank the community and research partners in the participatory and Indigenous mapping projects discussed in this chapter: Hilton and Marie Kelley of Community In-Power and Development Association; Miho Kim of DataCenter; Farrell Cunningham *yatam* and Lorena Gorbet of the Maidu Culture and Development Group; Ron Morales of Honey Lake Maidu; Beverly Ogle of Tásmam Koyóm Cultural Foundation; Eli Moore of the Othering and Belonging Institute; Corrina Gould and Johnella LaRose of Sogorea Te Land Trust; John Sullivan of University of Texas Medical Branch; and Chief Caleen Sisk and Michael (Pomtahatot) Preston of the Winnemem Wintu Tribe.

References

Aaronson, D., Hartley, D.A. and Mazumder, B. (2020) *The Effects of the 1930s HOLC 'Redlining' Maps* Chicago: Federal Reserve Bank of Chicago.

Allen, D., Lawhon, M. and Pierce, J. (2019) "Placing Race: On the Resonance of Place with Black Geographies" *Progress in Human Geography* 43 (6) pp.1001–1019.

Asselin, H. and Basile, S.I. (2018) "Concrete Ways to Decolonize Research" *ACME: An International Journal for Critical Geographies* 17 (3) pp.643–650.

Ayala, G., Maty, S., Cravey, A. and Webb, L. (2005) "Mapping Social and Environmental Influences on Health" In Israel, B., Eng, E. and Schultz, A. (Eds) *Methods in Community-Based Participatory Research for Health* San Francisco: Jossey-Bass, pp.188–209.

Bauer Jr, W.J. (2016) *California through Native Eyes: Reclaiming History* Seattle: University of Washington Press.

Blenkinsop, S. and Fettes, M. (2020) "Land, Language and Listening: The Transformations that can Flow from Acknowledging Indigenous Land" *Journal of Philosophy of Education* 54 (4) pp.1033–1046.

Bullard, R.D. (1994) *Unequal Protection: Environmental Justice and Communities of Color* San Francisco: Sierra Club Books.

Butler, T.T. (2018) "Black Girl Cartography: Black Girlhood and Place-making in Education Research" *Review of Research in Education* 42 (1) pp.28–45.

Cajete, G. (2000) *Native Science: Natural Laws of Interdependence* Santa Fe, NM: Clear Light Publishers.

Campos-Delgado, A. (2018) "Counter-mapping Migration: Irregular Migrants' Stories through Cognitive Mapping" *Mobilities* 13 (4) pp.488–504.

Cebul, B. (2020) "Tearing Down Black America" *Boston Review: A Political and Literary Forum* (22nd July) Available at: https://www.bostonreview.net/articles/brent-cebul-tearing-down-black-america/.

Collins, P.H. (2002) *Black Feminist Thought: Knowledge, Consciousness, and the Politics of Empowerment* London and New York: Routledge.

Corburn, J. (2017) "Concepts for Studying Urban Environmental Justice" *Current Environmental Health Reports* 4 (1) pp.61–67.

Craib, R.B. (2000) "Cartography and Power in the Conquest and Creation of New Spain" *Latin American Research Review* 35 (1) pp.7–36.

Craib, R.B. (2017) "Cartography and Decolonization" In Akerman, J.R. (Ed.) *Decolonizing the Map: Cartography from Colony to Nation* Chicago: University of Chicago Press, pp.11–71.

Dadigan, M. (2014) "Stop Damming Indians: Dancing Against the Shasta Dam Raise" *Indian Country Today* (18th September).

DataCenter (2015) *Our Voices, Our Land: A Guide to Community-Based Mapping of Indigenous Stories* Oakland, CA: DataCenter.

Day, I. (2015) "Being or Nothingness: Indigeneity, Antiblackness, and Settler Colonial Critique" *Critical Ethnic Studies* 1 (2) pp.102–121.

De Jager, A., Tewson, A., Ludlow, B. and Boydell, K. (2016) "Embodied Ways of Storying the Self: A Systematic Review of Body-mapping" *Forum: Qualitative Social Research* 17 (2) pp.1–31.

Estes, N. (2019) *Our History Is the Future: Standing Rock Versus the Dakota Access Pipeline, and the Long Tradition of Indigenous Resistance* London and New York: Verso.

Fanon, F. (1959) *A Dying Colonialism* New York: Grove/Atlantic, Inc.

Fanon, F. (1961) *The Wretched of the Earth* New York: Grove/Atlantic, Inc.

Farnham, A. (2007) "'Their Sleep is to be Desecrated': California's Central Valley Project and the Wintu people of Northern California, 1938–1943" *Ethnic Studies Review* 30 (1) pp.135–165.

Fox, J., Suryanata, K., Hershock, P. and Pramono, A.H. (2006) "Mapping Power: Ironic Effects of Spatial Information Technology" *Participatory Learning and Action* 54 (1) pp.98–105.

Fullilove, M.T. (2001) "Root Shock: The Consequences of African American Dispossession" *Journal of Urban Health* 78 (1) pp.72–80.

Garzón-Galvis, C., Wong, M., Madrigal, D., Olmedo, L., Brown, M. and English, P. (2019) "Advancing Environmental Health Literacy through Community-engaged Research and Popular Education" In Finn, S. and O'Fallon, L.R. (Eds) *Environmental Health Literacy* pp.97–134 New York: Springer.

Gilio-Whitaker, D. (2019) *As Long as Grass Grows: The Indigenous Fight for Environmental Justice, from Colonization to Standing Rock* Boston, MA: Beacon Press.

Gilmore, R.W. (2002) "Fatal Couplings of Power and Difference: Notes on Racism and Geography" *The Professional Geographer* 54 (1) pp.15–24.

Goeman, M. (2013) *Mark My Words: Native Women Mapping our Nations* Minneapolis: University of Minnesota Press.

Harris, C.I. (1993) "Whiteness as Property" *Harvard Law Review* 106 (8) pp.1707–1791.

Harris, T.M. (2016) "From PGIS to Participatory Deep Mapping and Spatial Storytelling: An Evolving Trajectory in Community Knowledge Representation in GIS" *The Cartographic Journal* 53 (4) pp.318–325.

Huang, G. and London, J.K. (2016) "Mapping in and out of 'Messes': An Adaptive, Participatory, and Transdisciplinary Approach to Assessing Cumulative Environmental Justice Impacts" *Landscape and Urban Planning* 154 pp.57–67.

Hunt, D. and Stevenson, S.A. (2017) "Decolonizing Geographies of Power: Indigenous Digital Counter-mapping Practices on Turtle Island" *Settler Colonial Studies* 7 (3) pp.372–392.

Johnson, J.T. (2012) "Place-based Learning and Knowing: Critical Pedagogies Grounded in Indigeneity" *GeoJournal* 77 (6) pp.829–836.

Johnson, J.T., Louis, R.P. and Pramono, A.H. (2005) "Facing the Future: Encouraging Critical Cartographic Literacies in Indigenous Communities" *ACME: An International Journal for Critical Geographies* 4 (1) pp.80–98.

Klasky, P. (2009) "The Salt Song Trail Map: The Sacred landscape of the Nuwuvi People" *News from Native California* (Winter) pp.8–10.

Kobayashi, A. and Peake, L. (1994) "Unnatural Discourse: 'Race' and Gender in Geography" *Gender, Place & Culture* 1 (2) pp.225–243.

Lawson, M. (2009) *Dammed Indians Revisited* Pierre, SD: South Dakota State Historical Society Press.

Lee, R., Ahtone, T., Pearce, M., Goodluck, K., McGhee, G., Leff, C., Lanpher, K. and Salinas, T. (2020) "Land-Grab Universities" *High Country News* (30th March).

Lerner, S. (2012) *Sacrifice Zones: The Front Lines of Toxic Chemical Exposure in the United States* Boston, MA: MIT Press.

Littman, D.M., Bender, K., Mollica, M., Erangey, J., Lucas, T. and Marvin, C. (2021) "Making Power Explicit: Using Values and Power Mapping to Guide Power-diverse Participatory Action Research Processes" *Journal of Community Psychology* 49 (2) pp.266–282.

Livingstone, D.N. (2010) "Cultural Politics and the Racial Cartographics of Human Origins" *Transactions of the Institute of British Geographers* 35 (2) pp.204–221.

Louis, R.P. (2007) "Can You Hear Us Now? Voices from the Margin: Using Indigenous Methodologies in Geographic Research" *Geographical Research* 45 (2) pp.130–139.

Lucchesi, A.H.E. (2018) "'Indians Don't Make Maps': Indigenous Cartographic Traditions and Innovations" *American Indian Culture and Research Journal* 42 (3) pp.11–26.

Lucchesi, A.H.E. (2019) "Mapping Geographies of Canadian Colonial Occupation: Pathway Analysis of Murdered Indigenous Women and Girls" *Gender, Place & Culture* 26 (6) pp.868–887.

Mahtani, M. (2014) "Toxic Geographies: Absences in Critical Race Thought and Practice in Social and Cultural Geography" *Social & Cultural Geography* 15 (4) pp.359–367.

Massey, D. and Denton, N.A. (1993) *American Apartheid: Segregation and the Making of the Underclass* Boston, MA: Harvard University Press.

McElroy, E. (2018) "Countermapping Displacement and Resistance in Alameda County with the Anti-Eviction Mapping Project" *American Quarterly* 70 (3) pp.601–604.

McKittrick, K. (2006) *Demonic Grounds: Black Women and the Cartographies of Struggle*. Minneapolis: University of Minnesota Press.

McTavish, A.K. (2010) *The Role of Critical Cartography in Environmental Justice: Land-use Conflict at Shasta Dam, California* (PhD thesis) San Francisco State University.

Middleton, E.R. (2010) "Seeking Spatial Representation: Reflections on Participatory Ethnohistorical GIS Mapping of Maidu Allotment Lands" *Ethnohistory* 57 (3) pp.363–387.

Middleton Manning, B.R. (2018) *Upstream: Trust Lands and Power on the Feather River* Tucson, AZ: University of Arizona Press.

Mignolo, W.D. (1992) "Putting the Americas on the Map (Geography and the Colonization of Space)" *Colonial Latin American Review* 1 (1–2) pp.25–63.

Mignolo, W.D. (2005) *The Idea of Latin America* Hoboken, NJ: Blackwell Publishing.

Mignolo, W.D. (2012) *Local Histories/Global Designs: Coloniality, Subaltern Knowledges, and Border Thinking* Princeton, NJ: Princeton University Press.

Minkler, M. (2004) "Ethical Challenges for the 'Outside' Researcher in Community-based Participatory Research" *Health Education & Behavior* 31 (6) pp.684–697.

Moore, E. and Garzón, C. (2010) "Social Cartography: The Art of Using Maps to Build Community Power" *Race, Poverty & the Environment* 17 (2) pp.66–67.

Morales, R. (2009) "Ko'domyeponi: The Worldmaker's Journey" (Map) Available at: http://www.honeylakemaidu.org/photos/maiduFINALlores.pdf.

Muhammad, M., Garzón, C., Reyes, A. and the West Oakland Environmental Indicators Project, (2017) "Understanding Contemporary Racism, Power and Privilege and their Impacts on CBPR" In Wallerstein, N., Duran, B., Oetzel, J. and Minkler, M. (Eds) *Community-Based Participatory Research for Health: Advancing Social and Health Equity* (3rd ed.) San Francisco: Jossey-Bass, pp.47–60.

Mundy, B.E. (1996) *The Mapping of New Spain: Indigenous Cartography and the Maps of the Relaciones Geográficas* Chicago: University of Chicago Press.

Nardone, A., Chiang, J. and Corburn, J. (2020) "Historic Redlining and Urban Health Today in US Cities" *Environmental Justice* 13 (4) pp.109–119.

National Community Reinvestment Coalition (NCRC) (2020) *Redlining and Neighborhood Health* Washington, DC: NCRC.

Nelson, R.K., Winling, L., Marciano, R. and Connolly, N. (2020) "Mapping Inequality: Redlining in New Deal America" In Nelson, R.K. and Ayers, E.L. (Eds) *American Panorama: An Atlas of United States History* Available at: https://dsl.richmond.edu/panorama/ Richmond, VA: University of Richmond.

Oliver, M.L. and Shapiro, T.M. (2006) *Black Wealth, White Wealth: A New Perspective on Racial Inequality* Abingdon: Taylor & Francis.

Omi, M. and Winant, H. (2014) *Racial Formation in the United States* London and New York: Routledge.

Ortiz, E., O'Boyle, K. and Smith, S. (2020) "The Next Wave of Statue Removals is Afoot. See Where They're Being Taken Down Across the U.S." *NBC News* (28th August). Available at: https://www.nbcnews.com/news/us-news/2020-next-wave-statue-removals-afoot-map-n1230506.

Peake, L. and Kobayashi, A. (2002) "Policies and Practices for an Antiracist Geography at the Millennium" *The Professional Geographer* 54 (1) pp.50–61.

Pearce, M. and Louis, R.P. (2008) "Mapping Indigenous Depth of Place" *American Indian Culture and Research Journal* 32 (3) pp.107–126.

Peet, R. (1985) "The Social Origins of Environmental Determinism" *Annals of the Association of American Geographers* 75 (3) pp.309–333.

Peluso, N.L. (1995) "Whose Woods are These? Counter-mapping Forest Territories in Kalimantan, Indonesia" *Antipode* 27 (4) pp.383–406.

Pualani Louis, R. (2017) *Kanaka Hawai'i Cartography* Corvallis, OR: Oregon State University Press.

Pulido, L. (2018) "Geographies of Race and Ethnicity III: Settler Colonialism and Nonnative People of Color" *Progress in Human Geography* 42 (2) pp.309–318.

Romo, F. and MacLachlan, M. (2020) "Mapping the Black Lives Matter Movement" Available at: https://blm-map.com/.

Rose-Redwood, R., Blu Barnd, N., Lucchesi, A.H.E., Dias, S. and Patrick, W. (2020) "Decolonizing the Map: Recentering Indigenous Mappings" *Cartographica* 55 (3) pp.151–162.

Sacred Land Film Project (SLFP) (2011) "Mapping Sacred Sites" (video) Available at: https://sacredland.org/mapping-sacred-sites/.

Sadd, J., Morello-Frosch, R., Pastor, M., Matsuoka, M., Prichard, M. and Carter, V. (2014) "The Truth, the Whole Truth, and Nothing But the Ground-truth: Methods to Advance Environmental Justice and Researcher-community Partnerships" *Health Education & Behavior* 41 (3) pp.281–290.

Senier, S. (2018) "Where a Bird's-eye View Shows More Concrete: Mapping Indigenous LA for Tribal Visibility and Reclamation" *American Quarterly* 70 (4) pp.941–948.

Smith, L.T. (1999) *Decolonizing Methodologies: Research and Indigenous Peoples* London: Zed Books Ltd.

Sovereign Bodies Institute (SBI) (2020) *Zuya Wicayu'onihan – Honoring Warrior Women: A Study on Missing and Murdered Indigenous Women and Girls in States Impacted by the Keystone XL Pipeline* San Francisco: SBI.

Sylvestre, P., Castleden, H., Martin, D. and McNally, M. (2018) "'Thank you very much… you can leave our community now': Geographies of Responsibility, Relational Ethics, Acts of Refusal, and the Conflicting Requirements of Academic Localities in Indigenous Research" *ACME* 17 (3) pp.750–779.

Tuck, E. and Yang, K.W. (2012) "Decolonization is Not a Metaphor" *Decolonization: Indigeneity, Education & Society* 1 (1) pp.1–40.

Walking Water (WW) (2019) "An Interview with Desirae Harp and Michael Preston (Pom) of Run4Salmon" Available at: https://walking-water.org/2019/09/01/run4salmon/.

Winlow, H. (2006) "Mapping Moral Geographies: WZ Ripley's Races of Europe and the United States" *Annals of the Association of American Geographers* 96 (1) pp.119–141.

Winlow, H. (2009) "Mapping the Contours of Race: Griffith Taylor's Zones and Strata Theory" *Geographical Research* 47 (4) pp.390–407.

Winlow, H. (2019) "Mapping, Race and Ethnicity" In Kobayashi, A. (Ed.) *International Encyclopedia of Human Geography* (2nd ed.) Amsterdam: Elsevier, pp.309–321.

Winnemem Wintu Tribe (WWT) (2017) *Winnemem Wintu Salmon Restoration Plan – McCloud River* Available at: https://cawaterlibrary.net/wp-content/uploads/2017/05/shasta_winnemem.pdf.

Yellowhead Institute (YI) (2019) *Land Back: A Yellowhead Institute Red Paper* Toronto: YI.

8
FEMINIST GEOGRAPHY AND GEOSPATIAL TECHNOLOGIES

Meghan Kelly

Feminist interventions have instrumentally shaped critical perspectives in mapping practice (Elwood and Leszczynski, 2018). Feminist theorists and geographers have long explored and uncovered power along with its ramifications across space (Rose, 1993). Drawing on these considerations, mapmakers have exposed power mechanisms and resisted power through personal (Lucchesi, 2019) and collective initiatives (Mappingback Network, n.d.). Feminists and feminist mapmakers recognize that all knowledge comes from someone somewhere (Haraway, 1988; Collins, 2009) and disrupt objective and universalizing approaches in mapping (Kelly, 2021). Geographers and mapmakers alike have recognized their own subjectivities and positionalities that are inherent in mapping (England, 1994; Ricker, 2017) and many graphically encode situated experiences in their maps (Pearce, 2008; Kelly, 2019). Feminist theory expands default knowledge systems to include emotion and everyday, embodied experiences as valued observations (Rose, 1993). Cartographers have embraced these personal geographies and mapped embodied experiences in provocative ways (Kwan, 2002b). Feminist mapmakers have sought ways to disrupt binaries altogether and transparently convey the impacts of binaries when disruption is not possible (Gieseking 2018; D'Ignazio and Klein 2020). Finally, feminists value labour, calling attention to participation across mapping processes and expanding documentation to include reflexive considerations of power, context, and labour practices (Kelly and Bosse, 2019, 2022). In sum, feminist theory and feminist geography create new and alternative vocabularies for mapping. These vocabularies are integral to geospatial technology and equitable, just futures.

Feminist mapping, past and present

Feminism and mapping are wide-ranging and dynamic terms that mean different things to different people and in different contexts. Western, white feminisms are frequently characterized as a series of waves (e.g., first, second, third wave) with shifting ideologies and agendas over time. Feminism writ large, however, has expansive histories in non-Western and non-white communities – histories that have been erased and remain in tension with white feminism today (Lewis and Mills, 2003; Mohanty, 2003; Cargle, 2018). In this chapter, I draw on Black feminist thought and intersectionality, which explores power in relationship to bodies and systems of privilege and oppression (Combahee River Collective, 1977; Crenshaw,

1989; Collins 2009; Nash, 2018). Intersectionality recognizes overlapping and interlocking systems of oppression that materialize at the site of the body. In other words, bodies can be marginalized and privileged across multiple axes of identity (e.g., Black, trans woman, white woman with a disability, or white, rich man). I use the second term – mapping – throughout this chapter in its broadest, most inclusive capacity. Mapping in this context encapsulates the sub-discipline of GIScience including GIS, cartography, and remote sensing, along with geospatial technologies. Further, the use of mapping strategically undisciplines the term, by recognizing maps and mapping practices outside the bounds of GIScience and further emphasizes mapping as processes (Bosse, 2020). As such, I defer to mapping unless otherwise noted by referenced mapmakers, authors, or projects.

Feminism is not new to mapping. At the 2000 American Association of Geographer's annual meeting, Regina Hagger, Susan Hanson, Mei Po Kwan, Sara McLafferty and Marianna Pavlovskaya gathered to explore the intersections of feminist geography and mapping in a paper session titled 'GIS: Gendered Science? Feminist method?' (Kwan 2002a). The two sub-areas, feminist geography and mapping, had been 'the most dynamic research areas in geography over the past decade' and yet, the two research thrusts 'remained stubbornly apart […] in divergent directions' (McLafferty, 2002: 263). The presumed epistemological incompatibility between feminist geography and GIS painted feminist theory as the antithesis to mapping and vice versa. The participants involved in the 2000 AAG session and the resulting 2022 publication series in *Gender, Place and Culture* refused such a diverging dichotomy, instead making space to explore the transformation of GIS and feminist possibilities of its use in research. Collectively, the authors opened new lines of geographic inquiry and alternative modes of knowledge production.

Despite the aforementioned work, and the work of many subsequent scholars and mappers, two related divergences continue to exist today – feminism/GIScience and feminist GIS/critical GIS. First, feminist perspectives in mapping remain niche amongst the broader GIScience community. Google Trends (Figure 8.1), for example, illustrates the ways in which feminist GIS has been digitally erased. In *Cartography and Geographic Information Science*

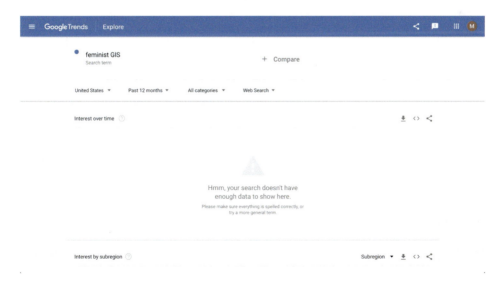

Figure 8.1 Google Trends results for the search term 'feminist GIS' procures 'Hmm, your search doesn't have enough data to show here'

(CaGIS), a premier journal for GIScience research, the word 'feminist GIS' has been used just 12 times since 1974. Second, feminist interventions in mapping have been enveloped and at times, appropriated, within a broader umbrella of 'critical GIS'. Elwood and Leszczynski (2018: 2) note 'critical GIS origin stories […] build from Derrida and Foucault, from Harley or Pickles, even though critical GIS emerged as a direct take-up of feminist critiques of scientific objectivity and vision'. The 12 uses of 'feminist GIS' in CaGIS pale in comparison to the 504 uses of 'critical GIS'. In an interview, Agnieszka Leszczynski and Nadine Schuurman (2017) discuss recent calls to 'revisit Critical GIS' and instead call for feminist action that builds 'new horizons' (Thatcher et al., 2016). Elwood and Leszczynski (2018: 2) further amplify new horizons in feminist mapping to include 'queer theory, Black geographies, postcolonial theory, black/queer code studies and more'. Like efforts in design justice, new horizons in feminist mapping make way for the 'worlds we need' (Costanza-Chock, 2020: 236).

Feminist mapping futures, or a gallery of possibilities

Today, feminist mapping is interdisciplinary, extending from geography and GIScience to fields like data science, design, and the digital humanities (D'Ignazio and Klein, 2016, 2020a, 2020b; Costanza-Chock, 2018, 2020; Elwood and Leszczynski, 2018). In their work on data feminism, Catherine D'Ignazio and Lauren Klein (2020a, 2020b) outline seven feminist principles and their applications within data science (Table 8.1). I draw on these principles to illustrate feminist mapping praxis across content (geospatial data), form (map design), and processes (mapping workflows) – three key sites for intervention within mapping. In the following sections, I reorganize and condense the feminist principles and define each

Table 8.1 This table collates D'Ignazio and Klein's (2020a) feminist principles for data feminism. These principles are reconfigured and applied to mapping contexts in the remainder of this entry

Feminist principles for data feminism	*Definition*
Examine power	Data feminism begins by analyzing how power operates in the world.
Challenge power	Data feminism commits to challenging unequal power structures and working toward justice.
Elevate emotion and embodiment	Data feminism teaches us to value multiple forms of knowledge, including the knowledge that comes from people as living, feeling bodies in the world.
Rethink binaries and hierarchies	Data feminism requires us to challenge the gender binary, along with other systems of counting and classification that perpetuate oppression.
Embrace pluralism	Data feminism insists that the most complete knowledge comes from synthesizing multiple perspectives, with priority given to local, Indigenous, and experiential ways of knowing.
Consider context	Data feminism asserts that data are not neutral or objective. They are the products of unequal social relations, and the context is essential for conducting accurate, ethical analysis.
Make labour visible	The work of data science, like all work in the world, is the work of many hands. Data feminism makes this labour visible so that it can be recognized and valued.

intervention within feminist geography and mapping. I then ground each feminist principle with existing projects.

Power

Black feminist thinkers conceptualize power as multi-dimensional systems – structural, disciplinary, hegemonic, and interpersonal layers – that differentially privilege some and oppress others (see Collins, 2009 for a matrix of domination). Further, Black feminist interventions acknowledge intersectionality or interlocking systems of oppression that cannot be disentangled (Combahee River Collective, 1977; Crenshaw, 1989; Collins, 2009). This includes racial and gendered categories, but also other axes of power like class, disability, education, ethnicity, religion, and so on, that might be experienced simultaneously. Intersectional power and systems of oppression are also embedded in content (geospatial data), form (map design), and process (mapping workflows). Geospatial technology and mapping engage with power in multiple directions. Data and maps carve out space for the interrogation of power including ways to reveal and analyse power structures and in/visibilities within mapping contexts. Further, feminist interventions actively 'commit to challenging unequal power structures and working towards justice' (D'Ignazio and Klein 2020a, 2020b: 49).

Power differentially determines who and what count. Missing datasets illustrate hierarchies within systems of oppression and demonstrate the ability to 'remove, hide, or obscure' certain lives that are often already marginalized (Onuoha, 2018). The *Missing and Murdered Indigenous Women Database* (MMIWD2) is one such example (Sovereign Bodies Institute, n.d.). In our data-saturated world, Indigenous women, girls, and two-spirit people have not been counted. To fill this absence, Annita Lucchesi and the Sovereign Bodies Institute devised a comprehensive database to log such geospatial data from 1900–present across colonial boundaries. MMIWD2 brings voice to those that have been erased by settler colonial violence and occupation. Lucchesi and the Sovereign Bodies Institute use this database to challenge power and 'transform data to action' (Sovereign Bodies Institute, n.d.).

Data analysis and map design also reveal systems of power. 'Land-Grab Universities' traces Indigenous lands stolen through the Morrill Act of 1862 that seeded the ongoing endowments of Land-Grant Universities across the United States. The authors reconstructed '10.7 million acres of lands taken from 250 tribes, bands, and communities through over 160 violence-backed land cessions' (Lee and Ahtone, 2020). This publicly available database and corresponding series of maps examine power by revealing the settler colonial footprints of US institutions that benefited from the dispossession of Indigenous lands. The authors' analyses challenge 'universities to re-evaluate the foundations of their success' and work toward decolonization and reconciliation (Lee and Ahtone, 2020).

Lastly, mapping processes and participation are critical sites for addressing power dynamics. 'Million Dollar Hoods' (n.d.) is a Los Angeles mapping project that exposes the power and impacts of mass incarceration on Black, Latinx, Indigenous, and other marginalized communities. The organization identified neighbourhoods with over a million dollars invested in local jails and argued for reinvestment in local communities. In terms of participation and process, 'Million Dollar Hoods' materializes the notion 'nothing about us without us', a phrase concretized within disability justice and centered with design justice frameworks (Hamraie, 2017; Costanza-Chock 2018, 2020). 'Million Dollar Hoods' disrupts power through community-led research by 'centering the voices of those directly impacted by policing', resulting in a grounded commitment towards mass decarceration and abolition.

Context and pluralism

Feminist thinkers and tinkerers question the notions of objectivity, neutrality, and universality that dominate Western perceptions of Science, technology, and mapping (Gieseking, 2018; Benjamin, 2019; Graziani and Shi, 2020). Feminists instead recognize situated knowledge or that all knowledge comes from somewhere and is therefore partial (Haraway, 1988). These partialities are political and are grounded in particular world views (Harding, 1986; Collins, 2009). Feminist objectivity in geospatial technology recognizes mapping as personal practice, which requires reflexivity or an examination of the workings of power that shape our understandings, subjectivities, and our position within systems of oppression as they relate to our work. More complete understandings can only be understood by bringing partialities together, which means embracing pluralism. In geographic and geospatial contexts, feminist practitioners have incorporated such interventions into understandings of data, design, and processes.

Mapping data from personal geographies is one way to map with situated knowledge in mind. Methods and calls for mapping personal and everyday geographies are not new, but they are ever important in understanding multiple worlds. Global lockdowns in response to the COVID-19 global pandemic undoubtably altered personal geographies. In an effort to document situated individuals and changing geographies amidst the crisis, *CityLab* (Bliss and Martin, 2022) put a call out for maps to better understand the transformation of private, public, and personal spaces. Mapmakers from around the world submitted remappings of their lived realities. Many of these maps detailed intimate geographies, reconfigured spaces, and relationships of the home. Others traced outdoor walking routes, new discoveries, and sounds in neighbourhoods or new, completely virtual realities. Personal maps like these introduce mapping practices that begin at the source, prioritizing first-hand, situated knowledge.

Perceived objectivity is often baked into maps and geospatial technologies opening sites for feminist intervention. Cartesian coordinate systems and top-down perspectives, for example, are default viewpoints in GIS technology. Struck by these limiting vantage points, Luke Bergmann and Nick Lally (2020) developed *Enfolding* as a 'geographic *imagination* system' (gis) to rethink spatial relationships, their partialities, and connections allowing for unique geographies that bend, fold, stretch, pinch, and tear (see Lally and Bergmann, 2021). The authors further rely on animations to undermine objectivity's reliance on static representations and illustrate the malleability and textures of lived experiences. Lowercase gis decentres assumptions built into conventional geospatial technology and subverts top-down, objective mapping practices in favour of alternative spatial theory, including situated knowledges and feminist critiques of objectivity.

A pluralistic approach to data and mapping brings partial, yet multidimensional perspectives together to capture nuance, difference, and connectedness. *Queering the Map* (n.d.; LaRochelle, 2021). harnesses such an approach in its aim 'to create a living archive of queer experience'. To do this, the project asks community members to geolocate queer moments, memories, and experiences collectively. Power stems from the overlapping partialities as you zoom and pan the map, building a collective memory greater than one single point on the map. Chandra Mohanty (2003) calls these intersecting perspectives moments of 'common difference' that challenge systems of oppression.

Pluralism and 'common difference' can also be a critical design strategy. In 'Collectively Mapping Borders', Meghan Kelly (2016) stitches sketches of Syrian border stories together based on interview excerpts. Each sketch provides one perspective and when brought together, the many border realities subvert standard border symbols. In a data feminism panel discussion (D'Ignazio and Klein, 2020b), Margaret Pearce further reflects on plurality within mapping processes in 'Land-Grab Universities' (mentioned above) noting 'there's an urgency in creating multiple entry points to the conversation to tell the story holistically'. For Pearce, this includes maps, but also text, photos, infographics, and providing the raw data for others to explore. Multiple entry points create space for multiple perspectives, multiple ways of knowing, and multiple moments for intervention.

Emotion and embodiment

Bodies and therefore emotions and embodiment are central to feminist thought. Feminist geographers challenge objectivity and mind/body dualisms by recognizing emotion and embodiment as ways of knowing. The body reflects the 'geography closest in' and understands place through lived experiences (Adrienne Rich in Valentine, 1999). Further, feminist geographers have highlighted everyday geographies and differential ways of experiencing the world (see Power above). Emotion has also been explored in the geospatial realm. Griffin and McQuoid (2012) have examined the collection of emotions as data attributes as well as the use of maps to depict emotive responses to place (i.e., cognitive maps of place) and to maps as artifacts. Caquard and Griffin (2018) extend such considerations to include 'the emotions that shape the mapping process and the map', shifting the focus from emotions and map artifacts to emotions amidst the mapping process.

From x/y coordinates attached to biometric data in fitness applications to cognitive mapping exercises and personal story maps, maps can be used to geolocate emotion (Kwan, 2008; Nold, 2009; Olmedo and Christmann, 2018). *New Not Normal* (n.d.) is community-generated map that geolocates emotive responses to the COVID-19 pandemic, collectively. *New Not Normal* calls individuals to take a deep breath, select a location on the map, select a colour associated with their emotion (blue = surprise, sadness = purple, trust = teal, and so on), and leave a message about their emotional status (e.g., 'VIRTUAL HUGS'). *New Not Normal* fosters a sense of community and resilience by sharing emotions with the map's community and by making emotive connections while 'planted in one place' (New Not Normal, n.d.).

Map design is a powerful mode of recognizing bodies, expressing emotion, and connecting to viewers in art, data journalism, and narrative mapping. Chelsea Nestel (n.d.), for example, captures 'geography closest in' in *Sum*, a one-to-one scale map of her body that uses georeferenced x-rays and collage to document experience. Margaret Pearce (2008) uses an emotive colour palette, frames, scale, and voice to highlight embodied experiences and emotions connected to place based on a French fur-trader's diary entries. Colour associations for emotion are notably dependent on culture and context. To convey the magnitude of 100,000 COVID-19 deaths in the United States, Lauren Tierney and Tim Meko – *Washington Post* data journalists – used a visual metaphor of light to symbolize each individual (Fisher, 2020). Compared to conventional symbolization techniques like points or polygons, beams of light humanized the 'peopled' data as well as the emotive toll of the pandemic. Emotion is a powerful design element that fosters connection, memorability, and empathy with audiences by recognizing human experience and alternative knowledge systems.

Bodies are at the centre of mapping processes, placemaking, and storytelling and yet the bodies of those closest to a given topic are too often left out of design processes. *Mapping Black Futures* is a digital project created by Black Futures Now Toronto (2020) that explores Black geographies and placemaking practices within Toronto. *Mapping Black Futures* is part digital archive, part resource, and part storytelling, created by and for non-binary, women, and Black youth. The interactive map geolocates important places (past and present) and further explores local resources, access to healthcare, and Black queer spaces. Lastly, *Mapping Black Futures* features an immersive, virtual reality map to engage more deeply with Black placemaking and storytelling. Such mapping centres Black bodies and amplifies emotion through spatial stories.

Binaries and hierarchies

Feminism today questions all binaries, not just the gender binary, as modes of categorization that artificially create difference and reinforce artificial hierarchies (D'Ignazio and Klein, 2020a). This includes, but is not exclusive to, binaries like man/woman, nature/culture, subject/object, reason/emotion, objectivity/subjectivities, and body/mind. The gender binary, for example, prioritizes masculine bodies while subjugating feminine bodies or any bodies of difference through systems of power and oppression. Feminist and queer theorists challenge such notions by revealing the social constructs of gender and sex while recognizing the lived realities of such inscriptions on bodies (Fausto-Sterling, 2000). Binaries, bins, and modes of classification are commonplace in mapping. They are frequently used to abstract and simplify the world into human- and computer-readable bites/bytes at the expense of nuance and messy, complex realities. Feminist mapmakers have sought ways to disrupt binaries altogether and make binaries visible when disruption is not possible.

Spatial data depend on binaries and categorization that are often presented as unquestionable facts. Bivens and Haimson (2016), for example, uncover the binary data structure for gender on Facebook that superficially presents users with a selection from one of thirty gender options. Yet, on the back end, gender options are reduced back into a binary (e.g., man/woman or male/female) to enable gendered advertisement. Further, OpenStreetMap (OSM) is largely applauded for crowdsourcing global data available to the public. Stephens' (2013) analysis found robust categorization for evening entertainment, yet a single category for 'childcare' revealed bias in data collection and classification practices. Geochicas (Yang *et al.*, 2019) is an international group actively working towards improving the structural bias, diversity, and quality of OSM data. Importantly, 'data can never speak for themselves' as classifications of people including racial categorization are deeply rooted in power relations (Muhammad, 2012).

Questions of binaries and classification are further evident in data analysis and map design. Thematic mapping relies on the data classification or the binning of data to uncover or hide patterns within the dataset. Data classification schemes (e.g., equal interval, quantiles, standard deviation) are chosen by the mapmaker and can be used to shape narratives and persuade audiences given their supposed 'trustworthiness'. Data distributions and resulting classification schemas require contextualization. Additionally, binaries are often default, unintended choices in map design. Open/closed border symbolization is one such example. Meghan Kelly (2019) examines the symbolization of nation-state borders represented in news maps, where open/closed borders are typically displayed as solid lines (closed) and dashed lines (open). In reality, the permeability of all borders fluctuates, and Kelly (2019) sought to develop a continuum for border porosity to better reflect the openness of borders. Further, Maria Rodó-de-Zarate (2014: 925) developed 'Relief Maps' to visualize intersectionality and 'disrupt homogenous categories while pointing towards the material consequences of oppression'. Such design strategies rethink and subvert binaries and classification within map design.

Given their prominence in mapping, it is important to contextualize binaries when they cannot be avoided. D'Ignazio and Klein (2020a) recommend clear communication and caution as to not reinforce stereotypes imposed by such classification. Further, there's a similar responsibility for the mapmaker to consider the role and power of visibility and invisibility in data and mapping. In *Underground R.R.*, for example, Kela Caldwell (n.d.) draws on Katherine McKittrick to understand the in/visibilities of Black diaspora geographies by intentionally revealing and concealing portions of the Underground Railroad. Caldwell's remapping challenges the priority of visuality and visual hierarchy in mapping by deliberately introducing invisibility into the map.

Labour

Feminist thinkers and geographers recognize the value of labour and expand labour definitions to include informal and often unrecognized labour practices (D'Ignazio and Klein, 2020a). The valuation of all forms of labour, including reproductive labour, emotional labour, care work, and other unseen labour, stems from long histories of feminist activism and labour rights. In the context of geospatial data and technology, 'a feminist approach to visualization can help to render visible the bodies that shape and care for data at every stage of the process' (D'Ignazio and Klein 2016). Labour visibility also coincides with the feminist practice of reflexivity (see 'Consider Context and Embrace Pluralism' above), which calls us to situate ourselves in relationship to power and our geospatial practices (Ricker, 2017). Labour is a key consideration in reflexive practice. D'Ignazio and Klein (2020a, 2020b) call for a visibility of both reflexive engagement and labour practices for accountability and equity.

Simply stated, metadata is data about data. While federal guidelines exist, metadata is not always consistent and is not accessible for all spatial datasets. Recent work on data biographies (Krause, 2017) and datasheets for datasets (Gebru *et al.*, 2020) expands conceptions of metadata to include labour practices as well as considerations of power. Krause (2017), for example, approaches data as a journalistic source raising important questions of who collected the data, what was collected, when and where collection took place, and finally, why the data are important. Ricker (2017) further calls for reflexivity throughout data lifecycles, particularly big data, that includes all forms of labour (collection, cleaning, storing, analysing, presenting, and so on).

Mapmakers are largely absent from resulting map artifacts. The mapmaker's presence and positionality as well as labour practices, however, can also be documented and embedded within the map in a variety of ways. Signifying voice and 'who's talking' in a map's annotations clearly differentiates when the mapper's voice is being used and when a voice is coming from another source (Pearce, 2008; Kelly, 2019). Further, Westerveld and Knowles (2018) use hand-drawn aesthetics in *I Was There, Places of Experience in the Holocaust* to illustrate the mapmaker's influence in the mapping of Holocaust stories. Reflexivity and feminist theory, more broadly, creates new possibilities for cartographic language and map design.

Lastly, reflexivity is imperative to accountability and transparent labour practices within mapping processes. Kelly and Bosse (2019, 2022) for example, call mappers to intentionally 'press pause' to engage with reflexive practice in mapping. Kelly and Bosse (2019, 2022) offer a feminist toolkit that includes written, audio, and visual tools and arenas for questioning as a starting point. They call attention to existing models like artist statements, mapmaker stories, live demos, and visual journaling that work toward reflexivity. 'Pressing pause', however, is a call to reflexive action. The phrase challenges mappers to adapt, expand, and repurpose tools for reflexive engagement that recognizes all forms of labour and care work in mapping.

Conclusion

In summary, feminist theory is fundamental to mapping, despite efforts to erase and minimize feminist efforts in GIScience and critical GIS. Perhaps most importantly, however, feminist mapping today considers 'new lines of flight' that expand with considerations of 'queer theory, critical race and postcolonial feminism, and Black geographies thought' (Elwood and Leszczynski, 2018: 7). Feminist perspectives in mapping and geospatial technology require interdisciplinary approaches and cross-boundary work. Drawing on feminist principles in data feminism, this entry serves as gallery of feminist work or a *gallery of possibilities* at the forefront of feminist mapping (Kelly and Bosse, 2023). Each example demonstrates one modality, one moment, and one possibility that can be learned from, questioned, adapted, and retooled. Taken collectively, this gallery of possibilities paves the way for a more equitable world.

References

Benjamin, R. (2019) *Race After Technology: Abolitionist Tools for the New Jim Code*. Cambridge: Polity Press.

Bergmann, L. and Lally, N. (2020) "For Geographical Imagination Systems" *Annals of the American Association of Geographers* 111 (1) pp.26–35 DOI: 10.1080/24694452.2020.1750941.

Bivens, R. and Haimson, O.L. (2016) "Baking Gender into Social Media Design: How Platforms Shape Categories for Users and Advertisers" *Social Media + Society* 2 (4) pp.1–12 DOI: 10.1177/2056305116672486.

Black Futures Now Toronto (2020) "Mapping Black Futures" *Toronto* Available at: https://mbf.blackfuturesnow.to/resources/ (Accessed: 16th July 2020).

Bliss, L. and Martin, J. (2022) "Covid Maps Reveal Personal Pandemic Landscapes" Available at: https://www.bloomberg.com/news/features/2022-01-25/homemade-covid-maps-navigate-life-around-the-world (Accessed: 10th February 2023).

Bosse, A. (2020) "Map Anyway" *Guerrilla Cartography: Atlas in a Day* Online: 16th May Available at: https://www.guerrillacartography.org/atlas-in-a-day (Accessed: 26th August 2020).

Caldwell, K. (n.d.) "Underground R.R." (Artwork).

Cargle, R.E. (2018) "When Feminism Is White Supremacy in Heels" *Harper's BAZAAR*. Available at: https://www.harpersbazaar.com/culture/politics/a22717725/what-is-toxic-white-feminism/ (Accessed: 26th August 2020).

Caquard, S. and Griffin, A. (2018) "Mapping Emotional Cartography" *Cartographic Perspectives* (91) pp.4–16 DOI: 10.14714/CP91.1551.

Collins, P.H. (2009) *Black Feminist Thought: Knowledge, Consciousness, and the Politics of Empowerment* (2nd ed.) New York: Routledge.

Combahee River Collective (1977) "The Combahee River Collective Statement" In Taylor, K-Y. (Ed.) *How We Get Free: Black Feminism and the Combahee River Collective* Chicago: Haymarket Books, pp.15–27.

Costanza-Chock, S. (2018) "Design Justice, A.I., and Escape from the Matrix of Domination" *Journal of Design and Science* DOI: 10.21428/96c8d426.

Costanza-Chock, S. (2020) *Design justice: Community-led Practices to Build the Worlds We Need* Cambridge: The MIT Press.

Crenshaw, K. (1989) "Demarginalizing the Intersection of Race and Sex: A Black Feminist Critique of Antidiscrimination Doctrine, Feminist Theory and Antiracist Politics" *University of Chicago Legal Forum* 1 (8) pp.139–167.

D'Ignazio, C. and Klein, L.F. (2016) "Feminist Data Visualization" *Workshop on Visualization for the Digital Humanities (VIS4DH)* Baltimore, MD, 23rd October 2016 IEEE.

D'Ignazio, C. and Klein, L.F. (2020a) *Data Feminism* Cambridge: The MIT Press.

D'Ignazio, C. and Klein, L.F. (2020b) "Conclusion: Now Let's Multiply!" *Data Feminism Reading Group* Online: 12th June Available at: http://datafeminism.io/blog/book/data-feminism-reading-group/ (Accessed: 20th August 2020).

Elwood, S. and Leszczynski, A. (2018) "Feminist Digital Geographies" *Gender, Place & Culture* 25 (5) pp.629–644 DOI: 10.1080/0966369X.2018.1465396.

England, K.V. (1994) "Getting Personal: Reflexivity, Positionality, and Feminist Research" *The Professional Geographer* 46 (1) pp.80–89.

Fausto-Sterling, A. (2000) *Sexing the Body: Gender Politics and the Construction of Sexuality*. New York: Basic Books.

Fisher, M. (2020) "U.S. Coronavirus Death Toll Surpasses 100,000, Exposing Nations Vulnerabilities" *Washington Post* (24th May 2020) Available at: https://www.washingtonpost.com/graphics/2020/national/100000-deaths-american-coronavirus/ (Accessed: 20th August 2020).

Gebru, T., Morgenstern, J., Vecchione, B., Vaughan, J.W., Wallach, H., Daumé, H. and Crawford, K. (2020) "Datasheets for Datasets" *arXiv* Available at: http://arxiv.org/abs/1803.09010 (Accessed: 29th May 2020).

Gieseking, J.J. (2018) "Size Matters to Lesbians, Too: Queer Feminist Interventions into the Scale of Big Data" *The Professional Geographer* 70 (1) pp.150–156.

Graziani, T. and Shi, M. (2020) "Data for Justice" *ACME: An International Journal for Critical Geographies* 19 (1) pp.397–412.

Griffin, A. and McQuoid, J. (2012) "At the Intersection of Maps and Emotion: The Challenge of Spatially Representing Experience" *Kartographische Nachrichten* 62 (6) pp.291–299.

Hamraie, A. (2017) *Building Access: Universal Design and the Politics of Disability* Minneapolis: University of Minnesota Press.

Haraway, D. (1988) "Situated Knowledges: The Science Question in Feminism and the Privilege of Partial Perspective" *Feminist Studies* 14 (3) pp.575–599 DOI: 10.2307/3178066.

Harding, S.G. (1986) *The Science Question in Feminism* Ithaca, NY: Cornell University Press.

Kelly, M. (2016) "Collectively Mapping Borders" *Cartographic Perspectives* 84 pp.31–39 DOI: 10.14714/CP84.1363.

Kelly, M. (2019) "Mapping Syrian Refugee Border Crossings: A Feminist Approach" *Cartographic Perspectives* (93) pp.34–64.

Kelly, M. (2021) "Mapping Bodies, Designing Feminist Icons" *GeoHumanities* 7 (2) pp.529–557.

Kelly, M. and Bosse, A. (2019) "Mapping and Positionality: A Call for Reflection and Action" *North American Cartographic Information Society* Tacoma, Washington, DC, 15th–18th October.

Kelly, M. and Bosse, A. (2023) "Pressing Pause: 'Doing' Feminist Mapping" *ACME: An International Journal for Critical Geographies* 21 (4) pp.399–415.

Krause, H. (2017) "Data Biographies: Getting to Know Your Data, Global Investigative Journalism Network" *Global Investigative Journalism Network* Available at: https://gijn.org/2017/03/27/data-biographies-getting-to-know-your-data/ (Accessed: 20th February 2020).

Kwan, M.P. (2002a) "Introduction: Feminist Geography and GIS" *Gender, Place & Culture* 9 (3) pp. 261–262 DOI: 10.1080/0966369022000003860.

Kwan, M.P. (2002b) "Is GIS for Women? Reflections on the Critical Discourse in the 1990s" *Gender, Place and Culture* 9 (3) pp.271–279.

Kwan, M.P. (2008) "From Oral Histories to Visual Narratives: Re-Presenting the Post-September 11 Experiences of the Muslim Women in the USA" *Social & Cultural Geography* 9 (6) pp.653–669.

Lally, N. and Bergmann, L. (2021) "*enfolding*: An Experimental Geographical Imagination System (gis)" In Kingsbury, P. and Secor, A. (Eds) *A Place More Void* Lincoln, NE: University of Nebraska Press, pp.167–180.

LaRochelle, L. (2021) "*Queering the Map*: On Designing Digital Queer Space" In Ramos, R. and Mowlabocus, S. (Eds) *Queer Sites in Global Contexts* Routledge: London, pp.133–147.

Lee, R. and Ahtone, T. (2020) "Land-Grab Universities" *High Country News* (30th March 2020) Available at: https://www.hcn.org/issues/52.4/indigenous-affairs-education-land-grab-universities (Accessed: 20th August 2020).

Lewis, R. and Mills, S. (Eds) (2003) *Feminist Postcolonial Theory* New York: Routledge.

Leszczynski, A. (2017) "Revisiting Cyborg GIS: A Conversation with Nadine Schuurman" *Society & Space* Available at: http://societyandspace.org/2017/05/02/revisiting-cyborg-gis-a-conversation-with-nadine-schuurman/ (Accessed: 11th April 2019).

Lucchesi, A.H. (2019) "Mapping Geographies of Canadian Colonial Occupation: Pathway Analysis of Murdered Indigenous Women and Girls" *Gender, Place & Culture* 26 (6) pp.868–887.

MappingBack Network (n.d.) "Indigenous Mapping, Extraction, & Alternative Representations" Available at: http://mappingback.org/home_en/ (Accessed: 31st January 2019).

McLafferty, S.L. (2002) "Mapping Women"s Worlds: Knowledge, Power and the Bounds of GIS" *Gender, Place & Culture* 9 (3) pp.263–269 DOI: 10.1080/0966369022000003879.

Million Dollar Hoods (n.d.) "Million Dollar Hoods" Available at: https://milliondollarhoods.pre.ss.ucla.edu/ (Accessed: 20th August 2020).

Mohanty, C.T. (2003) *Feminism Without Borders: Decolonizing Theory, Practicing Solidarity*. Durham, NC: Duke University Press.

Muhammad, K. (2012) "Khalil Muhammad on Facing Our Racial Past" *BillMoyers.com* Available at: https://billmoyers.com/segment/khalil-muhammad-on-facing-our-racial-past/ (Accessed: 20th August 2020).

Nash, J.C. (2018) *Black Feminism Reimagined: After Intersectionality*. Durham, NC: Duke University Press.

Nestel, C. (n.d.) "Sum" (Artwork) Available at: http://nestelmaps.azurewebsites.net/Content/images/BodyMap-Web.jpg (Accessed: 31st January 2019).

New Not Normal (n.d.) "New Not Normal" Available at: http://newnotnormal.ukai.ca (Accessed: 16th July 2020).

Nold, C. (2009) *Emotional Cartography*. Available at: http://emotionalcartography.net/ (Accessed: 3rd March 2019).

Olmedo, É. and Christmann, M. (2018) "Perform the Map: Using Map-Score Experiences to Write and Reenact Places" *Cartographic Perspectives* 91 pp.63–80 DOI: 10.14714/CP91.1486.

Onuoha, M. (2018) "Missing Datasets" Available at: https://github.com/MimiOnuoha/missing-datasets (Accessed: 20th August 2020).

Pearce, M.W. (2008) "Framing the Days: Place and Narrative in Cartography" *Cartography and Geographic Information Science* 35 (1) pp.17–32.

Queering the Map (n.d.) "Queering the Map" Available at: https://www.queeringthemap.com/ (Accessed: 7th April 2020).

Ricker, B. (2017) "Reflexivity, Positionality and Rigor in the Context of Big Data Research" In Thatcher, J., Shears, A. and Eckert, J. (Eds) *Thinking Big Data in Geography: New Regimes, New Research* Iowa City: University of Iowa Press, pp.96–118.Rodó-de-Zárate, M. (2014) "Developing Geographies of Intersectionality with Relief Maps: Reflections from Youth Research in Manresa, Catalonia" *Gender, Place & Culture* 21 (8) pp.925–944 DOI: 10.1080/0966369X.2013.817974.

Rose, G. (1993) *Feminism & Geography: The Limits of Geographical Knowledge*. Minneapolis: University of Minnesota Press.

Sovereign Bodies Institute (n.d.) "MMIWG2 Database" *Sovereign Bodies Institute* Available at: https://www.sovereign-bodies.org/mmiw-database (Accessed: 20th August 2020).

Stephens, M. (2013) "Gender and the GeoWeb: Divisions in the Production of User-Generated Cartographic Information" *GeoJournal* 78 (6) pp.981–996 DOI: 10.1007/s10708-013-9492-z.

Thatcher, J., Bergmann, L., Ricker, B., Rose-Redwood, R., O'Sullivan, D., Barnes, T.J., Barnesmoore, L.R., Beltz Imaoka, L., Burns, R., Cinnamon, J., Dalton, C.M., Davis, C., Dunn, S., Harvey, F., Jung, J.-K., Kersten, E., Knigge, L., Lally, N., Lin, W., Mahmoudi, D., Martin, M., Payne, W., Sheikh, A., Shelton, T., Sheppard, E., Strother, C.W., Tarr, A., Wilson, M.W. and Young, J.C. (2016) "Revisiting Critical GIS" *Environment and Planning A: Economy and Space* 48 (5) pp.815–824 DOI: 10.1177/0308518X15622208.

Valentine, G. (1999) "A Corporeal Geography of Consumption" *Environment and Planning D: Society and Space* 17 (3) pp.329–351 DOI: 10.1068/d170329.

Westerveld, L. and Knowles, A.K. (2018) "I Was There" *Visionsxarto* Available at: https://visionscarto.net/i-was-there (Accessed: 27th July 2020).

Yang, S., Jacquin, C. and González, M. (2019) "Geochicas: Helping Women Find their Place on the Map" *The Mapillary Blog* Available at: https://blog.mapillary.com/update/2019/05/28/putting-women-on-the-map-with-geochicas.html (Accessed: 20th August 2020).

9
MAPPING THE SUBALTERN

Penelope Anthias

While mapmaking has long been a tool of the powerful, today maps are being used by a variety of actors to challenge dominant geographical representations, resist dispossession, and make visible alternative spatial imaginaries, land-based practices, and territorial claims. Such initiatives are often referred to as 'counter-mapping'.[1] This chapter considers the historical evolution, varied methodologies and real-world applications of counter-mapping, and reflects on its possibilities, ambivalences, and limits. While the empowering effects of 'mapping the subaltern' are often taken for granted, critical accounts question the ability of maps to transform dominant colonial relations of knowledge, space and power. The role of powerful institutions and corporations in promoting, financing and providing technological infrastructure for mapping projects raises further questions about whose interests are ultimately served. Nevertheless, the diversity of counter-mapping projects and approaches defies simple generalization. While mapping is never power-free, it represents an important field for contemporary struggles over representation, rights and resources – and one in which diverse subaltern actors are increasingly seeking to intervene.

The chapter is structured as follows. It begins by introducing the concept of 'the subaltern' and its relevance to mapping. It then provides a brief overview of the recent history of counter-mapping, focusing on participatory mapping projects with Indigenous peoples. This is followed by a critical reflection on the possibilities and ambivalences of mapping as a tool for subaltern empowerment and decolonial struggle. The chapter concludes by discussing some alternative approaches to mapping the subaltern, which incorporate a broader range of participants, objectives and methodologies.

Mapping the subaltern

The term *subaltern* was used by Antonio Gramsci (1973) to refer to social groups whose interests and perspectives were not represented within dominant economic, social, and cultural institutions.[2] To achieve freedom, Gramsci argued that such groups must develop a consciousness independent from that of the ruling class. The concept was subsequently taken up by postcolonial historians in the Subaltern Studies Group, who used it to describe groups within colonized societies whose perspectives had been excluded from both colonial and elite nationalist historiography. Through their writings, these scholars sought to recover the historical agency of subaltern subjects.[3] However, in her famous essay 'Can the Subaltern Speak?', Gayatri Spivak (1978) questioned whether the subaltern can ever achieve representation within dominant discourse.

As this makes clear, the concept of the subaltern is intimately tied to the question of representation. The subaltern is not a fixed, bounded and mappable subject, but rather refers to a position of exclusion and marginalization within in a dominant representational field. If historiography is one such field, then the problem of subaltern representation is equally relevant to cartography. Maps have long been considered an instrument of the powerful, as well as a form of power in their own right. The 'power of maps' lies partly in their capacity to present themselves as objective and value-free, obscuring ethnocentric and class-based hierarchies of representation (Harley, 1989). The role of maps in European imperialism is well recognized (Craib, 2000; Akerman, 2009). In imperial cultures, maps helped demarcate racialized boundaries between colonizer and colonized (Saïd, 1994: 48), while in colonial contexts the cartographic representation of non-European peoples' lands as 'empty' or 'idle' was used to justify their violent appropriation (Razack, 2002; Bhandar, 2018). Such erasures persist in settler and postcolonial states, where the abstract space of state maps effaces Indigenous geographies and the violent origins of state sovereignty (Sparke, 2005; Radcliffe, 2011). Maps also play an important role in rendering land investible for capital (Li, 2014), driving new processes of accumulation by dispossession.

However, 'counter-mapping' seeks to use Western cartographic techniques to challenge dominant geographical representations of land relations (Hodgson and Schroeder, 2002: 80). It aims to 'counter earlier colonial and neo-colonial attempts to cartographically dispossess communities from their lands and resources' (Pualani Louis *et al.*, cited in Hunt and Stevenson, 2017: 376) and constitute 'the grounds for the recognition of indigenous peoples' [...] rights to territory, self-determination, and self-government' (Bryan and Wood, 2015: xvi). This raises a series of critical questions. To what extent can subaltern perspectives be represented using Western mapping techniques? What are the objectives of mapping the subaltern and whose interests are ultimately served? What power relations are involved in the production, ownership or circulation of maps and mapping technologies? And, to what extent does mapping have the capacity to transform material geographies of land control or advance subaltern agendas for self-determination? These questions will be considered below. First, I will provide a brief overview of the historical evolution of counter-mapping, focusing primarily on Indigenous mapping projects.

A brief history of (Indigenous) counter-mapping

While Indigenous peoples[4] have a long history of engagement with modern maps,[5] the origins of counter-mapping can be traced back to Indigenous mapping projects in Canada and Alaska in the 1950s and 1960s, which focused on documenting land use and occupancy for the purpose of negotiating aboriginal rights (see Chapin *et al.*, 2005; Bryan and Wood, 2015). A key method was 'map biographies', which charted the subsistence practices of individuals spatially and through time. Hugh Brody describes the process:

> hunters, trappers, fishermen, and berry pickers mapped out all the land they had ever used in their lifetimes, encircling hunting areas species by species, marking gathering

Mapping the subaltern

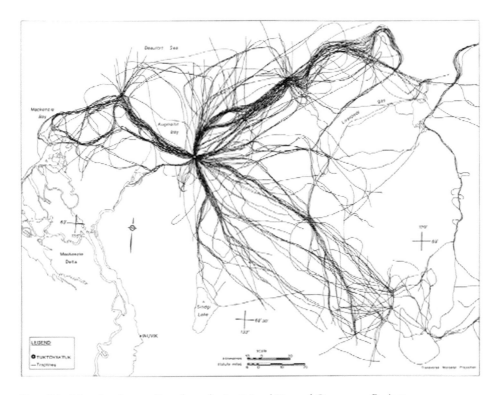

Figure 9.1 Map showing traplines from the Inuit Land Use and Occupancy Project
Source: Freeman, M. (Ed.) (1976) "Inuit Land Use and Occupancy Project Report Volume One: Land Use and Occupancy Ottawa: Supply and Services.

locations and camping sites – everything their life on the land had entailed that could be marked on a map.

(*Cited in Bryan and Wood, 2015: 60*)

Influential examples of this method include the Inuit Land Use and Occupancy Project (see Figure 9.1), Cree maps of harvest locations and Dene maps of traplines (Bryan and Wood, 2015: Chapter 4). In each case, maps produced played an important role in negotiations with the Canadian state over indigenous land and resource rights, although with ambivalent outcomes. Similar methods were subsequently adopted in other regions; for example, in the 1970s and 1980s cultural ecologist Bernard Nietschmann worked with Miskito people in eastern Nicaragua to produce maps of land use and occupancy that were used as a tool for territorial claims.

However, it was in the 1990s that Indigenous and community mapping was taken up more widely in diverse regions of Asia, Africa and the Americas. This was partly due to new mapping technologies of GPS (global positioning systems), GIS (geographic information systems), and remote sensing. These technologies were integrated with participatory techniques from Participatory Rural Appraisal and Participatory Action Research, leading to methodologies that often combined oral testimonies and map elicitation exercises with the GPS mapping of territorial boundaries (Hodgson and Schroeder, 2002; Chapin et al., 2005). The term 'counter-mapping' was coined by Nancy Peluso (1995) to describe forest dwellers' efforts to map their lands in Kalimantan, Indonesia in response to state and industry mapping efforts that sought to open these areas up to logging and mineral exploration. Latin America in particular saw a proliferation of counter-mapping during the 1990s, which was often linked to Indigenous claims for collective territorial rights (Offen, 2003; Bryan, 2012). Compared to earlier maps of indigenous land use and occupancy, maps from this period often show a more direct engagement with and contestation of state cartography (Figure 9.2) – although this could also place limits on what could be represented (discussed below).

It is worth considering the context for this explosion of Indigenous mapping projects in the 1990s. One factor was the intensification of capitalist processes of dispossession, as a result of neoliberal state reforms, which opened up new territories for investment by transnational mining, timber and agri-business companies, often in geographically

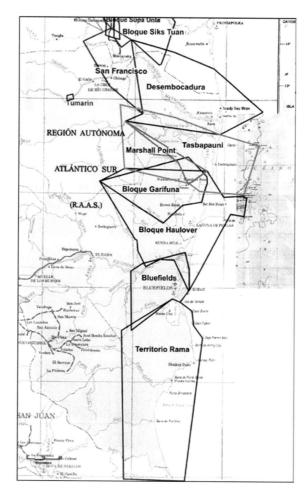

Figure 9.2 Outline map of black and indigenous communal land claims along Nicaragua's Caribbean coast, which make up the Autonomous Region of South Atlantic (Región Autónoma Atlántico Sur). Produced in a project funded by the World Bank, the map aimed to demonstrate that there is no 'state land' along this coastline

Source: Gordon et al. (2003: 371).

isolated regions where people lacked formal land rights. Another influence was the development of Indigenous rights in international law. The International Labour Organization's Convention 169, passed in 1989, called for recognition of Indigenous peoples' rights to territory, fuelling hopes of Indigenous that, once mapped, such rights would be respected and formalized (Anthias, 2018). Many mapping projects were also driven by interest in community-led natural resource management (Hodgson and Schroeder, 2002) and in Indigenous peoples' role in biodiversity conservation (Poole, 1995; Anthias and Radcliffe, 2015).

If the 1990s were something of a heyday for counter-mapping, then the first decades of the twentieth century have seen new developments and directions. As GIS and other spatial technologies have become more widely and cheaply available, communities have increasingly taken up these technologies themselves, relying less heavily on outsiders. Some Indigenous peoples have integrated newer technologies such as drones into their mapping efforts – often to make visible the impacts of other actors within their territories. For example, Mena *et al.* (2020: 411) describe how communities in the Ecuadorian Amazon are using a combination of drones, mobile phone apps and a geoportal to map the socio-environmental impacts of oil extraction in their territory, including information about new oil spills, oil pits containing disposed petroleum and highly toxic polluted waters and drilling muds.

There has also been a renewed global policy drive for mapping and formalizing community land rights, spearheaded by state and corporate funded platforms like the Rights and Resources Initiative, the Tenure Facility and the Land Portal. Amidst growing concern around climate change, the mapping and titling of collective territories is increasingly promoted as a means of reducing carbon emissions from deforestation (Ospina, 2018). Recent years have similarly seen a proliferation of new geo-spatial datasets on Indigenous and community land rights.[6] While some have expressed concerns about the (neo)colonial use of such datasets (Mann and Daly, 2019), social movements and activists are also using geospatial data to gain visibility and drive collective action (Milan and Gutierrez, 2018).

Many examples of counter-mapping are not covered in the above summary. Some of these alternative approaches will be presented below. First, the next section considers the achievements, ambivalences and limits of the kinds of counter-mapping projects discussed so far.

Debates and ambivalences

The possibilities and limits of counter-mapping have been subject to considerable debate. Many scholars emphasize the empowering effects of the mapping process, which can reinforce a cultural politics of identity and belonging following histories of colonial erasure (Gordon et al., 2003; Offen, 2003; Stocks, 2003). Indigenous maps present a powerful challenge to the exclusions of postcolonial geography, challenging the invisibility of Indigenous populations in the nation's cartography (Sparke, 2005; Radcliffe, 2011). Another common argument is summed up in the dictum 'map or be mapped'; that is, if Indigenous peoples do not produce their own maps, they will become victim to maps of their territories produced by others. As Rocheleau notes, maps can 'buy time and create space against the onslaught of mass eviction, wholesale land alienation and widespread despoliation by "outside" interests' (cited in Hodgson and Schroeder, 2002: 81). This remains a powerful incentive for many counter-mapping projects today, as extractive industry frontiers continue to dispossess local populations of land and resources (Tilley, 2020).

However, scholars have also raised critical questions about the ability of maps to represent subaltern geographies or advance broader decolonial struggles for territory and self-determination. As Wainwright and Bryan (2009: 156) note, the cartographic representation of Indigenous peoples 'is always already conditioned by unequal relations of social power: property, citizenship, territoriality, legal norms, the nation-state system, and so forth'. Standard base maps used in counter-mapping take the state's territory as a given, obscuring the history of Indigenous dispossession on which it is predicated. As such, by assuming the conventions of cartography, Indigenous peoples paradoxically collude in a performance of state sovereignty in the same breath that they assert their claims through a language of pre-colonial (ancestral) rights (Watson, 2002; Sparke, 2005; Bhandar, 2011).

Critiques of counter-mapping also highlight how modern cartography's focus on fixing boundaries in abstract space negates alternative territorialities, including those based on fluid or overlapping boundaries and relational understandings of space and belonging.[7] As Rundstrom (cited in Johnson *et al.*, 2006: 87) observed,

> [the] prevailing Cartesian-Newtonian […] epistemology does not prize key characteristics of indigenous thinking, including; the principle of the ubiquity of relatedness; non-anthropocentricity; a cyclical concept of time; a more synthetic than analytic view of the construction of geographical knowledge; non-binary thinking; the idea that facts cannot be dissociated from values; that precise ambiguity exists and can be advantageous; an emphasis on oral performance and other non-inscriptive means of representation; and the presence of morality in all actions.

As well as failing to represent subaltern geographies, the mapping process may disrupt Indigenous land use practices (Peluso, 1995). This is something I have observed in my own research in the Bolivian Chaco, where boundaries created during the mapping process led to conflict between neighbouring Guaraní communities (Anthias, 2018: Chapter 4). Mapping projects may also create particular understandings of 'community', effacing alternative forms of belonging (Leemann, 2019).

Recent scholarship has linked these concerns to a consideration of the racialized knowledge practices implicated in efforts to fix boundaries of identity and space (Bryan, 2009). Based on her research with Miskito and Garifuna communities on the Atlantic coast of Honduras, Mollett (2013) argues that the emancipatory possibilities of counter-mapping are constrained by an emphasis on legibility, which produces racializing effects (see also Bhandar, 2011).

There is also the question of whose knowledge is represented in efforts to map the subaltern. Women, minorities and other vulnerable or disenfranchised groups are particularly likely to be marginalized (Hodgson and Schroeder, 2002: 161), while elderly men in positions of leadership are often seen as the bearers of knowledge about territory. In many contexts, non-Indigenous activists, lawyers, NGOs and academics have played central roles in 'translating' Indigenous territorial knowledges into maps (Anthias, 2018). Given that these actors are rarely fluent in Indigenous languages, this mediates Indigenous participation in the mapping process, as people who only speak the Indigenous language (which may include many older people, women or those who have less formal education) can find that their knowledges of territory are overlooked by mapmakers.[8]

Scholars have also pointed to the power asymmetries inherent in subaltern engagements with digital technologies, with some arguing that GIS represents a form of 'epistemological assimilation' (Rundstrom, 1995). There are also growing concerns about how digital technologies give companies like Google 'potentially tremendous power […] to reach into pockets of Indigenous resistance' (Hunt and Stevenson, 2017: 382; McGurk and Caquard, 2020). Moreover, digital technologies remain dependent on material infrastructures that require the continuing dispossession and environmental degradation of Indigenous lands.

This links to a broader set of issues regarding the uses and ownership of Indigenous maps. That is, will cartographic data be held and utilized by local people, or could it serve the interests of more powerful outsiders? A particularly troubling example in this regard is the Bowman expeditions, a series of participatory mapping projects with Indigenous peoples funded by the US Army's Foreign Military Studies Office (Wainwright, 2013; Bryan and Wood, 2015).[9]

Some scholars have located Indigenous mapping within a 'neoliberal multicultural' agenda aimed at providing legibility and legal security to investors, while containing more radical Indigenous political projects (Hale, 2011; Rivera Cusicanqui, 2012). The World Bank's role in financing the mapping and titling of Indigenous territories in Latin America fuelled such concerns (Bryan, 2012). In many contexts, the mapping of Indigenous territories has not prevented the proliferation and socio-environmental impacts of extractive industry in such spaces, nor has it reversed postcolonial forms of racialized dispossession (see Anthias 2018, 2019a). This gap between cartographic recognition and material land control can be a source of intense frustration for Indigenous peoples, while generating false perceptions of Indigenous land control among other actors. In short, it cannot be assumed that putting subalterns 'on the map' will further struggles for resource control and self-determination. Rather, mapping can become part of a broader colonial politics of recognition (Coulthard, 2014).

As these critical perspectives demonstrate, the question of mapping the subaltern is fraught with ambivalence. While counter-mapping can pose a representational challenge to colonial geographies and support local processes of identity-formation, it cannot escape existing relations of knowledge and power. It is also important to note that, historically, 'much of the writing [on Indigenous counter-mapping] available has been produced by non-indigenous people, with academics and, most recently, GIS specialists in the lead; therefore, the indigenous view is often incompletely represented' (Chapin *et al.*, 2005: 620). The challenge of subaltern representation therefore lies not only in mapping processes, but also in their documentation and analysis.

In this regard, it is important to note the important and growing contributions of Indigenous scholars and cartographers to these debates. Jay Johnson, Renee Louis and Albertus Pramono (2006) highlight the importance of developing a 'critical cartographic consciousness' among Indigenous peoples and activists engaged in mapmaking in light of the ambivalences discussed above. Indigenous scholars have also drawn attention to a rich history of Indigenous cartographic traditions that centre place-based Indigenous knowledge and world-making practices over legibility within Western mapping conventions. A recent special issue on 'Decolonizing the Map' (Rose-Redwood *et al.*, 2020) provides illustrative examples (see also Pearce and Louis, 2008; Louis, 2017; Lucchesi, 2018). This work points us towards a broader conceptualization of counter-mapping than early definitions, which focus on the use of Western cartographic technologies to map subaltern territorial claims. In the final section of this chapter, I consider some alternative approaches and examples.

Alternative approaches

Counter-mapping need not be limited to the use of traditional cartographic techniques but can refer to a wide array of representational practices. As noted above, Indigenous peoples have their own cartographic traditions, which may, for example, involve songs, rituals and stories (Johnson et al., 2006). Hunt and Stevenson draw on Emily Cameron's (2015) work in the Arctic to argue that 'such illustrations may be viewed as Indigenous counter-mapping in their own right' (Hunt and Stevenson, 2017: 378), as they are a means through which Indigenous peoples resist colonial cartographies and structure material and imaginative landscapes on their own terms. From this perspective, a broad range of Indigenous practices might be considered to be forms of counter-mapping.[10]

Indigenous peoples have also used audio-visual media to narrate their presence on the land. The Karrabing film collective, formed by anthropologist Elizabeth Povinelli and her Indigenous collaborators in Australia, have used film to challenge the way Indigenous Australians are (mis)represented in mainstream settler colonial narratives (Lea and Povinelli, 2018). The films show the entanglement of colonial and Indigenous elements in particular landscapes – for example, the effects of mining contamination and relations with Dreamings. The effect is to make visible Indigenous presence amidst settler colonial violence, without reducing this to a mapping exercise that fixes difference temporally and geographically.

Other counter-mapping projects have focused on the mapping of mobilities. Amalia Campos-Delgado (2018) worked with Central American irregular migrants in transit through Mexico on their journey to the United States, to produce cognitive maps of their journeys. These maps include legends and scales not found on conventional state maps – including unnamed river crossings, locations of kidnapping and places of refuge. Through her analysis, Campos-Delgado seeks to expose how bordering processes are lived, perceived and defied by migrants.

In urban contexts, counter-mapping projects have been used to draw attention to spatial (often racialized) inequalities faced by particular communities (Buzzeli and Veenstra, 2007). For example, activists in the South Bronx used GIS databases to document connections between waste transfer stations and disproportionately high rates of asthma and other airborne diseases (Maantay, 2007; see Figure 9.4). At a broader scale, the Environmental Justice Atlas (notanatlas.org) maps ecological conflicts around the world to make visible protests against environmental pollution and degradation. In the context of the COVID-19 pandemic, interactive maps have been used to highlight racialized inequalities in infection rates and deaths, which might otherwise remain hidden (Oppel et al., 2020).

Counter-mapping may also include material acts of transforming place names and other geographical markers. Hunt and Stevenson (2017) describe a project that sought to restore Indigenous place names to streets, avenues, roads, paths and trails in Toronto. In the context of the June 2020 Black Lives Matter protests in the United Kingdom, Scottish activists replaced the names of streets in Glasgow that commemorated wealthy slave-owning merchants with placards of black campaigners, slaves and those who died in police custody. Such actions 'map the subaltern' by transforming the physical landscape to make visible histories that have been silenced by dominant historical and geographical representations.

In a very different context, Vega et al. (2022) describe how Munduruku Indigenous practices of 'autodemarcation' in the Brazilian Amazon combine participatory mapping of territorial boundaries with physical acts of visually signposting, patrolling and policing these boundaries as a means of challenging incursions from loggers, miners and other outsiders. Here, mapping is embedded in a series of embodied and material practices. The authors argue that such practices provide an important arena for 'intersubjective recognition' between neighbouring forest dweller groups; as such, 'autodemarcation moves firmly beyond a struggle for cartographic representation towards decolonial ways of knowing and being' (p.39).

Mapping the subaltern

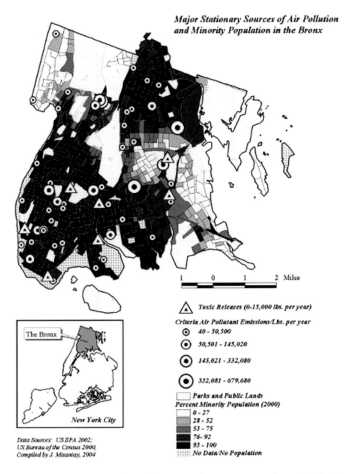

Figure 9.3 Major stationary sources of air pollution and minority population in the Bronx
Data Source: US EPA (2002); US Bureau of the Census (2000). Source: Mantaay (2007).

Other projects have used archival maps and documents to uncover histories of belonging and land rights that have been overwritten by settler colonial cartographies. The Palestine Open Maps project (palopenmaps.org) presents survey maps of Palestine from the British Mandate period (1922–1948) alongside census data, photographs and oral histories to convey 'layered visual stories that bring to life absent and hidden geographies'. By overlaying these data with more recent Israeli maps, the website enables users to visualize how Arab villages were displaced over time by Israeli settlements (Figure 9.5).

Counter-mapping can also be used as a critical research methodology to interrogate the erasures of existing cartographies. In my work with Guaraní communities in the Bolivian Chaco, I used map elicitation exercises, focus groups and a GPS-guided walk as tools for collective reflection on the erasures and power effects of state maps of Indigenous territories (Anthias, 2019b). The aim here was not to render subaltern geographies legible, but to reflect on the disjunctures between lived geographies and cartographic boundaries – including those created by previous counter-mapping initiatives. This highlights how mapping can be used as a self-reflexive and deconstructive tool, which seeks to make visible rather than cover up the knowledge-power relations involved in mapping the subaltern.

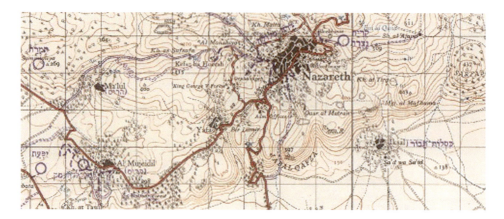

Figure 9.4 Israeli 1958 overprint (purple) on a British 1946 map, marking the destruction of Palestinian villages including Ma'lul and the new Jewish-Israeli settlements built on their lands

Source: Palestine Open Maps (https://palopenmaps.org/about).

Conclusion

As this chapter makes clear, mapping has become an important tool for subaltern groups in diverse regions of the world. While some mapping initiatives are undertaken by communities themselves, they often involve relationships with more powerful outsiders – from local NGOs and activists to global financial institutions and tech companies. The objectives of counter-mapping are also diverse, ranging from territorial governance to legal claims for land rights to efforts to communicate subaltern geographies to broader publics. While there are many positive examples, it is clear that mapping should not be treated uncritically as a tool for subaltern empowerment. There is a need for reflexivity regarding both the relations of knowledge and power involved in the mapping process, and the implications for broader struggles for land control and self-determination. Finally, it is important to keep in mind that most of the existing literature on counter-mapping has been produced by activists and scholars, rather than by Indigenous or other subaltern authors. If we accept that counter-mapping can include diverse representation practices that contest or refuse colonial cartographies, then most examples of counter-mapping surely remain undocumented. As such, Spivak's (1978) question 'Can the subaltern speak?' remains as relevant as ever.

Notes

1 Other related terms used include 'participatory mapping', 'participatory land use mapping', 'participatory resource mapping', 'community mapping', 'ethnocartography', 'auto-demarcation' and '"ancestral domain delimitation' (see Chapin *et al*., 2005).
2 Originating from the Latin *subalternus*, the term 'subaltern', meaning 'of a lower rank or position', was previously used to denote military rank and hierarchy.
3 See Loomba (1998) for an overview.
4 The term *Indigenous peoples* broadly refers broadly to peoples whose lands were colonized by Europeans and who remain marginalized within dominant settler colonial or postcolonial states. The term emerged in the mid-twentieth century in the context of international legal debates on Indigenous rights. While Indigeneity has become an important political identity, it must be placed in the context of colonial racial discourse and shifting national and global approaches to governing populations and territory (see Simpson, 2007; Radcliffe, 2017).

5 For example, Columbus's voyages relied extensively on Indigenous knowledge to navigate the Caribbean, while Captain Cook's efforts to devise a scientific means of calculating longitude were corrected by information from Polynesian navigators (Bryan, 2009).
6 The Land Portal's new Geoportal (see https://geoportal.landportal.org) is one such example.
7 See, for example, Peluso (1995), Hodgson and Schoeder (2002), Watson (2002), Johnson et al. (2006), Wainwright and Bryan (2009), Radcliffe (2011), Mollet (2013), and Tilley (2020).
8 This is not always the case; for example, Johnson et al. (2006) describe how well trained Indigenous cartographers have been using their education within the Cartesian/Newtonian cartographic epistemology to translate Indigenous geographic knowledge for Western audiences, producing autoethnographic cartographies.
9 The first of these expeditions, the México Indígena project, mapped Indigenous lands in Mexico as part of an effort to build a nation-wide GIS of property rights. This was informed by the US military's broader 'human terrain' strategy, that was developed in the context of the wars in Iraq and Afghanistan and echoes a longer history of mapping as a tool for US expansionism and counterinsurgency (Wainwright, 2013).
10 The term 'counter-mapping' may also be used metaphorically to describe a range of social movement practices that seek to expose or challenge dominant power relations, although examples of this go beyond the scope of this chapter.

References

Akerman, J. (Ed.) (2009) *The Imperial Map: Cartography and the Mastery of Empire* Chicago: University of Chicago Press.

Anthias P. (2018) *Limits to Decolonization: Indigeneity, Territory, and Hydrocarbon Politics in the Bolivian Chaco* Ithaca, NY: Cornell University Press.

Anthias, P. (2019a.) "Rethinking Territory and Property in Indigenous Land Claims" *Geoforum* 119 pp.268–278.

Anthias, P. (2019b) "Ambivalent Cartographies: Exploring the Legacies of Indigenous Land Titling through Participatory Mapping" *Critique of Anthropology* 39 (2) pp.222–242.

Anthias, P. and Radcliffe, S.A. (2015) "The Ethno-Environmental Fix and Its Limits: Indigenous Land Titling and the Production of Not-Quite-Neoliberal Natures in Bolivia" *Geoforum* 64 pp.257–269.

Bhandar, B. (2011) "Plasticity and Post-Colonial Recognition: 'Owning, Knowing, and Being'" *Law Critique* 22 pp.227–249.

Bhandar, B. (2018) *The Colonial Lives of Property: Law, Land, and Racial Regimes of Ownership* Durham, NC and London: Duke University Press.

Bryan, J. (2009) "Where Would We Be Without Them? Knowledge, Space and Power in Indigenous Politics" *Futures* 41 pp.24–32.

Bryan, J. (2012) "Rethinking Territory: Social Justice and Neoliberalism in Latin America's Territorial Turn" *Geography Compass* 6 pp.215–226.

Bryan, J. and Wood, D. (2015) *Weaponizing Maps: Indigenous Peoples and Counterinsurgency in the Americas* New York: Guilford Press.

Buzzeli, M. and Veenstra, G. (Eds) (2007) Part Special Issue: Environmental Justice, Population Health, Critical Theory and GIS *Health and Place* 13 (1) pp.1–298.

Cameron, E. (2015) *Far Off Metal River: Inuit Lands, Settler Stories, and the Making of the Contemporary Arctic* Vancouver: UBC Press.

Campos-Delgado, C. (2018) "Counter-Mapping Migration: Irregular Migrants' Stories through Cognitive Mapping" *Mobilities* 13 (4) pp.488–504.

Chapin, M., Lamb, Z. and Threlkeld, B. (2005) "Mapping Indigenous Lands" *Annual Review of Anthropology* 34 pp.619–638.

Coulthard, G.S. (2014) *Red Skins, White Masks: Rejecting the Colonial Politics of Recognition* Minneapolis: University of Minnesota Press.

Craib, R. (2000) "Cartography and Power in the Conquest and Creation of New Spain" *Latin American Research Review* 35 (1) pp.7–36.

Gordon, E.T., Gurdián, G.C. and C. R. Hale, C.R. (2003) "Rights, Resources, and the Social Memory of Struggle: Reflections and Black Community Land Rights on Nicaragua's Atlantic Coast" *Human Organization* 62 pp.369–381.

Gramsci, A. (1973) *Selections from the Prison Notebooks of Antonio Gramsci* (translated and edited by Hoard, Q and Nowell Smith, G.) New York: International Publishers.

Hale, C. (2011) "Resistencia para qué? Territory, Autonomy, and Neoliberal Entanglements in the 'Empty Spaces' of Central America" *Economy and Society* 40 pp.184–210.

Harley, J.B. (1989) "Deconstructing the Map" *Cartographica* 26 (2) pp.1–20.

Hodgson, D.L. and Schroeder, R.A. (2002) "Dilemmas of Counter-Mapping Community Resources in Tanzania" *Development and Change* 33 pp.79–100.

Hunt, D. and Stevenson, S.A. (2017) "Decolonizing Geographies of Power: Indigenous Digital Counter-mapping Practices on Turtle Island" *Settler Colonial Studies* 7 (3) pp.372–392.

Johnson, J.T., Louis, R.P. and Pramono, A.H. (2006) "Facing the Future: Encouraging Critical Cartographic Literacies in Indigenous Communities" *ACME* 4 pp.80–98.

Lea, T. and Povinelli, E.A. (2018) "Karrabing: An Essay in Keywords" *Visual Anthropology Review* 34 pp.36–46.

Leemann, E. (2019) "Who Is the Community? Governing Territory through the Making of 'Indigenous Communities' in Cambodia" *Geoforum* 119 pp.238–250.

Li, T.M. (2010) "Indigeneity, Capitalism, and the Management of Dispossession" *Current Anthropology* 51 (3) pp.385–414.

Li, T.M. (2014) "What Is Land? Assembling a Resource for Global Investment" *Transactions of the Institute of British Geographers* 39 (4) pp.589–602.

Louis, R.P (with Aunty Moana Kahele) (2017) *Kanaka Hawai'i Cartography: Hula, Navigation, and Oratory* Corvallis, OR: Oregon State University Press.

Lucchesi, A.H. (2018) "'Indians Don't Make Maps': Indigenous Cartographic Traditions and Innovations" *American Indian Culture and Research Journal* 42 (3) pp.11–26.

Loomba, A. (2015) *Colonialism-Postcolonialism* Oxford and New York: Routledge.

Maantay J. (2007) "Asthma and Air Pollution in the Bronx: Methodological and Data Considerations in Using GIS for Environmental Justice and Health Research" *Health and Place* 13 (1) pp.32–56.

Mann, M. and Daly, A. (2019) "(Big) Data and the North-in-South: Australia's Informational Imperialism and Digital Colonialism" *Television & New Media* 20 (4) pp.379–395.

McGurk, T.J. and Caquard, S. (2020) "To What Extent Can Online Mapping be Decolonial? A Journey throughout Indigenous Cartography in Canada" *The Canadian Geographer/Le Géographe canadien* 64 pp.49–64.

Mena, C.F., Arsel, M., Pellegrini, L., Orta-Martinez, M., Fajardo, P., Chavez, E., Guevara, A. and Espín, P. (2020) "Community-Based Monitoring of Oil Extraction: Lessons Learned in the Ecuadorian Amazon" *Society & Natural Resources* 33 (3) pp.406–417.

Milan, S. and Gutierrez, M. (2018) "Technopolitics in the Age of Big Data" In Caballero, F. and Gravante, T. (Eds) *Networks, Movements and Technopolitics in Latin America: Global Transformations in Media and Communication Research* Cham, Switzerland: Palgrave Macmillan, pp.95–109.

Mollett, S. (2013) "Mapping Deception: The Politics of Mapping Miskito and Garifuna Space in Honduras" *Annals of the American Association of Geographers* 103 (5) pp.1227–1241.

Offen, K. (2003) "Narrating Place and Identity, or Mapping Miskitu Land Claims in Northeastern Nicaragua" *Human Organization* 62 pp.382–392.

Oppel Jr., R.A., Gebeloff, R., Lai, K.K.R., Wright, W. and Smith, M. (2020) "The Fullest Look Yet at the Racial Inequity of Coronavirus" *New York Times* (5th July) Available at: https://www.nytimes.com/interactive/2020/07/05/us/coronavirus-latinos-african-americans-cdc-data.html.

Ospina, C. (2018) *Reducing Carbon Emissions through Indigenous Land Titles* New York: Climate Institute.

Peluso, N. (1995) "Whose Woods Are These? Counter-mapping Forest Territories in Kalimantan, Indonesia" *Antipode* 27 pp.383–406.

Poole, P. (1995) "Indigenous Peoples, Mapping and Biodiversity Conservation: An Analysis of Current Activities and Opportunities for Applying Geomatics Technologies" *Peoples, Forest Reefs Program Discussion Paper Series* Washington, DC: World Wildlife Fund.

Radcliffe, S.A. (2011) "Third Space, Abstract Space, and Coloniality: National and Subaltern Cartography in Ecuador" In Teverson, A. and Upstone, S. (Eds) *Postcolonial Spaces: The Politics of Place in Contemporary Culture* London: Palgrave, pp.129–145.

Radcliffe, S.A. (2017) "Geography and Indigeneity I: Indigeneity, Coloniality and Knowledge" *Progress in Human Geography* 41 (2): pp.220–229.

Razack, S. (Ed.) (2002) *Race, Space, and the Law: Unmapping a White Settler Society* Toronto: Between the Lines.

Rivera Cusicanqui, S. (2012) "Ch'ixinakax Utxiwa: A Reflection on the Practices and Discourses of Decolonization" *South Atlantic Quarterly* 111 (1) pp.95–109.

Rose-Redwood, R., Barnd, N.B., Lucchesi, A.H., Dias, S. and Patrick, W. (2020) "Decolonizing the Map: Recentering Indigenous Mappings" *Cartographica* 55 (3) pp.151–162.

Rundstrom, R.A. (1995) "GIS, Indigenous Peoples, and Epistemological Diversity" *Cartography and Geographic Information Systems* 22 (1) pp.45–57.

Saïd, E. (1994) *Culture and Imperialism* New York: Vintage Books.

Simpson, A. (2007) "Ethnographic Refusal: Indigeneity, 'Voice' and Colonial Citizenship" *Junctures: The Journal for Thematic Dialogue* 9 pp.67–80.

Sparke, M. (2005) *In the Space of Theory: Postfoundational Geographies of the Nation-State* Minneapolis: University of Minnesota Press.

Spivak, G.C. (1978) "Can the Subaltern Speak?" In Nelson, C. and Grossberg, L. (Eds) *Marxism and the Interpretation of Culture* Urbana, IL: University of Illinois Press, pp.271–313.

Stocks, A. (2003) "Mapping Dreams in Nicaragua's Bosawas Reserve" *Human Organization* 62 pp.344–356.

Tilley, L. (2020) "'The Impulse is Cartographic': Counter-mapping Indonesia's Resource Frontiers in the Context of Coloniality" *Antipode* 52 (5) pp.1434–1454.

Vega, A., Fraser, J., Torres, M. and Loures, R. (2022) "Those Who Live Like Us: Autodemarcations and the Co-becoming of Indigenous and Beiradeiros on the Upper Tapajós River, Brazilian Amazonia" *Geoforum* 129 pp.39-48 DOI: 10.1016/j.geoforum.2022.01.003. Wainwright, J. (2013) *Geopiracy: Oaxaca, Militant Empiricism, and Geographical Thought*. New York: Palgrave Macmillan.

Wainwright, J. and J. Bryan (2009) "Cartography, Territory, Property: Postcolonial Reflections on Indigenous Counter-mapping in Nicaragua and Belize" *Cultural Geographies* 16 (2) pp.153–178.

Watson, I. (2002) "Buried Alive" *Law Critique* 13 pp.253–269.

10
GEOSPATIAL TECHNOLOGIES AND RURAL AND INDIGENOUS SPATIAL KNOWLEDGES

María Belén Noroña

Rural and Indigenous spatial knowledges

In the last three decades, geographic information systems (GIS), global positioning systems (GPS), remote sensing, and Internet mapping services have emphasized the importance of rural and Indigenous knowledges (RIK) in the production of higher-quality datasets and scientific advancement (Sillitoe, 2007; Laituri, 2011). Rural and Indigenous (RI) spatial knowledges are typically diverse, non-hegemonic ontologies and epistemologies that resist, differ, and interact with mainstream understandings of place and space (Rundstrom, 1995; Descola, 2004; Sillitoe, 2007).

While some RI communities are associated with rural areas, land, and natural resources (Descola, 2004; Anderson, 2013; Pearce, 2018), other communities have livelihoods that rely on more than rural areas, including urban, service, and technological sectors (Escobar, 2008). Such diversity generates important conversations between mainstream and non-western understandings of place and space by moving between urbanities and peripheries (Escobar, 2008). RIK provides alternative ways of understanding place, space, and wayfinding that have major implications for the use of geospatial knowledge and technology.

Except for uncontacted tribes, such as those inhabiting parts of the Amazon rainforest (Brackelaire, 2006), RIK is produced and reproduced in opposition to, in relation to, and in conversation with western spatial knowledge, science, and technology (Dei, 2000; Dove et al., 2009). RI knowers engage with dominant science and technology not only to distance themselves from it, but also to legitimize their claims within hegemonic economic and political systems that operate using western science and technology (Castree, 2004; Escobar, 2008).

Scholars have called RI methods for understanding place, space, and wayfinding *process cartography* (Rundstrom, 1995; Pearce and Louis, 2008), *multiple cartography* (Johnson et al., 2005: 85), and *Indigenous cartography* (Woodward and Lewis, 1998). These terms emphasize the importance that RIK assigns to place, subjectivity, human and non-human relationships, and the importance of social processes and temporalities in place making. Throughout this chapter, I will use the term 'process cartography' to describe the incredibly varied ways in which RI communities understand and represent place, space, and wayfinding.

Process cartography may include some or all of the following characteristics: it presents time as cyclical rather than linear (Farris, 1995), it assumes non-anthropocentric views where other species and resources – animals and rivers, for example – have an ontological status similar to that of humans, and finally, process cartography operates through non-binary thinking where distinctions between culture and nature are blurred (Rundstrom, 1995; Descola, 2004).

A key aspect of process cartography is that cultural values and spirituality often cannot be separated from material aspects of place and space. For example, geological features such as mountains are considered to have agency, and the histories of these entities closely shape human morals (Roberts, 2012; De la Cadena, 2015). RI communities have traditionally maintained alternative understandings and practices of place and space while also finding ways to exist within the confines of western science and cartography. De la Cadena (2015), for example, details how highland Indigenous peoples of Peru maintain sacred and material relationships with their landscapes, especially mountains, while simultaneously using western discourse to negotiate with the state for resource management of these landscapes. Anthropomorphizing the landscape and assigning agency to it places RI communities at a disadvantage in state-led resource management processes. Therefore, Indigenous people begin to strategically treat the same mountains that they otherwise venerate as natural resources with commercial and ecological value that can be read on maps.

Another particularity of process cartography is that sometimes people make no ontological distinction between nature and culture. For instance, some Maya-Xinca and Aymara communities understand their bodies as a first territory of habitation that is highly dependent on a larger territory that allows everyone's material survival (Cabnal, 2010). Thus, relations between the humans, their territory, and non-human entities such as natural resources are interrelated and interdependent. Indigenous intellectuals such as Lorena Cabnal refer to these relationships and their existence as *territorio-cuerpo-tierra* or territory-body-land (2010: 23). Thus, in contrast to cartographic representations of space, the territory is viewed and represented as a body or fabric that weaves together relations with other territorial beings (Echeverri, 2004).

Process cartography sometimes emphasizes performances such as recitations, dances, and songs among others as representational elements (Turnbull, 2000; Johnson *et al.*, 2005; Pearce and Louis, 2008; Hirt, 2012). For example, Roberts (2012: 403) explains that the Polynesian Maori in New Zealand use cognitive maps called *whakapapas*. These cognitive maps form through narratives and music that consist of stories about the genealogy of things, places, values, morals, and specific practices associated to food production, consumption, and social relations (Roberts, 2012).

Hirt (2012) provides another example of process cartography in which the Mapuche shamans (wise elders and/or healers) in Chile use a traditional practice of dreaming to communicate with spirits to learn about strategic places. In Hirt's research, this type of wayfinding identifies places within a specific Indigenous territory and confirms the community's occupancy of territory over time. Hirt (2012) documents other scholars who compare Mapuche dreaming techniques to contemporary GPS devices. Rather than precision, dreaming allows shamans to learn about problematic realities elsewhere in ways that are similar to telekinesis. All these methods of organizing the world correspond to cognitive and perceptual mapping of the known environment, usually passed from generation to generation (McKenna *et al.*, 2008).

Compared to western GIS, RI wayfinding techniques use environmental landmarks, noting of constellations, and the movement of the sun and moon to develop cognitive mental maps for practical applications rather than to achieve precision and objectivity. For example, Turnbull (2000) and Feinberg *et al.* (2003) explain that Pacific Islanders historically used navigation techniques, known as star compasses, in which they memorized the rising and setting of stars and constellations. This knowledge allowed them to identify their positions in relation to local islands and landmarks, and to determine their direction of travel while at sea. Wind direction, wave patterns, cloud formation, and even human instinct also aided in Indigenous sea navigation (Feinberg *et al.*, 2003; Feinberg and Genz, 2012).

Process cartography and its wayfinding techniques have often proven to be more precise than conventional western methods. For instance, people living near the equatorial Andes have combined astrological and cultural knowledge of the landscape to orient themselves and administer territories since pre-Hispanic times. In Ecuador, locals have identified landmarks such as mountains to mark the sun's movement throughout the year. From the equinoctial line at 0 degrees and zero minutes, it is possible to observe how the sun moves north and south throughout the year (Cobo, 2017). Indigenous people observed the shadows produced by mountains, and ceremonial sites built for observation and located on the equinoctial line, to orient themselves in time and space. This also allowed them to plan for agriculture and arrange government and spirituality around the sun's movement. Indigenous people in the Andes were aware of the equator's exact location long before the French geodesical commission measured latitude 0 to establish the shape of the Earth in the eighteenth century (Cobo, 2017). We can see how the sun in in Figure 10.1 illuminates the Catequilla hill, located on the equinoctial line during the June solstice. The last spot to be illuminated on the right corresponds to an Indigenous archaeological remain built for such observation.

Oher examples include Indigenous Pacific Islander navigators (Turnbull, 2000) and rural Irish fishermen (McKenna *et al.*, 2008) that use landmarks and constellations when navigating; others memorize the location of sea beds and reefs in remote areas of the Solomon Islands (Feinberg *et al.*, 2003) to locate themselves. Such techniques are not considered reliable in western society because these methods are aimed at fulfilling practical needs and work only in specific contexts (Feinberg *et al.*, 2003). Even if they are precise, these methods are not scalable and cannot be used across different sets of data (Turnbull, 2000; Agarwal, 2005), making their incorporation or representation in geospatial technologies challenging. RI wayfinding usually represents distance by using temporal duration units such as the number of nights needed to travel from point A to point B, or observations that can be seen only during specific times of the year, whereas western maps use temporal and mathematical units (McKenna *et al.*, 2008). RI locational techniques are highly dependent on local landscapes and practices and are not applicable to broader applications from a western perspective.

Because of these differences, western science does not consider process cartography to be generalizable (Turnbull, 2000). Still, many scholars suggest that rural and Indigenous ways of knowing should be recognized as knowledge in its own right and be understood as having equal status when compared to western knowledge (Smith, 1999; Lucchesi, 2018). Such comparisons can raise questions about the validity of RIK (Feinberg *et al.*, 2003) by reemphasizing the hegemony of western science and technology in defining what has the legitimacy to exist and to be represented (Turnbull, 2000).

Figure 10.1 Pictures by Cristobal Cobo, director of the Quitsato research project. Catequilla hill near the city of Quito during June's solstice (21st June 2003)

Moreover, literature on the use and inclusion of RIK in geospatial technologies relies on narratives and representations that have been written in western terms (Johnson *et al.*, 2005; McKenna *et al.*, 2008; Wood, 2010). Additionally, data and technologies are mostly controlled by western knowers and users (Chambers *et al.*, 2004; McMahon *et al.*, 2017). Thus, many scholars have suggested that the production of knowledge through geospatial technologies may be incompatible with the knowledge and practices of some RI communities and might even endanger the survival of process cartographies (Chambers *et al.*, 2004; Reid and Sieber, 2019).

Rural and Indigenous knowledge and its inclusion in geospatial technologies

The production of georeferenced maps has been the most common geospatial application that incorporates RIK of place and space. These maps have been used as legal evidence to bolster Indigenous claims over traditional territories and resources (Herlihy and Knapp, 2003; Chambers *et al.*, 2004). Mapping Indigenous territories to secure land tenure began as early as the 1960s among British Columbia and Alaska's Native Nations (Chapin *et al.*, 2005).

Later, combined efforts of academics, activists, and Latin American Indigenous people to secure land and resource access led to counter-mapping practices (Peluso, 1995). These practices included participatory research mapping methodologies (PRM). PRM seeks Indigenous input in building databases for georeferenced maps to empower Indigenous land claims (Herlihy, 2003; Chapin *et al.*, 2005). This type of collaboration improved the quality of gathered data because it included the participation of those who best knew the territory and environment while also enhancing understanding of Indigenous social relations and land use (Herlihy and Knapp, 2003).

Mapping of RI territories through participatory methods became mainstream in the 1990s when multilateral institutions such as the United Nations, the US Agency for International Development, and the World Bank funded worldwide mapping initiatives (Bryan and Wood 2015). Although these initiatives were justified by the need to empower Indigenous claims over territory and to strengthen Indigenous voices politically, maps and data have also facilitated multilateral and state-based interventions in conservation, development, production, and modernization (Wood, 2010; Laituri, 2011; Bryan and Wood, 2015).

The methods of participatory mapping and participatory GIS (PGIS) have become more sophisticated as technologies have evolved. For example, efforts to include alternative understandings of space, place, and wayfinding have led to the development of cybercartographies. These GIS and GPS technologies combine maps, data, and descriptive functions that allow the inclusion of storytelling and other qualitative data into a user-friendly Web interface (Taylor, 2005, 2014). Among many existing applications, an important cybercartography tool geared specifically to RI users is the Nunaliit and the Native-land Atlas. Nunaliit concentrates on RI participation and empowerment through the development of databases and in the simplification of Web-based tools to encourage RI participation in the construction of datasets (Caquard *et al.*, 2009; Eisner *et al.*, 2012; Anonby *et al.*, 2018). The Native-land Atlas emphasizes Indigenous views of their territories over nation-states cartographies. Different from other software like Esri Story maps, Nunaliit and the Native-land Atlas have developed tools and features that allow alternative processual knowledge to be stored and represented, empowering Indigenous logics and knowledge (Hayes and Taylor, 2019; McGurk and Caquard, 2020).

The field has made tremendous progress towards including RIK in the analysis of georeferenced maps, satellite images, and 3D modelling technology, among others. Interest is driven by an increasing awareness of RI knowledge as important for better understanding climate and landscape change, along with social vulnerability and risk reduction (Alexander *et al.*, 2011; Pearce, 2018). In addition, there is renewed interest in finding alternatives to existing resource management frameworks using ecological Indigenous knowledges (McCall and Minang, 2005; Veland *et al.*, 2013).

These studies are particularly important when RI concerns match those of science (Eisner *et al.*, 2012; Pearce, 2018); for example, when analysing Alaskan landscape change that integrates Indigenous narratives (Eisner *et al.*, 2012), or when comparing climate change patterns with Indigenous perceptions of cooling and warming temperatures (Alexander *et al.*, 2011; Sheremata *et al.*, 2016). These tools have also been vital for improving disaster management and mitigation approaches by incorporating RI environmental knowledge. In these cases, RIK has served to anticipate landscape change based on experience and observation, improving models for local resilience (Lauer and Aswani, 2010; Veland *et al.*, 2013).

RIK has also been important in the assessment of resource management projects that leverage GIS and 3D modelling technology. These technologies have been used to develop models that anticipate different scenarios of landscape change (Lewis and Sheppard, 2006; Pearce and Louis, 2008). To make Indigenous concerns visible to state, non-governmental, and corporate institutions, it is common for Indigenous communities to contact universities, research tanks, development institutions, and private experts to gain technical support and expertise in legitimizing land and resource claims through science and geospatial technologies (Radcliffe, 2010; Thill, 2018; Noroña, 2020).

Criticism

When done ethically, GIS mapping, with or without the participation of rural and Indigenous peoples, has in many cases proven to be an important tool for legitimatizing Indigenous claims and empowering process cartographies. However, the involvement of Western academics and professionals in mapping projects has also been highly criticized because this work frequently implies colonial and paternalistic relations where data collection and mapping are portrayed as a benevolent intervention at the service of RI communities (Louis, 2007; Sletto, 2009a).

It is well documented that maps, GIS databases, and remote sensing have been used to privatize communitarian land, introduce forced modernization, identify resources key to corporate interests, implement governance models that do not serve RI interests and concerns, and even facilitate military interventions (Johnson *et al.*, 2005; Eisner *et al.*, 2012; Bryan and Wood, 2015; Reid and Sieber, 2020). Maps and geospatial technologies have been blamed for advancing colonization, state supremacy, and for furthering neoliberal agendas (Wood, 2010). For example, Bryan and Wood (2015) exposed how the US military funded US academic institutions to perfect participatory mapping approaches in Central America, where RIK about territory and terrain was shared with the military for counterinsurgency purposes.

Cartographers and scholars such as Johnson *et al.* (2005), Chambers *et al.* (2004), Pearce and Louis (2008), Johnson (2011), Olson *et al.* (2016), and McMahon *et al.* (2017), among others, have found that RIK is not well represented in geospatial technologies. RI awareness of space and place is frequently decontextualized, distorted, and assimilated into western logics (Rundstrom, 1995). For instance, to achieve cartographic accuracy in a map, GIS uses mathematical logic to define the nature of reality. It organizes reality by establishing domains, and within those domains there are classes and properties that define all data in discrete ways (Reid and Sieber, 2020). Conversely, RI understanding of place and space frequently resembles a woven fabric or meshwork (Descola, 2004; Hirt, 2012; Roberts, 2012) where some or all of its components are interdependent and understandings of place and space are processual and relational (Johnson et al., 2005; Sletto, 2009a). Thus, western cartography often fails to capture RI ways of representing reality and space (Rundstrom, 1995), challenging the survival of process cartography.

Many Indigenous groups and rural dwellers do not use borders to organize space because land and resources are either customarily shared, or borders are seen as porous membranes that allow the exchange of people and natural resources. When Indigenous communities present claims over territory to nation-states as ancestral shared territory, governmental frameworks fracture the land and privilege state, corporate, conservational, and private spatial logics (Sawyer, 2004; Ybarra, 2018). For example, private and state actors make claims over natural resources, such as oil and gas, which not only fractures the land but also emboldens internal ethnic conflicts within communities who feel forced to engage with mainstream spatial logics to maintain access to their traditional territories (Anthias, 2018; Noroña 2022). Rural community members that have shared flexible borders may then feel compelled to deny access to other users with whom they have traditionally shared rivers, lakes, or agricultural land (Sawyer, 2004, Sletto, 2009b).

In general, state governmentality has been aided by mainstream cartography and GIS technologies to administrate and govern territories, preventing RI communities from securing land and resources in their own terms and changing people's relationship with their territory (Scott, 1998; Johnson et al., 2005; Sletto, 2009a, 2009b; Wainwright and Bryan, 2009).

Additionally, GIS and GPS commonly misrepresent alternative knowledge affecting RI lives through policy making (McMahon et al., 2017). For example, rural community members that are not mapped as Indigenous by state authorities in the Amazon of Ecuador cannot access state services such as access to piped water or agricultural improvement programs (Noroña, 2020). This misrepresentation is the result of epistemological differences between how state officials and Indigenous nations define who is Indigenous and who is not, and how Indigenous relationships to resources are established in official maps and records.

Finally, the imbalanced power dynamics between western and RIK are exacerbated when rural and Indigenous people have no other option than to strategically adopt, or even embrace, the use of western science and geospatial technology to make their concerns and needs visible to western society (Thill, 2018; Reid and Sieber, 2020; Noroña, 2022). Adoption of these technologies is constrained by power dynamics that cause RI communities to have difficulties achieving geospatial literacy and access to technology such as the Internet, computers, and software (Chambers et al., 2004; Johnson et al., 2005; Laituri, 2011). For example, even when rural and Indigenous people participate in collecting GPS data for mapping purposes, they are often unable to interpret or use the final map (Wainwright and Bryan, 2009).

Geospatial technologies and efforts to represent space and place in rural and Indigenous terms

Scholars working at the intersection of geospatial technologies and process cartography recognize the limitations that geospatial technologies have in: (a) representing RIK in its own terms (Louis, 2007; McMahon et al., 2017), (b) producing geospatial tools and information that serve RI needs ethically while also advancing western scientific knowledge (Eisner et al., 2012; Reid and Sieber, 2020), and (c) encouraging RI participation in accessing and controlling geospatial technologies through literacy (Chambers et al., 2004; Laituri, 2011).

Scholars involved in the production of cartography, atlases, and other geospatial technologies continue to design tools sensitive to RI conceptualizations of place and space. GIS mapping projects involving rural and Indigenous knowers have developed several tools that allow representation of ecological, cultural, and relational aspects of processual knowledge as suggested by Kitchin *et al.* (2013) and Olson *et al.* (2016), including cybercartography software like the Nunaliit framework discussed earlier. These tools may include multimedia formats that allow for the storage of oral tradition, actual voice recordings of native languages, and video, among others (Caquard *et al.*, 2009; Anonby *et al.*, 2018).

Geospatial technology advocates argue that despite their limitations and inherent power dynamics, GIS, 3D modelling, and multimedia formats allow for better representations of process cartography (Pearce and Louis, 2008; Laituri, 2011). Advocates argue that GIS allows for information storage in multiple layers, each with different levels of detail, sensitivity, and confidentiality (Chambers *et al.*, 2004). Pearce and Louis (2008) have experimented with 3D modelling to bring a sense of positionality to map viewers, thus avoiding the idea of an objectified truth in a map where the viewer is detached from place. Others, like Lewis and Sheppard (2006) and Rambaldi *et al.* (2007), have been successful by using 3D visualizations of different landscape change scenarios in the implementation of resource management plans with First and Indigenous Nations.

Still, RI communities have very little control over these technologies and their technological literacy is generally low (Olson *et al.*, 2016), making them highly reliant on western expertise (McGurk and Caquard, 2020). Lack of literacy and control becomes even more pressing in a society whose reality increasingly operates through visual media and digital technology (Lukinbeal, 2014; Intahchomphoo, 2018). Indeed, geospatial literacy governs a user's ability to identify these technologies, evaluate them, and effectively use the technology by combining analysis and praxis (Lukinbeal, 2014).

Geographers and cartographers such as Johnson (2011), McMahon *et al.* (2017), and Lucchesi (2018) suggest the use of geospatial literacy programs as a means to decrease the imbalance of power dynamics between western and rural and Indigenous knowers. They add that literacy should include knowledge of the history and role of maps and other geospatial technologies in the processes of colonization (Johnson *et al.*, 2005), and strategic knowledge not only of how to collect and represent data, but also of how to re-represent and reclaim the nature of digital narratives (McMahon and Smith, 2017).

Efforts to achieve geospatial literacy among rural and Indigenous people are still far from complete; however, it is possible to find important efforts in the non-profit, business, and corporate sectors. The Native-Land.ca is a non-profit based in Canada and is governed by a majority Indigenous board of directors. This organization works closely with communities and grassroots organizations to map Indigenous territories tackling issues of inclusion, representation, and accuracy, among others (native-land.ca). Other projects include the Indigenous Mapping Workshop (IMW) sponsored by Esri, the Firelight Group, Universities, and Indigenous organizations in Colombia, Peru, Canada, and Australia. The IMW has around six to ten workshops per year in different parts of the world, reaching out to hundreds of rural and Indigenous users annually (indigenousmaps.com).

Other literacy efforts are financed by multilateral institutions like the Interamerican Development Bank who recently partnered with Google Earth Outreach, the Amazonian Conservation Team, Central American Universities, and grassroots organizations in Panamá to train Indigenous community members to map their own territories using Google Earth (Cotacachi, 2020). Although these efforts are important, and Indigenous people feel empowered, geospatial literacy is ever evolving and requires continuous education. As Nanki Wampankit, an Indigenous Ecuadorian cartographer, explains, Indigenous people remain dependent on external organizations for technological updates (4th December 2017: pers. comm.). According to

Wampankit, Indigenous cartographers are both critical and cautious of the involvement of multilateral institutions, corporate, and political interests in furthering Indigenous ability to produce georeferenced information due to its potential misuse (4th December 2017: pers. comm.).

Literacy opportunities in the formal education system are minimal among RI communities due to structural inequalities consistent with colonial legacies worldwide (United Nations, 2009). For example, Native American enrollment per capita in US colleges and higher education is lower than white, Asian, Pacific Islander and African American populations (Espinosa *et al.*, 2019). Although there are Indigenous specialists in cartography and other geospatial technologies mediating work with non-western knowers, Indigenous faculty in higher education are still underrepresented (Brayboy *et al.*, 2015).

In order to decrease power dynamics between western and process cartography, scholars interested in decolonizing geospatial technologies have suggested that western cartographers need to recognize Indigenous knowledge of space and place as cartography in its own right (Turnbull 2000; Johnson *et al.*, 2005). This would require cartographers to invest in learning about alternative spatial ontologies before producing maps (Johnson *et al.*, 2005). Scholars like Louis (2007) and Sundberg (2014) suggest that collaboration in geography should be a process of joint production rather than a process in which western knowers use Indigenous knowledge as empirics. Similarly, Reid and Sieber (2020) suggest that Indigenous people should be directly involved in shaping geospatial technologies themselves, instead of being treated as data sources or as technology consumers.

Although the need for such recognition has been widely acknowledged, little progress has been made to take process cartography seriously (Turnbull, 2000; Johnson *et al.*, 2005; Reid and Sieber, 2020). Additionally, RIK has proven to be of great importance in aiding society to solve pressing problems such as climate change. It is therefore incumbent on scholars and practitioners to not only invite rural and Indigenous knowers into the conversation, but to listen and allow these alternative knowledges to speak in their own terms.

References

Agarwal, P. (2005) "Ontological Considerations in GIScience" *International Journal of Geographical Information Science* 19 (5) pp.501–536 DOI: 10.1080/13658810500032321.

Alexander, C., Bynum, N., Johnson, E., King, U., Mustonen, T., Neofotis, P. and Vicarelli, M. (2011) "Linking Indigenous and Scientific Knowledge of Climate Change" *BioScience* 61 (6) pp.477–484 DOI: 10.1525/bio.2011.61.6.10.

Anderson, M.K. (2013) *Tending the Wild: Native American Knowledge and the Management of California's Natural Resources* Berkeley and Los Angeles: University of California Press.

Anonby, E., Murasugi, K. and Domínguez, M. (2018) "Mapping Language and Land with the Nunaliit Atlas Framework: Past, Present and Future" *Proceedings of FEL XXII (2018): Endangered Languages and the Land: Mapping Landscapes of Multilingualism* Available at: http://www.elpublishing.org/PID/4009.

Anthias, P. (2018) *Limits to Decolonization: Indigeneity, Territory, and Hydrocarbon Politics in the Bolivian Chaco* Ithaca, NY: Cornell University Press.

Brackelaire, V. (2006) "Situación de los últimos pueblos indígenas aislados en América latina (Bolivia, Brasil, Colombia, Ecuador, Paraguay, Perú, Venezuela)" *Diagnóstico regional para facilitar estrategias de protección V. Brackelaire* Brasilia DF 69.

Brayboy, B.M.J., Solyom, J.A. and Castagno, A.E. (2015) "Indigenous Peoples in Higher Education" *Journal of American Indian Education* 54 (1) pp.154–186 Available at: https://www.jstor.org/stable/10.5749/jamerindieduc.54.1.0154.

Bryan, J. and Wood, D. (2015) *Weaponizing Maps: Indigenous Peoples and Counterinsurgency in the Americas* New York: Guilford Press.

Cabnal, L. (2010) "Feminismos Diversos: el feminismo comunitario" Las Segovias: ACSUR Las Segovias Available at: https://porunavidavivible.files.wordpress.com/2012/09/feminismos-comunitario-lorena-cabnal.

pdf.Caquard, S., Pyne, S., Igloliorte, H., Mierins, K., Hayes, A. and Taylor, D.R.F. (2009) "A 'Living' Atlas for Geospatial Storytelling: The Cybercartographic Atlas of Indigenous Perspectives and Knowledge of the Great Lakes Region" *Cartographica* 44 (2) pp.83–100 DOI: 10.3138/carto.44.2.83.

Castree, N. (2004) "Differential Geographies: Place, Indigenous Rights and 'Local' Resources" *Political Geography* 23 (2) pp.133–167 DOI: 10.1016/j.polgeo.2003.09.010.

Chambers, K., Corbett, J., Keller, C. and Wood, C. (2004) "Indigenous Knowledge, Mapping, and GIS: A Diffusion of Innovation Perspective" *Cartographica* 39 (3) pp.19–31 DOI: 10.3138/N752-N693-180T-N843.

Chapin, M., Lamb, Z. and Threlkeld, B. (2005) "Mapping Indigenous Lands" *Annual Review of Anthropology* 34 pp.619–638 DOI: 10.1146/annurev.anthro.34.081804.120429.

Cobo, C. (2017) "Catequilla y los Discos Líticos, evidencia de la Astronomía Antigua en los Andes Ecuatoriales" *Revista de Topografía Azimut* 8 pp.41–62 Available at: https://revistas.udistrital.edu.co/index.php/azimut/article/view/8211/12750.

Cotacachi, D. (2020) "2020 Inclusivo: Tecnología Accessible para los Pueblos Indígenas" BID-Gender and Diversity Blog Available at: https://blogs.iadb.org/igualdad/es/2020-inclusivo-tecnologia-accesible-para-los-pueblos-indigenas/ Accessed: 6th July 2020.

De la Cadena, M. (2015) *Earth Beings: Ecologies of Practice Across Andean Worlds* Durham, NC: Duke University Press.

Dei, G.J.S. (2000) Rethinking the Role of Indigenous Knowledges in the Academy" *International Journal of Inclusive Education* 4 (2) pp.111–132 DOI: 10.1080/136031100284849.

Descola, P. (2004). "Ecology as Cosmological Analysis" In Suralles, A. and García, P. (Eds) *The Land Within: Indigenous Territory and the Perception of Environment* Copenhagen: IWGIA, pp.22–35.

Dove, M. R., Smith, D.S., Campos, M.T., Mathews, A.S., Rademacher, A., Rhee, S. and Yoder, L.M. (2009) "Globalization and the Construction of Western and Non-western Knowledge" In Sillitoe, P. (Ed.) *Local Science vs. Global Science: Approaches to Indigenous Knowledge in International Development* Oxford: Berghahn Books, pp.129–154.

Echeverri, J.Á. (2004) "Territory as Body and Territory as Nature: Intercultural Dialogue" In Suralles, A. and García, P. (Eds) *The Land Within: Indigenous Territory and the Perception of Environment* Copenhagen: IWGIA,, pp.259–276.

Eisner, W.R., Jelacic, J., Cuomo, C.J., Kim, C., Hinkel, K.M. and Del Alba, D. (2012) "Producing an Indigenous Knowledge Web GIS for Arctic Alaska Communities: Challenges, Successes, and Lessons Learned" *Transactions in GIS* 16 (1) pp.17–37 DOI: 10.1111/j.1467–9671.2011.01291.x.

Escobar, A. (2008) *Territories of Difference: Place, Movements, Life, Redes* Durham, NC: Duke University Press.

Espinosa, L., Turk, J., Taylor, M. and Chessman, H.M. (2019) *Race and Ethnicity in Higher Education: A Status Report* Washington, DC: American Council of Education and the Andrew Mellon Foundation Available at: https://www.equityinhighered.org.

Farriss, N.M. (1995) "Remembering the Future, Anticipating the Past: History, Time, and Cosmology among the Maya of Yucatan" In Hughes, D. and Trautmann, T. (Eds) *Time: Histories and Ethnologies* Ann Arbor, MI: University of Michigan Press, pp.107–138.

Feinberg, R., Dymon, U.J., Paiaki, P., Rangituteki, P., Nukuriaki, P. and Rollins, M. (2003) "Drawing the Coral Heads: Mental Mapping and its Physical Representation in a Polynesian Community" *The Cartographic Journal* 40 (3) pp.243–253 DOI: 10.1179/000870403225012943.

Feinberg, R. and Genz, J. (2012) "Limitations of Language for Conveying Navigational Knowledge: Way-Finding in the Southeastern Solomon Islands" *American Anthropologist* 114 (2) pp.336–350 DOI: 10.1111/j.1548-1433.2012.01429.x.

Hayes, A. and Taylor, D.R.F. (2019) "Developments in the Nunaliit Cybercartographic Atlas Framework" In Taylor, D.R.F., Anonby, E. and Murasugi, K. (Eds) *Further Developments in the Theory and Practice of Cybercartography: International Dimensions and Language Mapping (Modern Cartography Series Vol.7)* Amsterdam: Academic Press, pp.205–218 DOI: 10.1016/B978-0-444-64193-9.00013-0.

Herlihy, P.H. (2003) "Participatory Research Mapping of Indigenous Lands in Darien, Panama" *Human Organization* 62 (4) pp.315–331 DOI: 0018–7259/03/040315-17.

Herlihy, P.H. and Knapp, G. (2003) "Maps of, by, and for the Peoples of Latin America" *Human Organization* 62 (4) pp.303–314 DOI: 0018–7259/03/040303-12.

Hirt, I. (2012) "Mapping Dreams/dreaming Maps: Bridging Indigenous and Western Geographical Knowledge" *Cartographica* 47 (2) pp.105–120 DOI: 10.3138/carto.47.2.105.

Intahchomphoo, C. (2018) "Indigenous Peoples, Social Media, and the Digital Divide: A Systematic Literature Review" *American Indian Culture and Research Journal* 42(4), pp.85–111 DOI: 10.17953/aicrj.42.4.intahchomphoo.

Johnson, J.T. (2011) "Indigenous Knowledges Driving Technological Innovation" *AAPI Nexus: Asian Americans & Pacific Islanders Policy, Practice and Community* 9 (1–2) pp.241–248.

Johnson, J.T., Louis, R.P. and Pramono, A.H. (2005) "Facing the Future: Encouraging Critical Cartographic Literacies in Indigenous Communities" *ACME* 4 (1) pp.80–98 Available at: https://www.acmejournal.org/index.php/acme/article/view/729.

Kitchin, R., Gleeson, J. and Dodge, M. (2013) "Unfolding Mapping Practices: A New Epistemology for Cartography" *Transactions of the Institute of British Geographers* 38 (3) pp.480–496 DOI: 10.1111/j.1475–5661.2012.00540.x.

Laituri, M. (2011) "Indigenous Peoples' Issues and Indigenous Uses of GIS" In Nyerges, T., Couclelis, L. and McMaster, R. (Eds) *The SAGE Handbook of GIS and Society* London: Sage, pp.202–221.

Lauer, M. and Aswani, S. (2010) "Indigenous Knowledge and Long-term Ecological Change: Detection, Interpretation, and Responses to Changing Ecological Conditions in Pacific Island Communities" *Environmental Management* 45 (5) pp.985–997 DOI: 10.1007/s00267-010-9471-9.

Lewis, J.L. and Sheppard, S.R. (2006) "Culture and Communication: Can Landscape Visualization Improve Forest Management Consultation with Indigenous Communities?" *Landscape and Urban Planning* 77 (3) pp.291–313 DOI: 10.1016/j.landurbplan.2005.04.004.

Louis, R.P. (2007) "Can You Hear Us Now? Voices from the Margin: Using Indigenous Methodologies in Geographic Research" *Geographical Research* 45 (2) pp.130–139 DOI: 10.1111/j.1745–5871.2007.00443.x.

Lucchesi, A.H.E. (2018) "Indians Don't Make Maps": Indigenous Cartographic Traditions and Innovations *American Indian Culture and Research Journal* 42 (3) pp.11–26 DOI: 10.17953/aicrj.42.3.lucchesi.

Lukinbeal, C. (2014) "Geographic Media Literacy" *Journal of Geography* 113 (2) pp.41–46 DOI: 10.1080/00221341.2013.846395.

McCall, M.K. and Minang, P.A. (2005) "Assessing Participatory GIS for Community-based Natural Resource Management: Claiming Community Forests in Cameroon" *The Geographical Journal* 171 (4) pp.340–356 DOI: 10.1111/j.1475–4959.2005.00173.x.

McGurk, T.J. and Caquard, S. (2020) "To What Extent Can Online Mapping be Decolonial? A Journey Throughout Indigenous Cartography in Canada" *The Canadian Geographer/Le Géographe Canadien* 64 (1) pp.49–64 DOI: 10.1111/cag.12602.

McKenna, J., Quinn, R.J. Donnelly, D.J. and Cooper, J.A.G. (2008) "Accurate Mental Maps as an Aspect of Local Ecological Knowledge (LEK): A Case Study from Lough Neagh, Northern Ireland" *Ecology and Society* 13 (1) pp.1–24 Available at: https://www.jstor.org/stable/10.2307/26267918.

McMahon, R., Smith, T.J. and Whiteduck, T. (2017) "Reclaiming Geospatial Data and GIS Design for Indigenous-led Telecommunications Policy Advocacy: A Process Discussion of Mapping Broadband Availability in Remote and Northern Regions of Canada" *Journal of Information Policy* 7 pp.423–449 DOI: 10.5325/jinfopoli.7.2017.0423.

Noroña, M.B. (2020) "Luchas en Red o Luchas Colectivas en la Amazonía del Ecuador: El Caso de Tzawata" *Journal of Latin American Geography* 19 (2) pp.191–217 DOI: 10.1353/lag.2020.0044.

Noroña, M.B. (2022) "Extractive Governmentality, Ethnic Territories, and Racial Imaginaries in the Northern Amazon of Ecuador" *Geoforum* 128 pp.46–56 DOI: 10.1016/j.geoforum.2021.11.023.

Olson, R., Hackett, J. and DeRoy, S. (2016) "Mapping the Digital Terrain: Towards Indigenous Geographic Information and Spatial Data Quality Indicators for Indigenous Knowledge and Traditional Land-use Data Collection" *The Cartographic Journal* 53 (4) pp.348–355 DOI: 10.1080/00087041.2016.1190146.

Pearce, M. and Louis, R. (2008) "Mapping Indigenous Depth of Place" *American Indian Culture and Research Journal* 32 (3) pp.107–126.

Pearce, T.C.L. (2018) "Incorporating Indigenous Knowledge in Research" In McLemon, R. and Gemenne, F. (Eds) *The Routledge Handbook of Environmental Displacement and Migration* New York: Taylor and Francis, pp.125–134.

Peluso, N. (1995) "Whose Woods Are These? Counter-mapping Forest Territories in Kalimantan, Indonesia" *Antipode* 27(4), pp.383–406 DOI: 10.1111/j.1467–8330.1995.tb00286.x.

Radcliffe, S.A. (2010) "Re-mapping the Nation: Cartography, Geographical Knowledge and Ecuadorean Multiculturalism" *Journal of Latin American Studies* 42 (2) pp.293–323 DOI: io.ioi7/S00222i6Xi0000453.

Rambaldi, G., Muchemi, J., Crawhall, N. and Monaci, L. (2007) "Through the Eyes of Hunter-Gatherers: Participatory 3D Modelling Among Ogiek Indigenous Peoples in Kenya" *Information Development* 23 (2–3) pp.113–128 DOI:10.1177/0266666907078592.

Reid, G. and Sieber, R. (2020) "Do Geospatial Ontologies Perpetuate Indigenous Assimilation?" *Progress in Human Geography* 44 (2) pp.216–234 DOI: 10.1177/0309132518824646.

Roberts, M. (2012) "Mind Maps of the Maori" *GeoJournal* 77 (6) pp.741–751 DOI 10.1007/s10708-010-9383-5.

Rundstrom, R.A. (1995) "GIS, Indigenous Peoples, and Epistemological Diversity" *Cartography and Geographic Information Systems* 22 (1) pp.45–57 DOI: 10.1559/152304095782540564.

Sawyer, S. (2004) *Crude Chronicles: Indigenous Politics, Multinational Oil, and Neoliberalism in Ecuador* Durham, NC: Duke University Press.

Scott, J.C. (1998) *Seeing Like a State: How Certain Schemes to Improve the Human Condition have Failed* New Haven, CT: Yale University Press.

Sheremata, M., Tsuji, L. and Gough, W.A. (2016) "Collaborative Uses of Geospatial Technology to Support Climate Change Adaptation in Indigenous Communities of the Circumpolar North" In Imperatore, I. and Pepe, A. (Eds) *Geospatial Technology: Environmental and Social Applications* London: InTechOpen, pp.197–215 DOI: 10.57772/64214.

Sillitoe, P. (2007) "Local Science vs. Global Science: An Overview" In Sillitoe, P. (Ed.) *Local Science vs. Global Science: Approaches to Indigenous Knowledge in International Development* Oxford: Berghahn Books, pp.1–22.

Sletto, B. (2009a) "Indigenous Cartographies" *Cultural Geographies* 16 (2) pp.147–152 DOI: 10.1177/1474474008101514.

Sletto, B. (2009b) "'Indigenous People Don't Have Boundaries': Reborderings, Fire Management, and Productions of Authenticities in Indigenous Landscapes" *Cultural Geographies* 16 (2) pp.253–277 DOI: 10.1177/1474474008101519.

Smith, L.T. (1999) *Decolonizing Methodologies: Research and Indigenous Peoples* New York: Zed Books.

Sundberg, J. (2014) "Decolonizing Posthumanist Geographies" *Cultural Geographies*, 21 (1) pp.33–47 DOI: 10.1177/1474474013486067.

Taylor, D.R.F. (2005) "The Theory and Practice of Cybercartography: An Introduction" In Taylor, D.R.F. and Lauriault, T. (Eds) *Cybercartography: Theory and Practice* Amsterdam: Elsevier, pp.1–13.

Taylor, D.R.F. (2014) "Developments in the Theory and Practice of Cybercartography: Applications and Indigenous Mapping" (2nd ed.) Amsterdam: Elsevier.

In Taylor, D.R.F., Anonby, E. and Murasugi, K. (Eds) *Further Developments in the Theory and Practice of Cybercartography: International Dimensions and Language Mapping (Modern Cartography Series Vol.7)* Amsterdam: Academic Press, pp.205–218.

Thill, Z. (2018) *Rights Holders, Stakeholders, and Scientists: A Political Ecology of Ambient Environmental Monitoring in Alberta, Canada* (PhD thesis) University of Oregon Available at: https://alliance-primo.hosted.exlibrisgroup.com/permalink/f/to8ro2/uo_scholars_bank1794/23767.

Turnbull, D. (2000) *Masons, Tricksters and Cartographers: Comparative Studies in the Sociology of Scientific and Indigenous Knowledge* London and New York: Taylor and Francis.

United Nations (2009) "State of the World's Indigenous Peoples" (Report ST/ESA/328) New York: Department of Economic and Social Affairs Available at: https://www.un.org/esa/socdev/unpfii/documents/SOWIP/en/SOWIP_web.pdf.

Veland, S., Howitt, R., Dominey-Howes, D., Thomalla, F. and Houston, D. (2013) Procedural Vulnerability: Understanding Environmental Change in a Remote Indigenous Community *Global Environmental Change* 23 (1) pp.314–326 DOI: 10.1016/j.gloenvcha.2012.10.009.

Wainwright, J. and Bryan, J. (2009) "Cartography, Territory, Property: Postcolonial Reflections on Indigenous Counter-mapping in Nicaragua and Belize" *Cultural Geographies* 16 (2) pp.153–178 DOI:10.1177/1474474008101515.

Wood, D. (2010) *Rethinking the Power of Maps* New York: Guilford Press.

Woodward, D. and Lewis, M. (Eds) (1998) *Cartography in the Traditional African, American, Arctic, Australian and Pacific Societies* In: Harley, J.B. (Ed.) *The History of Cartography (Volume 2 Book 3)* Chicago: University of Chicago Press.

Ybarra, M. (2018) *Green Wars: Conservation and Decolonization in the Maya Forest* Oakland, CA: University of California Press.

11
SOCIAL CONSTRUCTIVISM AND GEOSPATIAL TECHNOLOGIES
Neogeography, big data, and deep maps

Barney Warf

Arguably the most significant transformation in the discipline of geography over the last half-century has been the rise of social constructivism, the discovery that everything – space included – is made, not given. Inspired by various versions of Marxist, feminist, poststructuralist, and postcolonial thought, geography has been radically reconstructed since the heyday of positivism. In human geography this has led to a sustained interest in social relations, power, and inequality, given rise to critical social theory. Yet social constructivism has spilled beyond the boundaries of human geography to influence other parts of the discipline as well. The analysis of human-environmental relations and biogeography, for example, has seen the rise of political ecology and concerns about the social construction of nature. Even physical geography has been touched by this turn (Inkpen and Wilson, 2004; Rhoads and Thorn, 2011).

Social constructivism has also penetrated into the domains of cartography and GIS, once the last enclaves of positivist thinking. Ever since the path-breaking work of Brian Harley (1989) and Denis Wood (1992), critical cartography has come to view maps as social discourses laden with power relations. Adopting a Foucauldian perspective in which knowledge and power are inseparable, this literature has aptly demonstrated that maps are embodiments of social relations, tools for advancing an interest that typically hide behind a façade of objectivity. As Wood (1992: 1) notes, 'power is ability to do work, and maps work'. Notably, their power lies in helping to bring into existence the world that they represent: discourses do not simply mirror social reality, but enter into its construction. This epistemological stance calls into question the conventional notion of representation as a purely technical act and recasts it as one intimately tied to social interests (i.e., with social origins and social consequences). Today, this line of thought is so widespread as to be unremarkable (Crampton, 2001; Crampton and Krygier, 2006).

Bouncing off the springboard provided by critical cartography, social constructivism has also engaged with the world of GIS. John Pickles' (1994) influential volume *Ground Truths* set the stage for a series of debates in which the origins of data, technology, and applications of GIS came to be viewed in social, not simply technical, terms. The volume noted the military origins of GIS, and situated it within the broader context of digitized, neoliberal capitalism, and the culture of hypermodernity. More recent works explored whether GIS should be viewed as a tool or science (Wright *et al.*, 1997). O'Sullivan (2006) opened the door for an explicitly critical GIS modelled after critical cartography. Others, such as Leszczynski (2009), forged linkages between poststructuralism, with its emphasis on the plurality of views and their inevitable ties to vested social interests, and GIS. A proliferation of different species of GIS – qualitative, public, participatory, feminist, queer, and so forth – emanated from the encounters between social theory and studies of geospatial technologies (Chrisman, 2005; Elwood, 2006; Aitken and Craine, 2009). In 'socializing the pixel and pixelizing the social' (Geoghegan *et al.*, 1998), constructivism even entered the rarefied domain of remote sensing (cf. Warf, 2012).

What is the trajectory of this body of scholarship in the contemporary moment? This chapter undertakes to address this issue in three steps. First, it explores the epistemological implications of neogeography, in which long-standing assumptions about truth must be seriously reconsidered. Second, it takes up the question of big data and its implications for geospatial technologies. Third, as a suggestion of strategy that ameliorates some issues of the former two topics, it turns to the matter of 'deep mapping', finely detailed spatial narratives that attempt to portray places in all their messy complexity, ambiguity, and contradictions. The point is not to summarize these three domains in detail, but to use them to explore how constructivism has reshaped our understanding of geospatial technologies. The conclusion ties together the major themes.

Neogeography and volunteered geographic data

As widely noted by many observers, the spread of interactive digital technologies – Web 2.0 – has put geospatial tools in the hands of a large number of people and users, giving rise to what is commonly called neogeography (Wilson and Graham, 2013; Leszczynski, 2014). The term of 'neogeography' was coined by Di-Ann Eisnor (2006), co-founder of Platial.com, who defined it as 'a diverse set of practices that operate outside, or alongside, or in the manner of, the practices of professional geographers'. By giving non-professionals and non-geographers access to Web mapping and spatial analytics online, neogeography is often heralded as inherently democratizing GIS (Rana and Joliveau, 2009; Byrne and Pickard, 2016), although Haklay (2013) argues correctly this view is simplistic and overstated. Neogeography has predictably unleashed a revolution in cartography, including a diverse array of map mashups, infographics and videographics, and geotagged and geolocated data (Haklay *et al.*, 2008; Batty *et al.*, 2010; Liu and Palen 2010; González 2019). In contrast to conventional spatial representations produced by the state or corporations, which may be expensive, neogeographic outputs are often free or very low cost. Rather than be updated periodically, they can be updated continuously.

At the heart of neogeography lies the issue of volunteered geographic information (VGI), or geolocated data collected, typically by unpaid workers, for a specific purpose (Sui *et al.*, 2012). For example, VGI was important in mobilizing aid after the 2009 Haitian earthquake (Zook *et al.*, 2010). VGI reflects the intersection of Web-based technologies and user-generated input and comprises a form of citizen science. However, one should note that crowdsourcing, VGI, and citizen science are not identical. Perhaps the best examples are Open Street Map, Bing Maps, Google Earth, and Wikimapia. Platforms such as GoogleMaps allow anyone to become a cartographer, much as YouTube allows anyone to become a film director. As Goodchild (2007: 214) notes, 'These are just a few examples of a phenomenon that has taken the world of geographic information by storm and has the potential to redefine the traditional roles of mapping agencies and companies'. Exactly what constitutes VGI is up for debate: for some, it must be generated by end-users; for others, it includes data produced by services such as Trip Advisor, Flickr, and Twitter. Elwood *et al.* (2012) herald VGI as a 'paradigmatic shift' in how geographic data are collected. On the one hand, VGI is widely celebrated as promoting participatory GIS (Elwood, 2008); on the other, it raises serious questions about its reliability, including its accuracy and degree of uncertainty (Flanagin and Metzger, 2008; Fonte et al., 2017).

Neogeography has markedly altered the philosophical terms for understanding how geographic knowledge is produced. Producers of VGI are not necessarily intent on generating random samples of data – the gold standard of traditional science – but rather seek to construct information tailored to their specific interests at hand. Can non-randomly sampled data be considered 'scientific'? The question brings to the fore concerns about the very foundations of knowledge and truth. For example, rather than the correspondence theory of truth, in which theoretical claims are buttressed or invalidated by appeals to the facts, VGI may lend itself to pragmatic and consensual views of truth (Warf and Sui, 2010). In this reading, which draws on the lineage established by John Dewey and William James, the lines between knower and known, subject and object, producer and consumer of knowledge become hazy and blurred. Arguably, there are multiple truths, which may be contradictory, that serve different interests. The democratization of geospatial technologies and VGI thus forces the discipline of geography and its practitioners to confront truth and knowledge as social, not merely technical, phenomena. Sieber and Haklay (2015) maintain that VGI calls for an epistemology that is self-conscious of its social origins and implications, that it cannot be studied without paying attention to how it is constructed and used.

Online, interactive mapping facilitates a bottom-up reconfiguration in how data are collected, transmitted, analysed, visualized, and utilized that differs considerably from traditional top-down models in which experts and government agencies dictate the criteria of data collection, analysis, applications, and standards of truth. In doing so, it calls into question the primacy of geographic expertise (Goodchild, 2007). In theory, VGI and neogeography draw upon the wisdom of the multitude, integrating the knowledge and experience of numerous individuals. VGI displaces the traditional 'expert' – the academic, the planner, the scientist – in favour of diverse groups of amateurs who may lack the traditional training involved in geospatial technologies. VGI thus runs the risk of low-quality data produced by naïve or poorly trained participants. Often it is accompanied by little metadata. As Foody et al. (2017: 6) note, 'Much VGI is collected opportunistically and is spatially biased, for instance by digital divides between urban and rural regions or between developed and developing countries'. Moreover, the producers of this type of geographic knowledge are also its consumers, or 'prosumers', blurring the boundaries between authors and audience, subjects, and objects (Budhathoki et al., 2008). However, it must be emphasized that not everyone is capable of generating VGI, leading to informational gaps and pockets of information poverty (Cinnamon and Schuurman, 2013; Graham et al., 2014). VGI is thus not simply a technical phenomenon, but very much a social and epistemological one.

Neogeography is one avenue through which social constructivism has penetrated the geosciences. Neogeographic spaces, generated through countless bottom-up interactions of users who are widely dispersed among many physical locations, are virtual, constantly changing, and often bear only tenuous linkages to material geographies. In this sense, neogeographic space is compatible with the Deleuze/Guattarian 'flat ontology', or spatialities that do not fall victim to the obfuscating effects of scale (Springett, 2015). Such a shift is vital to wean geosciences from their traditional dependence on absolute, Cartesian space, which carries the implicit assumption that space is given, not made. Rather, it opens up a world of spaces that are folded, origami-like, a reflection of how the intense time-space compression of globalization had altered traditional understandings of distance and proximity.

However, a critical perspective on neogeography also reveals that it is not without problems of its own. Inequalities in access to geospatial technologies and the knowledge about how to collect and upload data reflect deep schisms in terms of class, income, and often ethnicity (Haklay, 2013). Popular celebrations of VGI thus may obscure the reality of digital divides: bluntly, VGI is a middle-class phenomenon rarely practiced by the poor, marginalized, and disenfranchised. This dimension of neogeography has received remarkably little attention in the literature (but see Pavlovskaya, 2016). This issue raises the vital question of whose voices are heard in the practice of neogeography, and whose are silenced? For neogeography to be truly democratic, as its advocates assert, it must include a broad range of subject positions, including the poor, minorities, rural areas, the elderly, the handicapped, and others whose geographic knowledge is rarely hear or seen.

Big data

A recent concern regarding geospatial technologies is the rise of big data, gargantuan datasets (often petabytes) constructed in fine detail that are often exhaustive in scope and often created in or near real time (Thatcher *et al.*, 2018). Typical users are Internet and telecommunications companies, retail trade giants, financial institutions, marketing and advertising firms, and governments. As Crawford *et al.* (2014: 10) note, the term big data 'has spread like kudzu in a few short years, ranging across a vast terrain that spans health care, astronomy, policing, city planning, and advertising'. Kitchin (2014b) asserts that big data is the key to smart urbanism, or cities structured around information and communications technologies, including smart homes, smart crime-fighting agencies, and smart traffic networks.

Big data may be produced by volunteers or automatically by digitized systems, such as procedures that monitor clicks on websites and track the locations of visitors, Facebook and Twitter accounts, WiFi routers, credit card readers, smart machines and devices that constitute the Internet of things (e.g., a Fitbit), barcode readers, and devices that monitor online transactions such as banking and e-commerce (van der Zee and Scholten, 2014). Some may be created by applications such as FourSquare that operate on GPS-enabled cell or mobile phones. All of these create highly detailed information about users' sociodemographic profiles and spatial behaviour. While the social sciences have traditionally been data poor, they now face the prospect of being overwhelmed by it. As Kwan (2016) writes, such data is overwhelmingly generated by computer algorithms, which greatly complicates distinctions among procedures, methods, and techniques. She argues (p.274) that there is a 'need for geographers to remain attentive to the omissions, exclusions, and marginalizing power of big data'. Different algorithms may lead to different results, and thus different research findings. In this sense, the rise of big data reflects the growing penetration of computer code into the interstices of everyday life (Dodge and Kitchin, 2005; Kitchin and Dodge, 2011).

As with neogeography, data issues arise with big data. The sheer volume of data in question makes quality assessment difficult. Moreover, it hides the fact that all data are theory-laden (i.e., collected for a purpose and whose significance is determined by theory). As Gitelman and Jackson (2013) put it, 'raw data is an oxymoron' and 'data are always already cooked'. But *how* big data cooks the data is open to debate; who is doing the cooking, why, and for whom?

Kitchin (2014a) argues that big data is a disruptive technology that has received little epistemological reflection. In the social sciences, it has elevated computational models to new levels of sophistication. Big data has led to more sophisticated, fine-grained models of mass behaviour in a variety of contexts. Data-driven models have considerable explanatory and predictive power. In the process, big data may re-elevate empiricism and positivism, two epistemologies that were overcome only with great effort. It carries the risk of ceding theory to data (Miller and Goodchild, 2015). The old objections to positivism are worth revisiting here: the emphasis on patterns, not processes; the studied neglect of social relations, power, and consciousness; the façade of objectivity and value-freedom; the privileging of quantitative over qualitative data; the reliance on absolute, Cartesian space; and the substitution of methodology for explanation.

Big data's widespread use is changing how the world lives, works, and thinks (Mayer-Schonberger and Cukier, 2013). It offers opportunities, challenges, and risks (Kitchin, 2013). Needless to say, for corporations, big data offers a wealth of information about consumers. Commercial big data typically concerns those with disposable incomes (i.e., the relatively well-educated and often white populations), and do not adequately represent the socially marginalized (Crampton et al., 2013). Although it is usually discussed in a celebratory manner, its implications for privacy are frightening (Armstrong and Ruggles, 2005; Crawford et al., 2014).

Indeed, the distinction between data that is readily volunteered and that which is collected surreptitiously about individuals is critical. Neogeography by definition relies on data that individuals willingly provide; the same is not true about big data. Some users of geospatial technologies may not be concerned about the data collected about them, or what Birchall (2016) calls 'shareveillance'. More often, however, data about people and entire populations is collected without their knowledge or even permission. Most data on people are typically gathered without their knowledge or consent, and are usually decontextualized. Such data, which comprise the digital selves of billions of people, is bought and sold by data and marketing companies and data brokers in a complex series of transactions that commodify identity. Vast quantities of data on credit cards can be hacked and stolen, as what happened with the Marriott Hotels in 2018 (500 million people affected), Equifax in 2017 (143 million), Adult Friend Finder in 2016 (412 million), and eBay in 2014 (145 million). Moreover, big data is hopelessly biased to people in the global North, often ignoring the vast multitudes in the global South whose disposable incomes are much lower (Arora, 2016), limiting its utility for international development (Taylor and Schroeder, 2014).

Authoritarian governments that closely censor the Internet can use it to surveil and monitor people, as is widely done in China today. Combining facial recognition and other biometric technologies with algorithms that assess behaviour has greatly enabled the surveillant state, giving the digital panopticon unprecedented powers (Berner et al., 2014; Sullivan, 2014). For example, the enormous trove of materials put online by Edward Snowden reveals just how closely even ostensibly democratic societies like the United States monitor their own populations. Thus, to counter the surveillance that big data makes possible, Mann (2017) argues for a human-centered sousveillance, one that foregrounds transparency and makes clear how data are collected, analysed, and used.

Just as the overly optimistic views of the Internet gave way to more realistic and more measured assessments when it became clear that cyberspace is not inherently emancipatory, so too has big data come to be viewed in terms of power and inequality. When viewed critically, it is not always clear that more data are always better. Big data forces to the fore the question of 'better for whom?'.

Deep mapping

One of the more interesting developments in the application of GIS has been its use by scholars in the digital humanities (e.g., Bodenhamer et al., 2010). Historians, in particular, have deployed the technology for a variety of purposes. It should not surprise us that this encounter was mutually transformative: GIS was changed, even as it changed, by the humanities in which it was utilized. One result was the rise of 'deep maps' (Bodenhamer et al., 2015). Given the various issues of power, exclusion, and surveillance that arise in the context of neogeography, deep maps offer something of a more democratic and inclusionary alternative.

Deep maps differ from conventional maps in more ways than one. In contrast to conventional maps, deep maps are finely detailed depictions of a place, including representations of its history, landscape, ecology, and the people who inhabit it (Biggs, 2010). If, as David Harvey (1996: 18) observes, 'maps are typically totalizing, usually two-dimensional, Cartesian, and very undialectical devices', deep maps strive to be multidimensional, dialectical, and to represent relational space in all of its complexity.

Traditionally, deep maps have been narrative in format, such as Heat-Moon's (1999) celebrated volume *PrairyErth,* an enormously detailed portrait of Chase County in the Flint Hills of Kansas that is often upheld as the paradigmatic example of this phenomenon. Gregory-Guider (2011) calls the book a 'psychogeographic cartography'. Pearson and Shanks (2001: 64) eloquently explain that

> the deep map attempts to record and represent the grain and patina of place through juxtapositions and interpenetrations of the historical and the contemporary, the political and the poetic, the discursive and the sensual; the conflation of oral testimony, anthology, memoir, biography, natural history and everything you might ever want to say about a place." Deep maps are as much narratives as they are spatial representations, helping to organize "thoughts off the map.
>
> *(Oxx et al., 2013)*

Digital deep maps often contain embedded links to texts, photographs, images, oral histories, and biographical life stories. They convey a sense of change over time, and can be read as much synchronically as diachronically. In this sense, they are palimpsests that portray multiple layers of culture and events as they are superimposed upon a locale (Fishkin, 2011). They are intended to be maximally accessible or used pedagogically.

Deep maps are not confined to the material or observable, but include the discursive and ideological dimensions of place; they may seek to include the emotional dimensions of sense of place, a long-standing topic of humanistic geography. They solicit memories and paralinguistic cues, trying to make the invisible visible (Bodenhamer, 2016). They are, in short, positioned between matter and meaning, the tangible and the intangible. In this sense, deep maps serve as a useful bridge between geospatial technologies and geography's ventures into once obscure topics such as the non-representational, affect, and that which cannot be put into language (Anderson, 2016).

Deep maps not only show qualitatively more information than do shallow ones, they are self-conscious as social and political entities, making explicit their origins, purpose, whose interests they serve and whose they do not, and what they represent and what they do not. McLucas (n.d.) writes that:

> Deep maps will not seek the authority and objectivity of conventional cartography. They will be politicized, passionate, and partisan. They will involve negotiation and contestation over who and what is represented and how. They will give rise to debate about the documentation and portrayal of people and places. [...] Deep maps will bring together the amateur and the professional, the artist and the scientist, the official and

the unofficial, the national and the local. [...] Deep maps will be unstable, fragile and temporary. They will be a conversation and not a statement.

In lieu of portraits of places as coherent, integrated, and tidy, deep maps offer a range of embodied perspectives often at odds with one another and celebrate the inchoate chaos that undermines any aspiration of imposing a single narrative structure on a complex and heterogeneous reality. They portray places in terms of everyday life, with all of its contradictions and complexities. To do so, they can draw on the vast resources of the Geospatial Semantic Web (Bodenhamer *et al.*, 2013).

Unlike conventional maps, deep maps offer multiple viewpoints simultaneously, and do not hide behind the façade of the value-free observer, or what Donna Haraway (1991) famously called the god-trick, the foundation of positivist epistemology. In creating a cartography of non-Cartesian space, deep mapping helps to problematize the role of the observer: both space and knowledge are presented in relational terms. In their quest to 'explore how digital tools and interfaces can support ambiguous, subjective, uncertain, imprecise, rich, experiential content alongside the highly structured data at which GIS systems excel', Ridge *et al.* (2013) uphold the notion of a 'greedy deep map'. Online, digital deep maps allow much more detail to be presented than do conventional maps; information can be updated regularly; and the potential audience is much larger than traditional forms. Rather than have an observer stare at Cartesian space, deep maps are sensual and immersive, tying producers and consumers of spatial information together in webs of meaning (Harris, 2016). Neogeographic deep maps accommodate a wide range of perspectives, experiences, and worldviews, allowing them to be brought into a creative tension with one another. Neogeographic deep mapping is an ontological counterpoint for poststructuralist epistemology in that it allows for multiple voices to be heard, leading to a cacophony of representations in which places are depicted and viewed through multiple lenses.

Digital deep maps have found a variety of applications, including the study of landscape heritage (Fitzjohn, 2009), travel writing (Hulme, 2008), spatial history (Gregory and Geddes, 2014), anthropology (Roberts, 2016), and literature (Bodenhamer, 2016; Taylor *et al.*, 2018). Place-based examples include studies of Colorado Springs (Harner *et al.*, 2017), the Great Plains (Maher, 2001), and literary tourism in the English Lake District (Donaldson *et al.*, 2015). Bailey and Biggs (2012) use deep maps to evoke the lives of the elderly in North Carolina. By allowing local input from the people who live there, digital deep maps greatly democratize the process of spatial storytelling.

Concluding thoughts

There can be no doubt that social constructivism has penetrated the once-rarified world of geospatial technologies. The consequences have been enormous. People who are not experts collect and upload data and make their own maps. Experts get displaced by ordinary citizens. Geography escapes the geographers. Deep maps allow communities to represent themselves in enormous detail, challenging the view of maps as views from above. Corporations and governments seize on big data to monitor the behaviour of millions.

The arrival of social constructivism appears to have passed through two channels, bottom-up and top-down. In the cases of neogeography and deep maps, countless users (or prosumers) create and analyse data for their own purposes, to better understand the world and their place in. Such an approach reflects the democratization of geospatial technologies and the ability of millions to use Web 2.0 technologies to create mashups that reflect their interests. In contrast, big data has taken a markedly different path, used by large corporations and authoritarian governments for surveillance purposes, be it marketing or political control. This contrast indicates that one should not automatically equate social constructivism with emancipatory purposes: geospatial tools can be used against people as well as for them.

Epistemologically, social constructivism has heightened attention to how geographic data are collected, by whom, and for what purposes. There is, of course, a long history in philosophy and the social sciences concerned with the politics of knowledge, much of which adopts an implicitly Foucauldian perspective. Suffice it to say here that the diffusion of the Internet, mobile or cell phones, tablets, and smart technologies has brought to the fore issues such as privacy and surveillance in a critical and timely manner.

Moreover, in contending with the messy politics of data and knowledge, GIS practitioners have also been confronted with changing views of space. Traditional understandings that relied on the Cartesian perspective – the flat ontology in which location is equated with distance, in which social relations are rendered invisible, and in which space is portrayed as given, not made – have given way to much more dynamic relational approaches that view space as folded, twisted, and full of wormholes (Massey, 2005). A relational perspective obligates practitioners of geospatial technologies to focus on linkages and interactions, movements and flows, to view space in terms of networks rather than surfaces (Warf, 2008).

GIS has long grappled with the vexing issue of ontology (Schuurman, 2006), or what we take to be real. In the process, they have come to terms with the limits of language: our world is always more complex than our language allows us to admit. This notion exemplifies Wittgenstein's (1921/1998, remark 6.62) famous dictum that 'the limits of my language [...] mean the limits of my world'. But theoreticians of GIS have paid much less attention to epistemology – how we know what we know. This must change, urgently. The nature of geospatial technologies has been profoundly changed by VGI, big data, deep mapping, drones, and various forms of participatory GIS. There have, of course, been several attempts to foster a dialogue between geospatial theoreticians and social theory (Leszczynski, 2009), an exciting body of work that creatively blurs the boundaries between quantitative and qualitative research. In the process, there can be no retreat into the ostensibly apolitical practices of mapping and modeling. Rather, practitioners must note how data are created, for whom and for what purpose, toward what ends they are put. Doing geography in a digitized age means being sensitive toward the politics of knowledge, and the power relations involved in its production and utilization. Geosciences can no longer hide behind pretense of being apolitical or value free. Human geography made this transition long ago; it is time for scholars and practitioners of geotechnologies to do the same.

References

Aitken, S. and Craine, J. (2009) "Affective Visual Geographies and GIScience" In Cope, M. and Elwood, S. (Eds) *Qualitative GIS: A Mixed Methods Approach* London: Sage, pp.139–155.

Anderson, B. (2016) *Taking-place: Non-representational Theories and Geography* London: Routledge.

Armstrong, M. and Ruggles, A. (2005) "Geographic Information Technologies and Personal Privacy" *Cartographica* 40 (4) pp.63–73.

Arora, P. (2016) "The Bottom of the Data Pyramid: Big Data and the Global South" *International Journal of Communication* 10 pp.1681–1699.

Bailey, J. and Biggs, I. (2012) "'Either Side of Delphy Bridge': A Deep Mapping Project Evoking and Engaging the Lives of Older Adults in Rural North Cornwall" *Journal of Rural Studies* 28 (4) pp.318–328.

Batty, M., Hudson-Smith, A., Milton, R and Crooks, A. (2010) "Map Mashups, Web 2.0 and the GIS Revolution" *Annals of Geographical Information Science* 16 (1) pp.1–13.

Berner, M., Graupner, E. and Maedche, A. (2014) "The Information Panopticon in the Big Data Era" *Journal of Organization Design* 3 (1) pp.14–19.

Biggs, I. (2010) "The Spaces of 'Deep Mapping': A Partial Account *Journal of Arts and Communities* 2 pp.5–25.

Birchall, C. (2016) "Shareveillance: Subjectivity between Open and Closed Data" *Big Data & Society* 3 (2) pp.1–12 DOI: 10.1177/2053951716663965..

Bodenhamer, D. (2016) "Making the Invisible Visible: Place, Spatial Stories and Deep Maps" In Cooper, D., Donaldson, C. and Murrieta-Flores, P. (Eds) *Literary Mapping in the Digital Age* London: Routledge, pp.225–238.

Bodenhamer, D., Corrigan, J. and and Harris, T. (2010) *The Spatial Humanities: GIS and the Future of Humanities Scholarship* Indianapolis: Indiana University Press.

Bodenhamer D., Corrigan, J. and Harris, T. (2013) "Deep Mapping and the Spatial Humanities" *International Journal of Humanities and Arts Computing* 7 pp.170–175.

Bodenhamer, D., Corrigan, J. and Harris, T. (Eds) (2015) *Deep Maps and Spatial Narratives* Bloomington, IN: Indiana University Press.

Budhathoki, N., Bruce, B. and Nedovic-Budic, Z. (2008) "Reconceptualizing the Role of the User of Spatial Data Infrastructures" *Geojournal* 72 (3) pp.149–160.

Byrne, D. and Pickard, A.J. (2016) "Neogeography and the Democratization of GIS: A Metasynthesis of Qualitative Research" *Information, Communication & Society* 19 (11) pp.1505–1522.

Chrisman, N. (2005) "Full Circle: More Than Just Ssocial Implications of GIS" *Cartographica* 40 (4) pp.23–35.

Cinnamon, J. and Schuurman, N. (2013) "Confronting the Data-divide in a Time of Spatial Turns and Volunteered Geographic Information" *GeoJournal* 78 (4) pp.657–674.

Crampton, J. (2001) "Maps as Social Constructions: Power, Communication and Visualization" *Progress in Human Geography* 25 (2) pp.235–252.

Crampton, J. and Krygier, J. (2006) "An Introduction to Critical Cartography" *ACME: An International E-Journal for Critical Geographies* 4 (1) pp.11–33.

Crampton, J., Graham, M., Poorthuis, A., Shelton, T., Stephens, M., Wilson, M. and Zook, M. (2013) "Beyond the Geotag: Situating 'Big Data' and Leveraging the Potential of the Geoweb" *Cartography and Geographic Information Science* 40 (2) pp.130–139.

Crawford, K., Gray, M. and Miltner, K. (2014) "Critiquing Big Data: Politics, Ethics, Epistemology" *International Journal of Communication* 8 pp.1663–1672.

Dodge, M. and Kitchin, R. (2005) "Code and the Transduction of Space" *Annals of the Association of American Geographers* 95 (1) pp.162–180.

Donaldson, C., Gregory, I. and Murrieta-Flores, P. (2015) "Mapping Wordsworthshire: A GIS Study of Literary Tourism in Victorian Lakeland" *Journal of Victorian Culture* 20 (3) pp.287–307.

Eisnor, D.-A. (2006) "What is Neogeography Anyway?" Available at: https://platial.typepad.com/news/2006/05/what_is_neogeog.html.

Elwood, S. (2006) "Critical Issues in Participatory GIS: Deconstructions, Reconstructions, and New Research Directions" *Transactions in GIS* 10(5) pp.693–708.

Elwood, S. (2008) "Volunteered Geographic Information: Future Research Directions Motivated by Critical, Participatory, and Feminist GIS" *GeoJournal* 72 (3–4) pp.173–183.

Elwood, S., Goodchild, M. and Sui., D. (2012) "Researching Volunteered Geographic Information: Spatial Data, Geographic Research, and New Social Practice" *Annals of the Association of American Geographers* 102 (3) pp.571–590.

Fishkin, S. (2011) "'Deep Maps': A Brief for Digital Palimpsest Mapping Projects (DPMPs, or 'Deep Maps')" *Journal of Transnational American Studies* 3 (2) pp.1–31.

Fitzjohn, M. (2009) "The Use of GIS in Landscape Heritage and Attitudes to Place: Digital Deep Maps" In Sørensen, M. and Carman, J. (Eds) *Heritage Studies: Methods and Approaches* London: Routledge, pp.255–270.

Flanagin, A. and Metzger, M. (2008) "The Credibility of Volunteered Geographic Information" *Geojournal* 72 pp.137–148.

Fonte, C., Antoniou, V., Bastin, L., Estima, J., Arsanjani, J., Bayas, J., See, L. and Vatseva. R. (2017) "Assessing VGI Data Quality" In Foody, G., See, L., Fritz, S., Mooney, P., Olteanu-Raimond, A., da Costa Fonte, C., Antoniou, V. and Fonte, C. (Eds) (2017) *Mapping and the Citizen Sensor* London: Ubiquity Press, pp.137–163.

Foody, G., See, L., Fritz, S., Mooney, P., Olteanu-Raimond, A., da Costa Fonte, C., Antoniou, V. and Fonte, C. (Eds) (2017) *Mapping and the Citizen Sensor* London: Ubiquity Press.

Geoghegan, J., Pritchard, L., Ogneva-Himmelberger, Y., Chowdhury, R., Sanderson, S. and Turner, B. (1998) "'Socializing the Pixel'and 'Pixelizing the Social' in Land-use and Land-cover Change" InLiverman, D., Moran, E., Rindfuss, R. and Stern, P. (Eds) *People and Pixels: Linking Remote Sensing and Social Science* Washington, DC: National Academy Press, pp.51–69.

Gitelman, L. and Jackson, V. (2013) "Introduction" In Gitelman, L. (Ed.) *'Raw Data' Is an Oxymoron* Cambridge, MA: MIT Press, pp.1–14.

González, J. (2019) "Visual and Spatial Thinking in the Neogeography Age" In Koutsopoulos, K, de Miguel González, R. and Donert, K. (Eds) *Geospatial Challenges in the 21st Century* Dordrecht: Springer, pp.369–383.

Goodchild, M. (2007) "Citizens as Sensors: The World of Volunteered Geography" *GeoJournal* 69 (4) pp.211–221.

Graham, M., Hogan, B., Straumann, R. and Medhat, A. (2014) "Uneven Geographies of User Generated Information: Patterns of Increasing Informational Poverty" *Annals of the Association of American Geographers* 104 (4) pp.746–764.

Gregory, I. and Geddes, A. (Eds). (2014) *Towards Spatial Humanities: Historical GIS and Spatial History* Bloomington, IN: Indiana University Press.

Gregory-Guider, C. (2011) "'Deep Maps': William Least Heat-Moon's Psychogeographic Cartographies" *Journal of the Imaginary and the Fantastic* 2 (4) pp.1–17.

Haklay, M. (2013) "Neogeography and the Delusion of Democratisation" *Environment and Planning A: Economy and Space* 45 (1) pp.55–69.

Haklay, M., Singleton, A. and Parker, C. (2008) "Web Mapping 2.0: The Neogeography of the Geoweb" *Geography Compass* 2 (6) pp.2011–2039.

Haraway, D. (1991) "Situated Knowledges: The Science Question in Feminism and the Privilege of Partial Perspective" In Haraway, D. (Ed.) *Simians, Cyborgs, and Women* New York: Routledge, pp.183–201

Harley, J. (1989) "Deconstructing the Map" *Cartographica* 26 (2) pp.1–20.

Harner, J., Knapp, K. and Davis-Witherow, L. (2017) "'The Story of Us': Place-making Through Public Interaction with Digital Geohumanities in Colorado Springs" *International Journal of Humanities and Arts Computing* 11 (1) pp.109–125.

Harris, T. (2016) "Deep Mapping and Sensual Immersive Geographies" In Richardson, D., Castree, N., Goodchild, M., Kobayashi, A., Liu, W. and Marston, R. (Eds) *International Encyclopedia of Geography: People, the Earth, Environment and Technology* New York: Wiley-Blackwell, pp.1–13.

Harvey, D. (1996) *Justice, Nature, and the Geography of Difference* Oxford: Blackwell.

Heat-Moon, W. (1999) *PrairyErth: A Deep Map* New York: Mariner Books.

Hulme, P. (2008) "Deep Maps: Travelling on the Spot" In Kuehn, J. and Smethurst, P. (Eds) *Travel Writing, Form, and Empire* London: Routledge, pp.142–157.

Inkpen, R. and Wilson, G. (2004) *Science, Philosophy and Physical Geography* London: Routledge.

Kitchin, R. (2013) "Big Data and Human Geography: Opportunities, Challenges and Risks" *Dialogues in Human Geography* 3 (3) pp.262–267.

Kitchin, R. (2014a) "Big Data, New Epistemologies and Paradigm Shifts" *Big Data & Society* 1 (1) pp.1–12.

Kitchin, R. (2014b) "The Real-time City? Big Data and Smart Urbanism" *Geojournal* 79 (1) pp.1–14.

Kitchin, R. and Dodge, M. (2011) *Code/Space: Software and Everyday Life* Cambridge, MA: MIT Press.

Kwan, M.P. (2016) "Algorithmic Geographies: Big Data, Algorithmic Uncertainty, and the Production of Geographic Knowledge" *Annals of the American Association of Geographers* 106 (2) pp.274–282.

Leszczynski, A. (2009) "Poststructuralism and GIS: Is there a 'Disconnect'?" *Environment and Planning D: Society and Space* 27 (4) pp.581–602.

Leszczynski, A. (2014) "On the Neo in Neogeography" *Annals of the Association of American Geographers* 104 (1) pp.60–79.

Liu, S. and Palen, L. (2010) "The New Cartographers: Crisis Map Mashups and the Emergence of Neogeographic Practice" *Cartography and Geographic Information Science* 37 (1) pp.69–90.

Maher, S. (2001) "Deep Mapping the Great Plains: Surveying the Literary Cartography of Place" *Western American Literature* 36 (1) pp.4–24.

Mann, S. (2017) "Big Data is a Big Lie Without Little Data: Humanistic Intelligence as a Human Right" *Big Data & Society* 4 (1) pp.1–10.

Massey, D. (2005) *For Space* London: Sage.

Mayer-Schonberger, V. and Cukier, K. (2013) *Big Data: A Revolution That Will Change How We Live, Work and Think* London: John Murray.

McLucas, C. (n.d.) "Deep Mapping" Available at: http://cliffordmclucas.info/deep-mapping.html.
Miller, H. and Goodchild, M. (2015) "Data-driven Geography" *GeoJournal* 80 (4) pp.449–461.
O'Sullivan, D. (2006) "Geographical Information Science: Critical GIS" *Progress in Human Geography* 30 (6) pp.783–791.
Oxx, K., Brimicombe, A. and Rush, J. (2013) "Envisioning Deep Maps: Exploring the Spatial Navigation Metaphor in Deep Mapping" *International Journal of Humanities and Arts Computing* 7 (1–2) pp.201–227.
Pavlovskaya, M. (2016) "Digital Place-making: Insights from Critical Cartography and GIS" In Travis, C. and von Lünen, A. (Eds) *The Digital Arts and Humanities: Neogeography, Social Media and Big Data Integrations and Applications* Dordrecht: Springer, pp.153–167.
Pearson, M. and Shanks, M. (2001) *Theatre/Archaeology* London: Routledge.
Pickles, J. (Ed.) (1994) *Ground Truth: The Social Implications of Geographic Information Systems* New York: Guilford Press.
Rana, S. and Joliveau, T. (2009) Neogeography: An Extension of Mainstream Geography for Everyone Made by Everyone?" *Journal of Location Based Services* 3 (2) pp.75–81.
Rhoads, B. and Thorn, C. (2011) "The Role and Character of Theory in Geomorpholog" In Gregory, K. and Goudie, S. (Eds) *The SAGE Handbook of Geomorphology* London: SAGE, pp.59–77.
Ridge, M., Lafreniere, D. and Nesbit, S. (2013) "Creating Deep Maps and Spatial Narratives Through Design" *International Journal of Humanities and Arts Computing* 7 (1–2) pp.176–89.
Roberts, L. (2016) "Deep Mapping and Spatial Anthropology" *Humanities* 5 (5) pp.1–7.
Schuurman, N. (2006) "Formalization Matters: Critical GIS and Ontology Research" *Annals of the Association of American Geographers* 96 (4) pp.726–739.
Sieber, R. and Haklay, M. (2015) "The Epistemology(s) of Volunteered Geographic Information: A Critique" *Geo: Geography and Environment* 2 (2) pp.122–136.
Springett, S. (2015) "Going Deeper or Flatter: Connecting Deep Mapping, Flat Ontologies and the Democratizing of Knowledge" *Humanities* 4 (4) pp.623–636.
Sui, D., Elwood, S. and Goodchild, M. (Eds) (2012) *Crowdsourcing Geographic Knowledge: Volunteered Geographic Information (VGI) in Theory and Practice* Dordrecht: Springer.
Sullivan, J. (2014) "Uncovering the Data Panopticon: The Urgent Need for Critical Scholarship in an Era of Corporate and Government Surveillance" *Political Economy of Communication* 1 (2) pp.89–94.
Taylor, J., Donaldson, C., Gregory, I. and Butler, J. (2018) "Mapping Digitally, Mapping Deep: Exploring Digital Literary Geographies" *Literary Geographies* 4 (1) pp.10–19.
Taylor, L. and Schroeder, R. (2014) "Is Bigger Better? The Emergence of Big Data as a Tool for International Development Policy" *GeoJournal* 80 (4) pp.1–16.
Thatcher, J., Shears, A. and Eckert, J. (Eds) (2018) *Thinking Big Data in Geography: New Regimes, New Research* Lincoln, NE: University of Nebraska Press.
van der Zee, E. and Scholten, H. (2014) "Spatial Dimensions of Big Data: Application of Geographical Concepts and Spatial Technology to the Internet of Things In Bessis, N. and Dobre, C. (Eds) *Big Data and Internet of Things: A Roadmap for Smart Environments* Dordrecht: Springer, pp.137–168.
Warf, B. (2008) "From surfaces to networks" In Warf, B. and Arias, S. (Eds) *The Spatial Turn: Interdisciplinary Perspectives* London: Routledge, pp.59–76.
Warf, B. (2012) "Dethroning the View from Above: Toward a Critical Social Analysis of Satellite Occularcentrism" In Parks, L. and Schwoch, J. (Eds) *Down to Earth* Piscataway, NJ: Rutgers University Press, pp.42–60.
Warf, B. and Sui, D. (2010) "From GIS to Neogeography: Ontological Implications and Theories of Truth" *Annals of GIScience* 26 (4) pp.197–209.
Wilson, M. and Graham, M. (2013) "Situating Neogeography" *Environment and Planning A: Economy and Space* 45 (1) pp.3–9.
Wittgenstein, L. (1921/1998) *Tractatus Logico-Philosophicus* New York: Dover Publications.
Wood, D. (1992) *The Power of Maps* New York: Guilford Press.
Wright, D., Goodchild, M. and Proctor, J. (1997) "GIS: Tool or Science? Demystifying the Persistent Ambiguity of GIS as 'Tool' versus 'Science'" *Annals of the Association of American Geographers* 87 (2) pp.346–362.
Zook, M., Graham, M., Shelton, T. and Gorman, S. (2010) "Volunteered Geographic Information and Crowdsourcing Disaster Relief: A Case Study of the Haitian Earthquake" *World Medical & Health Policy* 2 (2) pp.7–33.

PART II

Understanding geospatial technologies

12
MOBILE MAPPING

Gordon Petrie

Mobile mapping is concerned with the process of acquiring 3D geospatial data from a mobile vehicle, device or person. Usually, a fairly complex system of instrumentation, comprising imaging and ranging devices integrated with a positioning and orientation sub-system, is needed to acquire this data, which is then used to compile digital maps, geo-referenced imagery products, three-dimensional models of terrain and/or cities and the data needed to populate geospatial and civil engineering databases.

Initially, this development began on a research basis at Ohio State University (Novak 1995, Bossler and Toth 1996, Toth and Grejner-Brzezinska, 2003, 2004) and the University of Calgary (Ellum and El-Sheimy, 2002; El-Sheimy, 2005) during the late 1980s and early 1990s with the collection of imagery using multiple video or digital cameras mounted on vans, with the geo-referencing of the resulting data being carried simultaneously using integrated GPS/IMU units. This data capture operation was then followed by the photogrammetric processing of the digital imagery on a highly automated basis to deal with the enormous number of images that were being acquired. This development gave rise to the establishment of a number of commercial companies that carried out mapping principally for road inventory purposes.

The development and the use of laser scanners and profilers on vehicular platforms came somewhat later, in the first years of the twenty-first century, with initial research carried out at the University of Tokyo and the Cartographic Institute of Catalonia. Again, this resulted a few years later in the establishment of commercial companies that either utilized laser scanners on their own for data collection or, more usually, in combination with imaging cameras. While initially these were often one-off systems developed in-house by the commercial companies themselves, very soon, mobile mapping systems were being offered by a wide range of system suppliers, including most of the largest and best-known suppliers of instrumentation and systems to the surveying and mapping community and to the civil engineering and construction industries (Petrie, 2010).

As the technology has developed, the individual components of mobile mapping systems have become ever smaller in size and lighter in weight. The result has been the development of compact, lightweight and portable systems that can be mounted on tricycles, trolleys or carts and on backpacks and frames. Now some can even be deployed in a hand-held operational mode.

Mobile mapping technologies

The basic hardware components of a typical mobile mapping system that acquires accurate 3D geospatial data comprise (i) *digital imaging devices*; (ii) *laser ranging profilers and/or scanners*; and (iii) a sub-system comprising a *GPS receiver, an inertial measuring unit (IMU) and an odometer* to provide a positioning (or geo-referencing) capability. These main components, as utilized in a vehicular system, are shown in Figure 12.1. The complexity of ensuring that all these various devices are linked together and integrated to work in a fully coordinated manner can be seen in Figure 12.2 which shows the network of cables and interface cards that are required for a system carrying out purely imaging (i.e., without laser scanning) or an IMU. The need for high-speed interfaces and the cabling required for the rapid transmission and storage of the measured data is especially important given the high-speed movement of the mapping vehicles and the relatively close proximity (some tens of metres) of the objects that are being imaged and ranged.

Imaging devices

The principal imaging devices that are used in mobile mapping systems are *small-format digital frame cameras* employing square or rectangular arrays of CCD or CMOS detectors as their imaging sensors. These cameras are manufactured in large numbers for industrial use or security purposes, which markedly reduces their cost. However, they do need to be calibrated if they are to be used for the photogrammetric measurements required for the generation of the 3D geospatial data. The typical specification of an individual camera for use in mobile mapping might require the output images to be 1k × 1k (= 1 megapixel) to 2k × 2k pixels (= 4 megapixels) in size and collected at a rate of 15 frames per second. Such cameras are available for example from Sony (Japan), Pelco and ImperX (USA) and Basler (Germany). Usually, several of these cameras will be included in a single mobile mapping system with each individual camera pointing in a different direction in accordance with the types of feature that are being mapped. Each camera will be enclosed in a special housing to protect it from rain, snow and dust and will be equipped with a heater to ensure its operation in bad weather.

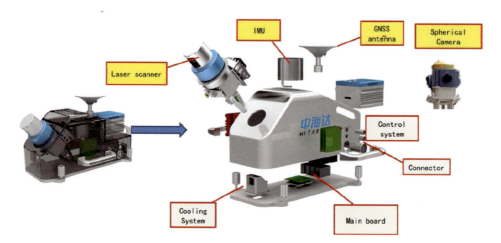

Figure 12.1 The principal hardware components of a vehicle-mounted mobile mapping system, including a multiple (spherical) camera unit, a laser scanner and a combined GNSS (GPS)/IMU sub-system

Source: Hi-Target Surveying Instrument Co.

Mobile mapping

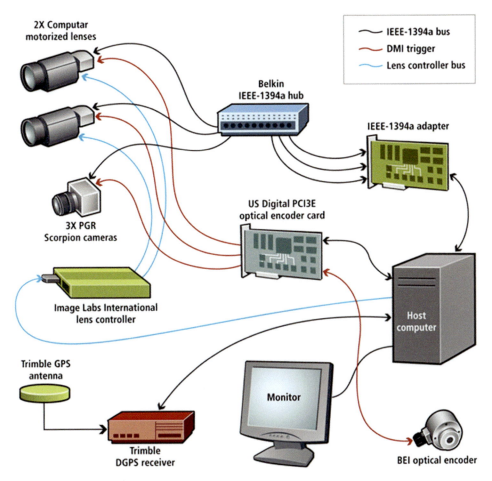

Figure 12.2 A diagram showing a system comprising imaging cameras, a GPS receiver and a measuring odometer (with its optical encoder) that is being linked together and controlled using high-speed IEEE1394 interface cards and cabling

Source: Point Grey Research.

However, the use of fully integrated *multiple camera units* instead of a set of individual cameras is now very common on mobile mapping vehicles. One of the best known of these is the Ladybug series (Figure 12.3) that is constructed by Point Grey Research based in Richmond, Canada, which is now owned by FLIR Systems located in Oregon, USA. These Ladybug camera units comprise six individual Sony CCD-based cameras, five of which are arranged concentrically,pointing outwards horizontally in a ring, while the sixth points vertically upwards. The Ladybug2 model images are 1024×768 pixels (= 0.8 megapixels) in size; the Ladybug 3 images are $1,600 \times 1,200$ pixels (= 2 megapixels); while the format size of the images collected by the current Ladybug 5+ model with its cameras equipped with CMOS arrays is $2,048 \times 2,464$ pixels (= 5 megapixels), with the images being exposed at a maximum rate of 15 images per second.

Still larger numbers of cameras have been utilized in the multi-camera units employed by *Google* in the collection of the images used in the Street View feature of its Google Maps and

Figure 12.3 The successive models of the Ladybug multiple camera units that are in widespread use on mobile mapping systems – (a) Ladybug2; (ii) Ladybug3; and Ladybug5+
Source: Point Grey Research, now FLIR Systems.

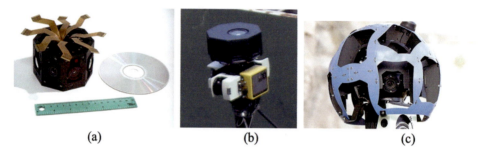

Figure 12.4 (a) A ring of eight cameras for use in a multi-camera system; (b) the integrated R5 nine-camera system that is mounted and operated on Google's Street View mapping cars; and (c) the R7 camera unit from Google with its 15 cameras providing panoramic coverage from each successive exposure position of the mapping platform
Source: (a) Elphel Inc. (c) Google.

Google Earth services. As with the Ladybug series, Google has steadily developed its own range of cameras (Figure 12.4). According to Anguelov *et al.* (2010), three successive generations of multi-camera units have been developed: (i) the R2 unit employing a ring of eight 11 megapixel CCD cameras; (ii) the R5 unit, again employing eight cameras, but equipped with CMOS instead of CCD arrays and having a ninth camera equipped with a fisheye lens placed above the main ring of cameras and pointing vertically upwards to order to image the upper levels of high buildings (Figure 12.4b); and (iii) the R7 units that used 15 of the same cameras employed in the R5 units to provide comprehensive panoramic coverage without the need to employ a fisheye lens (Figure 12.4c). Each of the R5 and R7 cameras is equipped with an electronic rolling shutter which allows the exposure of each row or line of the image in portrait orientation at a very slightly different time – which is beneficial both from the geometric point of view as well as the vehicle's motion. The individual cameras used in the Google multi-camera units are reported to have been supplied by Elphel Inc., which is based in Salt Lake City, Utah. In 2017, Google introduced a new unit that features just seven cameras (down from 15), each being fitted with 20 megapixel sensor arrays and mounted within a much smaller ball shield. Reportedly the new unit also plays host to two additional cameras that take still HD frame photos, as well as the required laser scanner(s).

Laser ranging, profiling and scanning devices

The main emphasis in terms of carrying out *ranging measurements* towards the objects that need to be located and mapped during mobile mapping operations is to use *2D laser profilers or scanners* for the task (Petrie and Toth, 2018). These devices can very rapidly measure the required distance and angular values to the objects lying within an individual 2D plane. Through the forward motion of the mobile vehicle or platform, a series of these 2D planes is measured in parallel at close intervals intersecting the road surfaces, kerbs, pavements, 'street furniture', buildings and vegetation that are present in the areas adjacent to the roads or tracks along which the mapping vehicle or platform is being driven. An important matter is the angle or orientation at which the 2D scanner is mounted relative to the vehicle and its direction of travel in order (i) to set the scan pattern over the ground and the surrounding objects and (ii) to obtain the maximum amount of useful data, such as the data needed to define the sides of buildings, from the scanner measurements (Elseberg et al., 2013).

By combining the measured range and angular data contained in this series of closely-spaced parallel 2D planes, a *three-dimensional model* of the surrounding area is created. To enable this 3D data to be transformed into the coordinate system being used by the geospatial database, the location of each new 2D plane needs to be accurately located (or georeferenced) by an integrated suite of positioning devices, normally comprising a GPS (or GNSS) receiver, an IMU and an odometer (distance measuring instrument). This suite will also be used to provide the location and coordinate data of the imaging cameras already described above.

A widely used example of the 2D scanners or profilers that are employed in mobile mapping is the LMS291 scanner (Figure 12.5) manufactured by the SICK AG company, which is based in Waldkirch, Germany. This instrument combines (i) a low-powered rapid-firing *laser rangefinder* using the time-of-flight (TOF) distance measuring principle with (ii) a *rotating mirror* whose angular directions are being measured continuously by a *rotary (angular) encoder*. Using this technology, the LMS291 scanner generates its range measuring pulses over a fan-shaped scanning angle of 180 degrees within its 2D scanning plane at ranges up to 80 m.

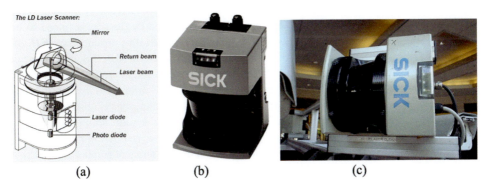

Figure 12.5 (a) Diagram showing the construction of a SICK 2D laser scanner including the laser diode and photo diode used as the transmitter and the receiver respectively for the range measurement and the rotary encoder used for the corresponding angular measurements; (b) a SICK LMS291 laser scanner; and (c) this SICK LMS291 scanner has been placed on its side to measure a vertical profile within its 2D plane to the side of the mapping vehicle

Source: SICK.

 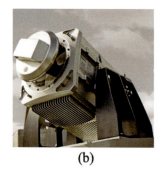 

(a) (b) (c)

Figure 12.6 (a) The RIEGL VQ-450 2D scanner showing its 'full-circle' scanning/profiling capability.; (b) The Lynx 2D scanner from Teledyne Optech, again with a 'full-circle' scanning capability; and (c) the Z+F Profiler 9012 scanner instrument, again allowing an unobstructed 2D profile scan over its full 360-degree rotation

Source: (a) RIEGL (b) Teledyne Optech (c) Zoller + Fröhlich.

with an accuracy of ±6 cm. These scanners have been used extensively, e.g., in the mobile mapping systems built by Google, Topcon and Mitsubishi.

Besides the SICK scanners, a number of *purpose-built TOF 2D laser scanners* are manufactured by specialist suppliers of laser-based measuring equipment to the surveying and mapping industry, such as *RIEGL* (Austria), Zoller + Fröhlich (Germany), Teledyne Optech (Canada) and Renishaw (UK). In general, these specialized 2D scanner units provide greater ranges, faster speeds and higher measuring accuracies than the SICK LMS291 model described above and are correspondingly more expensive. Thus, they tend to be used for those applications that demand high geometric accuracies. In the case of the VQ-450 model produced by *RIEGL* (Figure 12.6a), it provides a 'full-circle' 360-degree coverage within its 2D scanning plane; a pulse measuring rate of 550 kHz; a scan rate of 200 Hz; and can measure ranges up to 800 m, while still achieving a ranging accuracy of ±1 cm. The Lynx SG 2D scanner (Figure 12.6b) produced by Teledyne Optech and the Zoller + Fröhlich Profiler 5012 (Figure 12.6c) both have similar performance envelopes.

As with the digital cameras discussed above, *multi-scanner units* have been produced for use in certain applications where the 2D planes being measured by the scanner need to be set at different angles. Examples are the VMX-1Ha twin-headed unit produced by *RIEGL* (Figure 12.7a (Rieger et al., 2010) and the Lynx SG-1 unit manufactured by Teledyne Optech (Figure 12.7b). However, a completely different approach to the use of multiple scanners has been taken by the Velodyne Lidar company based in San Jose, California. The original Velodyne HDL-64E High-Definition Lidar (light detection and ranging) (Figure 12.8a) was based on the simultaneous use of 64 individual laser rangefinder units spinning around a common axis and covering a 26.8-degree vertical spread with a *series of 2D profiles* stacked on top of one another. Since then, Velodyne has introduced a series of smaller models that are smaller in size and lighter in weight, equipped with 32 laser rangefinder units (Model HDL-32E) and 16 units (VLP-16) (Figure 12.8b). A still lighter version of this last model, called the Puck LITE, was introduced in 2016. These various Velodyne models with their simultaneous measurements of multiple 2D profiles at differing angles of elevation and providing different densities of data have been adopted for mobile mapping purpose by many of the system suppliers and by a number of the service providers such as Google and Apple Maps who operate large fleets of mobile mapping vehicles.

Mobile mapping

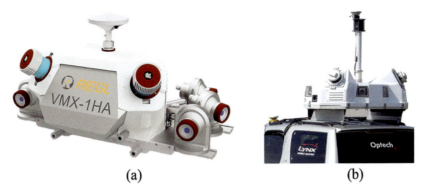

Figure 12.7 (a) A twin-headed VMX-1HA 2D scanner unit with a pair of 'full circle' scanners from RIEGL. Note also the accompanying imaging cameras mounted around the scanner unit and the GPS antenna mounted on top of the unit; (b) a Teledyne Optech Lynx SG scanner unit equipped with twin 'full circle' scanners

Source: (a) RIEGL (b) Teledyne Optech.

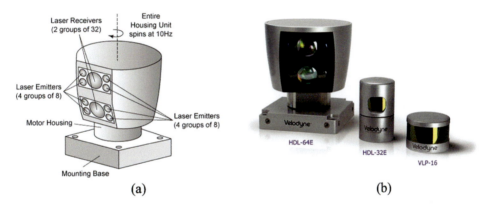

Figure 12.8 (a) Diagram showing the main external features of the Velodyne HDL-64E spinning laser scanner; (b) showing the comparative sizes of the HDL-64E, HDL-32E and VLP-16 multi-laser spinning lasers

Source: Velodyne.

Positioning (geo-referencing) devices

The primary device that is employed for the measurement of the position of the platform with its imaging cameras and laser scanners forming a mobile mapping system is a *GPS receiver*. The GPS satellite constellation comprises 24 satellites that are very precisely tracked by a world-wide network of ground stations so that the positions in space of each individual satellite are very precisely known at all times during the day and night. Similar constellations of satellites are operated by the European Union (Galileo) and Russia (GLONASS). Since 2018, these are known collectively as the *Global Navigation Satellite System (GNSS)*. Each satellite in the constellation continuously emits coded signals. The time when a specific signal is transmitted from a particular satellite is compared with the time when it arrives at the receiver. The precisely measured time difference can then be converted to the range (R) or distance between the satellite at its known position in space and the receiver mounted on the mobile platform. The GPS receiver can receive simultaneously

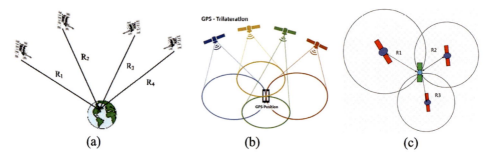

Figure 12.9 (a) The principle of location by the simultaneous measurement of ranges (R1–R4) from satellites at known positions in space to the GPS receiver mounted on the mobile mapping platform on the Earth; (b) showing how the range circles from four satellites intersect to define a unique position on the Earth; and (c) in fact, the minimum requirement to fix the location (position) of the platform is three ranges. The fourth range can act as a check on the position that has been determined and permits the measurement and determination of altitude (elevation)

the signals from a number of satellites that are in view at a given moment of time. If three such ranges can be received, then the precise position of the platform can be determined. The addition of a fourth range provides redundancy and a check on the positional value (Figure 12.9).

Arising from the needs for high positional accuracy of the imaging devices and scanners that are used in mobile mapping operations, almost invariably the GPS receivers that are used in combination with them are of the *dual-frequency* type. The use of the signals emitted at different frequencies by each individual satellite helps the combat and correct for the errors resulting from the refractions and delays caused by the ionosphere on their paths between the satellite and the receiver. However, a further technique to improve the positional accuracy of the measured data is to employ the *differential GPS (DGPS) method* by which the data collected at a local base station is used to compute corrections that can then be used to correct the data that has been measured by the receiver mounted on the mobile mapping platform (Figure 12.10). However often the mappers will make use of a commercial global correction service, such as those operated by the Fugro surveying and Navcom navigational organizations, or the data from a national monitoring network such as CORS (Continuously Operating Reference Stations) in the United States. where the National Geodetic Survey supplies correction data free-of-charge. A number of mobile mapping systems also feature a second GPS with its receiving antenna placed at a known distance from the primary GPS receiver. The difference in the positions given by the two receivers gives an accurate measurement of the platform's heading.

A special problem with GPS positioning systems concerns their application to the mapping of urban areas where so much mobile mapping takes place. In particular, the interruptions to the GPS satellite signals by high buildings and overhanging vegetation – creating the so-called 'urban canyons' – result in poor satellite configurations or even a complete loss of signal. Besides which, the reflections of signals from nearby buildings gives rise to *multi-path effects* that further reduce the positional accuracy of the GPS or GNSS measurements. Similar difficulties can be experienced in very mountainous areas. A further limitation is that the measurement of position using a GPS receiver can only be made after a certain time interval has elapsed – i.e., it has a low refresh rate – which causes problems when the receiver is mounted on a rapidly moving mobile mapping platform. These deficiencies have resulted, in most cases, in the GPS measurements being supplemented by the additional positional data supplied by inertial measuring units (IMUs) and odometers (distance measuring instruments or DMIs).

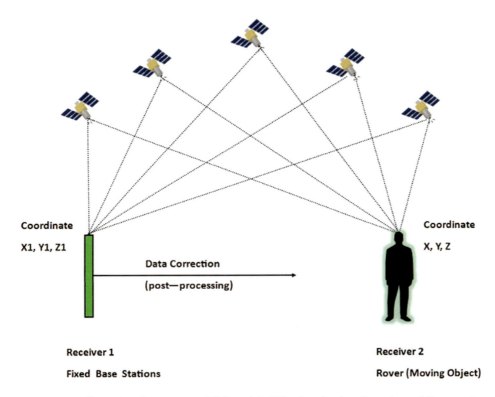

Figure 12.10 Illustrating the concept of differential GPS using the data from the mobile or roving mapping system in conjunction with that being continuously recorded at a fixed base station

An *IMU* measures the linear accelerations (using a three-axes accelerometer) and the rotational velocities (using a three-axes gyroscope) that are being experienced by the mobile platform (Figure 12.11). These measured values can then be integrated numerically by an embedded micro-controller unit to obtain the 3D position and orientation of the unit. The update frequency of the IMU is at least one order higher than that of the GPS receiver, which means that it can provide measurements of the position of a mobile mapping platform in between those provided by the GPS receiver. However, the positional values measured by the IMU are subject to drift with time leading to a loss of accuracy. Thus, the two technologies are, to a considerable extent, complementary to one another – with the GPS receiver providing more accurate data over a wider time interval, while the IMU gives more frequent updates with a lower accuracy. The integration of the two technologies (usually through the use of a Kalman filter) results in either a closely coupled or loosely coupled integrated GPS/IMU sub-system.

Regarding the IMUs that are commonly used in the combined GPS/IMU sub-system forming part of the overall mobile mapping system, three main types can be identified: (i) those utilizing *ring laser gyros*, which offer the highest accuracy data but are the most expensive to manufacture; (ii) those employing *fibre-optic gyros* (Figure 12.11) which are less expensive, but still offer a very acceptable accuracy; and (iii) those gyros based on *MEMS (Micro Electro-Mechanical Systems)* technology, which are the least expensive but still give an acceptable accuracy for many applications.

Figure 12.11 An early example of the Litton LN200 IMU (now being manufactured by Northrop Grumman) containing closed-loop fibre-optic gyros and solid-state silicon accelerometers

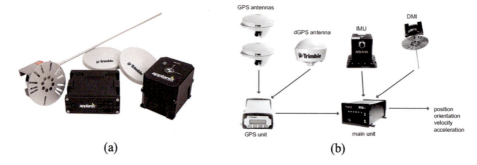

Figure 12.12 (a) The main components of an Applanix POS/LV GPS/IMU sub-system used in a mobile mapping system comprising (i) the IMU (at far right); (ii) the twin antennas of the Trimble GPS receiver (in the centre, at the back); the black electronics controller (in the centre, at the front); and the odometer with its wheel encoder and its attached rod carrying its cable to the controller (at far left); (b) a block diagram showing the connections between the various components of the Applanix POS/LV sub-system

Source: Applanix.

As mentioned above, the GPS/IMU sub-system is further supplemented by an *odometer* or distance measuring instrument. This comprises an optical angular encoder with an attached data transmission cable mounted inside a hollow rod that is fitted to the rear wheel of the mapping vehicle. The data generated by the combination of the IMU and the odometer allow positioning to be maintained during the period when GPS signals cannot be received. An example of the integration of all these various components into a fully operational sub-system is shown in Figure 12.12 for an Applanix POS/LV unit.

Mobile mapping systems

Although fully integrated systems utilizing all the components described above in the section 'Imaging devices' of this account were developed on a one-off basis by research establishments and commercial operating companies in the early days of mobile mapping, currently such systems are now available on a commercial off-the-shelf (COTS) basis from

Mobile mapping

a large number of systems suppliers to the surveying and mapping community. The major suppliers include Leica Geosystems (Switzerland) with its Pegasus range (Duffy 2014, 2015), Trimble (USA) with its TX series, Topcon (Japan) with its IP-S series; Mitsubishi (Japan) with its MMS range, Teledyne Optech (Canada) with its various Lynx and Maverick models; IGI (Germany) with its Street Mapper systems; etc. Besides these major suppliers, there are quite a number of much smaller specialist suppliers of complete mobile mapping systems. Representative examples of these integrated systems showing the arrangement of the various imaging, scanning and positioning elements of a complete system are given in Figure 12.13.

With the ever increasing trend towards the miniaturization of the individual components that are used in mobile mapping systems and the ready availability of less expensive commercial off-the-shelf (COTS) components, there is a current trend towards *more compact systems* – see Figure 12.14 showing the Topcon IP-S3, Optech Maverick and Siteco systems,

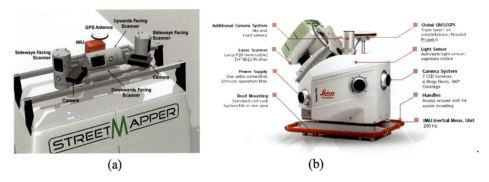

Figure 12.13 (a) An early Street Mapper mobile mapping system from IGI showing the various imaging, laser scanning and positioning elements that make up the complete system; (b) a more modern Pegasus Two system from Leica Geosystems with many of the measuring elements encased in a protective pod. In this particular example, the laser scanner is a Leica P20 terrestrial 3D scanner that has its azimuthal motion locked to allow the instrument to be used as a 2D scanner/profiler

Source: (a) 3D Laser Mapping. (b) Leica Geosystems.

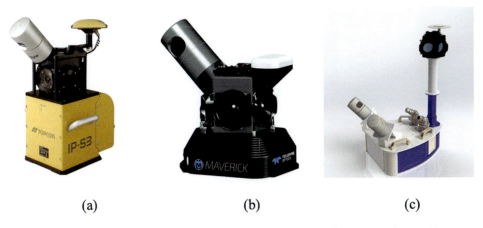

Figure 12.14 Examples of compact mobile mapping systems, each utilizing a Ladybug multi-camera unit; a Velodyne laser scanner; and a GPS receiver, from (a) Topcon; (b) Teledyne Optech; and (c) Siteco

all of which employ Point Grey Ladybug multi-cameras, Velodyne HDL-32E scanners and GPS receivers, together with their own electronic and controller units to form a complete mobile mapping system. This drive towards miniaturization has also resulted in the development of a number of *wearable or portable backpack systems* using these components (Ellum and El-Sheimy, 2000, Blaser et al., 2018), which allow the acquisition of geospatial data in areas where there is no vehicular access (Figure 12.15). This development has been taken a step further with the recent development of *hand-portable mobile systems* such as the Zeb range (Figure 12.16a) originally developed by CSIRO in Australia (Zlot et al., 2012) and now produced by GeoSLAM, based in Nottingham in the UK. The Zeb range is based on the use of laser scanners either from Velodyne (with 100 m range) or from Hokuyo (with 30 m range). A backpack model, called the Zeb Discovery (Figure 12.16b), which combines a Hokuyo 2D laser scanner with a Pulsar four camera unit fitted with fisheye lenses built by NC Tech in Edinburgh, Scotland has also been introduced.

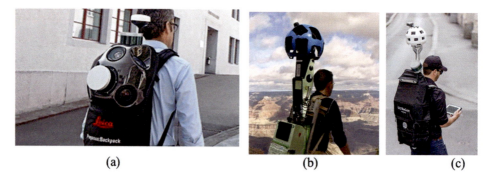

Figure 12.15   Examples of portable backpack mobile mapping systems from (a) Leica Geosystems (Pegasus Backpack); (b) Google Maps (Trekker); and (c) Vexcel Imaging (Panther), each featuring (i) a multi-camera unit; (ii) 2D laser scanners; and (iii) a GPS receiver

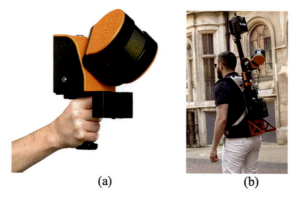

Figure 12.16   (a) A GeoSLAM Zeb hand-portable laser scanner for close-range mapping measurements; (b) a GeoSLAM Zeb Discovery backpack mapping unit combining a laser scanner; a cartographic grade GPS for accurate positional location; and a four camera Pulsar unit from NC Tech providing 360-degree imaging and scanner coverage

Source: GeoSLAM.

Mobile mapping operations, applications and service providers

One of the principal applications of mobile mapping systems at the present time is to generate geospatial data for use in *road navigation and cartography*. Closely associated with this activity is *street-level imaging* which is used for the inspection of the built environment and its infrastructure for assessment, planning, archival and personal viewing purposes. This type of data can also be used to help in the construction of *three-dimensional models* of cities and towns in conjunction with the data acquired through airborne digital mapping operations. Yet another important application of mobile mapping is the *detailed inspection of highways, railways, airports, harbours and waterways for engineering purposes* with a view to their management and maintenance. Obviously quite different specifications apply to each of these different activities, in particular, to the specific content that has to be provided and its positional accuracy. For example, the detailed content and the accuracy requirements of the data needed for highway engineering purposes will be far higher than those needed for the production of road atlases. In turn, these matters affect the systems that are employed and on their operational use and their cost.

Mobile mapping for road navigation and cartography

Mobile mapping for the purposes of road navigation and cartographic purposes is mostly carried out by large organizations working on a regional, national or international scale. The mobile mapping activities of two of the largest companies in this particular field – TomTom and Here Technologies – have been selected as examples and are outlined in the following sections.

TomTom

TomTom is a large Dutch company with 5,000 employees and an annual revenue of around 700 million Euros that has its headquarters in Amsterdam but conducts its mapping operations on an international scale. The company began to develop its mapping activities in the last few years of the twentieth century with its software products involving route planning and car navigation, the latter involving the use of a GPS receiver. With an ever-increasing range of products, in 2009, TomTom acquired TeleAtlas, which was a supplier of digital map data to a large network of users besides TomTom. The TeleAtlas company had a large data centre in Ghent, Belgium from which it directed its European mapping operations – which were mainly executed using camper vans (painted in an eye-catching bright orange colour (Figure 12.17a) that allowed the crews to work in remote and less populated areas (Petrie 2010). Another data centre was located in Lebanon, New Hampshire from which it directed its operations in North America. TeleAtlas had already entered the mobile mapping field in 2004 using technology developed by a Polish company and it had built up large fleets of mapping vehicles both in Europe and in North America. Smaller fleets operated in southeast Asia, in Taiwan, Thailand and Singapore. The initial processing and analysis of the raw data acquired by these vehicles was carried out partly in Poland and partly in New Delhi, India, where the large processing centre was operated on TeleAtlas's behalf by an Indian company, Infotech Enterprises.

Since its takeover by TomTom, the fleets of mapping vehicles have continued to grow. They are equipped with the usual combination of digital cameras, 2D laser scanners, a GPS receiver, an inertial sensor and odometers. Typical of the current vehicles are those shown in

Figure 12.17 (a) One of the original camper vans operated as a mobile mapping vehicle by TeleAtlas; note the imaging cameras at the front and on the side of the van. The logo on the side of van is a flying goose on whose back is a boy who, in this way, gets his view of the world – as recounted in a Swedish children's book; (b) a modern fleet of TomTom mobile mapping vehicles based in Poland, each equipped with a multiple camera unit, SICK 2D laser scanners and a GPS receiver

Source: TomTom.

Figure 12.18 (a) TomTom maps are heavily used by drivers in vehicles fitted with satnav devices (left) and cell phones (right); (b) a TomTom map of Manhattan, New York being used for traffic planning

Source: TomTom.

Figure 12.17b, which are equipped with multi-camera units and SICK 2D laser scanners. Once the data has been collected by the mobile mapping system, it is downloaded into the TomTom database. This data is supplemented by the data that is being collected by the satnav devices of TomTom's customers and by local specialists employed by TomTom to collect data in problem areas, as well as the changes that are being reported by customers. The giant TomTom database is maintained by its Cartopia map editing system. Once the changes have been made and verified, they are sent out to update the customers' navigation and map devices (Figure 12.18).

Since 2012, TomTom has also been one of the main suppliers of map data to Apple for use in its map products. From 2015 onwards, TomTom has also been a major supplier of map

Mobile mapping

data to the Uber transportation networking company. In recent years, TomTom has also sold its map data and its navigation and traffic solutions to various car manufacturers, including Volkswagen, Daimler (Mercedes-Benz), Toyota, and so on. It has also partnered with Bosch, the German electronics firm, to develop various map-based products for autonomous driving. Further specialist navigation products have been developed for use by the owners of caravan and camper vans, trucks and motorcycles.

Here Technologies

Here Technologies is an even larger mapping company (currently with 9,000 employees and an annual revenue of US$2 billion) with a world-wide operational spread that is also headquartered in Amsterdam. However, its origins are completely different to those of TomTom. Its origins can be traced back to the Navigation Technologies Corporation from 1987 (which was founded initially as Karlin & Collins in 1985), before being renamed as Navteq in 2004. Under these various titles, Navteq built up a large business as a leading supplier of automotive map data. Its other foundation was the Finnish electronics company, Nokia, which began to develop its Nokia Maps product (also called Ovi Maps for a period) in 2001. This map data, much of which was supplied by Navteq, could be downloaded into the company's smart phones (Figure 12.20a). In 2007, Nokia acquired Navteq, though at first, Navteq and Nokia Maps remained as separate divisions of the parent Nokia Corporation. In 2011, the two divisions were amalgamated to bring together the mapping, navigation and other location-based services under a single management. In 2012, the combined organization was re-branded as 'Here'. In 2015, Here was sold to a consortium of three German car manufacturers – Audi (part of the Volkswagen Group), BMW and Daimler (Mercedes Benz). Since then, several other large companies, including Intel, Bosch, Continental and Mitsubishi have acquired substantial shares in the company. The company has data centres in a number of countries with much of the processing being carried out in India, with over 4,000 employees and a large data processing centre located in Navi Mumbai.

Initially, Navteq employed vehicles for mapping in a simple way using a DGPS receiver to continuously plot its location with the revision being recorded by the crew on a tablet computer and as audio files (Figure 12.19a). In 2008, Navteq introduced a new fleet of mapping

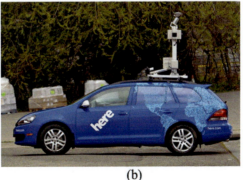

(a) (b)

Figure 12.19 (a) An early mapping car being used for map revision by Navteq. Note the GPS antenna on top of the car hat is being used for positioning purposes. The revision data is being recorded on a tablet computer and by an audio recording device; (b) a later Here mapping vehicle complete with a multi-camera unit, a laser scanner and a GPS receiver

Source: Navteq/Here.

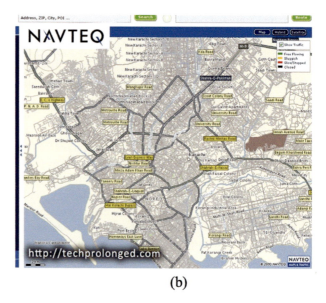

(a) (b)

Figure 12.20 (a) An Ovi Maps product being used for navigation in Paris; (b) a Navteq map of Karachi, Pakistan

Source: Here.

vehicles equipped with a set of imaging cameras. In 2010, Navteq finally began to use laser scanners as well as cameras for data collection with the adoption of the Topcon IP-S2 system using a Velodyne scanner. Many of its newer mapping vehicles also used the digital frame cameras that had been developed by Earthmine, a mobile mapping company based in Berkeley, California, which had been purchased by Nokia in 2012 (Figure 12.19b).

Besides supplying data to a large number of car manufacturers for incorporation in their integrated in-car navigation systems. Here sells or licences its map data to Garmin for use in its numerous GPS-based products developed for the automotive, aviation and marine industries; and to Internet service providers such as Facebook and Yahoo Maps. In 2009, Here also began the supply of map data and street-level imagery to Microsoft for incorporation in its Bing Maps service; however, this partnership ended in 2014. Here has also been the supplier of map data to large enterprises such as Amazon and Oracle as well as government agencies in many countries that use the data as the basis of their GIS services (Figure 12.20b).

Mobile mapping for imaging services

A large number of companies that provide street view services featuring 360-degree panoramic images have come into existence around the World over the last 20 years. Examples are *CycloMedia*, a pioneering company based in the Netherlands, whose charged street view services also extend into Germany, Belgium and the Scandinavian countries; and *Yandex*, which carries out similar activities in Russia, Ukraine, Belarus, Kazakhstan and Turkey. However, the majority of those companies supplying street view images are relatively small, both in size and in terms of their coverage, especially when compared with the giant international companies such as Google, Apple Maps and Bing Maps that also offer this type of product and service on an international scale – the first two of which are those whose activities will be discussed in this section.

Google Maps and Street View

Google Maps is a hugely popular web-based map and geospatial imaging service, based on rectified aerial photography for most urban areas and on satellite imagery for more remote areas. Its conventional line maps that are available to users via the Internet are mainly derived from these imageries along with additional image and data that has been purchased or leased from local suppliers. However, this basic geospatial data is supplemented by the information derived from Google's extensive mobile mapping activity which has resulted in its map-based products for route planning and GPS-based navigation which are extensively used on smartphones. According to press reports, overall Google has 1,100 employees working full-time on its mapping activities and products with a further 6,000 people employed by contractors carrying out mapping-related work for the company.

Google's Street View is a special feature of the Google Maps and Google Earth services that can also be accessed over the Internet. The software provides access (without charge) to the images that have been acquired by its mobile mapping systems, usually at 10–20 m intervals, along the streets of many cities, especially in the more highly developed countries of the World – for example in the United States, Western Europe, Japan and Australasia. Other large countries with partial coverage include Russia, India and Brazil. However other parts of the World, including many countries in Africa and the Middle East, have poor coverage (Figure 12.21).

Although initially, in 2007, the Street View images were acquired by external contractors, since 2008, Google has been generating the required image, scanner and positional data using its own fleets of mobile mapping vehicles; various press reports give estimates that between 250 and 400 such vehicles are being operated by Google world-wide. These vehicles are equipped with Google's own unique high-resolution multi-camera units and the various 2D laser scanners from SICK and Velodyne that have been outlined above, in combination with DGPS/IMU positioning sub-systems supplied by Topcon and the required wheel encoders (Figure 12.22a).

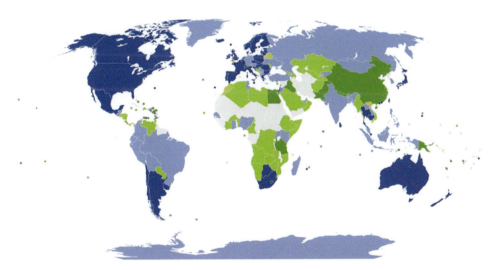

Figure 12.21 Google Street View coverage. Dark Blue – Countries with mostly full coverage; Light Blue – Countries with partial coverage; Dark Green – Countries with certain selected items or areas only; Light Green – Countries with views of private businesses only

Source: Wikipedia.

 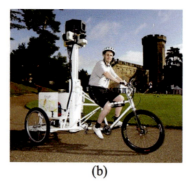 

(a) (b) (c)

Figure 12.22 (a) A current model of the Google Street View mapping vehicle equipped with a seven-camera imaging unit mounted on its mast, below which are mounted two Velodyne laser scanners; (b) a pedal powered tricycle equipped with the same suite of sensors as a Google mapping car, as seen at Warwick Castle in England; and (c) two different models of the Google Trekker backpack system, that on the left featuring the older R7 unit with its 15 cameras, while that on the right has the current lighter unit with seven cameras

Source: Google.

Besides which, Google also developed versions of its systems that could be mounted on tricycles (Figure 12.22b), backpacks (Figure 12.22c) and small boats for the collection of data in areas such as university campuses, tracks, paths, etc., where motorized vehicles cannot be deployed.

The 2D laser scan data and its associated positioning data are used first to form a 3D model or mesh to both sides and to the front of the street, on to which the individual photographic images are fitted (Madrigal 2012). However, arising from privacy concerns, the blurring of persons' faces and car licence plates is undertaken in a prior operation for each image using automated procedures, supplemented by visual inspection as a check (Frome *et al.*, 2009). The sets of multiple images with their 360-degree coverage allow the formation of continuous panoramic views along the streets and roads being covered by the mapping vehicles. Much of the detailed data processing and manual editing is carried out in India, reportedly in centres located in Hyderabad and Bangalore. Besides the measured data acquired using its mapping vehicles, Google receives large numbers of corrections and updates from the users of its Street View services on a daily basis and these need to be verified on the ground before being added to its database. To ensure a reasonably rapid response to requests from users of the service, Street View servers are located in a network of data centres distributed around the World.

Apple Maps and Look Around

Apple Maps is the web mapping service that has been developed by Apple Inc. for use with the various operating systems that are used with its computers, tablets and smartphones (Figure 12.23a). The service was first introduced using data that had been sourced by independent providers such as TomTom and Digital Globe. However, in 2015, Apple announced that it had started to operate a fleet of mapping vehicles to supplement this material. These vehicles were fitted with various types of cameras and laser scanners (Figure 12.23b), the acquired data being used to construct and revise the service's map data with an emphasis on road navigation and traffic information. Reportedly the required data processing to produce the final map products is being carried out at a large centre in Hyderabad in partnership

Mobile mapping

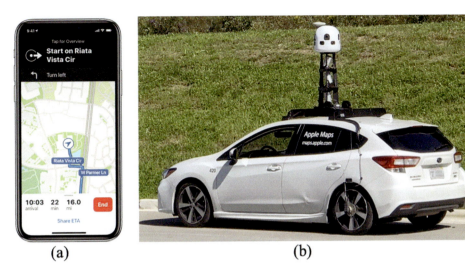

(a) (b)

Figure 12.23   (a) An Apple Maps display with road navigation directions on an Apple iPhone; (b) an Apple Maps mapping vehicle equipped with a multiple camera unit and three Velodyne laser scanners that is engaged in the collection of image and laser scan data

Source: (a) Apple Inc. (b) Wikipedia.

with an Indian company, RMSI, with 4,000 people being employed on these activities. The Apple Maps coverage is now on an international scale. In June 2019, Apple introduced its *Look Around* service by which the user can view 360° street-level imagery with smooth transitions as the scene is navigated along much the same lines as Google's well known Street View service. At the present time (in January 2020) this service is only available for a few major cities in the United States.

Mobile mapping for infrastructure

Many commercial survey and mapping companies have invested in mobile mapping systems with a view to undertaking surveys of infrastructure, principally for engineering applications. This widespread activity is mostly concerned with the 3D mapping of the corridors containing road, rail and marine transport arteries. Such surveys require much higher requirements (i) for positional and elevation accuracy – at the single cm. level – and (ii) for image resolution, than those that apply to the road navigation, cartographic and street view applications that have been discussed in the preceding sections of this account.

With regard to *road infrastructure applications*, the use of mobile mapping systems allow the details of the road surface and the adjacent pavements, kerbs, ditches and vegetation to be captured, together with the associated road signs, street furniture, power and telephone poles and overhead wiring to create highly accurate 3D digital models and databases for engineering, maintenance, planning and CAD purposes (Figure 12.24a and b). Detailed assessments of bridge structures and of possible routes for wide loads, including overhead clearances, can also be made. The detailed detection and location of cracks and potholes in the surfaces of both roads and pavements can also be carried out. Indeed, some survey companies, such as Fugro (Netherlands), Mandli (USA) and Siteco (Italy) (Figure 12.25) have added downward pointing high-resolution cameras, laser scanners and even ground penetrating radar (GPR) to certain of their systems specifically for the purpose.

165

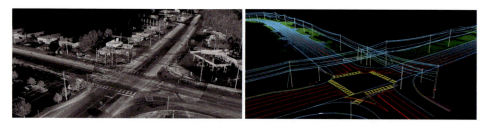

Figure 12.24 (a) Data depicting the highway infrastructure at a busy road junction in Atlanta, Georgia, USA (b) Showing the detail of the road markings, crossings, poles and overhead wires that has been extracted from these data

Source: Sanborn Map Company.

Figure 12.25 This Siteco Road-Scanner mobile mapping system is equipped with additional high-definition cameras that are used to image the surface of the road and/or its pavements. A powerful lighting system can be added for its use in night time surveys of airport runways, taxi ways and service roads

Source: Siteco.

Mobile mapping systems can be used in a similar manner for *railway and tramway infrastructure surveys* with the mounting of the system either on a railway maintenance railcar (Figure 12.26a,c) or using a normal mapping vehicle such as a pick-up truck with its wheels being kept on the rails of the track using a re-railer device (Figure 12.26b). As with road surveys, the use of such equipment results in much more rapid surveys being undertaken, while minimizing the disruption to traffic and the danger to surveyors that attends manual instrumental surveys of the track (Kremer and Grimm, 2012). The resulting image, scanner and positional data allows automated or visual track inspection to be implemented, since it provides details of overhead wires and supporting poles, signals, switches, clearances, crossing points and obstructions as well as the all-important measurements of the width between rails and between the individual supporting sleepers. Needless to say, surveys of railway tunnels become thoroughly practical to execute with mobile mapping systems, the use of alternative terrestrial survey techniques being time-consuming, costly and quite dangerous.

In a similar manner, mobile mapping vehicles are used to undertake the detailed mapping of airports and their surrounding infrastructure, including runways, terminals, support facilities and structures and lighting. When mounted on a suitable boat or marine craft (Figure 12.27a), mobile mapping systems can also undertake the detailed mapping of

Mobile mapping

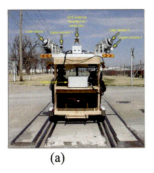  

(a) (b) (c)

Figure 12.26 (a) A railroad speeder (track maintenance car) equipped with a Teledyne Optech Lynx mapping system undertaking the mapping of a railroad in Oklahoma; (b) a Rail Mapper vehicle equipped with a Z+F Profiler 9012 2D laser scanner, a multi-camera unit and a GPS receiver; and (c) a Z+F Profiler 6007 scanner system mounted on the front of a train that is carrying out a detailed survey of the railroad track and its fittings in New York

Source: (a) Teledyne Optech. (b) IGI. (c) Zoller + Fröhlich.

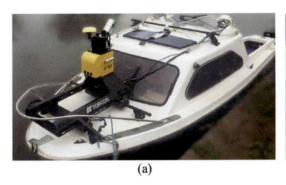 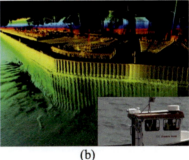

(a) (b)

Figure 12.27 (a) A survey launch equipped with a Topcon IP-S3 mobile mapping system; (b) a model showing the land and underwater features of a harbour area carried out using a Renishaw Dynascan mobile mapping system (see bottom right corner) combined with a multi-beam sonar

Source: (a) PLA. (b) Merrett Surveys.

harbours, rivers and canals, including the walls, banks, bridges, vegetation as well as the beacons, buoys, etc., used for navigation purposes. Often this data will be combined with the data from underwater multi-beam sonar systems, resulting in a high-resolution dataset both above and below the waterline (Figure 12.27b).

Mobile mapping for three-dimensional city modelling

3D City Models comprise digital virtual models of large urban areas that attempt to represent in all three dimensions, the roads, buildings, transport infrastructure, parks, woods and open spaces that are present in the area, together with the details of the terrain on which they are constructed. Given the enormous numbers of these features that exist in such areas and their complex relationship with one another, these 3D models are notoriously difficult to construct and the acquisition of the required data can be enormously costly to execute. To help cope with these problems, the CityGML standard data model that has been adopted widely, defines different levels of detail to be included in such a model –

- LOD 0: 2.5D footprints only;
- LOD 1: Buildings are represented by block models only (usually extruded footprints);
- LOD 2: Building models are included with standardized roof structures;
- LOD 3: Detailed (architectural) building models are included;
- LOD 4: LOD 3 building models that are supplemented with interior features.

The scale and resolution of the imagery and scanned data will depend greatly on the specific LOD level that is being aimed at. In general terms, most commonly, the basic data acquisition for 3D City Models is carried out using aerial photogrammetric mapping methodology, based on the use of aerial photography and airborne laser scanning. Since mobile mapping operations are confined to the narrow corridors of roads, streets, railways, rivers and canals and cannot measure roof structures, they can only play a secondary role in 3D City Modelling. Nevertheless, mobile mapping plays an important part in the implementation of the more detailed (higher level) LOD models through its supply of detailed facades, street furniture, and so on. (Cavegn and Haala, 2016) (Figure 12.28a).

An early example is the 3D model of the Historic Peninsula of Istanbul (Baz *et al.*, 2008; Dursun *et al.*, 2008; Kersten *et al.*, 2009) – which is a UNESCO World Heritage site covering an area of 1,500 ha and 48,000 buildings mostly located in very narrow streets. The roof mapping of the buildings was carried out using both analogue and digital aerial photography. Initially the mapping of the facades was being carried using static terrestrial laser scanning (TLS) methods. However, productivity was low, but it increased greatly when mobile mapping technology was introduced and largely replaced the TLS methodology. Thereafter the two datasets (aerial and terrestrial) were combined to form the base for the 3D modelling, successively at LOD 1 and 2 levels (Figure 12.28b). Later the most interesting features, such as the landmark buildings much visited by tourists, were modelled at LOD3 level. Since then, a number of cities or, more often, parts of cities world-wide have been mapped and modelled in a similar manner with mobile mapping techniques playing a secondary but important part in the process.

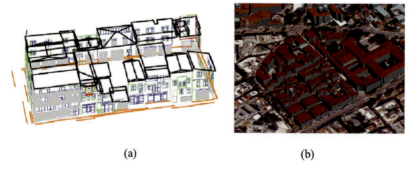

(a)   (b)

Figure 12.28 (a) The building facades that are acquired through mobile mapping need the roof patterns acquired by airborne imaging in order to create a 3D model of this block of properties; (b) a small part of the 3D City Model of the Historic Peninsula of Istanbul. The data have been collected at LOD2

Source: Bimtas, Istanbul.

Conclusion

Mobile mapping based on vehicle-based digital imaging and laser scanning has developed rapidly over the last 20 years largely on the back of the technological developments in GPS, digital imaging, laser scanning and inertial systems that took place at the very end of the twentieth century. Within a short period of time, it has given rise to and underpins road navigation and street imaging services and databases that are now viewed as being an integral part of life in the more developed parts of the World. The result has been the growth of mapping organizations such as those working for TomTom, Here Technologies, Google and Apple that are larger than any national mapping agency. At the same time, mobile mapping methodology also provides vital information about the road, rail and marine transport infrastructure and its maintenance. However, because of its confines to the mapping of narrow transport corridors and networks, the methodology has played little part in national mapping programmes where aerial mapping methods still dominate (Stoter *et al.*, 2016). For the future, if indeed the present development of autonomous (driverless) cars is successful and they come to be widely used, then there will be a need for the very detailed and accurate mapping of extensive road and street networks, in which case, mobile mapping will undoubtedly need to be implemented on a massive scale and in a very detailed manner (Walker, 2018).

References

Anguelov, D., Dulong, C., Filip, D., Frueh, C., Lafon, S., Lyon, R., Ogale, A., Vincent, L. and Weaver, J. (2010) "Google Street View: Capturing the World at Street Level" *Computer* 43 (6) pp.32–38.

Baz, I., Kersten, Th., Buyuksalih, G. and Jacobsen, K. (2008) "Documentation of Istanbul Historic Peninsula by Static and Mobile Terrestrial Laser Scanning" *International Archives of Photogrammetry, Remote Sensing and Spatial Information Sciences* 37 (Part B5) ISPRS Congress, Beijing, 3rd to 11th July.

Blaser, S., Cavegn, S. and Nebiker, S. (2018) "Development of a Portable High Performance Mobile Mapping System Using the Robot Operating System" *ISPRS Annals of the Photogrammetry, Remote Sensing and Spatial Information Sciences* IV-1 2018 ISPRS TC I Mid-term Symposium "Innovative Sensing – From Sensors to Methods and Applications" Karlsruhe, Germany, 10th to 12th October.

Bossler, J. and Toth, C. (1996) "Feature Positioning Accuracy in Mobile Mapping: Results Obtained by the GPSVan™" *International Archives of Photogrammetry and Remote Sensing* ISPRS Comm. IV 31 (Part B4) pp.139–142.

Cavegn, S., and Haala, N. (2016) "Image-Based Mobile Mapping for 3D Urban Data Capture" *Photogrammetric Engineering & Remote Sensing* 82 (12) pp.925–933.

Dursun, S. Sagir, D., Buyuksalih, G., Buhur, S., Kersten, Th. and Jacobsen, K. (2008) "3D City Modelling of Istanbul Historic Peninsula by Combination of Aerial Images and Terrestrial Laser Scanning Data" *4th EARSel Workshop on Remote Sensing for Developing Countries/GISDECO 8* Istanbul, Turkey, 4th to 7th June.

Duffy, L. (2014) "Can One Mobile Mapping System Do It All?" *Lidar Magazine* 4 (4) pp.38–42.

Duffy, L. (2015) "Exploring New Revenue Streams with Mobile Mapping" *Lidar Magazine* 5 (7) pp.50–53.

Ellum, C. and El-Sheimy, N. (2000) "The Development of a Backpack Mobile Mapping System" *International Archives of Photogrammetry and Remote Sensing* 33 (Part B2) pp.184–191.

Ellum, C. and El-Sheimy, N. (2002) "Land-Based Mobile Mapping Systems" *Photogrammetric Engineering and Remote Sensing* 68 (6) pp.13, 15–17, 28.

Elseberg, J., Borrmann, D. and Nuchter, A. (2013) "A Study of Scan Patterns for Mobile Mapping" *International Archives of Photogrammetry, Remote Sensing and Spatial Information Sciences* Vol. Xl-7/W2 pp.75–80.

El-Sheimy, N. (2005) "An Overview of Mobile Mapping Systems" *Proceedings of the FIG Working Week, 2005, and GSDI-8 – From Pharaohs to Geoinformatics* FIG/GSDI, Cairo, Egypt, 16th to 21st April.

Frome, A., Cheung, G., Abdulkader, A., Zennaro, M., Wu, B., Bissacco, A., Adam, H., Neven, H. and Vincent, L. (2009) "Large-scale Privacy Protection in Google Street View" *12th IEEE International Conference on Computer Vision* Kyoto, Japan.

Kersten, Th., Buyuksalih, G., Baz, I. and Jacobsen, K. (2009) "Documentation of Istanbul Historic Peninsula by Kinematic Terrestrial Laser Scanning" *The Photogrammetric Record* 24 (126) pp.122–138.

Kremer, J. and Grimm A. (2012) "The Rail Mapper – A Dedicated Mobile Lidar Mapping System for Railway Networks" *International Archives of the Photogrammetry, Remote Sensing and Spatial Information Sciences* 39 (B5), pp.477–482 *22nd ISPRS Congress* Melbourne, Australia, 25th August to 1st September.

Madrigal, A. (2012) "How Google Builds Its Maps – and What It Means for the Future of Everything" *The Atlantic* (6th September). Available at: https://www.theatlantic.com/technology/archive/2012/09/how-google-builds-its-maps-and-what-it-means-for-the-future-of-everything/261913/.

Novak, K. (1995) "Mobile Mapping Technology for GIS Data Collection" *Photogrammetric Engineering and Remote Sensing* 61 (5) pp.493–501.

Petrie, G. (2010) "Mobile Mapping Systems: An Introduction to the Technology" *GEOInformatics* 13 (1) pp.32–43.

Petrie, G. and Toth, C. (2018) "Dynamic Terrestrial Laser Scanners" In Shan, J. and Toth, C. (Eds) *Topographic Laser Ranging and Scanning: Principles and Processing* (2nd ed.) London: CRC Press, pp.62–88.

Rieger, P., Studnicka, N., Pfennigbauer, M. and Ulrich, A. (2010) "Advances in Mobile Laser Scanning Data Acquisition" *FIG Congress* Sydney, Australia, 11th to 16th April.

Stoter, J., Vallet, B., Lithen, T., Pla, M., Wozniak, P., Kellenberger, T., Streilein, A., Ilves, R. and Ledoux, H. (2016) "State-of-the-Art of 3D National Mapping" *EuroSDR Special Session at ISPRS Conference "Innovative Technologies and Methodologies for NMCAs"* Prague, 16th July.

Toth, C. and Grejner-Brzezinska, D. (2003) "Driving the Line: Multi-sensor Monitoring for Mobile Mapping" *GPS World* 14 (3) pp.16–22.

Toth, C. and Grejner-Brzezinska, D. (2004) "Re-Defining the Paradigm of Modern Mobile Mapping: An Automated High-Precision Road Centerline Mapping System" *Photogrammetric Engineering and Remote Sensing* 70 (1) pp.685–694.

Walker, A.S. (2018) "Autonomous Vehicles Operational Thanks to Lidar" *Lidar Magazine* 8 (5) pp. 20–26.

Zlot, R., Bosse, M., Wark. T., Flick, P. and Duff, E. (2012) "CSIRO: Moving Mobile Mapping Indoors" *Lidar Magazine* 2 (4) pp.2–5.

13
AIRBORNE AND GROUND-BASED LASER SCANNING

Mathias Lemmens

Since the mid-1990s, laser scanning has matured into a mapping technology routinely used in a variety of applications such as 3D modelling of cities, power stations and factories; power line mapping; dike and dune inspection for flood prevention; and monitoring of forests, open pit mines, construction sites, tunnels and glaciers, just to mention a few. Laser scanners acquire point clouds of many millions or even billions of points during one survey. They are active sensors that emit laser beams for measuring the distances (ranges) to objects; 'active' means that the sensors themselves emit electromagnetic (EM) energy.

The laser scanner primarily measures: (1) range – distance from the sensor to the first surface hit by the laser beam; (2) scan angle; and (3) intensity of the return. The ranges are the key asset for computing 3D coordinates of object points in a preferred reference frame. A high geometric fidelity, which is a necessity to obtain high-quality geoinformation, is obtained through accompanying the laser scanners operating outdoors with a GNSS (Global Navigation Satellite System) positioning device and an inertial measurement unit (IMU). They enable to determine the 3D position coordinates accurately as well as the three angles defining the orientation of the sensors in 3D space. In order to further improve the quality and to transform the 3D coordinates to a common reference frame, registration of the diverse scans and ground control points (GCP) are required.

Laser scanning technology is usually referred to as Lidar (light detection and ranging). Lidar sensors can be mounted on a wide variety of platforms, including vehicles, the human back and manned and unmanned aircraft and helicopters. This chapter first elaborates on basics of laser light and point clouds, then continues with georeferencing, followed by airborne Lidar and ground-based Lidar. Finally, the point capturing characteristics of airborne laser scanning are compared with those of mobile laser scanners.

Basics

A laser creates a narrow, intense beam of coherent and monochromatic light. Compared to other light sources (Siegman, 1986; Wilson and Hawkes, 1987; Koechner, 1992; Silfvast, 1996), laser beams have:

- high spectral purity;
- high intensity (i.e., they contain a lot of energy); and

DOI: 10.4324/9780367855765-16

- small beam divergence (i.e., the width of the beam and thus its footprint on the object is small.)

Furthermore, as they are monochromatic, together with the small beam divergence and high energy, this allows laser beams to bridge distances of several kilometres across the atmosphere. On the flip side of these beneficial properties, laser beams can harm human skin and eyes, and so laser systems are therefore subject to safety regulations.

Interaction with objects

A Lidar sensor emits EM energy in the visible green part or in the near-infrared (NIR) part of the EM spectrum. The most commonly used is the NIR portion, which is invisible to the human eye. When EM energy meets objects, including particles in the atmosphere, three types of interaction can occur: the energy may be (partly) reflected, absorbed or transmitted. The degree of reflection, absorption and transmission depends on wavelength and object characteristics (Figure 13.1).

In the visible part of the EM spectrum, the human eye perceives reflected light as colour. For example, humans see leaves of deciduous trees as green in spring and summer. However, the reflection of near-infrared (NIR) light – which is invisible for the human eye – on vegetation is stronger than the reflection of green. NIR, used in topographic or land Lidar systems, is absorbed by water, resulting in data gaps (Figure 13.2).

Green passes the water surface and reflects on the bottom of the water body. Therefore, green laser pulses are used in bathymetric Lidar, including mapping sea, river and lake bottoms. Some manufacturers combine green laser and NIR laser in one and the same device. Such devices can be used to monitor sand transport from dunes and adjacent seabed by recording time series. The transmission of laser beams is impeded by water droplets in the atmosphere (clouds).

Measurement principles

After emission, a receiver records the EM energy that an object reflects in the direction of the sensor. The main goal is to determine the range. When the position and orientation of the sensor as well as the scan angle is known the 3D coordinates of points in a reference frame can be calculated from the ranges (Figure 13.3).

Exploiting lasers as surveying sensors is based on three principles:

1 *Phase-shift*: laser beams are amplitude modulated; signals of varying wavelength are superimposed on the laser carrier beam.

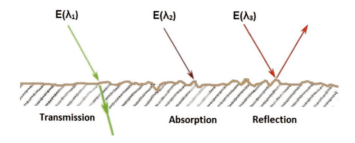

Figure 13.1 Whether an incident EM beam is (partly) reflected, absorbed or transmitted depends on its wavelength

Airborne and ground-based laser scanning

Figure 13.2 Part of the highly detailed digital terrain model (DTM) of the Netherlands available as open data; the data gaps are caused by ground filtering which removes vegetation, buildings, cars and other above-ground objects, and no return on water bodies due to the use of NIR laser

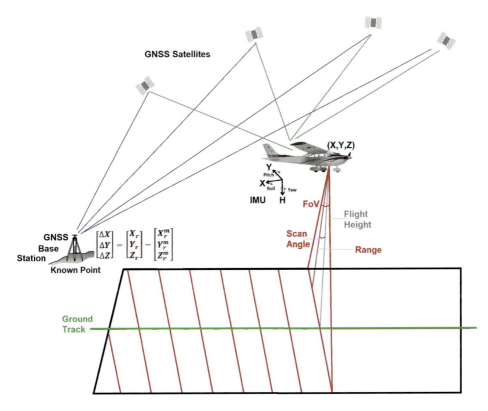

Figure 13.3 During an airborne Lidar survey GNSS and IMU continuously measure the six parameters of the exterior orientation enabling automatic calculation of the 3D coordinates of ground points from range and scan angle; the scan pattern of parallel lines is typical for a rotating mirror

Source: Lemmens (2023).

2   *Pulse measurements* or time-of-flight (ToF): pulses are emitted and their travel time forth and back from the object is measured ($\Delta t = t_1 - t_0$); the range is calculated from multiplying Δt with the speed of light, and dividing the outcome by two.
3   *Triangulation-based*: this method aims at creating point clouds for short-range applications and small objects.

Phase-shift and ToF systems are both used in ground-based systems that operate outdoors. The range precision of phase-shift scanners is in the order of a few millimetres. They produce higher point densities compared to ToF scanners. Pulse scanners can bridge longer distances than phase-shift scanners – which is why they are used in airborne Lidar systems – but the rate of laser beam emission and range accuracy are lower. Triangulation-based systems do not measure distances but angles. Compared to ToF and phase-shift systems they have a simple concept, low price, high acquisition speed and high precision (up to 0.01 mm). They can capture complex shapes of small objects in great detail, but the range is limited to a maximum of 5 metres.

2.5D and 3D point clouds

The basic product of a laser scanning survey is a set of data points, each with three position coordinates (X, Y, Z) represented in a reference frame. They provide the *where* component of each point. The *what* component is determined by attributes and class labels. The attributes can be reflection values of a laser beam (intensity return), or the red, green and blue (RGB) values in a colour image (Figure 13.4).

The result is a description of the scene in both geometric and thematic terms. The automated class labelling of point clouds is far from a solved problem and a topic that is extensively investigated by researchers all over the world. Since digital elevation models (DEM) are an important and critical data source for many GIS analysts and geo-scientists, it is appropriate to distinguish 2.5D point clouds from 3D point clouds. A DEM is a digital 2.5D

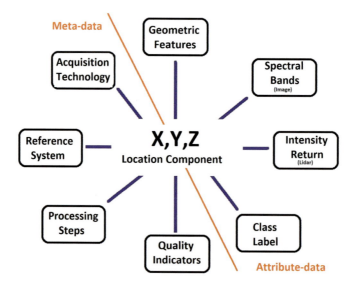

Figure 13.4   Diagram representing the nucleus of a point cloud consisting of three location coordinates, measured and derived attributes, and meta-data

Source: Lemmens (2023).

representation of the surface of the Earth, either bare ground or with objects such as vegetation and buildings, bridges and other constructions on top of it. It is a collection of planar coordinates (X, Y). Each X, Y location is assigned one elevation value, which is often considered an attribute. Often GIS analysts and geo-scientists need a bare ground DEM, a so-called digital terrain model (DTM), to calculate the water run-off capacity of a watershed, for example. The DTM represents the ground surface without buildings, vegetations and other above ground objects. However, an ALS captures points on the umbrella surface, including roads, roofs, foliage, other non-moving objects called the digital surface model (DSM). Additionally, a raw ALS point cloud contains points returned from cars, cattle, humans and other moving objects. To be usable as a DTM, all these points must be removed. Since there are millions of them, manual removal of non-ground points is practically impossible. Many algorithms based on a variety of mathematical models have been developed since the mid-1990s to automate the ground filtering process. Figure 13.2 shows the gaps in a bare ground DEM where buildings, cars and vegetation have been removed by a combination of automated and manual ground filtering. Figure 13.5 shows an example of applying a morphological ground filtering approach to a dense airborne Lidar DSM of an urban area.

However, automated ground filtering approaches are not faultless. Often 10% or more of the points are mis-labelled. Points that belong to the ground surface are labelled non-ground (false negatives) and non-ground points as ground points (false positives). Using ground truth (i.e., reference values collected by a human operator) the correctly classified points (Ng and Nn), false positives (Nng) and false negatives (Ngn) can be calculated and stored in a cross-matrix (Table 13.1). The cross-matrix indicates the suitability of an automated ground filtering approach for the terrain types used.

Manual editing of the output of automated ground filtering remains necessary to obtain high-quality DTMs. Unlike 2.5D point clouds, 3D point clouds aim to represent the entire 3D space through three coordinates: each planar position (X, Y) can be assigned multiple Z-coordinates.

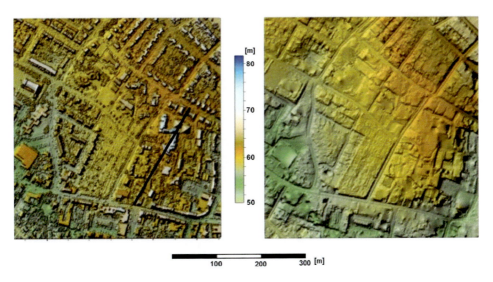

Figure 13.5 Application of a geodesic morphological ground filtering method on the National Elevation Model of an urban landscape type in the Netherlands; the left part depicts the DSM and the right part the DTM after ground filtering

Source: Li et al. (2017).

Table 13.1 Cross-matrix confronting a small fraction of the points in the filtered point cloud with ground truth

	Filtered point cloud		
Ground Truth	Ground Points	Non-ground Points	
Ground Points	Ng	Ngn	Ng + Ngn
Non-ground Points	Nng	Nn	Nn + Nng
	Ng + Nng	Nn + Ngn	N = Ng+Nn+Nng+Ngn

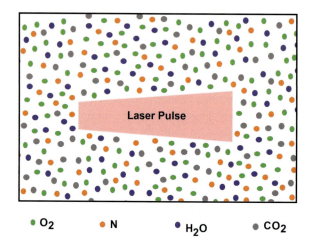

Figure 13.6 During propagating through the atmosphere at approximately the speed of light, the laser beam diverges and its energy is weakened by scattering and absorption by air molecules

Source: Lemmens (2023).

Beam divergence

The terms point and point cloud can cause confusion. The points are acquired by physical devices operating in a physical environment. Particles in the atmosphere cause beam divergence (BD) (Figure 13.6).

The laser beam will thus have a circularly shaped extension – footprint – and is not a point in the mathematical sense, having no dimension. The larger the distance, the larger the footprint. If the BD is 0.8 mrad the diameter of the footprint is 80 cm at a distance of 1 km. With modern laser scanners the user can often adapt the footprint within certain limits. Edges of buildings, roads and so on are preferably captured as sharp as possible (Xu *et al.*, 2008). This can be achieved through small footprints. In forest areas with high canopy, an enlargement of the footprint results in better penetration of the foliage so that the pulse can reach the ground (Hopkinson, 2007), which is beneficial for creating ground DEMs. However, a large BD means that the pulse contains less energy per unit of area, reducing the likelihood that the ground return will be detected by the laser scanner. That is why foresters prefer small footprints of airborne Lidar. When planning a Lidar flight, the relationship between pulse strength and footprint size must be taken into account.

Georeferencing

The points in a point cloud are rivetted in a reference frame by three values that each represent a dimension (X, Y, Z) and measured using metric units, such as metre, kilometre and so on. These three numbers relate to a particular reference frame, of which there are many types in use. They can be grouped into two fundamental categories: those based on a geometrical view of the Earth and those based on a geophysical view. The geometrical view has resulted in ellipsoidal coordinate systems and map projections, such as the Universal Transverse Mercator (UTM) coordinate system. The geophysical view, in which the Earth's gravity field plays a major role, has resulted in a representation in the form of the geoid (Figure 13.7).

Proper water drainage and flood prevention measures require that dike constructors, dune managers, drainage technicians and other water managers and engineers are sure that water flows from high to low. The heights above the ellipsoid – a geometric construct – do not determine the water flow, which is instead determined by gravitational forces. Consequently, to be practical, elevation values must refer to the gravity field of the Earth. The associated reference frame is an equipotential surface called the geoid.

The accurate georeferencing of point clouds, preferably automated, is one of the most challenging parts of the production phase. A pre-stage of geo-referencing is registration, the purpose of which is to bring two or more point clouds (or photogrammetric images) taken at different times, and/or from different sensors, and/or from different viewpoints into one local reference system. To gain further insight into the georeferencing process, the section 'Terrestrial laser scanning (TLS)' provides the pipeline to bring TLS point clouds into a common reference frame. The process of georeferencing and registration of laser point clouds is described in detail in Toth (2009) and Lichti and Skaloud (2010); and see also Shan and Toth (2018) and Lemmens (2023).

Airborne laser scanning (ALS)

Airborne laser scanning (ALS), often also referred as Airborne Lidar, has made great advances and has become a mainstream remote sensing technology. Together with aerial photogrammetry, ALS has become a major data source for creating DSMs, DTMs, 3D city models and 3D digital landscape models. An important facility to recording a strip is a scanning mirror, which distributes the pulses across the flight track. Most systems use an oscillating, rotating or nutating mirror. The scanning mechanism largely determines the scan pattern and thus the distribution of points on the ground (Figure 13.8).

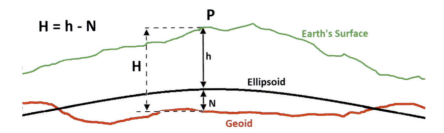

Figure 13.7 Relationship between height above ellipsoid (h) and height above geoid (H), with geoid undulation N = h-H

Source: Lemmens (2023).

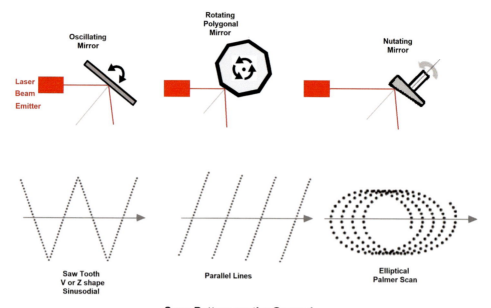

Figure 13.8 Various scanning mechanisms used in airborne Lidar systems (top) and the resulting scan patterns on the ground

Source: Lemmens (2023).

The scan angle range (e.g., 37.5°) determines the field of view (FoV), which is 75° in this example (see Figure 13.3). Flight height and FoV together determine the width of the strip.

The ranges are obtained from measuring the travel time of laser pulses forth and back. Their transformation to 3D coordinates in a reference frame requires – at the moment of firing the pulse – measurement of: (1) the 3D coordinates of the position of the laser sensor and (2) its orientation in 3D space defined by three rotation angles. In addition to the six parameters of position and orientation, the scan angle has to be measured. Initially, most ALS detected the first returns. Forestry became an important field of application and foresters found that pulses not only reflected off the leaves of trees (first return) but could also reach the ground through gaps between the leaves, enabling the generation of DTMs. The difference between the last and first returns provides a measure on the height of the tree and ultimately the forest stand as a whole. For deriving more information, returns between the first and last return were also measured, resulting in full waveform digitization (FWD). The principles of multiple return recording and FWD digitization are sketched in Figure 13.9.

Along with forestry, the mapping of aboveground power line corridors was one of the first ALS applications. Trees can grow so high that encroachment from wires and towers can cause short circuits. When long corridors pass through rugged, wooded areas where ground surveys are cumbersome, ALS offers an alternative to ground surveys.

Until 2015, the use of ALS was mainly confined to manned flights. Lidar sensors used to be heavy – in the order of tens of kilograms – and consume a lot of power, which constrains their use on small platforms. In parallel with the increasing miniaturization of unmanned airborne systems (UAS), airborne Lidar sensors became less heavy and bulky (Figure 13.10). Some weigh less than one kilogram. Lidar from UAS is gaining popularity

Airborne and ground-based laser scanning

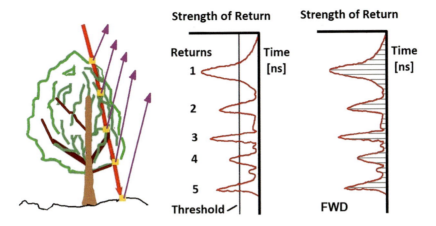

Figure 13.9 Principles for capturing of multiple returns and full-waveform digitization (FWD) of a tree

Source: Lemmens (2023).

Figure 13.10 Miniaturization of laser scanners allows UAS Lidar; top shows two examples (left: Velodyne Lidar Puck VLP-16, right: RIEGLVUX-1); bottom shows YellowScan Mapper in a UAS survey

for obtaining top viewpoint clouds of small sites, including factories, power plants, archaeological sites, office buildings and open pit mines.

Ground-based laser scanning

Ground-based Lidar may be operated in a static way, that is, the platform – usually a tripod – remains in a fixed, stationary position (Figure 13.11). This is called terrestrial laser scanning (TLS). The laser scanner can also be mounted on a car, van, boat, trolley or other platform moving on the ground, including the human back and a stick held by a human hand. In these dynamic cases, ground-based Lidar is called mobile laser scanning (MLS), or, when cameras are also part of the system, it is called mobile mapping system (MMS).

Scan mechanism

To act as a measurement device, the beams must be distributed across the scene in a regular, controlled manner. The dispersion can be done by an oscillating or rotating mirror, or by a solid-state device, which has no mechanical parts. The angular velocity of revolving mirrors may show anomalies, reducing the location precision. Therefore, regular calibration of the laser unit is required, which adds to surveying costs. The most common systems use revolving mirrors, which is the focus here.

Regular dispersion is achieved by a constant angular velocity of the mirror. The laser beams must be distributed in two directions, which are more or less perpendicular to each other. For TLS, these will usually be the vertical and horizontal direction. The dispersion across the vertical direction is achieved through a rotating or oscillating mirror. The regular distribution in horizontal direction is done by a second revolving mirror placed orthogonally to the first mirror or by rotating the TLS instrument stepwise around its vertical axis. Also in most MLS systems, the laser beam distribution is implemented in one direction by a revolving mirror. The movement of the platform causes the distribution in the other direction.

Figure 13.11 Three commercially available terrestrial laser scanners, mounted on tripods, at work for tunnel inspection (left), on a construction site (centre), and for glacier monitoring in Central Chile

Sources: Individual manufacturers as listed; courtesy right image: Geocom, Alexis Caro.

Point density

Unlike images, the scene is not captured at an instantaneous moment defined by the exposure time. In principle, the laser beams are emitted and recorded one after the other. In today's systems, multiple laser beams are usually emitted simultaneously, for example 32, 64 or 128 concurrent beams, and these sensors are called multi-beam Lidar systems (Figure 13.12).

The scanning mechanism may cause artefacts in dynamic scenes. For example, when pedestrians or cyclists cross the scene, they may not be recognized as people but as spikes. If such artefacts are undesirable, the surveyor must close the scene area to all traffic, which may require permission from the local authorities. In addition to the point accuracy, which is at millimetre level for phase-shift systems and at the (sub-)centimetre level for pulse-based systems, point density is another important quality indicator. Because the laser beams are distributed over 3D space at constant angles and/or increments, the point density decreases with increasing distance of the object from the laser scanner. When the distance increases from 10 m to 20 m – i.e., the distance becomes two times greater – the point density decreases approximately four times. If the distance increases from 10 m to 100 m – i.e., the distance increases ten times – this number decreases to 100 times. The point density thus decreases more or less quadratically with distance. In addition, the point density decreases with distance because the strength of the return decreases with distance; the return will be too weak to provide a reliable range, or no return will be received.

Figure 13.12 Intensity image of a TLS scan of Delft's historic centre where passing pedestrians and cyclists cause spikes and occlusion. Courtesy of Delft University of Technology, The Netherlands

Terrestrial laser scanning (TLS)

Terrestrial laser scanning has become a preferred surveying technique for many applications. The high accuracy paired with measurement speed and high point density has boosted its popularity. A TLS system is often equipped with an on-board calibrated camera, which can simultaneously generate textured 3D models. When objects are large, such as open pit mines, multiple set-up positions are needed and the resulting multiple scans must be transformed into a single local reference frame. This computational step, called registration, requires corresponding (tie) points in the scan overlap. Usually the geometric transformation model (GTM) is based on a rigid body. Subsequently, the local system must be transformed into a common reference frame to allow combination with other geodatasets. This process, called georeferencing, requires ground control points (GCP), of which the 3D coordinates refer to the common reference frame, using a seven-parameter 3D similarity transformation. Both registration and georeferencing are conducted based on least squares adjustment (LSA) using more GCPs than necessary to calculate the transformation parameters. Redundancy adds to precision and the ability to detect outliers.

The identification of tie points and GCPs can be done by placing signalized targets, called markers, in the scene. Also (parts of) scene objects, called landmarks, which stand out strongly against the background can be used, and their detection can be done manually or automatically. The 3D coordinates of GCPs can be measured with a total station or a GNSS receiver. Instead of measuring the coordinates of markers or landmarks, the coordinates of the TLS set-up positions can be measured. The signalized set-up points are measured with a total station or GNSS receiver. Permanent signalization of set-up positions is feasible when the scene has to be captured regularly. The coordinates of the TLS position can also be measured by mounting a GNSS receiver on top of the instrument (Figure 13.11, right).

TLS thus measures 3D polar coordinates (r, α, β): range, vertical angle and horizontal angle. The polar coordinates refer to the frame of the TLS device. Their georeferencing to X, Y, Z coordinates in a reference frame requires three steps:

1 Transformation of (r, α, β) coordinates to x, y, z coordinates (Figure 13.13)
2 Registration: bringing the TLS centred x, y, z, coordinates in a local reference system using tie points
3 Georeferencing: transformation of the local reference system to a common reference frame

The first step is done through the geometric transformation model (GTM):

$$\begin{pmatrix} x \\ y \\ x \end{pmatrix} = r \begin{pmatrix} \cos\alpha\cos\beta \\ \sin\alpha\cos\beta \\ \sin\beta \end{pmatrix}$$

The second and third steps are performed via a 3D similarity transformation, also called a Helmert or seven-parameter transformation, with the general form:

$$\begin{bmatrix} X \\ Y \\ Z \end{bmatrix}^G = \lambda \mathbf{R} \begin{bmatrix} X \\ Y \\ Z \end{bmatrix}^L + \begin{bmatrix} X_0 \\ Y_0 \\ Z_0 \end{bmatrix}$$

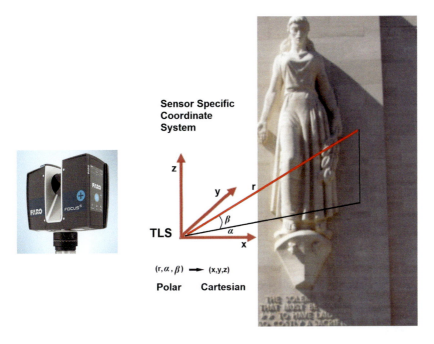

Figure 13.13 Relationship between TLS centred polar coordinates and Cartesian coordinates

The superscripts G and L refer to the two different 3D Cartesian reference frames; λ is the scale factor, which is one in the registration part; R is a 3 × 3 matrix containing sine and cosine functions of the three rotation parameters; and X_0, Y_0, Z_0 are the three translation (shift) parameters.

Mobile laser scanning (MLS)

Around the year 2003, MLS systems using a car or van as platform became operational for surveying and 3D mapping of road scenes. This technology has undergone a steady and rapid development ever since. In contrast to TLS, the platform carrying the system is continuously on the move. As with ALS, registration and georeferencing requires continuous measurement of the position and orientation of the Lidar sensors. To acquire laser points, images and positioning and orientation data, an MLS integrates one or more laser scanners, one or more digital cameras and a positioning and orientation system (POS) that integrates a GNSS receiver and IMU (Figure 13.14).

The main device for measuring orientation is the IMU. Often, odometers (wheel rotation counters), which can be used to calculate speed and distance travelled, complement the POS part. The offset between IMU and GNSS antenna must be measured through a calibration procedure, usually performed by the manufacturer. Vibrations during the survey, shocks from holes in the road surface and sudden deceleration cause mutual displacement of the sensors and other disturbances. Regular recalibration is a necessity to ensure accurate measurements at all times.

In cities, GNSS signals will often be blocked during an MLS survey due to tall buildings (urban canyons), tunnels, treetops and noise barriers. Satellite outage or low signal reception deteriorates the positioning precision of GNSS. The main remedy is to use the IMU by

Figure 13.14 Basic system components of a mobile mapping system, the components of which are rigidly assembled in a rack and mounted on a car; the system shown is the Trimble MX-9

Source photo: The manufacturer; description of the components: author.

Figure 13.15 NavVis 3D Mapping Trolley equipped with laser scanners, cameras and a control and data recording unit for indoor 3D mapping (left); Leica Pegasus backpack equipped with GNSS receiver, five cameras and two laser scanners mounted on a carbon-fibre chassis (centre); and Handheld GeoSLAM ZEB-REVO mobile 3D scanner

Source: Left: image source Trolley: NavVis; indication of system components by author. Centre: Leica Geosystems; indication of system components by author.

integration of the accelerations twice, where the first integration leads to velocity, the second to positioning. However, the precision of IMU positioning rapidly degrades with time due to drift, so that only short GNSS outages can be bridged. The apparatus used to combine GNSS and IMU measurements is based on Kalman filtering (see Brown and Hwang, 1997; Grewal and Andrews, 2000; Kim, 2011). Another remedy is to use GCPs distributed along the survey trajectory. The two remedies are usually combined to increase reliability and precision. Placing and measuring GCPs with a total station or a GNSS receiver is both tedious and labour-intensive.

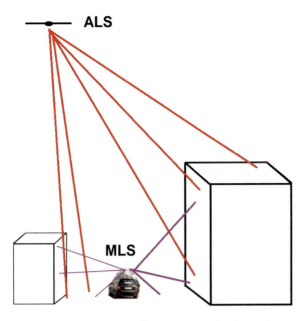

Figure 13.16 Comparison on data collection characteristics of airborne laser scanning (ALS) and mobile laser scanning (MLS)

Manufacturers are increasingly producing small MLS systems that can be mounted on a trolley, placed in a backpack or attached to a pole for handheld laser scanning (Figure 13.15). Indoors, GNSS positioning of the MLS is severely hampered or impeded by the blocking of the GNSS satellite signals. The solution is found in simultaneously determining the location of the platform and creating a map exploiting on-board sensor data. Simultaneous localization and mapping (SLAM) solutions for the 3D mapping of spaces rely on sensors, including odometers, IMU and laser rangers. A position paper and tutorial on SLAM is provided by Cadena *et al.* (2016).

Comparing ALS and MLS

Airborne and mobile laser scanning platforms are both dynamic systems equipped with similar ranging, positioning and orientation sensors. Georeferencing their point clouds is similar. Both can capture extended scenes in a relatively short time. However, they have different data capture abilities due to different viewpoints (perspectives) and distance to objects. To obtain insight in the abilities of ALS and MLS we conducted a comparison on their data capture characteristics (Figure 13.16).

The distance of ALS to the scene is large. The height above ground level (AGL) is usually 1 km or higher. As a result, point density is almost the same everywhere and details at decimetre level are not recognizable. Depending on flight tracks, flight directions and strip overlaps, ALS pulses may hit facades and other vertical objects at one side of the road but miss them on the other. ALS has a top view and thus hits building roofs, which remain invisible for MLS scans. The distance of MLS to objects is much shorter. The distance to road surfaces is a few metres to tens of metres. Fine details such as cracks can be captured. Pilot studies have demonstrated that combining MLS point clouds with orthoimagery (ground sample distance (GSD) 1 cm) allowed the mapping of ruts and potholes with a depth of 2 cm and larger (Schwarzbach, 2014). Cracks can even be automatically detected (Chen and Li, 2016). Road markings can also be easily mapped (Ye *et al.*, 2020). The point density decreases with increasing range. For MLS, the ranges vary widely and the point densities vary accordingly. Only the first object in the line of sight of a laser pulse can be captured, except for tree foliage. So only roadside objects are visible in an MLS point cloud; most other objects are occluded. This means that (part of) an object is invisible due to the presence of another object in the sensor's line of sight.

Mobile laser scanning is particularly suited for mapping road furniture, including road signs, lampposts and road markings, and mapping road pavement damages and building facades. The latter makes MLS, equipped with cameras, suitable for completing 3D city models obtained from ALS and (oblique) photogrammetric images. When creating the 3D city model of Istanbul, Baskaraca *et al.* (2019) recognized that to increase the level of detail and to cover facades with image textures, these should be recorded with laser scanners and cameras mounted on a car and – for narrow streets – on a backpack.

Concluding remarks

Laser scanning is a 3D mapping technology with an almost infinite number of applications. Users can purchase their own equipment for laser point cloud collection, which can be beneficial when regular surveys are required. Laser point cloud collection can also be outsourced to specialized surveying companies. This is a practical solution when large areas need to be captured. Increasingly (processed) point cloud data is available as open data, such as the DTM shown in Figure 13.2. The rate at which laser beams are fired continues to soar and for several commercial systems the emission rate has reached over 1 million beams per second. It can be expected that the beam emission rate will continue to rise gradually over time. The number of professionals appreciating the point clouds produced will increase accordingly.

All geo-sciences and geo-management tasks can benefit from laser scanning. The greater the benefit, the more craftsmanship required. Lemmens (2023) provides in-depth treatise on airborne and ground-based Lidar, including georeferencing, ground filtering, feature extraction and applications.

References

Baskaraca, P., Buyuksalih, G. and Rahman, A.A. (2019) "3D City Modelling of Istanbul – Issues, Challenges and Limitations" *GIM International* 33 (1) pp 16–19.

Brown, R.G. and Hwang, P.Y.C. (1997) *Introduction to Random Signals and Applied Kalman Filtering* (3rd ed.) New York: Wiley.

Cadena, C., Carlone, L., Carrillo, H., Latif, Y., Scaramuzza, D., Neira, J., Reid, I. andLeonard, J.L. (2016) "Past, Present, and Future of Simultaneous Localization and Mapping: Towards the Robust-perception Age" *IEEE Transactions on Robotics* 32 (6) pp.1309–1332.

Chen, X. and Li, J. (2016) "A Feasibility Study on Use of Generic Mobile Laser Scanning System for Detecting Asphalt Pavement Cracks" *The International Archives of the Photogrammetry, Remote Sensing and Spatial Information Sciences* XLI-B1 pp.545–549.

Grewal, M.S. and Andrews, A.P. (2000) *Kalman Filtering: Theory and Practice* (2nd ed.) New York: Wiley.

Hopkinson, C. (2007) "The Influence of Flying Altitude and Beam Divergence on Canopy Penetration and Laser Pulse Return Distribution Characteristics" *Canadian Journal of Remote Sensing* 33 (4) pp.312–324.

Kim, Ph. (2011) *Kalman Filter for Beginners: With MATLAB Examples* North Charleston, SC: Createspace Independent Publishing Platform.

Koechner, W. (1992) *Solid-state Laser Engineering* (3rd ed.) Berlin: Springer.

Lemmens, M. (2023) *Introduction to Pointcloudmetry – Point Clouds from Laser Scanning and Photogrammetry* Dunbeath, Scotland: Whittles Publishing.

Li, Y., Yong, B., van Oosterom, P., Lemmens, M., Wu, H., Ren, L., Zheng, M. and Zhou, J. (2017) "Airborne LiDAR Data Filtering Based on Geodesic Transformations of Mathematical Morphology" *Remote Sensing* 9 (11) 1104 pp.1–21.

Lichti, D. and Skaloud, J. (2010) "Registration and Calibration" In Vosselman, G. and Maas, H-G. (Eds) *Airborne and Terrestrial Laser Scanning* Dunbeath, Scotland: Whittles Publishing, pp 83–133.

Schwarzbach, F. (2014) "Road Maintenance with MMS – Accurate and Detailed 3D Models Using Mobile Laser Scanning" *GIM International* 28 (7) pp.34–37.

Shan, J. and Toth, C.K. (Eds) (2018) *Topographic Laser Ranging and Scanning: Principles and Processing* (2nd ed.) London: CRC Press.

Siegman, A.E. (1986) *Lasers* Mill Valley, CA: University Science Books.

Silfvast, W.T. (1996) *Laser Fundamentals* Cambridge: Cambridge University Press.

Toth, C.K. (2009) "Strip Adjustment and Registration" In Shan, J. and Toth, C.K. (Eds) *Topographic Laser Ranging and Scanning: Principles and Processing* London: Taylor & Francis, pp 235–268.

Wilson, J. and Hawkes, J.F.B. (1987) *Lasers: Principles and Applications (International Series in Optoelectronics)* Upper Saddle River, NJ: Prentice Hall.

Xu, B., Li, F., Keshu, Z. and Lin, Z. (2008) "Laser Footprint Size and Pointing Precision Analysis for Lidar Systems" *The International Archives of the Photogrammetry, Remote Sensing and Spatial Information Sciences*, XXXVII-B1 pp.331–336.

Ye, C., Li, J. Jiang, H., Zhao, H., Ma, L. and Chapman, M. (2020) "Semi-automated Generation of Road Transition Lines Using Mobile Laser Scanning Data" *IEEE Transactions on Intelligent Transportation Systems* 21 (5) pp.1877–1890.

14
DRONES AND UNMANNED AERIAL VEHICLES (UAVs)

Faine Greenwood

A 'drone' is an unmanned aerial vehicle that is guided by a remote-control system or with the assistance of a computer. This seemingly conceptually simple object – an airplane without a pilot – has become loaded with considerable social, political, and technical meaning since the first true drones (or UAVs) took to the skies in the early twentieth century. The arrival of inexpensive civilian drone technology has intensified social, political, and academic debate over what drones are, the information they collect, and where they fit into our societies. In the 2010s and 2020s, drones and their rise have been depicted in the public imagination as physical, visible representations of technological overreach and continuous surveillance (Liang and Lee, 2017). This chapter, which will focus primarily on civilian, or non-military drones, will briefly summarize how drones are defined, where they came from, how they work, and what they are used for today.

Defining drones

The popular, informal term 'drone' is used to describe an extremely broad category of unmanned vehicles, encompassing both miniscule drugstore toys and 12-ton military surveillance aircraft. The more formal terms 'unmanned aerial vehicle' (UAV), 'unmanned aerial system' (UAS), and remotely piloted aircraft (RPAS) are more often used by industry professionals, in government, and in academia, often somewhat interchangeably. The words used to describe unmanned aerial vehicles are of considerable importance to those who work with the technology, who have long reported frustration with the public's military associations with the word 'drone' (Clothier et al., 2015). Today, there are still no broadly agreed-upon terms that differentiate military from civilian drones. While most people are familiar with the distinction between a battleship and a canoe under the category of 'boats', there is no such public consciousness around drones. This lack of clarity around the differences between military drones and consumer drones poses considerable problems for the industry, as it attempts to foster public acceptance of the technology (Schulzke, 2018).

Military drones are, perhaps most importantly, aircraft that are not available for purchase by consumers: they are aircraft designed with military uses in mind. Some of the most widely known – and feared – military drones fit into this category, such as the immense General Atomics MQ-9 Reaper. The aforementioned MQ-9 Reaper has a range of 1,000 nautical miles, weighs 2,223 kilograms (when empty), and can cruise at a speed of 200 knots; it carries four Air-to-Ground (AGM)-114 Hellfire missiles (United States Air Force, 2015).

Some sources refer to drones that carry aircraft ordnance as UCAVs, or 'unmanned combat air vehicles' to more clearly differentiate them from military UAVs that only carry surveillance equipment (Gilli and Gilli, 2015). While the Trump administration cancelled the Obama-era policy of tracking US government drone strike deaths in 2019, a number of non-government organizations collect figures on the death toll (Dilanian and Kube, 2019). As of February 2020, the Bureau of Investigative Journalism estimated that US drones have killed between 8,845 and 16,794 people in Afghanistan, Somalia, Yemen, and Pakistan, with an estimated civilian death toll ranging from 910 to 2,181. The US military also occasionally uses large military drones for surveillance within the country; in May 2020, Customs and Border Protection used a Predator drone to monitor police brutality protests in Minneapolis (McKinnon and Hackman, 2020).

Civilian drones are, as the title implies, small, short-range drones designed for civilian purposes and are either sold directly to consumers, or are built by hobbyists from component electronic parts. The world's most popular consumer drone models, such as the Chinese manufacturer DJI's camera-equipped, foldable, '"Mavic' quadcopter series falls into this category. It also encompasses more sophisticated drones designed for professionals, such as the multispectral-camera equipped Sensefly eBee X, a fixed wing drone that costs $19,800 and is largely marketed as a mapping tool to surveyors, farmers, and other professionals (SenseFly, 2020).

There are, indisputably, grey areas between civilian and consumer drones. Both militaries and armed non-state actors alike have used consumer-intended drones in the recent past, including the US and Israeli militaries and ISIS. In the case of the US military, concerns over the security of Chinese-made drone systems prompted the US Department of Defense to ban the use of consumer, commercial drones by the military in 2018 (Gibbons-Neff, 2017; Newman, 2017). In some cases, drones designed for military purposes have been used to support humanitarian efforts, drawing criticism from aid workers concerned these technologies will sow confusion and harm the public's perception of aid worker neutrality.

Ultimately, these issues revolve around a single question: is a civilian drone *more* of an inherently military-oriented object than other consumer electronics that can also be appropriated for military purposes, such as mobile phones and digital cameras? The civilian drone industry argue that their technology is not inherently more menacing or privacy-violating than an iPhone or a Jeep is; whether the public will come around to this way of thinking remains to be seen.

How drones work

The most important feature that sets a drone apart from other aircraft is its ability to fly autonomously – that is, to operate without a human pilot in the cockpit. Today's long-range military drones rely upon complex computerized systems and long-range communications that enable them to function with minimal latency from thousands of miles away from their operators (Sayler, 2015). Meanwhile, small, civilian drones are, in essence, model aircraft that have been fitted with computers and small sensors, which became more widely available after the rise of the smartphone in the 2010s. Modern drones all have some onboard computing capability, via the flight controller, that permit a limited form of autonomy above and beyond the creation of flight routes. This includes features that allow the drone to avoid obstacles and follow moving objects (both to a rather limited extent) and to land without direct human input.

The civilian drone industry splits the technology into two categories: multirotors (or helicopter-style drones). and fixed-wings (which resemble a traditional model airplane). These aircraft use computerized flight controllers as a central 'brain', which controls the functions of the aircraft's other electrical components, including propeller speed and aircraft direction.

Drones are usually equipped with a global positioning system (GPS) receiver that permits it to know where it is in space. The GPS works with the flight controller to enable drone pilots to design pre-determined flight routes the drone is capable of automatically following. Most mid-range consumer drones today are marketed primarily as tools for collecting data, most commonly photographs and video: they carry cameras beneath them using a stabilization tool known as a 'gimbal'.

A brief history of drones

Ideas and concepts that can be linked to the drones of today date back to pre-modern times. Literature from ancient Chinese and Greek sources describes inventors who crafted 'mechanical birds' capable of moving on their own, as well as kites and balloons that could deliver messages (Olshin, 2019). The notion of using a secondary object or vessel separate from the body to 'see' beyond the limits of the naked eye is not a new idea. The Norse god Odin, as one example, had two ravens that he deployed as spies, and a number of ancient fairytales refer to the existence of objects that allow their owner to virtually 'view' anything that is happening in the world.

The invention of the hot air balloon in the sixteenth century ushered in the era of human flight, and led to newfound interest in the possibilities of the aerial or cartographic gaze (Specht and Feigenbaum, 2018). The French artist and balloonist Nadar is credited with taking the first photograph from a balloon in 1858, although, sadly, the image does not survive today. The earliest surviving aerial photograph was taken by J.W. Black in a balloon over Boston in 1860, and the Union Army used balloons to survey Confederate positions during the Civil War (Hauser, 2012).

The first true unmanned aerial vehicles can be tracked to the early twentieth century. The notion of an aircraft capable of flight without placing a human being at risk was attractive from the outset of the aviation era, and inventors began to grapple with the problem soon after the Wright Brother's first flight at Kitty Hawk in 1903. In 1914, aviator and inventor Lawrence Sperry successfully demonstrated the first gyroscopic autopilot, a device that permitted an aircraft to maintain a stable altitude and compass heading (Davenport, 1978) (Figure 14.1).

During the First World War, American, British, and German researchers began to experiment with radio-controlled, unmanned aircraft, inspired in large part by the inventions of Nikola Tesla, who publicly demonstrated a wireless radio-controlled boat (which he dubbed the 'Teleautomaton') at a demonstration in 1898 (Marincic and Budimir, 2008). In January 1916, Archibald Low of the UK Royal Flying Corps demonstrated a radio-controlled aircraft that could be both filled with explosives and used as an aerial target. During the same year, US Navy inventors introduced the similarly intended 'Hewitt-Sperry Automatic Airplane' (Whitmore, 2016; Benchoff, 2019).

In 1933, the Royal Air Force successfully tested a remote-controlled version of the de Havilland Tiger Moth aircraft known as the 'Queen Bee', which it used for target practice in exercises against the Home Fleet. This aircraft's name and its consistent, droning sound likely inspired the first use of the word 'drone' to describe an unmanned aerial vehicle (Keane and Carr, 2013).

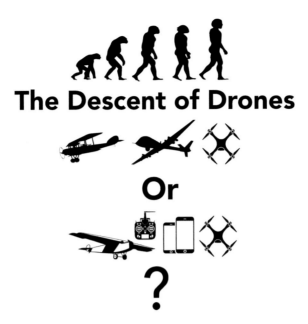

Figure 14.1 The descent of drones

The first drones that brought together both autonomous flight and remote sensing technology date to 1938, when the US Navy began talks with the Radio Corporation of America (RCA) regarding the potential use of television technology as a means of providing drone pilots with a real-time view from their aircraft (Clark, 1999).

By 1941, British actor Reginald Denny's Radioplane company was supplying the US Army Air Corps with flying aerial targets. That year, German inventors began to refine the self-guided, pulse-jet powered aircraft that would eventually become the infamous V-1 flying bomb (Zaloga, 2011). The US military continued to experiment with the use of drone technology for surveillance and decoy purposes after the close of the Second World War; camera-carrying surveillance drones and decoy drones were used during both the Korean and Vietnam Wars. During the Yom Kippur war in 1973, the Israeli military used UAVs for reconnaissance and as decoys against Egyptian and Syrian forces (Sloggett, 2014).

Israeli inventor Abraham Karem's GNAT 750 surveillance UAS was used by the CIA during peacekeeping operations in Bosnia and Herzegovina from 1993 to 1994. The GNAT was soon followed by an improved version dubbed the General Atomics RQ-1 Predator, which boasted a live-satellite link. The US military used the Predator in operations over Kosovo in 1999 and Afghanistan in 2000 (Connor, 2018). By autumn of 2001, the US military had deployed the first weaponized, armed long-endurance Predator drone – equipped with Hellfire missiles – over Afghanistan, ushering in a new era of 'hunter killer' drone technology. In 2007, the Predator was joined by the General Atomics MQ-9 Reaper, a larger aircraft meant to supplant the Predator on the battlefield (United States Air Force, 2015). As of March 2020, an estimated 102 nations have active military drone programs (Gettinger, 2020a).

Contrary to popular belief, small consumer drones are not the *direct* conceptual or intellectual 'descendants' of long-range, weaponized military drones like the Predator or the Reaper; their origins are more directly traceable to the model aviation hobby.

While inventors have tinkered with model aircraft for hundreds of years, the hobby's formalization dates to the mid-nineteenth century in Europe, at the height of the ballooning craze. In 1871, Alphonse Penaud displayed his 'Planophore' at a Royal Aeronautical Society event – the first heavier-than-air model aircraft, powered by a rubber-band operated propeller (Gibbs-Smith, 2003). Nineteenth-century hobbyists also experimented with aerial photography, using unmanned balloons, kites, and even pigeons as lifting apparatus.

By the 1930s, modellers were building gas-powered model aircraft capable of free flight (Mueller, 2009). The first flight by a radio-controlled model aircraft likely took place at 1936 competition in Germany, although this achievement is less well known than that of twin brothers William and Walter Good, who demonstrated their gas-powered, remote-controlled aircraft at a college science fair in January 1937 (Mueller, 2009).

Electric-powered RC aircraft were introduced in 1957 by Great Britain's H.J. Taplin, while West Germany's Dieter Schluter demonstrated the first remote-controlled model helicopter in 1968. Enthusiasts continued to develop more sophisticated flight control systems for hobby helicopters in the following decades, concepts that would eventually lead to the computerized control systems used in the small drones of today (Cai *et al.*, 2010).

Filmmakers began to use camera-carrying model helicopters in the 1980s: the decade also saw the first use of remote-controlled, computerized helicopters for agriculture in Japan, with the Yamaha R-MAX aircraft. In 1989, Japanese toymakers released what was likely the first known toy quadcopter, the Keyence Gyrosaucer II E-570. While it was only capable of flying indoors for two to three minutes, the design presaged the dominant look and features of multirotor drones today (Darack, 2017).

In the 1990s, Japan used Yamaha R-MAX helicopters to take photographs of volcanic eruptions, in one of the first known uses of drone technology for civilian disaster response purposes. In 1991, the first 'international aerial robotics competition' for unmanned systems took place at the Georgia Institute of Technology, an event that has continued to the present (Cai *et al.*, 2010).

By the late 1990s, some model aircraft builders had begun to experiment with using television technology to stream video from their model aircraft back to screens on the ground, an early version of the high-latency 'first-person video' or FPV technology used on most drones today (Rivera, 1990). A number of hobbyists, including the future founders of now-ascendant DJI, developed better multirotor designs, largely selling their creations to hobbyists and specialized filmmakers. Researchers from CRASAR used a fixed-wing drone to capture photographs and video of destruction from Hurricane Katrina in 2005, although US government regulators quickly stepped in to ban the use of drones in US airspace until further notice (Murphy & Baraniuk, 2016).

During the 2000s, some police departments in the United States and in Europe began to acquire small camera-carrying drones, although regulations, public distrust, and technical challenges limited their practical use by law enforcement (Finn, 2011). Parrot's AR. Drone in 2010 was released as a toy: it carried a camera and could be flown for limited distances via a linked Apple iPhone app. Unlike the model aircraft that preceded it, it required no skill to fly, and could shoot (rather grainy) photographs and videos out of the box.

At the beginning of 2013, DJI released the Phantom drone, a white quadcopter that was much more capable than Parrot's AR Drone model – and could also be flown out of the box by the inexperienced. The DJI Phantom drone could carry a GoPro camera, and purchasers quickly began publishing impressive videos shot with the devices to YouTube and other online platforms. Reporters used DJI Phantom drones (alongside other consumer and home-built models) to shoot video of news events later that year, including mass protests in

Thailand and scenes of devastating after Typhoon Haiyan in the Philippines. In late 2013, Amazon CEO Jeff Bezos released a video claiming the company would soon use drones to deliver packages to customer's doors, introducing drone delivery to the public imagination.

As of 2020, consumer drones have become an almost-mainstream technology. By March 2020, the US Federal Aviation Administration reported a total of 1,563,263 registered drones (both commercial and recreational), while the UK Civil Aviation Authority estimated there were 90,000 drone users in the country in late 2019 (Federal Aviation Administration, 2020; Loughran, 2019). Many nations have introduced comprehensive laws around drones, and are now considering the problem of 'UTM' or 'unmanned traffic management' – a set of technologies and regulations that will fully integrate drones into the air traffic management systems that govern the movements of manned aircraft today.

What drones do

Civilian drones are widely used in many industries and sectors around the world. The vast majority of practical drone uses today center around using the drone to carry sensors capable of collecting some form of *information*, or data – most typically (though by no means exclusively). photographs or video. In this respect, a drone is simply a flying camera, or a mobile platform for a given type of sensor. A number of popular consumer drones, such as the DJI Matrice, can be used with different sensors: the user simply has to take one off and put another one on.

The vast majority of consumer, civilian drones today are capable of taking photographs and video using standard electric optical (EO) sensors. The quality of these sensors differs dramatically, from grainy images shot by $50 drones that can be purchased from toymakers, to ultra-high resolution, 4K still photographs and images captured by costlier models.

Drones that carry multispectral sensors, capable of capturing imagery from across the electromagnetic spectrum, are most widely used in agriculture, as farmers derive considerable value from using multispectral imagery to capture detailed information about crop health and moisture: they can repeat these flights daily or multiple times a day if they so wish, giving them highly up-to-date information on their farms. Archaeologists use multispectral equipped drones to search for hard-to-see indicators of hidden structures beneath the ground, while environmental researchers use it to identify plant species and to quantify plant health.

Light detection and ranging technology, or Lidar, is widely used on manned aircraft: smaller, lighter sensors that can be used with drones are beginning to come to market, although they are still extremely expensive. These sensors use lasers to very accurately capture spatial information on the ground beneath it; a Lidar sensor mounted on a small drone can operate in tight and constricted areas that a Lidar-carrying helicopter or aircraft could not function in.

Drone-collected imagery can be stitched together and geographically referenced to produce maps, in a process called photogrammetry. Today, most consumer drones carry onboard GPS receivers, which assign each photograph a specific spatial location. Drone pilots who wish to make a map can use flight-planning software that gives the drone a specific, tightly controlled flight plan, which usually resembles a lawnmower or grid pattern. As the drone flies this pattern, it takes a photograph every few seconds, ensuring that all photographs adequately overlap with one another.

A number of companies now offer software that is capable of processing these photographs into orthomosiacs (geographically accurate pictures or maps), often using cloud-based computing. There are also free, open-source software packages available. These software packages resize the overlapping photographs taken by the drone and 'stitches' them together, eventually resulting in a high-resolution photographic map.

While delivery drones have captured the public imagination, the use of drone technology for these purposes remains experimental. As of this writing, most drone delivery experiments today are collaborations between private companies and governments, who afford drone-makers special privileges to operate cargo-carrying drones in their national airspace. In one example, UNICEF has collaborated with the government of Malawi to open a drone testing corridor in the country, which private companies and researchers can apply to use.

The Silicon Valley-based Zipline company has come closest to creating an operational drone delivery system with its network of medical-supply delivery drones in Rwanda: investors had valued the company at $1.2 billion as of May 2019 (Kolodny, 2019). Zipline intends to bring the technology to the United States in the near future, in what co-founder Keller Rinaudo described as a 'trend of seeing technology developed in other countries and brought here' (Krouse, 2019).

Who uses drones

Drones have been widely adopted by many sectors around the world, and new use-cases are constantly emerging. Mapmakers and geographers have swiftly adopted the technology, using it as an inexpensive means to capture geographically accurate and extremely high-resolution imagery of the Earth without the cost or constraints of satellite imagery and manned aerial photography. GIS specialists use drone imagery to support a wide range of mapping operations, from agricultural health analysis to disaster response image classification, while GIS software makers like Esri now smoothly integrate drone data into their products.

Drone imagery is now widely used to support participatory mapping exercises, such as those carried out by the Humanitarian OpenStreetMap Team during a disaster, which once relied upon satellite imagery or aerial photographs. Indigenous rights groups in a number of nations, including Indonesia, Panama, and Guyana, have used drone mapping to document their land holdings and to defend their land rights against interlopers (Smith, 2017). Humanitarian aid workers and disaster responders increasingly rely upon drone technology as an inexpensive and swift means of gathering up-to-date information on crisis situations, planning post-disaster recovery efforts, and evaluating potential disaster risks (Greenwood and Joseph, 2020).

Construction workers and infrastructure specialists use drones for a wide variety of monitoring and surveillance tasks, including monitoring the progress of construction sites over time and creating detailed elevation maps and models of prospective project areas. Transportation authorities now regularly use drones to inspect roadways and bridges, while power and communication utilities use them to inspect power lines, telephone lines, and masts while minimizing risk to human inspectors (Lercel *et al.*, 2018).

Drones are widely used by professional photographers and videographers. Sweeping drone shots are an increasingly ubiquitous sight in major movies, while freelance photographers use drones to capture impressive shots of weddings and events; many social media influencers regularly rely upon drones for impressive travel photographs. Journalists were early adopters of drone technology, and drone shots have become a regular components of news broadcasts: they are a popular framing tool for mass protests, natural disasters, and other stories that require a clear sense of scale.

Academic researchers and scientists have widely adopted drones to facilitate their work. Duke University's Marine Robotics and Remote Sensing program uses small drones to capture a wide range of data, including counting seal populations, measuring the size of whales, and monitoring beach geomorphology (Seymour et al., 2017). Ecologists use drones to monitor air pollution, map plant cover, and to identify key habitat characteristics for study species. Archaeologists regularly use drones to comprehensively map dig sites and to identify new ones; in 2018, Peruvian researchers identified 50 new examples of the Nazca lines using drone photography.

Wildlife protection programs also use drones monitor illegal poaching. Researchers have also experimented with using drones to impact animal behaviour: researchers are attempting to use drones to train South African rhinoceros to avoid border areas that may expose them to poaching.

Drones are also widely used for entertainment, such as filmed drone races, in which pilots fly tiny and exceedingly fast drones through a series of obstacles with the assistance of FPV or first-person view cameras, which use high-latency video feeds and goggles to permit the pilot to 'see' through the drone's eyes. Since the mid-2010s, drone racing has moved from a somewhat underground activity to a growing industry; televised competitions for large prize purses now take place around the world.

Civilian police forces are increasingly adopting drone technology around the world, largely as a means of collecting situational intelligence. Data from March 2020 found that 1,578 state and local public safety agencies in the United States-owned drones, a significant increase from ownership rates just a few years prior (Gettinger, 2020b). During the 2020 COVID-19 pandemic, police forces in the United States, Italy, India, and China experimented with using loudspeaker-equipped DJI Mavic drones to convey public health information to the public; some forces also attempted to use drones to monitor social distancing practices (with uncertain levels of success) (WeRobotics, 2020).

Regulating and managing drones

The rise of consumer, civilian drones in the 2000s and 2010s prompted nations around the world to develop regulatory systems that encompassed them. One drone-industry source that monitors relevant laws around the world found in October 2019 that 64% of countries in the world have introduced some form of drone law, while about 9% of the world's countries ban the use of the technology completely (Ducowitz, 2019). Most nations' drone regulatory systems set a maximum flight altitude; the United States, the UK, and Canada set this altitude at 400 feet. Most heavily restrict drone flight in close proximity airports or other sensitive areas, and require that pilots maintain 'line of sight' with their aircraft at all times.

Some international organizations are developing overarching technical and regulatory standards for civilian drone operations. The International Civil Aviation Organization (ICAO) is currently developing a set of Model UAS Regulations, which are intended to 'offer a template for Member states to adopt or supplement their existing UAS regulations' (ICAO, 2020). The Joint Authorities for Rule-making on Unmanned Systems (JARUS), an international group of aviation safety experts, works 'to recommend a single set of technical, safety and operational requirements for the certification and safe integration of Unmanned Aircraft Systems (UAS)'. One of its key projects is the SORA (Specific Operations Risk Assessment) document, which provides guidance for national aviation authorities who wish to evaluate the safety of specific types of drone operations (Date, 2017).

Unmanned traffic management (UTM) systems are under development around the world, and are intended to incorporate drones into existing manned air traffic management systems. In 2019, ICAO introduced a new guidance document geared towards states hoping to develop their own UTM systems, with the goal of harmonizing their global development (ICAO, 2019). In late 2019, the United States released a final draft of its new digital ID standards, which will (when implemented) require all drones to be integrated into an interconnected UTM network (Federal Aviation Administration, 2019).

Drone debates: privacy and surveillance

Perhaps the central debate around drones – which concerns both military and civilian sectors – relate to matters of surveillance and privacy. Long range military drones are capable of flying for days on end, and can follow an individual of interest – often without their notice – for long periods, engaging in a practice of so-called 'lethal surveillance' (Kindervater, 2016). Surveillance concerns are also linked to smaller, consumer drones. Fear of an unobtrusive 'eye in the sky', capable of quietly and persistently monitoring an individual's day-to-day activities, have been prominent in public and media discourse since civilian drones first became widespread in the 2010s, in wording that implied that their capabilities were similar to those of far larger military drones.

Civilian drones are not as capable of long-range surveillance as the armed drones used by militaries, a fact that may not be widely known outside of the industry itself – leading to further confusion over their actual capabilities. A battery-powered multirotor drone is usually capable of hovering persistently in one place for approximately 30 minutes, while fixed wing drones are often capable of flight for about an hour. While these smaller aircraft can capture high-resolution images of people or objects on the ground, they cannot quietly monitor a given area for many hours at a time, as long-range surveillance drones like the General Atomics Reaper can do.

Limited public opinion research has been conducted to date centered on non-military drones, and the research that *does* exist is centered on the United States and Europe. A recent US research project found respondents opinion on drone use was heavily dependent on what the drone was being used for: search and rescue drone missions were much more acceptable than recreational drone use (Zwickle *et al.*, 2018).

Although privacy issues are a prominent feature in public debate and discourse around drones, many national and international regulators appear reluctant to take strong positions on the matter. In the United States, the Federal Aviation Administration has declined to create or produce national laws that explicitly address privacy – although the agency still maintains that it has exclusive authority over crafting legislation that addresses airspace, and has moved to strike down local laws that are directed at drone privacy protection in the recent past (Roberson and Nowak, 2017).

In the European Union, the General Data Protection Rule (GDPR) governs the collection and use of drone-collected data, although the meaningful enforcement of these rules remains somewhat in doubt. There is still little regulatory guidance – either on the national or international level – devoted specifically to setting standards around regulating data collection from drones, or that specifically address privacy concerns. Although international aviation organization ICAO is in the process of developing safety standards for civilian unmanned aircraft, it has explicitly avoided issuing public recommendations on drone data privacy.

'Killing by remote control'

The rise of armed, long-range military drones – capable of killing at long distances – have prompted fierce legal and ethical debate since their introduction in 2001. Critics have noted the use of long-range drones could lend warfare a detached, 'video game'-like quality, dehumanizing combatants and making it easier for soldiers to kill other human beings (Gregory, 2011; Wall and Monahan, 2011; Allinson, 2015). A number of advocacy organizations, such as the Campaign to Stop Killer Robots, call for the introduction of international covenants or laws that would outlaw the use of autonomous weapons systems in conflict (Manjikian, 2017).

At the same time, some argue that the technology's ability to 'surgically strike' makes it more humane than older means of aerial bombardment, and that there may even be, in this sense, an ethical duty to use them (Gross, 2015). Others note that remote killing is not actually a *new* development, and that current debates around armed drones focus too much on the technology's novelty, confusing the conversation and failing to consider relevant ethical and regulatory history (Carvin, 2015).

This ethical concern over military drones has been compounded by recent improvements in artificial intelligence technology, and an accompanying increase in public concern over the prospect of aircraft capable of killing without a human 'in the loop' (Liu, 2012). In 2018, employees of Google, Amazon.com, and Microsoft protested their company's collaboration with the US government on developing AI systems intended to improve the accuracy of drone strikes (Shane and Wakabayashi, 2018).

The use of AI technologies for facial recognition is a topic of great public concern, as numerous governments and companies have begun to experiment with its use in real-world settings. At the time of writing, there is no clear evidence that facial recognition technology has been successfully used with either military or civilian drones in a real-world setting. The fact that a drone sensor is up in the air and is likely to be constantly moving represents a considerable computational challenge.

International humanitarian law and differentiation

Some research indicates that one factor behind public fear of drones is their inscrutability – in other words, it is very difficult to determine what a drone is doing, why it is doing it, and who it is doing it for (Bajde *et al.*, n.d.). In the absence of systems that enable people on the ground to access information about a drone's identity or intentions, people must come up with their own conclusions about *who* (or *what*, if we take fears over artificial intelligence into account) is looking at them.

In May 2018, the Israeli military used a specially modified DJI S1000 drone – usually used by filmmakers to carry heavy cameras – to drop tear gas canisters over the border into Gaza onto protesters gathered there (Greenwood, 2018). The Israeli military's use of this consumer-intended drone raises a number of ethical questions. It is conventional under international humanitarian law that military aircraft carry identifiable markings, and it could be argued the unmarked Israeli tear-gas drone violated this rule. The fact the drone was a modified version of a model commonly used by civilian photographers is also worrisome: could this have fostered confusion over what it was being used for, and by whom?

During the spring of 2018, Palestinian reporter Yaser Murtaja was shot and killed by Israeli Defense Forces (IDF) soldiers during border protests. Murtaja had used drones to shoot video for his reporting work in the past, although there was no evidence that he was piloting a drone at the time of his death (Greenwood, 2018). Israeli Defense Minister Avigdor Lieberman speculated Murtaja may have been killed because it was unclear if his drone was dangerous or not – inaccurately attributing his death to a case of drone mistaken identity (Kubovich and Zikri, 2018).

Aid organizations have, partially due to these problems, largely avoided using drones as part of their activities in conflict. The UTM systems being developed around the world represent perhaps the most realistic future solution to the problem of drone mistaken identity.

Democratization of the aerial view

Civilian drones have arguably *democratized* the aerial view – or, some might argue in a more negative light, have democratized the practice of surveillance. Prior to the rise of inexpensive consumer and 'DIY' camera drones, access to the aerial view was restricted to a relatively small, elite group of people with the financial resources and time to fly a camera-carrying manned aircraft or to capture and analyse satellite imagery. While the rise of free online satellite services like Google Earth enabled more people to look at the planet from above via the 'cartographic gaze', these systems did not permit their users to dictate where the satellite's camera would be pointed: they remained passive *consumers* of the imagery, not creators of it (Specht and Feigenbaum, 2018). Civilian drones have, for the first time in history, made it relatively easy for the untrained to inexpensively produce extremely high-resolution orthomosaic maps of the Earth's surface, outside of the traditional professional domains of cartography and surveying – an important power shift with implications that we have only just begun to understand.

While many controversies around drones centre around fears of the technology being used in exploitative ways by entities or individuals who already hold power, there are numerous examples of drones being used by less powerful entities to challenge existing power structures. A widely publicized example of this took place during the 2016 protests of the Dakota Area Pipeline in 2016, where Native American activists used drones to capture and disseminate video of police violence against them. Authorities responded aggressively to the activists' use of drones: the FAA placed a restricted flight area around the site, legally barring drone flights, and activist drones were shot down (Smith, 2017; Tuck, 2018). Activist drone pilots were faced with criminal charges for their drone flights, although the charges were eventually dropped.

The new, increasingly restrictive drone laws that many nations have introduced in the interest of safety may also reduce the appeal – or usefulness – of drone technology for activists, journalists, and other groups. For example, the US Remote ID rule, as presently written, will require that a drone pilot's physical location be made publicly available; a measure that marginalized people may find too dangerous to cope with. In the near future, these transformative uses of drones may become increasingly rare, reverting access to the drone-enabled cartographic view towards the powerful – or, perhaps, these drone pilots will again move away from using premade consumer drones, and back towards building their own aircraft from electronic parts that even highly authoritarian states will be hard-pressed to effectively ban.

Conclusion

Drone technology is in a state of transition. These devices have shifted from obscurity to ubiquity in just a few short years, forcing societies to scramble to develop new law and behavioural norms around their use. The rise of inexpensive and easy-to-use civilian drones has placed the power of aerial observation into the hands of a previously unimaginably large number of people – a phenomenon whose practical and ethical implications remain poorly understood. The brief, so-called 'Wild West' period of civilian drone experimentation that began in earnest in the early 2010s is now over, as the public becomes more accustomed to drones and as they become more aggressively regulated by governments. Drone users now must determine how their technology transitions from novelty to a practical part of daily life. What is clear now, at least, is that drones are not going anywhere. They are the compelling realization of an ancient idea: what could we do if we could fly without our bodies, view the Earth from above as casually as the birds do?

References

Allinson, J. (2015) "The Necropolitics of Drones" *International Political Sociology* 9 (2) pp.113–127 DOI: 10.1111/ips.12086.

Bajde, D., Hojer Bruun, M., Sommer, J. and Waltorp, K. (n.d.) *General Public's Privacy Concerns Regarding Drone Use in Residential and Public Areas* Odense, Denmark: University of Southern Denmark; Aalborg University.

Benchoff, B. (2019) "The Evolution Of Drones: A Brief History Of 'Drone'" *AMA Flight School* Available at: https://www.amaflightschool.org/DRONEHISTORY.

Cai, G., Lin, F., Chen, B.M. and Lee, T.H. (2010) "Development of Fully Functional Miniature Unmanned Rotorcraft Systems *Proceedings of the 29th Chinese Control Conference* pp.32–40.

Carvin, S. (2015) "Getting Drones Wrong" *The International Journal of Human Rights* 19 (2) pp.127–141 DOI: 10.1080/13642987.2014.991212.

Clark, R.M. (1999) *Uninhabitated Combat Aerial Vehicles: Airpower by the People, for the People, but Not with the People* Montgomery, AL: Air University, Maxwell Air Force Base..

Clothier, R.A., Greer, D.A., Greer, D.G. and Mehta, A.M. (2015) "Risk Perception and the Public Acceptance of Drones" *Risk Analysis* 35 (6) pp.1167–1183 DOI: 10.1111/risa.12330.

Connor, R. (2018) "The Predator, a Drone that Transformed Military Combat" *National Air and Space Museum* Available at: https://airandspace.si.edu/stories/editorial/predator-drone-transformed-military-combat.

Darack, E. (2017) "A Brief History of Quadrotors" *Air and Space* (May) Available at: https://www.airspacemag.com/daily-planet/brief-history-quadrotors-180963372/.

Date, E. (2017) "JARUS Guidelines on Specific Operations Risk Assessment (SORA)" *Joint Authorities for Rulemaking of Unmanned Systems* Available at: http://jarus-rpas.org/sites/jarus-rpas.org/files/jar_doc_06_jarus_sora_v2.0.pdf.

Davenport, W.W. (1978) *Gyro!: The Life and Times of Lawrence Sperry* New York: Scribner.

Dilanian, B.K. and Kube, C. (2019) "Trump Cancels Obama Policy of Reporting Drone Strike Deaths" *NBC News* (6th March) Available at: https://www.nbcnews.com/politics/donald-trump/trump-cancels-obama-policy-reporting-drone-strike-deaths-n980156.

Ducowitz, Z. (2019) "No Flying Allowed: The 15 Countries Where Drones Are Banned" UAV Coach (25th Febrary) Available at: https://uavcoach.com/drone-bans/.

Federal Aviation Administration (2019) "Remote Identification of Unmanned Aircraft" *Federal Register* 86 (10) pp.4390–4513.

Federal Aviation Administration (2020) "UAS by the Numbers" Available at: https://www.faa.gov/uas/resources/by_the_numbers/.

Finn, P. (2011) "Privacy Issues Hover Over Police Drone Use" *The Washington Post* (23rd January) Available at: https://www.washingtonpost.com/national/privacy-issues-hover-over-police-drone-use/2011/01/22/ABEw0uD_story.html.

Gettinger, D. (2020a) "Drone Databook Update: March 2020" *Centre for the Study of the Drone* pp.1–30 Available at: https://dronecenter.bard.edu/files/2020/03/CSD-Databook-Update-March-2020.pdf.

Gettinger, D. (2020b) *Public Safety Drones* (3rd ed.) Annandale-on-Hudson, NY: Center for the Study of the Drone at Bard College Available at: https://dronecenter.bard.edu/files/2020/03/CSD-Public-Safety-Drones-3rd-Edition-Web.pdf.

Gibbons-Neff, T. (2017) "ISIS Drones are Attacking U.S. Troops and Disrupting Airstrikes in Raqqa" *The Washington Post* (14th June) Available at: https://www.washingtonpost.com/news/checkpoint/wp/2017/06/14/isis-drones-are-attacking-u-s-troops-and-disrupting-airstrikes-in-raqqa-officials-say/?utm_term=.2ca7e3e4f62b.

Gibbs-Smith, C.H. (2003) *Aviation: An Historical Survey from its Origins to the End of the Second World War* Available at: http://eds.a.ebscohost.com/eds/detail/detail?vid=1&sid=25f291a7-fd8d-44de-a28b-622cd0cd7865%40sessionmgr4010&bdata=JnNpdGU9ZWRzLWxpdmUmc2NvcGU9c2l0ZQ%3D%3D#AN=mts.b1494158&db=cat00263a.

Gilli, A. and Gilli, M. (2015) "The Diffusion of Drone Warfare? Industrial, Organizational and Infrastructural Constraints: Military Innovations and the Ecosystem Challenge. *Security Studies* (March) DOI: 10.2139/ssrn.2425750.

Greenwood, F. (2018) "Drones Don't Wear Uniforms. They Should" *Foreign Policy* (22nd May) Available at: https://foreignpolicy.com/2018/05/22/drones-dont-wear-uniforms-they-should/.

Greenwood, F. and Joseph, D. (2020) "Aid from the Air: A Review of Drone Use in the RCRC Global Network" *American Red Cross* Available at: https://americanredcross.github.io/rcrc-drones/.

Gregory, D. (2011) "From a View to a Kill: Drones and Late Modern War" *Theory, Culture & Society* 28 (7–8) pp.188–215 DOI: 10.1177/0263276411423027.

Gross, O. (2015) "The New Way of War: Is There a Duty to Use Drones?" *Florida Law Review* 67 (1) pp.1–72.

Hauser, K. (2012) "Photography and Flight" *History of Photography* 36 (4) pp.463–465 DOI: 10.1080/03087298.2012.712268.

ICAO (2019) "Unmanned Aircraft Systems Traffic Management (UTM) – A Common Framework with Core Principles for Global Harmonization" Available at: https://www.icao.int/safety/UA/Documents/UTM-Framework.en.alltext.pdf.

ICAO (2020) "Model UAS Regulations" Available at: https://www.icao.int/safety/UA/Pages/ICAO-Model-UAS-Regulations.aspx.

Keane, J.F. and Carr, S. S. (2013) "A Brief History of Early Unmanned Aircraft" *Johns Hopkins APL Technical Digest (Applied Physics Laboratory)* 32 (3) pp.558–571 DOI: 10.4321/S1134-80462013000300002.

Kindervater, K.H. (2016) "The Emergence of Lethal Surveillance: Watching and Killing in the History of Drone Technology" *Security Dialogue* 47 (3) pp.223–238 DOI: 10.1177/0967010615616011.

Kolodny, L. (2019, May 17) "Zipline, Which Delivers Lifesaving Medical Supplies By Drone Medical Delivery Drone, Now Valued At $1.2 Billion" *CNBC* (7th May) Available at: https://www.cnbc.com/2019/05/17/zipline-medical-delivery-drone-start-up-now-valued-at-1point2-billion.html%0Ahttps://www.cnbc.com/2019/05/17/zipline-medical-delivery-drone-start-up-now-valued-at-1point2-billion.html?fbclid=IwAR03bSurSQ0yUi-0tR6LncKNNJWhSa.

Kubovich, Y. and Zikri, A. (2018) "Lieberman Accused Palestinian Photographer Killed in Gaza of Operating Drone Despite Lack of Evidence" *Haaretz* (9th April) pp.1–8.

Krouse, S. (2019) "Humanitarian Drones are Coming to the U.S., says Zipline CEO" *The Wall Street Journal* (23rd October) Available at: https://www.wsj.com/livecoverage/wsjtechlive/card/ifULQzOc91K7XSV1KRzN.

Lercel, D., Steckel, R. and Pestka, J. (2018) "Unmanned Aircraft Systems: An Overview of Strategies and Opportunities for Missouri" Jefferson City, MO: Missouri Department of Transport Available at: http://www.modot.org/services/or/byDate.htm.%0Ahttps://library.modot.mo.gov/RDT/reports/TR201808/cmr18-009.pdf.

Liang, Y. and Lee, S. A. (2017) "Fear of Autonomous Robots and Artificial Intelligence: Evidence from National Representative Data with Probability Sampling" *International Journal of Social Robotics* 9 pp.379–384 DOI: 10.1007/s12369-017-0401-3.

Liu, H. (2012) "Categorization and Legality of Autonomous and Remote Weapons Systems" *International Review of the Red Cross* 94 (886) pp.627–652 DOI: 10.1017/S181638311300012X.

Loughran, J. (2019) "50,000 Drone Owners Face £1,000 Fine as Registration Deadline Looms" *Engineering and Technology* (27th November) Available at:https://eandt.theiet.org/content/articles/2019/11/50-000-drone-owners-face-1-000-fine-as-registration-deadline-looms.

Manjikian, M. (2017) "A Typology of Arguments about Drone Ethics" *Army War College Monographs* 288 Available at: https://press.armywarcollege.edu/monographs/288.

Marincic, A. and Budimir, D. (2008) "Tesla's Multi-frequency Wireless Radio Controlled Vessel" *IEEE History of Telecommunications Conference HISTELCON 2008* pp.24–27 DOI: 10.1109/HISTELCON.2008.4668708.

McKinnon, J.D. and Hackman, M. (2020)) "Drone Surveillance of Protests Comes Under Fire" *The Wall Street Journal* (10th June) Available at: https://www.wsj.com/articles/drone-surveillance-of-protests-comes-under-fire-11591789477.

Mueller, T.J. (2009) "On the Birth of Micro Air Vehicles" *International Journal of Micro Air Vehicles* 1 (1) pp.1–12.

Murphy, R. and Baraniuk, C. (2016) "Robot Rescue? We're Ready to Roll" *New Scientist* 230 (3067) pp.28–29.

Newman, L. (2017) "The Army Grounds its DJI Drones Over Security Concerns" *Wired* (7th August) Available at: https://www.wired.com/story/army-dji-drone-ban/.

Olshin, B.B. (2019) *Lost Knowledge: The Concept of Vanished Technologies and Other Human Histories* Leiden: Koninklijke Brill.

Rivera, J. (1990) "R/C Helicopter ATV" *73 Amateur Radio* (August) pp.57–58.

Roberson, J.E. and Nowak, J.M. (2017) "Local Drone Law Struck Down by Federal Preemption" *Holland & Knight Aviation Law Blog* (27th September) Available at: https://www.hklaw.com/en/insights/publications/2017/09/local-drone-law-struck-down-by-federal-preemption.

Sayler, K. (2015) "A World of Proliferated Drones: A Technology Primer" *Center for a New American Security* (10th June) Available at: https://www.cnas.org/publications/reports/a-world-of-proliferated-drones-a-technology-primer.

Schulzke, M. (2018) "Drone Proliferation and the Challenge of Regulating Dual-Use Technologies" *International Studies Review* 21 (3) pp.1–21 DOI: 10.1093/isr/viy047.

SenseFly (2020) "Introducing the senseFly eBee X With MicaSense RedEdge-MX, A Seamless Dual Solution for Accurate & Efficient Crop Analysis" *sUASNews* (13th March) Available at: https://www.suasnews.com/2019/03/introducing-the-sensefly-ebee-x-with-micasense-rededge-mx-a-seamless-dual-solution-for-accurate-and-efficient-crop-analysis/.

Seymour, A.C., Dale, J., Hammill, M., Halpin, P.N. and Johnston, D.W. (2017) "Automated Detection and Enumeration of Marine Wildlife Using Unmanned Aircraft Systems (UAS) and Thermal Imagery" *Scientific Reports* 7 (45217) pp.1–10 DOI: 10.1038/srep45127.

Shane, S. and Wakabayashi, D. (2018) "'The Business of War'": Google Employees Protest Work for the Pentagon" *The New York Times* (4th April) Available at: https://www.nytimes.com/2018/04/04/technology/google-letter-ceo-pentagon-project.html.

Sloggett, D. (2014) *Drone Warfare* Barnsley: Pen & Sword Aviation.

Smith, K.N. (2017) "Indigenous People are Deploying Drones to Preserve Land and Traditions" *Discover* (11th December) Available at: https://www.discovermagazine.com/environment/indigenous-people-are-deploying-drones-to-preserve-land-and-traditions.

Specht, D. and Feigenbaum, A. (2018) "From Cartographic Gaze to Contestatory Cartographies" In Bargués-Pedreny, P., Chandler, D. and Simon, E. (Eds) *Mapping and Politics in the Digital Age* London: Routledge, Chapter 2.

Tuck, S. (2018) "Drone Vision and Protest" *Photographies* 11 (2–3) pp.169–175 DOI: 10.1080/17540763.2018.1445020.

United States Air Force (2015) "MQ-9 Reaper" Available at: https://www.af.mil/About-Us/Fact-Sheets/Display/Article/104470/mq-9-reaper/.

Wall, T. and Monahan, T. (2011) "Surveillance and Violence from Afar: The Politics of Drones and Liminal Securityscapes" *Theoretical Criminology* 15 (3) pp.239–254 DOI: 10.1177/1362480610396650.

WeRobotics (2020) *Drones and the Coronavirus: Do These Applications Make Sense?* (9th April) Available at: https://blog.werobotics.org/2020/04/09/drones-coronavirus-no-sense/.

Whitmore, B.A. (2016) *Evolution of Unmanned Aerial Warfare: A Historical Look at Remote Airpower – A Case Study in Innovation* (Master of Military Art and Science thesis) US Army Command and General Staff College, Fort Leavenworth, KS Available at: https://apps.dtic.mil/sti/pdfs/AD1020384.pdf.

Zaloga, S.J. (2011) *V-1 Flying Bomb 1942–52: Hitler's Infamous 'Doodlebug'* London: Bloomsbury Publishing.

Zwickle, A., Farber, H.B. and Hamm, J.A. (2018) "Comparing Public Concern and Support for Drone Regulation to the Current Legal Framework" *Behavioral Sciences & the Law* (April), 1–16. https://doi.org/10.1002/bsl.2357

15
AIRBORNE PHOTOGRAMMETRIC MAPPING

Gordon Petrie

By far the largest amount of base data that is being held in most geospatial databases has been acquired by or derived from airborne cameras and laser scanners mounted on *manned airborne platforms,* with the support of the appropriate photogrammetric technologies. Besides the traditional vector map data that can be extracted from the resulting aerial photographs and the stereo-models that are formed from them, data from these airborne sensors can also produce newer geospatial products such as orthophotographs and 3D surface or terrain models. Although other platforms and technologies have been introduced for data collection over the last 50 years, airborne digital mapping still holds sway as the dominant method of acquiring the base topographic data on which geospatial databases are built.

The alternative *spaceborne remote sensing techniques* are still largely confined to the acquisition of data for environmental mapping and monitoring on the one hand and for the gathering of military geospatial intelligence data on the other. Of course, spaceborne remote sensing data could be used for the mapping and provision of geospatial base data at all but the largest scales and highest resolutions. However, it is seldom employed in this role due to the greater ease and flexibility and the lower cost of carrying out these operations using airborne photogrammetric mapping technologies and procedures. In actual practice, most of the mapping that is carried out by non-military agencies using spaceborne imagery is thematic in type and at much smaller scales and lower resolution values than are required for topographic, engineering or cadastral mapping.

At the other end of the operational spectrum and scale range, in recent years, *unmanned aerial vehicles (UAVs)* have been introduced and have been welcomed by field scientists carrying out mapping for their research projects over limited areas of terrain. However, given the severe restrictions governing the operation of UAVs, such as the maximum allowable flying height, needs for line-of-sight operation, severe weight restrictions; extensive exclusion zones, and so on, their wider use to cover large areas of terrain for geospatial data collection and mapping purposes (e.g., at a regional or national level) still remains impractical. The area of mapping where the introduction of UAVs has been felt most strongly has been surveys of limited areas of terrain, especially for engineering and construction purposes, where they can offer strong competition to the ground surveying and mapping operations of traditional land surveyors, with the operators of UAV platforms utilizing similar photogrammetric technologies and procedures to those that are used with the data acquired from manned airborne platforms. So much so that many land surveying practices have now invested in the appropriate UAV and photogrammetric technologies.

Airborne non-digital mapping technologies

Much of the original aerial photographic camera technology and the accompanying photogrammetric instrumentation that was needed to extract map information from the resulting photographic images was developed during the 1920s and 1930s. However, it was not until after the Second World War that this particular combination of technologies began to be adopted widely both by national mapping agencies and by commercial survey and mapping companies.

Airborne film cameras

From 1950 until the end of the twentieth century, the airborne data acquisition side was completely dominated by large-format aerial film cameras. These cameras were manufactured primarily by Wild Heerbrugg (Switzerland) (Figure 15.1a), Zeiss Oberkochen (West Germany) and Carl Zeiss Jena (East Germany). The size of the images being acquired on film by these single-lens cameras for mapping purposes was standardized at 23 × 23 cm, but they could be fitted with different lenses having a wide range of focal lengths to provide aerial photography of different scales and resolution values and different geometries from a wide range of flying heights. The simple geometry of the single-lens film camera exposing a vertical photograph resulting in a frame image of the ground is shown in Figure 15.1b. The aerial photographic coverage of a larger area of terrain than that covered by a single photograph or a stereo-pair was acquired by flying a series of parallel strips with a 60% forward overlap of the frame images in each strip so that a continuous series of stereo-models of the terrain could be formed and measured. Each of the adjacent strips overlapped on its neighbour by 20% so that there was no gap in the ground coverage should there be some deviation from the planned flight line pattern and coverage (Figure 15.1c). The stereo-models of the terrain that were formed by the overlapping frame photographs (Figure 15.2a) acted as the basic material for the extraction and compilation of the vector line data and contours from which topographic maps were produced.

Analogue and analytical photogrammetric instrumentation

On the photogrammetric side, initially the extraction of the required map data from the aerial photographs was almost invariably carried out by human operators employing analogue stereo-plotting instruments (Figure 15.2b). These were based on optical, optical-mechanical or mechanical projection principles to generate 3D stereo-models, which were measured visually in great detail by a trained photogrammetrist to generate the vector-based graphical map detail (roads, railways, rivers, built-up areas, woods, and so on) and the spot heights and contours required to construct a topographic map. However, from the 1970s onwards, these instruments were often equipped with rotary or linear encoders within the 3D stereo-model space, so that they could generate digital 3D data that was suitable for entry into a computer-based CAD system. From the 1980s onwards, with the development of small high-speed computers, the analogue stereo-plotting instruments began to be replaced by *analytical plotters*, in which the optical or mechanical projection system of the analogue instrument was replaced by a purely mathematical solution implemented and driven by the

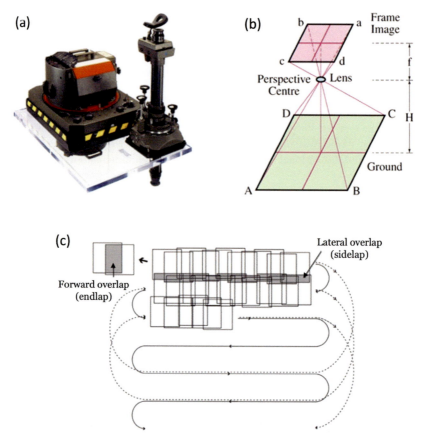

Figure 15.1 (a) A Wild RC30 aerial film camera with its tracking and navigation sight at front. This camera exposes frame images with a 23 × 23 cm format; (b) the projection of the ground through the perspective centre (the camera lens) results in a frame image being exposed in the focal plane of the aerial camera; and (c) a block of aerial photography comprising several strips of overlapping photographs and showing both the forward (60%) and lateral (20%) overlaps

Source: (a) Leica Geosystems.

computer in real-time (Petrie, 1990). However, although the photogrammetric solution was mathematical and digital, the aerial photos of the stereo-pair taken by the film camera still remained in hard-copy form for the operator to view in 3D (Figure 15.2c).

During the late 1980s, a wholly *digital photogrammetric workstation* (DPW) was introduced with the required digital images being produced by digitizing the images recorded on the aerial films using a film scanner (Petrie, 1997). As the cost of these DPWs dropped and with the advent of airborne digital cameras generating digital images after the year 2000, gradually the analogue and analytical stereo-plotters employing hard-copy photographs have disappeared almost entirely from the photogrammetric mapping community. However, it is important for the present geospatial community to understand something about these instruments and their capabilities and limitations – especially in terms of the accuracy of the final product – since the map data that was produced (in graphical form) on a vast scale by national mapping agencies using these instruments during the second half of the twentieth

Airborne photogrammetric mapping

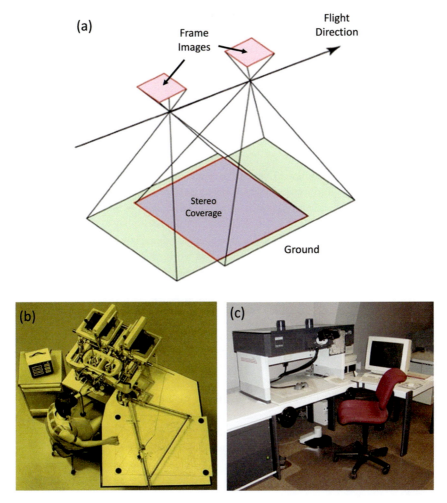

Figure 15.2 (a) The aerial photographs that overlap in the flight direction provide stereo-coverage of the ground and allow the formation of a stereo-model of the terrain on which measurements can be made to extract the information required for the production of topographic maps; (b) a Kern PG-2 analogue stereo-plotting instrument. At the back of the instrument are the pair of overlapping aerial photographs needed to form a stereo-model. In the centre of the instrument are the pair of mechanical rods that replicate the geometry of the intersecting rays from the two overlapping photographs for every specific point in the stereo-model. At the right side of the instrument is the plotting table on which the measured data is recorded graphically, complete with a pantograph to implement changes in the scale at which the final map data will be plotted. The operator has a binocular viewing system to view the stereo-model; and (c) Zeiss Planicomp analytical plotter. The stereo-pair of photographs in the form of film diapositives are mounted within the dark coloured box at the top of the instrument. However, the analogue mechanical representation of the intersecting rays shown in Figure 15.2b has been replaced by a mathematical representation that is being acted upon in real time by the instrument's computer and software which are being controlled by an operator using a trackball or similar device that allows 3D measurements to be undertaken on the stereo-model. The viewing of the 3D stereo-model by the operator is again made using a binocular viewing system

Source: (b) Hexagon. (c) Cardinal Systems.

century still forms the base of substantial parts of the geospatial databases that are in use today. The map detail and the contours that were measured and recorded in these photogrammetrically produced graphical maps were digitized later in huge campaigns conducted in the years around the turn of this century. For many countries in Western Europe and North America, much of this graphical-to-digital conversion work was outsourced and carried out in Eastern Europe and India (Petrie, 2005).

Airborne digital mapping technologies

The two primary data acquisition technologies that are used in airborne digital mapping are digital cameras and laser scanners. These are supported by various secondary technologies, primarily GPS receivers and inertial measuring units (IMUs) that are used for the in-flight measurement of position and altitude both for navigation purposes and for various post-flight photogrammetric operations. These latter operations are no longer being carried out using purpose-built optically and mechanically based instrumentation, but mathematically and digitally using suitably programmed computers and workstations.

Airborne digital cameras

For the purposes of this chapter, the airborne digital cameras that are used for the acquisition of basic geospatial image data will be discussed on the basis of the format size of their imaging arrays – small, medium and large – since each category tends to have its own particular set of applications. As appropriate to the situation existing in 2020, small-format cameras are defined here as having images up to 50 megapixels in size; the images produced by medium-format cameras range from 50 to 150 megapixels in size; while large-format cameras produce images greater in size than 150 megapixels. For those readers who may wish to read about digital airborne camera technology and operations in greater depth and detail than will be attempted here, reference may be made to the books by Graham and Koh (2002) and by Sandau (2010), although obviously there have been considerable advances in the technology since these books were published.

Large-format cameras

The first large-format airborne digital cameras that were designed to replace the standard aerial film cameras were introduced during the period 2000 to 2003. However, it took a few years of development before they began to be accepted, after which they have been quite widely and rapidly adopted as replacements for the standard film cameras, at least in the more highly developed countries. In 2010, the first large-format digital camera utilizing a single area array of detectors to produce a discrete panchromatic frame image was introduced in the shape of the Intergraph (now Leica Geosystems) DMC-II model (Petrie 2010). Prior to this, its predecessor, the DMC-I model (Hinz et al., 2000), had used four individual medium-format imaging arrays generating four separate tilted images that required further processing and stitching together to produce a single large-format image. By contrast, the DMC-II camera could generate individual frame images that were 250 megapixels in size using a single area array. The DMC-III (Figure 15.3a), which was introduced in 2015, generates panchromatic images that are 390 megapixels in size. The current model (2022) is the DMC-4, with almost a thousand megapixels. The UltraCam series of models from Vexcel Imaging, which were first introduced in 2003 (Leberl and Gruber, 2003), offer comparable

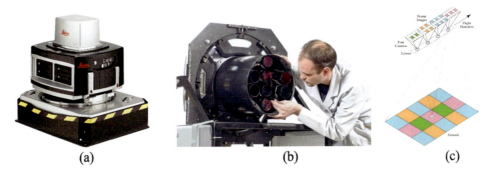

(a)   (b)   (c)

Figure 15.3 (a) A Leica DMC III large-format aerial digital camera mounted on a PAV gyro-controlled mount; and (b) a Vexcel UltraCam Eagle Mk. 3 large-format camera showing the multiple lenses of the design. The middle four lenses (placed in a line from top to bottom) expose the four sets of arrays that are used to construct the high-resolution panchromatic image. The two pairs of lenses on either side of the central four lenses expose the separate RGB and NIR spectral channels, the resulting image data serving to colourize the central panchromatic image and produce colour and false-colour images; and (c) a diagram showing how the large-format panchromatic frame image produced by the Vexcel UltraCam camera is built up from the 15 smaller medium-format images exposed using four separate lenses, all from a single exposure station in the air as the aircraft flies forward. The resulting 15 images are stitched together to form a single large-format panchromatic frame image

Source: (a) Leica Geosystems. (b) Wild Carinthia. (c) Vexcel Imaging.

performances in terms of their image sizes while still retaining the use of multiple medium-format arrays whose images have to be stitched together to form the final single image (Figure 15.3c). The UltraCam Eagle Mk. 3 (Figure 15.3b) generates panchromatic images that are 450 megapixels in size. The current model is the UltraCam Eagle 4.1, with over 500 megapixels. Both the DMC and UltraCam ranges feature an additional set of four auxiliary lenses that simultaneously collect additional frame image data in the red, green, blue (RGB) and near infra-red (NIR) spectral channels on reduced size (medium-format) arrays. These additional sets of image data are used to 'colourize' the large-format panchromatic images to produce colour and false-colour images. In the UK, for example, models from these two-camera series (DMC and UltraCam) are in widespread use on manned aircraft for the image data capture of large areas both by the national mapping agency (Ordnance Survey) and by three of the leading commercial suppliers of airborne digital imagery (Blue Sky International, GetMapping and Infoterra). The resulting imagery and the data that is extracted from it are the basis for the 2D and 3D datasets that are currently being used to form the base data of most geospatial databases and to update them on a regular basis.

Besides the large-format frame cameras described above, there also exists another type of large-format imager in the form of the *pushbroom line scanner*. Instead of imaging the terrain in a series of overlapping frame images from which 3D stereo-models of the terrain are formed, a pushbroom scanner images the terrain as a continuous strip image using a single linear array of detectors pointing vertically downwards to expose the image. In order to acquire stereo imagery, overlapping strip images are exposed simultaneously using additional backward and forward pointing linear arrays. Thus, this type of device is often described as a three-line pushbroom scanner. This particular imaging arrangement is widely used in high-resolution satellites, which of course operate in the very different environment and circumstances of space instead of the Earth's atmosphere and at a very different altitude (250–700

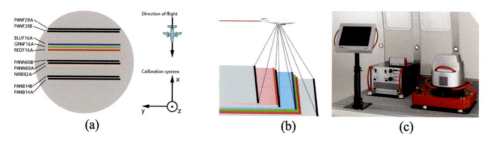

Figure 15.4 (a) The arrangement of the linear arrays in the focal plane of the ADS pushbroom scanner. There are three pairs of linear arrays that point in the forward, nadir and backward directions respectively to generate the basic overlapping panchromatic strip images, while the three linear arrays that record the corresponding RGB images are offset as a group from the nadir position. The remaining single linear array images the NIR radiation alongside the pair of panchromatic linear arrays in the nadir position; (b) a perspective view of the imaging geometry of the ADS pushbroom scanner as it generates its continuous strip coverage of the terrain; and (c) a Leica ADS80 laser scanner with its control electronics cabinet and display monitor

Source: Leica Geosystems.

km) and speed (29,000 km/hour). The re-development of this concept as an airborne system was undertaken by Leica Geosystems with the aid of the German Space Agency (DLR) and has resulted in a series of Leica ADS models (Sandau, 2010). As with the frame cameras, the three main panchromatic images captured by the ADS pushbroom scanner require the acquisition of additional data from the red, green, blue (RGB) and near infra-red (NIR) spectral channels in order to produce colour and false-colour imagery. This is carried out using an additional group of four linear arrays, one for each of the four required (RGB+NIR) spectral channels (Figure 15.4a).

The resulting imagery produced by the ADS pushbroom scanners has a completely different geometry to that of a frame camera – since each individual line in the image is exposed from a different position in the air (Figure 15.4b), whereas the whole of an individual image exposed by a frame camera is captured from a single position in the air (Figure 15.1b). Thus, a completely different photogrammetric solution and process needs to be applied to the pushbroom scanner imagery to produce the final geospatial data in the ground coordinate system. Besides which, since the position in the air of every line in the image needs to be known, it is necessary to include a GPS/IMU sub-system as an integral part of the overall system. A gyro-stabilized mount is a further requirement to ensure that there are no gaps or overlaps between adjacent lines in the image. Leica ADS pushbroom line scanners (Figure 15.4c) have been used extensively in certain countries for wide area image coverage (e.g. for the National Agriculture Imagery Program (NAIP) in the United States) In the UK, Infoterra, which forms part of the GEO-Information Division of the Astrium aerospace manufacturer, operates an ADS pushbroom scanner and another example has been used quite extensively in the Republic of Ireland by the Ordnance Survey of Ireland (OSI).

Medium-format cameras

Besides these large-format digital cameras, there are numerous models of medium-format digital cameras that are in widespread use for geospatial data collection. Examples include those from the two leading suppliers of large-format cameras – Leica with its series of RCD30 cameras and Vexcel with its Falcon model. However, there are also several smaller

Airborne photogrammetric mapping

(a) (b) (c)

Figure 15.5 (a) An IGI DigiCam medium-format camera installed in the grey Wild gyro-stabilized mount with an IGI AeroControl GPS/IMU sub-system and controller contained in the red box mounted on the small shelf directly above the camera; (b) at lower right, is the latest Phase One iXM-RS150 medium-format camera, while at upper left, the camera has been installed in the yellow Somag Compact Stabilization Mount CSM 40 with an Applanix POS/AV GPS/IMU sub-system mounted on a supporting metal frame directly above the camera; and (c) a Teledyne Optech CM-11K medium-format camera producing 86 megapixel images

Source: (a) IGI. (b) Phase One. (c) Teledyne Optech.

independent suppliers of medium-format airborne digital cameras; prominent among these are Phase One (Denmark) and IGI (Germany) (Figure 15.5a). Teledyne Optech in Canada also build medium-format digital cameras, but chiefly in the form of modules to fit into their airborne laser scanner systems (Figure 15.5c). In general terms, the image sizes of most cameras falling within the medium-format category range currently between 60 and 100 megapixels, although Phase One has recently introduced a new model that produces images that are 150 megapixels in size (Figure 15.5b). An important point of distinction is that all of these medium-format cameras have a single lens and use a single square or rectangular array of CCD or CMOS detectors having a mosaic filter and either a Bayer or some other interpolation scheme to generate colour or false colour images directly – instead of requiring and employing the supplementary set of lenses and arrays of detectors to record the required RGB and NIR data that are deployed in large-format airborne cameras.

In general, these medium-format cameras are being used for acquisition of high-quality imagery over smaller areas than the wide area coverage of the large-format cameras. They are also much used for 'corridor mapping' along transport infrastructure networks such as roads, railways, canals and other water courses where they are frequently operated in tandem with airborne laser scanners. Their lower cost and their smaller size and lighter weight, which allows them to be mounted and operated on manned light aircraft and helicopters, are other reasons for their increasing popularity. As will be seen later in this account, they also form the bases for a number of multiple camera systems such as oblique imaging systems and multi-spectral camera systems, where they are being deployed in various different configurations, all of which need to be taken into account in the subsequent photogrammetric data processing.

Small-format cameras

In sharp contrast to the large- and medium-format cameras described in the previous sections, almost all the airborne digital cameras falling within the small-format camera category are based on DSLR (Digital Single-Lens Reflex) cameras that are produced in quite large numbers aimed mainly at the professional photography market. For airborne

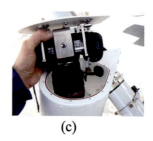

(a) (b) (c)

Figure 15.6 (a) A GeoTechnologies' SF-DMC small-format digital camera mounted on an anti-vibration mount equipped with shock absorbers and accompanied by its electronic control box; (b) a Canon EOS small-format camera equipped with a Canon GP-E2 GPS receiver for the measurement of positioning data; and (c) a Canon EOS 5DS camera being inserted into an FAA-approved pod to be mounted on the side or the undercarriage of a Cessna light aircraft

Source: (a) GeoTechnologies. (b) Canon. (c) Sky Imaging and Mapping Data.

use, they will often be fitted to a simple anti-vibration mount equipped with shock absorbers (Figure 15.6a), together with (i) an electronic unit that controls the timing and exposure of the image; and (ii) an inexpensive GPS receiver that can be used for flight navigation and geo-referencing purposes (Figure 16.6b). Usually, the system will be interfaced to a small laptop computer that is used both for flight planning and navigation purposes and for the storage of the digital images acquired in-flight. In some cases, a simple, relatively inexpensive, lower performance inertial measurement unit (IMU) may be used to produce the orientation (tilt) data needed to help carry out the rectification or ortho-rectification of the images. Like the medium-format cameras, the small-format cameras utilize a single lens with an area array of charge coupled device (CCD) or complementary metal oxide semiconductor (CMOS) detectors equipped with mosaic filters and using a Bayer interpolation procedure to generate colour and false-colour frame imagery. This category is dominated by two major Japanese companies, Canon with its EOS series and Nikon with its Dxxx range. Most of the models in these two series produce images in the range of 20–30 megapixels. However, recently introduced models from both manufacturers feature imaging arrays up to 50 megapixels in size. Examples of these Canon and Nikon cameras have been fitted to light aircraft such as the widely used Cessna 150, 172 and 182 models (Figure 15.6c) and to helicopters to acquire both vertical and oblique photography usually of specific sites or of relatively small areas of terrain. In this respect, the relatively low cost and light weight of the cameras are decisive factors in their use for these applications. As with the medium-format cameras, these small-format cameras also form the basis of a number of multiple camera systems that have come into widespread use for the acquisition of airborne geospatial image data and will be discussed next.

Multiple frame cameras generating oblique images

A major development in image data collection for geospatial purposes in recent years has been that of *multiple frame camera systems,* which are designed specifically to acquire systematic aerial oblique photography of substantial areas of terrain to generate image and elevation data that will populate geospatial databases. A large number of such systems have been developed by various system suppliers utilizing both small- and medium-format frame cameras that can

Airborne photogrammetric mapping

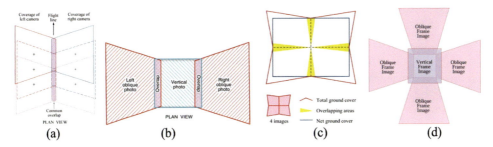

Figure 15.7 Diagrams showing the ground coverage of frame photography acquired by (a) twin (fan); (b) triple (fan); (c) quadruple (block); and (d) pentagonal (Maltese Cross) multi-camera systems

vary in number between two and thirteen in a wide variety of configurations. These include fan, block, Maltese Cross and concentric configurations (Petrie, 2009a).

i   In the case of the *fan configuration*, this commonly consists of from two to five small- or medium-format cameras that are set out in a line with tilted optical axes to expose simultaneously a series of overlapping oblique images in the cross-track direction to ensure greater coverage (a wider swath) of the terrain on either side of the flight line (Figure 15.7a and b).

ii With regard to the *block configuration*, usually this involves setting a group of four or six cameras in a square or rectangular formation with their optical axes pointing outwards and with simultaneous exposures of their images to increase the overall ground area that can be covered from a single exposure station, typically for mapping or surveillance applications (Figure 15.7c).

iii As for the distinctive five-camera '*Maltese Cross*' *configuration*, this comprises a combination of a single near-vertical (nadir) and four oblique pointing cameras, with the latter group having two of the cameras pointing obliquely in the forward and backward directions along the flight track and the other pair pointing obliquely in opposite directions cross-track at right angles to the flight line (Figure 15.7d). Essentially this means that the optical axes of the four oblique cameras point in the four cardinal directions relative to the flight line.

iv In the case of the *concentric (or star-type) configuration*, typically this comprises nine cameras with a single central near-vertical (nadir) pointing camera surrounded by eight oblique pointing cameras, each with its optical axis pointing outwards from the centre at 45 degrees in the azimuth direction to each of its neighbours (Figure 15.9b). However, it is possible to implement this configuration with other numbers of cameras, as indeed has been done recently with 13 cameras (Figure 15.9c).

While vertical aerial photography is still the standard method of acquiring imagery for topographic mapping purposes, systematic oblique photographic coverage using multiple digital cameras is becoming increasingly common, especially over urban areas (Remondino and Gerke, 2015). On the one hand, the development of the technique has allowed wide area coverage to be achieved using lower-cost, small- and medium-format cameras. However, more importantly, there is much more interest and value being placed on the availability of multiple oblique views of urban areas at fairly high (steep) vertical angles. In particular, this type of photography is perceived as readily providing invaluable information about the characteristics

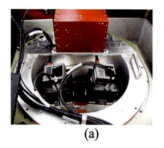  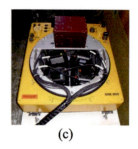

(a) (b) (c)

Figure 15.8 (a) An IGI Dual-DigiCAM system with the twin medium-format cameras tilted so that their optical axes are pointing obliquely to either side of the flight line in the cross-track direction – with an AEROcontrol GPS/IMU and controller unit mounted on the shelf above the cameras; (b) the DLR-3k three-camera system makes use of a fan of three Canon EOS small-format digital cameras with one firing vertically and two obliquely to provide wide cross-track coverage; and (c) an IGI Quattro-DigiCAM system with its four closely-coupled DigiCAM medium-format cameras that have been configured to produce a single composite large-format frame image – after the rectification and stitching together of the four oblique images has been undertaken

Source: (a) IGI. (b) DLR. (c) IGI.

of buildings and other structures that are difficult to obtain with classical near-vertical photography. Furthermore, the detailed interpretation of the highly oblique imagery, resulting in a much-improved identification of the objects that are present on the ground, is much more easily carried out by non-professional users of aerial photography – since it is closer to what they perceive in their normal daily life. This has led to its widespread adoption by the emergency (police, fire and ambulance) services in many countries, as well as by the urban and landscape planners and architects who are the more obvious users of this type of visual geospatial information. However, until now, this type of systematic oblique aerial photography has only been adopted to a limited extent by national mapping agencies (Remondino *et al.*, 2016).

The German systems supplier, IGI, offers a wide range of multiple camera systems for the acquisition of systematic oblique photography (Figure 15.8) in the shape of its Dual-, Triple-, Quattro- and Penta-DigiCAM systems. These are based on its single-lens medium-format cameras fitted with mosaic filters to produce colour and false-colour images, which are combined with its own in-house produced GPS/IMU and CCNS flight navigation subsystems (Petrie 2009b). Another very prominent supplier of five-camera systems producing "Maltese Cross" imagery, employing Canon EOS small-format cameras, has been the Dutch systems supplier Track'Air and its American Lead'Air associate company with its MIDAS (Multi-Image Digital Acquisition System) system, with over one hundred of these having been sold world-wide (Figure 15.9a). Moreover, Track 'Air has also developed and sold both nine- and thirteen-camera systems using a concentric arrangement of the cameras and lenses (Figures 15.9b and c). Following the successful introduction of these systems, the two major suppliers of large- and medium-format airborne camera systems, Leica Geosystems and Vexcel Imaging, have both entered the field with their RCD30 Oblique (Leica) and Osprey (Vexcel) five-camera systems that also produce 'Maltese Cross' patterned imagery.

Multiple frame cameras generating multi-spectral imagery

A substantial number of multi-camera systems have also been developed for the acquisition of *multi-spectral imagery,* especially in the United States, where there is a substantial demand from

Airborne photogrammetric mapping

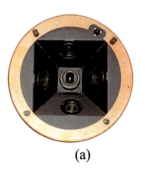 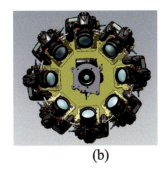 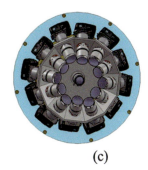

(a) (b) (c)

Figure 15.9 The layout of various small-format multi-camera systems developed by Track'Air and designed to acquire systematic oblique digital aerial photography: (a) a MIDAS five-camera system producing 'Maltese Cross' imagery; (b) an Octoblique nine-camera system exposing one vertical and eight oblique images; and (c) a Dodecablique 13 camera system with a single vertical camera surrounded by 12 oblique pointing cameras

Source: Track'Air.

 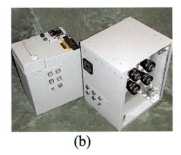 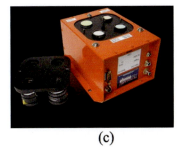

(a) (b) (c)

Figure 15.10 Multi-spectral camera systems from (a) GeoVantage; (b) Airborne Data Systems; and (c) Tetracam, all using four small-format digital cameras with parallel optical axes to produce separate red, blue, green and near infra-red images

the agricultural industries, forest enterprises and environmental agencies. Most of these systems are based on the use of readily available very small-format cameras, on grounds of their low cost, small size and light weight. Most of these systems employ four nadir-pointing monochrome cameras with parallel optical axes using filters to produce separate images in the classic red, green, blue (RGB) and near infra-red (NIR) spectral channels, supported by simple MEMS-based GPS/IMU sub-systems (Figure 15.10). The resulting compact size and light weight of these systems allows them to be used on light aircraft, which are readily available for hire in the United States. The imagery is usually acquired from low altitudes and over relatively small areas of terrain. Examples of companies that are producing these systems are GeoVantage based in Massachusetts; Airborne Data Systems (ADS) based in Minnesota (Petrie and Walker, 2007); and Tetracam in California (Petrie, 2013). In fact, GeoVantage is also a service provider, operating over a large part of the United States as well as being a system constructor and supplier.

Airborne laser scanning technologies

Taking a broad view, airborne laser scanning technologies can be divided into two broad categories: (i) those that are designed to carry out topographic mapping operations; and (ii) those that are designed primarily to carry out bathymetric mapping and charting operations, although the latter will often include some topographic mapping of coastal or shoreline areas.

Airborne topographic laser scanners

Airborne laser scanners have evolved to become a mainstream mapping technology over a relatively short period of time. Their predecessors were the laser altimeters and profilers that were developed for use in aircraft, principally by NASA, for scientific research purposes during the 1980s and 1990s. At that time, the continuous accurate positioning of the aircraft undertaking the measurements was a problem that could only be solved using a network of microwave ranging systems and instruments that was both elaborate and expensive to implement. The first successful laser scanners were constructed in the mid-1990s in parallel with the successful development of GPS/IMU units that could determine the successive positions, altitudes and attitudes of the scanner system as it carried out its scanning and ranging measurements. After an initial rather cautious acceptance of the technology, since 2005, it has come to be widely accepted and implemented for the measurement of terrain elevation and bathymetric depth values and now forms a major element in the current airborne mapping scene (Petrie and Toth, 2018).

The overall concept of the airborne laser scanner (Figure 15.11a) is that (i) the position, height and attitude of the airborne platform (together with that of the scanner that is mounted on it) are being measured continuously in-flight by an on-board GPS/IMU or GNSS/IMU unit with specific reference to a nearby GPS ground base station or a wide area correction service such as the satellite-based OmniSTAR. (ii) Simultaneously a dense series of ranges and the corresponding scan angles from the platform to the ground are being measured successively and very rapidly across the terrain in the cross-track direction by the laser rangefinder and by the angular encoder that is attached to the scanning mechanism. (iii) Combining these two sets of measurements (from (i) and (ii)) results in the determination of a line of elevation

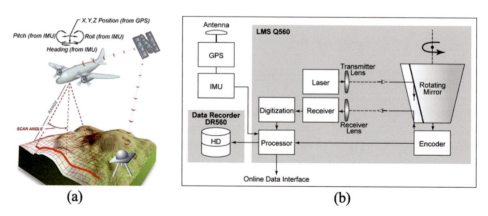

Figure 15.11 (a) Diagram showing the overall principle of airborne laser scanning, with (i) the position, height and attitude of the scanner being measured by the GPS/IMU sub-system on board the aircraft, while (ii) the slant range and scan angle values between the aircraft and the ground are being measured simultaneously by the laser rangefinder and the angular encoder attached to the scanning mechanism, resulting (iii) in a profile of measured ground elevation values (in red) in the cross-track direction; (b) this diagram shows the relationship between the main hardware components of an airborne laser scanner system – using the RIEGL LMS-Q560 scanner as the example

Source: (a) Leica Geosystems; re-drawn by Mike Shand (b) RIEGL; redrawn by M. Shand.

Airborne photogrammetric mapping

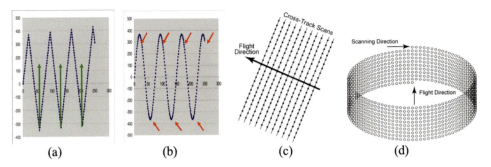

Figure 15.12 Ground measuring patterns – (a) the saw-tooth scan pattern; (b) the sinusoidal pattern using oscillating (bi-directional) mirrors to produce the scanning pattern; (c) the raster scan pattern produced using a continuously rotating (uni-directional) optical polygon; and (d) the elliptical pattern produced using a nutating mirror or prism in the so-called Palmer scan

Source: (b) Teledyne Optech. (c) RIEGL. (d) Leica Geosystems.

values at known positions (with X, Y, Z coordinates) forming a profile across the terrain in the cross-track direction (Figure 15.11a). The successive series of these measured profiles that are acquired in parallel as the airborne platform flies forward forms a digital terrain model or 3D point cloud of the terrain area that has been scanned. Besides the measurement of the slant range values using a very precise clock, the scanner detectors will also measure the intensity (or energy) value of the returned pulse. However, this latter information is often very 'noisy' and difficult to interpret or to utilize. Until now, these intensity values appear to be of limited interest to most users of airborne laser scan data, whose interest and attention is usually focused on the positional and elevation data that is provided by the airborne laser scanner.

It is important to realize that, in practice, quite a number of different measuring patterns over the terrain can result from airborne laser scanning depending on the particular type of scanning mechanism that is employed by the scanning system (Petrie, 2011a). In turn, this has its effect on the pattern and the density of the geospatial data that are lodged in the database.

(i) If an *oscillating (bi-directional) mirror* is used to carry out the scanning, then it results overall in either a Z-shaped (or saw-toothed) or a similar sinusoidal pattern of measurement over the ground (Figures 15.12a and b). This type of mechanism is widely used in the airborne scanners constructed by two of the leading system suppliers, Teledyne Optech and Leica Geosystems. (ii) If a *continuously spinning (uni-directional) polygon or mirror* is used to carry out the scanning of the terrain, then the resulting measuring pattern over the ground is a set of parallel lines at nearly right angles to the flight line. The constant rotational speed of the polygon means that there is no repetitive acceleration or deceleration of the mirror or polygon as is experienced with the previous type of scan being implemented using an oscillating mirror. This particular regular raster-like pattern (Figure 15.12c) is that adopted by another leading system supplier, RIEGL. (iii) The use of a *nutating mirror or prism* as the basis of the scanning mechanism was first developed by NASA for use on its pioneering laser scanner instruments and has also been adopted later by various developers of scanner systems in Sweden for use in both land (topographic) and bathymetric laser scanning instruments and systems. The scanning pattern that results from the use of this so-called Palmer scan is elliptical in shape (Figure 15.12d). As mentioned, it has been used in airborne topographic laser scanners by AHAB (later Leica-AHAB) and TopEye in Sweden and has been implemented on a number of current models of airborne bathymetric laser scanner.

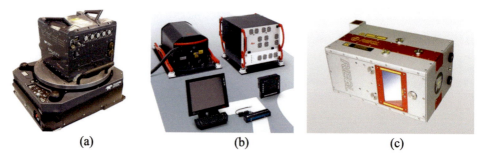

Figure 15.13 Powerful high-performance airborne topographic laser scanners capable of conducting mapping operations from high altitudes: (a) Teledyne Optech Galaxy mounted on a Somag GSM gyro-controlled mount; (b) Leica Geosystems ALS80 scanner at top left with its controller box at top right and the displays for the scanner operator and the pilot at front; and (c) RIEGL LMS-Q780i scanner

Source: (a) Teledyne Optech. (b) Leica Geosystems. (c) RIEGL.

At the high end, both in terms of cost and performance, are various models from the three leading suppliers of airborne laser scanners and systems – Teledyne Optech (ALTM Galaxy) (Figure 15.13a), Leica Geosystems (ALS) (Figure 15.13b) and *RIEGL* (LMS-Qxxx) (Figure 15.13c). In rough terms, all of these instruments are capable of acquiring 3D elevation data from flying heights of around 5 km, while the maximum pulse repetition frequency (PRF) of their measurements is usually from around 500 to 1,000 kHz and with a maximum scan rate of 100 Hz. In order to achieve such high-performance levels, twin measuring streams of pulses are generated either through the use of two laser rangefinders or by employing a beam splitter to split the stream of pulses being generated by a single laser rangefinder. Another aid to performance was the introduction by all three major laser system suppliers of the technique of having *multiple pulses* measured in the air simultaneously within a single profile scan. This feature is called 'multiple-pulses-in-the-air' (MPiA) by Leica Geosystems; 'continuous multipulse' (CMP) by Optech; and 'multiple time around' (MTA) by *RIEGL*. Irrespective of these differences in the name, their common adoption of this particular technique means that the laser rangefinder can fire a new pulse towards the ground without having to wait for the arrival of the reflection of the previous pulse at the instrument. Thus, more than one measuring cycle can be taking place at any specific moment of time. Besides these high-altitude models, all three of these principal manufacturers also offer a range of models that are designed to operate specifically from lower altitudes (e.g., for corridor mapping). For reasons of eye safety, these will often be fitted with a laser rangefinder that operates at the wavelength of 1,540 nm in the short wave infra-red (SWIR) part of the spectrum instead of the 1,046 nm wavelength value in the near infra-red (NIR) that is normally used.

Since the position, altitude and the orientation of the laser scanner in the air needs to be known at all times during its operation, a GPS/IMU sub-system is included as an integral part of the overall system. Furthermore, since the monochrome image that is produced from the intensity values generated by the reflected laser ranging pulses from the ground is 'noisy', almost invariably the laser scanner will be fitted with a medium-format digital camera that can generate much higher quality images in terms of their resolution and texture as well as having the desired colour content. The camera and the laser scanner are usually mounted

Airborne photogrammetric mapping

(a)

(b)

(c)

Figure 15.14 (a) Teledyne Optech's compact Orion low-altitude laser scanner and CS medium-format camera mounted rigidly together on a specially built base plate; (b) Leica CityMapper integrated system showing the five windows of the camera system acquiring one nadir and four oblique images, located around the periphery of the camera pod and the window of the laser scanner at the central position of the case; and (c) RIEGL VQ-1560i-II integrated system with the windows of the twin laser scanners at lower left and right respectively within the system case, while the window of the Phase One medium-format camera is located at top centre

Source: (a) Teledyne Optech. (b) Leica Geosystems. (c) RIEGL.

rigidly together on a common base plate or mount (Figure 15.14a). The spatial relationship of the two devices is then determined very exactly through measurement during a prior calibration procedure. For operational use, the two devices are fairly closely integrated and will normally share the flight management and control sub-systems, together with the shared GPS/IMU sub-system.

These developments have led to an ever-closer integration of laser scanners and digital cameras and the recent introduction of so-called "hybrid systems" such as the 'TerrainMapper' and the 'CityMapper' by Leica Geosystems. The former of these two systems combines a high-performance laser scanner having a 2 MHz pulse repetition frequency (PRF) mounted together with an RCD30 medium-format (80 megapixel) camera in a single case and is designed specifically to implement wide-area mapping with the image and scanner data being acquired from relatively high flight altitudes. The latter system (Figure 15.14b) has a set of five medium-format (80 megapixel) RCD30 cameras, comprising a single vertical (nadir) pointing and four oblique pointing cameras to generate 'Maltese Cross' imagery, which are operated in conjunction with a Hyperion laser scanner unit for the mapping of urban areas, again mounted closely together within a pod or case. A CityMapper unit is already in operation in the UK in the hands of Blue Sky International. In a somewhat similar development to that of the TerrainMapper, *RIEGL* has introduced its latest VQ-1560-II model (Figure 15.14c) for high-altitude operations fitted with twin 2 MHz laser scanners and a Phase One 150 megapixel medium-format camera, again a highly integrated system designed for wide-area mapping.

Airborne bathymetric laser scanners

As with the airborne topographic laser scanner systems, their bathymetric equivalents began as relatively simple profiling systems that were developed by NASA and various naval research establishments in the United States, Sweden and Australia during the 1980s. Again, it was only in the mid-1990s with the simultaneous development of suitable scanning mechanisms for use with laser rangefinders and the availability of the fully operational GPS system that fully capable bathymetric systems came into use by naval hydrographic agencies in these three countries. Following the considerable developments that have taken place since then, in 2020, airborne bathymetric laser scanner systems became available for both deep-water and shallow-water operations for sale from a number of commercial system suppliers. Besides the official naval hydrographic agencies that are charged with the task of undertaking nautical charting, there are a number of private operators who offer bathymetric survey and charting services on a commercial basis (Petrie, 2011b).

The basic principle of measuring the depth of the seabed (or lakebed) below the sea (or lake) surface using an airborne laser scanner normally involves the use of two laser rangefinders emitting their ranging pulses simultaneously at different wavelengths – in the NIR and green parts of the electro-magnetic spectrum. The pulsed NIR radiation is reflected back from the surface of the water, while the pulse of green radiation passes into and through the water column and is reflected by the seabed back towards the laser rangefinder (Figure 15.15a). The reflected radiation at both wavelengths is received by the detectors of the rangefinder and the elapsed time between the emitted and received pulses is measured for both pulses, converted to range values and compared to provide a depth value. The depth that can be measured is restricted to between 2 to 70 m depending on the particular system that is being used and the clarity or turbidity of the water column through which the green laser pulse is travelling. The cross-track scanning mechanism that is placed in front of the laser rangefinder points forward at a shallow angle and provides the coverage of a swath of the sea surface and the corresponding area of the seabed below (Figure 15.15b). The position, altitude and orientation of the laser scanner are measured continuously by the GPS/IMU sub-system that is mounted onboard the airborne platform.

The manufacture and supply of airborne bathymetric laser scanners is largely in the hands of two of the same three suppliers as their topographic equivalents. Teledyne Optech was at the core of the original development of the systems that were built for NASA and various US naval and maritime agencies through its construction of the successive models in the SHOALS series built for surveys in deeper water. Its current model is the CZMIL system built for use in the US Coastal Mapping Program. This features a single laser rangefinder that is capable, through its use of a frequency doubling technique, of generating both the required NIR and green pulses simultaneously. The depth data that is generated by the scanner is supplemented by the images generated by a hyperspectral pushbroom line scanner. The system is now also available commercially to other customers as the CZMIL Nova (Figure 15.16a). A second system that is also capable of scanning operations in deeper water is the Hawk Eye system, originally developed by Saab in Sweden and then developed further by AHAB, a company formed by former Saab employees before being taken over by Leica Geosystems in 2013. The current model offered by Leica is the Hawk Eye III (Figure 15.16b), which operates along much the same lines as the Optech series but features three laser rangefinders – a topographic scanner for use over land, a shallow-water system and the deeper-water system. It also features the elliptical scan pattern of the Palmer scanning system. Besides the systems from these two well-known system suppliers, it is also worth noting the LADS (laser airborne depth sounder) series was developed originally for the Royal Australian Navy. It has since been taken over by the Fugro surveying organization, which has operated the systems commercially on a world-wide basis, including the development of new models in the series (Figure 15.16c). The development of these different systems has made a dramatic difference to the collection of bathymetric data in coastal waters since their productivity is many times that of a hydrographic survey ship, especially when such a survey is being conducted along

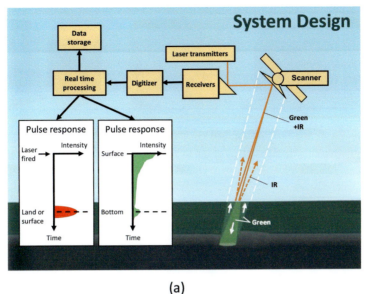

(a)

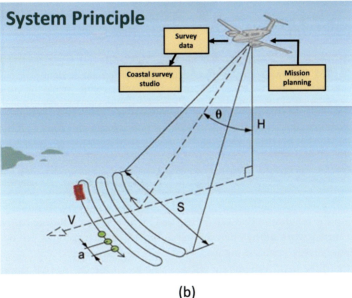

(b)

Figure 15.15 (a) The overall concept and design of an airborne bathymetric laser scanner showing how the pulses from the NIR laser rangefinder are reflected back from the sea surface – whereas the pulses from the green laser rangefinder penetrate the water column and are reflected back from the seabed; (b) the operating principle of the Hawkeye airborne bathymetric scanning system showing the forward oblique pointing of the rangefinder and the resulting scanning pattern over the sea surface

Source: Leica-AHAB.

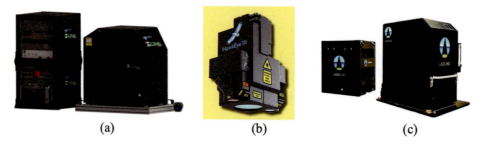

Figure 15.16 Powerful bathymetric laser scanners for use over deeper water: (a) Teledyne Optech CZMIL Nova; (b) Leica Geosystems HawkEye III; and (c) Fugro LADS HD

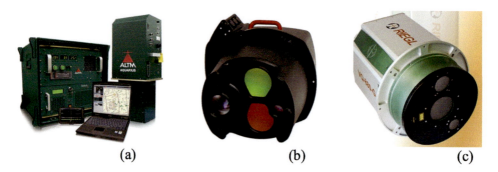

Figure 15.17 Bathymetric laser scanners for operation over shallower water: (a) Teledyne Optech ALTM Aquarius with the laser rangefinder and scanner at right rear, and the control electronics cabinet at left rear. The operator's laptop computer and the small pilot display monitor are at the front; (b) Leica Chiroptera II with the two larger optical windows for the green and near infra-red laser rangefinders, while the smaller window at left is that for the accompanying RCD30 medium-format digital camera and (c) RIEGL VQ-880-G with the optical windows for the green and NIR laser scanners separated by that for the digital camera

Source: (a) Teledyne Optech. (b) Leica Geosystems. (c) RIEGL.

rocky coastlines where navigation is complex and perilous. Of course, beyond the 70 or 80 m depth, which is the maximum that can be reached by the current airborne bathymetric laser scanners, the survey work in the open sea and in the deeper oceans must still be undertaken by hydrographic survey vessels using underwater sonar technology.

In 2011, the three main systems suppliers – Teledyne Optech, AHAB (soon to become Leica-AHAB) and *RIEGL* – all decided independently to introduce systems that were aimed principally at surveys of inland waters such as lakes and reservoirs and inshore coastal waters with shallow depths (Petrie, 2011b). In general terms, all of these systems feature a single green laser rangefinder operating from a low altitude with a short pulse duration resulting in markedly lower power requirements. In turn, this allows more frequent measurements of depth producing a very high data density and a high spatial accuracy, albeit with a restricted water penetration – about 20 m. It also allows the laser scanners to meet the eye safety standards when operated at a low altitude over a populated inland or coastal area. The systems that fall into this shallow water survey category include the Teledyne Optech Aquarius (Figure 15.17a), the Leica Chiroptera (Figure 15.17b) and the *RIEGL* VQ-820-G and VQ-880-G systems (Figure 15.17c).

Generating geospatial data

Having acquired the basic raw airborne image and laser scan (Lidar) data, the next matter to be discussed is its conversion into the 2D and 3D data that is needed to populate geospatial data bases.

Photogrammetric processing

Since the raw images that have been acquired by the airborne digital cameras contain substantial displacements due to the presence of terrain relief and aircraft tilts, these displacements first need to be removed over the whole of each image. After which, the resulting displacement-free images or the data derived from them needs to be placed in the coordinate system (e.g., geographical latitude and longitude or the national grid) that is being utilized in the database. The most usual method of doing so is (i) to form a 3D stereo-model of the terrain from the overlapping aerial photographs; and then (ii) to place this in the required coordinate system. After which, the required 2D or 3D data is extracted most often either (i) as a conventional vector line map with elevation contours; or (ii) as an orthophotograph and a matrix of elevation values, usually known as a 'point cloud' or, more formally, as a digital terrain model (DTM) or a digital elevation model (DEM). With all of these geospatial products, each stereo-model requires a set of control points to ensure its accurate placement (geo-referencing) in the ground coordinate system. These points are normally provided through an aerial triangulation procedure in which a block of stereo-models is first tied together through the linking of common points in the overlaps between adjacent models. Usually, these linkages are carried out using a highly automated image matching procedure. Once the block of photographs (and stereo-models) have been tied rigidly together (Figure 15.18a), then the block as a whole is fitted on to a set of ground control points located mostly around the periphery of the block, whose coordinate positions and elevation have already been established by ground survey measurement, usually carried out using GPS receivers. This procedure is much aided by the availability of GPS/IMU data providing positional and elevation data, which indeed is usually available from most flights undertaken with airborne digital cameras.

I   For the construction of a vector line map by a photogrammetrist, a digital photogrammetric workstation (DPW) is required. Instead of measuring hard-copy photographs via an optical or mechanical model of the terrain in an analogue or analytical stereo-plotting

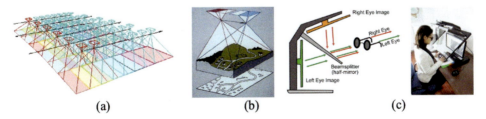

Figure 15.18   (a) Block of stereo-models linked together by aerial triangulation; (b) compilation of the vector line data required for map production is undertaken in a DPW through the orthogonal projection of the measured features from a single stereo-model on to the datum plane; and (c) diagram of a stereo-viewing system to form the visual stereo-model in a DPW that can be viewed and measured in 3D by the photogrammetric operator

instrument, as outlined in the introduction to this chapter, the stereo-model formed by the digital images in the DPW is a purely digital and mathematical model, albeit supplemented by a 3D stereo-viewing sub-system to allow the selection, stereo-measurement and compilation of the required map data to be undertaken by a human operator (Figures 15.18b and c). The final compiled and edited data will be output directly into standard CAD software packages such as AutoCAD or MicroStation or a dedicated cartographic or GIS package such as ArcGIS. While software routines will be available to help speed up the process, obviously much of the work still involves manual operations that need to be carried out visually by the human operator. Moreover, the compilation of the map is normally followed by a detailed field completion procedure carried out on the ground by a land surveyor or topographer. This means that overall, the production of vector line maps is a comparatively lengthy and costly operation – which is why its use has declined in recent years, though certainly it has not disappeared. Furthermore, when this type of product is required, the detailed photogrammetric work is often outsourced to India or to a country in south-east Asia, where labour costs are lower than in Western Europe or North America.

II The alternative to the classic compilation of vector map data using DPWs is the use of highly automated photogrammetric routines to generate *digital terrain models* and *orthophotographs*. With this methodology, after the aerial triangulation procedure has been completed and all the stereo-models have been established within the ground coordinate system, the measurement of each individual stereo-model in the block is carried out wholly automatically using dense multi-image matching techniques to generate the position (X, Y) and elevation (Z) value for each pixel in the model, together with its intensity and colour values. The final output from this procedure for each pixel in the stereo-model is (i) the orthophotograph in which all the tilt and relief displacements have been removed from the aerial photo images and (ii) the corresponding matrix of elevation values that has been formed to produce the DTM (Figure 15.19a). Adjacent orthophoto images can then be stitched together to produce seamless *orthophotomosaics*. The use of this highly automated procedure means of course that the final orthophoto output image or mosaic has been produced wholly automatically without the selection

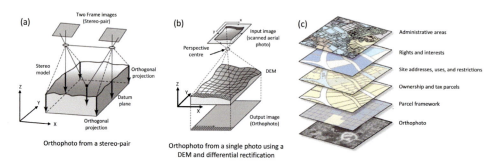

Figure 15.19 (a) The orthophotograph is formed by the orthogonal projection of the 3D stereo-image data contained in the stereo-model on to the datum plane; (b) an alternative method to generate an orthophotograph is to project a single photograph on to the DTM derived from airborne laser scanning and again project the image data orthogonally on to the datum plane and (c) diagram showing how the orthophoto acts as the base layer for all the other thematic map layers containing other data themes within the geospatial database

Source: Esri.

and interpretation of features that has been carried out by the operator during the alternative vector map compilation procedure. Thus, the interpretive processes that are required for feature extraction are thrown on to the user of the orthophotograph. Nevertheless, nowadays the purely image-based orthophotograph often forms the base positional data on which the rest of the geospatial database will be built (Figure 15.19c). The considerably increased speed with which this orthophoto data becomes available is seen as being very desirable and highly advantageous to many users. This has also led to the repeated aerial digital photographic coverage of many areas every year or two because of the relative ease with which the geospatial base data can be revised and updated through the use of the rapidly produced orthophotographs – hence their popularity at the present time.

III It will be seen that, so far, in this account, there has been no mention of laser scanning and indeed the production of (i) vector line data and contours on the one hand and (ii) orthophotographs and a DTM on the other hand can be carried out purely from the image data that has been captured by the airborne digital camera. However, as an alternative procedure, the positional and elevation data captured by the airborne laser scanner can readily be processed to form first a strip and then a block of elevation data covering the whole of the area of ground that needs to be modelled, resulting in a digital terrain model (DTM). In which case, an orthophotograph can be generated from the aerial digital photographs of the area in the same way as before (Figure 15.19b). There is much debate currently within the photogrammetric community as to which of these two processes – purely photogrammetric image matching or laser scanning – is the better from the points of view of the density and the accuracy of the resulting DTM datasets, as well as the speed with which they can be generated (Leberl *et al.*, 2010). The additional very substantial investment that is needed to purchase an airborne laser scanner is balanced by the fact that many hundreds of such laser scanners have already been purchased and are being operated by commercial service providers alongside their airborne digital cameras in order to implement this third procedure (III). However, there are many other service providers that rely entirely on the image matching approach without the need for a laser scanner.

Photogrammetric software developments

The role of recent developments in software, especially the marked advances in digital image matching (Haala, 2009, 2014; Remondino *et al.*, 2014), have been crucial in bringing about the very substantial degree of automation that is now available for the photogrammetric processing of airborne digital imagery. Nearly all the leading photogrammetric system suppliers now offer comprehensive data processing packages that can implement a near continuous and highly automated workflow. Taking Trimble as an example, its Inpho software suite includes modules that implement automated aerial triangulation (Match-AT and inBLOCK); feature extraction and vector line data capture (Summit Evolution); automated digital elevation extraction (Match-T); DTM editing (DTMaster); Lidar data processing (LP Master); DTM management, contouring, etc. (SCOP++); orthophoto generation (OrthoMaster); and orthophoto mosaicking (OrthoVista). Leica Geosystems offers a similar range of modules in the form of its HxMap suite, as does Vexcel Imaging with its UltraMap suite. Similar packages are available from specialist photogrammetric software houses such as Racurs (Russia) and SimActive (Canada). The concurrent acquisition of vertical (nadir) image and laser scan (Lidar) data is already a common occurrence, so software has already been developed to handle this particular combination of data. However, considerable efforts are now being made to ensure that photogrammetric software suites can also handle the images taken by the multiple camera systems producing systematic nadir and oblique aerial photographs (Jacobsen, 2008; Gerke et al., 2016) in combination with laser scan (Lidar) data – as integrated, for example, in the Leica CityMapper system – given their future potential in urban mapping and modelling (Toschi et al., 2018).

Conclusion

Airborne photogrammetric mapping based on the twin technologies (and synergies) of digital cameras and laser scanners is now ubiquitous in the more highly developed parts of the world and is the biggest supplier of the accurate base data on which all geospatial databases depend. The large-format digital cameras, which are the staple airborne imaging sensors of national mapping agencies and the large commercial mapping companies, are used to undertake the mapping of large areas of terrain on a regional and national scale, often producing this wide-area coverage on a repetitive basis. Moreover, there is extensive use of medium-format digital cameras for corridor mapping and for the repetitive coverage of relatively small areas, being used, for example, for monitoring purposes in agriculture and forestry. Often these large- and medium-format cameras are complemented by laser scanners that can generate the elevation data required to form the DTMs that can then be used for the rapid production of orthophotographs from the imagery. Other simpler, lighter solutions that are based on the use of small-format cameras and can be mounted in low-cost light aircraft are well suited to applications such as coverage of very small areas and specific sites where cost is a major consideration and high accuracy of the final product is not needed – though this does bring them into competition with UAVs that can also use these small-format cameras to satisfy these requirements.

A notable trend in recent years has been the development of multi-camera imaging systems, based on multiple medium- and small-format cameras, to undertake systematic oblique aerial photographic coverage, especially over urban areas with an emphasis on the implementation of 3D visualization and modelling. The most recent technological development has been that of hybrid systems with the close integration of cameras and laser scanners as a combined unit in a single pod; it seems inevitable that more of these systems will come into operation in the coming years. So far, the photogrammetric techniques that are needed to process the data from all these very different imaging and scanning technologies and configurations appear to have kept pace with the many technological developments in airborne data capture. This has come about, most notably, through the development of highly efficient image matching techniques that form the basis of the automation that is needed to satisfy the never-ending demands for the ever-quicker supply of 2D and 3D geospatial data from airborne digital mapping operations.

References

Gerke, M., Nex, F., Remondino, F., Jacobsen, K., Kremer, J., Karel, W., Hu, H. and Ostrowski, W. (2016) "Orientation of Oblique Airborne Image Sets – Experiences from the ISPRS/EUROSDR Benchmark on Multi-Platform Photogrammetry" *International Archives of the Photogrammetry, Remote Sensing and Spatial Information Sciences* 41 (PartB1) pp.185–191.

Graham, R. and Koh, A. (2002) *Digital Aerial Survey: Theory and Practice* Latheronwheel, Scotland: Whittles Publishing.

Haala, N. (2009) "Comeback of Digital Image Matching" In Fritsch, D. (Ed.) *Proceedings Photogrammetric Week* Heidelberg: Wichmann-Verlag, pp.289–301.

Haala, N. (2014) "Dense Image Matching Final Report" *(EuroSDR Official Publication)* 64 pp.115–145.

Hinz, A., Dörstel, C. and Heier, H. (2000) "Digital Modular Camera: System Concept and Data Processing Workflow" *International Archives of Photogrammetry and Remote Sensing* 33 (Part B2) pp.164–171.

Jacobsen, K. (2008) "Geometry of Vertical and Oblique Image Combinations" *Proceedings of the 28th EARSel Symposium* Istanbul, Turkey Available at: https://www.ipi.uni-hannover.de/fileadmin/ipi/publications/KJ_oblique.pdf.

Leberl, F. and Gruber, M. (2003) "Flying the New Large Format Digital Aerial Camera UltraCam-D" In Fritsch, D. (Ed.) *Proceedings Photogrammetric Week* Heidelberg: Wichmann-Verlag, pp.67–76.

Leberl, F., Irschara, A., Pock, T., Meixner, P., Gruber, M., Scholz, S. and Wiechert, A. (2010) "Point Clouds: Lidar versus 3D Vision" *Photogrammetric Engineering & Remote Sensing* 76 (10) pp.1123–1134.

Petrie, G. (1990) "Developments in Analytical Instrumentation" *ISPRS Journal of Photogrammetry and Remote Sensing* 45 (2) pp.61–89.

Petrie, G. (1997) "Developments in Digital Photogrammetric Systems for Topographic Mapping Applications" *ITC Journal* 2 pp.121–135.

Petrie, G. (2005) "Outsourcing and Offshoring: Hot Topics in the Mapping Field: Especially Important in Major English-Speaking Countries" *GEOInformatics* 8 (8) pp.8–15.

Petrie, G. (2009a) "Systematic Oblique Aerial Photography Using Multiple Digital Frame Cameras" *Photogrammetric Engineering and Remote Sensing* 75 (2) pp.102–107.

Petrie, G. (2009b) "The IGI DigiCAM Range: Modular Cameras; Multiple Configurations" *GEOInformatics* 12 (7) pp.60–63.

Petrie, G. (2010) "The Intergraph DMC II Camera Range: New Large-Format Airborne Digital Frame Cameras" *GEOInformatics* 13 (1) pp.8–11.

Petrie, G. (2011a) "Airborne Topographic Laser Scanners: Current Developments in the Technology" *GEOInformatics* 14 (1) pp.34–44.

Petrie, G. (2011b) "Airborne Bathymetric Laser Scanners: A New Generation is Being Introduced" *GEOInformatics* 14 (8) pp.18–24.

Petrie, G. (2013) "Tetracam's Range of Airborne Products: Cameras & UAVs for Agriculture, Forestry and Vegetation Applications" *GEOInformatics* 16 (3) pp.26–31.

Petrie, G. and Toth, C. (2018) "Airborne and Spaceborne Laser Profilers and Scanners" In Shan, J. and Toth, C. (Eds) *Topographic Laser Ranging and Scanning: Principles and Processing* (2nd ed.) Boca Raton, FL: CRC Press, pp.89–157.

Petrie, G. and Walker, A.S. (2007) "Airborne Digital Imaging Technology: A New Overview" *The Photogrammetric Record* 22 (119) pp.1–23.

Remondino, F. and Gerke, M. (2015) "Oblique Aerial Imagery – A Review" In Fritsch, D. (Ed.) *Proceedings Photogrammetric Week*, pp.75–83.

Remondino, F., Spera, M.G., Nocerino, E., Menna, F. and Nex, F. (2014) "State of the Art in High Density Image Matching" *The Photogrammetric Record* 29 (146) pp.144–166.

Remondino, F., Toschi, I., Gerke, M., Nex, F., Holland, D., McGill, A., Talaya Lopez, J. and Magarinos, A. (2016) "Oblique Aerial Imagery for NMA – Some Best Practices" *International Archives of the Photogrammetry, Remote Sensing and Spatial Information Sciences* 41 (Part W4) pp.639–645.

Sandau, R. (2010) *Digital Airborne Camera: Introduction and Technology* Dordrecht: Springer.

Toschi, I., Remondino, F., Rothe, R. and Klimek, R. (2018) "Combining Airborne Oblique Camera and Lidar Sensors: Investigations and New Perspectives" *International Archives of the Photogrammetry, Remote Sensing and Spatial Information Sciences* 42 (1) pp.437–444.

16
DIGITAL ELEVATION MODELS (DEMs)

Oluibukun Gbenga Ajayi

Introduction – definitions of digital elevation models

Accurate description, depiction and representation of the topography or configuration of the surface of the Earth in three dimensions (3D) is very important to many branches or fields of the Earth sciences. The simplest and most commonly adopted form of representing or depicting the terrain in 3D is the digital elevation model (DEM). The simplicity of its data structure (Olivera *et al.*, 2002) means that it is also described as the data structure used for storing topographic information more generally (Manuel, 2004), in which elevation values of the depicted terrain can be extracted by simple interpolation.

A DEM can also be described as a 3D raster image that contains the elevation of points of interest on the ground or Earth's surface, which are represented by pixel values, referenced to the mean sea level (MSL) (Ajayi *et al.*, 2017). It contains the elevation of points on a surface above the MSL and effectively models the spatial configuration or the topography of a region of interest on the Earth's surface. Digital elevation models are spatially or geographically referenced 3D datasets that continuously and effectively depict the topography and landforms of an area with the aid of elevation values. They are georeferenced pictorial representations of the relief surface whose elevation values are known. Digital elevation models are also digital imageries in which each matrix point has a value that corresponds to its altitude above sea level (Francesco *et al.*, 2011), and are basically created from elevation data.

These models can be represented either as a raster or as a vector. The raster representation of a DEM is often referred to as the 'computed or secondary DEM', while the vector representation of a DEM is called the 'measured or primary DEM' (Gandhi and Sarkar, 2016), which is also known as a Triangular Irregular Network (TIN). The raster DEM contains files with unique elevation values or data for each raster cell (Sitabi, 2015).

Whereas a DEM consists of a matrix or an array of numbers, it is usually rendered and presented in a visually understandable form, so that the spatial information it conveys can be easily extracted by the users. The most common form of this visualization is the assignment of shades and pseudo colours that will express the elevation differences between points in a pronounced and noticeable way. Also, vertical exaggeration is sometimes used, especially in oblique visualization where the terrain's visual image is reconstructed to enable oblique viewing, so that subtle differences in elevation can be clearly noticeable and made more pronounced. However, too much vertical exaggeration may mislead viewers about the true landscape of the surface by misrepresenting it (Morrison, 1992).

The term digital elevation model is often synonymously or interchangeably used with the terms digital terrain model (DTM) and digital surface model (DSM). While these three terminologies are used to describe the three most commonly implemented geospatial models used to depict various types of continuous surface in 3D, there are slight differences between them. A DEM depicts the bare surface of the Earth without any natural and man-made (artificial or built) features, whereas a DTM is a mathematical representation of the spatial distribution of terrain characteristics, and it is more of an augmentation of the DEM by including features that form the natural terrain of the scene. Digital elevation models can be generated by interpolating the DTM, whereas DEMs cannot be interpolated to generate DTMs. Meanwhile, a DSM is a raster image in which the elevation of the ground above the MSL, together with several other features (both natural and artificial features) present, are represented by unique pixel values.

A brief history of digital elevation models

According to Doyle (1978), the term 'digital terrain model' originated from the work conducted by Charles Miller at Massachusetts Institute of Technology between 1955 and 1960 (Miller, 1957; Miller and Laflamme, 1958). The objective of this research was to expedite highway design by digital computation based on photogrammetrically acquired terrain data where the data was acquired using a jury rig attached to Kelsh plotters. Specifically, digital terrain modelling, which is an extension of digital elevation modelling, emerged in photogrammetry in the mid-1950s as a new field of research (Rosenberg, 1955) known as geomorphometry, which is the quantitative study of topography (Pike, 2000), and became the primary source of topographic information. At this time, DTMs were used for the design of railways and highways (Miller, 1957) and for the production of relief maps using milling machines that were computer controlled (Florinsky, 2011). Subsequent improvements led to advancements in geophysical, computer and space technologies to ensure the transition from conventional geomorphometry to digital terrain modelling (Florinsky, 2011).

As a term, 'DEM' was introduced in the 1970s, and was originally meant only for raster representations. It was introduced mainly to distinguish this simplest form of relief modelling from other complex means of representing digital surfaces. See Pike *et al.* (2009) and Florinsky (2011) for further details of the historical evolution of digital elevation modelling.

Data sources and methods of DEM generation

Digital elevation models are produced through a process that involves the interpolation of known elevation data of points on the Earth's surface. These points are sampled at horizontal intervals that are regularly spaced and that are derived from different spatially referenced data sources. These data sources may be classified into three major categories: (1) ground-based data sources, which use conventional ground surveying techniques and, more recently, the terrestrial laser scanner (TLS); (2) space or satellite-based data sources; and (3) aerial or airborne data sources. Specifically, early techniques of the generation of DEMs involved direct digital interpolation of contour maps produced from topographic surveying using conventional ground surveying techniques or from irregularly spaced 3D points that were collected from field surveys. Due to advances in surveying technology, additional methods such as aerial photogrammetric techniques, synthetic aperture radar (SAR) interferometry, laser altimetry, and others have been introduced as viable data sources for DEM generation (Manuel, 2004; Yue *et al.*, 2015).

Ground-based data sources

Digital elevation models are generated from 3D spatial data (x, y, z values) following topographic survey. Traditionally, these data are obtained with the aid of a theodolite accompanied with distance measuring instrument and a level, and more recently, a total station, global navigation satellite system (GNSS) receivers and terrestrial laser scanners (TLSs), which use the principles of light detection and ranging (Lidar) to give millimetre accuracy. Different contour interpolation algorithms have been developed for DEM generation using ground surveying techniques. Reports of the performance and robustness evaluation of these interpolation models are provided by Kuta *et al.* (2018) and Alcaras *et al.* (2019).

The Lidar mapping procedure generates a set of crown structural variables, based on the intensity and ranges of the return of individual pulses or the full waveform's characterization (Alonzo *et al.*, 2014). The laser pulses that travel to the surface are rapidly transmitted, and as soon as the pulses contact the targeted surface, the signals are returned back to the sensor. The returned pulses are then converted from photons to electrical impulses, which are collected by a high-speed data recorder. The time intervals for the transmission are derived and converted to distance based on positional information obtained from ground or aircraft GPS receivers and the onboard inertial measurement unit (IMU). This process leads to the recording of data in the form of a 'point cloud' that is interpolated for DEM generation. Although very accurate products can be derived from conventional ground-based data techniques, their applicability is often limited, particularly in areas with inaccessible or dangerous terrain.

Satellite-based data sources

Various satellite-based sensors that are used for the acquisition of data can be used to produce DEMs. The first known satellite to provide stereoscopic images for DEM generation over vast areas of the Earth's surface was the Satellite Pour l'Observation de la Terre (SPOT) satellite, which was launched by France in 1986 (Elkhrachy, 2018). Today, DEMs derived from spaceborne synthetic aperture radar (SAR) offer the most complete coverage of the Earth's surface (Patel *et al.*, 2016).

The generation of DEMs from SAR comprises two major methods: interferometry SAR (InSAR) and stereo SAR. The InSAR technique measures the distance between the illumination target and the radar sensor by utilizing the phase information from the backscatter (received) signal (Thuan and Lindenschmidt, 2017). Each SAR interferogram is generated from the precise co-registration of two SAR images of a particular scene (Huang *et al.*, 2004), cross-multiplying the first SAR image with the complex conjugate of the other (Hanssen, 2001; Uppuluri and Jost, 2006). The processing of stereo SAR images involves three popular models: (1) the equivalent line central projection model, which is based on photogrammetric theory (Huang *et al.*, 2004; Ajayi and Palmer, 2020); (2) the model of range and Doppler equations; and (3) the parallax and elevation relation model.

In 2007, SAR imaging witnessed the introduction of TerraSAR-X (TSX), a second-generation SAR satellite that offers a spatial resolution of up to 1 m (Sefercik *et al.*, 2020), whereas its twin satellite, TanDEM-X (TDX), was launched in the year 2010 and enabled the generation of a global DEM from 2014 (see Soergel et al. (2013) and Sefercik et al., 2020).

While a comparatively good accuracy of about 1–10 m is obtainable from InSAR-based DEMs (Gruber et al., 2012), especially from single-pass systems like the TanDEM-X and the Shuttle Radar Topography Mission (SRTM) (Thuan and Lindenschmidt, 2017), models that are generated with this system are often subjected to compromises caused by the gaps that occur due to radar shadow and layover. These effects are more pronounced in extreme conditions such as highly mountainous terrain where little or no information can be extracted from the gapped areas (Hoja et al., 2006). Also, repeat-pass interferometry (e.g., ERS-ENVISAT, RADARSAT and ALOS-PALSAR) suffers from temporal decorrelation, which adversely affects the accuracy of estimated elevations, particularly for dynamic land surfaces, such as vegetation and snow or ice-covered areas (Rott, 2009).

In order to minimize these inherent limitations and to improve the accuracy of the acquired surface and elevation models, the single-pass systems TSX and TanDEM-X (TDX) were synchronized in a helix geometry. The introduction of TDX bistatic InSAR pairs made it possible to generate accurate global DEMs by the elimination of the atmospheric decorrelation caused by data acquisition during repeat passes. The German Aerospace Center (DLR, 2017) affirmed that the expected absolute and relative accuracies of the TDX global digital models are 6 m and 2 m respectively.

The SRTM, which produces one arc-second (approximately 30-metre) resolution DEMs, is a combined venture of NASA's Jet Propulsion Laboratory (JPL), the German and Italian Space Agencies and the National Geospatial-Intelligence Agency (NGA). About 12 terabytes of data covering over 80% of the Earth's landmass had been collected by the mission by February, 2000 (Kolar et al., 2007). The SRTM DEM data is reported to have a vertical error of less than ±16 m (Miliaresis and Paraschou, 2005) and is distributed via several public and private agencies at three spatial resolutions, though it is primarily housed on the USGS Earth Explorer online data repository (http://earthexplorer.usgs.gov).

Other open-source satellite-based global DEM data sources include the Advanced Spaceborne Thermal Emission and Reflection Radiometer (ASTER), a joint effort of Japan and NASA, which provides 90-metrr resolution DEMs globally and 30-metre resolution DEMs for the United States, which can also be freely accessed and downloaded from the USGS Earth Explorer website. Another satellite-based DEM is the ALOS World 3D, which offers a DSM of 30-metre resolution, captured by the Japan Aerospace Exploration Agency (JAXA). This uses the Advanced Land Observing Satellite (ALOS) based on stereo mapping from PRISM (panchromatic remote-sensing instrument for stereo mapping) principles. It is judged to offer the most precise data available for global elevation, and can be downloaded from the JAXA Global ALOS portal. The comparative performance of open-source satellite-based DEMs are discussed by Patel et al. (2016), Thuan and Lindenschmidt (2017), Elkhrachy (2018) and Jalal et al. (2020). A flowchart summarizing the step-by-step procedure of extracting ALOS phased array band Synthetic Aperture Radar (ALOS PALSAR) DEM, SRTM DEM and TanDEM-X data is presented in Figure 16.1, with further details recorded by Jalal et al. (2020).

Airborne or aerial data sources

Airborne data sources can be broadly classified into Airborne Laser Scanning (ALS) and aerial photogrammetry (manned and unmanned aerial systems).

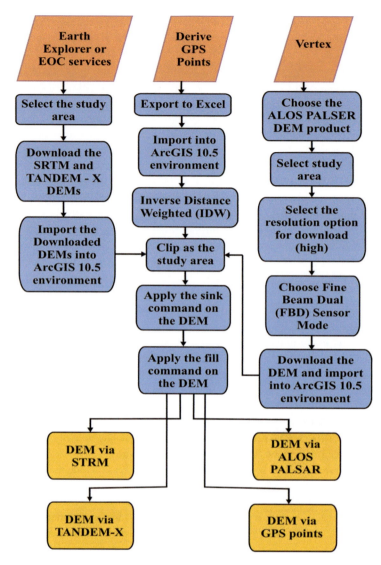

Figure 16.1 Flowchart summarizing the step-by-step procedure of extracting ALOS Phased Array band Synthetic Aperture Radar (ALOS PALSAR) DEM, SRTM DEM, and TanDEM-X

Airborne laser scanning (ALS)

This method determines the elevation of points by estimating the time taken for the laser pulses released or fired from a laser scanner (on an aircraft) to reflect off the targeted surface and return back to the scanner (Olsen et al., 2012; Doyle and Woodroffe, 2015). The laser scanner collects a cloud of laser range measurements, using direct georeferencing techniques, which are used for the computation of the 3D coordinates of the surveyed area and are then interpolated for the generation of a DEM. Lidar can measure the topography of a surface rapidly and accurately (to centimetre precision) and can be deployed for the survey of landscapes that are not easily accessible (Reddy et al., 2015). Lidar mapping can be expensive to perform (Nathalie et al., 2016), costs are decreasing and national datasets are often released to the public. In the UK, Lidar-generated DTM, DSM and point clouds are available from the EDINA Digimap service and can be downloaded from their website (https://digimap.edina.ac.uk/) with an institution-based subscription (Ajayi and Palmer, 2020).

Aerial photogrammetry (manned and unmanned systems)

Aircraft are the conventional platforms used for aerial surveys. Cameras attached to the bottom of an aircraft are flown to a predefined flight height, from where the photographs are captured with appropriate overlaps (forward and sideways) to ensure accurate registration or mosaicking, 3D stereoscopic viewing and the generation of DEMs (Ajayi et al., 2017).

Recently, miniaturized, lightweight unmanned aerial vehicles (UAVs), commonly referred to as drones, are fast gaining attention in digital photogrammetry and for the generation of DEMs. Photogrammetry using structure from motion (SfM) technology and multi-view stereo (MVS) for DEM generation has received much attention and acceptance due to its many advantages from a UAV platform. These include: (1) automation of the entire photogrammetric process; (2) high repeatability of the survey, which affords flexible and user-defined temporal resolution; (3) very low operating costs; (4) high accuracy with the possibility of obtaining aerial photographs with centimetric resolution; and (5) easily modifiable platforms for obtaining different types of remote sensing data (Gonçalves and Henriques, 2015; Bhardwaj et al., 2016). Unmanned aerial vehicles are also more flexible and easier to control when compared to manned aircraft and satellite remote sensing (Candiago et al., 2015), which makes them a low-cost alternative for accurate large-scale or wide-area topographic mapping and 3D recording of ground information (Ajayi et al., 2018).

The process of DEM generation using UAVs involves the acquisition of overlapping pairs of images of the area of interest and the processing of the acquired images into map products or deliverables. Different types of commercially available software packages have been developed for the processing of UAV or drone images for DEM generation. Some of these packages are the Agisoft photoscan (Metashape), DroneDeploy, IMAGINE Photogrammetry, AirGon, PrecisionMapper, Propeller, Trimble Business Center Photogrammetry Module, 3DF Zephyr, Pix4D Mapper, SimActive Correlator3D, and Open Drone Map Photogrammetry. The image processing procedure that leads to the generation of DEMs from UAVs involves interior, absolute and relative orientations (Ajayi et al., 2017; Ajayi, 2019). Based on SfM algorithms, which use the SGM and scale-invariant feature transform algorithm (Lowe 2004; Westoby et al., 2012), several control points (or tie points) are extracted from the UAV-acquired overlapping image pairs, which are then interpolated for DEM generation. The aim of SfM is to accurately recover the parameters of the camera, pose estimates and sparse 3D scene geometry from 2D overlapping image pairs (Hartley and Zisserman, 2004), taken from different orientations and locations in order to accurately reconstruct the photographed scene (Ajayi and Palmer, 2020). The accuracy of UAV-generated DEMs is often improved with the integration of ground control points (GCPs). The step-by-step process of generating DEMs from drone-acquired nadir images using Agisoft photoscan is presented in Figure 16.2.

Applications and quality of DEMs

There has been an unprecedented increase in the demands and uses of accurate DEMs within the last decade, due to their multivariate applications that cut across different disciplines, such as the built environment, urban planning and mapping (Ajayi et al., 2018; Nwilo et al.,

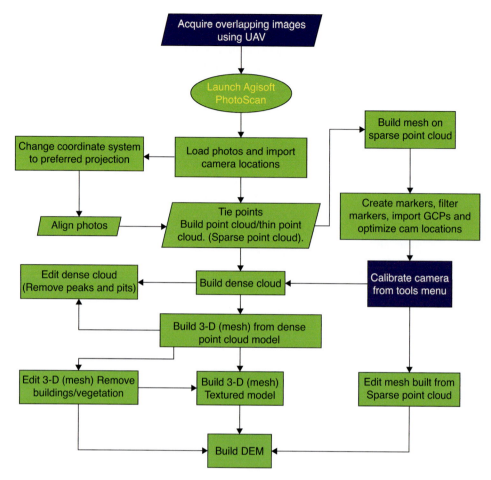

Figure 16.2 The step-by-step process of generating DEM from drone-acquired nadir images using Agisoft photoscan

2020), civil and environmental engineering (Li *et al.*, 2018), geological and geophysical studies (Ricchetti, 2001), agriculture (Ajayi *et al.*, 2018), disaster, hazards and environmental monitoring and/or analysis (Demirkesen *et al.*, 2007; Tsai *et al.*, 2010), and many others.

Specifically, some of the common uses of DEMs as highlighted by Balasubramanian (2017) include the extraction of terrain parameters for geomorphological applications, water flow modelling for hydrological studies, modelling soil wetness, composition of relief maps, image rectification, rendering of 3D visualizations, reduction or terrain correction of gravity data, 3D flight planning, geographic information systems (GIS), Earth observation and satellite navigation, engineering and infrastructure design, line-of-sight analysis, intelligent transportation systems, precision agriculture, as well as surface and terrain analysis.. Coveney and Roberts (2017) also highlighted some applications of digital elevation models, especially for flood risk modelling.

The quality of a DEM depends on its vertical and horizontal accuracy. Accuracy of a global or world DEM depends on the grid size used. Some of the factors that influence the quality and accuracy of a generated DEM and its derived products include the sampling density, methods of collection of the interpolated elevation data, choice of interpolation algorithm, pixel size or grid resolution, terrain roughness, and terrain analysis algorithm (Hanuphab et al., 2012; Balasubramanian, 2017). The achievable accuracy of DEMs based on spaceborne imagery mainly depends on the contrast and resolution of the images and the height-to-base relation (Pecherska, 2013).

Future outlook and concluding remarks

This chapter has provided a definition of the digital elevation model and an overview of its emergence as a technique for understanding and analysing the Earth's surface. It introduced some methods of their construction and the data sources used (ground-based, spaceborne and airborne), and explained how DEMs, DSMs, and DTMs are differentiated. Furthermore, some of the applications and uses of DEMs were highlighted and the chapter indicated how the quality of DEMs may be compared and measured.

Going forward, the already broad and vast market for DEMs stands to grow further, and the accuracy and resolution of DEMs will also keep improving with technological advancements. To keep up with demand, he launch of more satellite missions and production of more airborne systems will enable the generation of higher-resolution DEMs to suit an ever-growing range of applications.

References

Ajayi, O.G. (2019) "Producing Accurate 3-D Models from Low Percentage Overlapping Images Acquired Using UAV" *Sensed* 73 pp.15–18.

Ajayi, O.G. and Palmer, M. (2020) "Modelling 3D Topography by Comparing Airborne LiDAR Data with Unmanned Aerial System (UAS) Photogrammetry Under Multiple Imaging Conditions" *Geoplanning: Journal of Geomatics and Planning* 6 (2) pp.122–138.

Ajayi, O.G., Palmer, M. and Salubi, A.A. (2018) "Modelling Farmland Topography for Suitable Site Selection of Dam Construction Using Unmanned Aerial Vehicle (UAV) Photogrammetry" *Remote Sensing Applications: Society and Environment* 11 pp.220–230 DOI: 10.1016/j.rsase.2018.07.007.

Ajayi, O.G., Salubi, A.A., Angbas, A.F. and Odigure, M.G. (2017) "Generation of Accurate Digital Elevation Models from UAV Acquired Low Percentage Overlapping Images" *International Journal of Remote Sensing* 8–10 (38) pp.3113–3134, DOI: 10.1080/01431161.2017.1285085.

Alcaras, E., Parente, C. and Vallario, A. (2019) "Comparison of Different Interpolation Methods for DEM Production" *International Journal of Advanced Trends in Computer Science and Engineering* 8 (4) pp.1654–1659 DOI: 10.30534/ijatcse/2019/91842019.

Alonzo, M., Bookhagen, B. and Roberts, D.A. (2004) "Urban Tree Species Mapping Using Hyperspectral and LiDAR Data Fusion" *Remote Sensing of Environment* 148 pp.70–83

Balasubramanian, A. (2017) "Digital Elevation Model (DEM) in GIS" (*Technical Report*) University of Mysore DOI: 10.13140/rg.2.2.23976.47369.

Bhardwaj, A., Sam, L., Akanksha, Martín-Torres, F.J. and Kumar, R. (2016) "UAVs as Remote Sensing Platform in Glaciology: Present Applications and Future Prospects" *Remote Sensing of Environment* 175 pp.196–204.

Candiago, S., Remondino, F., De Giglio, M., Dubbini, M. and Gattelli, M. (2015) "Evaluating Multispectral Images and Vegetation Indices for Precision Farming Applications from UAV Images" *Remote Sensing* 7(4) pp.4026–4047.

Coveney, S. and Roberts, K. (2017) "Lightweight UAV Digital Elevation Models and Orthoimagery for Environmental Applications: Data Accuracy Evaluation and Potential for River Flood Risk Modelling" *International Journal of Remote Sensing* 38 (8–10) pp.3159–3180 DOI: 10.1080/01431161.2017.1292074.

Demirkesen, A.C., Evrendilek, F., Berberoglu, S. and Kilic, S. (2007) "Coastal Flood Risk Analysis Using Landsat-7 ETM+ Imagery and SRTM DEM: A Case Study of Izmir, Turkey" *Environmental Monitoring and Assessment* 131 pp.293–300 DOI 10.1007/s10661-006-9476-2.

DLR (2017) "German Aerospace Center TanDEM-X information page" Available at: https://directory.eoportal.org/web/eoportal/satellite-missions/t/tandem-x.

Doyle, F.J. (1978) "Digital Terrain Models: An Overview" *Photogrammetric Engineering and Remote Sensing* 44 (12) pp.1481–1485.

Doyle, T. and Woodroffe, C.D. (2015) "A Method to Investigate Foredune Morphodynamics and Ecology Using Airborne LiDAR in SE Australia" *24th NSW Coastal Conference* Club Forster, New South Wales, 11th–13th November.

Elkhrachy, I. (2018) "Vertical Accuracy Assessment for SRTM and ASTER Digital Elevation Models: A Case Study of Najran City, Saudi Arabia" *Ain Shams Engineering Journal* 9 pp.1807–1817 DOI: 10.1016/j.asej.2017.01.007.

Florinsky, I. (2011) *Digital Terrain Modelling: A Brief Historical Overview in Digital Terrain Analysis in Soil Science and Geology* (2nd ed.) Amsterdam: Elsevier.

Francesco, D., Domenico, G. and Antonello, C. (2011) "Nature and Aims of Geomorphological Mapping" In Griffiths, J.S., Smith, M.J. and Paron, P. (Eds) *Geomorphological Mapping Methods and Applications* (Volume 15) Amsterdam: Elsevier, pp.39–73.

Gandhi, S.M. and Sarkar, B.C. (2016) *Essentials of Mineral Exploration and Evaluation* Amsterdam: Elsevier Science, pp. 53–79.

Gonçalves, J.A. and Henriques, R. (2015) "UAV Photogrammetry for Topographic Monitoring of Coastal Areas" *ISPRS Journal of Photogrammetry and Remote Sensing* 104 pp.101–111.

Gruber, A., Wessel, B., Huber, M. and Roth, A. (2012) "Operational TanDEM-X DEM Calibration and First Validation Results" *ISPRS Journal of Photogrammetry and Remote Sensing* 73 pp.39–49.

Hanssen, R.F. (2001) *Radar Interferometry: Data Interpretation and Error Analysis* (Volume 2) New York: Springer Science & Business Media, Kluwer.

Hanuphab, T., Suwanprasit, C. and Srichai, N. (2012) "Effects of DEM Resolution and Source on Hydrological Modelling" *33rd Asian Conference on Remote Sensing* Pattaya, Thailand: 26th–30th November, pp.26–30.

Hartley, R.I. and Zisserman, A. (2004) *Multiple View Geometry in Computer Vision* (2nd ed.) Cambridge: Cambridge University Press.

Hoja, D., Reinartz, P. and Schroeder, M. (2006) "Comparison of DEM Generation and Combination Methods Using High Resolution Optical Stereo Imagery and Interferometric SAR Data" *Proceedings of the ISPRS Commission I Symposium* 36 (1) pp.1–6

Huang, G.M., Guo, J.K., Zhao, Z., Xiao, Z., Qiu, C.P., Pang, L. and Wang, Z.Y. (2004) "DEM Generation from Stereo SAR Images Based on Polynomial Rectification and Height Displacement" *Geoscience and Remote Sensing Symposium* 6 pp.4227–4230 DOI: 10.1109/IGARSS.2004.1370068.

Jalal, S.J., Tajul, A.M., Taher, H.A., Ami, H.M.D., Wan, A.W.A. and Jwan, M.E. (2020) "Optimizing the Global Digital Elevation Models (GDEMs) and Accuracy of Derived DEMs from GPS Points for Iraq's Mountainous Areas" *Geodesy and Geodynamics* 11 (5) pp.338–349 DOI: 10.1016/j.geog.2020.06.004.

Kolar, J., Kjems, E., Bodum, L. and Sorensen, E. M. (2007) "Global Surface Model Using Spaceshuttle Radar Topographic Mission Dataset and Global Indexing Grid" *Strategic Integration of Surveying Services FIG Working Week* Hong Kong SAR, China, 13th–17th May, pp.1–13.

Kuta, A.A., Ajayi, O.G., Osunde, T.J., Ibrahim, P.O., Dada, D.O. and Awwal, A.A. (2018) "Investigation of the Robustness of Different Contour Interpolation Models for the Generation of Contour Map and Digital Elevation Models. Contemporary Issues and Sustainable Practices in the Built Environment" *School of Environmental Technology Conference (SETIC)* Minna, Nigeria, 10th–12th April: SET-FUTMinna, pp.1527–1541.

Li, L., Yang, J., Chuan-Yao, L., Chua, C.T., Wang, Y., Zhao, K. and Yun-Ta, W., Liu, P.L-F., Switzer, A.D., Mok, K.M., Wang, P. and Peng, D. (2018) "Field Survey of Typhoon Hato (2017) and a Comparison with Storm Surge Modeling in Macau" *Natural Hazards and Earth System Sciences* 18 pp.3167–3178 DOI: 10.5194/nhess-18-3167-2018.

Lowe, D. (2004) "Distinctive Image Features from Scale-Invariant Key Points" *International Journal of Computer Vision* 60 (2) pp.91–110 DOI: 10.1023/B:VISI.0000029664.99615.94.

Manuel, P. (2004) "Influence of DEM Interpolation Methods in Drainage Analysis" *GIS Hydro 04* Austin, Texas.

Miliaresis, G.C. and Paraschou, C.V.E. (2005) "Vertical Accuracy of the SRTM DTED Level 1 of Crete" *International Journal of Applied Earth Observation and Geoinformation* 7 (1) pp.49–59.

Miller, C.L. (1957) "The Spatial Model Concept of Photogrammetry" *Photogrammetric Engineering* 23 (1) pp.31–35.

Miller, C.L. and Laflamme, R.A. (1958) "The Digital Terrain Model-Theory and Application" *Photogrammetric Engineering and Remote Sensing* 34 (3) pp.433–442.

Morrison, D. (1992) "'Flat-Venus Society' Organizes" *EOS* 73 (9) p.99 DOI: 10.1029/91EO00076.

Nathalie, L., Bastien, M., Benoît, G., Frédéric, P. and Xavier, B. (2016) "Monitoring the Topography of a Dynamic Tidal Inlet Using UAV Imagery" *Remote Sensing* 8 (5) pp.1–18 DOI: 10.3390/rs8050387.

Nwilo, P.C., Okolie, C.J., Onyegbula, J.C., Abolaji, O.E., Orji, M.J., Daramola, O.E. and Arungwa, I.D (2020) "Vertical Accuracy Assessment of 20-metre SPOT DEM Using Ground Control Points from Lagos and FCT, Nigeria" *Journal of Engineering Research* 25 (2) pp.153–164.

Olivera, F., Furnans, J., Maidment, D., Djokic, D. and Ye, Z. (2002) "Drainage Systems" In Maidment, D.R. (Ed.) *Arc Hydro: GIS for Water Resources* Redlands, CA: Esri Press, pp.55–86.

Olsen, M.J., Young, A.P. and Ashford, S.A. (2012) "TopCAT-topographical Compartment Analysis Tool to Analyze Seacliff and Beach Change in GIS" *Computers & Geosciences* 45 pp.284–292.

Patel, A., Katiyar, S.K. and Prasad, V. (2016) "Performances Evaluation of Different Open Source DEM Using Differential Global Positioning System (DGPS)" *The Egyptian Journal of Remote Sensing and Space Sciences* 19 pp.7–16 DOI: 10.1016/j.ejrs.2015.12.004.

Pecherska, J. (2013) "A System Dynamics Model for Basin Level Forecasting on the Basis of a Digital Elevation Model" *Information Technology and Management Science* 16 (1) pp.101–105 DOI: 10.2478/itms-2013-0016.

Peralvo, M. (2004) "Influence of DEM Interpolation Methods in Drainage Analysis" *GIS Hydro 04* Austin, TX.

Pike, R.J. (2000) "Geomorphometry-Diversity in Quantitative Surface Analysis" *Progress in Physical Geography: Earth and Environment* 24 (1) pp.1–20 DOI: 10.1177/030913330002400101.

Pike, R.J., Evans, I.S. and Hengl, T. (2009) "Geomorphometry: A Brief Guide" In: Hengl, T. and Reuter, H.I. (Eds) *Geomorphometry – Concepts, Software, Applications*, (Volume 33) Amsterdam: Elsevier, pp.3–30.

Reddy, A.D., Hawbaker, T.J., Wurster, F., Zhu, Z., Ward, S., Newcomb, D. and Murray, R. (2015) "Quantifying Soil Carbon Loss and Uncertainty from a Peatland Wildfire Using Multi-temporal LiDAR" *Remote Sensing of Environment* 170 pp.306–316.

Ricchetti, E. (2001) "Visible-infrared and Radar Imagery Fusion for Geological Application: A New Approach Using DEM and Sun-illumination Model *International Journal of Remote Sensing* 22 (11) pp.2219–2230 DOI: 10.1080/713860801.

Rosenberg, P. (1955) "Information Theory and Electronic Photogrammetry" *Photogrammetric Engineering* 21 (4) pp.543–555.

Rott, H. (2009) "Advances in Interferometric Synthetic Aperture Radar (InSAR) in Earth System Science" *Progress in Physical Geography* 33 (6) pp.769–791.

Sefercik, U.G., Buyuksalih, G. and Atalay, C. (2020) "DSM Generation with Bistatic TanDEM-X InSAR Pairs and Quality Validation in Inclined Topographies and Various Land Cover Classes" *Arabian Journal of Geosciences* 13(560) pp.1–15 DOI: 10.1007/s12517-020-05602-5.

Sitabi, A. (2015) "DEM Analysis – The Many Uses and Derivatives of a Digital Elevation Model" *Earth Data Analysis Center, University of New Mexico* Available at: https://rgis.unm.edu/dem_analysis/.

Soergel, U., Jacobsen, K. and Schack, L. (2013) "The TanDEM-X Mission: Data Collection and Deliverables" In Fritsch, D. (Ed.) *Proceedings of Photogrammetric Week '13* Berlin and Offenbach: Wichmann/VDE Verlag, pp.193–203

Thuan, C. and Lindenschmidt, K.E. (2017) "Comparison and Validation of Digital Elevation Models Derived from InSAR for a Flat Inland Delta in the High Latitudes of Northern Canada" *Canadian Journal of Remote Sensing* 43 (2) pp.109–123 DOI: 10.1080/07038992.2017.1286936.

Tsai, F., Hwang, J-H., Chen, L-C. and Lin, T-H. (2010) "Post-disaster Assessment of Landslides in Southern Taiwan After 2009 Typhoon Morakot Using Remote Sensing and Spatial Analysis" *Natural Hazards and Earth System Sciences* 10 pp.2179–2190 DOI: 10.5194/nhess-10-2179-2010.

Uppuluri, A.V. and Jost, R.J. (2006) "An Application of the SAR Image Processing Toolkit: InSAR" *IEEE Conference on Radar* Verona, NY, 24th–27th April DOI: 10.1109/RADAR.2006.1631826.

Westoby, M.J., Brasington, J., Glasser, N.F., Hambrey, M.J. and Reynolds, J.M. (2012) "'Structure-from-Motion' Photogrammetry: A Low-Cost, Effective Tool for Geoscience Applications" *Geomorphology* 179 pp.300–314 DOI: 10.1016/j.geomorph.2012.08.021.

Yue, L., Shen, H., Yuan, Q. and Zhang, L. (2015) "Fusion of Multi-scale DEMs Using a Regularized Super-resolution Method" *International Journal of Geographic Information Science* 29 (12) pp.2095–2120.

17
EXTENDED REALITY (XR)

Łukasz Halik and Alexander J. Kent

Introduction

Interest in extended reality (XR) technologies has grown over the last few years, with some estimates of their market value predicting a rise from $31 billion in 2021 to $300 billion by 2024 (Alsop, 2021). The impact of the global pandemic COVID-19 has facilitated this trend, particularly with the possibilities of remote accessibility and experience that these technologies offer, and the rate of innovation is likely to accelerate. This chapter provides an overview of XR technologies and their geospatial applications. It defines the key terminology of XR and identifies the principles that have shaped the advancement of its technologies, before discussing their applications and their wider impact on society. Firstly, however, it defines XR and introduces those technologies that together form its scope.

Definitions

Extended reality (XR) is a broad term that refers to real-and-virtual environments created by computer technology and involving human-machine interaction. The term 'extended reality' derives from the possibility of extending our lived experience with geographic spaces through well-designed, compelling and meaningful ways of accessing information (Çöltekin et al., 2020b). As predominantly digital technologies that are varyingly immersive and/or interactive, XR includes augmented reality (AR), virtual reality (VR), mixed reality (MR), as well as their combinations or inter-dependent technologies, since the boundaries that distinguish between them are blurred. Consequently, XR technologies represent a wide spectrum of experience, which may be described as spanning from the completely real to the completely virtual (Figure 17.1).

Although the terminology has evolved since Milgram and Kishino's (1994) virtuality continuum was devised, their diagram provides a useful starting point for understanding the key technologies of XR and their interdependence. Between the opposing ends of the completely real and completely virtual environments, it distinguishes two technologies: augmented reality (AR) and augmented virtuality (AV). The term 'augmented reality' was introduced by Caudell and Mizell (1992) to describe their apparatus of see-through head-mounted displays for aircraft manufacturing. While AR remains in use, 'augmented

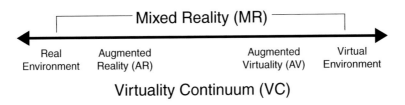

Figure 17.1 Simplified representation of a 'virtuality continuum' (redrawn from Milgram and Kishino, 1994)

virtuality' (AV) has been superseded by virtual reality (VR), which was popularized by Jaron Lanier (co-founder of the Visual Programming Lab) in the mid-1980s.

Nevertheless, the defining characteristics of these technologies have remained largely consistent. Azuma (1997) described AR as a technology that combines real and virtual images (both can be seen at the same time); that allows virtual content to be interacted with in real-time; and that this content is registered in 3D so that the virtual objects appear to be fixed in space. By contrast, virtual reality (VR) is a display and control technology that provides an interactive, multi-sensory, computer-generated, three-dimensional virtual environment (VE) to a user. A virtual environment has been defined by Slocum *et al.* (2008: 521) as 'a 3D computer-based simulation of a real or imagined environment that users are able to navigate through and interact with'. Çöltekin *et al.* (2020a: 258) have more recently described the ideal VR system as providing humans with experiences that 'stimulate all the senses and that are indistinguishable from those that could be real'.

Therefore, AR and VR technologies offer different user experiences. The distinguishing feature of AR is its capacity to display a real-world object's attribute information (or virtual objects associated with a real version that may be hidden at first sight) in front of the user, such as via a smartphone (Figure 17.2). This allows the possibility of increasing the density of information available, such as terrain data, to enhance the spatial awareness of the user. The environmental information provided by VR (Figure 17.3) allows users to experience a more comprehensive simulation by immersing them into past, present or planned real places (or fictitious ones), often with the use of a head-mounted display (HMD). Virtual reality may be used to train users to respond to certain situations that occur in real life, especially in places that are inaccessible due to time, cost or danger.

The technologies represented by mixed reality (MR) in Milgram and Kishino's (1994) diagram arose from developments in AR, rather than from a combination of AR and AV (VR) capabilities. Consumer devices, particularly smartphones, gradually became equipped with sensors that could distinguish between objects that were closer and further away, allowing digital content to be blended with the real-world environment (Figure 17.4). According to Çöltekin *et al.* (2020b), MR occurs when environments contain spatially registered (i.e., real-time georeferenced) virtual objects in the real world and those virtual objects can be obscured by real objects located closer to the observer. Hence, the user experience of MR incorporates elements of the real and virtual environments based on those elements' actual locations and the user's visual perspective of reality.

Not surprisingly, these technologies have developed and diversified since Milgram and Kishino's (1994) continuum was devised. The availability of XR to an increasingly wider consumer base has accelerated its development and broadened its range of applications. Consequently, the definitions and scope of XR technologies have evolved beyond academia.

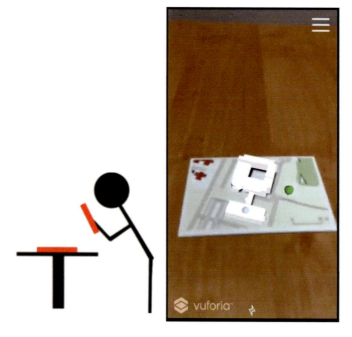

Figure 17.2   An example of an augmented reality (AR) user interface on a smartphone and its method of user interaction. In this case, a QR (quick response) code on a traditional business card is scanned via the camera on a smartphone, which links to a website that generates a 3D model of a building on the screen of the smartphone and superimposes this on the card. Unlike VR, other elements of the real world (i.e., the desk on which the card rests) are visible on the screen simultaneously and remain unaffected

Source: Authors.

Figure 17.3   An example of a virtual reality (VR) user interface via an HMD (head-mounted display) and its method of user interaction. In this arrangement, the user is seated and controls their movement in the virtual environment through a hand-held controller. Unlike AR (augmented reality), no aspect of the real world is visible

Source: Authors.

Extended reality (XR)

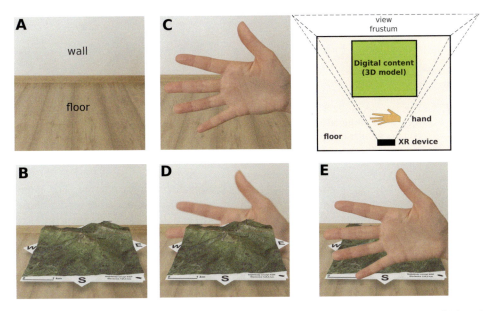

Figure 17.4 An example of the difference between augmented reality (AR) (images B and D) and mixed reality (MR) (images B and E)

Source: Authors.

In an attempt to standardize understanding of XR technologies amongst producers and consumers, the Consumer Technology Association (2020) offers the following definitions:

- *augmented reality* (AR) overlays digitally created content into the user's real-world environment. AR experiences can range from informational text overlaid on objects or locations to interactive photorealistic virtual objects. AR differs from Mixed Reality in that AR objects (e.g., graphics, sounds) are superimposed on, and not integrated into, the user's environment.
- *virtual reality* (VR) is a fully immersive user environment affecting or altering the sensory input(s) (e.g., sight, sound, touch and smell) and can allow interaction with those sensory inputs by the user's engagement with the virtual world. Typically, but not exclusively, the interaction is via a head-mounted display, use of spatial or other audio and/or motion controllers (with or without tactile input or feedback).
- *mixed reality* (MR) seamlessly blends a user's real-world environment with digitally created content, where both environments coexist to create a hybrid experience. In MR, the virtual objects behave in all aspects as if they are present in the real world – e.g., they are occluded by physical objects, their lighting is consistent with the actual light sources in the environment, [and] they sound as though they are in the same space as the user. As the user interacts with the real and virtual objects, the virtual objects will reflect the changes in the environment as would any real object in the same space.

In short, these definitions focus on how the user experience differs: they either see digital content as overlaid onto their real-world environment (AR); or they are fully immersed in a digital environment (VR); or they experience a hybrid environment that combines elements of both (MR). As these technologies have become more consumer-driven than their

Figure 17.5 Level of detail (LOD) for buildings according to the OGC CityGML 2.0 standard, which was proposed by the Open Geospatial Consortium in 2012 and remains the most popular standard adopted for building 3D city models

Source: Authors.

predecessors, consumer-oriented definitions have become increasingly relevant. The extension of lived experience offered by these technologies also incorporates other modes of user engagement, such as audio, haptic, smell and touch as well as sight. However, since the latter monopolizes our attention (up to 70% according to Heilig, 1992), this chapter focuses on the visualization and display of XR technologies and their applications.

Key concepts

Extended reality technologies rely on two fundamental principles: immersion and interactivity. Immersion is the extent to which technology forms a vivid illusion of reality as sensed by the participant. This combines the illusion of movement of the observer and what they observe with the illusion of depth. The goal of immersion is to create a sense of presence; that is, what Slater and Wilbur (1997) have called 'an invariant sense of being there'. Hence, the observer is conscious of being within a virtual environment, but their awareness of this – as a computer-generated space – is minimized. Hence, the user believes that they are situated inside a virtual environment, firstly, through a personal presence (the extent to which a person feels like he or she is part of the virtual environment); secondly, through a social presence (the extent to which other beings, whether living or synthetic, co-exist within the virtual environment); and thirdly, through an environmental presence (the extent to which the virtual environment acknowledges and reacts to the user).

As XR technologies have developed, their capacity to provide a more immersive sense of illusion has increased. This ranges from creating the illusion of depth with stereoscopy (where each eye is presented with a different image of a scene) to presenting different images and changing the view of these relative to the observer. Similarly, as the Level of Detail (LOD) in digital environments has increased, this has enhanced the user's sense of immersion (Figure 17.5). The classification of hardware (devices), human behaviour (movement) and geovisualization (data dimensions) in Figure 17.6 illustrates the various methods for constructing the user's sense of illusion in XR technologies.

In addition to immersion, XR technologies have developed to offer higher levels of interactivity. As the rate of human-computer interaction evolved through the 1960s to emulate the speed of person-to-person conversation, the introduction of graphical user interfaces in the 1970s and 1980s brought greater accessibility, which, with higher connectivity between computers and portability of devices, particularly smartphones, increased the level of human-computer interaction further (Gaines, 2019). This trend is set to continue, since the demand for better interaction hardware and software is higher than ever (Doolani et al., 2020).

In XR, interactivity concerns how users are able to perform real-time actions (whether their input is from a keyboard, joystick or by making hand gestures) to which the simulated

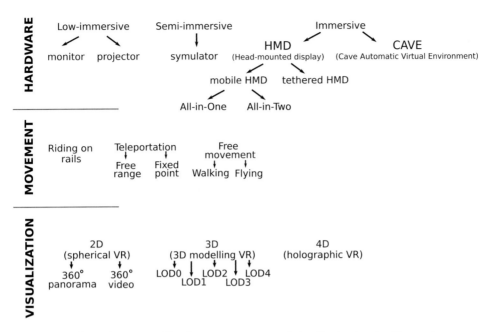

Figure 17.6 The various types of hardware, movement, and visualization involved in creating a sense of illusion with XR technologies

Source: Authors.

environment reacts and responds. The concepts of immersion and interactivity are therefore closely related within XR, since a greater level of interactivity increases the user's sense of environmental presence. The development of XR technologies that maximize the capabilities of hardware and software to combine immersion and interactivity will continue to shape the evolution of the user experience.

Technological approaches

Augmented reality

AR devices employ different technologies that may be distinguished by how they display virtual content (Figure 17.7). Specifically, these two methods are Video Pass-Through and Optical See-Through. The first of these simulates stereopsis (depth perception) by using a head-mounted display (HMD) and a camera to capture a video image of the surroundings. The virtual content is then blended with the video image of the real world and displayed on a small screen within the HMD, close to the user's eyes. By contrast, the Optical See-Through method displays the virtual content on a semi-transparent mirror or glass that is placed in front of the user's eyes, allowing them to see the real world through this apparatus.

Amongst the most significant recent developments in AR technology are the incorporation of biometric sensors (for eye-tracking) in HMD devices and the introduction of controller-less interaction with digital objects (hand tracking). This includes, for example, Microsoft's Hololens 2, which was released in autumn 2019. In the future, the development of AR glasses (smart glasses) as a street-wearable accessory is likely to reduce the demand for, and use of, handheld screens.

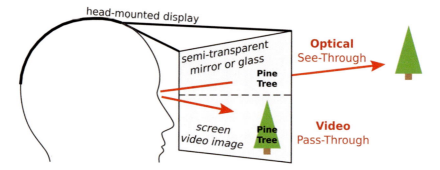

Figure 17.7 The different AR display technologies, using video or optical methods
Source: Authors.

Virtual reality

Virtual reality technologies work by directly coupling the viewer's position with the image shown on their display (Kraak, 2002), usually on a desktop monitor (low-immersive), simulator (semi-immersive) or head-mounted display (HMD) (immersive). The user's sense of immersion is increased when they are surrounded by the display, either through the use of an HMD or on walls surrounding the user in a room-sized cube, the so-called cave automatic virtual environment (CAVE). This sense is enhanced further by greater interactivity, especially when the user's motor actions are linked directly to the virtual environment (Cruz-Neira *et al.*, 1992).

The most common and affordable VR technology is currently HMD-based VR. This is capable of delivering an immersive and immediate sensory experience of simulated environments (which may or may not resemble reality). The two most important components of HMD hardware are its display and lenses. The display shows two images, one for each eye, whereas the lenses (placed close to the eyes) are used to alter the user's field of view. In addition, a gyroscope (which measures yaw, pitch and roll) and an accelerometer (which measures the speed of the user's head movement) are often incorporated within HMD apparatus.

As with developments in AR, the most important innovation in recent HMD VR technology is the incorporation of biometric sensors. These include eye-tracking, such as the FOVE 0 (2016), heart-rate and pupil dilation monitors and face camera, such as the HP Reverb G2 Omnicept edition (2021) and controller-less interaction with digital objects via hand-tracking, such as the Oculus Quest 2 (2019). These enable the return of a greater level of feedback for measuring the effects of external stimuli on the user, enhancing the capacity for the virtual environment to respond to their actions.

A successful VR system requires hardware, software and user to work well together. Advances in hardware require up-to-date software to function properly, whereas the user experience has to be comfortable and beneficial. Indeed, some systems, such as the Google Glass project, were abandoned because the user experience they offered was unsatisfactory (largely due to the imaging device being located in the far corner of the right-hand lens). The availability of an 'ecosystem' of compatible applications (apps) extends the versatility of hardware and provides greater choice for the user (the Oculus VR headset, for example, is supported by numerous games and other apps).

Most XR experiences (apps) are created using native software development kits (SDKs), such as Android, iOS or cross-platform game engines (e.g., Unity3D or Unreal Engine).

These apps need to be downloaded from proprietary online app stores such as Google Play or App Store and installed directly on the device. In the near future, more VR/AR experiences may be created using WebXR standards. One of these is the WebXR Device API (application programming interface), which provides access to input and output capabilities commonly associated with VR and AR devices. This allows a user to develop and host VR and AR experiences on the Web, so that a Web browser is a gateway to those experiences. The combination of the internet with WebXR standards may herald a wave of Web-based 3D spaces in which people gather and collaborate, where moving between these 3D spaces would be as easy as clicking on a link.

As XR technologies become more accessible, current methods of interaction will also evolve. A recent trend is the equipping of VR goggles with a set of cameras that enable the use of video pass-through technology. These cameras record images in increasingly higher quality, providing more realistic images. Although the use of video pass-through technology is currently a cheaper option and more commonly used in consumer VR devices, typical VR goggles are likely to become mixed reality devices as pass-through technology develops (e.g., the Varjo XR-3).

XR applications and their impact on society

The technologies encompassed by XR support a wide range of applications, from gaming to firefighting. As an extension of the visualization of environments presented by maps, they are particularly useful for experiencing situations remotely, such as by emergency services or military personnel. Empirical studies have demonstrated the benefits of 'learning by doing' on memory recall and performance, proving the effectiveness of using XR technologies for training in the use of complex operations, such as in manufacturing (Doolani *et al.*, 2020). Similarly, the gradual incorporation of XR technologies in education settings will accelerate as devices become cheaper and more accessible.

The possibility of merging and superimposing 2D or 3D content with or over real-world objects afforded by AR/MR is especially useful where the user's experience of the real world is enhanced – but not interrupted – by the provision of supplemental information. These range from simple navigation aids to the in situ identification of plant species using artificial intelligence (AI). Additionally, the capacity for AR/MR to realize 3D data, such as that of a mountain range triggered by geolocation or a QR code, virtually eliminates the need to use physical materials, including plastic, for the interpretation of complex landscapes.

It has long been recognized that VR offers considerable potential for simulating the exploration of environments that would otherwise be physically inaccessible to the user due to location, scale, time or danger (McGreevy, 1993). As the quality of its image rendition has improved, VR has come to offer the most immersive experience of remote places or spaces and with the benefits of true-to-life scale (1:1). There are significant advantages offered by the combination of accurate and comprehensive topographic data with immersive VR capabilities, particularly for real-world user-groups and applications that require such datasets, such as emergency services (police, paramedic, fire and rescue), organizations engaged in disaster relief, urban planning, transport, architecture and utility companies (Halik and Kent, 2021). Furthermore, VR will be the next-generation platform for social media, facilitating social interactions through purpose-built VR platforms, such as VRChat, Rec Room, Mozilla Hubs and Meta Horizon Worlds.

Although XR technologies have a far-reaching impact on many aspects of society, perceptions have not always been positive. In popular culture, for example, the use of AR in *The*

Terminator (1984) as the dynamic visual interface of a killer-cyborg, presents an incarnation of XR as a destructive, 'alien' technology that constitutes a threat to humanity. Similarly, films such as *Tron* (1982), *The Lawnmower Man* (1992) and *The Matrix* (1999) have provided a frame of reference for the concept of VR that has also tended to present the technology in a negative light (e.g., as a means of exerting the power to control a population), and therefore as a technology to be feared. Nevertheless, that power is also recognized by initiatives to build resilience in vulnerable communities. The UN film *Clouds Over Sidra* (2015), which features the Za'atari Refugee Camp in Jordan that is home to over 80,000 Syrians fleeing war and violence, uses the medium of VR to generate greater empathy and new perspectives on people living in conditions of great vulnerability, and has been twice as effective in raising funds (UN SDG Action Campaign, n.d.).

The harvesting of biometric data through XR technologies presents questions over the ethics of their use. Although these technologies offer more scope for bringing people together in a 3D space, the capturing of eye-tracking data to indicate attention hotspots and pupil dilation data to indicate excitement or boredom, for example, could be used for neuromarketing campaigns that target advertisements according to user behaviour.

More generally, the question of whether user behaviour in VR can affect social schema has wider ethical significance. Since there are fewer consequences if real-world rules are not obeyed in VR, whether this leads to the erosion of rule-keeping in the real world merits further investigation. Certainly, users appear to apply real-world schema in VR, such as not walking through walls (Halik and Kent, 2021). With AR, the lower level of user engagement with digital content suggests fewer ethical concerns, although the gathering, sharing and display of personal data – for instance the use of face-recognition technology to access information about individuals – raises deeper questions. Other issues, such as VR sickness (with symptoms similar to motion sickness), the requirement for geovisualizations to be converted to a cartesian coordinate system and the need for intermediate software (e.g., for 3D modelling) present technical, if not ethical, challenges to be overcome.

Conclusion

Since their infancy, XR technologies have sought to extend and enhance the scope of human experience. The rapid pace of their development – encompassing the full gamut from AR to VR – is set to accelerate as devices become more accessible, affordable and applicable. In addition, the evolution from single-purpose to multi-purpose virtual environments (which may be considered as a 'canvas' that allows various potential applications to be developed) will increase the versatility and diversity of XR technologies. As the ethical dimensions of these technologies, particularly VR, are explored more fully with their widespread implementation, it is likely that regulation will be introduced to protect users and their data.

References

Alsop, T. (2021) "Extended Reality (XR): AR, VR, and MR – Statistics & Facts" Available at: https://www.statista.com/topics/6072/extended-reality-xr/ (Accessed: 15th September 2021).

Azuma, R. (1997) "A Survey of Augmented Reality" Presence: Teleoperators and Virtual Environments 6 (4) pp.355–385.

Caudell, T.P. and Mizell, D.W. (1992) "Augmented Reality: An Application of Heads-Up Display Technology to Manual Manufacturing Processes" *Proceedings of the Twenty-Fifth Hawaii International Conference on System Sciences* pp.659–669 DOI: 10.1109/HICSS.1992.183317.

Çöltekin A., Griffin, A.L., Slingsby, A., Robinson, A.C., Christophe, S., Rautenbach, V., Chen, M., Pettit, C. and Klippel, A. (2020a) "Geospatial Information Visualization and Extended Reality Displays" In Guo, H., Goodchild, M.F. and Annoni, A. (Eds) *Manual of Digital Earth* Singapore: Springer, pp.229–277 DOI: 10.1007/978-981-32-9915-3_7.

Çöltekin, A., Lochhead, I., Madden, M., Christophe, S., Devaux, A., Pettit, C., Lock, O., Shukla, S., Herman, L., Stachoň, Z., Kubíček, P., Snopková, D., Bernardes, S. and Hedley, N. (2020b) "Extended Reality in Spatial Sciences: A Review of Research Challenges and Future Directions" *International Journal of Geo-Information* 9 (439) pp.1–29 DOI: 10.3390/ijgi9070439.

Consumer Technology Association (2020) *CTA Standard: Definitions and Characteristics of Augmented and Virtual Reality Technologies (CTA-2069-A)* Available at: https://cta.tech/standards (Accessed: 26th July 2021).

Cruz-Neira, C., Sandin, D.J., DeFanti, T.A., Kenyon, R.V. and Hart, J.C. (1992) "The CAVE: Audio Visual Experience Automatic Virtual Environment" *Communications of the ACM* 35 (6) pp.64–72.

Doolani, S., Wessels, C., Kanal, V., Sevastopoulos, C., Jaiswal, A., Nambiappan, H. and Makedon, F. (2020) "A Review of Extended Reality (XR) Technologies for Manufacturing Training" *Technologies* 8 (77) pp.1–20 DOI: 10.3390/technologies8040077.

Gaines, B.R. (2019) "From Facilitating Interactivity to Managing Hyperconnectivity: 50 Years of Human–Computer Studies" *International Journal of Human-Computer Studies* 131 pp.4–22 DOI: 10.1016/j.ijhcs.2019.05.007.

Halik, Ł. and Kent, A.J. (2021) "Measuring User Preferences and Behaviour in a Topographic Immersive Virtual Environment (TopoIVE) of 2D and 3D Urban Topographic Data" *International Journal of Digital Earth* 14 (12) pp.1835–1867 DOI: 10.1080/17538947.2021.1984595.

Heilig, M. (1992) "El Cine del Futuro: The Cinema of the Future" *Presence: Teleoperators and Virtual Environments* 1 (3) pp.279–294 DOI: 10.1162/pres.1992.1.3.279.

Kraak, M-J. (2002) "Visual Exploration of Virtual Environments" In Fisher, P. and Unwin, D. (Eds) *Virtual Reality in Geography* London and New York: Taylor & Francis, pp.58–67.

McGreevy, M.W. (1993) "Virtual Reality and Planetary Exploration" In Wexelbrat, A. (Ed.) *Virtual Reality: Applications and Explorations* Boston, MA: Academic Press Professional, pp.163–198.

Milgram, P. and Kishino, F. (1994) "Taxonomy of Mixed Reality Visual Displays" *IEICE Transactions on Information and Systems* 77 pp.1321–1329.

Slater, M. and Wilbur, S. (1997) "A Framework for Immersive Virtual Environments (FIVE): Speculations on the Role of Presence in Virtual Environments" *Presence: Teleoperators and Virtual Environments* 6 (6) pp.603–616.

Slocum, T.A., McMaster, R.M., Kessler, F.C., Howard, H.H. and McMaster, R.B. (2008) *Thematic Cartography and Geographic Visualization* (3rd ed.) Upper Saddle River, NJ: Prentice Hall.

UN SDG Action Campaign (n.d.) "Syrian Refugee Crisis - Clouds over Sidra" Available at: http://unvr.sdgactioncampaign.org/cloudsoversidra/#.YLoB0_kzZEY (Accessed: 26th July 2021).

18
FREE AND OPEN SOURCE SOFTWARE FOR GEOSPATIAL APPLICATIONS (FOSS4G)

Rafael Moreno-Sanchez and Maria Antonia Brovelli

What is FOSS4G?

The concepts of Free and Open Source Software (FOSS) and FOSS for Geospatial Applications (FOSS4G) refer to software, code libraries, and platforms that are created, developed, maintained, and distributed following the philosophies of 'Free Software' and 'Open Source Software'. These two philosophies have similarities and are related. 'Free' refers to freedom, not to the absence of cost. Free Software provides the freedom to run, copy, distribute, study, change, and improve the software (GNU, 2020). 'Open Source' refers to the software that not only provides access to its source code, but also complies with ten criteria listed in the Open Source Initiative (https://opensource.org/osd). Stallman (2009) discusses the similarities, complementarity, and consequences of the differences between these two philosophies.

FOSS4G is also immersed in the broader philosophy of 'Openness'. Openness promotes transparency and no-cost unrestricted access to data, information, knowledge, or technologies with emphasis on collaborative development, management, and decision-making (Moreno-Sanchez, 2018). Compliance with the principles of Openness is often denoted by the use of the word 'Open', and it is used in many and diverse fields of science, technology, government, education, and the arts. The following are closely related and relevant to the area of FOSS4G: Open Science, Open Education, Open Hardware, Open Innovation, Open Data, and Open Standards. These areas of Openness complement and reinforce each other and there are significant synergies and positive impacts when there is concurrent adherence to all of them (Coetzee *et al.*, 2020). Because of their relevance to the development and use of FOSS4G, a brief overview of Open Standards and Open Data are presented next.

There are numerous studies aimed at understanding the motivations of individuals and organizations that contribute to FOSS/FOSS4G projects (e.g., Ke and Zhang, 2009; Baytiyeh and Pfaffman, 2010; Andersen-Gott *et al.*, 2012). Each FOSS4G software is developed and maintained by its user and developer communities, which are distributed around the world. Some communities are immersed in government organizations (e.g., http://step.esa.int/main/community/), while others are more heterogeneous and independent (e.g., https://grass.osgeo.org/support/community/). In both cases, what enables this distributed development approach is compliance with Open Standards and the coordination and oversight provided by organizations such as the Open Source Geospatial Foundation (OSGeo, see https://www.osgeo.org/) and the support of the communities of users and developers of each FOSS4G (e.g., QGIS, see https://www.qgis.org/en/site/forusers/support.html). Also, organizations such as GeoForAll (https://www.osgeo.org/initiatives/geo-for-all/) and the OSGeo Education Initiative (https://wiki.osgeo.org/wiki/Edu_current_initiatives) promote communication and cooperation between FOSS4G researchers and academic institutions working on their development, applications, and geospatial education using FOSS4G.

Open Standards are technical documents that detail interfaces or encodings for geospatial data, software, or online services development. Open Standards are freely and publicly available, non-discriminatory, free of licence fees, vendor neutral, data neutral, and agreed upon by a formal member-based consensus process (OGC, 2020). When developers working independently create software or online services that comply with Open Standards, the resulting products can seamlessly exchange data and work together without further debugging or modifications (i.e. they are interoperable; Percivall, 2010). The leading organizations working on the creation of Open Standards for geospatial data, software, and technologies are the Open Geospatial Consortium (OGC; https://www.opengeospatial.org/) and the International Organization for Standardization (ISO; https://committee.iso.org/home/tc211). The benefits of interoperability of geospatial software and data are numerous and important (Hamilton, 2005; Percivall, 2010). Among them are (Percivall, 2010): improved sharing, integration, and reuse of data resulting in efficiencies and decreased costs; ability to choose the best tool or software for the job from multiple sources reducing the risk of being locked with a single vendor; ability to have diverse software pieces of a solution work together seamlessly.

Open Data is data that can be freely used, re-used, and distributed by anyone, subject only, at most, to the requirement of attributing its source (Open Data Handbook, 2020). Geospatial Open Data applies the principles of 'Free' and 'Open' to geospatial data allowing communities to collaborate on a data product (OSGeo, 2020). There are political, social, economic, and technical/operational areas in which Open Data could create benefits (e.g., self-empowerment of citizens; stimulation of innovation; reduced data collection redundancy). At the same time, given the nature and speed of innovation, it is difficult to predict exactly where and how some of these benefits will be realized. Discussions and lists of the benefits, barriers to adoption, and myths of Open Data are presented by Janssen *et al.* (2012). Also, some studies point to potential risks of Open Data such as the possibility of privacy violations or risks to the security of critical infrastructure and fraud (e.g., publication of inaccurate or improper data) (Kucera and Chlapek, 2014). Regardless, Open Data supports (The White House, 2013, 2016; Lathrop and Ruma, 2010; Kitchin, 2014): Interoperability of data and information; data reuse and integration; transparency and democratic control; diverse stakeholder participation; self-empowerment; innovation; measurement of impact of activities and policies; creation of new knowledge by combining data sources and partners into large data volumes; and improvement of efficiency or effectiveness of government services.

A brief history of FOSS4G

The history of FOSS goes back to the early days of the computer industry in the 1950s. At that time closed/proprietary software was rare or nonexistent. Early programmers in this decade such as those that formed the SHARE Program Library Agency worked in academia and industry research labs and developed a 'hacker' culture of openly sharing and collaborating to improve their code and advance computer science (Raymond, 1999). The importance, need, and principles of the FOSS movement re-emerged strongly in the 1980s (e.g., creation of the Free Software Foundation in 1985) as a response to the progressive closing of source code and privatization of software. Today some of the most reliable and influential software in the computer industry are FOSS (e.g., Unix operating system; Apache web-server). An in-depth analysis of the key historical developments of the FOSS movement is provided by Tozzi (2017).

More specifically, FOSS4G has a history dating back to the late 1970s. The OSGeo provides a detailed timeline for the historical evolution of some of the most important FOSS4G initiatives (see https://wiki.osgeo.org/wiki/Open_Source_GIS_History). The Map Overlay and Statistical System (MOSS) was designed in 1977 (Reed, 2004) and GRASS GIS (https://grass.osgeo.org/) has been in continuous development since 1982 (Mitasova and Neteler, 2004; Neteler, 2013). Both were early examples of geographic information systems (GIS).

In the 1990s, the growth of the Internet and of the World Wide Web had major impacts on the growth of FOSS4G. Communication and collaboration became easier and the need for coordination and creation of Open Standards became more necessary. The Open Geospatial Consortium (OGC; https://www.ogc.org/) was founded in 1994, with the mission of creating – through a consensus-based collaborative process – Open Standards that facilitate the creation of geospatial software and technologies. In 1995, the web mapping software UNM MapServer was created, allowing use of the WWW for accessing and distributing maps and geospatial functionality (see https://www.osgeo.org/projects/mapserver/). In 1998, the development of the GDAL/OGR software library began (see https://www.osgeo.org/projects/gdal/) with the purpose of reading, writing, and translating between raster and vector geospatial data formats.

Starting in the 2000s the number and capabilities of FOSS4G operating on local computers or using the WWW as a platform grew very quickly. One of the currently most popular desktop GIS systems QGIS started in 2002 together with others like gvSIG in 2003 and UDig in 2004. In 2001, PostGIS (https://postgis.net/) was released as an extension to the powerful and popular Database Management System (DBMS) PostgreSQL (https://www.postgresql.org/) enabling it to store, query, and analyse spatial data in the DBMS. In the same year, another powerful web mapping software, GeoServer, was released.

In 2004, the web map service OpenStreetMap (OSM) was created, which is the equivalent of the private/closed web service Google Maps with the added advantage that any individual or organization can contribute to the data provided through OSM (i.e., through what is known as crowdsourcing). The rapid growth of FOSS4G created the need for an organization to coordinate the development and foster the use of high-quality FOSS4G; in 2006 the OSGeo Foundation (https://www.osgeo.org/) was created for this purpose. Today this organization supports several of the leading FOSS4G for addressing diverse geospatial information needs (see https://www.osgeo.org/projects/).

The current state of FOSS4G

Today, FOSS4G offers at least one mature, capable, and reliable software, code library, or development platform for almost every geospatial technology area (e.g., desktop GIS; web-GIS; spatially enabled DBMS; big data; remote sensing; real-time data; web services) and geospatial information need (e.g., geospatial data collection, storage, management, analysis, and distribution) (Moreno-Sanchez, 2012; Brovelli *et al.*, 2017; Coetzee *et al.*, 2020). Currently, the OSGeo-Live DVD (see http://live.osgeo.org/en/index.html) functions as a portal to access and download the most comprehensive list of popular high-quality FOSS4G.

Several studies have presented how FOSS4G is being used in different contexts and applications: Jolma et al. (2008a), Sanz-Salinas and Montesinos-Lajara (2009), Steiniger and Bocher (2009), Garbin and Fisher (2010); Steiniger and Hunter (2012); Horvat (2013), Brovelli et al. (2017), and Coetzee et al. (2020). As seen in these reports, the FOSS4G software ecosystem is large and diverse. Some FOSS4G are used together to create software stacks (i.e., a set of individual software that complement each other) to seamlessly address the whole geospatial information workflow from data collection, storage, analysis, tools development, and information delivery in printed or digital form (see Jolma et al., 2008b for FOSS4G stack examples). The decision about which FOSS4G or stack is best depends on the specific socio-cultural, financial, institutional, and technological context in which the software is to be used.

FOSS4G is used by governments, private companies, universities, leading research centers, non-profit organizations, and users' communities in developed and in developing countries to address their needs for geospatial data collection, management, analysis, and distribution, as well as for geospatial education and research. Brovelli et al. (2017) and Coetzee et al. (2020) provide numerous web links to resources and references to studies and reports of how FOSS4G is being used to create large-scale, sophisticated, mission-critical systems and applications developed in diverse technological, institutional, socio-cultural, and economic contexts around the globe.

FOSS/FOSS4G have been declared essential for the developing world (Naronha 2003; Camara and Fonseca, 2007; Sowe, 2011; Choi et al., 2016) because they are developed in a participatory-distributed way that includes local efforts, thus allowing countries to develop their own technology and reducing dependency on imports. They assist in closing the digital divide between developed and developing nations by lowering access barriers (e.g., software and training costs) to sophisticated geospatial software, data, and technologies. They support the creation of National Spatial Data Infrastructures. High-income countries (e.g., in Europe and North America) also have growing interest in FOSS/FOSS4G (White House, 2013; Coetzee et al., 2020) because they boost innovation and collaboration through their culture of open sharing and distributed software development. They enable ease of access to geospatial software and training thus facilitating the distribution of geospatial information resulting in an increase in the value and use of this information. They address some national security and technology independence concerns when relying on foreign closed/proprietary software whose source code is not available for review and modification. It is interesting to note that the largest concentration of the international network of laboratories working on FOSS4G development, education, and research (GeoForAll) is found in Europe (see https://www.osgeo.org/initiatives/geo-for-all/). The GeoForAll mission is to make geospatial education, data, and capabilities available to everyone that needs them around the world.

Today, learning and adopting FOSS4G in diverse contexts and organizations is becoming easier and more common. There are numerous materials (i.e., books, online tutorials, application examples, and code repositories) and resources (i.e., regular meetings of local communities of users, organizations such as OSGeo and GeoForAll, university education programs, consulting companies offering training and technical support services, and large FOSS4G conferences) to support from entry-level users with basic needs to large sophisticated mission-critical deployments in the private sector, government, research, and education arenas. A great resource is Github (https://github.com/) that is used by FOSS4G developers and users as code repository, platform for software development, and learning resource for developers (e.g., GRASS GIS https://github.com/OSGeo/grass).

Findings of studies on the strengths, weaknesses, barriers, or concerns for the use of FOSS/FOSS4G for specific purposes and contexts vary (Erlich and Aviv, 2007; Ven et al., 2008; Ransbotham, 2010; Kuechler et al., 2013; Mount and Fernandes, 2013; Gupta 2018). Regardless, there is ample evidence that FOSS/FOSS4G are capable, reliable, and provide healthy competition for private/closed software while offering opportunities for mutual benefit and complementarity. Major geospatial private/closed software companies make use and support connections to the most popular FOSS4G software (e.g., EsriArcGIS connection to and use of PostgreSQL for geodatabases (see https://desktop.arcgis.com/en/arcmap/latest/manage-data/gdbs-in-postgresql/get-started-gdb-in-postgresql.htm). These companies also recognize the importance and value of adhering to the principles of Openness and have initiatives to support it and contribute to Open Source software efforts (see, for example,https://www.esri.com/en-us/arcgis/open-vision/overview; https://www.esri.com/en-us/arcgis/open-vision/standards/open-source). The combined use of FOSS/FOSS4G and private/closed software to create hybrid information technology infrastructures offers organizations more flexibility to address their geospatial information needs in their specific technological, financial, and socio-cultural contexts (Carahsoft, 2016; Ingold, 2017).

FOSS4G software selection and adoption

Selecting the best individual FOSS4G or FOSS4G stack to address specific geospatial information needs in specific contexts might seem overwhelming given the number and diversity of the options available in the FOSS4G software ecosystem. Often the challenge starts with a lack of awareness of the existence, capabilities, and success stories of these software. However, there are numerous resources and studies that can guide potential users in their evaluation, selection, and adoption of FOSS4G to address their needs; most of them are listed in Brovelli et al. (2017) and Coetzee et al. (2020). The OSGeo Foundation has compiled the best-in-class FOSS4G and provides a helpful interface to assist potential users in their software selection (see https://www.osgeo.org/choose-a-project/).

FOSS/FOSS4G should be evaluated on a par with their private/closed counterparts. Common factors that must be considered in a software evaluation whether it is private/closed or FOSS are (Woods and Guliani, 2005; Chen et al., 2010; Brovelli et al., 2017): technical features, reliability, ease of use, documentation, technical support, customizability and extensibility, costs of training, total cost of ownership, support and maintenance, and management requirements (e.g., budget, in-house personnel expertise, long-term maintainability). Besides these criteria, the evaluation of FOSS4G must include the following criteria (Moreno-Sanchez, 2012; Brovelli et al., 2017): Maturity of the software, code library, or platform (i.e., history, quality of the code, technical capabilities); size, diversity, and geographical distribution of the community of users; size of the developers community; level of activity and contributions of the user and developer communities; quality of documentation; number, quality, and ease of access to tutorials and learning resources; use and compliance with Open Specifications; software has linkages with other FOSS/FOSS4G and has been reviewed and passed through an incubator process carried out by the OSGeo (see http://old.www.osgeo.org/incubator); and software has strong positive OSGeo software metrics (see https://live.osgeo.org/en/metrics.html). Also, Steiniger and Hunter (2013) present an extensive list of criteria that can be used to evaluate FOSS4G prioritizing them based on the software intended final use (business, research, or teaching).

There are numerous studies on the facilitators and barriers for the adoption of FOSS/FOSS4G for diverse purposes in organizations with varied financial, technological, and institutional contexts. The most common inhibitors and facilitators affecting adoption are discussed among others by Hauge *et al.* (2010), Li *et al.* (2011), Macredie and Mijinyawa (2011), Qu *et al.* (2011), Buffett (2014), and Brovelli *et al.* (2017). Some of the inhibitors are lack of awareness of software existence, capabilities, and success stories; concerns regarding service and support; uncertainty regarding total cost of ownership; lack of in-house technical knowledge of the FOSS/FOSS4G; degree of difficulty to integrate with current information technology infrastructure; concerns regarding intellectual property and licensing; concerns regarding software security; current favorable or mandatory arrangements with private/close software vendors; and staff fears of distancing from popular commercial private/closed software. On the other hand, among the facilitators to adoption there are: FOSS/FOSS4G technological and no-cost benefits outweigh its perceived disadvantages or concerns; availability of personnel knowledgeable on FOSS4G; existence of a FOSS4G promoter in the organization; support from top management for FOSS use; limited financial resources force the consideration of FOSS4G; FOSS4G can help bridge organizations' or countries' technological and financial disparities and facilitate cooperation. Wheeler (2015) discusses some of the most common concerns and negative myths about adopting FOSS (e.g., reliability, performance, scalability, security, and total cost of ownership). Through references to research and usage statistics, he shows that these concerns are unfounded.

Importance and relevance of FOSS4G to society and current challenges

Current social, economic, and environmental challenges are growing beyond traditional spatial, temporal, and jurisdictional scales and boundaries. Local and regional issues must be analysed and addressed considering the interactions of multiple socio-economic and environmental processes that span larger geographical areas, longer time spans, and multiple socio-political/economic/institutional jurisdictions. Conversely, macro-level (i.e., national or global) policies and actions must consider their short-term and mid-term socio-economic and environmental impacts at the local/regional level. This requires the inclusion of multiple stakeholders (i.e., governments, private companies, universities, scientists, policy makers, and general public) with very diverse geographical information needs and socio-cultural, economic, technological, and institutional contexts. This complexity requires the inclusion of the best scientific and local/traditional knowledge, as well as the use of diverse best-in-class geospatial software and technologies that are appropriate for each context and capable of addressing each need for information and decision support. FOSS4G and private/closed software solutions can be used exclusively or in hybrid information technology infrastructures that include pieces of both. This means more options, more flexibility, and improved capability to better fit the large diversity of contexts, stakeholders, and geospatial information needs that need to be addressed.

Current multi-scale, multi-dimensional challenges require governments, companies, scientists, policy makers, and citizens willing to work together in a transparent, collaborative, and democratic way facilitating access and distribution of data, information, knowledge, and technologies. These are the characteristics that define the development, distribution, and application of FOSS4G, as well as the members of their communities of developers. Some examples of FOSS/FOSS4G projects working on current challenges are: The Global Open Data for Agricultural and Nutrition (GODAN; https://www.godan.info/); the diverse projects carried out by the Humanitarian OpenStreetMap Team (HOT; https://www.hotosm.org/); the United Nations Open GIS Initiative (http://unopengis.org/unopengis/main/main.php);

Open Climate Workbench (https://climate.apache.org/); and the diverse research and education projects by the GeoForAll international network of FOSS4G labs (https://www.osgeo.org/initiatives/geo-for-all/).

FOSS4G not only offers best-in-class software, but also because of its adherence to the philosophy of Openness it has multiple important benefits besides the reasons previously listed for their adoption in developing and developed countries. FOSS/FOSS4G supports (Ramachandra and Kumar, 2009; Brovelli et al., 2017; Coetzee et al., 2020): development of local capacities (human and technological); increased communication, networking, collaboration, and synergies between organizations and individuals around the world; increased creativity and innovation; improved access, use, modification and sharing of geospatial science and education; lower barriers to the access, distribution, and use of geospatial data and technologies; increased engagement and empowerment of diverse stakeholders; democratization of geospatial information and technologies; and reduced risk of being locked in with a single software vendor or solution. These features coincide with essential steps required to move toward Sustainable Development in developed and developing countries such as (Harris, 2009; Becker, 2014; Mulligan, 2014): increase resilience, learning capacity, and adaptability; reduce hazards, risks, and vulnerability; develop local capacities; improve self-reliance; provide equal access; strengthen social networks; enhance communication and cooperation; and promote democracy and stability.

FOSS4G and their communities of users and developers are mature, capable, and ready to tackle the mission-critical tasks necessary to convert data into actionable information and knowledge for the management and solution of the most pressing issues of our time, from the local to the global, from individuals to societies, and from the short-term to the long-term. As important, if not more, as offering high-quality software, FOSS4G offers passionate and knowledgeable communities of users and developers distributed around the world with a frame of mind that emphasizes values that we must remember and apply more than ever in our time: openness, communication, collaboration, equal access, empowerment of stakeholders, democracy, and altruism.

References

Andersen-Gott, M., Ghinea, G. and Bygstad, B. (2012) "Why Do Commercial Companies Contribute to Open Source Software?" *International Journal of Information Management* 32 (2) pp.106–117 DOI: 10.1016/j.ijinfomgt.2011.10.003

Baytiyeh, H. and Pfaffman, J. (2010) "Open Source Software: A Community of Altruists" *Computers in Human Behavior* 26 (6) pp.1345–1354 DOI: 10.1016/j.chb.2010.04.008.

Becker, P. (2014) *Sustainability Science: Managing Risk and Resilience for Sustainable Development* Amsterdam: Elsevier.

Brovelli, M.A., Minghini, M., Moreno-Sanchez, R. and Oliveira, R. (2017) "Free and Open Source Software for Geospatial Applications (FOSS4G) to Support Future Earth" *International Journal of Digital Earth* 10 (4) pp.386–404 DOI: 10.1080/17538947.2016.1196505.

Buffett, B. (2014) "Factors Influencing Open Source Software Adoption in Public Sector National and International Statistical Organizations" *Meeting on the Management of Statistical Information Systems (MSIS 2014)* Dublin, Ireland and Manila, Philippines (Volume 114) Available at: https://www.unece.org/fileadmin/DAM/stats/documents/ece/ces/ge.50/2014/Topic_1_UNESCO.pdf (Accessed: 13th March 2020).

Camara, G. and Fonseca, F. (2007) "Information Policies and Open Source Software in Developing Countries *Journal of the American Society for Information Science and Ttechnology* 58 (1) pp.121–132 DOI: 10.1002/asi.20444.

Carahsoft (2016) "The Benefits of Hybrid GIS" Available at: https://www.carahsoft.com/community/-the-benefits-of-hybrid-gis (Accessed: 24th March 2020).

Chen, D., Shams, S., Carmona-Moreno, C. and Leone, A. (2010) "Assessment of Open Source GIS Software for Water Resources Management in Developing Countries" *Journal of Hydro-Environment Research* 4 (3) pp.253–264 DOI: 10.1016/j.jher.2010.04.017.

Choi, J., Hwang, M.H., Kim, H. and Ahn, J. (2016) "What Drives Developing Countries to Select Free Open Source Software for National Spatial Data Infrastructure?" *Spatial Information Research* 24 (5) pp.545–553 DOI: 10.1007/s41324-016-0051-9.

Coetzee, S., Ivánová, I., Mitasova, H. and Brovelli, M.A. (2020) "Open Geospatial Software and Data: A Review of the Current State and A Perspective into the Future" *ISPRS International Journal of Geo-Information* 9 (2) p.90 DOI: 10.3390/ijgi9020090.

Erlich, Z. and Aviv, R. (2007) "Open Source Software: Strengths and Weaknesses" In St. Amant, K. and Still, B. (Eds) *Handbook of Research on Open Source Software: Technological, Economic and Social Perspectives* Hershey, PA: IGI Global, pp.184–196.

Garbin, D.A. and Fisher, J.L. (2010) "Open Source for Enterprise Geographic Information Systems" *IT Professional* 12 (6) pp.38–45 DOI: 10.1109/mitp.2010.152.

GNU (2020) "What is Free Software?" Available at: http://www.gnu.org/philosophy/free-sw.html (Accessed: 21st February 2020).

Gupta, D. (2018) "Adopting Free and Open Source Software (FOSS) in Education" *Journal of Educational Technology* 14 (4) pp.53–60.

Hamilton, B.A. (2005) "Geospatial Interoperability Return on Investment Study" *National Aeronautics and Space Administration (NASA), Geospatial Interoperability Office* Available at: http://lasp.colorado.edu/media/projects/egy/files/ROI_Study.pdf (Accessed: 24th February 2020).

Harris, J.M. (2009) "Basic Principles of Sustainable Development" In Kamaljit, S. and Seidler, R. (Eds) *Dimensions of Sustainable Development* (*Encyclopedia of Life Support Systems Volume 1*) Oxford: UNESCO, pp.21–41.

Hauge, O., Cruzes, D.S., Conradi, R., Velle, K.S. and Skarpenes, T.A. (2010) "Risks and Risk Mitigation in Open Source Software Adoption: Bridging the Gap Between Literature and Practice" In Agerfalk, P., Boldyreff, C., Gonzalez-Barahona, J.M., Madey, G.R. and Noll, J. (Eds) *Open Source Software: New Horizons* Berlin: Springer, pp.105–118.

Horvat, Z. (2013) "Building Spatial Data Infrastructure Using Free and Open Source Software" *SDI DAYS* Available at: https://bib.irb.hr/datoteka/648040.Hecimovic_Cetl_Ed_Proceedings_SDI_DAYS_2013.pdf#page=145 (Accessed: 2nd March 2020).

Ingold, T. (2017) "How to Implement a Hybrid Architecture for Open Source Geographic Information Systems" Available at: https://thenewstack.io/implement-hybrid-architecture-open-source-geographic-information-systems/ (Accessed: 24th March 2020).

Janssen, M., Charalabidis, Y. and Zuiderwijk, A. (2012) "Benefits, Adoption Barriers and Myths of Open Data and Open Government" *Information Systems Management* 29 (4) pp.258–268 DOI: 10.1080/10580530.2012.716740.

Jolma, A., Ames, D. P., Horning, N., Mitasova, H., Neteler, M., Racicot, A. and Sutton, T. (2008a) "Free and Open Source Geospatial Tools for Environmental Modelling and Management" *Developments in Integrated Environmental Assessment* 3 pp.163–180

Jolma, A., Ames, D.P., Horning, N., Mitásová, H., Neteler, M., Racicot, A. and Sutton, T. (2008b) "Environmental Modeling Using Open Source Tools" Available at: http://fdo.osgeo.org/files/viscom/library/jolma_2006a.pdf (Accessed: 9th March 2020).

Ke, W. and Zhang, P. (2009) "Motivations in Open Source Software Communities: The Mediating Role of Effort Intensity and Goal Commitment" *International Journal of Electronic Commerce* 13 (4) pp.39–66 DOI: 10.2753/jec1086-4415130403.

Kitchin, R. (2014) *The Data Revolution: Big Data, Open Ddata, Data Infrastructures and Their Consequences* London: Sage.

Kucera, J. and Chlapek, D. (2014) "Benefits and Risks of Open Government Data" *Journal of Systems Integration* 5 (1) pp.30–41 DOI: 10.20470/jsi.v5i1.185.

Kuechler, V., Jensen, C. and Bryant, D. (2013) "Misconceptions and Barriers to Adoption of FOSS in the US Energy Industry" In Petrinja, E., Succi, G., El Ioini, N. and Sillitti, A. (Eds) *Open Source Software: Quality Verification* (*OSS 2013. IFIP Advances in Information and Communication Technology Volume 404*) Berlin, Heidelberg: Springer, pp.232–244.

Lathrop, D. and Ruma, L. (2010) *Open Government: Collaboration, Transparency, and Participation in Practice* Sebastopol, CA: O'Reilly Media, Inc.

Li, Y., Tan, C.H., Xu, H. and Teo, H.H. (2011) "Open Source Software Adoption: Motivations of Adopters and Motivations of Non-adopters" *ACM SIGMIS Database* 42 (2) pp.76–94 DOI: 10.1145/1989098.1989103.

Macredie, R.D. and Mijinyawa, K. (2011) "A Theory-Grounded Framework of Open Source Software Adoption in SMEs" *European Journal of Information Systems* 20 (2) pp.237–250 DOI: 10.1057/ejis.2010.60.

Mitasova, H. and Neteler, M. (2004) "GRASS as Open Source Free Software GIS: Accomplishments and Perspectives" *Transactions in GIS* 8 (2) pp.145–154 DOI: 10.1111/j.1467-9671.2004.00172.x.

Moreno-Sanchez, R. (2012) "Free and Open Source Software for Geospatial Applications (FOSS4G): A Mature Alternative in the Geospatial Technologies Arena" *Transactions in GIS* 16 (2) pp.81–88 DOI: 10.1111/j.1467-9671.2012.01314.x.

Moreno-Sanchez, R. (2018) "FC-35 – Openness" In Wilson, J.P. (Ed.) *The Geographic Information Science & Technology Body of Knowledge* (4th Quarter 2017 ed.) University Consortium for Geographic Information Science DOI: 10.22224/gistbok/2018.1.5.

Mount, M.P. and Fernandes, K. (2013) "Adoption of Free and Open Source Software Within High-velocity Firms" *Behaviour & Information Technology* 32 (3) pp.231–246 DOI: 10.1080/0144929x.2011.596995.

Mulligan, M. (2014) *An Introduction to Sustainability: Environmental, Social and Personal Perspectives* New York: Routledge. Naronha, F. (2003) "Developing Countries Gain from Free/Open-Source Software" *Linus Journal* (20th May) Available at: https://www.linuxjournal.com/article/6884 (Accessed: 27th February 2020).

Neteler, M.G. (2013) "Scaling up Globally: 30 Years of FOSS4G Development" (Keynote) *FOSS4G Central and Eastern Europe 2013* Bucharest, Romania: 16th–20th June Available at: http://openpub.fmach.it/bitstream/10449/22195/10/neteler2013_keynote_scaling_up_foss4g_cee2013%20(3).pdf (Accessed: 27th February 2020).

OGC (Open Geospatial Consortium) (2020) "OGC Standards and Supporting Documents" Available at: https://www.opengeospatial.org/standards (Accessed: 21st February 2020).

Open Data Handbook (2020) "What is Open Data?" and "Why Open Data" *Open Knowledge Foundation* Available at: https://opendatahandbook.org/guide/en/what-is-open-data/ and https://opendatahandbook.org/guide/en/why-open-data/ (Accessed: 21st February 2020).

OSGeo (Open Source Geospatial Foundation) (2020) "What is Open Data?" Available at: https://www.osgeo.org/about/what-is-open-data/ (Accessed: 21st February 2020).

Percivall, G. (2010) "The Application of Open Standards to Enhance the Interoperability of Geoscience Information" *International Journal of Digital Earth* 3 (Supplement 1) pp.14–30 DOI: 10.1080/17538941003792751.

Qu, W.G., Yang, Z. and Wang, Z. (2011) "Multi-level framework of Open Source Software adoption" *Journal of Business Research* 64 (9) pp.997–1003 DOI: 10.1016/j.jbusres.2010.11.023.

Ramachandra, T.V. and Kumar, U. (2009) "FOSS for Geoinformatics (FOSS4G)" *Discussion meeting: Open Source GIS in India* (*CiSTUP Technical Report 2*) Available at: https://www.researchgate.net/profile/T_V_Ramachandra/publication/257030284_Open_Source_GIS_in_India_Present_Scenario/links/02e7e5243f6755065d000000/Open-Source-GIS-in-India-Present-Scenario.pdf#page=5 (Accessed: 14th March 2020).

Ransbotham, S. (2010) "An Empirical Analysis of Exploitation Attempts Based on Vulnerabilities in Open Source Software" *Workshop on the Economics of Information Security 2010* Available at: https://www.econinfosec.org/archive/weis2010/papers/session6/weis2010_ransbotham.pdf (Accessed: 13th March 2020).

Raymond, E.S. (1999) "A Brief History of Hackerdom" In *Open Sources: Voices from the Open Source Revolution* Sebastopol, CA: O'Reilly Media, pp.19–30.

Reed, C. (2004) "MOSS: A Historical Perspective" Available at: https://www.scribd.com/document/4606038/2004-Article-by-Carl-Reed-MOSS-A-Historical-perspective (Accessed: 28th February 2020).

Sanz-Salinas, J.C. and Montesinos-Lajara, M. (2009) "Current Panorama of the FOSS4G Ecosystem" *Upgrade* 10 (2) pp.43–51.

Sowe, S. (2011) "Free and Open Source Software in sub-Saharan Africa" *United Nations University* Available at: https://unu.edu/publications/articles/free-and-open-source-software-in-sub-saharan-africa.html (Accessed: 10th March 2020).

Stallman, R. (2009) "Viewpoint Why 'Open Source' Misses the Point of Free Software" *Communications of the ACM* 52 (6) pp.31–33 DOI: 10.1145/1516046.1516058.

Steiniger, S. and Bocher, E. (2009) "An Overview on Current Free and Open Source Desktop GIS Developments" *International Journal of Geographical Information Science* 23 (10) pp.1345–1370 DOI: 10.1080/13658810802634956.

Steiniger, S. and Hunter, A.J. (2012) "Free and Open Source GIS Software for Building a Spatial Data Infrastructure" In Bocher, E. and Neteler, M (Eds) *Geospatial Free and Open Source Software in the 21st Century* Berlin and Heidelberg: Springer, pp.247–261.

Steiniger, S. and Hunter, A.J. (2013) "The 2012 Free and Open Source GIS Software Map – A Guide to Facilitate Research, Development, and Adoption" *Computers, Environment and Urban Systems* 39 pp.136–150 DOI: 10.1016/j.compenvurbsys.2012.10.003.

Tozzi, C. (2017) *"For Fun and Profit: A History of the Free and Open Source Software Revolution"* Cambridge, MA: MIT Press.

Ven, K., Verelst, J. and Mannaert, H. (2008) "Should You Adopt Open Source Software?" *IEEE Software* 25 (3) pp.54–59 DOI: 10.1109/ms.2008.73.

The White House, President Barack Obama (2013) "Open Government Initiative" Available at: https://obamawhitehouse.archives.gov/open (Accessed: 10th March 2020).

The White House, President Barack Obama (2016) "Open Data: Empowering Americans to Make Data-driven Decisions" Available at: https://obamawhitehouse.archives.gov/blog/2016/02/05/open-data-empowering-americans-make-data-driven-decisions (Accessed: 21st February 2020).

Wheeler, D.A. (2015) "Why Open Source Software / Free Software (OSS/FS, FLOSS, or FOSS)? Look at the Numbers!" Available at: https://dwheeler.com/oss_fs_why.html (Accessed: 10th March 2020).

Woods, D. and Guliani, G. (2005) *Open Source for the Enterprise: Managing Risks, Reaping Rewards* Sebastopol, CA: O'Reilly Media, Inc.

19
APIS, CODING AND LANGUAGE FOR GEOSPATIAL TECHNOLOGIES

Oliver O'Brien

Application programming interface (API) frameworks

Application programming interfaces come in a wide variety of forms, and their level of access can vary from using graphical user interfaces that completely hide the code connection to the service providing the mapping, through using specially crafted URLs to return an image, to function calls to libraries that are compiled directly into a software application. In this section we present, from the highest (easiest/quickest) to lowest level, the range of APIs available to access geospatial technology, focusing principally on Web mapping. These are summarized in Table 19.1 as follows.

Ecosystems

Examples: Mapbox Studio, CARTO Builder, Esri ArcGIS Online, Tableau, Google My Maps, HERE Studio

Geospatial ecosystems are the easiest APIs to use, for developing Web maps, because no programming or scripting knowledge, or webspace, is required. Instead, the APIs are contained within Web applications on the provider's own websites, to which the user's spatial information and configuration is also transferred and resides. The servers also typically contain the base content such as background maps (many based on custom renders of OpenStreetMap data), aerial imagery and routing intelligence.

Ecosystems aim to be a complete solution to the online mapping needs of businesses and for personal projects, with data management and format converting, cartography selection and customization, data analysis and even simple GIS-style spatial operations. However, the capabilities of each ecosystem are limited to what has been made available by the provider themselves – they may not have functionality allowing the advanced user the ability to add custom functionality with code or scripting. Ecosystems generally operate on a 'freemium' business model where limited functionality and bandwidth is available for free, with additional options and resources available on payment of a recurring fee. The platforms are also vulnerable to policy changes and API deprecation by the provider. For example, in late 2019, Google, deprecated its Fusion Tables tool, which allowed for an easy spreadsheet-based interface into creating point maps with popup attribute windows.

API wrappers

Examples: Google Maps Embed API, Bing Maps Embed a Map

API wrappers provide a GUI or simple documentation to allow a custom map to be created, for placement on a website or in an app created by the user. An interactive map component can be obtained, often in the form of simple HTML tags for 'iframe' embedding in another website, rather than requiring scripting or programming to achieve this. These are simple to set up but are still subject to operator bandwidth restrictions, and also require webspace owned and managed by the user.

Static Map Image APIs

Examples: Google Maps Static API, OSM Static Maps, HERE Map Image API, Apple Maps Web Snapshots

Static map image services provide a non-scripting way to allow a bespoke image, for placement on the user's website. This can be as simple as specifying a map size, zoom level, and map centre longitude/latitude, as URL (uniform resource locator) parameters (also known as a 'REST' API) upon which a simple image containing the map is returned. These again are subject to operator quotas. Static map APIs wrappers are widely used by large third-party websites that do not provide their own mapping platform, such as Strava or TripAdvisor, because of more generous quotas. Clicking on a static map will typically then redirect to an interactive equivalent which are more bandwidth intensive.

Map Tile APIs

Examples: HERE Map Tile API, HERE Vector Tile API, Bing Maps Imagery REST Service

The 'XYZ' standard for defining image tiles can be supplied though a REST API. The tiles are defined simply with a URL that includes three positional parameters (X and Y the tile coordinates, and Z the zoom level). Some of the major providers, such as Google, no longer provide XYZ tiles through a REST API because there is less control of the 'brand' – as including the provider logo on every tile is likely undesirable for the consumer.

These map tiles are separate from, and can be layered into a Managed API or Open Framework. This means it is technically possible, for example, to include Microsoft Bing map tiles in a HERE Maps Javascript API, though this is a breach of Microsoft's terms and conditions.

Managed APIs

Examples: Google Maps JavaScript API, HERE Maps API for JavaScript, Microsoft Bing Maps V8 Web Control, Apple MapKit JS

Managed APIs provide structured programmatic access to tools and assets (maps and data) on the providers' servers, allowing richly featured custom mapping applications to be created. The most well-known managed API is the Google Maps JavaScript API, which kick-started 'Web Mapping 2.0', or neocartography, in 2007 (Haklay et al., 2008).It was formally released as a response to 'map mashups' incorporating and customizing the JavaScript code used on the then-new Google Maps website. The bulk of the code remains on the server, and is accessed through documented APIs. Quotas and usage restrictions apply, but users can typically pull in assets and services from other sites, and incorporate local data and imagery without it needing to be placed on the providers' server. These APIs remain extremely popular and have continued to be refined, giving Web application developers access to richer mapping such as street-level photo imagery and 3D buildings. The cartography supplied can often be adapted by configuring API calls appropriately, such as re-colouring roads or omitting labels.

Open frameworks

Examples: OpenLayers, Leaflet, D3, RStudio extensions, Uber Deck/Kepler

In a contrast to the above examples, which are all dependent on normally commercial mapping providers, open frameworks download the complete codebase to the users' server (or a third-party content delivery network) and then document and provide API access to the local code, while allowing for the possibility of refining the code beyond API-level access, if needed by the user. Open Frameworks seek to provide the user with more flexibility and reliability than APIs dependent on a commercial provider, as they are not necessarily dependent on external resources.

Open Frameworks are typically based on JavaScript, as this is the most widely available scripting language available in most people's Web browsers, and JavaScript has seen considerable development over the last decade, significantly optimizing its use for display large volumes of data and content such as 'slippy' Web maps. However, some utilize other languages and then wrap the output in JavaScript for the Web. An example is RStudio; its output, which includes mapping libraries detailed in this chapter, can be wrapped into a website for display in a Web browser.

Spatial servers

Examples: MapServer, GeoServer, Tilecache, Tilestache

Spatial servers predate neocartography and generally involve configuration and programming that is more advanced. Public spatial server installations can be useful for obtaining mapping data or imagery, using REST APIs. The most common spatial servers are MapServer, which uses the C++ programming language, and GeoServer, which is based on Java. These utilize standards for accessing the data, such as WMS for imagery and WFS for vector map features.

Server programming

Examples: R libraries, Java libraries, Mapnik (Python/C++)

Despite the rise of increasingly powerful smartphones and heavily optimized Web browser rending engines, allowing much of the processing to be done locally by the user's device, server-side services remain useful for, in particular, producing raster imagery from complex spatial data sources and serving image tiles. Server programming can make use of APIs that package together spatial-related functions into code that can be accessed via function calls.

For example, Mapnik, a C++ library with additional Python and Node bindings is widely used to produce such imagery; for example, the Stamen Design 'Watercolour' map, which incorporates a number of textures difficult to reproduce with vectors (Watson, 2012).

Table 19.1 Summary of different kinds of API frameworks

Type	Eco-systems	API wrap-pers	Static map APIs	Map tile APIs	Managed APIs	Open frameworks	Spatial servers	Server programming
Programming required	No	No	No	No	No	No	No	Yes
Scripting required	No	No	No	Yes	No	Yes	Yes	No
Web space required	No	Yes	Yes	Yes*	Yes	Yes	Yes	Yes
Quota based/ freemium	Yes	Yes	Yes	Yes	Yes	No	No	No
Quick to implement	No	Yes	Yes	Yes	No	No	No	No
Dependent on third-party availability	Yes	Yes	Yes	Yes	Yes	No	No	No
Server administration required	No	No	No	No	No	No	Yes	Yes

* Unless an ecosystem is used.

Toolkits

While API Frameworks are mainly focused on Web mapping (which may include a Web view within a smartphone app), toolkits (or SDKs – software development kits) provide a deeper level of integration with the native software used within mobile platforms, including Android and iOS. Toolkits are created and provided to developers by the platforms (Google for Android and Apple for iOS) and so tend to be tied to their mapping products (Google Maps and Apple Maps, respectively). Google also provides Maps SDK for iOS and some third parties do provide toolkits too, such as the HERE SDK for Android/iOS.

Google's Maps SDK for Android or iOS give access to Google Maps API functionality at a similar level to the JavaScript API for websites, such as map imagery, points of interest (POIs) and routing, directly into their app. Apple's MapKit similarly allows iOS developers similar access to Apple Maps from within their app. More advanced capabilities, such as access to ground-camera imagery, or indoor mapping, are also often available to SDK users, often as 'beta' level functionality, which is still being refined by the supplier.

The toolkits work in conjunction with geolocation capabilities that are present in most smartphones and many other devices – these typically augment GPS receiver-based positioning with proprietary database-driven location data such as Wi-Fi hotspot and home broadband router locations, mobile phone tower IDs and signal strengths, public Bluetooth tags, accelerometer and compass extrapolation and potentially known common locations and associated temporal patterns by the logged-in user. These capabilities are abstracted away so that if an application requires current location data from the device, it can access it without needing to use the map toolkit.

Apple's geolocation framework is called Core Location and is available across the manufacturer's set of hardware. Google's is called the Geolocation API and is a Web service.

Spatial libraries for popular programming languages

JavaScript/Node.js

JavaScript has evolved into a language that powers client-side interactivity in Web applications in increasingly sophisticated ways, more recently being extended to server-side functionality with solutions such as Node.js, which allow the programmer to retain the lightweight nature of JavaScript and take advantage of its many rich libraries. There are therefore a number of spatial-focused projects that extend JavaScript, allowing for incorporation of spatial operations into a JavaScript-based toolchain.

Turf.js is one of the best-known examples of a JavaScript spatial library. It includes regular spatial operations that would conventionally be found in a standalone GIS, such as joins, transforms and distance calculations, bringing the power of 'proper' spatial manipulation natively to the Web browser or Web server. Like many very new spatial libraries, it only uses the WGS84 coordinate reference system and GeoJSON as its data format.

JavaScript Topology Suite (JSTS) is a fuller JavaScript library that processes geometry that is defined by the Simple Features Specification for SQL. It is a conversion of the Java Topology Suite (JTS) that reflects JavaScript's increased dominance as the Web application language of choice over the older Java platform. Like Turf, it can be used either on the client-side or on a Node.js server. Its range of topological operations include point cloud buffering and unions, which are, for example, used on the client-side on Bike Share Map's Numbers tabs to determine spatial footprints (O'Brien, 2023). It can be directly integrated into the OpenLayers open framework detailed above (OpenLayers, 2023), although the two frameworks have different geometrical objects so a parser is used to convert between the two.

Python

PySAL (Python Spatial Analysis Library) is a key library for managing spatial data in this popular language, which is widely used for data manipulation and processing.

Shapely allows for sophisticated manipulation of spatial data, such as Delauney triangulation, cascaded unions and interior hole removal of polygons.

R

R has greatly increased in popularity in recent years, as it has expanded beyond its academic foundation into the general data science community, with many examples of detailed and powerful libraries (known as packages in R parlance) written to work with it.

The 'ggplot2' package is widely used in R to create graphs and maps. While not designed for geospatial manipulation itself, its combination with the 'sf' package (simple features) allows for spatial analysis to be performed and visualized.

The 'tmap' package is similar to ggplot2, but is specifically designed for thematic map production. Like ggplot2, it uses the sf package to manage the geospatial data.

R also has a Leaflet library allowing for easy integration of this Web mapping framework, and for spatial data managed by R to be quickly and efficiently displayed in a "slippy map" framework familiar to users in a broader context. Leaflet maps can also be directly viewed within the RStudio interactive development environment for R, allowing for rapid processing and analysis of spatial data.

Base libraries

A number of base libraries or geospatial toolsets are very widely used by the full-feature libraries detailed so far, as well as directly in full applications such as QGIS and ArcGIS. The base libraries are typically written in lower-level languages such as C++ and heavily optimized resulting in fast and powerful spatial granular operations, such as reprojection or format conversion. Typically, the popular programming languages, such as JavaScript, Python and R, will have their own libraries that provide bindings to these base libraries, allowing their use for more sophisticated operations by the libraries described above.

Three significant base libraries, widely used in other GIS software and other libraries, are GDAL, PROJ and GEOS. Bindings include, for instance, rgdal, which allows R libraries and direct user scripts to access GDAL that reliably converts a wide variety of spatial file formats as well as carry out core spatial operations. Proj4js is a JavaScript version of PROJ, which spatially defines the main coordinate reference systems (CRS) in use around the world, allowing, in this case, JavaScript packages such as OpenLayers to project between different CRS. Shapely is based on a library called GEOS, which is itself based on JTS.

Programmatic integration with GIS software

Major desktop GIS software solutions, such as QGIS and ArcGIS, have introduced programmatic ways to interact and control the software, from running spatial operations on a batch basis, to directly manipulating, hiding and altering graphical user elements in the software itself.

QGIS

The QGIS Python API, also known as pyqgis, allows any software user or advanced QGIS user to access most functions and elements of QGIS by scripting in Python, including creating layers, manipulating them, and producing a finished map. This means standard GIS operations can be carried out without needing to manipulate the software using keyboard and mouse as normal, but retaining access to the software's rich feature set. The API can also build custom plugins for use in the QGIS software by users, and can alter the functionality of QGIS such as loading in a standard set of input files on launch, and even create a standalone limited version of QGIS set up to do a simple task such as fetching geospatial data from a specific URL and producing print-ready maps in a standard way.

Python code can be executed directly from a console built in to QGIS, or from an editor also built into the software, which allows for script management.

QGIS also has a similar C++ API.

ArcGIS

Esri's ArcGIS similarly has a Python API, which allows for programmatic operation of the software, automating repetitive spatial analysis while retaining the power and familiarity of the desktop GIS, or documenting and performing reproduceable research in a medium such as Jupyter notebooks. It can also be used to perform technical operations such as provisioning users with licences for using the software.

ArcGIS also has a JavaScript API. This is more focused on producing standalone Web applications that can take some of ArcGIS's spatial functionality, encapsulate it in JavaScript functions and allow users of the Web applications to manipulate spatially data provided to them by the app, or their own data. Increasingly, ArcGIS is extending its heavyweight GIS functionality to Web-based spatial frameworks, as other spatial frameworks look to strengthen their GIS capabilities. Increasingly, options are merging. ArcGIS, which started out as a standalone desktop application, is strengthening its Web capabilities, while in the other direction, Web projects that started visualizing spatial data very simply are building out more sophisticated GIS options.

Conclusion

The growth of Web maps, smartphone apps and standalone software products that offer geospatial services relies upon the development of a range of tools, scripts and programming languages, which allow access to external sources of geospatial data or incorporate them for analysis and display to the user. As the scope of geospatial services widens, the integration of programming languages (such as JavaScript, Python and R) with GIS software is set to increase, enhancing its functionality and application across many sectors in society.

References

Haklay, M., Singleton, A. and Parker, C. (2008) "Web Mapping 2.0: The Neogeography of the GeoWeb" *Geography Compass* 2 pp.2011–2039 DOI: 10.1111/j.1749-8198.2008.00167.x.

O'Brien, O. (2023) "Bike Share Map: London" Available at: https://bikesharemap.com/london/.

OpenLayers (2023) "JSTS Integration" Available at: https://openlayers.org/en/latest/examples/jsts.html.

Watson, Z. (2012) "Watercolor Process" Available at: https://hi.stamen.com/watercolor-process-3dd5135861fe.

20
SPATIAL ANALYSIS AND MODELLING

Timofey Samsonov

Spatial analysis and modelling are the core components of geographical information science (or GIScience). In their discourse on the topic, de Smith *et al.* (2018: 22) conclude that spatial analysis is essentially 'the subset of [analytical] techniques that are applicable when, as a minimum, data can be referenced on a two-dimensional frame and relate to terrestrial activities'. It is essential that the results of spatial analysis depend on location, otherwise the phenomena under study does not exhibit spatially varying properties, and a simple non-spatial analysis can be used instead. A time series of carbon dioxide (CO_2) concentrations measured by an automatic station is defined along a single temporal dimension, which is not spatial. However, if a multitude of these stations performs measurements at different locations, their time series begin to characterize the spatiotemporal field of CO_2 concentration, and the context expands automatically from non-spatial to spatial followed by addition of at least two spatial dimensions and a dramatic increase in the complexity of analysis.

Fundamentally, spatial modelling is a two-sided notion. First, spatial modelling concerns spatial data models, which are essentially the rails needed for the implementation of spatial analysis workflows. Spatial data models allow the representation of geographic phenomena as formalized data structures such as vectors and rasters, which are designed for analysis by computer. Second, spatial modelling is the process for constructing spatial data models that reflect the results of spatial analysis operations. For example, the vector spatial model for an air measurement station is a *point* feature, which has a pair of coordinates and a value of (for example, CO_2) concentration. With a series of these points it is possible to interpolate the field of CO_2 concentrations and represent it as a raster data model. In this case, a spatial interpolation is a spatial analysis operation that transforms one spatial data model into another.

This chapter presents a broad range of spatial analysis and modelling methods. Since these components of GIScience are closely interrelated, there is no dedicated focus on the differences between them; instead, the primary focus is on their common foundations and high-level applications related to geographical problem-solving. The suggested foundations include conceptualization, data modelling, spatial relations and basic operations, which are explained consecutively in the sections that follow.

Conceptualization

Conceptualization of the problem under the study is the first step of spatial analysis and modelling. The *Cambridge Dictionary* (2023) defines conceptualization as 'the act or process of forming an idea or principle in your mind'. In GIScience, the term has strong roots in artificial intelligence, where it describes the procedure of defining the objects presumed or hypothesized to exist in the real world and their relationships (Genesereth and Nilsson, 1987).

As an example, imagine the task of selecting the optimum location for a new underground metro station in a large city, such as London or Moscow. There are many actors in geographic space that influence the decision: the existing network of metro lines with a number of stations that already serve some parts of the city; residential and office buildings that generate human traffic; spatially varying geological, ground, vegetation and surface water conditions that limit the areas of construction; land properties that induce similar spatial limitations; electricity, sewerage and other underground communications that must be relocated if they coincide with the areas to be excavated; and many others.

Conceptualization allows the GIS scientist to reveal and relate all the objects needed to solve the problem, and not to omit important aspects for consideration. The entities resulting from conceptualization, however, are not always discrete objects with well-defined boundaries in space. Sometimes they are continuously changing fields, induced either by existing objects or by the physical processes acting in space. One of such fields is the elevation of the Earth's surface. In the case of the metro station construction, this field defines the 3D geometry of the tunnel to a great extent, since the line depth is defined in relation to its elevation. Conceptualization also involves the determination of possible relations and processes that appear in a system of entities, which together define the context of the problem.

Data modelling

After the problem is conceptualized in terms of related entities, their representation models must be selected. Discrete objects are modelled as *features*, while the fields that exist at any point in space are modelled as *coverages*. Sometimes, the relations between discrete entities are so strong that they begin to constitute a *network*, and GIScience provides specialized high-level data structures for modelling such phenomena. There are also specialized data models for representing the data on the Earth's sphere, such as *discrete global grid systems*, which are also briefly covered in this chapter.

A feature is basically an abstraction of real-world phenomena that depends on the scale of observation. Abstraction principles in GIScience are taken from cartography, where all phenomena are traditionally represented as a set of *points*, *lines* and *polygons* on a map (Robinson et al., 1995; Kraak and Ormeling, 2021). Decisions over which feature type is appropriate largely depends on the *scale* of the mapping process, since the scale defines the relationship between distances on a map and the actual distances on the Earth's surface. On a small-scale map (e.g., one that depicts an entire continent on an atlas page or smartphone), it is impossible to show all cities or mountains with their physical boundaries with any accuracy or detail, since they are too small to perceive visually. Consequently, these objects are shown as points on a map, and these points are actually much larger than the corresponding objects. Similarly, rivers are depicted graphically as lines, and only countries are shown as polygons in their actual boundaries, since their relative size allows this. On the contrary, large-scale maps or plans depict both the city boundaries and the rivers as polygonal objects that occupy a certain amount of space.

Spatial analysis and modelling

The principles of feature abstraction in spatial analysis and modelling are similar to cartographic principles, and the basic forms used to represent the object are points, lines and polygons (Burrough *et al.*, 2015; de Smith *et al.*, 2018). However, rather than being dependent on a particular scale, digital feature type selection now depends on the *level of detail* (LOD) that is sufficient to solve the problem. Earlier in this chapter, residential and office buildings were described as generators of human traffic. On a map of the whole city, it is quite reasonable to represent these objects as points, since the actual building geometry does not influence the distribution of human traffic at that scale of problem solving. Similarly, underground electricity cables can be effectively represented as lines. However, large hydrographic objects (such as the River Thames in the example of London) still need to be represented as polygons, since their size remains significant in relation to the whole metropolitan area.

While acknowledging cartography for lending these basic forms, GIScience explores the object representation problem in more depth, since it adopts appropriate feature types for handling spatial data in different ways. Most commonly, features are represented as so-called *simple features* (SF), a standardized way to model discrete objects in space (OGC, 2010). A simple feature consists of the components called *attributes*: a spatial attribute, which describes where on the Earth the feature is located, and a set of non-spatial attributes that describe all other properties of the object. A spatial attribute of a simple feature is composed of points, which can be two-, three-, or four-dimensional. Whereas the first two dimensions are commonly dedicated to projected or geographic coordinates, the remaining two can be used to represent elevation, time, distance or some other supplementary attribute which positions the point in space-time. All simple feature geometries consist of points, and there are 17 feature types in total according to the Open Geospatial Consortium (OGC, 2010). However, only seven of these (POINT, LINESTRING, POLYGON, MULTIPOINT, MULTILINESTRING, MULTIPOLYGON and GEOMETRYCOLLECTION) are commonly used in spatial analysis (Pebesma, 2018) and are summarized in Figure 20.1a.

The non-spatial attributes of a feature are numerical, text, Boolean, date/time and other types of variables that describe the entity. For example, a land property polygon may have an *integer* type attribute with a unique identifier of the land property, a *text* type attribute with the name of the owner and a *date* type attribute with the date of the property's registration.

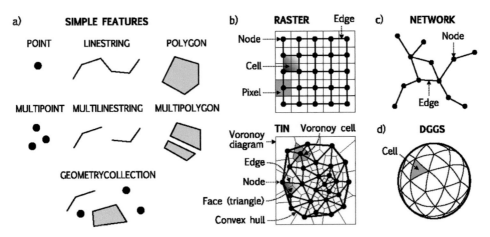

Figure 20.1 Basic spatial data models for objects (a), fields (b), networks (c) and global distributions on a sphere (d)

A collection of simple features constitutes a spatial dataset, which is usually stored as a table with one geometry column (spatial attribute) and a multitude of non-spatial columns.

Multi-geometries emerge when the logical object consists of multiple physical objects. For example, a MULTIPOLYGON feature type is commonly used to represent countries in spatial databases. Many coastal countries have islands, and their entire area can be effectively processed using this geometry type. MULTILINESTRING can be used to represent a river that is interrupted by a lake (another POLYGON object). MULTIPOINTs are rarely used to represent some geographical entities, but are often generated as a result of feature-type conversion operations (which is explained below). For example, if a GIS user needs to extract all points from a state boundary (usually a MULTILINESTRING), the resulting dataset will have a MULTIPOINT feature type. Finally, GEOMETRYCOLLECTION is almost exclusively a supplementary feature collector, and is commonly needed to store the results of overlay operations irrespective of their feature types (see geometry operations below). The remaining ten feature types defined in OGC SF standard deal with various types of curves and surfaces. There is an increasingly popular opensource GeoPackage format (OGC, 2017) that implements the SF standard and is supported by modern GIS software (such as QGIS, ArcGIS). More technical details on simple features and how they can actually be implemented in software are provided by Pebesma (2018).

These fields are abstractions of spatial variables, such as terrain elevation, temperature, distance to the state border, surface reflectance or land cover type. Fields can be continuous (smooth) or discontinuous (abruptly changing). For example, the land cover type changes abruptly when a forest abuts a water body. Still, most fields are smooth (e.g., temperature), which means that by moving over a small distance, the value of the variable also changes by a small amount. Fields differ from objects in that the field exists and can be measured at *every point* within its domain and is allowed to vary between any two locations. In contrast, an object is discrete and has a strictly defined boundary, and it is presumed that all the attributes of the object within are constant. In other words, fields are *variable*-oriented, not *object*-oriented concepts.

Since a field can be measured at any point in space, representing its complete spatial variation requires an infinite set of values, irrespective of the size of the area within which the field is investigated. A set of point locations comprising a 1×1 km geographic area is equivalent to a set of locations comprising a $1,000 \times 1,000$ km area, since these locations can be unambiguously related one-to-one. Since every value requires a storage resource to maintain and computing power to process, it is impossible to work with spatial fields in their entirety. To deal with the problem, the field must be *discretized* (i.e., represented by a limited number of sampled observations) and then *reconstructed* using a specialized data structure that approximates the complete distribution of the variable by interpolation between samples. This data structure is called a *coverage*. A technical definition of the coverage regards this data structure as a feature that acts as a function to return values from its range for any direct position within its spatiotemporal domain (Baumann et al., 2017).

Most commonly, the spatial fields are discretized using matrix sampling and the continuous field is then reconstructed by means of a *raster* coverage data model (Figure 20.1b). The sample values are written to raster *nodes,* which are essentially the point locations arranged in rows and columns. Each node has eight neighbours, except for the boundary nodes. Based on the nodes, the reconstruction of the continuous field is made by means of pixels and cells. The *pixel* is the regular and (usually) rectangular area associated with each raster node (see the grey square in Figure 20.1b). Importantly, the value of the variable is considered constant within the pixel, and a system of pixels covers the whole modelled area without any gaps.

When the arbitrary point is queried within the raster, the GIS will return the value of the underlying pixel, which corresponds to the so-called *nearest neighbour* interpolation (where the nearest node is the nearest neighbour).

Reconstruction of the smooth spatial field based on the raster nodes is also possible. To achieve this, the raster is interpreted as a system of *cells*. Each cell is the area between four neighbouring nodes. When a *bilinear* interpolation is used, the value of the field at any point within the cell is obtained by reconstructing the two-dimensional function based on four values of its nodes (see the cell square with the gradient fill in Figure 20.1b). The resulting surface is smooth inside the cell, but it is linear on the cell's edges, and results in non-smooth behaviour across its edges. A truly smooth surface is reconstructed by *bicubic* interpolation, which covers nine neighbouring cells and therefore requires 16 node values to fit the surface.

It is important to note that pixel-based interpretation of the raster is conceptually related to how the sampling is performed, which reflects the peculiarities of spatial data generation (Stasch *et al.*, 2014; Scheider *et al.*, 2016). In most cases, the values are *aggregated* over the pixel area, which is caused by the limited precision of the survey techniques used, the necessity to remove the noise from the results, or just that the data are generated by aggregation of a more detailed raster. For example, the sea surface temperature (SST) sensed by the MODIS satellite is actually *averaged* over the pixel area, and the set of such averaged temperatures constitutes the temperature raster. Therefore, raster nodes are actually abstractions needed to reconstruct the smooth surface, and generally it is not reliable to assume that the value is sampled exactly at each node's location.

There are two major types of raster data structure from a coordinate system (CS) perspective: geographical and projected. Geographical rasters are reconstructed from samples on a sphere or ellipsoid, which are equally spaced in spherical coordinates: longitude and latitude. On the contrary, projected rasters are sampled from projected spatial data and are defined in metric coordinates: X and Y. Geographical rasters are a somewhat unofficial standard for representing the global or semiglobal fields, which cover the whole Earth's surface or its terrestrial/oceanic component. A typical example is a digital elevation model (DEM), such as GEBCO (GEBCO Bathymetric Compilation Group, 2020) or GMTED2010 (Danielson and Gesch, 2011). Projected rasters are more commonly used for limited areas. Although raster nodes are usually equidistant at the stage of raster generation, they are not required to be such in later stages of processing, which may result in rectilinear or curvilinear raster geometries in some cases. One of these cases is the transformation of the raster coordinate system, which can be done through either warping or projection. Warping creates the new regular raster in the resulting coordinate system and in most cases there is no one-to-one relationship between the source and the resulting raster nodes. When the raster is reprojected, its nodes are treated as point features and are precisely transformed into the new coordinate system. The resulting raster has the same matrix of values as the source raster, and there is a strict one-to-one relationship between the source and resulting nodes. However, resulting nodes are not necessarily equally spaced, which may result in a curvilinear raster geometry. Since raster analysis methods assume a regular spacing between nodes, a true projection is rarely performed in GIS, and a warping is applied by default.

If, for some reason, matrix sampling is ineffective, then the coverage can be built based on irregular sampling. This technique is often used for modelling terrain: elevation samples are extracted more densely in the rugged terrain and are sparser in plain areas. Irregular samples are most commonly triangulated, resulting in a triangulated irregular network (TIN) (Peucker *et al.*, 1976). The samples are written to triangulation *nodes*, which are connected using the *edges*. Each three neighbouring nodes contribute to a triangle called a TIN *face*, and the spatial extent of the TIN is limited by a *convex hull* (Figure 20.1b). Having a set of triangles, the continuous field is reconstructed in a way similar to the raster method. For nearest-neighbour interpolation, the so-called Voronoy diagram is reconstructed from the TIN, and the resulting Voronoy cells act similarly to raster pixels (see the grey polygon in Figure 20.1b). A piecewise linear surface is reconstructed by fitting a plane that goes through three triangle points (see the TIN face with the gradient fill in Figure 20.1b). A smooth surface can be generated by using the Voronoy-based natural neighbour interpolation (Sibson, 1981) or fifth order polynomials that require an extended set of neighbouring faces (Akima, 1978). It is worth noting that the OGC Simple Feature Access standard envisages TIN feature type support (OGC, 2010).

Network data structures (Figure 20.1c), sometimes referred to as *graphs*, are used for representing the phenomena which comprise the set of spatially connected entities. Each entity is represented by node or vertex, while connections between entities are modelled as edges. Hence, the terminology is similar to that of the field models. Basically, the network is a collection of points joined together in pairs by lines (Newman, 2018), such as a map of airline routes that connect cities. Formally, a simple LINESTRING representing the river can also be considered to be a network, since it is essentially a set of point coordinates in which every two consecutive points are connected by an edge. However, network data structures are more complex, since more than two edges may be adjacent to each node, and each network edge has individual significance and properties (such as direction) during the analysis. This property is quite useful for modelling hydrographic, transportation, infrastructure and mobility systems (Barthélemy, 2011).

Discrete global grid systems (DGGS) are specific data structures that tesselate the surface of the Earth into equally-shaped grid units, usually spherical triangles (Figure 20.1d) or hexagons, which constitute a hierarchy (Sahr *et al.*, 2003). While DGGSs are quite useful for spatial data indexing, their main advancement over the regularly spaced raster or TIN coverages is that DGGS nodes are defined on a sphere, and therefore do not inherit distortions that are introduced by the map projection. In comparison with geographical rasters, DGGSs have equally sized elements and are safer to process due to the absence of edge effects along the prime meridian and poles. They can also be used to model features (like rasters), and their hierarchical nature allows automatic generalization of data (Dutton, 1999). The applications of DGGS for spatial analysis and modelling are still quite limited, but their potential is very strong due to the possibility of seeking to 'calculate on a round planet' (Chrisman, 2017). There are basic computational frameworks for using the DGGS for geographical analysis, which require further development (Lin *et al.*, 2018; Robertson *et al.*, 2020).

Spatial relations

Spatial relations are so fundamental and purely geographic that they underpin nearly every spatial analysis operation. They allow *spatial queries* to be performed (i.e., by extracting the spatial data based on its location). There are three commonly used types of spatial relations: metric, directional and topological (Li, 2006), as represented in Figure 20.2.

Metric relations (Figure 20.2a) are essentially distances. By calculating the distance, we are able to answer how far or close the objects are from each other. An example of a spatial query based on a metric relation is: 'Select all weather stations located no further than 50 kilometres from the specified location'. Depending on the purpose and coordinate system, the distance can be calculated differently. The Euclidean (straight-line) distance is a standard for projected data. However, applying this distance to the data in geographic coordinates

Spatial analysis and modelling

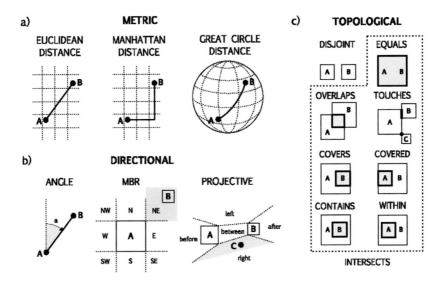

Figure 20.2 Spatial relations (examples)

(degrees/radians) is meaningless, and the great circle or geodesic distance must be used instead. The Manhattan distance is calculated as a sum of absolute coordinate differences and is sometimes used as a rough approximation of the real travelling distance along the city's street network (which is more realistic than Euclidean for rectangular street arrangements).

Directional relations (Figure 20.2b) establish the relative position of the target object in relation to one or more axes defined by one or more source objects. Most often, the direction is expressed in terms of an angle, and there are derivative directional systems such as cone-based (Peuquet and Ci-Xiang, 1987). If the axis going through the source object points towards North, then the angle is called the azimuth. Using the azimuth, a military officer can find the targets using the following spatial query: 'Select all objects located in an azimuth interval between 45 and 55 degrees'. However, the angles are not universal, and they can even be undefined in some computational contexts. Imagine a task to classify all suburban areas in terms of their directional relation to the city: southern suburbs, northwestern suburbs, and so on. Both the city and the suburbs are large and modelled as POLYGON features. What are the source and target points between which the angle should be calculated? If the exact angle is not needed, then the minimum bounding rectangle (MBR) matrix approach (Goyal and Egenhofer, 2001) can be used to subdivide the whole area into nine tiles: the central tile for the city itself, and the remaining eight sectors for classified directions. The resulting direction will be defined by the tile in which the suburban area is located. There is also a projective directional model that allows defining the direction to the third object C in relation to the system of two source objects A and B (Clementini and Billen, 2006), which can be quite useful in analysing movement data.

Topological relations (Figure 20.2c) describe the relative position of objects in terms of their intersections. In this case, two spatial features (A and B) are either disjoint or intersecting. The variety of intersection cases is large, depending on the feature types and their actual positions and shapes. Most commonly, each feature is divided into three components: its internal area ($A°$), boundary (∂A) and external area (A^-). If two features are under consideration, then there are nine possible intersection combinations such as internal A to boundary B, external A to external B, and so on. This model is called a *Dimensionally Extended 9-Intersection*

Model or *DE-9IM* (Egenhofer and Franzosa, 1991; Clementini *et al.*, 1993). Some of the possible intersection types are illustrated in Figure 20.2c. For example, the OVERLAPS intersection type emerges when all nine intersection combinations are realized, whereas the TOUCHES intersection type emerges when the inner area of each object does not intersect with the inner area or boundary of another object, while all other intersections are realized.

Topological relations are very powerful and are often used for spatial queries. For example, it is possible to perform the following spatial query based on topology: 'Select all forest areas that meet the coastline and are located inside the natural reserve'.

Relations can be combined to perform complex spatial queries. Finding the object that is located within a specified distance and inside/outside a polygon is a common task for spatial analysts. It is also worth noting that the number of spatial relations that we use in everyday life is much wider than those that can be easily and strictly defined. Such relations are sometimes quite challenging to be formalized. One example is the recently developed SURROUNDS relation (Worboys and Duckham, 2021).

Basic operations

Despite the large variety of spatial analysis and modelling methods, all are based on a limited number of basic operations:

- spatial operations;
- non-spatial operations;
- grid operations; and
- data model transformations.

Spatial operations

Spatial operations are used to manipulate object coordinates, and many of them are essentially computational geometry algorithms applied to spatial data (Berg *et al.*, 2008). The variety of such algorithms is quite large, and they are useful both for data maintenance and for research purposes. Several commonly used examples are illustrated in Figure 20.3.

- *Construction* (Figure 20.3a) is used to build feature geometries from coordinates, addresses or from another features. For example, linear GPS (Global Positioning System) tracks along a field boundary can be converted into a polygon. Addresses are converted to coordinates and then point features via *geocoding*.
- *Subsetting* and *type conversion* (Figure 20.3b) allow the extraction of feature components and transforming them to another feature type. It is possible to extract all vertices of a polygon as point features or to keep only the first and the last points of a linestring.
- *Bounding geometries* (Figure 20.3c) are built to obtain the simplified representation of the area covered by the feature or the set of features. Most commonly, rectangles, convex and concave shapes are used for this purpose. For example, when given a set of points representing plant observations, it is possible to derive the approximate species distribution using a concave shape (although more precise distribution modelling requires a more complex approach).
- *Skeletons* and *medial axes* (Figure 20.3d) are the results of collapsing the geometry to a representative geometry of a lower dimension. In particular, centroids and skeletons are used to find the locations for labelling the polygonal features (e.g., to label a sea strait).

Spatial analysis and modelling

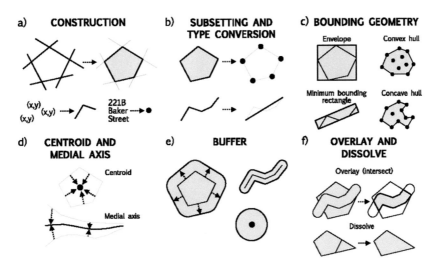

Figure 20.3 Spatial operations (examples)

- *Buffering* (Figure 20.3e) offsets the boundary of the object to a specified distance. Buffers are commonly used to visualize the locations affected by distance-based spatial queries. For example, it is possible to draw a 100-metre-wide buffer zone around an underground metro exit, or a 200-metre-wide water protection zone along a river. However, buffers are as valuable as polygonal objects in cases when further processing is required (e.g., if a water protection zone should be intersected with land property parcel). Buffers can be negative for polygonal objects, which means that the polygon is shrunk towards its medial axis, forming a 'setback'.
- *Overlay* and *dissolve* (Figure 20.3f) operations are used to combine geometries. The previous example of intersecting a water protection zone with a land property parcel is the result of an overlay operation between these two objects. Dissolving is used to remove the boundaries between the adjacent polygons. For example, it is possible to derive the polygon of the country by dissolving the subordinate administrative units.

Non-spatial operations

Non-spatial operations are applied to the variables (non-spatial attributes) associated with features and coverages. Some of the most common operations are as follows (see Figure 20.4):

- *(Re)calculation* (Figure 20.4a) is the most widely used non-spatial operation, which stands for calculating the attribute values based on the values of the attribute itself and/or the values of other attributes. Using attribute calculation, it is possible to transform ratios of 0…1 value range into percentages of 0…100 value range.
- *Aggregation* (Figure 20.4b) is needed when the value for one spatial location is calculated based on the values from multiple spatial locations. Hence, some grouping must be established before applying the aggregation. The type of aggregation operator depends heavily on the type of the variable (Stasch et al., 2014). Spatially extensive (absolute) variables must be summed, while the intensive (relative) variables can only be averaged. Total population and population density are examples of extensive and intensive

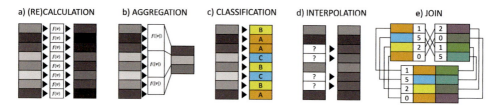

Figure 20.4 Non-spatial operations (examples)

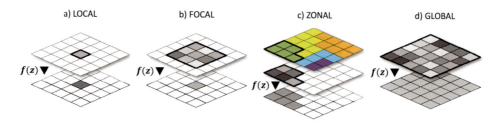

Figure 20.5 Grid operations

variables. Still, aggregation is not limited by summing or averaging, and actually any meaningful statistical transformation can be applied as well.
- *Classification* (Figure 20.4c) is used to reduce the variety of values and can be applied to both numerical and categorical variables. Space imagery is usually processed by classification, which allows transforming the surface reflectance values into the meaningful categories, such as land cover classes.
- *Interpolation* (Figure 20.4d) is used to fill gaps in the data and can be applied if some non-spatial object ordering exists. This operation is quite typical for restoring time series data such as river discharge values related to one location.
- *Join* (Figure 20.4e) is used to merge two tables based on common row identifiers. For example, mapping based on census data usually requires joining the statistical data to the dataset of administrative units. This is possible if each record in both tables contains the unique identifier of the corresponding administrative unit.

Grid operations

Operations on gridded data structures, such as rasters, are usually expressed using the conceptual framework of map algebra (Tomlin, 2012). Map algebra is the formalized approach to solving a wide range of spatial analysis problems using four types of operations: local, focal, zonal and global (Figure 20.5).

- *Local* analysis (Figure 20.5a) operates at one location at a time. When the raster of ratios (0…1) should be rescaled to the raster of percentages (0…100), this is essentially the local operation, since the value at each raster node can be rescaled independently of others. Similarly, arithmetic calculations on multiple rasters (often performed in GIS packages by so-called 'raster calculators') are also local. For example, the spatial differences in climate change can be calculated by subtracting the average temperature rasters for different years and obtaining the difference raster.

Spatial analysis and modelling

- *Focal* analysis (Figure 20.5b) is performed within the neighbourhood of the location. The neighbourhood can be fixed and extended. A fixed neighbourhood is the template that is centred on the calculated pixel and is used to extract the values for calculation. Most commonly, this is a square matrix with odd number of elements (3×3, 7×7, and so on). Extracted values are aggregated or interpolated to obtain the resulting value. Gaussian smoothing or calculation of elevation derivatives are typical focal analysis operations. Extended neighbourhood analysis is not limited by any template and is allowed to grow until some condition is satisfied. For example, a Euclidean allocation assigns each raster pixel the identifier of the closest source (a marked pixel), and to achieve this, the neighbourhoods of the sources are growing until they touch.
- *Zonal* analysis (Figure 20.5c) requires two spatial datasets: one with zones and one with values to be aggregated. The calculations are somewhat similar to the focal case, but instead of defining the fixed neighbourhood or allowing it to grow, the shape of the neighbourhood is defined by a supplementary zone raster. Another difference from focal analysis is that all pixels that belong to the zone receive the same aggregated value. A possible application of zonal analysis is to calculate the mean terrain slope for each elevation interval, which can be useful for planning a recreational site.
- *Global* analysis (Figure 20.5d) is probably the most simple and straightforward amongst all map algebra variants. A global analysis operation takes all raster values, aggregates them in some way, and the resulting value is assigned to *all* pixels of the output raster. The operation can be useful for obtaining a supplementary raster that is used as a weighting coefficient in local calculations. For example, a weighted average of two rasters can be achieved by multiplying them by corresponding constant rasters obtained by the global operation.

Data model transformations

Data model transformations are applied in various cases when the initial data model is not detailed enough for analytical or representational purposes. Some of the commonly used transformations are illustrated in Figure 20.6.

- *Vector-raster* (Figure 20.6a) and *raster-vector* (Figure 20.6b) *conversions*. If spatial analysis involves multiple datasets (some of which are raster and others are vector), it is often required to unify the data model and convert either vector data into raster format or vice versa.

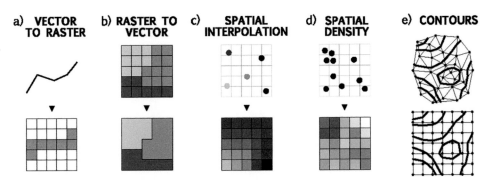

Figure 20.6 Data model transformations (examples)

- *Spatial interpolation* (Figure 20.6c) is one of the most frequently used model transformations during which the discrete point samples are turned into the raster surface, which covers the modelling domain completely. For example, noise-level measurements in a city can be interpolated to analyse the noise distribution at a scale covering the whole city.
- *Spatial density* of objects can be estimated as a continuous field (Figure 20.6d), but the method of density estimation differs significantly from interpolation approaches (Baddeley et al., 2016). Visualizations of spatial density fields are informally called *heatmaps*, and are useful for analysing the concentration of objects. A typical application is to visualize the density of geotagged Instagram photos to show which places are popular tourist destinations.
- *Contouring* is the transformation of the field model into the lines of a constant value (Figure 20.6e). Though contours do not usually correspond to some geographical entities, they are useful in cartographic applications, and in combination with other spatial analysis techniques. In particular, a zero-degree temperature contour can be extracted from a climatic raster to subdivide the territory into the areas of negative and positive temperatures, which is useful from a water management perspective.

Applications

The basic approaches explained above lay the foundation for high-level methods of spatial analysis and modelling that tackle a diverse variety of research investigations. The questions under study are always primary and define the selection of the most appropriate solution. In many cases, the same problem can be looked at from different points of view, and multiple solutions exist. Therefore, it is helpful to build a picture of spatial analysis and modelling techniques from the perspective of their use, and so the following subsections explain some of the more common applications.

Visualization and mapping

It is virtually impossible to imagine any rigorous geographical research being conducted without a map. Statisticians often work with purely one-dimensional data represented by a time series, but they still use graphical approaches to understand their data more effectively. Geographical approaches begin when data have locations that can be identified by their spatial coordinates. The multidimensional nature of spatial data makes it difficult to understand in tabular form, not least because the unique mathematical ordering is only possible in one-dimensional space. Our vision is what helps us to order and understand complex geographic information, and maps are the essential tools that facilitate this process (MacEachren, 1995). Visual analytics of spatial data is valuable at all stages of geographical analysis: at the beginning where it helps to understand the initial state; in the middle, when decisions must be made how to direct the following processing; and at the end, when the final results have to be communicated effectively (Andrienko and Andrienko, 1999).

The visualization of spatial data and automated mapping are highly technical processes that require a multitude of spatial data manipulations to visualize information. The generation of one-to-many distributive flow maps (Figure 20.7) is based on the network tree-like model between endpoints (Sun, 2018), whereas the placement of supplementary contour lines for terrain representation requires an estimation of width and centrality of a free space between regular contours, as completed by map algebra expressions in raster mode (Samsonov et al., 2019).

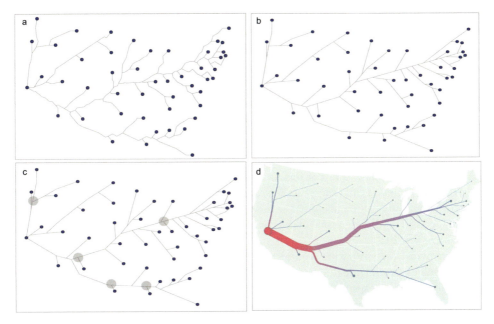

Figure 20.7 Application of spatial analysis and modelling to visualization and mapping: population migration to California in 2000 US Census: (a) initial approximate Steiner tree; (b) after simplification and straightening; (c) after angle regulation; and (d) final rendering (Sun, 2018)

Searching and joining

Spatial analysis workflows often require the searching of data subsets that are located in specified areas. This is performed through spatial queries, as explained earlier. However, a raw spatial query may take an excessive amount of time if the dataset from which the data are extracted is large. Imagine extracting a small portion of forest polygons located within a 5×5 km square from a dataset that covers a country that is $1{,}000 \times 500$ km in size and contains millions of similar polygons. Depending on the power of the computer, the task will take many seconds or minutes to solve if one-at-a-time checking of each polygon is performed.

To resolve issues like these, spatial databases support a specific data structure, called a *spatial index,* which can be associated with each feature collection (Shekhar and Chawla, 2003; Samet, 2006). Spatial indices are auxiliary data and usually consist of virtual rectangles arranged in a grid or in hierarchical tree-like structures. When the spatial index is built, the features or other index rectangles (in case of hierarchical trees) located within each rectangle are determined. Later, when a spatial query is executed against the feature collection, it is first evaluated on a spatial index, and the objects are extracted only for those index rectangles that intersect the queried area. Such a strategy may increase the speed of spatial queries by several orders of magnitude.

One of the important applications of spatial queries is merging two datasets via *spatial join* – a subcategory of the join operation that is based on relations between feature geometries. For example, we can enrich the dataset of traffic accident points with the name of a city district where the accident occurred. For this, district polygons are joined to accident points using the WITHIN relation. Spatial joins are commonly supplemented with *aggregation.* If we substitute the roles of points and polygons in the last example, it is possible to count the number of accidents in each city district using the CONTAINS spatial relation and the COUNT aggregation function.

Recognition and enrichment

Data preparation often takes the largest part of research time. Typically, the researcher has to identify *what* geographical entity is located at specific place, or, inversely, *where* the entity under study is located. If the data are spatiotemporal, the *where* predicate may also be supplemented with the *when* statement. The classical example of spatial recognition is interpretation of satellite imagery, which helps to extract the extent of different land cover types (Buchhorn et al., 2020). For this, non-spatial operations such as classification, as well as spatial operations such as segmentation, are both valuable. Similarly, the main routes of pedestrian mobility can be identified as medial axes from mobile tracking data, and routes can be specified for different time of the day, since mobile data are naturally spatio-temporal (Lee and Sener, 2020).

In many cases, identification requires a preliminary analysis of the data using the analysis of morphology of individual features or their distribution and grouping. Recognition can be one of the stages of spatial data enrichment, when unknown properties are inferred about objects. For example, Hecht et al. (2015) used pattern recognition and machine learning techniques to identify the building types that were absent in a database (Figure 20.8).

Morphometry and shapes

Morphology aims at learning the character of object's geometry and providing its qualitative interpretation. GIScience supplies the morphological characterization of spatial phenomena with two quantitative instruments: morphometric analysis and shape analysis. Morphometry provides the basic measures for describing object's geometry, or variables, such as area, perimeter, curvature, density of elements, and many others. Morphometric analysis of the three-dimensional Earth surface via digital elevation models (DEMs) is usually referred to as *geomorphometry* (Hengl and Reuter, 2009; Florinsky, 2016). Geomorphometric analysis is commonly performed on raster elevation models using the focal map algebra approach. A

Figure 20.8 Application of spatial analysis and modelling to object recognition and data enrichment: input data (left) and classified building footprints (right) for a small section of the city of Krefeld (Hecht et al., 2015)

quadratic surface is fitted to elevations in a 3 ×3 moving window, and then its derivatives are calculated, based on which numerous morphometric indices can be derived (Evans, 1980; Zevenbergen and Thorne, 1987). Based on the derivatives, numerous useful topographic variables can be calculated, such as slope, aspect and curvature.

Shape analysis is closely related to morphometry and shape characterization is often based on morphometric indices. However, it is more abstract, since size-related morphometric indices are not used, and the overall shape characterization can be objectively achieved using multiple indices (Zhang and Lu, 2004). Generally, shape is the only geometrical information that remains when location, scale and rotational effects are removed from an object (Dryden and Mardia, 2016). Shape is therefore a very important property of a geographical object, since it can provide clues about its origins. For example, a rectangular object shape is typical for buildings and is used to extract them automatically from remotely sensed imagery (Jin and Davis, 2005).

Distribution and structure

Spatial distribution is a key concept in geography. It can be applied both to objects (i.e., how objects are arranged in space) and to fields (i.e., how a spatial variable changes between locations). Since distribution is a statistical concept, spatial distributions are investigated by *spatial statistics* (Ripley, 1981; Cressie, 1993). This aims to describe the main properties of distribution under study and then building an associated model. In particular, when the distribution of objects is focused, it can be approached by *point pattern analysis*, which is one of the branches of spatial statistics (Diggle, 2014; Baddeley *et al.*, 2016). Since the exact distribution cannot be fully described, point patterns are usually investigated in terms of first-order (density) and second-order (distances) characteristics. An example of density-based point pattern analysis is shown in Figure 20.9, which depicts the spatial density of tornadoes over Northern Eurasia (Chernokulsky *et al.*, 2020).

By contrast, the properties of variable distribution over space are investigated by *geostatistics* (Matheron, 1962; Chiles and Delfiner, 2012). The core concept of geostatistics is a variogram, which is the variance of the difference between values as a function of the distance between points at which these values are sampled. A typical variogram shows that the variance increases with distance and this continues until a certain distance (called the *range*) after which there is no dependency (unless there is some bias in the data). The shape of the variogram is determined by the variogram model, which is then used to calculate the interpolation weights of the source points – a procedure known as *kriging* (Cressie, 1993). When the field is categorical, such as the land cover type, its spatial structure can be analysed. Structural analysis is well developed in the field of landscape ecology, where numerous indices have been developed to characterize the mosaic of land cover at patch, class and landscape level (Hesselbarth *et al.*, 2019).

Grouping and polygons

Whereas point pattern analysis is focused mainly on general point distributions and their statistical properties, grouping analysis aims to reveal exact groups of phenomena in space-time. Most commonly, this is achieved by *clustering*, which, essentially, is a technique of unsupervised classification. If applied to spatial coordinates (planar or geographic), it can reveal the groupings of phenomena. Depending on the exact arrangements of phenomena into groups, different clustering methods may be more effective, but all of them are based on

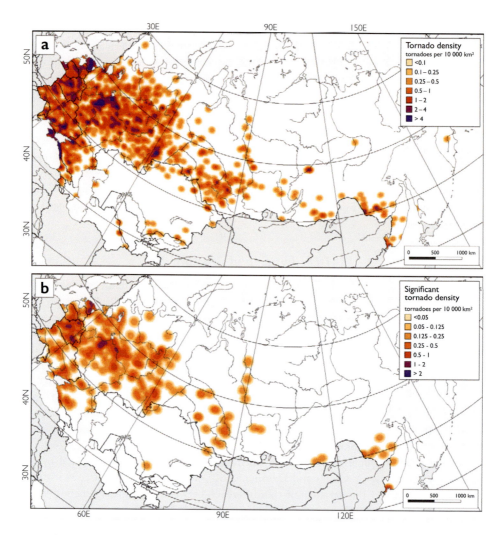

Figure 20.9 Application of spatial analysis and modelling to study spatial distribution: density of tornado cases per 10,000 km² for the 1900–2016 period for (a) all tornadoes over land and (b) significant tornadoes over land (Chernokulsky *et al.*, 2020)

calculation of distances between phenomena. For isometric and simple configurations with convex shapes, simplistic approaches like K-means (MacQueen, 1967) or ISODATA (Ball and Hall, 1965) are sufficient, while clusters of more complex shapes are recognized more robustly through density-based clustering algorithms such as DBSCAN (Ester *et al.*, 1996) and OPTICS (Ankerst *et al.*, 1999).

Grouping is an essential operation in cartographic generalization. For example, the automated generalization of buildings on topographic maps is performed using typification, when a group of buildings is identified for replacement by a smaller number of representative features (He *et al.*, 2018). It is sometimes desirable not to group the objects, but to delineate the area covered by them. This can be informative if the spatial analysis workflow requires having the point coverage as a polygonal feature. Convex clusters can be effectively described by convex shapes, while in general, various concave hulls such as α-shape (Edelsbrunner *et al.*, 1983) or *χ*-shape (Duckham *et al.*, 2008) can be used instead. For example, Bu *et al.* (2021) used *α*-shape to delineate the area of activity of elderly people in Shanghai based on smartphone-based GPS tracking points (Figure 20.10).

Influence and dominance

Since all geographical entities are located in physical and/or social space, they inevitably fall into the sphere of influence of their neighbourhood. This space is not only about the central point and dimensional vectors, but also about the matter that fills it. When we speak about social space, it is the myriad of artificial actors that fill this space with various forms of communication. For example, in a multicultural city, the peculiarities of local social habits can be influenced by primary language, religion and other attributes of cultural identity. Using the spatial analysis of geotagged social media posts, it is possible to infer characteristic properties of local communities (Lansley and Longley, 2016). A large city is a powerful attractor of daily migration, and its influence can be assessed from analysing ticket transactions on public transport stops.

Similarly, influence is spread in a physical space, too. In particular, industrial and transportation pollutants are spread over significant distances. Using gridded data structures, it is possible to model such a spread and to map the resulting distributions (Lateb *et al.*, 2016). Another example of influence is visibility. It is important for human navigation to reveal and map the most visible and dominant orienteering objects, either in urban or natural landscapes. This can be achieved through viewshed analysis or more advanced techniques such as scale-space transformation (Witkin, 1984) used by Rocca *et al.* (2017) to label mountain peaks (Figure 20.11).

Combinations and optimum solutions

Some geographical research problems seek the *best location* or the set of best locations. When the problem is conceptualized, the result reveals the spatial variables that influence the decision. After being analysed together, their values allow the best location to be found, which is most commonly a compromise of the optimal combination of values. This type of problem is a subcategory of multicriteria decision analysis or MCA (Pereira and Duckstein, 1993; Malczewski, 2006). A classic example is defining the best site for a construction, such as a new wind

Figure 20.10 Application of spatial analysis and modelling to derive polygons: space of older adults' daily activities: (a) concave shape; (b) activity space (Bu *et al.*, 2021)

or solar farm. There must be a sufficient power source: wind or sunlight; the geological, terrain and environmental conditions must be safe and favourable for construction; and the site should be located in the vicinity of the consumers of the power generated, such as industry, residential and office buildings, otherwise there will be unacceptable losses of power supply.

Since such optimization problems often include some field values (such as distance), a raster model is commonly used, with the corresponding method sometimes called a *weighted overlay*. Each spatial variable is represented as a raster layer and then classified into an ordinal scale in terms of its suitability. Therefore, the classes of suitability for each factor involved in the decision are known for each location. The subsequent overlay is essentially the local map algebra operation that aggregates (e.g., sums) the class values from multiple rasters to calculate the resulting suitability value, optionally with weights applied to each variable. After setting the minimum suitability value, it is possible to reclassify the raster into a binary (1/0) scale and then vectorize the contiguous areas with pixel values equal to 1. A specific group of problems that refer to spatial optimization are also network analysis problems. Spatial optimization, which combines the network analysis with multiple-criteria decision making, is widely used in geomarketing for locating the best locations for facilities (retail stores, pharmacies, and so on) (Cliquet, 2006).

Dynamics and evolution

The possible changes of spatial objects can be classified in terms of *internal structure* (uniform/evolving), *geometry* (rigid/elastic) and *movement* (stationary/moving) (Goodchild et al., 2007). The changes of internal structure are related to inner content of the object, which is either described implicitly by a set of non-spatial attributes, or is modelled explicitly through a subordinate features or coverages. For example, a change of the crops' structure in an experimental agricultural field can be modelled implicitly by multitemporal attributes of a field polygon, or explicitly by a multitemporal raster obtained by hyperspectral imagery

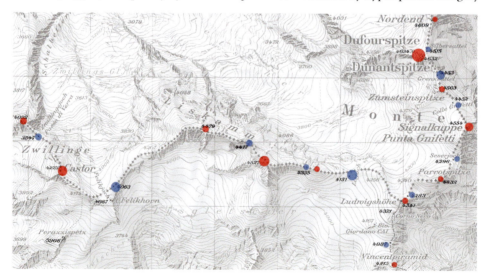

Figure 20.11 Application of spatial analysis and modelling to quantify dominance: spot height locations selected by the continuous scale-space method; red: summits; blue: mountain passes. Circle size indicates importance values (Rocca et al., 2017). Background: 1:50,000 map by swisstopo

and linked to the polygon in a database. A change in geometry can be modelled by multiple representations of the feature geometry, which reflect the changes in its *boundary* not related to the object moving.

Sometimes it is not easy to discriminate geometry changes from feature movements, but movement results in the global displacement of all object points, while changes of geometry are local. For example, an oil slick on the water surface is moving and changing its shape simultaneously. A shrinkage of a mountain glacier due to global warming is another example of geometry dynamics that can be reconstructed from multitemporal space imagery and then modelled by multiple feature representation (Petrakov *et al.*, 2016). Fields can also be classified as static or dynamic, which is true both for numerical and categorical fields. A classic example of characterizing a dynamic categorical field is a so-called change detection, which is performed on remotely sensed data. With two interpreted remote sensing images, it is possible to perform a local map algebra expression to reveal the difference between them. For example, a tree cover change can be mapped in such a way (Hansen *et al.*, 2013). Evolution of spatial phenomena can also be modelled by geosimulation, genetic algorithms and other computational approaches (de Smith *et al.*, 2018). One of the widely used methods from this category is the raster-based *cellular automata* approach that is commonly applied to model the development and growth of cities (Li and Yeh, 2000). An example of such modelling combined with analysis of remote sensing data for Delhi urban area is represented in Figure 20.12 (Tripathy and Kumar, 2019).

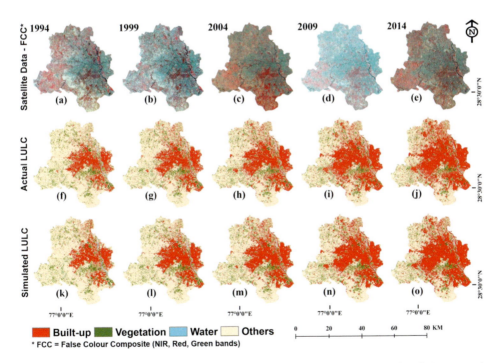

Figure 20.12 Application of spatial analysis and modelling to explore dynamics and evolution: growth of Delhi's urban area learned from space imagery and simulated by cellular automata. LANDSAT satellite images (a–e) (obtained from USGS), satellite-based land use/land cover (LULC) maps (f–j) and simulated LULC map (k–o) of Delhi for the year 1994 (a, f, k), 1999 (b, g, l), 2004 (c, h, m), 2009 (d, i, n) and 2014 (e, j, o) (Tripathy and Kumar, 2019)

Movement and trajectories

Movement is a change of position, and it is unusual if a physical object moves only along a straight line. Even aircraft do not fly along pure geodesics continuously and follow more complex paths determined by atmospheric conditions, state boundaries, runway orientation, and many other factors. Geographical space is characterized by the perpetual movement of its inhabitants, be it the water in a river system or people in a city. Although object movement may appear to conform to an established network, it does not have to emerge in such a restricted way. Animals can move quite randomly, and their actual movements can be motivated by problem-solving (e.g., hunting) or playing.

While network analysis is more focused on theoretical properties of networks and their elements, computational movement analysis (Laube, 2014) and the visual analytics of movement (Andrienko et al., 2013) deal with actual movement data. These data are usually registered either by the Langrangian approach, where the trajectory of each moving object (e.g., a car or a bird) is recorded in space-time, or the Eulerian approach, in which the actual movement is inferred from the volume of the matter (e.g., a volume of water or a number of pedestrians) that passes through a set of observation sites. There are computational models that allow restoring the actual trajectories from volumes (Tobler, 1981; Wang et al., 2015), although they are inevitably based on some theoretical or empirical assumptions, and most of the movement analysis is performed based on trajectories (Zheng, 2015).

The processing and interpretation of movement data is a relatively new direction of spatial analysis and modelling, and it is evolving quickly (Dodge et al., 2020). Trajectory information requires the development of specialized approaches that overcome the limited temporal capabilities of standard spatial data models and allow the processing of big datasets associated with movement (Graser et al., 2021). An example of valuable trajectory analysis is represented in Figure 20.13, which shows the dangerous intersections between trajectory tubes of neighbouring airports as recognized by using raw trajectory data (Murça et al., 2018).

Networks and accessibility

A system of connections between geographical entities is commonly modelled as a network. The essence of connection is usually the movement of an entity, such as electricity or water, or transportation between places. Additionally, information and social links can connect geographically distributed entities, such as social media posts. Networks are powerful abstractions that allow the analysis of the system of these connections as a whole and describing its subsets and atomic components, i.e., nodes and edges (Newman, 2018). While computational movement analysis is focused on actual movement data, spatial network analysis is focused on theoretical properties of an established network system that makes this connection possible (e.g., a transportation network) or, inversely, is generated by connections (e.g., social networks). Network analysis is based on topological relations between nodes and edges, which, alongside edge and node properties, determine the possibility of building a *path* between a selected pair of nodes. A shortest path determines the network distance between selected nodes. Since geographic space creates obstacles to movement due to its heterogeneity, some edges become difficult to traverse. Therefore, network nodes are characterized by different degrees of *accessibility* (i.e., how easy it is to reach the node in terms of the time, distance or some other measure) (Brans et al., 1981). A line of constant accessibility bounds the area that can be reached from the node within the selected time or the distance limit (which is frequently referred to as the *service area*).

There are many other applications of spatial analysis, such as facility location, vehicle routing or assessing the importance of particular edges in a network structure (usually through 'betweenness' measures). If the movement is allowed through any point in space, it is possible to combine the spatial field and network models and establish the movement directions between neighbouring raster or TIN nodes. This is how the hydrological analysis of digital elevation models is performed, and this type of analysis allows the extraction of potential stream networks, watersheds and many other hydrology-related features (O'Callaghan and Mark, 1984; Freitas et al., 2016).

Co-location and correlation

Often, geographical phenomena behave similarly in space. This observation can be true for objects, fields, networks, and their combinations. The primary research hypothesis for this case is that the presence (or value) of one phenomenon defines the presence or the value of the second phenomenon, or both are influenced by the unknown others. If the problem under investigation is about objects that are frequently found together, this is usually modelled as spatial association or co-location (Huang et al., 2004). For example, petrol stations are co-located with road networks, whereas the density of street food locations may increase in the vicinity of railway stations. The latter can be interpreted both in terms of objects (street food is co-located with the stations), and fields (the density field is correlated with the distance field). In a similar way, the price of real estate is commonly correlated with the distance to the city centre. Vegetation and soils are co-located, and every species of vegetation has its most favourable soil type. This fundamental observation can be used to reveal the soil type and its properties from satellite imagery if the direct observation of soil is not possible.

Co-location dependencies can be quantized using vector or raster *overlay*. In the case of vector data, the objects are intersected and the table of their combinations can be used to calculate the association between object types. If intersections are areal or linear, their area or length can be used as weightings. If a raster overlay is used, two variables (e.g., density and distance) are interpolated on the same raster grid, and the values at each raster node are extracted to form two samples. Correlation between these samples can be calculated using standard statistical methods. Correlated phenomena can be related by non-spatial regression when the values of one (dependent) variable are expressed as a function of the values of a set of other (independent) variables. Regressions can be used as a part of spatial interpolation where they determine a spatial drift, i.e., the deterministic component of spatial variation (Cressie, 1993; Hengl et al., 2004).

Dependency and autocorrelation

According to Tobler's 'First Law of Geography', everything is related to everything else, but near things are more related than distant things' (Tobler, 1970). From a statistical perspective, this means that if a series of observations of variables at two randomly selected points are made, thus forming two samples, then the dependency between the derived samples will be stronger for point pairs separated by shorter distances. Usually, the dependency is formalized by correlation, and since the samples are derived from the same spatial variable, it is *spatial autocorrelation*. A temperature field is an example of a variable with high spatial autocorrelation. Knowing that a variable is highly autocorrelated may be interesting per se, but gaining meaning requires regression analysis. As mentioned earlier, co-location and correlation analysis commonly ends with building a regression model. If the residuals of the regression are spatially autocorrelated, this means that a simple regression model does not explain the full variation of the dependent variable appropriately. To treat the effect, it is possible to build a few types of spatially explicit regression models, from which two are most commonly used: a *spatial lag model* and a *spatial error model* (Anselin, 1988; LeSage, 2008). Both types are most commonly applied to lattice data (e.g., data measured at some fixed sites) and require identification of the neighbouring areal units and quantizing the strength of their neighbouring relation through a spatial weighted matrix. The main power of these models is that they allow more reliable predictions to be made for spatially autocorrelated variables, including cases when independent and/or dependent variables are not known for some areal units (Goulard *et al.*, 2017).

Heterogeneity and non-stationarity

Both spatial and non-spatial dependencies may vary in space, which is called *spatial heterogeneity* (Anselin and Griffith, 1988). From a statistical perspective, this means that the regression model that is fitted between variables will change its coefficients. Since this behaviour can be caused by changing the statistical properties of the variables, it is often called *non-stationarity*. Non-stationarity of relations between variables can be handled by geographically weighted regression or GWR (Fotheringham *et al.*, 2002). This is based on a simple idea that the fitting of the regression model does not have to be global, and it can be done in the vicinity of a selected point, similarly to what is done in the LOESS method in statistical graphics (Cleveland, 1979). For example, it is possible to build a GWR model that handles the spatial variation of dependency between real estate prices and the various factors that determine it, such as population density, migration index, unemployment rate, air pollution, and others (Cellmer *et al.*, 2020). Such maps show clearly the heterogeneous nature of this dependency (Figure 20.14).

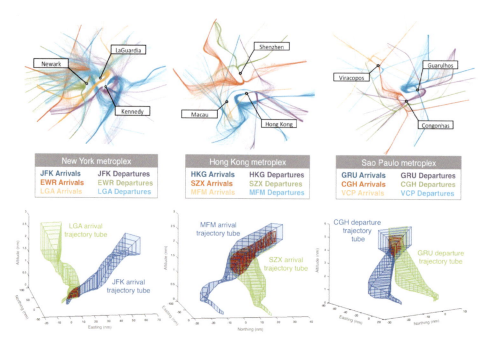

Figure 20.13   Application of spatial analysis and modelling to analyse movement and trajectories: arrival and departure trajectories for one day of terminal area operations in the New York, Hong Kong and Sao Paulo metroplexes (top). Examples of trajectory tube intersections (bottom) (Murça *et al.*, 2018)

Spatial analysis and modelling

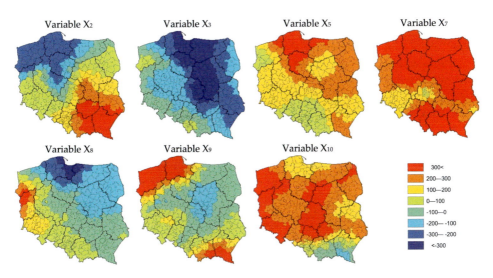

Figure 20.14 Application of spatial analysis and modelling to study spatial heterogeneity: spatial distribution of the parameters at local variables in the geographically weighted regression model built for average unit flat price (Cellmer et al., 2020)

Spatial heterogeneity may also complicate the mechanisms of autocorrelation. The localized effects of non-stationarity – as well as their influence on the global picture of spatial dependence – can be learned by indicators of spatial association (LISA) (Anselin, 1995). For example, Eugenio-Martin et al. (2019) have shown that *Airbnb* spatial correlation (co-location) with incumbent hotels depends on the kind of tourism, and nature-based tourism (as opposed to beach tourism) areas have not shown any sign of marked spatial correlation.

Hierarchy and multiscaling

Multiscaling, or the presence of multiple organizational levels, is a quintessential property of natural and social phenomena (Levin, 1992; McMaster and Sheppard, 2004), and is therefore of particular interest in spatial analysis and modelling (Zhang et al., 2014). Each organizational level can be characterized typically by a specific spatial and temporal scale of observed events and processes. Revealing such levels is one of the most intricate problems of geographical analysis. A mismatch between the granularity of the phenomena and the spatial configuration of sampling areal units leads to the so-called *modifiable areal unit problem* (MAUP), a commonly recognized analytical issue in GIScience (Openshaw and Taylor, 1979).

Multiscale representation is also very important in mapping, where it helps to show the most relevant spatial features in each scale of map visualization. Numerous algorithms for spatial data and map generalization are developed for this purpose (Li, 2006; Mackaness and Ruas, 2007; Burghardt et al., 2014). A search for characteristic scales in spatial data can be achieved by various kinds of transformations, including spatial and non-spatial. Generally, each of the previously considered applications of spatial analysis has its own multiscale problem. For example, a GWR method can be adapted to catch the characteristic scale of spatial heterogeneity (Fotheringham et al., 2017), while multiscale spatiotemporal organization of object groups can be inferred from hierarchical clustering (Lamb et al., 2020).

Conclusion

Spatial analysis and modelling use a diverse number of theoretical and computational approaches to answer geographical questions. Among them, conceptualization and data modelling allow geographical space to be built and represented in ways that are suitable for processing with spatial algorithms. Spatial relations and basic operations in spatial and non-spatial domains are the low-level components that form the basis of high-level methods of spatial analysis and modelling. These methods, such as multicriteria analysis, network analysis, spatial regression, computational movement analysis, and many others, serve numerous applications. Some of these require the combined use of multiple methods, which is only natural for a highly diverse landscape of geographical problems.

References

Akima, H. (1978) "A Method of Bivariate Interpolation and Smooth Surface Fitting for Irregularly Distributed Data Points" *ACM Transactions on Mathematical Software* 4 2) pp.148–159 DOI: 10.1145/355780.355786.

Andrienko, G., Andrienko, N., Bak, P., Keim, D. and Wrobel, S. (2013) *Visual Analytics of Movement* Berlin, Heidelberg: Springer.

Andrienko, G. and Andrienko, N. (1999) "Interactive Maps for Visual Data Exploration" *International Journal of Geographical Information Science* 13 (4) pp.355–374 DOI: 10.1080/136588199241247.

Ankerst, M., Breunig, M.M., Kriegel, H-P. and Sander, J. (1999) "OPTICS: Ordering Points to Identify the Clustering Structure" *Proceedings of the 1999 ACM SIGMOD International Conference on Management of Data - SIGMOD '99* pp.49–60 DOI: 10.1145/304182.304187.

Anselin, L. (1988) *Spatial Econometrics: Methods and Models* Dordrecht: Springer.

Anselin, L. (1995) "Local Indicators of Spatial Association-LISA" *Geographical Analysis* 27 (2) pp.93–115 DOI: 10.1111/j.1538-4632.1995.tb00338.x.

Anselin, L. and Griffith, D. A. (1988) "Do Spatial Effects Really Matter in Regression-analysis" *Papers of the Regional Science Association* 65 (1) pp.11–34 DOI: 10.1111/j.1435–5597.1988.tb01155.x.

Baddeley, A., Rubak, E. and Turner, R. (2016) *Spatial Point Patterns: Methodology and Applications with R* Boca Raton, FL: CRC Press.

Ball, G.H. and Hall, D.J. (1965) "ISODATA, a Novel Method of Data Analysis and Pattern Classification" (*Technical Report AD699616*) Menlo Park, CA: Stanford Research Institute DOI: 10.1016/0031-3203(92)90114-X.

Barthélemy, M. (2011) "Spatial Networks" *Physics Reports* 499 (1–3) pp.1–101 DOI: 10.1016/j.physrep.2010.11.002.

Baumann, P., Hirschorn, E. and Masó, J. (Eds) (2017) *OGC Coverage Implementation Schema* Open Geospatial Consortium Available at: http://www.opengis.net/doc/IS/cis/1.1 (Accessed: 21st February 2021)

Berg, M., Cheong, O., Kreveld, M. and Overmars, M. (2008) *Computational Geometry: Algorithms and Applications* (3rd ed.) Berlin: Springer.

Brans, J.P., Engelen, G. and Hubert, L. (1981) "Accessibility to a Road Network: Definitions and Applications" *The Journal of the Operational Research Society* 32 (8) pp.653–673 DOI: 10.1057/jors.1981.133.

Bu, J., Yin, J., Yu, Y. and Zhan, Y. (2021) "Identifying the Daily Activity Spaces of Older Adults Living in a High-density Urban Area: A Study Using the Smartphone-based Global Positioning System Trajectory in Shanghai" *Sustainability* 13 (9) 5003 DOI: 10.3390/su13095003.

Buchhorn, M., Lesiv, M., Tsendbazar, N-E., Herold, M., Bertels, L. and Smets, B. (2020) "Copernicus Global Land Cover Layers – Collection 2" *Remote Sensing* 12 (6) 1044 DOI: 10.3390/rs12061044.

Burghardt, D., Duchene, C. and Mackaness, W. (Eds) (2014) *Abstracting Geographic Information in a Data Rich World: Methodologies and Applications of Map Generalization* Berlin, Heidelberg: Springer.

Burrough, P.A., McDonnell, R. and Lloyd, C.D. (2015) *Principles of Geographical Information Systems* (3rd ed.) Oxford, New York: Oxford University Press.

Cambridge Dictionary (2023) "Conceptualization" Available at: https://dictionary.cambridge.org/dictionary/english/conceptualization (Accessed: 15th March 2023).

Cellmer, R., Cichulska, A. and Bełej, M. (2020) "Spatial Analysis of Housing Prices and Market Activity with the Geographically Weighted Regression" *ISPRS International Journal of Geo-Information* 9 (6) 380 DOI: 10.3390/ijgi9060380.

Chernokulsky, A., Kurgansky, M., Mokhov, I., Shikhov, A., Azhigov, I., Selezneva, E., Zakharchenko, D., Antonescu, B. and Kühne, T. (2020) "Tornadoes in Northern Eurasia: From the Middle Age to the Information Era" *Monthly Weather Review* 148 (8) pp.3081–3110 DOI: 10.1175/MWR-D-19-0251.1.

Chiles, J-P. and Delfiner, P. (2012) *Geostatistics: Modelling Spatial Uncertainty* (2nd ed.) Hoboken, NJ: John Wiley & Sons.

Chrisman, N.R. (2017) "Calculating on a Round Planet" *International Journal of Geographical Information Science* 31 (4) pp.637–657 DOI: 10.1080/13658816.2016.1215466.

Clementini, E. and Billen, R. (2006) "Modelling and Computing Ternary Projective Relations Between Regions" *IEEE Transactions on Knowledge and Data Engineering* 18 (6) pp.799–814 DOI: 10.1109/TKDE.2006.102.

Clementini, E., Di Felice, P. and van Oosterom, P. (1993) "A Small Set of Formal Topological Relationships Suitable for End-User Interaction" In Abel, D. and Chin Ooi, B. (Eds) *Advances in Spatial Databases* Berlin, Heidelberg: Springer pp.277–295 DOI: 10.1007/3-540-56869-7_16.

Cleveland, W.S. (1979) "Robust Locally Weighted Regression and Smoothing Scatterplots" *Journal of the American Statistical Association* 74 (368) pp.829–836 DOI: 10.2307/2286407.

Cliquet, G. (Ed.) (2006) *Geomarketing: Methods and Strategies in Spatial Marketing* Newport Beach, CA: ISTE USA.

Cressie, N. (1993) *Statistics for Spatial Data* (2nd ed.) Hoboken, NJ: John Wiley & Sons.

Danielson, J.J. and Gesch, D.B. (2011) "Global Multi-resolution Terrain Elevation Data 2010 (GMTED2010)" (*Open File Report 2011-1073*) DOI: 10.3133/ofr20111073.

de Smith, M.J., Goodchild, M.F. and Longley, P.A. (2018) *Geospatial Analysis: A Comprehensive Guide to Principles, Techniques and Software Tools* (6th ed.) Winchelsea: The Winchelsea Press.

Diggle, P. (2014) *Statistical Analysis of Spatial and Spatio-temporal Point Patterns* (3rd ed.) Boca Raton, FL: CRC Press.

Dodge, S., Gao, S., Tomko, M. and Weibel, R. (2020) "Progress in Computational Movement Analysis – Towards Movement Data Science" *International Journal of Geographical Information Science*, 34 (12) pp.2395–2400 DOI: 10.1080/13658816.2020.1784425.

Dryden, K.V. and Mardia, I.L. (2016) *Statistical Shape Analysis with Applications in R* (2nd ed.) Chichester: John Wiley & Sons.

Duckham, M., Kulik, L., Worboys, M. and Galton, A. (2008) "Efficient Generation of Simple Polygons for Characterizing the Shape of a Set of Points in the Plane" *Pattern Recognition* 41 (10) pp. 3224–3236 DOI: 10.1016/j.patcog.2008.03.023.

Dutton, G.H. (1999) *A Hierarchical Coordinate System for Geoprocessing and Cartography* Berlin, Heidelberg: Springer DOI: 10.1007/BFb0011617.

Edelsbrunner, H., Kirkpatrick, D. and Seidel, R. (1983) "On the Shape of a Set of Points in the Plane" *IEEE Transactions on Information Theory* 29 (4) pp.551–559 DOI: 10.1109/TIT.1983.1056714.

Egenhofer, M.J. and Franzosa, R.D. (1991) "Point-set Topological Spatial Relations" *International Journal of Geographical Information Systems*, 5 (2) pp.161–174 DOI: 10.1080/02693799108927841.

Ester, M. Kriegel, H-P., Sander, J. and Xu, X. (1996) "A Density-based Algorithm for Discovering Clusters in Large Spatial Databases with Noise" *Proceedings of the 2nd International Conference on Knowledge Discovery and Data Mining. 2nd International Conference on Knowledge Discovery and Data Mining* Portland, OR: AAAI Press, pp.226–231.

Eugenio-Martin, J.L., Cazorla-Artiles, J.M. and González-Martel, C. (2019) "On the Determinants of Airbnb Location and its Spatial Distribution" *Tourism Economics*, 25(8) pp.1224–1244 DOI: 10.1177/1354816618825415.

Evans, I.S. (1980) "An Integrated System of Terrain Analysis and Slope Mapping" *Zeitschrift fur Geomorphologie* 36 pp.274–295.

Florinsky, I.V. (2016) *Digital Terrain Analysis in Soil Science and Geology* (2nd ed.) Amsterdam: Elsevier.

Fotheringham, A.S., Yang, W. and Kang, W. (2017) "Multiscale Geographically Weighted Regression (MGWR)" *Annals of the American Association of Geographers*, 107 (6) pp.1247–1265 DOI: 10.1080/24694452.2017.1352480.

Fotheringham, S.A., Brunsdon, C. and Charlton, M. (2002) *Geographically Weighted Regression: The Analysis of Spatially Varying Relationships* Chichester: Wiley DOI: 10.1007/978-3-642-03647-7.

Freitas, H., Freitas, C., Rosim, S. and Oliveira, J. (2016) "Drainage Networks and Watersheds Delineation Derived from TIN-Based Digital Elevation Models" *Computers and Geosciences* 92 pp.21–37 DOI: 10.1016/j.cageo.2016.04.003.

GEBCO Bathymetric Compilation Group 2020 (2020) "The GEBCO_2020 Grid – A Continuous Terrain Model of the Global Oceans and Land" British Oceanographic Data Centre, National Oceanography Centre, NERC, UK Available at: https://www.bodc.ac.uk/data/published_data_library/catalogue/10.5285/a29c5465-b138-234d-e053-6c86abc040b9/ (Accessed: 15th March 2023).

Genesereth, M.R. and Nilsson, N.J. (1987) *Logical Foundations of Artificial Intelligence* Los Altos, CA: Morgan Kaufmann.

Goodchild, M.F., Yuan, M. and Cova, T.J. (2007) "Towards a General Theory of Geographic Representation in GIS" *International Journal of Geographical Information Science* 21 (3) pp.239–260 DOI: 10.1080/13658810600965271.

Goulard, M., Laurent, T. and Thomas-Agnan, C. (2017) "About Predictions in Spatial Autoregressive Models: Optimal and Almost Optimal Strategies" *Spatial Economic Analysis* 12 (2–3) pp.304–325 DOI: 10.1080/17421772.2017.1300679.

Goyal, R.K. and Egenhofer, M.J. (2001) "Similarity of cardinal directions" In Jensen, C.S., Schneider, M., Seeger, B. and Tsotras, V.J. (Eds) *Advances in Spatial and Temporal Databases* Berlin, Heidelberg: Springer, pp.36–55 DOI: 10.1007/3-540-47724-1_3.

Graser, A., Dragaschnig, M. and Koller, H. (2021) "Exploratory Analysis of Massive Movement Data" In Werner, M. and Chiang, Y-Y. (Eds) *Handbook of Big Geospatial Data* Cham, Switzerland: Springer, pp.285–319 DOI: 10.1007/978-3-030-55462-0_12.

Hansen, M.C., Potapov, P.V., Moore, R., Hancher, M., Turubanova, S.A., Tyukavina, A., Thau, D., Stehman, S.V., Goetz, S.J., Loveland, T.R., Kommareddy, A., Egorov, A., Chini, L., Justice, C.O. and Townshend, J.R.G. (2013) "High-resolution Global Maps of 21st-Century Forest Cover Change" *Science* 342 (6160) pp.850–853 DOI: 10.1126/science.1244693.

He, X., Zhang, X. and Xin, Q. (2018) "Recognition of Building Group Patterns in Topographic Maps Based on Graph Partitioning and Random Forest" *ISPRS Journal of Photogrammetry and Remote Sensing* 136 pp.26–40 DOI: 10.1016/j.isprsjprs.2017.12.001.

Hecht, R., Meinel, G. and Buchroithner, M. (2015) "Automatic Identification of Building Types Based on Topographic Databases – a Comparison of Different Data Sources" *International Journal of Cartography* 1 (1) pp.18–31 DOI: 10.1080/23729333.2015.1055644.

Hengl, T., Heuvelink, G.B.M. and Stein, A. (2004) "A Generic Framework for Spatial Prediction of Soil Variables Based on Regression-kriging" *Geoderma* 120 (1–2) pp.75–93 DOI: 10.1016/j.geoderma.2003.08.018.

Hengl, T. and Reuter, H.I. (Eds) (2009) *Geomorphometry: Concepts, Software, Applications* Amsterdam: Elsevier.

Hesselbarth, M.H.K., Sciaini, M., With, K.A., Wiegand, K. and Nowosad, J. (2019) "Landscapemetrics: An Open-Source R Tool to Calculate Landscape Metrics" *Ecography* 42 (10) pp.1648–1657 DOI: 10.1111/ecog.04617.

Huang, Y., Shekhar, S. and Xiong, H. (2004) "Discovering Colocation Patterns from Spatial Data Sets: A General Approach" *IEEE Transactions on Knowledge and Data Engineering* 16 (12) pp.1472–1485 DOI: 10.1109/TKDE.2004.90.

Jin, X. and Davis, C. H. (2005) "Automated Building Extraction from High-Resolution Satellite Imagery in Urban Areas Using Structural, Contextual, and Spectral Information" *EURASIP Journal on Advances in Signal Processing* 2005 (14) 745309 DOI: 10.1155/ASP.2005.2196.

Kraak, M-J. and Ormeling, F. (2021) *Cartography: Visualization of Geospatial Data* (4th ed.) Boca Raton, FL: CRC Press.

Lamb, D., Downs, J. and Reader, S. (2020) "Space-time Hierarchical Clustering for Identifying Clusters in Spatiotemporal Point Data" *ISPRS International Journal of Geo-Information* 9 (2) 85 DOI: 10.3390/ijgi9020085.

Lansley, G. and Longley, P.A. (2016) "The Geography of Twitter Topics in London" *Computers, Environment and Urban Systems* 58 pp.85–96 DOI: 10.1016/j.compenvurbsys.2016.04.002.

Lateb, M., Meroney, R.N., Yataghene, M., Fellouah, H., Saleh, F. and Boufadel, M.C. (2016) "On the Use of Numerical Modelling for Near-Field Pollutant Dispersion in Urban Environments – A Review" *Environmental Pollution* 208 pp.271–283 DOI: 10.1016/j.envpol.2015.07.039.

Laube, P. (2014) *Computational Movement Analysis* Cham, Switzerland: Springer.

Lee, K. and Sener, I.N. (2020) "Emerging Data for Pedestrian and Bicycle Monitoring: Sources and Applications" *Transportation Research Interdisciplinary Perspectives* 4 100095 DOI: 10.1016/j.trip.2020.100095.

LeSage, J.P. (2008) "An Introduction to Spatial Econometrics" *Revue d'économie industrielle* 123 pp.19–44.

Levin, S.A. (1992) "The Problem of Pattern and Scale in Ecology" *Ecology* 73 (6) pp.1943–1967.

Li, X. and Yeh, A.G-O. (2000) "Modelling Sustainable Urban Development by the Integration of Constrained Cellular Automata and GIS" *International Journal of Geographical Information Science* 14 (2) pp.131–152 DOI: 10.1080/136588100240886.

Li, Z. (2006) *Algorithmic Foundation of Multi-Scale Spatial Representation* Boca Raton, FL: CRC Press.

Lin, B., Zhou, L., Xu, D., Zhu, A-X. and Lu, G. (2018) "A Discrete Global Grid System for Earth System Modelling" *International Journal of Geographical Information Science* 32 (4) pp.711–737 DOI: 10.1080/13658816.2017.1391389.

MacEachren, A.M. (1995) *How Maps Work: Representation, Visualization, and Design* New York: Guilford Press.

Mackaness, W.A. and Ruas, A. (Eds) (2007) *Generalisation of Geographic Information: Cartographic Modelling and Applications* Amsterdam: Elsevier.

MacQueen, J. (1967) "Some Methods for Classification and Analysis of Multivariate Observations" *Proceedings of the Fifth Berkeley Symposium on Mathematical Statistics and Probability* Berkeley, CA, pp.281–296.

Malczewski, J. (2006) "GIS-based Multicriteria Decision Analysis: A Survey of the Literature" *International Journal of Geographical Information Science* 20 (7) pp.703–726 DOI: 10.1080/13658810600661508.

Matheron, G. (1962) *Traité de géostatistique appliquée* Paris: Technip.

McMaster, R.B. and Sheppard, E. (2004) "Introduction: Scale and Geographic Inquiry" In Sheppard, E. and McMaster, R.B. (Eds) *Scale and Geographic Inquiry* Malden, MA: Blackwell Publishing, pp. 1–22 DOI: 10.1002/9780470999141.ch1.

Murça, M.C.R., Hansman, R.J., Li, L. and Ren, P. (2018) "Flight Trajectory Data Analytics for Characterization of Air Traffic Flows: A Comparative Analysis of Terminal Area Operations Between New York, Hong Kong and Sao Paulo" *Transportation Research Part C: Emerging Technologies* 97 pp. 324–347 DOI: 10.1016/j.trc.2018.10.021.

Newman, M.E.J. (2018) *Networks* (2nd ed.) Oxford, New York: Oxford University Press.

O'Callaghan, J.F. and Mark, D.M. (1984) "The Extraction of Drainage Networks from Digital Elevation Data" *Computer Vision, Graphics, and Image Processing* 28 (3) pp.323–344 DOI: 10.1016/S0734-189X(84)80011-0.

OGC (Open Geospatial Consortium) (2010) "Simple Feature Access – Part 1: Common Architecture" *OGC Implementation Standard for Geographic Information* Available at: https://www.ogc.org/standard/sfa/ (Accessed: 15th March 2023).

OGC (Open Geospatial Consortium) (2017) "OGC® GeoPackage Encoding Standard (OGC 12-128r14)" *(Version 1.2)* Available at: https://portal.opengeospatial.org/files/12-128r14 (Accessed: 15th March 2023).

Openshaw, S. and Taylor, P.J. (1979) "A Million or So Correlation Coefficients: Three Experiments on the Modifiable Areal Unit Problem" In Wrigley, N. (Ed.) *Statistical Applications in the Spatial Sciences* London: Pion, pp.127–144.

Pebesma, E. (2018) "Simple features for R: Standardized support for spatial vector data" *The R Journal* 10 (1) pp.439–446 DOI: 10.32614/RJ-2018-009.

Pereira, J.M.C. and Duckstein, L. (1993) "A Multiple Criteria Decision-Making Approach to GIS-Based Land Suitability Evaluation" *International Journal of Geographical Information Systems* 7 (5) pp. 407–424 DOI: 10.1080/02693799308901971.

Petrakov, D., Shpuntova, A., Aleinikov, A., Kääb, A., Kutuzov, S., Lavrentiev, I., Stoffel, M., Tutubalina, O. and Usubaliev, R. (2016) "Accelerated Glacier Shrinkage in the Ak-Shyirak Massif, Inner Tien Shan, During 2003–2013" *Science of the Total Environment* 562 pp.364–378 DOI: 10.1016/j.scitotenv.2016.03.162.

Peucker, T.K., Fowler, R.J., Little, J.J. and Mark, D.M. (1976) "Digital Representation of Three-Dimensional Surfaces by Triangulated Irregular Networks" (TIN) (*Technical Report 10* [Revised]) Burnaby, British Columbia: Simon Fraser University.

Peuquet, D.J. and Ci-Xiang, Z. (1987) "An Algorithm to Determine the Directional Relationship Between Arbitrarily-shaped Polygons in the Plane" *Pattern Recognition* 20 (1) pp.65–74 DOI: 10.1016/0031-3203(87)90018-5.

Ripley, B.D. (1981) *Spatial Statistics* John Wiley & Sons.

Robertson, C., Chaudhuri, C., Hojati, M. and Roberts, S.A. (2020) "An Integrated Environmental Analytics System (IDEAS) Based on a DGGS" *ISPRS Journal of Photogrammetry and Remote Sensing* 162 pp.214–228 DOI: 10.1016/j.isprsjprs.2020.02.009.

Robinson, A.H., Morrison, J.L., Muehrcke, P.C., Kimerling, A.J. and Guptill, S.C. (1995) *Elements of Cartography* New York: John Wiley & Sons.

Rocca, L., Jenny, B. and Puppo, E. (2017) "A Continuous Scale-Space Method for the Automated Placement of Spot Heights on Maps" *Computers & Geosciences* 109 pp.216–227 DOI: 10.1016/j.cageo.2017.09.003.

Sahr, K., White, D. and Kimerling, A.J. (2003) "Geodesic Discrete Global Grid Systems" *Cartography and Geographic Information Science* 30 (2) pp.121–134 DOI: 10.1559/152304003100011090.

Samet, H. (2006) *Foundations of Multidimensional and Metric Data Structures* Amsterdam: Elsevier and Morgan Kaufmann Publishers.

Samsonov, T., Koshel, S., Walther, D. and Jenny, B. (2019) "Automated Placement of Supplementary Contour Lines" *International Journal of Geographical Information Science* 33 (10) pp.2072–2093 DOI: 10.1080/13658816.2019.1610965.

Scheider, S., Gräler, B., Pebesma, E. and Stasch, C. (2016) "Modelling Spatiotemporal Information Generation" *International Journal of Geographical Information Science* 30 (10) pp.1–29 DOI: 10.1080/13658816.2016.1151520.

Shekhar, S. and Chawla, S. (2003) *Spatial Databases: A Tour* Upper Saddle River, NJ: Prentice Hall.

Sibson, R. (1981) "A Brief Description of Natural Neighbour Interpolation" In Barnett, V. (Ed.) *Interpreting Multivariate Data*. Chichester: John Wiley & Sons, pp.21–36.

Stasch, C. Scheider, S., Pebesma, E. and Kuhn, W. (2014) "Meaningful Spatial Prediction and Aggregation" *Environmental Modelling and Software* 51 pp.149–165 DOI: 10.1016/j.envsoft.2013.09.006.

Sun, S. (2018) "A Spatial One-to-many Flow Layout Algorithm Using Triangulation, Approximate Steiner Trees, and Path Smoothing" *Cartography and Geographic Information Science* pp.1–17 DOI: 10.1080/15230406.2018.1437359.

Tobler, W.R. (1970) "A Computer Movie Simulating Urban Growth in the Detroit Region" *Economic Geography* 46 pp.234–270 DOI: 10.2307/143141.

Tobler, W.R. (1981) "A Model of Geographical Movement" *Geographical Analysis* 13 (1) pp.1–20 DOI: 10.1111/j.1538-4632.1981.tb00711.x.

Tomlin, D. (2012) *GIS and Cartographic Modelling* (2nd ed.) Redlands, CA: Esri Press.

Tripathy, P. and Kumar, A. (2019) "Monitoring and Modelling Spatio-temporal Urban Growth of Delhi Using Cellular Automata and Geoinformatics" *Cities* 90 pp.52–63 DOI: 10.1016/j.cities.2019.01.021.

Wang, J., Duckham, M. and Worboys, M. (2015) "A Framework for Models of Movement in Geographic Space" *International Journal of Geographical Information Science* 30 (5) pp.970–992 DOI: 10.1080/13658816.2015.1078466.

Witkin, A. (1984) "Scale-space Filtering: A New Approach to Multi-Scale Description" *Proceedings of ICASSP '84: IEEE International Conference on Acoustics, Speech, and Signal Processing* San Diego, CA: Institute of Electrical and Electronics Engineers, pp.150–153 DOI: 10.1109/ICASSP.1984.1172729.

Worboys, M. and Duckham, M. (2021) "Qualitative-geometric 'Surrounds' Relations Between Disjoint Regions" *International Journal of Geographical Information Science* 35 (5) pp.1032–1063 DOI: 10.1080/13658816.2020.1859513.

Zevenbergen, L.W. and Thorne, C.R. (1987) "Quantitative Analysis of Land Surface Topography" *Earth Surface Processes and Landforms* 12 (1) pp.47–56 DOI: 10.1002/esp.3290120107.

Zhang, D. and Lu, G. (2004) "Review of Shape Representation and Description Techniques" *Pattern Recognition* 37 (1) pp.1–19 DOI: 10.1016/j.patcog.2003.07.008.

Zhang, J., Atkinson, P.M. and Goodchild, M.F. (2014) *Scale in Spatial Information and Analysis* Boca Raton, FL: CRC Press.

Zheng, Y. (2015) "Trajectory Data Mining: An Overview" *ACM Transactions on Intelligent Systems and Technology* 6 (3) pp.1–41 DOI: 10.1145/2743025.

21
THE GEOVISUALIZATION OF BIG DATA

Nick Bearman

This chapter will discuss a range of aspects of big data, including what it is, the different tools and techniques we can use to analyse and visualize it and the ethical considerations we need to take when thinking about visualizing and working with big data. We will also discuss practical methods for understanding the ethical considerations required. These are relevant to all types of big data, but the vast majority of big data contain some geospatial element. Geospatial data brings some additional considerations, which will also be discussed in this chapter.

What is big data?

Big data is[1] a term that came to the fore in the early 2010s for which there is no formal definition. It is typically defined as datasets that are larger than 'we are used to working with', although, of course, the size of the data 'we are used to working with' has increased. Colloquially, when big data first came to the fore, it was often thought that if your data could not fit into an Excel worksheet (i.e., 65,535 rows and 256 columns[2]) it was considered as 'big data'. Typically, working with big data requires specific software and techniques, including managing and cataloguing the data, as well as analysing or creating visualizations.

Big data is usually secondary data, in that it has been collected for one specific purpose, but then is used for another purpose. An example of this is loyalty card data, which is initially collected by a retailer to better understand and market to their customers (Cortiñas *et al.*, 2008). Loyalty card data can also be used for a wide range of other applications, such as identifying stores customers choose to use relative to their home address (Primerano *et al.*, 2008), understanding purchases of over-the-counter pain medication (Davies *et al.*, 2018), and detecting unexpected correlations of product purchases such as an increase in buying nappies with an increase in buying beer (Lloyd *et al.*, 2018).[3]

Big data is also usually unstructured, in that it does not have a well-defined data model, and/or contains large chunks of text that require some initial processing before analysis can take place. For example, with loyalty card data, product details (e.g., own brand 16 × 500 mg paracetamol) may or may not be easily grouped with other types of paracetamol (own brand and branded), other pain-killers (ibuprofen/aspirin/codeine) or other over-the-counter medicine. Commonly, big data will often require significant re-working before any analysis can be performed.

Additionally, there is a key geographical element to this data. For example, with loyalty card data, we usually have home addresses of customers and the location of the store where they bought their products. This can bring some insightful analysis by looking at the relationship between these two, for example customers who use a click and collect service to pick up products from stores on their commute (Davies et al., 2019).

We can also use a customer's home address to link them to their (most likely) geodemographic – a picture of where similar people are based - which can help a retailer group similar types of customer together to better understand their behaviour and buying habits (see section on summarizing and simplifying data for more details on geodemographics).

Whichever type of big data we are working with, it is very important to establish the research question. Even if the analytical process begins with some exploratory data analysis and no formal hypothesis testing, it is still good to have a clear idea of what we are looking for. Having a clear research question will help to focus on the parts of the data that are useful and the extraneous information that is not relevant to be filtered out. This is relevant both to the analysis stage and to the visualization stage of the research process.

Working with big data

Many of the tools used to work with big geospatial data are still under active development. Whereas a number of useful tools are available for working with big data in general (e.g., Apache Hadoop, Apache Spark, MongoDB, Tableau), few of these are explicitly designed for use with geospatial data. Many of the concepts within this area are key, for example the big data concept of moving the processing code to the data, rather than moving the data to the code – a key principle in the Apache Hadoop software, including in the data processing approach of MapReduce.[4]

The active development of these tools means that working with these types of data usually requires a variety of 'data-science' skills, including database configuration and management, scripting/programming (e.g., R or Python) and version control (e.g., Git). There is a clear need for social scientists (including geographers) to possess these skills alongside their traditional disciplinary expertise (Bearman, 2021). Otherwise, those without Social Science skills will create tools that may not be as useful for Social Science research (Singleton, 2014).

Many good Geographical Information Science or Geographic Data Science master's degrees[5] include both technical skills (e.g., programming, databases, big data analytics, data mining, Web maps, cartography) and Social Science skills (e.g., survey analysis, human geography, health, spatial inequalities, and ethics). Many of these programmes and the modules within contain a mixture of both aspects, i.e., what the technology can do and how we can apply it ethically.

Summarizing and simplifying data

A common approach to working with large datasets is to create an index or a classification of some type. There are a number of approaches, including multi-dimensional indices and k-means clustering. The idea is to take a wide range of input data that have a conceptual link to the factors we are trying to summarize, and create an index or series of classifications (sometimes with a hierarchical element) to summarize the data.

One approach is to create a multi-dimensional index, where multiple datasets are brought together, weighted and ranked in order to produce an index (Bearman, 2020). The most well-known of these is the IMD (index of multiple deprivation), which brings together data on education, occupation, income, housing, health, crime and access to services (Noble et al., 2006; Smith et al., 2015). It is widely used in central and local government planning in the UK, and ranks small areas (LSOAs)[6] in order from most deprived to least deprived. The same technique has been used to develop AHAH (Access to Healthy Assets & Hazards), an index measuring how healthy neighbourhoods are, including the retail environment (access to fast food outlets, pubs, off-licences, tobacconists, gambling outlets), health services (access to GPs, hospitals, pharmacies, dentists, leisure services), the physical environment (blue space, green space – active, green space – passive) and air quality (nitrogen dioxide, particulate matter, sulphur dioxide) (Green et al., 2018).

A more advanced approach is k-means clustering, where similar data is gathered together. This is commonly used for geodemographic classifications, where data is prepared and weighted. The final stage is the clustering process, where an iterative process (k-means clustering) is used to find the best set of categories for this data, and these categories are often hierarchical, with four to five top level categories, and 15–20 sub-categories (Alexiou et al., 2018).

Geodemographics often uses this approach to take a wide range of variables and try to group people together. This is often used by retail companies to better understand their customers. A number of generalized commercial classifications are available (such as Experian and Mosaic) as well as open classifications such as OAC (Output Area Classification).[7] There are also bespoke classifications for specific applications, such as the Internet User Classification[8] (Alexiou et al., 2018).

Summarizing the data in this way allows it to be displayed much more easily, by just showing the classification rather than all of the source data. Indexes and geodemographics can be represented on a map, which can work well. They can also be shown non-spatially, which is particularly useful if the user is focused on a specific local authority. For example, it is possible to highlight which LSOAs in the local authority are in the top decile (top 10%) deprived of all LSOAs in the country and show links with deprivation and a range of other data by combining a graph of the IMD decile and the other variable or dataset.

Visualizing big data

Whichever tools and analysis techniques we use with our data, whether or not it is 'big data', researchers need to think about how they are going to present it. One of the key things to think about is to establish the main aims of the analysis. For a research project, these will form the research questions. For a different type of project, the same principle applies: what are you trying to find out from the data? Once you are clear on what you want to find out, you can then think about how you are going to present the findings.

There are many different techniques for visualizing big geospatial data. Given that it is geospatial, the first thought might be to create some type of map, although maps are not always the best way of representing spatial data. For example, if we are interested in the relationship between two groups of information (one of which happens to be data aggregated spatially), then a scatterplot may be much more effective than a map.

The type of approach depends on the type of data to be summarized, and what it is intended to reveal. For information related to populations, or people, then a cartogram might be a suitable approach, i.e., where a geographic map is resized to represent the population. This could then be used to show a variety of data related to population, (e.g., income per capita, education, literacy, cases of disease, and so on). Hennig (2013, 2019) describes the process in detail and illustrates how cartograms are a visually powerful tool. There are many further examples of these on the Worldmapper[9] website (Figure 21.1).

For data that includes a temporal element, as well as a spatial one, then a space time cube can be a useful representation method, providing much more information than just a map (see Figure 21.2). This often works well with a small number (up to 5) trajectories that move through space and time. For more data than this, some grouping may be required.

Figure 21.1 Cartograms are a really useful way of showing population related data. A cartogram is resized to show a specific variable; in this case, population in 2020 (Hennig, 2019)

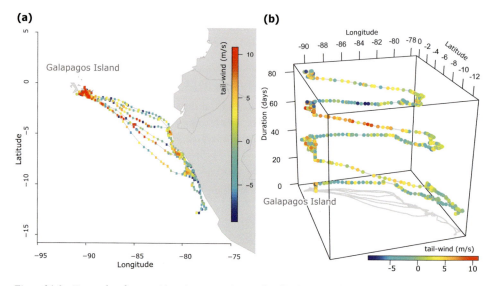

Figure 21.2 Example of a map (a) and a space-time cube (b) showing the GPS trajectory of a Galapagos Albatross (Dodge and Noi, 2021). Note that on the map, the spatial pattern is clear, but the temporal aspect is not. In the space-time cube, the flight patterns of the albatross are clear, showing its outward and return journeys

How is big data impacting on society?

The use of big data can have a massive impact on society, both good and bad. The impact will depend on how the data is used and different organizations that make use of big data will influence public opinion. Very often this comes down to trust (i.e., do we trust the organization to look after our data securely and only use it for appropriate uses?). This will vary between organizations. For example, our view of whether we trust the government with our data will probably be very different to whether we trust a social media company, such as Facebook or Twitter, with our data.

When working with data (including geographic information), researchers have a responsibility for the security of the data they are using. This is both a legal responsibility (through GDPR) as well as an ethical responsibility; all ethical reviews will include a section on data security. Researchers are required to follow appropriate guidance to ensure that the data they are working with does not fall into the wrong hands. The specifics of this will, of course, vary depending on what information the dataset contains, and most universities and public institutions have ethical guidelines for researchers to follow. This is an opportunity to consider the data and its use, whether that use is ethical and whether the same or similar research aims could be achieved with a less-sensitive dataset.

Ethical standards have often been led by medical fields such as health and psychology rather than geospatial sciences. This has changed significantly since 2018 with the introduction of GDPR, and the prevalence of geospatial elements within big data.

What is GDPR?

The General Data Protection Regulation (GDPR) was implemented in 2018 by the European Union. One of the key aspects of GDPR is that the user (data subject) has control over what their data is used for, and needs to provide informed consent for their data to be used in this way. GDPR also provides for users to be able to correct data about them if it is wrong. The regulation is still relatively recent and has yet to be fully tested in the courts, although there have been some cases where technology giants have been fined substantially for non-compliance (see, for example, Hern, 2019).

One of the fundamental principles of GDPR is that data should be collected for a specific purpose, and that the end user is aware of and agrees to this same purpose. Therefore, after the data is collected, it can only be used for this specific purpose. When working with data collected under GDPR, it is important to be aware of why it was collected (i.e., was according to its specific purpose).

Big data will always be secondary data, and the intended use of this data may clash with the requirement from GDPR for data to be used for a specific purpose. Sometimes this purpose may be quite general (e.g., 'to be used in research to help us understand customers more effectively'/'to be used in research to help us understand the impact of COVID-19 on ethnic minorities'/'to be used in research to understand how the public make use of health facilities'), or may be very specific. Whatever the case, researchers should always be aware of the specific purpose for the collection and use of data.

Ethical considerations

The increase in the use of big data also raises the need to consider the ethics of its use. This is particularly relevant when working with data that refers to individual people, whether they are identifiable (e.g., via name, date of birth) or non-identifiable. It is important to always be aware of how this information could be combined with other information and whether our use of the data is disclosive. Although GDPR is a general piece of legislation for collecting and working with personal data, it is highly relevant to big data, and was drafted with big data in mind. It is not specific to working with geographical data, however, and there are other guidelines for working with spatial data.

The UK Statistics Authority (2021) have published 'Ethical Considerations in the Use of Geospatial Data for Research and Statistics', which outlines a range of issues to consider when publishing spatial data. Their 16-point checklist offers a good starting point for analysing and publishing spatial data.

The Locus Charter[10] is a relatively new great framework, and one of the first to address these issues explicitly taking into account geographical data. It was developed by EthicalGeo and Benchmark and provides an initial framework to consider ethical issues. It contains a series of ten key principles (see Chapter 48 on the Locus Charter in this *Handbook* for more details).

The World Wide Web Consortium (W3C, 2021) has also published their Interest Group Note on 'The Responsible Use of Spatial Data', and the Royal Statistical Society (Royal Statistical Society, 2019) have led the way with their 'Guide for Ethical Data Science'. These cover a range of similar themes from slightly different points of view.

These frameworks are the first attempts to address these issues by taking geographical data explicitly into account. However, they lack some specifics on how they should actually be implemented when working with spatial data. As Specht (2021) comments, the Locus Charter is a good starting point, but some of the concepts need more detail. For example, the concept of 'minimize data' is helpful (keep no more data than the minimum required to do your analysis), but exactly who decides on the minimum data required is not stipulated.

There are still many unknowns about ethics in big geospatial data, including standard working practices and what is and what is not acceptable. Currently, showing that ethical implications have been considered is enough, depending on the exact use of the data. It will be interesting to see how privacy laws – and public expectations – develop in this regard.

Secure data infrastructures

One approach often used when working with big data that are commercially or personally sensitive is secure data storage. In this case, access to the data is controlled, often according to ISO 27001, which outlines the key requirements for systems working with sensitive data and requirements for keeping the data secure. There are many systems available for academic research in the UK, including the ONS Secure Research Service, UK Data Service Secure Lab, CDRC Secure Labs programme and many others.

Confidential or sensitive big data (e.g., electoral roll, loyalty card data, information on bankruptcies), this is managed in a secure environment. Access is controlled to specific users, who cannot export data from the secure environment without going through a series of output checks. In addition, specific information may be redacted to reduce data sensitivity (e.g., removing people's names, replacing date of birth with year of birth). Which methods are required depends on the data being managed, the purpose of its storage and the results of the DPIA (Data Protection Impact Assessment) that helps to inform the steps required.

A secure environment would typically comprise a special set of computers that might need to be accessed physically in a room or laboratory, or accessed over a VPN (virtual private network) or remote access environment. The user would need to complete a range of training before they are given access to the environment. Their research proposal would also be reviewed and would only be allowed to access the data for this research. The environment would give them a range of tools to analyse the data (including potentially geospatial and/or big data tools) and they would not be permitted to remove any data or analysis from the environment. This would need to go through a disclosure checking procedure, where two (or more) people would verify that the data is not disclosive and meets the specific requirements of the research project and the terms under which the data is shared. Subsequently, then the data would be released to the end user.

Conclusion

There is no one solution to visualizing big, geo-data. The best way depends on the data itself, and what we are trying to show. Much like cartography, there is a mixture of science and art in choosing the best visualization method; there are many good examples available, so these should be used as an inspiration and to see what see what works best for the data.

There are a number of common techniques that are often used with big data, with a combination of technological solutions to handle big data and data reduction techniques. Additionally, there is a range of ethical considerations when working with big data that contains personal or sensitive information (which is the vast majority of big data) and a number of conceptual, procedural and technical solutions available to address these. From a geospatial point of view, the Locus Charter is beginning to address some of the procedural issues. The technical solutions include secure data infrastructures, and for best practice, both of these need to be considered in combination.

Big data, particularly with a spatial element, can offer amazing insights into many different areas, with substantial research potential. However, with great power comes great responsibility; researchers need to ensure that they manage and work with big data within the law and within the ethical frameworks that exist within our academic areas so that researchers can continue to have access to these invaluable, powerful and very detailed datasets.

Notes

1 Formally, the word 'data' is the plural form of the singular term 'datum', so one should write 'data are', but in the majority of uses it is written 'data is', which is the standard adopted in this chapter.
2 This was the limit imposed by the XLS file format, which was superseded by XLSX in 2003, which can handle 1,048,576 rows and 16,384 columns.
3 Remembering that correlation ≠ causation!
4 See this question on StackExchange for an explanation: https://stackoverflow.com/questions/40601991/.
5 For example, see University of Liverpool's Geographic Data Science or UCL's Social and Geographic Data Science courses (https://www.liverpool.ac.uk/study/postgraduate-taught/taught/geographic-datascience-msc/module-details/ and https://www.geog.ucl.ac.uk/study/graduate-taught/msc-social-and-geographic-data-science).
6 Lower layer Super Output Areas, a common geography used with census data in the UK, containing about 1,500 people.
7 https://data.cdrc.ac.uk/dataset/output-area-classification-2011.
8 https://data.cdrc.ac.uk/dataset/internet-user-classification.
9 https://worldmapper.org/.
10 Locus is the Latin word for 'place' https://ethicalgeo.org/locus-charter/.

References

Alexiou, A., Riddlesden, D., Singleton, A., Longley, P. and Cheshire, J. (2018) "The Geography of Online Retail Behaviour" *Consumer Data Research* London: UCL Press, pp.96–109.

Bearman, N. (2020) "Advanced GIS Methods Training: AHAH and Multi-Dimensional Indices | CDRC Data" Aavailable at: https://data.cdrc.ac.uk/dataset/advanced-gis-methods-training-ahah-and-multi-dimensional-indices (Accessed: 31st May 2021).

Bearman, N. (2021) *GIS: Research Methods* London: Bloomsbury Publishing.

Cortiñas, M., Elorz, M. and Múgica, J.M. (2008) "The Use of Loyalty-cards Databases: Differences in Regular Price and Discount Sensitivity in the Brand Choice Decision Between Card and Non-card Holders" *Journal of Retailing and Consumer Services* 15 (1) pp.52–62.

Davies, A., Dolega, L. and Arribas-Bel, D. (2019) "Buy Online Collect In-store: Exploring Grocery Click&Collect Using a National Case Study" *International Journal of Retail & Distribution Management* 47 (3) pp.278–291.

Davies, A., Green, M.A. and Singleton, A.D. (2018) "Using Machine Learning to Investigate Self-medication Purchasing in England via High Street Retailer Loyalty Card Data" *PLoS One* 13 (11) e0207523 pp.1–14 DOI: 10.1371/journal.pone.0207523.

Dodge, S. and Noi, E. (2021) "Mapping Trajectories and Flows: Facilitating a Human-centered Approach to Movement Data Analytics" *Cartography and Geographic Information Science* 48 (4) pp.353–375.

Green, M.A., Daras, K., Davies, A., Barr, B. and Singleton, A. (2018) "Developing an Openly Accessible Multi-Dimensional Small Area Index of 'Access to Healthy Assets and Hazards' for Great Britain, 2016" *Health & Place* 54 pp.11–19.

Hennig, B.D. (2013) *Rediscovering the World: Map Transformations of Human and Physical Space* Berlin and Heidelberg: Springer DOI: 10.1007/978-3-642-34848-8.

Hennig, B.D. (2019) "Remapping Geography: Using Cartograms to Change Our View of the World" *Geography* 104 (2) pp.71–80.

Hern, A. (2019) "Google Fined Record £44m by French Data Protection Watchdog" *The Guardian* (21st January) Available at: https://www.theguardian.com/technology/2019/jan/21/google-fined-record-44m-by-french-data-protection-watchdog (Accessed: 31st May 2021).

Lloyd, A., Cheshire, J., Squires, M., Longley, P. and Singleton, A. (2018) "The Provenance of Customer Loyalty Card Data" *Consumer Data Research* London: UCL Press, pp.28–39.

Noble, M., Wright, G., Smith, G. and Dibben, C. (2006) "Measuring Multiple Deprivation at the Small-Area Level" *Environment and Planning A: Economy and Space* 38 (1) pp.169–185.

Primerano, F., Taylor, M.A.P., Pitaksringkarn, L. and Tisato, P. (2008) "Defining and Understanding Trip Chaining Behaviour" *Transportation* 35 (1) pp.55–72.

Royal Statistical Society (2019) "A Guide for Ethical Data Science" Available at: https://rss.org.uk/RSS/media/News-and-publications/Publications/Reports%20and%20guides/A-Guide-for-Ethical-Data-Science-Final-Oct-2019.pdf (Accessed: 31st May 2021).

Singleton, A. (2014) "Learning to Code" *Geographical Magazine* Available at: http://geographical.co.uk/opinion/item/278-learning-to-code (Accessed: 13th November 2019).

Smith, T., Noble, M., Noble, S., Wright, G., McLennan, D. and Plunkett, E. (2015) "English Indices of Deprivation 2015" (*Technical Report*) Available at: https://www.gov.uk/government/publications/-english-indices-of-deprivation-2015-technical-report (Accessed: 31st May 2021).

Specht, D. (2021) "Is the Locus Charter Enough to Rein in the Power of Tech Companies?" *Geography Directions* (29th April) Available at: https://blog.geographydirections.com/2021/04/29/is-the-locus-charter-enough-to-rein-in-the-power-of-tech-companies/ (Accessed: 31st May 2021).

UK Statistics Authority (2021) "Ethical Considerations in the Use of Geospatial Data for Research and Statistics" *UK Statistics Authority* Available at: https://uksa.statisticsauthority.gov.uk/publication/-ethical-considerations-in-the-use-of-geospatial-data-for-research-and-statistics/ (Accessed: 31st May 2021).

W3C (2021) "The Responsible Use of Spatial Data" Available at: https://www.w3.org/TR/responsible-use-spatial/ (Accessed: 31st May 2021).

22
MACHINE LEARNING AND GEOSPATIAL TECHNOLOGIES

Izabela Karsznia

Contemporary cartography involves interdisciplinary studies from the environmental, social, and computer science domains, including geographical data exploration and analysis. Cartographic generalization is a crucial element of map design, and is a very complex process that requires consideration of the multifaceted character of objects. In the past few years, research on automated generalization has become enormously complex with increasing applications of data science principles and machine learning-based frameworks (Feng *et al.*, 2019). At the same time, research on map generalization has almost exclusively focused on maps and spatial data at large scales, for instance, from 1:10,000 to 1:50,000 (Stoter *et al.*, 2014). Since basic spatial databases with nationwide coverage have typically been collected at the scales of 1:10,000, 1:25,000, or 1:50,000, the need for their automated generalization at medium and small scales has become urgent.

Unfortunately, neither coherent standards nor homogeneous principles of small-scale map generalization have been elaborated so far. As a result, in the current production of small-scale topographic maps, cartographic generalization is primarily carried out manually in interactive systems and thus relies on the fundamental decisions made by a cartographer. The output of the process is therefore prone to bias and subjectivity. Consequently, maps of the same small scale may differ considerably among different authors. From this point of view, after a wave of research on large-scale map generalization, the automation of generalization procedures at medium and small scales is necessary, highly innovative, and challenging.

Machine learning (ML) has recently been recognized as a promising generalization technique and even cited by some researchers as a new paradigm for cartographic generalization (Touya *et al.*, 2019). The great potential of ML for knowledge acquisition was recognized more than 20 years ago (Weibel *et al.*, 1995). Nevertheless, limited computing power remained a significant obstacle in making full use of those ideas. Recently, the potential of ML methods has been explored, for example, to enrich data with implicit structures and relations (Touya and Dumont, 2017), to acquire procedural knowledge to orchestrate and parametrize generalization algorithms (Zhou and Li, 2017), and to evaluate generalized maps (Harric *et al.*, 2015).

The research described in this chapter follows this paradigm shift in the map generalization process as it aims to propose and compare automated ML methods for settlement generalization. Promising results have been achieved for Polish datasets (Karsznia and Weibel,

2018; Karsznia and Sielicka, 2020), and therefore the motivation is to learn from this experience. The purpose of this chapter is to introduce the idea and basic algorithms of ML in the context of cartographic generalization. The goal is not to provide exhaustive information, but to discuss the possibilities of ML application in this field. Specifically, in this chapter we discuss the constraints and opportunities of selected ML-based model application in automatic settlements selection for small-scale maps.

Machine learning

Machine learning (ML) is a subfield of artificial intelligence (AI) that addresses the possibility of constructing computer programs and algorithms that automatically learn and improve with experience (Mitchell, 1997). Based on input data and correct examples, a machine can explore knowledge (new information) and draw conclusions based on the regularities detected in the data. The task of the machine is to find patterns in the data. Machine learning is connected with computer programming to optimize a performance criterion using example data or experience. Given a model defined by certain parameters, learning is the execution of a computer program to optimize the parameters of the model using a separate set of training data. The model may be predictive so as to make calculations about the future, or descriptive to gain knowledge from data, or both (Alpaydin, 2010).

The application of ML can be found in many areas related to vision, speech recognition, robotics, and other fields. With advances in computer technology, it is currently possible to store and process large amounts of data, as well as access it from distant locations over a computer network. The application of ML methods to large database analysis is called 'data mining'. This term comes from the analogy of a large volume of material being extracted from a mine, which leads to exploring a small amount of precious material, while in data mining a large volume of data is processed to construct a model with valuable use and reasonable accuracy (Alpaydin, 2010). To be intelligent, a system should have the ability to learn and explore. If the system can learn and explore new knowledge, the system designer need not foresee and provide solutions for all possible situations as the system will automatically find them.

Machine learning types

There are four types of ML: supervised, unsupervised, reinforcement, and evolutionary learning (Marshland, 2015). Supervised learning, also called predictive learning, assumes that the training set of examples with correct responses (also called targets or labels) is provided. Based on this training set, the algorithm or ML-based model generalizes (decides) to respond correctly to all possible inputs. This learning type is also commonly called learning from examples (Murphy, 2012; Marshland, 2015). It means that while building the ML model, we have gathered the examples of proper and correct responses (answers to the defined problem) and the ML model learns how to find correct responses using these examples. Cases where the aim is to assign each input example to one of a finite number of discrete categories are called classification problems. On the other hand, in cases where the desired output consists of one or more continuous variables the task is called regression (Bishop, 2006).

In unsupervised learning, also called descriptive learning, the correct responses are not provided, but instead the algorithm or ML-based model tries to find patterns in the data (Murphy, 2012). The algorithm looks for similarities between inputs so that the inputs that have some common features are categorized together. In some cases, the aim in unsupervised learning is to discover groups of similar examples within the data, and this is called clustering. If we want to determine the distribution of data within the input space, we deal with density estimation. Furthermore, if we try to project the data from a high-dimensional space down to two or three dimensions, the purpose is visualization (Bishop, 2006).

The third learning type, called reinforcement learning, can be seen as a learning type between supervised and unsupervised learning (Sutton and Barto, 1998). In this learning type, an algorithm is given information where the response is wrong; however, it does not get examples of correct responses. The algorithm's task is to explore and try various solutions in a trial-and-error approach until it reaches the right decision. Reinforcement learning is a trade-off between exploration, in which the algorithm tries out new kinds of actions to see how effective they are, and exploitation, in which it makes use of actions that are known to be successful (Bishop, 2006).

The evolutionary learning type is related to biological evolution, where organisms adapt and evolve to improve their survival rate. Evolutionary learning concerns models where certain aspects of natural evolution have been proposed as learning methods to improve the performance of computer programs (Back 1996; Bishop, 2006). The most commonly applied ML type is supervised learning and the research described in this chapter uses a supervised ML approach.

Selected machine learning models

The following are the supervised learning models that were used in the research concerning the automatic selection of settlements presented in this chapter.

Decision tree

A decision tree is a hierarchical data structure. The idea of a decision tree is to break a decision problem down into a set of choices about each feature in turn, starting at the root (base) of the tree and progressing down to the leaves, where we receive the classification decision. A decision tree is a tree-like collection of nodes intended to create a decision on values associated with a class or an estimate of a numerical target value. Each node represents a splitting rule for one specific attribute. For classification, this rule separates values belonging to different classes; for regression, it separates them to optimally reduce the error for the selected parameter criterion. The building of new nodes is repeated until stopping criteria are met. A decision for the class label attribute is determined depending on the majority of examples that reached this leaf during generation. At the same time, an estimation for a numerical value is obtained by averaging the values in a leaf (RapidMiner, 2019). The root of the decision tree indicates the most important feature in the process, while decisions are made based on terminal leaves.

Decision trees often do not deliver the best performance in the supervised learning process; however, they have an important advantage. They are straightforward to understand, and can even be turned into a set of if-then rules suitable for use in a rule-based system (Bishop, 2006).

Genetic algorithms

A genetic algorithm (GA) is a search heuristic that mimics the process of natural evolution. This heuristic is routinely used to generate useful solutions for optimization and search problems. Genetic algorithms belong to the larger class of evolutionary algorithms, which generate solutions using techniques inspired by natural evolution such as inheritance, mutation, selection, and crossover. Genetic algorithms are a search heuristic that helps to optimize the selection of the most relevant features in the classification model or select the most relevant attributes of the given dataset. When using genetic algorithms for feature selection, 'mutation' means switching features on and off and 'crossover' means interchanging used features (RapidMiner, 2019).

Different implementations of genetic algorithms vary in their details; however, they typically share a similar scheme. The algorithm operates by iteratively updating a pool of hypotheses, called the population. On each iteration, all members of the population are evaluated according to some fitness function. For instance, if the task is to learn a strategy for selecting settlements, the fitness function could be defined as the number of cases where a particular settlement was selected or omitted by both the GA and by an experienced cartographer in the current population. A new population is then generated by probabilistically selecting the fittest individuals from the current population. Some of these selected individuals are carried forward intact into the next generation population. Others are used as the basis for creating new offspring individuals by applying genetic operations such as crossover and mutation (Mitchell, 1997). Even if the GA does not find an optimal solution, it often succeeds in finding a solution with high fitness. Genetic algorithms have been applied to many optimization problems within and outside of ML.

Random forest models

In recent years, the popularity of random forest models has grown significantly. The idea of using random forest models is mostly based on the fact that if a single tree works well, then many trees (forest) should be even more effective. A random forest model functions by training many trees with the randomness being used in a twofold way. First, in creating a forest, particular trees are trained on slightly different data by taking bootstrap samples from the dataset for each tree. Secondly, at each node a random subset of the features is given to the tree, and the tree can only pick from that subset rather than from the whole set.

Deep learning

Deep learning (DL), also called deep structured learning, belongs to the family of ML methods based on artificial neural networks (ANNs) with representation learning. DL architectures, such as deep neural networks, deep belief networks, recurrent neural networks, and convolutional neural networks, have been applied in many fields from computer vision, speech recognition, and social network filtering.

Artificial neural networks were inspired by information processing and distributed communication nodes in biological systems. The adjective 'deep' comes from the use of multiple layers in the network. The idea of deep learning is that initial analysis can only deal with simple tasks. Learners at different levels can produce higher-order correlations of the data so that eventually the whole system can learn more complicated functions.

Cartographic generalization

Cartographic generalization is an essential element of map design that concerns the selection and simplification of geographic information in order to present a map at a smaller scale. It is a complex decision-making process that requires taking into account the underlying geography of the objects being generalized. Essential aspects of cartographic generalization constitute modelling and abstracting geographic information while maintaining and emphasizing important relations and patterns existing in the data (Mackaness et al., 2017). To automate this process, the relevant cartographic knowledge needs to be gathered, made explicit, and applied in the program supporting automated generalization. In theory, automated generalization should simulate processes of traditional manual map generalization in order to meet cartographic quality and correctness. However, in manual map design, different cartographers will produce different solutions for generalized maps, even if the same large-scale map is provided for generalization. Thus, it is not easy to describe the generalization procedure of a particular map object using a single algorithm (Yan, 2019).

This chapter specifically considers the selection of objects, which is the initial and crucial task to undertake in cartographic generalization. Selection affects the quantity of visual information since it removes part of the map content (Stanislawski et al., 2014). Selection, also called elimination or pruning, constitutes one of the generalization operators. A generalization operator can be seen as a conceptual descriptor for the type of spatial or attribute operation to be achieved on a set of data (Regnauld and McMaster, 2007). Generalization operators are implemented by generalization algorithms or models. For a given operator, several implementations (algorithms or models) may exist (McMaster and Shea, 1992; Bereuter and Weibel, 2013). The goal of this operator is to select essential objects and omit irrelevant objects or object classes. Thus, selection deals with deciding which objects we will show on the map at a smaller scale. This decision is not straightforward, especially at small scales. Although we always select the most relevant and important objects, importance or relevance can be measured differently. For example, if the task is to select the five most important settlements, these may not be the largest ones, since if a settlement is small but isolated from other settlements it may be selected in order to maintain the relative settlement density in an area.

One of the first selection algorithms was 'Radical Law' as proposed by Töpfer and Pillewizer (1966). Their proposed formula calculates the number of objects to be retained on the generalized map, taking into consideration the number of initial objects as well as the source and target scales. However, the formula is now commonly used instead to evaluate selection results, as it does not take into account the spatial patterns and density of the source data. Radical Law informs us how many objects should be selected; however, it does not let us know which ones to select.

There have also been proposed algorithms that use some semantic and spatial characteristics of a set of objects. The goal of object set generalization is to reduce the number of objects in a set while maintaining the similarity among the objects before and after generalization (Yan, 2019). For settlement selection, a settlement-spacing ratio algorithm was proposed by Langan and Poicker (1986). Another algorithm, called circle growth, was introduced by Van Kreveld et al. (1995). Other solutions include an online algorithm for point clustering thematic feature simplification (Burghardt et al., 2004), a dot map simplification algorithm (De Berg et al., 2004), and the Voronoi-based algorithm (Yan and Weibel, 2008). A somewhat different algorithm was proposed by Samsonov and Krivosheina (2012), where both geometric and thematic information were considered.

The limitation of these algorithms is that they usually consider only very basic semantics of settlements, such as population or administrative status. In recently proposed solutions that use algorithms based on ML, specifically decision trees in settlement selection for small-scale maps, Karsznia and Weibel (2018) and Karsznia and Sielicka (2020) argue that more thorough settlement semantic and spatial attributes should be taken into account in order to fully reflect settlement importance. Studies in urban geography also suggest that other factors, such as a settlement's touristic, cultural, or education functions and local density differentiation or settlement size structure should also be considered (Carol, 1960; Smith, 1965; Batty, 2006).

Machine learning in cartographic generalization

Machine learning has recently gained importance as an effective method of cartographic knowledge exploration and a way of automating the generalization process for various types of objects. The first attempts at applying ML models in cartographic generalization concerned the exploration of rule sets (Weibel et al., 1995; Plazanet et al., 1998). At the same time, further research concentrated on extending the initial generalization rule set by exploring new, important generalization rules with ML usage (Mustière et al., 2000; Duchêne et al., 2005).

Machine learning models have also been tested for road and building generalization for large-scale maps. Mustière (1998) used ML to identify the optimal sequence of the generalization operators for roads, whereas Lagrange et al. (2000) tested neural networks in line smoothing. The application of neural networks to road line classification was also investigated by Balboa and López (2009), with Zhou and Li (2017) verifying the suitability of various ML techniques in road selection. In another study, Lee et al. (2017) used ML techniques, namely naive Bayes, DT, k-nearest neighbour, and support vector machine, to perform the elimination and aggregation of buildings. The best classification result was achieved with the use of the decision tree algorithm, especially at the initial steps of building elimination. Recently, Sester et al. (2018) and Feng et al. (2019) proposed the application of deep learning for the task of generalization buildings.

Promising results of automatic settlement selection for small-scale maps were also achieved by Karsznia and Weibel (2018) and Karsznia and Sielicka (2020). To improve the settlement selection process, they used data enrichment and ML, especially decision trees and genetic algorithms. As a result, they explored new variables and generalization rules that are important in the settlement selection process. To examine this topic in more detail, the next section of this chapter focuses on a research project that was designed to use ML models to improve settlement selection in small-scale maps.

Using machine learning models in settlement selection for small-scale maps

The development of automatic selection models supports the process of small-scale map design, making it faster, cost-effective, and more objective. The challenge is not only to make the selection process automatic in order to support human map designers but to achieve the automatic results that are optimal (i.e., as similar as possible to manual approaches to map design). Because manual map design is seen as a pattern of correct selection, manually designed small-scale maps are used for the learning process and verification of the results.

The example discussed here addresses automatic settlement selection in small-scale maps at the range of 1:250,000–1:500,000. An automatic settlement selection with the use of decision tree (DT), decision tree supported with genetic algorithms (DT-GA), random forest (RF), and deep learning (DL) models was conducted, followed by the thorough evaluation of the results achieved. The work was carried out in two steps: validation against the selection status acquired from the atlas map (taken as the reference for evaluation) and comparison of performance statistics across the different models. As a reference and pattern for the correct selection of settlements, an atlas map at a reduced scale of 1:500,000 (GGK, 1993–1997) was used. This research therefore aims to contribute towards extending the toolbox for small-scale mapping.

Materials and methods

The source data comprises the settlement layer contained in the General Geographic Object Database (GGOD) at the detail level of 1:250,000. The GGOD is the database covering the whole of Poland, which, as with other countries, is divided into a hierarchy of administrative units. The first level of administrative units constitutes voivodships and the second level is composed of districts (there are 16 voivodships and 314 districts in Poland). In this case study, 16 Polish districts were used as test areas (5% of all districts in the country) and the districts were differentiated in terms of their population density, settlement density, and settlement size structure (Figure 22.1).

The structure of the GGOD is relatively similar to the general databases of other countries. The GGOD is rich in geometry, but poor in semantics. Thus, to make the database an appropriate source for cartographic generalization, the settlement layer was enriched with additional semantic and spatial attributes (named variables). In order to take the additional variables into account, the source database was enhanced by information from the large-scale Topographic Objects Database (BDOT10k). To automatically enrich the GGOD source database, basic ArcGIS software tools were used in the form of a Python script. Points-of-interest (POIs) information was used, particularly for buildings with specific functions in the BDOT10k database (the enrichment process is described in detail by Karsznia and Weibel, 2018 and by Karsznia and Sielicka, 2020).

The list of considered variables includes 18 semantic and 15 spatial variables. The semantic variables include: Administrative status, Cultural function, Educational function, Trading function, Industrial function, Monumental, Sacred, Accommodation, Communication and finance, Health, Other, Number of railway stops within the settlement area, Number

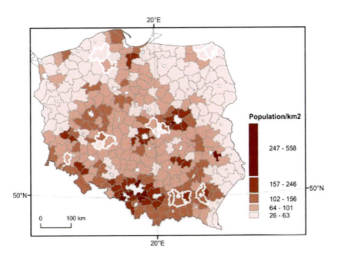

Figure 22.1 District groups considered in the research outlined in white (author's work based on GGOD data)

of airports within the settlement area, Number of ports within the settlement area, Number of district rank roads crossing the settlement border, Total number of communication nodes within the settlement area, and Number of crossings of higher category roads within the settlement area. The spatial variables include: Administrative area of the settlement, Built-up area of the settlement, Residential area of the settlement, Service and commercial area of the settlement, Industrial and storage area of the settlement, Population density of residential areas, population, Density of settlements in the district, Density of settlements calculated using a rectangular grid, Density of settlements calculated using a hexagonal grid, Population density calculated using a rectangular grid, Population density calculated using a hexagonal grid, Population density of districts, Voronoi area, and Distance to the nearest neighbour. The atlas map at the reduced scale of 1:500,000 was used as a reference (GGK, 1993–1997).

The methodology comprises four main stages: (1) selection of districts; (2) data enrichment with the use of the information contained in the BDOT10k considering the new variables; (3) ML model construction; and (4) verification and comparison of the research results. To accommodate cartographic knowledge into the automatic selection process, an additional attribute was added to the defined spatial and semantic variables. This variable concerns the status of the settlement on the atlas reference map at the scale of 1:500,000 and indicates if the map designer would select a particular settlement in the manual map design process. With this variable, some cartographic knowledge has been added to the selection process. Using all of the variables, including the status of the settlement on the atlas map, the selection of the settlement was formulated as a binary classification problem with two labels (i.e., selected or omitted).

Subsequently, four parallel selection processes with the use of the selected ML models were designed. The models were built and executed in RapidMiner Studio. The selection processes consisted of developing and applying decision trees supported with genetic algorithms (DT-GA), decision tree (DT), random forest (RF), and deep learning (DL) models. For the learning, DT-GA and DT models were selected in particular. These models reveal the variables that influence the decision process, since the variables appear explicitly on the tree and also reveal the stages of the decision process by working down the branches of the tree (see Figures 22.4 and 22.5). However, decision trees are generally known to not provide the best performance results. Thus, random forest and deep learning models were included and recognized as both robust and promising. An evaluation of the results was carried out in two steps: validation against the selection status acquired from the atlas map (taken as reference for evaluation), and the comparison of performance statistics across the different ML models.

The performance of machine learning models

The performance achieved with particular ML models as compared to the atlas map can be found in Table 22.1. The overall accuracy describes the similarity of the automatic selection results to the manually generalized atlas map. Accuracy is also termed precision or positive predictive value, and it constitutes the fraction of correctly classified settlements (as selected and omitted) among all settlements considered.

To validate the selection results on the maps, we selected two district group examples. The first example represents the districts characterizing high population and settlement density (Figure 22.2), while the second example represents districts with low population density and small settlement density (Figure 22.3). These examples were selected because they represent different district groups and because they show a range of different performances.

Table 22.1 ML models performance for all districts (author's work)

Learning model	Overall accuracy in %
Random forest (RF)	83.27
Deep learning (DL)	82.76
Decision tree with genetic algorithms (DT_GA)	81.69
Decision tree (DT)	78.67

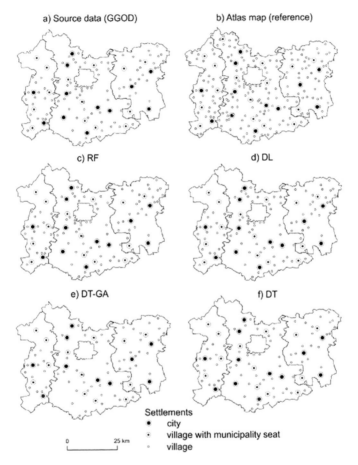

Figure 22.2   Maps of Tarnowski, Dębicki, and Brzeski districts: (a) source data from GGD; (b) atlas map used as evaluation reference; (c) RF; (d) DL; (e) DT_GA; and (f) DT (author's work)

The maps present automatic settlements selection results for two example district groups, all considered ML models, and the selection taken from the reference atlas map. The decision tree-based models (DT and DT_GA) also generated decision trees. The decision tree for all districts acquired in the DT_GA learning model is presented in Figure 22.4, while the decision tree for all districts resulting from the DT model is shown in Figure 22.5.

This case study example showcases a methodology that uses various ML models to make settlement selection as a subprocess of map generalization more effective, holistic, and aware of semantic and spatial context. The goal was to find the solutions that would be optimal

Izabela Karsznia

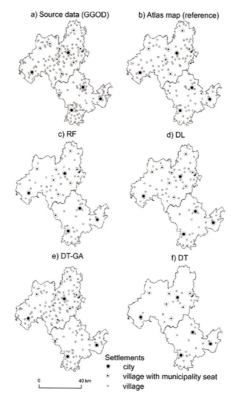

Figure 22.3 Maps of Chojnicki and Bytowski districts: (a) source data from GGD; (b) atlas map used as evaluation reference; (c) RF; (d) DL; (e) DT_GA; and (f) DT (author's work)

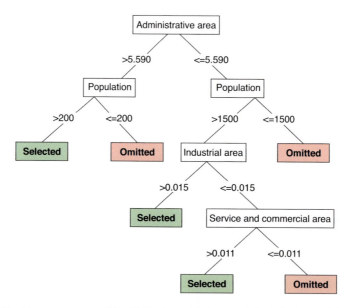

Figure 22.4 Decision tree generated for all districts with the use of the DT_GA model (author's work)

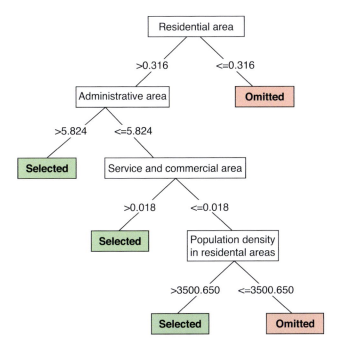

Figure 22.5 Decision tree generated for all districts with the use of the DT model (author's work)

in the sense that results will be as close as possible to the manual map design. Thus, the outcomes of automatic selection models have been compared to the atlas map taken as a reference.

Looking at Table 22.1, the best performing models are RF (83.27%) and DL (82.76%), then DT_GA (81.69%) followed by DT (78.67%). The difference between the best and least performing model is 4.6%, which indicates a performance improvement of 4.6% with the use of different ML models. Here, the RF and DL models provide the results that are closest to the manual atlas map.

The drawback of DL and RF models in comparison to DT and DT_GA models is that when DL is used, this does not provide an intuitive and straightforward description of the decision process. Instead, there is just the decision itself with no explanation, and for RF there may be dozens of trees, which makes it difficult to understand the variables that influence the decision process or the stages of the decision process most strongly.

On the other hand, using DT or DT_GA does not yield high performance results, but does generate holistic decision trees for each model and all considered districts (Figures 22.4 and 22.5), with the decision trees developed in DT_GA and DT models consisting of four decisive and easily understood steps.

The essential variables are placed at the root of the tree. For DT_GA it is *Administrative area*, whereas for DT it is *Residential area*. The next three decisive steps for DT_GA model are made based on the *Population, Industrial area,* and *Service and commercial area* variables, which are placed at the leaves of the tree. In the case of the tree developed using the DT model, the next three decisive steps are made based on the following variables: *Administrative area, Service and commercial area,* and *Population density in residential areas*. Looking at the DT_GA and DT models, the variables concerning population, population density, and different area of settlement seem to be most important in the decision process.

Finally, when considering the maps presenting the selection results of all source data and reference atlas map for the Tarnowski, Dębicki, and Brzeski districts, the RF and DL models performed better than DT_GA and DT, especially in terms of maintaining settlement density (Figure 22.2). However, the differences among the particular models were not as evident as in the case of the results obtained for Chojnicki and Bytowski districts (Figure 22.3). In these cases, the RF and DL results are very similar to each other and closer to the reference atlas map than the DT and DT_GA results, where DT selects too many settlements and DT_GA selects too few settlements compared to the reference atlas map.

Conclusions

This chapter has aimed to clarify ML concepts as well as types and selected models while providing an example of automatic settlement generalization that was driven by the motivation to fill the knowledge gap in the selection algorithms for small-scale maps. This application of ML to cartographic generalization, was designed to include significant and holistic variables as well as various ML models. The ML model results obtained for all districts and presented for the example of two district groups showed that different variables were crucial for selection depending on the region. The obtained selection accuracy in each tested model, understood as a similarity to the settlement selection on atlas reference map, was very high, ranging from 78% (DT) to nearly 84% (RF). The goal of this research case study was to automatically achieve results that would be nearly equivalent to manual map design, and these results were achieved with the use of RF, DL, and DT_GA models. Moreover, DT, in combination with optimized feature selection using GA, namely the DT_GA model, performed similarly to the RF and DL models.

The solutions presented in this chapter are a further step towards the full automation of the selection process for small-scale maps. Other classification algorithms, such as SVM (support vector machines) or different neural network models, may generate better classification results. However, one has to be aware that none of these algorithms, including DL and RF models, deliver results that can be easily interpreted and transformed into human-readable results – which is of key importance when the ultimate aim is to support rule formation for cartographic practice.

Acknowledgements

This research was funded by the National Science Centre, Poland, grant number UMO-2020/37/B/HS4/02605, "Improving Settlement and Road Network Design for Maps of Small Scales Using Artificial Intelligence and Graph Theory".

References

Alpaydin, E. (2010) *Introduction to Machine Learning* Cambridge: MIT Press.
Back, T. (1996) *Evolutionary Algorithms in Theory and Practice* Oxford: Oxford University Press.
Balboa, J.L.G and López, F.J.A. (2009) "Sinuosity Pattern Recognition of Road Features for Segmentation Purposes in Cartographic Generalization" *Pattern Recognition* 42 (9) pp.2150–2159.
Batty, M. (2006) "Hierarchy in Cities and City Systems" In Pumain, D. (Ed.) *Hierarchy in Natural and Social Sciences*. Dordrecht: Springer, pp.143–168 DOI: 10.1007/1-4020-4127-6_7.
Bereuter, P. and Weibel, R. (2013) "Real-time Generalization of Point Data in Mobile and Web Mapping Using Quadtrees" *Cartography and Geographic Information Science* 40 (4) pp.271–281 DOI: 10.1080/15230406.2013.779779.
Berg, M. de, Bose, P., Cheong, O. and Morin, P. (2004) "On Simplifying Dot Maps" *Computational Geometry: Theory and Applications* 27 (1) pp.43–62.

Bishop C.M. (2006) *Pattern Recognition and Machine Learning* New York: Springer.

Burghardt, D., Purves, R.S. and Edwardes, A.J. (2004) "Techniques for on-the-Fly Generalization of Thematic Point Data Using Hierarchical Data Structures" *Proceedings of the GIS Research UK 12th Annual Conference* Norwich, 28th–30th April.

Carol, H. (1960) "The Hierarchy of Central Functions Within the City" *Annals of the Association of American Geographers* 50 (4) pp.419–438 DOI: 10.1111/j.1467-8306.1960.tb00359.x.

Duchêne, C., Dadou, M. and Ruas, A. (2005) "Helping the Capture of Expert Knowledge to Support Generalization" *ICA Workshop on Generalization and Multiple Representation* A Coruña, Spain Available at: http://tinyurl.com/plsum7c.

Feng, Y., Thiemann, F. and Sester, M. (2019) "Learning Cartographic Building Generalization with Deep Convolutional Neural Network" *International Journal of Geo-Information* 8 (6) pp.1–20 DOI: 10.3390/ijgi8060258.

GGK (1993–1997) *Atlas Rzeczypospolitej Polskiej Mapa 1:500 000* Warsaw: Head Office of Geodesy and Cartography.

Harrie, L., Stigmar, H. and Djordjevic, M. (2015) "Analytical Estimation of Map Readability" *ISPRS International Journal of Geo-Information* 4 (2) pp.418–446.

Karsznia, I. and Sielicka, K. (2020) "When Traditional Selection Fails: How to Improve Settlement Selection for Small-scale Maps Using Machine Learning" *ISPRS International Journal of Geo-Information* 9 (4) pp.1–18 DOI: 10.3390/ijgi9040230.

Karsznia, I. and Weibel, R. (2018) "Improving Settlement Selection for Small-scale Maps Using Data Enrichment and Machine Learning" *Cartography and Geographic Information Science* 45 (2) pp. 111–127 DOI: 10.1080/15230406.2016.1274237.

Lagrange, F., Landras, B. and Mustiere, S. (2000) "Machine Learning Techniques for Determining Parameters of Cartographic Generalization Algorithms" *International Archives of Photogrammetry and Remote Sensing* 33 (Part B4) pp.718–725.

Langan, C. and Poicker, T. (1986) "Integration of Name Selection and Name Placement" *Proceedings of the 2nd International Symposium on Spatial Data Handling* pp.50–64.

Lee, J., Jang, H., Yang, J. and Kiyun, Y. (2017) "Machine Learning Classification of Buildings for Map Generalization" *International Journal of Geo-Information* 6 (10) pp.1–15 DOI: 10.3390/ijgi6100309.

Mackaness, W.A., Burghardt, D. and Duchêne, C. (2017) "Map Generalization" In Richardson, D., Castree, N., Goodchild, M.F., Kobayashi, A., Liu, W. and Marston, R.A. (Eds) *The International Encyclopedia of Geography* New York: John Wiley & Sons DOI: 10.1002/9781118786352.wbieg1015.

Marshland S. (2015) *Machine Learning: An Algorithm Perspective* Boca Raton, FL: CRC Press.

McMaster, R.B. and Shea, K.S. (1992) *Generalization in Digital Cartography* Washington, DC: Association of American Geographers.

Mitchell, T.M. (1997) *Machine Learning* New York: McGraw-Hill.

Murphy K.P. (2012) *Machine Learning: A Probabilistic Perspective* Cambridge, MA: MIT Press.

Mustière, S. (1998) "GALBE: Adaptive Generalization: The Need for an Adaptive Process for Automated Generalization, an Example on Roads" *Proceedings of the 1st GIS'PlaNet Conference* Lisbon, Portugal, 22nd July.

Mustière, S., Zucker, J-D. and Saitta, L. (2000) "An Abstraction-Based Machine Learning Approach to Cartographic Generalization" *Proceedings of the 9th International Symposium on Spatial Data Handling* Beijing: Study Group on Geographical Information Science of the International Geographical Union, pp.50–63.

Plazanet, C., Bigolin, N.M. and Ruas, A. (1998) "Experiments with Learning Techniques for Spatial Model Enrichment and Line Generalization" *Geoinformatica International* 2 (4) pp.315–333 DOI: 10.1023/A:1009753320636.

RapidMiner (2019) "RapidMiner 9 Operator Reference Manual" Available at: http://docs.rapidminer.com/studio/operators/rapid miner-studio-operator-reference.pdf.

Regnauld, N. and McMaster R.B. (2007) "A Synoptic View of Generalization Operators" In Mackaness, W.A., Ruas, A. and Sarjakoski, L.T. (Eds) *Generalization of Geographic Information: Cartographic Modelling and Applications* Amsterdam: Elsevier, pp.37–66. Sester, M., Feng, Y. and Thiemann, F. (2018) "Building Generalization Using Deep Learning" *International Archives of the Photogrammetry, Remote Sensing and Spatial Information Sciences* 42 (4) pp.565–572.

Samsonov, T. and Krivosheina, A. (2012) "Joint Generalization of City Points and Road Network for Small-Scale Mapping" *Proceedings of the 7th International Conference on Geographic Information Science (GIScience 2012)* Columbus, OH, 18th–21st September.

Smith, R.H.T. (1965) "Method and Purpose in Functional Town Classification" *Annals of the Association of American Geographers* 55 (3) pp.539–548 DOI: 10.1111/j.1467–8306.1965. tb00534.x.

Stanislawski, L.V., Buttenfield, B.P., Bereuter, P., Savino, S. and Brewer, C.A. (2014) "Generalization Operators" In Burghardt, D., Duchêne, C. and Mackaness, W. (Eds) *Abstracting Geographic Information in a Data Rich World* Cham, Switzerland: Springer, pp.157–195.

Stoter, J., Post, M., van Altena, V., Nijhuis, R. and Bruns, B. (2014) "Fully Automated Generalization of a 1:50k Map from 1:10k Data" *Cartography and Geographic Information Science* 41 (1) pp.1–13 DOI: 10.1080/15230406.2013.824637.

Sutton, R.S. and Barto A.G. (1998) *Reinforcement Learning: An Introduction* Cambridge, MA: MIT Press.

Topfer, F. and Pillewizer, W. (1966) "The Principles of Selection" *The Cartographic Journal* 3 (1) pp.10-16 DOI: 10.1179/caj.1966.3.1.10.

Touya, G. and Dumont, M. (2017) "Progressive Block Graying and Landmarks Enhancing as Intermediate Representations Between Buildings and Urban Areas" *Proceedings of the 20th ICA Workshop on Generalization and Multiple Representation* Washington, DC Available at: https://kartographie.geo.tu-dresden.de/downloads/ica-gen/workshop2017/genemr2017_paper_1.pdf.

Touya, G., Zhang, X. and Lokhat, I. (2019) "Is Deep Learning the New Agent for Map Generalization?" *International Journal of Cartography* 5 (2–3) pp.142–157 DOI: 10.1080/23729333.2019.1613071.

Van Kreveld, M., Van Oostrum, R. and Snoeyink, J. (1995) "Efficient Settlement Selection for Interactive Display" *Proceedings of AutoCarto 12* Bethesda, MD.

Weibel, R., Keller, S. and Reichenbacher, T. (1995) "Overcoming the Knowledge Acquisition Bottleneck in Map Generalization: The Role of Interactive Systems and Computational Intelligence" In *Lecture Notes in Computer Science, Proceedings COSIT '95 (Volume 988)* Berlin: Springer, pp.139–156 DOI: 10.1007/3-540-60392-1_10.

Yan, H. (2019) *Description Approaches and Automated Generalization Algorithms for Groups of Map Objects* Singapore: Springer DOI: 10.1007/978-981-13-3678-2.

Yan, H. and Weibel, R. (2008) "An Algorithm for Point Cluster Generalization Based on the Voronoi Diagram" *Computers & Geosciences* 34 pp.939–954 DOI: 10.1016/j.cageo.2007.07.008.

Zhou, Q. and Li, Z. (2017) "A Comparative Study of Various Supervised Learning Approaches to Selective Omission in a Road Network" *The Cartographic Journal* 54 (3) pp.254–264.

23
ARTIFICIAL INTELLIGENCE FOR GEOSPATIAL APPLICATIONS

Vit Vozenilek

The term 'artificial intelligence' (AI) was coined in 1956 (Smith, 2006), but it has become more popular thanks to increased data volumes, advanced algorithms, and computing power and storage improvements. Early AI research in the 1950s explored topics like problem-solving and symbolic methods. In the 1960s, the US Department of Defense took an interest in this type of computing and began training computers to imitate basic human reasoning. More recently, AI has come to be understood as the ability of a machine or a computer program to think, act, and learn like a human mind. Today, AI is used across many different disciplines and industries.

Artificial intelligence

Artificial intelligence is intelligence demonstrated by machines, unlike human intelligence, which is developed by the human mind. The term may also be applied to any machine that possesses features associated with a human mind, such as learning and problem-solving. The goals of artificial intelligence include learning, reasoning, and perception.

AI has become a buzzword that symbolizes the next stage of innovative technological transformation and how the industries of the future will be driven. Using intelligent algorithms, data classification, and intelligent predictive analysis, AI has applications in many sectors. However, since their inception, AI machines have caused concerns that they would become so highly developed that people would no longer be able to control them, e.g., that they would invade people's privacy and be armed.

Despite the many potential applications of AI, it is not easy to provide a succinct definition. The Organization for Economic Cooperation and Development (OECD, 2019) defines an AI system as a machine-based system that can make predictions, recommendations, or decisions influencing real or virtual environments for a given set of human-defined objectives.

AI technologies already permeate many aspects of our lives, both directly and indirectly. In recent years, a combination of huge datasets, increased computing power, and the availability of advanced learning algorithms have enabled rapid advances in what AI can do. They include considerable breakthroughs in the ability of AI programs to recognize images and sounds, solve complex problems, translate languages, power autonomous robots, and analyse the sentiment of human speech (Figure 23.1). As a result of these advances, tasks that

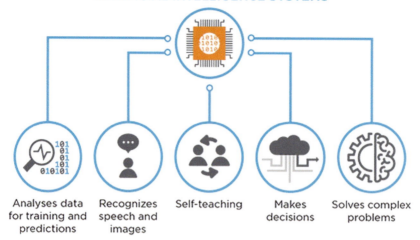

Figure 23.1 AI systems can analyse, recognize, make decisions, and solve problems without human intervention (Piovesan, 2019)

previously required human cognition to perform are becoming automated through businesses and organizations.

The recent explosion of interest in AI is largely due to three factors: the availability of large datasets; increases in parallel computer processing power; and the availability of improved machine-learning algorithms that anyone can use, often through open-source sharing. These factors have allowed increasingly complex problems to be solved more quickly and at a lower cost.

Artificial intelligence is continuously growing to profit many different scientific disciplines and industries. Machines are wired using cross-disciplinary approaches based on mathematics, computer science, linguistics, psychology, and many others. Algorithms play a significant role in AI, from simple algorithms for performing basic applications, to complex algorithms that characterize strong AI.

Weak and Strong AI

A key aspect of AI is its ability to rationalize and act with the best opportunities to obtain specific aims. Machine learning, which is a subset of AI, refers to the concept that computer programs can routinely learn from and adapt to new data without being assisted by humans. Deep learning techniques enable this automatic learning by engaging huge amounts of unstructured data such as text, images, or video. In general, however, a distinction is made between Weak and Strong AI.

Weak AI

Weak AI (also called narrow AI) embodies a system designed to carry out one job. Weak AI is limited to a specific or narrow area and simulates human cognition (Long *et al.*, 2020). As a result, it can benefit society by automating time-consuming tasks and analysing data in ways that humans sometimes cannot.

All currently existing AI systems are Weak AI. These AI systems do not have general intelligence; they have specific intelligence. Weak AI systems include video games, such as chess, or navigation from A to B. Weak AI helps to turn big data into usable information by detecting patterns and making predictions. Examples include Facebook's newsfeed, Amazon's suggested purchases, and Apple's Siri (the iPhone technology that answers users' spoken questions). Email spam filters are another example of Weak AI – a computer uses an algorithm to learn which messages are likely to be spam.

Besides its limited capabilities, some of the problems with Weak AI include the possibility to cause harm if a system malfunctions or ceases to operate. For example, consider an autonomous car that underestimates the location of an oncoming vehicle and causes a fatal collision. The system can also cause harm if the system is used by someone who wishes to cause harm, such as a terrorist who uses an autonomous car to deploy explosives in a crowded area.

Strong AI

Strong AI (also called True Intelligence or Artificial General Intelligence) conducts tasks that are considered human-like. Strong AI is – at present – theoretical; intelligent machines that can successfully perform any intellectual task that a human can do. Strong AI tend to be more complicated systems that are programmed to control situations and solve problems in the absence of human supervision or input. Significant features of Strong AI involve the ability to reason, solve puzzles, make judgments, plan, learn, and communicate. A Strong AI system should also have consciousness, objective thoughts, self-awareness, sentience, and sapience.

Some experts predict that Strong AI may be developed by 2030 or 2045, while others foresee its development within the next century, yet there are some who do not see the development of Strong AI as a possibility at all. Some theorists argue that a machine with strong AI should go through the same development process as a human, starting with a naive mind and developing an adult mind through learning. It should interact with the world and learn from it, obtaining its common sense and language.

Whereas Weak AI merely simulates human cognition, Strong AI would have human cognition. A single Strong AI system could theoretically handle the same problems that a single human could. Weak AI may replace many low- and medium-skilled workers, although Strong AI might be used to replace certain categories of highly skilled workers (Figure 23.2).

The possibility of developing Strong AI comes with serious concerns. Some people fear that if Strong AI becomes a reality, AI may become more intelligent than humans. The idea is that Strong AI will be so intelligent that it can alter itself and pursue its own goals without human intervention, possibly in harmful ways. Another major concern is that AI will increasingly take jobs away from people, resulting in high unemployment – even for knowledge-intensive roles. However, just as the Industrial Revolution dramatically changed the types of jobs that workers performed, an AI Revolution could result not in massive unemployment but a massive employment shift. Hence, Strong AI could have a significant positive impact on society by increasing productivity and wealth, and humans could perform jobs that we cannot yet imagine.

AI technologies

The family of technologies currently referred to as AI includes many different approaches. Historically, rule-based or symbolic AI, such as Expert Systems, was the leading paradigm

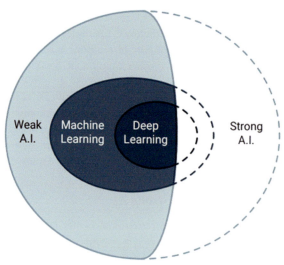

Figure 23.2 Machine learning and deep learning related to Weak and Strong AI (Komodo Technologies)

for AI. Other popular AI approaches included planning and scheduling, knowledge representation, and reasoning. Today, AI has many sub-fields and advanced technologies, including the following (with new technologies being developed constantly):

- Artificial neural networks – simulating the working of neurons in the brain;
- Natural language processing – aiming to produce computer systems that can understand, translate, and communicate in human languages;
- Theorem provers – allowing computers to solve mathematical problems and to discover new mathematical concepts;
- Genetic algorithms – solving problems by a loose analogy with biological evolution by natural selection;
- Knowledge-based systems – encoding human expert knowledge in such a way a computer can reason with it;
- Case-based reasoning – simulating how humans reason from past experience;
- Robotics – focusing on the construction of intelligent robots that adapt to their environment; and
- Vision – focusing on tasks such as face recognition.

Neural networks

Warren McCulloch and Walter Pitts conceived the first neural network 80 years ago. Their seminal paper (McCulloch and Pitts, 1943) explained how neurons may operate and how they modelled their ideas by forming a simple neural network using electrical circuits. Research in AI accelerated over the next 30 years, with Fukushima (1975) developing the first true multilayered neural network.

The original aim of the neural network approach was to build a computational system that could solve problems like a human brain. However, over time, researchers moved their

Artificial intelligence for geospatial applications

focus to using neural networks to match specific tasks, which deviated from the earlier narrowly biological approach. Neural networks have since developed to support distinct tasks, including computer vision, speech recognition, machine translation, social network filtering, playing board and video games, and medical diagnosis.

Neural networks are designed to help people to solve complex problems in real-life situations. They can learn and model the relationships between inputs and outputs that are complex, make generalizations and inferences, uncover hidden relationships, patterns, and predictions, and model very unstable data (such as climate time series data) and variances needed to predict exceptional events (such as earthquake detection) (SAS Institute, 2022).

A neural network includes an input layer, a hidden layer, and an output layer. Nodes connect the layers, and these connections form a 'network' – the neural network – of interconnected nodes (Figure 23.3).

Data are consumed into a neural network through the input layer, which communicates with the hidden layers (Figure 23.1). Processing appears in the hidden layers through a system of weighted connections. Nodes in the hidden layer then relate data from the input layer with coefficients and assign proper weights to inputs. Then the input-weight products are counted. The sum is passed through a node's activation function, which decides how a signal must progress further through the network to influence the final output. Finally, the hidden layers link to the output layer, which is where the outputs are recovered.

A node is designed after a neuron in a human brain. Similar in behaviour to neurons, nodes are stimulated when there are appropriate stimuli or input. This stimulation spans throughout the network, creating a response to the stimuli (output). The connections between artificial neurons operate as simple synapses, enabling signals to be conducted from one to another. Signals across layers travel from the first input to the last output layer and are processed along the way. When posed with a request or problem to solve, the neurons run mathematical calculations to decide if there is enough information to pass on the information to the next neuron. They read all the data and to determine where the strongest relationships exist. As the number of hidden layers within the neural network increases, more

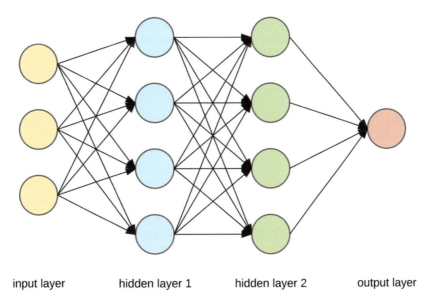

Figure 23.3 Principle of neural networks

complex neural networks are created. Learning architectures move simple neural networks to the next level and the computer can learn on its own by recognizing patterns in many layers of processing (Horak et al., 2011).

The SAS Institute (2022) distinguishes between different kinds of neural networks, each having advantages and disadvantages, depending upon their use:

- Convolutional neural networks (CNNs) contain five types of layers: input, convolution, pooling, fully connected, and output. Each layer has a specific purpose, like summarizing, connecting, or activating. CNNs have popularized image classification and object detection.
- Recurrent neural networks (RNNs) use sequential information such as time-stamped data from a sensor device or a spoken sentence, composed of a sequence of terms. Unlike traditional neural networks, all inputs to recurrent neural networks are not independent of each other, and the output for each element depends on the computations of its primary elements. RNNs are used in forecasting and time series applications, sentiment analysis, and other text applications.
- Feedforward neural networks, in which each perceptron in one layer is connected to every perceptron from the next layer. Information is fed forward from one layer to the next in the forward direction only. There are no feedback loops.
- Autoencoder neural networks are used to create abstractions called encoders, created from a given set of inputs. Although similar to more traditional neural networks, autoencoders seek to model the inputs themselves, and therefore the method is considered unsupervised. The premise of autoencoders is to desensitize the irrelevant and sensitize the relevant. As layers are added, further abstractions are formulated at higher layers (those closest to the point at which a decoder layer is introduced). Linear or nonlinear classifiers can then use these abstractions.

Machine learning

Machine learning is a subset of AI that often uses statistical techniques to give computers the ability to learn with data – creating models to process, interpret, and respond – without being explicitly programmed with a predefined set of rules. Machine learning is about making systems capable of learning and improving themselves without being programmed again and again. This is possible by developing new computer programs that enable analysis of massive quantities of data.

While AI is the broad science of imitating human abilities, machine learning is a particular subgroup of AI that trains a machine how to learn. Reviving interest in machine learning is due to increasing capacities and diversities of available data, computational processing that is cheaper and more powerful, and proper data storage.

There are many machine learning techniques, each with specific capabilities and applications, and machine learning is becoming an essential component of spatial analysis in GIS (geographical information systems). These tools and algorithms have been applied to geoprocessing tools to solve problems in three broad categories (Singh, 2019): (i) with classification, vector machine algorithms create land-cover classification layers; (ii) clustering supports process large quantities of input point data, identify the meaningful clusters within them, and separate them from the sparse noise; and (iii) prediction algorithms, such as geographically weighted regression, to give the ability to model spatially varying relationships. These methods work well in several areas, and their results are interpretable, but they require experts to identify or introduce the factors affecting the outcome attempting to be predicted (Marjanovic *et al.*, 2011).

Deep learning

Deep learning is one of the foundations of AI and the current interest in deep learning is due in part to the buzz surrounding AI. Deep learning techniques have improved the ability to classify, recognize, detect, and describe. Deep learning is used to classify images, recognize speech, detect objects, and describe content. Human-to-machine interfaces have evolved too, with the mouse and the keyboard being replaced with haptic gestures, swipe, touch, and natural language ushering in a renewed interest in AI and deep learning.

Many developments are currently advancing deep learning:

- Algorithmic developments have expanded the performance of deep learning methods;
- Much more data are available to develop neural networks with multiple deep layers, including streaming data from the Internet of Things, textual data from social media, and investigative transcripts;
- New machine learning procedures have advanced the efficiency of models;
- New aspects of neural networks have been received that fit well for applications such as text translation and image classification; and
- Higher computational power due to computational progress of distributed cloud computing and graphics processing units.

A lot of computational power is demanded to solve deep learning difficulties due to the iterative basis of deep learning algorithms, their intricacy as the number of layers increase, and the large volumes of data required to train the networks. On the other hand, the nature of deep learning methods – their ability to coherently enhance and adapt to improvements in the underlying information pattern – presents a great possibility to add more dynamic behaviour into analytics.

More comprehensive personalization of customer analytics and advancing accuracy and performance in applications where neural networks have been utilized for a long time are big opportunities for deep learning. Through better algorithms and more computing power, machine learning shifts deep learning.

The new market focus of deep learning techniques lies within applications of cognitive computing. There is also considerable potential in some of the more common analytics applications, such as time series analysis. Another possibility is to be more productive and streamlined in performing existing analytical procedures (Figure 23.4). Compared to the conventional techniques, recent deep neural networks in speech-to-text transcription tasks can reduce the word error rate by more than 10%. Performance gains and time savings are mostly found in:

- Speech recognition – Microsoft XBox, Skype, Google Now, and Apple Siri are already applying deep learning in their systems to distinguish between human speech and voice patterns.
- Image recognition – automatic image captioning and scene description has become essential in law enforcement investigations for recognizing criminal activity from

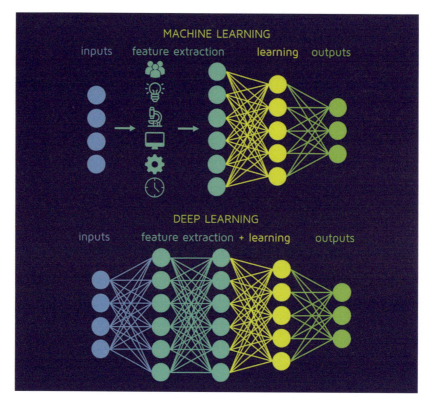

Figure 23.4 Differences between machine and deep learning (QuantDare.com)

thousands of images submitted by observers in crowded areas where a crime has happened. Autonomous cars will also profit from image recognition through the use of 360-degree cameras.
- Natural language processing – neural networks, a fundamental component of deep learning, have been used to treat and analyse written text for many years. Text mining can be used to discover patterns in (for example) customers' complaints, physicians' notes, or news reports.
- Recommendation Systems – Amazon and Netflix have familiarized recommendation systems with a good possibility of understanding what the consumer might purchase next, based on earlier behaviour. Deep learning improves advice in complex environments such as music interests or clothing preferences across multiple platforms.

For example, Esri collaborates with NVIDIA to use deep learning to automate the manually intensive process of generating complex 3D building models from aerial Lidar data. Moreover, in operating with satellite imagery, Esri applies deep learning in producing digital maps by automatically extracting road networks and building footprints. Imagine utilizing a trained deep learning model on a large geographic territory and producing a map indicating all the roads in the region, then having the power to generate driving navigation instructions using this detected road network (Figure 23.5). Such an example can be particularly helpful for low-income countries without high-quality digital maps, or for mapping regions where newer developments have been built.

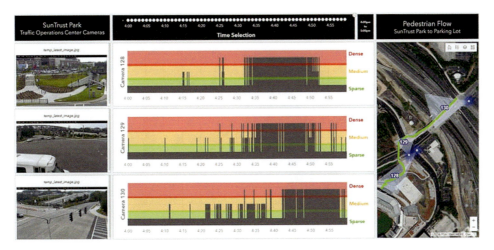

Figure 23.5 Esri applies AI in image classification to conduct traffic and pedestrian movement planning in Cobb County, Georgia (Singh, 2019)

Geospatial AI

Geospatial AI (also Geo.AI, or geospatial intelligence, GEOINT), which lies at the intersection of geospatial data and AI, is the new challenge of technological innovation and promotes the transformation of whole business industries. Geospatial AI is machine learning based on geographic components. It is combining geography, location, and AI. Geospatial AI is crucial to enterprises and governments, varying from weather centres, national laboratories, defence agencies, healthcare, agriculture, insurance, transportation, and many more.

Machine and deep learning offer promising approaches for analysing geospatial data, allowing a practical method for distinguishing features or objects in detailed satellite imagery. Today, numerous governmental and commercial organizations use deep learning techniques and geospatial data from satellites to manifest geospatial AI's potential (Vopham *et al.*, 2018).

Geospatial deep learning can instantly distinguish time and space features from geospatial data and efficiently build complex features. Combining AI with GIS provides considerable opportunities that were previously unavailable. Artificial intelligence, machine learning, and deep learning can, for example, increase crop yield through precision agriculture, understand crime patterns, and predict when the next big storm will hit (and being better equipped to handle it) (Figure 23.6).

Three trends lie behind the rise of geospatial AI (Hahn, 2019): an increased availability of data from satellites and remote sensors (geospatial big data); the progress of AI (especially machine and deep learning); and the availability of massive computational power (supercomputing).

Geospatial big data

Most Earth observation data consist of very detailed imagery and time-series data in large file sizes. With growing availability, geospatial data are being used by organizations to explore the ability to analyse the location of terrestrial objects and obtain actionable intelligence (Duan *et al.*, 2019). Since 2020, there has been considerable reduction in the cost of launching satellites, falling from an average of €16,445/kg in 2000 to €2,418/kg in 2018 for

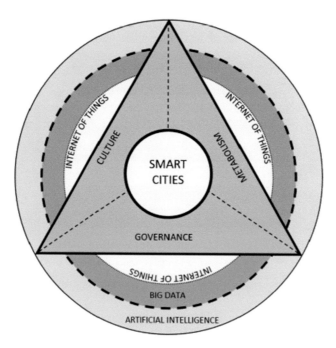

Figure 23.6 A proposed framework for the integration of AI and big data in smart cities to ensure livability (Allam and Dhunny, 2019)

sending a satellite into low Earth orbit. In 2018, Euroconsult predicted (Hahn, 2019) that over 7,000 small satellites would be launched over the next ten years, raising the number of Earth's observation satellites from 540 to over 1,400.

The emergence of low-cost nanosatellites, satellite constellations, and newer sensor technologies, such as synthetic aperture radar and hyperspectral imaging, is creating a growing collection of geospatial big data (Robinson et al., 2017). The increase in remote sensing capabilities has given rise to the Earth observation data market, which Northern Sky Research (2018) predicts will reach $6.9 billion by 2027, doubling its 2017 size. In Europe alone, the market for Earth observation data is projected to increase in 2021 from its 2016 value of €639 million, rising to €1.26 billion, according to Technavio (2017).

Supercomputing

Supercomputing is a system that provides the massive computing and storage resources required for timely AI application development and delivery. Processing and analysing growing volumes of data, maintaining complex processing, and allowing distributed training methods all need increasingly powerful and proficient computing architectures.

Geospatial AI is a supercomputing problem since supercomputers are the only machines that have the capabilities organizations need to embrace the coming geospatial AI wave. The digital world is doubling in size every two years and AI applications grow on massive datasets.

Supercomputers are closely integrated, deeply scalable, zero-waste architectures that propose the right technology for each task to facilitate maximum application productivity and reduce computational limits. They stand out at ingesting, moving, and processing massive volumes of data. Today, systems are being purpose-built. Everything from the processors to the software allows diverse AI and enterprise workflows to operate in parallel on a single system. Ultimately, the computing power provided by supercomputers makes it possible to train larger deep learning neural networks using larger geospatial training sets in less time.

Geospatial AI is currently an important research direction. Geospatial AI users can solve various GIS application problems such as spatial clustering, spatial classification, and spatial regression based on geospatial machine learning. However, most studies mainly focus on specific application scenarios and rarely involve the research and exploration of Geospatial AI.

Geospatial AI and GIS

Using Geospatial AI to improve and develop the new generation of GIS is a powerful method for solving current issues with GIS systems (Vozenilek, 2009). Geospatial AI originally realized the computer-vision extraction of spatial information, such as from satellite images (Vozenilek, 2015) and AI technologies, such as solutions for voice recognition and language processing, can be introduced to provide further empowerment. However, Geospatial AI is still in the Weak AI stage and is far from the Strong AI stage. Accordingly, the technology represented by Strong AI is also a relevant course for the future development of Geospatial AI.

Geospatial AI systems refer to the application of AI technology to enhance the intelligence of GIS software (Guanfu, 2020), including AI Attribute Collecting, AI Survey & Mapping, AI Cartography, and AI interaction (Figure 23.7). AI Attribute Collecting can help users intelligently classify and identify multi-source targets such as video images. AI Survey & Mapping can provide lower cost and more convenient indoor mapping solutions. AI Cartography saves users from the tedious process of manual mapping, and the style transfer from image to map can proceed through a simple operation. AI interaction includes rich application interaction of using voice, haptic gestures, and so on.

To serve a variety of GIS applications, Geospatial AI needs to be deeply integrated with various forms of GIS software, including Component GIS, Desktop GIS, Server GIS, in order to build a Geospatial AI product architecture jointly. For example, ArcGIS Pro includes tools for supporting data training for deep learning and has been improved for deploying trained models for feature extraction and classification. ArcGIS Image Server (in the ArcGIS

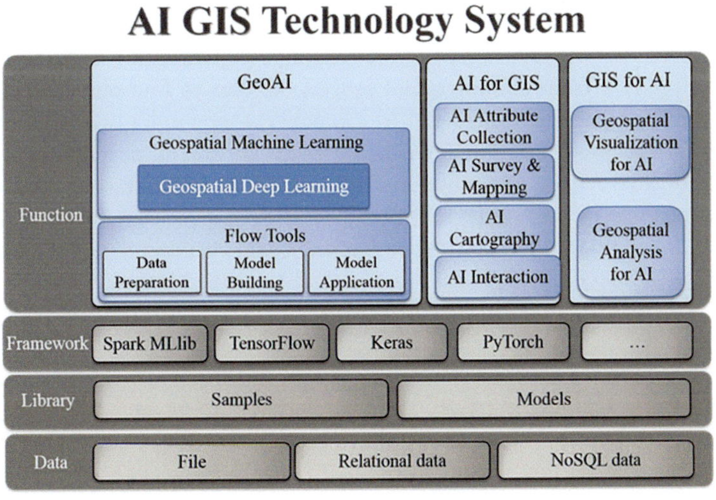

Figure 23.7 AI SIS Technology System according to Gunafu (2020)

Enterprise 10.7) has related capabilities to use deep learning models by leveraging distributed computing. The arcgis.learn module in the ArcGIS API for Python permits GIS analysts and data scientists to train deep learning models with an intuitive API. ArcGIS Notebooks provides a ready-to-use environment for training deep learning models. ArcGIS integrates built-in Python raster functions for object detection and classification using CNTK, Keras, PyTorch, fast.ai, and TensorFlow.

Geospatial AI advantages and limits

Geospatial AI brings significant advantages thanks to linking spatial data and AI. Since AI avoids hours of hand-operated effort needed to analyse GIS data, by automating thousands of repetitive procedures like classification or clustering, AI can minimize processing time. As the solutions are AI-powered, they are accurate and reduce human error. AI algorithms can consider new geographic data as and when it is received, 'learn' from these patterns, and generate suitable outputs. They can also run 24/7.

Several significant technical factors have encouraged the rapid growth of geospatial analysis in the field of AI to make AI more robust and ensure accelerated growth (Hahn, 2019):

- More Computing Power, Cloud Storage – Geospatial AI generally requires high computing power due to the volume of geodata. Recent researchers noticed that an approximate minimum computation requirement for training a deep neural network on a dataset of 1.28 million images would be on the order of an ExaFLOP. At a supercomputing level, in 1993, supercomputers could perform 124.5 GigaFLOPS. In 2020, this number stood at 415.53 PetaFLOPS. This means that Geospatial AI algorithms can run faster and more efficiently on today's computers. Another advance is that of cloud computing. As satellite imagery requires enormous storage space, the cloud is a logical solution to host data and even the programs that manipulate these.
- Better Algorithms with Geospatial Analysis – AI provides better results than other analysis techniques. Progress in computer vision has made it possible to obtain credible intelligence from satellite imagery using AI techniques such as deep learning. The increased computing capacity has ensured that more efficient and complex AI algorithms can be run on massive geospatial datasets.

Geospatial AI has various applications. Ride-sharing companies like Uber or Lyft can receive similar feedback from customers and process the data to find out the density of cars and the availability of drivers. Geospatial AI can plug the gaps in logistics and supply chains and gather more accurate location information that can streamline product delivery and save time. Geospatial AI can also substantially improvise planning, resource allocation, and decision-making – predicting surges in demand and supply, identifying the prospects of high and low margins, multiplying supply chain efficiency, and optimizing service delivery.

The future of AI

Artificial intelligence is a modern-day technique that is making our machines smarter than before. It is not about adding more programs to the software or reprogramming them, but is about improving the capabilities of a machine so that it can learn, analyse, memorize, and make the best decisions in the circumstances.

AI-assisted environmental monitoring expands the capacity for organizations and companies to review, monitor, and understand environmental data (Allam and Dhunny, 2019).

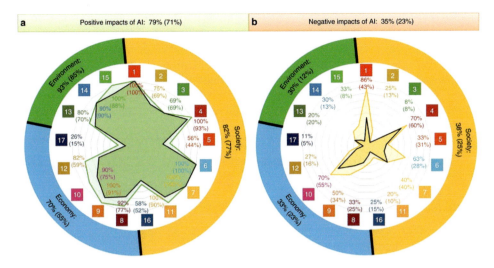

Figure 23.8 The role of AI in achieving the Sustainable Development Goals (Vinuesa *et al.*, 2020)

These systems combine video, photographic, or satellite monitoring systems with AI analysis to recognize changes in an environment (Figure 23.8).

Recent advances in AI models can generate written content that is indistinguishable from that by a human. For example, current AI writing technologies can draft email subject lines, finish the sentences of a human writer, and produce full passages of text based on a short sample. Other current and potential uses include drafting news articles, song lyrics, advertising copy, and assisting authors with their writing. However, concerns have been mentioned about the potential abuse of models that can write like humans, including the quick spread of disinformation online.

The website www.climatechange.ai was launched in June 2019 following the paper "Tackling Climate Change with Machine Learning" that was co-authored by some of the world's leading international academics and technology-industry researchers. The paper shows how machine learning techniques can reduce greenhouse gas emissions and help society adapt to a changing climate.

Ethical questions and implications

The vast power of AI also brings tremendous opportunities for exploitation and neglect. Some current ethical concerns include:

- Data siloing and resource inequity – much machine learning and geospatial work depends on open datasets. Governments have commonly provided large sources of free data for research. Not all nations, however, have the same technical capabilities or infrastructure. This disparity is often reinforced through the 'brain-drain' – when experts emigrate to seek professional opportunities. The increasing value of data further exacerbates such trends as a market commodity.
- Privacy – anonymizing patient data is key to their use in aggregated studies. The Internet of Things, social media algorithms, and the use of remote sensing also raise serious questions of privacy and consent.

- Diversity and equity – neither the technology nor the medical industry is known for their track records on diversity in race, gender, or disability. The connection between bias in engineers and bias in models and algorithms is well-documented. A lack of diversity exists in the datasets used to train machine learning models.

The Future Today Institute (2019) published a comprehensive report detailing international trends in AI, which provides a glimpse of more future innovations soon to arrive (Box 23.1).

Box 23.1 Future Today Institute AI trends 2019

1. Consumer-Grade AI Applications
2. Ubiquitous Digital Assistants
3. A Bigger Role for Ambient Interfaces
4. Deep Linking Everywhere
5. Proliferation of Franken-algorithms
6. Deployable AI Versions of You
7. Ongoing Bias in AI
8. AI Bias Leads to Societal Problems
9. Making AI Explain Itself
10. Accountability and Trust
11. AI Hiding its Own Data
12. Undocumented AI Accidents on the Rise
13. The AI Cloud
14. Serverless Computing
15. New Kinds of Liability Insurance for AI
16. Generating Virtual Environments from Short Videos
17. AI Spoofing
18. Ambient Surveillance
19. Proprietary, Home Grown AI Languages
20. AI Chipsets
21. Marketplaces for AI Algorithms
22. Even More Consolidation in AI
23. Real-Time Machine Learning
24. Natural Language Understanding (NLU)
25. Machine Reading Comprehension (MRC)
26. Natural Language Generation (NLG)
27. Generative Algorithms for Voice, Sound and Video
28. Real-Time Context in Machine Learning
29. General Reinforcement Learning Algorithm
30. Machine Image Completion
31. Hybrid Human-Computer Vision Analysis
32. Predictive Machine Vision
33. Much Faster Deep Learning
34. Reinforcement Learning and Hierarchical RL
35. Continuous Learning

> 36. Multitask Learning
> 37. Generative Adversarial Networks (GANs)
> 38. New Generative Modelling Techniques
> 39. Capsule Networks
> 40. Probabilistic Programming Languages
> 41. Automated Machine Learning (AutoML)
> 42. Customized Machine Learning
> 43. AI for the Creative Process
> 44. Bots
>
> Source: Future Today Institute, shared under CC BY-NC-SA 4.0 International Licence

Artificial intelligence solutions can either be built in-house using custom solutions or frameworks or deployed using 'infrastructure as a service' technologies provided by large AI companies. Key players in this space include Google Cloud, Microsoft Azure, Amazon AWS, and IBM Watson. Subscriptions to these services allow organizations to upload and manage their data in the cloud and deploy a range of ready-to-use AI solutions such as machine vision or natural language processing across the data. In addition, these services can provide popular open-source deep learning frameworks, efficient AI development tools, and powerful servers.

Many enterprise software companies also offer AI as a service solution that comes pre-integrated with the companies' other services. In addition, many enterprise database servers are increasingly incorporating embedded machine learning – this includes Oracle, Microsoft SQL Server, and MarkLogic.

Deloitte (2022) note that established vendors are not the only organizations supplying AI tools and services, and it is not yet transparent what the dominant vendor model will be. Several highly innovative startups are developing various tools and solutions, and this proliferation will likely continue until the consolidation of AI tools and infrastructure markets and standards begin to emerge.

For a program called 'AI for Earth', Microsoft makes cloud and AI tools available to those working to protect the environment. Users can apply for grants to help them label datasets, which are then made public and used to train AI models with sustainability goals in mind. Alternatively, grants are awarded to those whose data are ready to analyse. For example, Microsoft has partnered with Chesapeake Conservancy to generate land cover data with one-metre resolution to enable precision conservation. Also, Microsoft support helps machine learning models to scour millions of images and thousands of hours of field recordings to detect species of interest to conservationists.

The potential applications for AI are endless. Technology enriches many different sectors and industries. Whether through image analysis, natural language processing, process automation, autonomous vehicles, or making better predictions, there is an infinite variety of opportunities for people to deploy AI.

References

Allam, Z. and Dhunny, Z.A. (2019) "On Big Data, Artificial Intelligence and Smart Cities" *Cities* 89 pp.80–91 DOI: 10.1016/j.cities.2019.01.032.

Deloitte (2022) "How the US Government Can Accelerate AI Entrepreneurship" *Deloitte Insights* (23rd August) Available at: https://www2.deloitte.com/us/en/insights/industry/public-sector/accelerating-entrepreneurship-in-artificial-intelligence.html.

Duan, Y., Edwards, J. and Dwivedi, Y. (2019) "Artificial Intelligence for Decision Making in the Era of Big Data – Evolution, Challenges and Research Agenda" *International Journal of Information Management* 48 pp.63–71 DOI: 10.1016/j.ijinfomgt.2019.01.

Fukushima, K. (1975) "Cognitron: A Self-organizing Multilayered Neural Network" *Biological Cybernetics* 20 pp.121–136.

Future Today Institute (2019) "2019 Tech Trends Today Annual Report" *What We See* Available at: https://futuretodayinstitute.com/2019-tech-trends.

Guanfu, S. (2020) "What is AI GIS (Artificial Intelligence GIS)?" *SuperMap* Available at: https://www.supermap.com/en-us/news/?82_2701.html.

Hahn, P. (2019) "Three Trends Driving the Geospatial AI Revolution" *The European Business Review* (23rd September) Available at: https://www.europeanbusinessreview.com/three-trends-driving-the-geospatial-ai-revolution/.

Horak, Z., Kudelka, M., Snasel, V. and Vozenilek, V. (2011) "Orthophoto Map Feature Extraction Based on Neural Networks" *CEUR Workshop Proceedings* 706 pp.216–225.

Long, Q., Ye, X. and Zhao, Q. (2020) "Artificial Intelligence and Automation in Valvular Heart Diseases" *Cardiology Journal* 27 (4) pp.404–420 DOI: 10.5603/CJ.a2020.0087.

Marjanovic, M., Kovacevic, M., Bajat, B., Vozenilek, V. and Marek, L. (2011) "Využití Klasifikačních Algoritmů Metod Strojového Učení pro Účely Prostorového Modelování" [Use of Classification Algorithm Methods of Machine Learning for the Purpose of Spatial Modeling] In Voženílek, V., Dvorský, J. and Húsek, D. (Eds) *Metody umělé Inteligence v Geoinformatice [Methods of Artificial Intelligence in Geomatics]* Olomouc, Czech Republic: Univerzita Palackého v Olomouci, pp.161–168.

McCullough, W.S. and Pitts, W. (1943) "A Logical Calculus of the Ideas Immanent in Nervous Activity" *Bulletin of Mathematical Biophysics* 5 pp.115–133.

OECD (2019) *Artificial Intelligence in Society* Paris: OECD DOI: 10.1787/eedfee77-en.

Northern Sky Research (2018) "NSR Report: Earth Observation Markets to Generate $54 Billion in Revenue by 2027" *NSR Press Releases* (22nd October) Available at: https://www.nsr.com/nsr-report-earth-observation-markets-to-generate-54-billion-revenue-by-2027/.

Piovesan, C.J. (2019) "Artificial Intelligence: Coming to a Store Near You" *Canadian Retailer Magazine* (14th January) Available at: https://www.retailcouncil.org/community/technology/artificial-intelligence-coming-to-a-store-near-you/.

Robinson, A.C., Demšar, U., Moore, A.B., Buckley, A., Jiang, B., Field, K., Kraak, M.-J., Camboimm S.P. and Sluter, C.R. (2017) "Geospatial Big Data and Cartography: Research Challenges and Opportunities for Making Maps That Matter" *International Journal of Cartography* 3 (Supplement 1) pp.32–60 DOI: 10.1080/23729333.2016.1278151.

SAS Institute (2022) "Artificial Neural Network: What They Are & Why They Matter" Available at: https://www.sas.com/en_gb/insights/analytics/neural-networks.html.

Singh, R. (2019) "Where Deep Learning Meets GIS" *ArcWatch* (June) Available at: https://www.esri.com/about/newsroom/arcwatch/where-deep-learning-meets-gis/.

Smith, C. (2006) "Introduction" In Smith, C., McGuire, B., Huang, T. and Yang, G. (Eds) *The History of Artificial Intelligence* Seattle, WA: University of Washington.

TechNavio (2017) "Global 3D Printing in Low-Cost Satellite Market 2017–2021" Available at: https://www.technavio.com/report/global-3d-printing-in-low-cost-satellite-market.

Vinuesa, R., Azizpour, H., Leite, I. Balaam, M., Dignum, V., Domisch, S., Felländer, A., Langhans, S.D., Tegmark, M. and Nerini, F.F. (2020) "The Role of Artificial Intelligence in Achieving the Sustainable Development Goals" *Nature Communications* 11 (233) pp.1–10 DOI: 10.1038/s41467-019-14108-y.

Vopham, T., Hart, J.E., Laden, F. and Chiang, Y.Y. (2018) "Emerging Trends in Geospatial Artificial Intelligence (geoAI): Potential Applications for Environmental Epidemiology" *Environmental Health* 17 (40) pp.1–6.

Vozenilek, V. (2009) "Artificial Intelligence and GIS: Mutual Meeting and Passing" *International Conference on Intelligent Networking and Collaborative Systems (INCOS 2009)* DOI: 10.1109/INCOS.2009.83.

Vozenilek, V. (2015) "Detection and Clustering of Features in Aerial Images by Neuron Network-Based Algorithm" *Proceedings of the 7th International Conference on Graphic and Image Processing (ICGIP 2015)* 98170V DOI: 10.1117/12.2228918.

PART III

Applications of geospatial technologies

24
LOCATION MATTERS
Trends in location-based services

Georg Gartner

Ultimately, humanity is concerned with 'knowing' and 'understanding', and thus seeks to gather information to make decisions, to plan or to be aware. This is true for any spatial context, and so knowledge about the 'where' of objects, phenomena or spatial relations is a fundamental pillar of human life and identity.

The production and use of maps address this concern. Maps are an attempt to communicate spatial information for expanding human abilities; to be informed about spatial objects and phenomena beyond our own perception. Maps have allowed humans to find their way, to plan routes, to fight wars, to trade, to depict relations and to tell stories. Although many of these functions can be derived without the dimensions of 'when' and 'where' in their usage, a very close connection of the location of the map user with the depicted area is often very important, as, for instance, in wayfinding and navigational tasks.

However, it is up to the skills and competencies of the map reader to make this connection and to position the depicted information with respect to the perceived reality, leaving room for misinterpretations and misuse. Only when recent technological developments allowed the position of the map user to adapt the map's content and style according to their location has this become possible. Such a personalization of communicating spatial information demands underpinning technologies to position a user; devices that are capable of positioning technologies and for data acquisition, transfer and display; and a network that connects mobile devices and servers. The rise of these respective technologies has taken place just recently at rapid speed and has allowed for personalized – thus location-based – map services. The potential of considering the location of mobile users expanded quickly beyond map use scenarios into a whole new field of applications and specific uses now known as location-based services (LBS). These allow for ubiquitous access to spatial information based on the location of the user. Hence, LBS have become popular and developed into a fundamental instrument of a mobile information society.

The potential of 'location': where we are

Recent years have seen rapid advances in LBS with the continuous evolution of mobile devices and telecommunication technologies. Location-based services are computer applications (specifically mobile computing applications) that provide information depending on the location of the device and the user, mostly through mobile portable devices (e.g., smartphones) and mobile networks. They became ever more popular not only in citywide outdoor environments, but also in shopping centres, museums, airports, large transport hubs, and many other indoor environments. Location-based services have been used by emergency services, tourism services, navigation guidance, intelligent transport services, entertainment (gaming) producers, assistive services, healthcare/fitness professionals, in social networking, and many others (Huang and Gao, 2018).

As an emerging field in cartography and GIScience, LBS bring many challenges to these disciplines. For example, when compared to users of traditional maps, LBS users are often the general public, whose knowledge, abilities, preferences and needs are diverse. Location-based services are also often used in a mobile environment, which requires LBS to be context-aware (i.e., adapted to the locational context of the user). Additionally, in LBS, the devices to render maps or communicate spatial information are diverse, ranging from smartphones and tablets to wearable devices (e.g., smartwatches, digital glasses) and public displays. All these changes – in users, map use contexts and devices – challenge the discipline of cartography to define and offer (improved or new) principles, rules and techniques that can be used to design usable LBS applications to meet users' information needs (Raper *et al.*, 2007; Huang and Gao, 2018).

Therefore, it is no surprise that LBS has developed into a very interdisciplinary field. It is influenced by technological developments and scientific research from various domains, with its common denominator being the idea of tailoring services based on the 'location' of the user. This is also reflected in the definition provided by Raper *et al.* (2007: 5):

> Location-based services are computer [typically, mobile device] applications that deliver information depending on the location of the device and user.

It can be argued that LBS are typical applications of ubiquitous computing that emphasize the importance of location-awareness (Gartner *et al.*, 2007). Several attempts at envisioning the range and direction of the potential of LBS have been offered since then (see Raper *et al.*, 2007; Huang and Gao, 2018). Generally, a path towards the evolution of such services, from technology-oriented instruments to human-centred embedded assistance tools that are immersive in daily life, becomes recognizable when looking at the evolution of LBS. The development of LBS is therefore driven by some fundamental ideals:

- **Ubiquitous availability ('Information is available anytime and anywhere')**
  Information is available (in terms of accessibility) wherever and whenever it is needed. It is not the user that needs to (physically) move to get access to information, but that the information is delivered to the user's location. Telecom infrastructure and mobile devices have made that vision a reality in a broad sense already, so that the accessibility of specific information is available anytime and anywhere on demand, brought to the location of the user.
- **Personalization ('Information is tailored to user's context and needs')**
  Information is not only made available and accessible, but is tailored to the user's context, constraints and needs, and thus is adapted in terms of personalization. This requires a modelling of the context as well as the ability to adapt the information content and to its depiction accordingly.
- **Selection ('Location is a key selector for which information is provided and how')**
  The location of the user is a key parameter for the selection of relevant information, since information about phenomena that are physically closer to the user might be more relevant than other information.

- **Popularity ('LBS are widespread and used daily')**
    Location-based services are often intended and developed for general public use. Multiplied by the availability of telecom infrastructure and the high penetration rate of mobile devices, services using these requirements and making use of the positioning functionalities of mobile devices are available for many purposes, including social media, entertainment, business, and are thus a part of daily life for many people.
- **Contribution ('LBS contributes to capture and analyse data according to everything-happens-somewhere')**
    Location-based services may be regarded as a platform for contributing to attempts to capture objects, phenomena and relations at various scales and levels of detail. As modern mobile devices can be regarded as sensor platforms, various data can be collected in relation to and for the benefit of LBS. All smart environment and digital twin concepts include the role of LBS as data collectors as well as interfaces for data acquisition.

Driven by these theoretical foundations, an ever-increasing number of services and software applications ('apps') have been produced, which serve several aims (Figure 24.1). They share, unlike other traditional means of communication spatial information such as GIS or Web mapping applications, an awareness of the locational context that users currently occupy, and the ability to adapt the contents and their presentation accordingly (Steiniger et al., 2006). In addition, they are sensitive to the mobile and highly dynamic environment in which they are used (Huang and Gao, 2018).

Subsequently, the potential of 'location' as a key parameter in information and communication services of all kinds will lead to a unique development of LBS applications, and open many research questions beyond current fields of research, which might eventually form a unique scientific discipline (Huang and Gao, 2018).

The rapid early advances in the enabling technologies of LBS, such as mobile devices and telecommunication, have therefore regarded the development of LBS as becoming a separate field of engineering. This was followed by an increasing demand in expanding LBS

Figure 24.1 Schematic illustration of location-based services as interface (by M. Schmidt)

and therefore the need for a differentiated and more holistic understanding of problems and consequences in developing and offering a range of LBS. This refers to the trend of expanding services from outdoors to indoors, to go beyond navigation systems and mobile guides to more diverse applications (e.g., healthcare, transportation, and gaming). In addition, new interface technologies (e.g., more powerful smartphones, smartwatches, digital glasses, and augmented reality devices) have emerged and there has been an increasing smartness of our environments and cities (e.g., with different kinds of sensors). Finally, we have seen more and more LBS entering into the daily lives of the general public, which greatly influences how people interact with each other and their behaviours in different environments. This brings many opportunities (e.g., for traffic management and urban planning) and challenges (e.g., privacy, ethical, and legal issues) to our environments and human societies.

From engineering to data science to human-centred approaches: how LBS evolves

In general, the development and success of LBS is underpinned by the availability of telecom infrastructure, mobile devices including respective sensors, (big) data, data infrastructure, data standards, standardized application programming interfaces, innovative human-computer-interaction and multimedia communication processing (Gartner, 2007; Meng et al., 2007; Krumm, 2009; Fellner et al., 2017).

Consequently, early research trajectories followed questions aligned with engineering, such as positioning, modelling, and the communication aspects of developing and adapting technology. This early era of LBS can be characterized as an *engineering discipline* (Gartner et al., 2007); trying to 'make it work', thus getting services built that use position information acquired by mobile devices. Areas of concern from the 'engineering era' are still highly relevant and ongoing in many ways as the underpinning technologies evolve.

Every LBS has basic architecture elements, dealing with positioning, modelling and communication tasks. With positioning, the aim is to calculate the position of a mobile device with adequate accuracy and reliability. Technologies such as global navigation satellite systems (GNSS) have become a de facto standard on contemporary smartphones and are widely used in outdoor environments. However, the determination of an accurate and reliable position is, despite recent advances in indoor positioning (Huang and Gartner, 2009; Wang et al., 2019), still challenging in dense urban environments, in mixed indoor/outdoor environments, and especially in indoor and underground environments.

Location-based services are meant to assist, support or inform users. In order to fulfil this aim, the modelling of the context of a user, including their location, characteristics and needs, is of concern, as this informs how the services adapt accordingly. So far, rather semantic concepts, such as 'place', are only rarely included in such models, thus hindering full integration of the Semantic Web and linked data approached with LBS (Feng and Liu, 2015; Jiang et al., 2016).

The task of providing relevant information via mobile devices to support decision making, to raise awareness, to inform planning or simply to inform users is limited and constrained by the device being used. As device characteristics are changing rapidly and new devices are introduced dynamically, research on this aspect is focusses on the question 'What information should be communicated to the user, and in which presentation forms?'. These forms can include several communication channels (visual, acoustic, haptic), the use of multimedia and interactivity, and might be expanded to newly emerging devices. Often maps play a central role in these services (Gartner et al., 2007).

As the engineering challenges on assembling LBS are being addressed, more services are being offered and consequently used (Chiang *et al.*, 2021). This has brought new opportunities as well as challenges, especially in the context of the data needed to feed LBS and the data produced by LBS. Consequently, developments and research surrounding the handling of data related to mobile activities led to the creation of specific applications directed towards specific applications (e.g., related to walking and biking). In this respect, the conceptualization of mobile activities, user needs and constraints in the context of these activities and their provision as a service became a concern. In this context, the solution to access or model the data needed to feed those concepts are relevant, as well as the use of the data being produced by the mobile activity (Baker, 2017). This stage of LBS development can be called a '*data-driven era*' (Gidofalvi, 2019; Graser *et al.*, 2019), since the focus of many activities was directed towards the many facets of data handling.

Mobile phone applications, based on the advances of positioning, modelling and communication), make use of LBS in various domains, going beyond classical navigation and wayfinding applications. Location-based services can be found in the context of domains such as gaming and entertainment, fleet management, social networks, healthcare, tourism, culture, collaboration tools, disaster management, marketing, and so on (UN-GGIM, 2013). It is not hard to predict that this list will get longer as 'location' – and information about location – become integrated parts of mobile devices in many ways, and because spatial data infrastructures, data standards and standardized application programming interfaces (APIs) have reached maturity.

A substantial amounts of data is generated every time a service is used. Analysing these data can be useful for gaining new insights on how people use, travel and interact with each other in the environment (Huang and Gao, 2018). Indeed, the analysis of these data, especially location-based tracking data, social media data, floating car data and crowdsourced geographic information, has developed into a significant focus for research, simply because of the vast amount of data being produced. Often, the associated analysis is motivated by gaining a better understanding of people's behaviour in different environments or enabling more innovative and personalized applications (Capineri *et al.*, 2018).

The consequent current research activities – as well as ongoing questions associated with engineering and data-driven approaches – also follow *human-centred concepts*, which start by understanding the user's needs, demands and questions, and trying to build LBS around them. In this respect, understanding human perception and cognitive abilities are relevant (Liao *et al.*, 2018), as well as our habits and the social context or impacts of LBS. As LBS have also become ubiquitously available, using these services leads to questions associated with privacy and ethical issues, especially on the 'risk of privacy breaches' and 'the possibility of increased monitoring leading to unwarranted surveillance by institutions and individuals' (Abbas *et al.*, 2014: 11).

In a collaborative attempt led by the Commission on Location-Based Services (lbs.icaci.org) of the International Cartographic Association (ICA), a cross-cutting research agenda was established, identifying 'the key research questions and challenges that are essential to meet the increasing societal demands of LBS' (Huang *et al.*, 2019). These research challenges include open questions related to positioning, modelling, communication, evaluation, applications, analysis of LBS-generated data, and social and behavioural implications of LBS (Figure 24.2); and therefore also questions related to engineering, to data science, as well as to human-centred holistic research.

The identified challenges are further differentiated into research questions related to (a) *ubiquitous positioning*, where – despite remarkable advances – reliable indoor positioning is

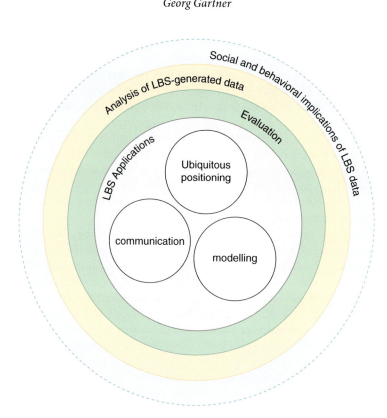

Figure 24.2 Research challenges of LBS (Huang *et al.*, 2019). The three circles clockwise from top: Ubiquitous positioning, modelling, communication. Labels on each ring, innermost first: LBS Applications, Evaluation, Analysis of LBS-generated data, Social and behavioral implications of LBS data

an issue and sensor fusion is a major research topic. Furthermore, it is important to address whether service interfaces to indoor positioning solutions can be standardized, just as the Global Positioning System (GPS) may be regarded as a de facto standard for outdoor environments. Finally, every application has different location accuracy requirements, especially with respect to position accuracy, frequency of access, reliability and latency, but there is little knowledge and evidence of guidelines for which requirements apply to LBS applications.

For open questions related to (b) *modelling aspects*, the research agenda identifies the challenge of the integration of indoor environments with LBS. For example, it might be favourable to be able to derive navigational instructions direct from a database of a building model (Fellner *et al.*, 2017) or to enrich automatically semantic descriptions of indoor environments, e.g., by using signs in wayfinding instructions (Wang *et al.*, 2017). Furthermore, context modelling remains an ongoing significant research challenge, due to the high complexity of this task.

The ongoing trend for tailoring LBS (as personalized services) has a close connection to gaining a better understanding of the nature, the conceptualization, and the development of an appropriate methodology to acquire and structure context parameters (Feng and Liu, 2015). These context parameters might include any information that can be used to characterize the situation of a person, place or object. Prominent context parameters can include information about the environment, the user, and technical, social, or temporal aspects of the history of the ongoing activity. Therefore, contemporary research activities need to pursue questions such as 'What kind of context information should be modelled by which methods?', 'How can context information acquired from different sources be integrated?', 'In which way can context information be efficiently retrieved and maintained?', and 'How can the dynamics of context be modelled in LBS?'.

Ultimately, context data are useful only if they are applied to tailor and personalize LBS (Li *et al.*, 2007). Such personalization can include the adaptation of the content being presented as well as a change of presentation style or functionalities, all based on the user's context. Open questions exist with respect to 'Which features can be adapted?', 'Which techniques and methods are needed for adaptation?', and 'When should the adaptation process happen?'. There is also no comprehensive knowledge on the acceptance of such personalized services, thus it is unclear how to balance a user's demands for transparency and control versus an efficient level of automatization.

When considering open challenges related to LBS from a human-centred perspective rather than from a technological focus, it is no surprise that users are often not alone when moving in space or referring to spatial information, and thus might aim for collaborative decision making and information retrieval. So far, services are developed primarily for independent users, neglecting the social and collaborative nature of many human activities. Designing services to support collaborative tasks and activities is therefore an open research task, especially when considering various type of collaboration, developing efficient mechanisms for enabling collaboration, resolving potential conflicts, and preserving privacy (Shokri *et al.*, 2011).

Users of LBS ultimately need to be informed, and thus need to be able to access the information the service is providing (based on the location, task and other parameters), in an accessible way. Ongoing research questions therefore evolve around (c) *communication aspects*, including the need to better understand 'How can relevant information be communicated in an optimal way to facilitate decision-making activities in space', as well as the continuous task of tackling the question 'How we can employ newly emerging mobile devices' (Figure 24.3). Additionally, as new devices might allow scope for a different set of communication channels, they raise the question of 'How can visual, acoustic and tactile communication methods be integrated meaningfully' (Thrash *et al.*, 2019).

It is often a characteristic of highly dynamic technology-driven fields that their development is driven in an iterative manner, i.e., by trying to get services built and 'making them work'. Thus, efforts towards developing a systematic evaluation of underpinning aspects, as well as overall parameters concerning issues like quality assessment of usability or usefulness are often missing. It is therefore essential to include (d) *evaluation methodology questions* on research efforts related to the development of LBS. These questions include aspects of building a holistic framework for evaluating LBS applications. This framework must systematically investigate all components and relationships within the User – Technology – Environment triangle, which might consist of user interfaces, user properties and skills, cognition, device

Figure 24.3 Accessing traces of historical information about European culture of archive collections through LBS (Verstockt *et al.*, 2019)

and service properties, environmental factors and social aspects. Additionally, how can lab-based and field-based evaluations be effectively combined to give important new insights? (Schneckenburger, 2013; Delikostidis *et al.*, 2015).

Once LBS are in use, they are collecting and generating data. These can include tracking data, social media data, data about how and when a service is requested and how it is used, which interactions are taking place, and which features are used. This is an area of research in respect to the (e) *analysis of LBS-generated data*, dealing with the aim of gaining better insights into the use, usage and users of LBS, which might be beneficial for informing future LBS developments. However, all collected data can also give insights into the habits, patterns, movements and characteristics of individuals, leading to questions of privacy. Research is therefore needed with respect to better understanding data models of LBS-generated data, as these data often have the characteristics of big data. In this context, better insights into appropriate metadata are needed to allow for full integration and participation in data infrastructures, Semantic Web and linked data structures. Further research questions include the development of appropriate analysis methods of LBS-generated data (Porras *et al.*, 2019).

Location-based services are meant to be used by humans. Often technologies have the potential to impact individual habits, patterns and abilities as well as influence or even change aspects of society. For example, the invention of the automobile has influenced how humans travel and society responded by producing a car-oriented infrastructure, changing cities and landscapes. The implications of an increasing use of newly emerging technologies is thus an important area of research. This applies to LBS as well, around (f) *societal and behavioural implications of LBS* research questions, such as 'How to LBS influence people's spatial abilities?' and 'How can we design LBS to facilitate people's activities and decision-making without harming their spatial abilities?' (Parush *et al.*, 2007; Huang *et al.*, 2012; Thrash *et al.*, 2019). Finally, privacy and ethical concerns need a systematic response from an interdisciplinary perspective, helping to understand which privacy and ethical issues exist, how they might differ over time or between cultures, and which technical and non-technical answers might lead to which reactions.

Research activities have also developed recently on the integration of LBS with the Semantic Web. Defining ontologies and making full use of relations via Linked Data standards may become an additional field within a future LBS research agenda (Kang, 2018; Hbeich and Roxin, 2020) (Figure 24.4).

Figure 24.4 Semantic LBS making use of an 'annotated' world (Wang *et al.*, 2019)

Living inside the map: where we are heading

When trying to envision a trajectory from the advances of LBS into future developments, the following trends are likely to emerge and continue.

- **Getting closer (Immersiveness)**

    Devices are developed with the intention of making them non-intrusive and allowing for immersive communication experiences. When trying to gain spatial information in the past, users needed to apply effort to acquire a map. Even in the digital era, users first needed to find a computer to get access to a digital map. It was only when such devices become mobile that interactive services became part of mobile activities. By monitoring the ongoing evolution of mobile devices of all kinds, it is possible to predict that devices might become even more 'natural', in terms of being immersive and non-intrusive. Examples include contact lenses, wearable devices, or augmented reality glasses with interfaces to LBS. Some predict the idea of brain-computer-interfaces becoming a vehicle for LBS in future as well (Liu *et al.*, 2005).

- **Getting embedded (Integration)**

    We are living in a digital era. Data are collected automatically about objects, phenomena, relations, activities and emotions at all sorts of different scales. The vast quantity of sensors available (including those in mobile devices) is permanently contributing to the ever-increasing amount of data being captured. Hence, concepts like smart cities and smart environments of ubiquitous computing have been influenced from these developments significantly. There is even the concept of the Digital Twin of the Earth (Bauer *et al.*, 2021), which asks for clear interfaces between the virtual environment and the real environment. In this respect, LBS can play a key role, allowing the actors (users) to be in a relational semantic connection with other semantic descriptions of the environment, other actors or other entities. Such semantic LBS can be seen as doorways into an annotated world, contributing to defining the annotated world (entities, relations and linked data) and interconnecting the physical and virtual world by simultaneously generating, harvesting and presenting data to humans or even machines, which might be analysed and processed through machine learning algorithms and iteratively further optimized by artificial intelligence methods.

- **Getting intersected (Intertwining)**

    The story of LBS is, in many ways, prototypical for technology-driven inventions with significant impact. It has become crucial to tackle aspects beyond simply 'making it work'. Location-based services are meant to assist humans in gaining spatial information, in supporting decision-making, and in building spatial awareness. Often the research challenges related to LBS include interdisciplinary characteristics. Therefore, more cross-disciplinary endeavours and intertwined research activities are anticipated, particularly at the intersection of geospatial science (e.g. GIScience, cartography and geodesy), information and communication technology (ICT, e.g., ubiquitous computing), human-computer interaction and data science, and the social sciences.

- **Getting user-controlled (Integrity)**

    The big developments we are all facing in our societies, including digitalization, automation, artificial intelligence and machine learning, have a tremendous impact on how we live, how we work, how we communicate, how we are entertained, and how we socialize. It is no surprise that such enormous influences can lead to a range of different human responses, including the wish for transparency, integrity and control. In the realm of LBS, all of this applies and thus it might be key for future developments of LBS, so that users can have trust and confidence in the services. These developments might include the introduction of transparent procedures and open methods, regulations governing the use of data, and the user has fuller control over what services shall be used and how.

Conclusion

Location-based services are becoming more ubiquitous in many aspects of daily life and have attracted significant research interest from various scientific disciplines. With the continuous evolution of communication technologies and mobile devices that underpin and support these services, rapid advances in LBS have been observed in the past few years. As summarized above, there have been several key evolutions of LBS research, as well as their motivations and aims, and some anticipated areas of research were outlined.

Humans have developed maps, and more recently, LBS, as tools for satisfying their need for spatial information. By using these tools, humans gain some advantages but they also face some challenges. As it is unlikely that the need for spatial information will go away, tools such as LBS, which allow for making use of spatial information in context with the full potential of 'location', will continue to be a key area of interest for our scientific endeavours.

References

Abbas, R., Michael, K. and Michael, M. (2014) "The Regulatory Considerations and Ethical Dilemmas of Location-Based Services (LBS): A Literature Review" *Information Technology & People* 27 (1) pp.2–20 DOI: 10.1108/ITP-12-2012-0156.

Baker, K. (2017) "Improving Navigation Services for Leisure Activities Exploiting the Full Potential of Route-sharing Communities and Their Crowd-based Sources" (*PhD thesis*)University of Ghent, Belgium.

Bauer, P., Stevens, B. and Hazeleger, W. (2021) "A Digital Twin of Earth for the Green Transition" *Nature Climate Change* 11 pp.80–83 DOI: 10.1038/s41558-021-00986-y.

Capineri, C., Huang, H. and Gartner, G. (2018) "Tracking Emotions in Urban Space: Two Experiments in Vienna and Siena" *Rivista Geografica Italiana* 125 (3) pp.273–288.

Chiang, K.W., Tsai, G.J. and Zeng, J.C. (2021) "Mobile Mapping Technologies" In Shi, W., Goodchild, M.F., Batty, M., Kwan, MP. and Zhang, A. (Eds) *Urban Informatics (Urban Book Series)* Singapore: Springer, pp.439–465 DOI: 10.1007/978-981-15-8983-6_25.

Delikostidis, I., Fritze, H., Fechner, T. and Kray, C. (2015) "Bridging the Gap Between Field- and Lab-Based User Studies for Location-Based Services" In Gartner, G. and Huang, H. (Eds) *Progress in Location-Based Services 2014* Cham, Switzerland: Springer, pp. 257–271 DOI: 10.1007/978-3-319-11879-6_18.

Fellner, I., Huang, H. and Gartner, G. (2017) "Turn Left after the WC, and Use the Lift to Go to the 2nd Floor-Generation of Landmark-Based Route Instructions for Indoor Navigation" *ISPRS International Journal of Geo-Information* 6 (6) 183 DOI: 10.3390/ijgi6060183.

Feng, J. and Liu, Y. (2015) "Intelligent Context-Aware and Adaptive Interface for Mobile LBS" *Computational Intelligence and Neuroscience* 2015: Article ID 489793 DOI: 10.1155/2015/489793.

Gartner G. (2007) "LBS and TeleCartography: About the Book" In Gartner G., Cartwright, W. and Peterson M.P. (Eds) *Location Based Services and TeleCartography (Lecture Notes in Geoinformation and Cartography)* Berlin, Heidelberg: Springer DOI: 10.1007/978-3-540-36728-4_1.

Gartner, G., Bennett, D. and Morita, T. (2007) "Towards Ubiquitous Cartography" *Cartography and Geographic Information Science* 34 (4) pp.247–257 DOI: 10.1559/152304007782382963.

Gidofalvi, G. (2019) "Trajectory and Mobility Based Services: A Research Agenda" *Adjunct Proceedings of the 15th International Conference on Location Based Services* Vienna, pp.123–128 DOI: 10.34726/lbs2019.60.

Gartner, G. and Huang, H. (Eds) (2019) *Adjunct Proceedings of the 15th International Conference on Location Based Services (LBS 2019)* Vienna, pp.11–13 DOI: 10.34726/lbs2019.60.

Graser, A., Schmidt, J., Dragaschnig, M. and Widhalm, P. (2019) "Data-driven Trajectory Prediction and Spatial Variability of Prediction Performance in Maritime Location Based Services" In Gartner, G. and Huang, H. (Eds) *Adjunct Proceedings of the 15th International Conference on Location Based Services (LBS 2019)* Vienna, pp.129–134 DOI: 10.34726/lbs2019.23.

Hbeich, E., Roxin, A. (2020) "Linking BIM and GIS Standard Ontologies with Linked Data" *Proceedings of the 8th Linked Data in Architecture and Construction Workshop (LDAC2020)* Dublin, pp.146–159.

Huang, H. and Gao, S. (2018) "Location-Based Services" In Wilson, J.P. (Ed.) *The Geographic Information Science & Technology Body of Knowledge* Ithaca, NY: University Consortium for Geographic Information Science (UCGIS).

Huang, H. and Gartner, G. (2009) "A Survey of Mobile Indoor Navigation Systems" In Gartner, G. and Ortag, F. (Eds) *Cartography in Central and Eastern Europe (Lecture Notes in Geoinformation and Cartography)* Berlin, Heidelberg: Springer, pp.305–319 DOI: 10.1007/978-3-642-03294-3_20.

Huang, H., Schmidt, M. and Gartner, G. (2012) "Spatial Knowledge Acquisition with Mobile Maps, Augmented Reality and Voice in the Context of GPS-Based Pedestrian Navigation: Results from a Field Test" *Cartography and Geographic Information Science* 39 (2) pp.107–116 DOI: 10.1559/15230406392107.

Jiang, L., Yue, P. and Guo, X. (2016) "Semantic Location-based Services" *IEEE International Geoscience and Remote Sensing Symposium (IGARSS)* pp.3606–3609 DOI: 10.1109/IGARSS.2016.7729934.

Kang, T. (2018) "Development of a Conceptual Mapping Standard to Link Building and Geospatial Information" *ISPRS International Journal of Geo-Information* 7 (5) 162 DOI: 10.3390/ijgi7050162.

Krumm, J. (2009) "A Survey of Computational Location Privacy" *Personal and Ubiquitous Computing* 13 (6) pp.391–399.

Li, F., Li, X. and Bian, F. (2007) "Autonomic LBS Based on Context: Preview" *International Conference on Wireless Communications, Networking and Mobile Computing* pp.3266–3269 DOI: 10.1109/WICOM.2007.809.

Liao, H., Dong, W., Huang, H., Gartner, G. and Liu, H. (2018) "Inferring User Tasks in Pedestrian Navigation from Eye Movement Data in Real-world Environments" *International Journal of Geographical Information Science* 33 (4) pp.739–763 DOI: 10.1080/13658816.2018.1482554.

Liu, H., Wang J. and Zheng, C. (2005) "Using Self-organizing Map for Mental Tasks Classification in Brain-Computer Interface" In Wang, J., Liao, X.F. and Yi, Z. (Eds) *Advances in Neural Networks* Berlin, Heidelberg: Springer DOI: 10.1007/11427445_53.

Meng, L., Zipf, A. and Stephan, W. (2007) *Map-Based Mobile Services: Interactivity and Usability* Berlin, Heidelberg: Springer.

Parush, A., Ahuvia, S. and Erev, I. (2007) "Degradation in Spatial Knowledge Acquisition When Using Automatic Navigation Systems" In Winter, S., Duckham, M., Kulik, L. and Kuipers, B. (Eds) *Spatial Information Theory*, pp.238–254 Berlin, Heidelberg: Springer.

Porras Bernárdez, F., Gartner, G., Van de Weghe, N. and Verstockt, S. (2019) "Finding Cultural Heritage Traces from Modern Social Media" *Abstracts of the International Cartographic Association* 1 (302) DOI: 10.5194/ica-abs-1-302-2019.

Raper, J., Gartner, G., Karimi, H. and Rizos, C. (2007) "A Critical Evaluation of Location Based Services and Their Potential" *Journal of Location Based Services* 1 pp.5–45.

Schneckenburger, J. (2013) "Überprüfung des Navigationsverhaltens in der DAVE" (*Master's thesis*) TU Wien.

Steiniger, S., Neun, M. and Edwardes, A. (2006) *Foundations of Location Based Services* Zurich: University of Zurich.

Thrash, T., Fabrikant, S., Brügger, A., Do, C., Huang, H., Richter, K., Lanini-Maggi, S., Bertel, S., Credé, S., Gartner, G. and Münzer, S. (2019) "The Future of Geographic Information Displays from GIScience, Cartographic, and Cognitive Science Perspectives" *Leibniz International Proceedings in Informatics* 142 pp.19:1–19.11 DOI: 10.4230/LIPIcs.COSIT.2019.19.

UN-GGIM (2013) "Future Trends in Geospatial Information Management: The Five to Ten Year Vision" Available at: https://ggim.un.org/future-trends/.

Verstockt, S., Melville, K., Dilawar, A., Porras-Bernardez, F., Gartner, G. and Van de Weghe, N. (2019) "EURECA – EUropean Region Enrichment in City Archives and Collections" *Proceedings of the 14th Conference Digital Approaches to Cartographic Heritage* Available at: http://cartography.web.auth.gr/ICA-Heritage/Thessaloniki2019.

Wang, W., Huang, H. and Gartner, G. (2017) "Considering Existing Indoor Navigational Aids in Navigation Services" *Proceedings of Workshops and Posters at the 13th International Conference on Spatial Information Theory (COSIT 2017)* pp.179–189.

Wang, W., Klettner, S., Gartner, G., Fian, T., Hauger, G., Angelini, A., Söllner, M., Florack, A., Skok, M. and Past, M. (2019) "Towards a User-oriented Indoor Navigation System in Railway Stations" In Gartner, G. and Huang, H. (Eds) *Adjunct Proceedings of the 15th International Conference on Location Based Services (LBS 2019)* Vienna, pp.63–70 DOI: 10.34726/lbs2019.58.

25
MAPPING BUILDINGS AND CITIES

Tomasz Templin

Introduction

From the beginning of the human culture, large population centres, or urban areas, allowed civilizations to develop. Recent studies suggest that the percentage of the human population living in cities will continue to grow (Batty, 2019). It is estimated that by the end of this century, more than 90% of the world's population will live in some form of the city. This makes the process of mapping of urbanised areas a fundamental issue, which has become the subject of many studies, projects, and publications (Musialski *et al.*, 2013; Wang *et al.*, 2018).

Mapping plays an important role in the process of reconstruction and modelling of 3D buildings and cities. It enables the creation of virtual 3D replicas by precisely defining the shape and appearance of objects. The importance of virtual 3D city models is growing every year. They are gaining popularity in industrial, consumer, entertainment, healthcare, education, and governmental applications. The role of models is becoming especially significant in municipal applications. The popularization of the idea of a smart city and digital twins significantly changes the way in which authorities approach data acquisition. Today's 3D building models are being combined with comprehensive geographic information system (GIS) data, along with Building Information Modelling (BIM) data. This brings the vision of a digital twin to life for both the inside and outside of buildings. Today, all over the world, local governments and municipal authorities are starting to roll out complex smart city/digital twin projects.

The dynamic growth of data acquisition techniques leads to changes in the methodology of measuring buildings, cities, or agglomerations. Emerging technologies like Lidar (light detection and ranging), UAS/UAV (Unmanned Aerial Systems/Unmanned Aerial Vehicles, more commonly known as drones), mobile mapping, or location intelligence are now widely available and boost 3D mapping of objects and structures.

Mapping buildings and cities can be based on a variety of different input data sources, including existing maps and databases (topographic/cadastre), street-level photos, satellite imagery, aerial images, or laser scans. Buildings are important human-made objects of 3D city models. Existing concepts for the representation of buildings are restricted to a specific level-of-quality. They range from the simplest structure like block models (polygonal) to roof-including models, architectural models, and indoor reality models.

Mapping buildings and cities is still under active research. The main purpose of this chapter is to review and summarise information that will help practitioners and researchers know the history and state-of-the-art methods used in this field. The scope of the text concentrates on the reconstruction of 3D geometric models of urban areas and individual buildings. To systematise the presented contents, a top-down approach has been proposed. The discussion starts with the least accurate measurement techniques, allowing to create coarse 3D models of vast areas. In the following sections, different methods are described presenting a methodology for modelling high-complexity 3D urban space with photorealistic, architectural 3D building models. At the end methods supporting semantics used in specialised solutions such as Building Information Modelling (BIM)/City Information Modelling (CIM) are described.

Mapping buildings – challenges and constraints

The measurement of buildings and creation of their 3D models require the selection of an appropriate measurement technique. Buildings located in dense urban environments are affected by a large number of obstacles or barriers preventing access or limiting the ability to obtain all necessary data to build a model of the object. It is often impossible to measure a building directly using a single measurement method. To obtain a complete dataset, it is necessary to collect data using different platforms (ground, air) and sensors (active or passive). Usually, there is a need to apply complementary methods in order to obtain a satisfactory result. The following constraints are among the most common (Figure 25.1):

- no easy access, narrow streets between buildings with many obstacles;
- obstacles in front of buildings, such as cars, trees, utility poles, pedestrians, or similar;
- items displayed outside buildings (advertising, billboards, others);
- electricity and telecommunication wires above ground;
- irregularly shaped facades, which require additional manual tasks to complete the measurements;
- construction materials used to build the structure (glass and similar materials absorbing the signal from measuring devices, so they are often impossible to measure in the visible range).

Automatic extraction of buildings from satellite or aerial images is a difficult task. As is the case with ground-based measurements, many factors make the identification of buildings

Figure 25.1 Figures illustrate common problems for mapping buildings from the ground (left – street-level photo) and air (right – aerial photo with building footprints). Courtesy of Head Office of Geodesy and Cartography in Poland, geoportal.gov.pl

problematic. Building tops in an urban area do not have similar shapes, sizes, and textures. The varied structure and shape of buildings makes it challenging to discover fully automatic detection methods for all types of objects. Another factor affecting the detection of buildings is the presence of obstacles created by surrounding objects, such as trees, tall buildings, etc. In addition, the contrast between the roof of a building and the surrounding region is usually low, which is an important problem for traditionally used segmentation methods that are based on spectral characteristics of the objects.

Level of details and standards

There are many approaches to classify methods of mapping buildings and urban areas. The type of classification depends on different factors, like application purpose (visualization, urban planning, smart city, BIMs/CIMs), level of automation, and expected Level of Detail (LOD). In most applications, the open standard (Open Geospatial Consortium (OGC) City Geography Markup Language (CityGML) Encoding Standard) is used for urban reconstruction based on LOD (Gröger et al., 2012). Other formats also commonly used for storing both geometric and semantic information are X3D, CityJSON, COLLADA, KML, or IFC (Nys et al., 2020).

CityGML is an open data model and XML-based (*eXtensible Markup Language*) format for the storage of virtual 3D city models. It is a frequently used international standard for spatial data exchange issued by the OGC and the International Standardization Organization (ISO) TC211. It contains a common definition of the basic entities, attributes, and relations of a 3D city model.

CityGML supports different LODs. To explain the consecutive concept of LOD, in Figure 25.2 a single CityGML building is presented from a coarse LOD0 representation up to a semantically rich and geometric-topologically sound LOD4 model, including the building interior.

The expected LOD makes it possible to adapt the methods of data acquisition and related methodology of representing the data to various application requirements. Additionally, LOD facilitates efficient data visualization and analysis. In CityGML, the same object can be represented in different LODs at the same time, which allows the analysis and visualization of the same object in relation to different resolution levels. Here is a short description of each of the consecutive LODs:

- LOD0 – regional/landscape applications (2.5 Digital Terrain Model, usually draped by orthophoto);
- LOD1 – city/region applications (block-models, without roof structures);
- LOD2 – city district/site model (geometry-models, differentiated roof structures);
- LOD3 – outside architectural model (architectural models, facade details);
- LOD4 – interior architectural model (detailed indoor models).

Some researchers propose an additional level of LOD. To enhance the attractiveness of the LOD1 model (realism of visualization), they suggest dividing the footprint of buildings into sections, so that it is possible to extrude these parts to different heights (Xiong et al., 2016). Other publications show limitations concerning the LOD definition from CityGML models and suggest expanding their number to improve modelling and data exchange (Biljecki et al., 2016).

According to the CityGML standard, each LOD has a different application. The lowest one (LOD1) defines coarse building models and is typically created using satellite and aerial

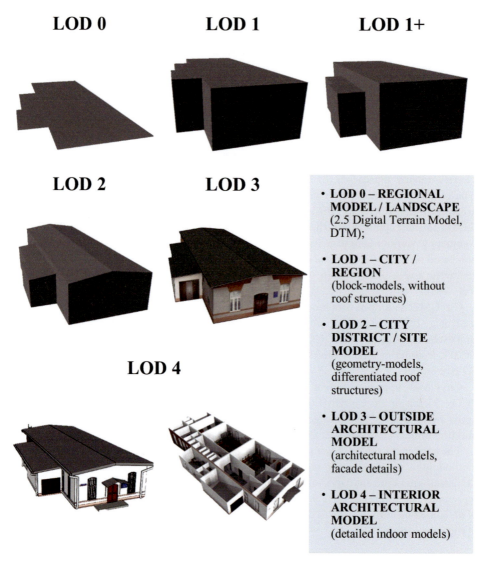

Figure 25.2 The five LODs of CityGML. The geometric details and semantic information changes with consecutive levels. The LOD1+ proposed by some authors was additionally presented in the figure

data technology. LOD2 has differentiated roof structures and thematically differentiated boundary surfaces. Starting from LOD3 a fusion of aerial and ground-based data is required. This level and the following one are already created based on reality capture methodology with detailed ground-based measurements or acquired using mobile aerial platforms, such as UAVs or small helicopters.

The OGC CityGML standard does not explicitly define the methods of automatic model generation for different LODs. In addition, the relationship between the different LODs is also missing. In practice, the creation of virtual 3D building or city models is based on a number of independent data sources. A short list of commonly used data is as follows (Döllner *et al.*, 2006):

- **Cadastral/topographic databases**
  The cadastral databases provide the official building outlines, boundaries of the land parcels, address of the building, as well as ownership information. These databases are official data sources, gathered by public institutions in accordance with existing laws and surveying standards. Although the databases usually do not contain 3D data, they are often used as a foundation for the construction of 3D buildings or city models.
- **Crowdsource or commercial data**
  Nowadays, global databases are available in addition to official, national sources. They are created both by commercial providers and services created using crowdsourcing and Volunteered Geographic Information, VGI (Sui et al., 2013). They offer easy access to maps, vector data, as well as street-level photos.
- **Digital terrain model**
  The DTM is a structure used as a reference surface for all geometric objects of a virtual 3D city model. Typically, a GRID or TIN model is used, supplemented by a thematic map or true orthophoto overlayed on it.
- **Satellite-based data sources**
  Satellite data is a widely used source of data for city mapping. Satellite images are commonly applied to identify and extract buildings and other man-made structures. The application areas of satellite-based data have been considerably increased with the availability of sub-metre resolution data from high-resolution Earth satellites, such as IKONOS, WorldView, and QUICKBIRD. In some cases, the Synthetic Aperture Radar (SAR) interferometry method is also used as a complementary instrument to compensate for the disadvantages of passive methods.
- **Aerial-based data sources**
  Aerial-based platforms are the most important data source for measuring buildings. The rapid development of measurement platforms such as light aeroplanes, helicopters, or drones has led to the popularization of data obtained using this method. They allow collecting data simultaneously from both passive and active sensors. It is a realization of the concept of reality capture that relies on the integration of data from multiple sensors to obtain more realistic models and create visualizations of the world 'as built'. Aerial laser scanners are more favourable measurement devices for the generation of rooftops and extraction of buildings. Aerial photography provides essential data for photorealistic visualization (e.g., land coverage images, roof textures, oblique photos for facades).
- **Ground-based data sources**
  For ground-based platforms, two solutions are used in practice to collect data. The first is short-range photogrammetry with single-image or multiple-image techniques. This method is particularly useful for extracting geometric details of buildings. The second one is based on the use of mobile platforms, in particular, mobile Lidar scanners, to obtain data of building facades. This method allows for quick data acquisition along designated paths. For measuring complex objects, like historic buildings or buildings with multiple facades, stop-and-go measurements are preferred using terrestrial Lidar and total station (TS) scanners.

A methodology for modelling cities

Three-dimensional city modelling is of growing interest in a variety of applications. Since the first release of Google Earth and Bing Maps Platform (previously Microsoft Virtual Earth) in 2005, mapping buildings and cities is becoming a regular topic in many disciplines (Liang et al.,

2018). The launching of these products on the market has become a milestone in the development of 3D city models. This has shown the potential of new technology and a wide range of applications, and as a consequence, has had a significant impact on both academia and industry.

Since then, the number of cities represented in 3D city models has been increasing rapidly. The majority is primarily related to the progress in the development of mapping techniques, accessibility of geospatial data, and ongoing automation. The methodology of creating a 3D city model depends mostly on the aim of its development. In most cases, models are generated for visualization purposes to provide a photorealistic representation of urban objects. This solution is based solely on graphic and geometric models without semantic information. The most recent implementations concentrate on incorporating into the model semantic data required for spatio-temporal analysis and modelling. Figure 25.3 presents the image of the city model at different LODs.

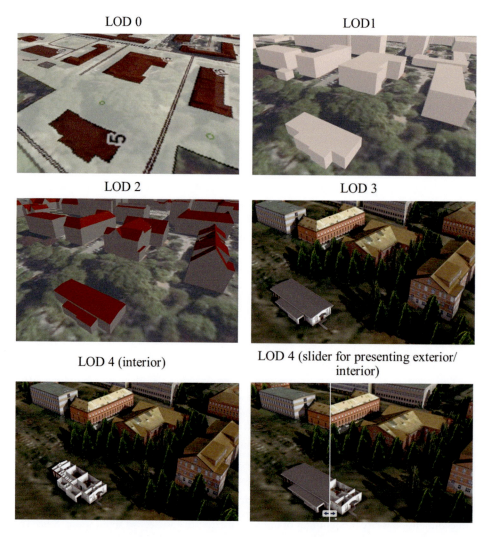

Figure 25.3 Comparison of virtual city models at various level of details (LOD). Courtesy of Head Office of Geodesy and Cartography in Poland, geoportal.gov.pl

Mapping buildings and cities

Different measurement techniques and modelling strategies can be used for consecutive LODs. Figure 25.4 illustrates the differences in the appearance of the sample building from the city model of Olsztyn. For each level, a list of possible measurement technologies that can be used at various stages of construction of building models is added. Due to the variety of possible measurement methods, or the use of hybrid solutions combining different methods, the list is not a closed one.

A brief description characterising the individual steps (building footprint extraction, identification and extraction of a rooftop, facade measurement, interior measurements) in the model building process at a specific level of detail is described in the following.

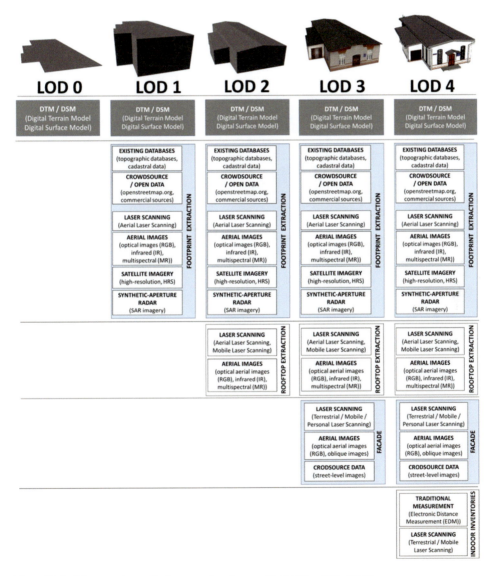

Figure 25.4   The mapping techniques used at the specific LODs of CityGML

Building footprint extraction

The building footprint is a crucial component in the mapping process. It is a requirement for basemaps preparation, 3D urban modelling, cadastre systems, transportation, and many other applications. Building outlines combined with the height of the buildings are essential parameters for the construction of models at LOD1/LOD1+ level. The schematic workflow for designing the LOD1 model with information about typically used data sources is presented in Figures 25.5 and 25.6, respectively.

Building outlines have been used for geodetic measurements for years. Traditional methods rely on surveying techniques to measure building corners. They create a precise representation of their shape in 2D. The results of the measurements are verified and stored in the cadastral/topographic databases, making them a reliable source of data, which is commonly used as a base layer to build a model at every level of accuracy defined in the OGC CityGML standard.

In addition to the classic measurements, building footprints can be extracted from both satellite and aerial missions. There are several ways to extract building outlines, with both

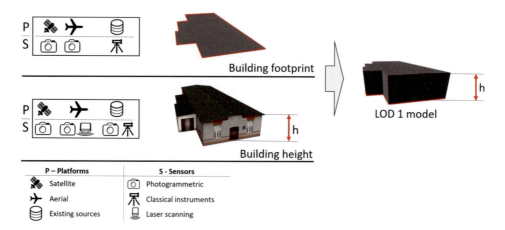

Figure 25.5 Data sources and platforms used for creating LOD1 building model

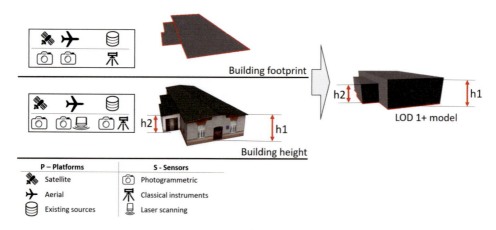

Figure 25.6 Data sources and platforms used for creating LOD1+ building model

active and passive measuring sensors. The primary source of data that is commonly used in practice is aerial images, one of the most available data sources. Their spatial resolution and optical characteristics allow to identify buildings and determine the edge of buildings. In practice, both images taken in the visible bands (RGB) and an infrared band (IR) along with multispectral bands are used to identify buildings.

Access to aerial images has increased significantly with the popularity of Mobile Mapping Systems (MMS). Built on aerial platforms, such as light aircraft, helicopters, or UAV/UAS, MMS allowed to accelerate the process of acquisition and reduce data acquisition costs. Due to the high resolution of the generated orthophotomaps (centimetre level), it is a valuable data source for the extraction of buildings. The introduction of sub-metre resolution data from high-resolution Earth satellites such as IKONOS, WorldView, and QUICKBIRD to the market has also made satellite images an excellent source of data. This high-resolution (HRS) imagery facilitates the identification of natural or human-made structures like roads, buildings, or even trees.

Traditionally in GIS, the building footprints are digitized using aerial and high-resolution satellite imagery. This process involves manual digitization by using tools to draw the outline of each building. However, it is a labour-intensive and time-consuming process. Therefore, most of the current works are focused on automating the process of creating the building outlines. Automatic methods use different techniques related to radiometric, geometric, edge detection, and object-based classification. They are common topics in the literature and are discussed by various researchers or commercial companies (Czerniawski and Leite, 2020).

The use of laser scanning from an airborne measurement platform is less applicable for the extraction of the building footprint. The possibilities of its use are largely defined by the density of the generated point cloud. Low-density point clouds (2 pts/m2), which are more readily available for numerous geographic areas, can be used to determine building footprints with a simple roof structure. For other cases, the dense point cloud (20–40 pts/m^2) must be utilised.

In case of creating models based on open data or crowdsourcing, databases containing ready-to-use datasets that include building footprints are often used. These data have recently become widely available through worldwide geospatial data-sharing efforts (e.g., OpenStreetMap, Microsoft Bing Map, Data.Gov in the US, Ordnance Survey in the UK, Head Office of Geodesy and Cartography in Poland, and local providers in other countries).

In addition to the building outline at the stage of construction of LOD1/LOD1+ models, it is necessary to determine the height of buildings. This is usually done using airborne light detection and ranging (Lidar) data. Lidar data are combined with building footprints to extract the height information. Two stages can be distinguished: (1) extracting Lidar points within the boundaries of the building outline; and (2) conducting a statistical analysis of the extracted points in order to obtain the height of buildings. As the rooftop area is usually cluttered by artefacts, like trees, chimney, elevator shafts, towers, antennas, and so on, extracted rooftop points are filtered at this stage, and irrelevant points are excluded.

Another commonly used method is the use of aerial or satellite imagery for height determination. The height of each building is calculated by the number of floors of each building. The number of floors is estimated from the satellite or aerial images. For each floor, a standard value (based on the type of building) is used to estimate the total height of a building.

Roof extraction

The rooftop extraction is one of the critical issues for constructing accurate 3D building models. Based on the rooftop, models at LOD2 and LOD3 are generated (Figure 25.7). At

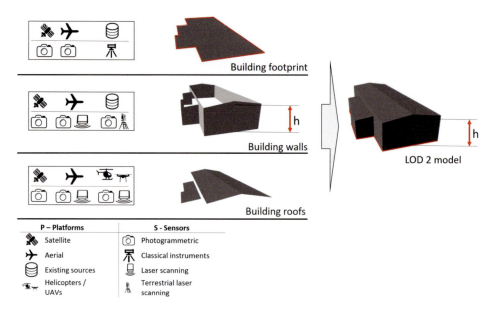

Figure 25.7 Data sources and platforms used for creating LOD2 building model

this moment, there is no consensus among the authors on the optimal method of obtaining data for detecting the rooftop of a building and automatically generating a shape-sensitive 3D model.

To date, most of the rooftop modelling methods focus on the reconstruction of the roof structure from the information captured by a single data acquisition mode. Traditionally, for this task, photogrammetric products, like aerial images and satellite imagery, are used. Many authors have reported the advantages of this type of solution, but also point out significant limitations. The most important are shadows, occlusions, texture problems, and variations in brightness and contrast (Jung and Sohn, 2019).

Some authors suggest aerial scanning as the best measurement method for roof extraction. There is a belief that Lidar should be the preferred method, as it is less dependent on external factors. However, its application is limited by the density of the aerial Lidar point cloud. Segmentation of the roof plane based on this data is a complex task since the point clouds data do not provide any information about connectivity and semantic characteristics of the scanned surfaces. Additional problems related to the use of laser scanning are (Gilani *et al.*, 2018): systematic and stochastic measurement inaccuracies, changing point density and irregular spatial distribution, holes in the point clouds and unnecessary objects resulting from occlusion by adjacent objects (e.g., clusters of vegetation), and the physical limitation of data acquisition sensors and multiple reflections (multipath effect). Moreover, the absorption of laser impulses by glass elements of roofs and reflection from vents, chimneys, and dormers are also mentioned.

The recent developments indicate the fusion of multiple data sources as the best method. This solution offers improved accuracy and eliminates the disadvantages of single data source methods. Many researchers have explored the advantages of combining information obtained from multiple sources. Due to their complementary properties, the aerial images and the Lidar are usually combined to achieve the goal of roof reconstruction in 3D. For point cloud segmentation, the RANdom SAmple Consensus (RANSAC), Hough Transform,

Mapping buildings and cities

region growing, model-driven tactic, support by aerial or satellite images in addition to Lidar data, or other approaches are often utilised for point set segmentation (Tarsha Kurdi and Awrangjeb, 2020).

In new approaches, time is also an important factor. The novel fusion methods integrate multi-sensor data to refine building rooftop models over time and focus on temporal changes. Smart cities require regular updates for 3D virtual representations of buildings and cities.

Building facade extraction

The need for rapid collection of data increases as local governments move to digitized 3D models for city planning. Designing detailed 3D models of buildings with facades is the further stage realised from the LOD3. The construction of models of the building at this stage requires combining data from multiple sources (Figure 25.8).

Airborne Lidar data and aerial images are highly suitable for extracting features, such as building footprints and roofs structures. The acquired point density from ALS (1–40 pts/m^2) is high enough for the rough extraction of building roof structures and outlines. However, building facades are more challenging to model due to nadir-type viewing direction. For that reason, the details of buildings, such as balconies and windows, are missed. Mapping of buildings at the highest level of detail demands access to more accurate and dense data.

The photorealistic 3D model of a building requires mapping of georeferenced image texture to 3D object geometry. This task is performed mostly based on oblique photos taken from the aerial-based platform. The oblique photos are made from different viewing

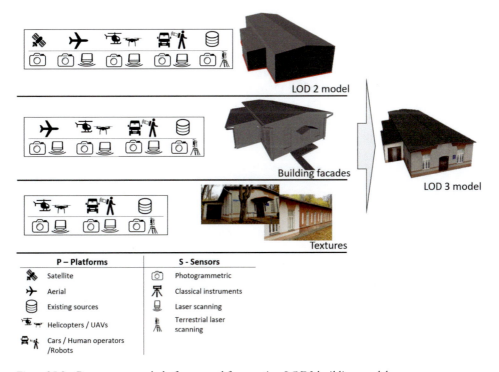

Figure 25.8 Data sources and platforms used for creating LOD3 building model

directions and provide the correct textures for building roofs or even facades (Kersten and Stallmann, 2012, Xiao *et al.*, 2012). However, occlusion and differences in viewpoint greatly perturb the bottom parts of buildings, leading to holes in geometry and visual effect of blurring on textures. To eliminate distortion, it is necessary to use street-level photos and integrate them with aerial photos to enhance the 3D model of a building.

Two measuring methods are usually chosen to obtain data from the terrestrial platforms. The first is short-range photogrammetry with single-image or multiple-image techniques. One of the main advantages of ground photogrammetry is the relatively low cost of data acquisition. This method is particularly useful for extracting geometric details of buildings. As an input, street-level photos are used, which are quite simple to obtain. In addition, already existing photographs available on the Internet (crowdsourcing) can also be used to reconstruct models automatically (Menzel *et al.*, 2016). The second approach is based on scanning methodology. It uses stationary laser scanners (terrestrial laser scanning, or TLS) or mobile mapping systems for dynamic measurements (Figure 25.9).

The main limitation of ground-based platforms are areas that are inaccessible for measuring directly using a terrestrial sensor, like rooftops and backs of buildings. Therefore, it is difficult to obtain a complete, sufficient dataset to reconstruct the building only from ground-level data acquisition systems. Therefore, terrestrial measurements are often supported by data obtained from other platforms. UAV/UAS are popular and widely used measurement platforms. Due to the ease of use, reduced cost, and accuracy of data obtained from multiple sensors (both passive and active), UAV/UAS are commonly used in the process of obtaining data for buildings modelling (Templin and Popielarczyk, 2020). In addition, they allow for taking measurements periodically in order to update the city model or detect temporal changes in its structure.

Indoor measurements

In recent years a growing need for up-to-date and detailed interior models has been observed. This type of model is required in many fields and is needed in various applications. These expectations are reflected in the OGC CityGML specification by the LOD4 model.

Creating a model on the LOD4 level requires data from the interior as well as the exterior of the building. Compared to outdoor measurements, the main problem with a survey of indoor environments is the lack of the GNSS signal. This necessitates the use of other

Figure 25.9 Visualization of typical point cloud acquired using the terrestrial and aerial (helicopter) mobile laser platform (orthophotomap on the left, classified point cloud (buildings – red colour, vegetation – green colour, overhead power line – purple colour) on the right). Courtesy of Vimap Sp. z o.o., https://www.vimap.pl

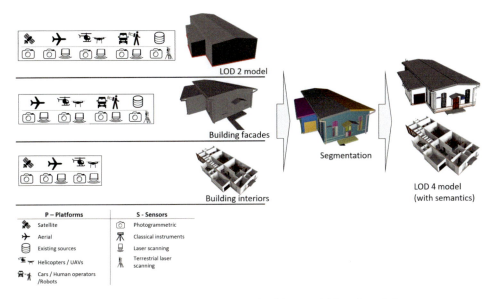

Figure 25.10 Methodology for preparing an LOD4 building model based on different measurement platform and sensors

positioning methods inside buildings to inventory the rooms and objects inside them. The dynamic growth of simultaneous localization and mapping (SLAM) algorithms and the miniaturization of mobile laser scanners makes them the leading equipment solution for indoor measurements (Figure 25.10).

Despite the increasing demand, the reconstruction of the interior of a building based on point clouds is still at an early stage, compared to outdoor scene reconstruction. Interior surveying generates specific problems that have not been observed in the solutions dedicated to exterior surveying. They are mainly related to the complex layout of buildings and the occurrence of a large number of objects such as furniture that causes clutter and occlusions.

In outdoor applications, the sensor is located using GNSS and the Inertial Measurement Unit (IMU). In indoor applications, the positioning process is based on the SLAM. The main advantages of indoor MMSs are primarily faster acquisition time, no need for target-based georeferencing, and less occlusion due to continuous scanning along the trajectory. Nevertheless, their nominal accuracy is still lower than that of static scanners.

Today, many commercial indoor mobile mapping platforms are available on the market. They vary in hardware algorithms and SLAM. Typical indoor mobile mapping systems include wheeled mobile systems, backpacked mobile systems, or handheld devices, depending on the platform in which they are carried. The primary product of the measurement is point cloud and trajectory over which the system moves during the measurement. They are used to create a 3D indoor model of the building with objects inside and 2D floorplans.

Conclusion

Data acquisition is an important stage of the modelling process. Currently most of the work is based on the synergy of multiple data sources. The fusion of data from multiple technologies/sensors helps to eliminate the limitations of a single type of data source and minimises errors. Moreover, it creates a possibility for automatic object extraction and 3D object modelling – and consequently, elimination of human intervention required to process data in order to avoid errors.

The deployment of the reality capture concept (and its further evolution as a real-time 'digital twin') accelerates the process of data acquisition by using various measuring platforms and sensors. This leads to higher accuracy, more realistic models, and creation of visualizations of the world 'as built'. With the increase in accuracy, level of detail, and completeness of the raw data, it is possible to create better and more faithful replicas of cities.

Better information can be used more effectively and not only for visualization purposes, but also to build complex BIM/CIM solutions. The multidimensionality of these solutions facilitates not only the time aspect but also other factors (costs, building life cycle control, facility, and asset management). Additionally, they form new technological platforms, which are the central point for the cooperation of specialists in various fields.

The use of artificial intelligence algorithms (machine learning, deep learning) to extract objects directly from raw data, point cloud, and quickly generated mesh models can be a breakthrough in the development of mapping of these objects. Engaging cloud computing in this process with increased computing power creates a good opportunity to realise the vision of automatic mapping of buildings and objects in the near future.

References

Batty, M. (2019) *Inventing Future Cities* Cambridge, MA: MIT Press.
Biljecki, F., Ledoux, H. and Stoter, J. (2016) "An Improved LOD Specification for 3D Building Models" *Computers, Environment and Urban Systems* 59 pp.25–37.
Czerniawski, T. and Leite, F. (2020) "Automated Digital Modeling of Existing Buildings: a Review of Visual Object Recognition Methods" *Automation in Construction* 113 pp.103–131.
Döllner, J., Baumann, K. and Buchholz, H. (2006) "Virtual 3D City Models as Foundation of Complex Urban Information Spaces" *Proceedings of CORP 2006 & Geomultimedia06*, Vienna, 13th–16th February, pp.107–112.
Gilani, S.A.N., Awrangjeb, M. and Lu, G. (2018) "Segmentation of Airborne Point Cloud Data for Automatic Building Roof Extraction" *GIScience and Remote Sensing* 55 (1) pp.63–89.
Gröger, G., Kolbe, T.H., Nagel, C. and Häfele, K-H. (Eds) (2012) "OpenGIS City Geography Markup Language (CityGML) encoding standard, Version 2.0.0" *OGC Document Reference: OGC 08-007r1* Available at: https://portal.ogc.org/files/?artifact_id=28802 (Accessed: 22nd February 2023).
Jung, J. and Sohn, G. (2019) "A Line-Based Progressive Refinement of 3D Rooftop Models Using Airborne Lidar Data With Single View Imagery" *ISPRS Journal of Photogrammetry and Remote Sensing* 149 pp.157–175.
Kersten, T.P. and Stallmann, D. (2012) "Automatic Texture Mapping of Architectural and Archaeological 3D Models" *ISPRS – International Archives of the Photogrammetry, Remote Sensing and Spatial Information Sciences* 39 (Part B5) pp.273–278.
Liang, J., Gong, J. and Li, W. (2018) "Applications and Impacts of Google Earth: a Decadal Review (2006–2016)" *ISPRS Journal of Photogrammetry and Remote Sensing* 146 pp.91–107.
Menzel, J.R., Middelberg, S., Trettner, P., Jonas, B. and Kobbelt, L. (2016) "City Reconstruction and Visualization from Public Data Sources" *Proceedings of the Eurographics Workshop on Urban Data Modelling and Visualisation* Goslar, Germany: Eurographics Association, pp.79–85.
Musialski, P., Wonka, P., Aliaga, D.G., Wimmer, M., Van Gool, L. and Purgathofer, W. (2013) "A Survey of Urban Reconstruction" *Computer Graphics Forum* 32 (6) pp.146–177.
Nys, G.-A., Poux, F. and Billen, R. (2020) "CityJSON building generation from airborne LiDAR 3D point clouds" *ISPRS International Journal of Geo-Information*, 9, (521) pp.1–19.
Sui, D., Elwood, S. and Goodchild, M. (2013) *Crowdsourcing Geographic Knowledge: Volunteered Geographic Information (VGI) in Theory and Practice* Dordrecht: Springer.
Tarsha Kurdi, F. and Awrangjeb, M. (2020) "Automatic Evaluation and Improvement of Roof Segments for Modelling Missing Details Using Lidar Data" *International Journal of Remote Sensing* 41 (12) pp.4700–4723.

Templin, T. and Popielarczyk, D. (2020) "The Use of Low-cost Unmanned Aerial Vehicles in the Process of Building Models for Cultural Tourism, 3D Web and Augmented/Mixed Reality Applications" *Sensors (Switzerland)* 20 (19) pp.1–26.

Wang, R., Peethambaran, J. and Chen, D. (2018) "LiDAR Point Clouds to 3-D Urban Models : A Review" *IEEE Journal of Selected Topics in Applied Earth Observations and Remote Sensing* 11 (2) pp.606–627.

Xiao, J., Gerke, M. and Vosselman, G. (2012) "Building Extraction from Oblique Airborne Imagery Based on Robust Façade Detection" *ISPRS Journal of Photogrammetry and Remote Sensing* 68 (1) pp.56–68.

Xiong, B., Elberink, S.O. and Vosselman, G. (2016) "Footprint Map Partitioning Using Airborne Laser Scanning Data" *ISPRS Annals of the Photogrammetry, Remote Sensing and Spatial Information Sciences* 3 (3) pp.241–247.

26
UNDERGROUND MAPPING

Aurel Saracin

Exact knowledge of the positions and routes of underground pipes and cables, which constitute the utility networks in localities, has become a necessity in the processes of design and completion of any type of construction. Usually, utility networks in the localities are located under the street (Figure 26.1). Thus, when rehabilitating a street or boulevard, it is considered to avoid the destruction of these existing networks if they are kept in operation, and, when they are partially replaced for the economic efficiency of the work, it is necessary to know as precisely as possible the route and the depth at which they are located. (Figure 26.2).

In the last 25–30 years, the branch of engineering that concerns with the detection and mapping of underground network elements has developed significantly through the evolution of underground techniques and technologies. At the same time it has been necessary to adopt standards and rules for the installation of new underground utility networks (regulated depths for the installation of water, sewerage or gas pipes, regulated minimum distances between the routes of different types of pipes or buried cables, and so on. For example, when installing a new non-metallic pipe (gas, water or sewer), a metal strip can be glued or placed on the pipe in question before the ditch of the pipe is blocked (Figure 26.3). Subsequently (over the years) in order to establish the exact route of the pipe, a transmitter can be connected to this metal strip (in a manhole), and a receiver device set on the appropriate frequency is used on the surface (Saracin, 2017).

When installing a new pipe or cable, electronic markers can be buried that become active when they receive signals from the surface at a preset frequency at installation, depending on the type of utility network and the standards adopted (Figure 26.4 and Table 26.1).

The extension of subway routes in major cities involves excavations in areas of subway stations with surface exits, which require the bypasses of sewerage routes, water supply, gas, and other utility networks that are buried under the roads of large cities.

An efficient and accurate investigation of these underground utility networks is provided by ground penetrating radar (GPR) systems, which are based on modern microwave transmission and reception technologies, automatic data acquisition software and specialized software for processing and interpreting this information.

Underground mapping

Figure 26.1 Types of underground utilities

Figure 26.2 Cluster of cables and underground pipes

Figure 26.3 Underground marking of the route of a pipe with metallic tape

Figure 26.4 Electronic markers for underground utilities

Table 26.1 Network type and frequency

Network type	Frequency
Power lines	145.7 kHz
Telecommunication lines	121.6 kHz
Water pipes	169.8 kHz
Sewer pipes	101.4 kHz
Gas pipeline	83.0 kHz

The possibility of correlating GPR data with the data provided by modern satellite positioning systems or total stations can lead to obtaining high-quality products for 3D modelling of the studied area, which can be easily integrated into a GIS (geographic information system). For a modern cadastre, spatial information and textual data on underground utility networks are very useful, allowing better estimation of the market value of land and buildings, all of which can be integrated into GIS. At the same time, utility operators can access, update and optimize the information using GIS, depending on the requirements of the users of these urban underground utility networks.

Operating principles of GPR systems

As a rule, most underground utility networks do not have underground markings to allow easy detection. Current legislation also provides for the protection of archaeological sites if they are found underground in areas where foundation pits for constructions are made. For these and related situations, it is recommended to use GPR systems, which are compact and relatively easy to use, and emit electromagnetic pulses in the ground, measuring the time (t) from emission to reception of the signal reflected by the existing object(s) in the soil at different depths. Knowing the wave propagation speed through the ground will make it possible to determine the depth to the target object in the ground (Manacorda *et al.*, 2009) (Figure 26.5). If the radar pulses cross different materials on their way to the 'target' the speed of these pulses will change, depending on the physical-chemical properties of the materials they traverse. Thus, materials with different dielectric properties will lead to different reflection and refraction rates of the respective waves.

Different antennas allow the emission of radar signals of different frequencies. Higher frequencies allow a higher resolution at the expense of depth of investigation. Conversely,

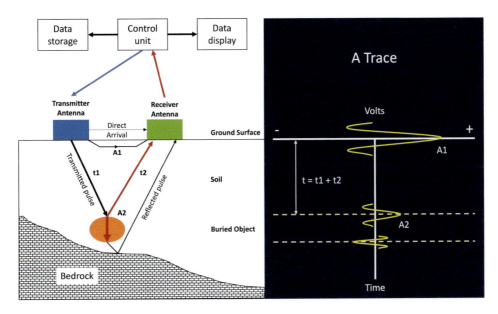

Figure 26.5 Principle of wave propagation in the soil

when the frequency is low, the depth of investigation increases, but the resolution decreases (Jaw and Hashim, 2013). The depth of investigation (d) is also controlled by the nature of the surveyed lands whose absorption characteristics (of radar waves) are variable:

$$d = \frac{t}{2} \times V_m$$

$$V_m = \frac{c}{\sqrt{\varepsilon_r}}$$

where c – is the speed of wave propagation in free space (3 × 108 m/s), V_m – is the speed of wave propagation in the ground, and ε_r – is the relative permittivity of the ground.

The type of GPR antenna and frequency chosen depends on the field of use, the depth at which we want to investigate or the location of the objects of interest and the size of these objects (Table 26.2). Depending on the requirements of a project, we can use simpler or more complex GPR systems. Thus, for the investigation of small areas, a single antenna system

Table 26.2 Type of GPR antenna and frequency

Working frequency (MHz)	Scanning depth (m)	Minimum size of scanned objects (cm)
80	14	25
160	10	15
450	6	6
760	3	4
1200	1.5	1.5
1600	1	1
2300	0.6	0.5

can be used on a stroller directed by an operator, with a paint device to mark the detected positions of a particular underground network and then determine the positions of the paint signs with the help of a total station, relative to a local coordinate system (EuroGPR Association, 1997) (Figure 26.6).

To cover larger investigation areas in a single pass, we can configure GPR systems with several antennas, on which is mounted global navigation satellite system (GNSS) receivers, allowing all utility networks on a street to be investigated, and after post-processing of the taken data and the correlation with the GNSS data, georeferencing of the underground utility networks within the digital maps of the locality can also be done (Figure 26.7).

Ground penetrating radar systems towed by vehicles (e.g., motorcycle, car, all-terrain vehicle) can be used for large projects at the level of a locality for example, and can travel on streets with speeds of 15–40 km/h. These can be STREAM (Subsurface Tomographic Radar Equipment for Assets Mapping) with antenna arrays, some with vertical polarization set on a certain frequency and others with horizontal polarization set on another frequency, so that it can detect underground utilities arranged parallel to the travel

Figure 26.6 GPR systems with a single antenna

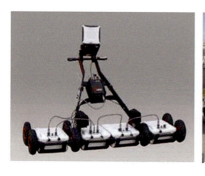

Figure 26.7 GPR systems with antenna arrays

direction and utilities arranged transversely on the travel direction and at different depths (Simi et al., 2010).

Such GPR systems can acquire multi-channel information (depending on the frequency setting in the array and the polarization of these antennas) that can be correlated and compatible with the positioning information provided by GNSS systems and a digital map previously loaded on the data acquisition computer.

Acquisition and viewing data in the field

For GPR mapping, the use of other sources of information, from existing maps and plans to the visual location of the layout of the networks, their depth, diameter and direction through the manholes, is necessary to have a complete set of geospatial information.

It is possible to estimate the size of the area to be investigated, the characteristics of the targets, the design of preliminary scanning routes ensuring accessibility (traffic interruption permits, the presence of parked cars, permission to access private premises, and so on), select the appropriate frequency, executing calibration paths and adjust the system for soil parameters (electrical conductivity and relative dielectric permittivity), after which the actual data acquisition is performed. The data recorded with the GPR system can be viewed in situ on the field computer and can resemble that shown in Figures 26.8–26.10.

In the case of street intersections, where there are several pipeline routes at different depths, it is recommended to design an investigation grid and the simultaneous use of low- and high-frequency antennas (Bell, 2014).

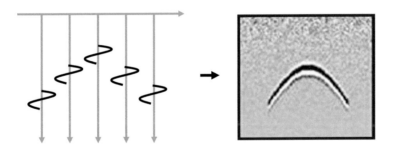

Figure 26.8 Locating a pipe perpendicular to the investigation route

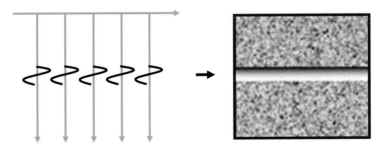

Figure 26.9 Locating a pipeline along the investigation route

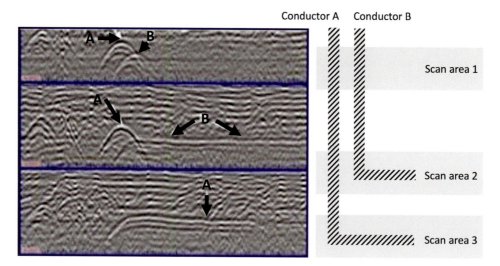

Figure 26.10 Underground investigation tapes

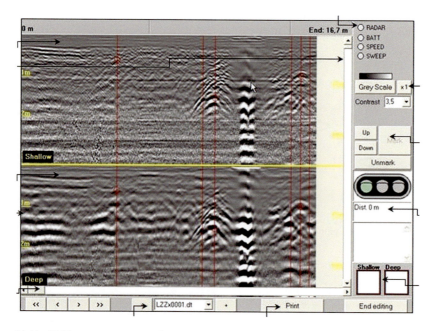

Figure 26.11 Field computer screen of a GPR system with two antennas

During data acquisition, both radar representations can be viewed on a field computer screen. When both antennas operate (e.g., 600 MHz and 200 MHz), the two sections are separated by a horizontal yellow line. To mark the presence of a parabola on the screen during purchase (generated by the presence of a pipe or cable), the GPR system moves backwards in the opposite direction of purchase, which temporarily suspends recording and produces a vertical red line on the display (Figure 26.11).

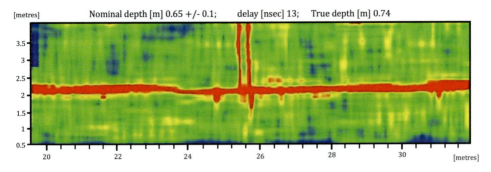

Figure 26.12 Pipelines visualization into a tomography (in red colour)

It is clear that these issues can have an impact on productivity with GPR, taking into account the effort of further processing and analysis of a large dataset, which can take at least twice as long as the time required to acquire data in the field. A recent innovation in field data acquisition and analysis software is the hyperbolic representation of reflected waves with automatic marking of their peaks (Falorni *et al.*, 2004; Windsor *et al.*, 2003), and the position of pipes can be easily recognized to obtain a 2D map of the scanned area. The Hyperbolus Automatic Detection Algorithm (HADA) allows a visualization of GPR data similar to tomography, and is easy to understand for the operator, who can change settings to obtain optimal tomography (Figure 26.12). The operator can mark the pipes detected on this tomography, since they are visible along with the intersections of underground networks on a single screenshot, thus reducing the time required for data analysis. These software tools can increase productivity by up to four times.

GPR data georeferencing

For small areas of georadar investigation, such as civil construction sites, where there is cartographic support of existing details, the surface elements of underground utility networks (manholes, vents, and so on) can be identified on the ground, establishing the positions of the georadar investigation profiles depending on the positions of the visible elements of the underground networks or in relation to the edges of the streets from around the investigated area (Figure 26.13).

The route positions of the underground utility networks identified with the GPR system can be marked on the ground. Subsequently, with a total station, the coordinates of the marked points can be determined, and they can be reported in CAD (computer-aided design) software and can be joined to the coded points corresponding to the routes of the underground utility networks, updating the initially existing cartographic support (Figure 26.14).

Additional information about these underground utilities (type of network and administrator or owner, manufacturing material, estimated diameter or thickness, depth, and so on), collected from the field, will be added later as attributes in the context of a GIS or if the information is to be integrated in BIM (Building Information Management).

A multi-channel GPR system, for assessing the condition of a road pavement or for locating an underground utility network under the road, is usually mounted on a vehicle that collects GPR data on a strip of 1.5–1.6 m, at speeds of 15–40 km/h. This high rate of data

Figure 26.13   Establishing the investigation routes according to the surface elements

Figure 26.14   Ways of marking on the ground the detected positions of the underground utilities

Figure 26.15   Direct georeferencing of GPR records using GPS

collection allows us to reduce data acquisition costs, to minimize the impact with car and pedestrian traffic.

This approach requires direct and precise georeferencing, in order to allow the correct overlapping of GPR investigation bands with the digital support of street routes. To this end, the GPR system is paired with a high-performance inertial GNSS system. The use of a dual-antenna GNSS receiver with an inertial measurement unit (IMU) and a distance measuring indicator (DMI) (odometer) provides complementary position solutions for the direct georeferencing of GPR record positions with centimetre accuracy (HEXAGON – IDS GeoRadar, 1980) (Figure 26.15). For complex projects, the GPR system can be integrated into a Mobile Mapping system that can create a 3D model of the locality, which also includes the underground utility networks (Figure 26.16).

Underground mapping

Figure 26.16 Integrated Mobile Mapping + GPR solutions

GPR data acquisition software

Manufacturers of GPR systems have also developed software for field data acquisition, with user-friendly interfaces. The operator can make some settings before starting work and can view the downloaded data stream. Most GPR systems require an initial calibration at the investigation location, when the antenna and software recognize the predominant soil characteristics and adjust the displayed images for optimal data collection (Giannopoulos, 2005).

This whole process is related to the emission, reflection and reception of signals, all of which are influenced by the propagation environment, interference and noise. A signal pulse is sufficient and the hardware-software interaction will recognize the duration of the reflected pulse and will adequately amplify the received signal to be correctly interpreted regarding the type of utilities investigated and the depth at which they are located (Figure 26.17).

The operator can enter their own values for the dielectric constant and can enter their own attenuation values in dB/m or can introduce advanced filtering algorithms, such as low-pass and high-pass FIR filters. Similar to focusing a camera, during the investigation the operator can increase or decrease the gain over a certain depth range, after identifying the

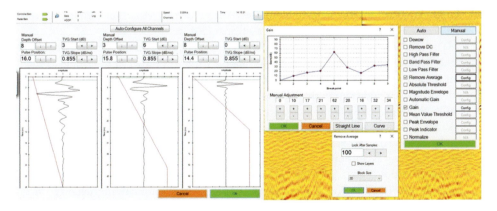

Figure 26.17 System calibration and parameter setting GPR investigation

Figure 26.18 Marking on the screen the types of underground networks according to the standards

target. The reading of the target position, depth and X and Y coordinates, can be displayed by marking on the screen, both in real time in the field and after the post-processing of data in the office (Figure 26.18).

The colours selected for the markings of the utility types are chosen according to the codes of the industrial standards of representation of the underground utilities on the map.

Some of the most common GPR data acquisition software packages include:

- MALÅ Object Mapper, created to interpret utility detection GPR data with MALÅ GPR equipment, is appreciated by its users for the simplicity and effectiveness of providing results. The software also supports Global Positioning System (GPS) data from a GPS receiver installed with the GPR system (GUIDELINE GEO AB (PUBL) MALÅ & ABEM, 2012).

  Object Mapper works with two main windows, a data view (GPR profiles) and a site map. If GPS positioning data are available and connected to the Internet, the sitemap will also include a Google Maps image of the project and the positions of the data profiles. Viewing the data and identifying each GPR profile on the sitemap will then help to answer whether markers align in a predictable way (Figure 26.19).

- ImpulseRadar ViewPoint is an Android data acquisition application (App) that controls the ImpulseRadar, CrossOver and PinPointR antennas. When installed on a suitable Android device (smartphone or tablet), ViewPoint allows wireless connectivity between the device and the GPR antenna (ImpulseRadar, 2015).

  The modern and intuitive user interface is designed to maximize productivity in the field, with features and functionalities to support daily workflows. It is possible to carry out linear or multi-line projects with full control over profile lengths and distances to suit the actual conditions of the location and objectives.

  Combining these systems with internal GPS and Google Map integration allows the possibility of seeing profile locations, survey paths and assigned markers in real-time, along with radar data (Figure 26.20).

- EKKO_Project is an optional modular software package. The basic package provides file organization, quick viewing of project information in a single window and automatic reporting. EKKO_Project works with.gpz files automatically exported from current GPR systems. A built-in basic structure allows all GPR data and auxiliary

Underground mapping

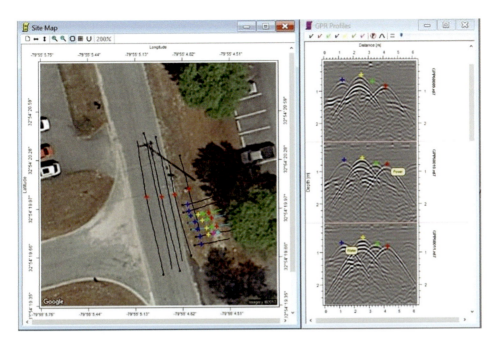

Figure 26.19 Screen windows in Object Mapper software

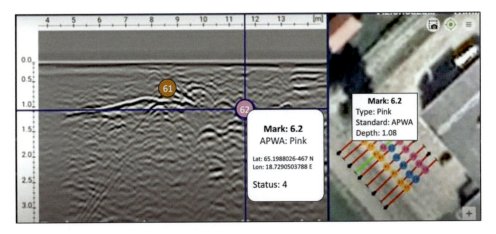

Figure 26.20 Screen windows from the ViewPoint application

files (GPS, topography, photos, notes, and so on) to be housed in a single.gpz file for a given project. A project file can contain a single GPR line or several hundred lines from a complex survey. Any available project information can be easily added to the. gpz file, providing a single point for all survey information (Sensors & Software Inc., 1988). The main user interface consists of a series of windows to provide an overview of the project (Figure 26.21). The optional modules of the EKKO_Project software package are: Interpretation Module, SliceView Module, SliceView 3D Visualization,

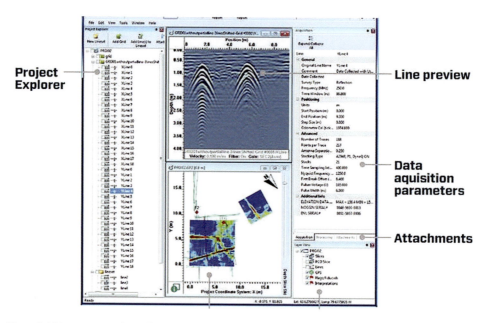

Figure 26.21 User interface of the EKKO Project software

Processing Module, Bridge Deck Condition Report Module and Pavement Structure Report Module.
- GPR-SLICE is a complete and unique software package that has a direct interface with raw field data, processes raw data and creates a multitude of possible image presentations. GPR-SLICE is for processing and creating images from reflection data that penetrate the ground (Ground Penetrating Radar Imaging Software, 1994). GPR-SLICE integrates radargram profiles taken from a survey grid to produce from the slices, maps of ground radar anomalies. Sliced maps can effectively show the size, shape, location, and depth of buried targets. Subtle reflective features that do not distinguish between adjacent radar diagram profiles can be imagined and detected on time slice maps created from 3D volume datasets. GPR-SLICE is compatible with all major GPR equipment manufacturers.

GPR-SLICE manages all the data densities surveyed to create the most comprehensive underground images. Many researchers have migrated to software due to the professional quality images that can be created in this software. It is designed to eliminate line noise and artifacts that show profile directions and pixelation noise from incomplete soil sampling. GPR-SLICE images help interpreters extract hidden information contained in noisy radar charts that would otherwise be lost and never revealed in raw data.

Civil engineers monitoring the infrastructure can use GPR-SLICE features that provide software support for 4D GPR monitoring. GPR monitoring is fast becoming an important market to reoccupy sites and determine the differences that have occurred over time. GPR monitoring of railway systems and other critical infrastructures is observed by the engineering community, because recently, the change in volume deduced by 4D measurements has become available, especially through GPR investigations (Figures 26.22–26.24).

Underground mapping

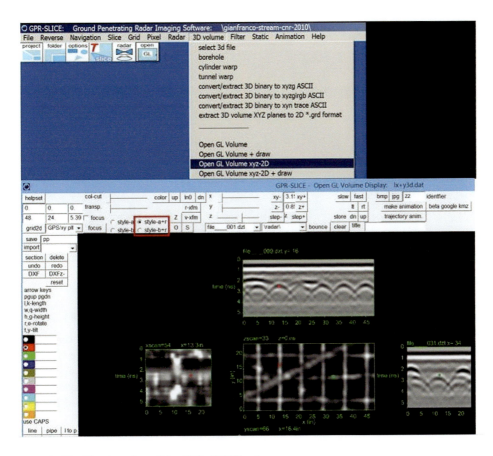

Figure 26.22   User interface of the GPR–SLICE software

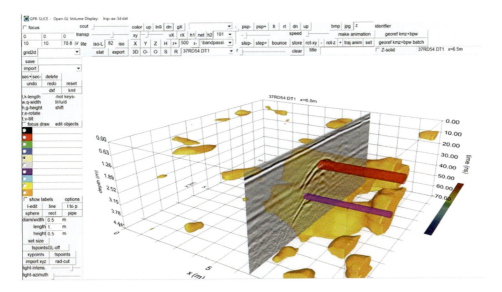

Figure 26.23   View sections with GPR–SLICE software

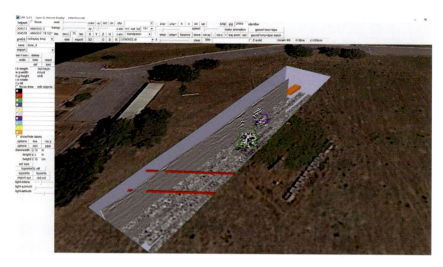

Figure 26.24   3D view with GPR-SLICE software

Software for post-processing GPR data

To obtain more accurate information regarding underground objects, further processing and better correlation with the georeferencing data of GPR records are needed. Some GPR data post-processing software has existed since the 1990s. These include the aforementioned GPR-SLICE software, which is compatible with all major manufacturers of GPR systems, including: Mala Guideline Geo, Impulse Radar, IDS GeoRadar, US Radar, 3D of Norway, Geoscanners, Leica, GeoTec Russia, 3D Radar Geoscope and Terravision USA.

The following are among the operations that can be performed with GPR-SLICE in post-processing (Ground Penetrating Radar Imaging Software, 1994) (Figure 26.25):

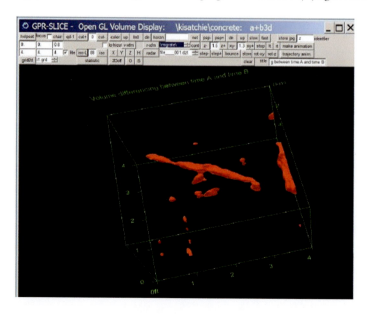

Figure 26.25   3D Visualization menu from GPR-SLICE software

- Analysis, filtering and decimation of GPS positioning data.
- Numerous operations can be performed in the menu to view radar charts, including: segmenting recordings into smaller portions, identifying incomplete radar charts or uncovered areas, combining two or more radar charts, noise filtering, identifying positions with reflection anomalies of the signal, identification of migration zones of the reflected signal and application of filters or signal amplifiers in unclear portions, correlation and integration of GPS data with GPR data and displaying a topographic map in the background. After all these steps, a correction can be made to establish where the soil surface is to proceed to mark objects of interest in the soil (if this was not done directly in the stage of taking data from the field).
- A 3D Visualization menu, to build semi-automatic objects in the basement (pipes, cables, gaps, and so on) with the possibility of exporting in DXF or other formats, or continue processing with the combination of two or more sets of data collected at different times in order to monitor the movements or changes in the position of underground elements.
- 3D analysis of data volumes can be performed by viewing any vertical, horizontal or oblique sections, with soil or soil elements allowing the entry of attributes regarding the material or dimensional characteristics, as well as the depth at which it is located, as information of primary interest.

All these results of processing with this software can be structured in layers, transformed into vector formats and then exported in a GIS.

Radar Studio – Post Processing GPR Software is a complete software suite for viewing and processing radar data. It has an extensively customizable user interface, with intuitive commands and functions. Users can turn the collected data into dynamic images that can be shared and stored for later retrieval, while providing many data processing options to improve their interpretation. Automatic settings allow users to instantly get usable, real-time updated 3D results to generate underground utilities in a digitally displayed map (USRADAR INC. SUBSURFACE IMAGING SYSTEMS, 2019). The final 3D modelling product can be exported in a wide variety of formats, from automatically generated reports to advanced CAD, GIS or Google Earth files.

For studies and visualization, this software can generate sections in different directions to find detailed characteristics of the underground objects researched, highlighting the importance and accuracy of data in complex projects (Figure 26.26).

GeoPointer X is a GPR post-processing software that combines in a single system the many aspects of a survey. These are the coordinates that come from the GPS receiver, the position on the map, images of the surroundings and many other useful information. While using GeoPointer X it is possible to choose which of the data windows to display. All data windows can be opened or only a few of them; this will not be reflected in the amount of data collected or in their quality. The only window that needs to be opened to store the data is to view the photo data (GEOSCANNERS AB Geophysical Survey Solutions, 1989). GeoPointer X brings new possibilities to the way geophysical data are analysed. Not only the data from the geophysical instrument are available, but also a whole set of data about the location of the survey. Using GeoPointer X in conjunction with trigger devices provides a real way to link ground-based radar data to a database that describes the site where the data were collected. In other words, by combining the geophysical data of the survey and the data stored in GeoPointer X it is possible to see both sides of the surface. Post analysis of the data will reveal the features available below the surface, along with an image of the surroundings above the surface. This related data gives us a better idea of the survey site and

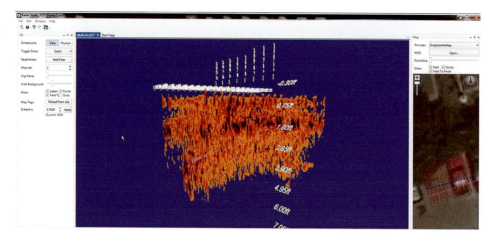

Figure 26.26 3D view with Radar Studio – Post Processing GPR Software

Figure 26.27 Information viewing windows with GeoPointer X Software

its position. The information collected may be used when writing reports or detailed descriptions of a site. There are several options for controlling the software. All data are stored in a single database, which makes the processing of the acquired information a simpler and more harmonious task (Figure 26.27).

Underground mapping

Figure 26.28 Instant 3D viewing with IQMaps application

IQMaps is a new post-processing software application for advanced GPR data analysis that provides a fast interface between user and GPR data. This allows processing, management and 3D visualization in real time. IQMaps offers a step-by-step approach to guide the user in performing the best and fastest data analysis with a customizable processing and analysis tool, both for experienced users and those who are not qualified for utility mapping, and for archaeological and environmental research and extending mapping to larger projects (Figure 26.28).

Leica DX Office Vision allows even inexperienced CAD users to digitally map the detected underground utilities and obtain professional 3D CAD drawings and view the detected underground utilities in a simple way. The intuitive interface allows users to filter, select, identify and annotate localized targets. With DX Office Vision, further processing for all GPR data does not require additional software (TOP GEOCART Leica Geosystems, 2013). The software was created to reduce post-processing time and eliminate all unnecessary steps to convert data or choose parameters. The software guides the user to create a reliable 3D map of the detected underground utilities (Figures 26.29 and 26.30).

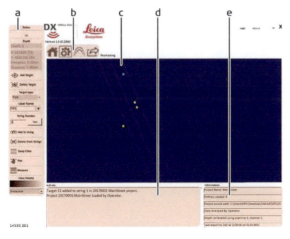

a) Main toolbar
b) Menu bar
c) Positioning plan view
d) Area for logging messages
e) Area for project Information

Figure 26.29 DX Office Vision software user interface

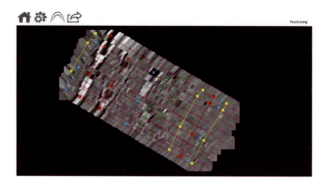

Figure 26.30 Post-processing window with DX Office Vision software

Utility networks and GIS

Utility, pipeline and telecommunications companies have used GIS to manage data on network assets for decades. After all, knowing the locations of assets, their conditions and their connections is fundamental to their management. But GIS data were often kept from the people who needed it most: field workers, directors, managers, service technicians and accountants. In this new world of limited resources and increasingly complex networks, companies need a new wave of GIS-based network management, with more functionality, increased flexibility and better access. Hence, Esri is building its 'Utility Network' on the ArcGIS platform (Esri Romania, 1999). The 'Utility Network' allows users to create, manage and share data on electrical, water, wastewater, gas, district heating and telecommunications assets. Esri will provide basic data models in some of these areas. Moreover, as part of the ArcGIS platform, the 'Utility Network' is available on any device, anytime, anywhere, and can easily manage billions of data items. Workers will be able to edit and track the path of network data in the field. The technology can meet modern data needs and facilitate the secure exchange of information with those who need it (Figures 26.31 and 26.32).

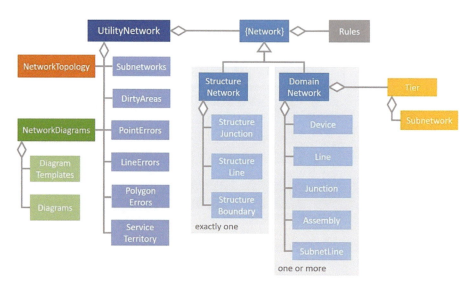

Figure 26.31 Information structure in 'Utility Network'

Underground mapping

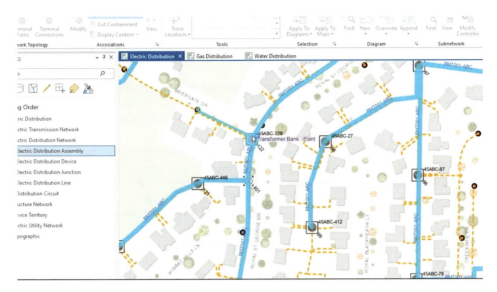

Figure 26.32 Graphical view in ArcGIS 'Utility Network'

The 'Utility Network' module of the ArcGIS platform offers:

- Flexibility: users can edit data through Web services.
- Active connectivity: users can exploit the behaviour of networks and can simulate the reaction of the type of utilitarian network if consumers or optimization elements are added.
- Attachment: users can model attachments to network elements, such as how a electrical transformer or telephone cable is attached to a pole.
- Quality: the technology uses built-in, industry-standard rules that protect against data entry errors by users. Editors will not be allowed to connect a high voltage cable to a low voltage one, for example.
- Modelling: users can specify where electricity, gas or water sources are located to facilitate network simulation. They can also model devices that have many connection points, such as complex switches and valves.
- Increased productivity: users have access to shortcuts, templates and simplified workflows in the software.
- View: Utility Network supports 3D.

The 'Utility Network' module works directly in ArcGIS and provides tools for extracting data from existing data sources and transforms the data to automatically conform to industry standard data models in the utility network.

Bentley Systems is a global provider of integrated digital software and digital cloud services for advanced design, construction and infrastructure, smart cities and a more connected regional infrastructure (Bentley Systems, 1984). 4D engineering representations enable new collaborative digital workflows to serve public works management and development planners and engineers for the benefit of city owners and administrators. Digital cloud services provide an intuitive and captivating 4D environment that converges the digital context and digital components of the infrastructure over asset life cycles. For infrastructure professionals, BIM and GIS are effective 4D digital components. Bentley Systems has extensive experience in geospatial technology (GIS) and BIM, municipal infrastructure applications, reality modelling and data management. Reality modelling using ContextCapture and Orbit GT to obtain 3D models provides a precise and real engineering context to support planning, design, construction and operations. Users of Bentley's open source applications (OpenBuildings, OpenSite, OpenRoads, OpenRail, OpenUtilities) can use this digital context to model new, improved buildings, roads, transit systems, tunnels, bridges, utilities and more.

Hence, 4D digital information becomes a common index, without the need to change data formats from existing source systems, including digital terrain models (DTM), images and GIS sources. Engineering models (from any BIM software) of buildings, streets, transit systems, utilities and other city infrastructure, both surface and basement, are semantically aligned and georeferenced to enhance the richness and relevance of digital data over time.

For the sustainability and resilience of cities, surface and basement surveys can be combined with engineering data to ensure asset performance, resilience and sustainability of investments over time. Using Bentley's open simulation applications during the life cycles of built assets, seismic resistance (STAAD), evacuation of people from subway stations, stadiums and other public places (LEGION and CUBE) can be assessed, the impact of events in case of floods can be determined (OpenFlows FLOOD), and we can be ensured the adequacy of the underground conditions for urban projects (PLAXIS, SoilVision) (Figure 26.33).

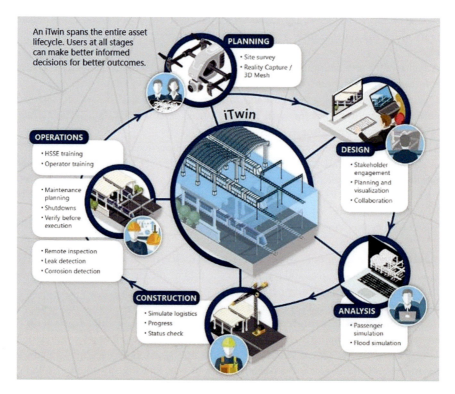

Figure 26.33 Geospatial data management with Bentley Systems

Towards a reliable map of underground utilities

A reliable map can be defined as a map with a sufficient amount of information and a high degree of quality to support urban planners, land managers and underground utility development engineers in their daily work practices. The most important characteristics of the quality of an underground utility map are (1) accuracy, (2) timeliness and (3) completeness.

The accuracy of spatial information is the degree of closeness with which the data match the real world. Accuracy is affected by inaccurate measurements, data ambiguity (e.g., depth values of buried items relative to a variable reference), completeness of data (e.g., omission of a vertical position) and data veracity (e.g., map features labelled 'How it is built').

Actuality is the degree to which the data on the map is updated and reflects the current state of the real world. In addition, for land planning and management purposes, it should include information on future projects related to underground utilities, planned works and diversions and ongoing works.

Data integrity or completeness is the degree to which all utilities that exist in the real world are represented on the map. A complete map contains all types of underground utilities, including abandoned utilities alongside functional ones (Figure 26.34).

A reliable map of underground utilities will contribute to an intelligent, resilient and sustainable system and will allow professionals (designers, administrators or plant engineers) to be informed and make effective decisions. Topographers and cartographers have the ability to produce accurate and reliable 3D information about underground utilities and can correctly integrate this data into GIS.

We need a digital image of the underground that covers many dimensions, from infrastructure to technical qualities and resources, making the underground a crucial realm in the city's sustainable and responsive future. The conversion of GIS data into augmented reality (AR), mixed reality (MR) and virtual reality (VR) is relevant in many fields, as well as in the field of underground utility networks. Augmented reality in real time to display underground utility infrastructure helps us quickly locate underground assets, using regular mobile devices or holographic displays, such as HoloLens.

The 'vGIS Utilities' application, designed by Esri GIS for municipalities and utility companies, turns traditional GIS data into augmented reality displays and holograms over what

Figure 26.34 Example map with all underground utilities

you see with the naked eye on the spot, where you have work to do. Water pipes, sewer systems, gases and power lines appear in your field of vision as a natural extension of the real world (vGIS Inc., 2020).

During large infrastructure projects, vGIS Utilities can help infrastructure project managers and field workers become aware of all underground infrastructure, preventing errors, delays and cost overruns. This app can help emergency response teams work quickly to identify underground infrastructure that should be avoided or stopped before any emergency excavation when responding to accidents, floods, water pipe ruptures and other critical incidents. vGIS amplifies the power of Microsoft HoloLens to visualize CAD and GIS data as holograms to create a natural extension of the real world. 'vGIS' integrates natively with the GIS and CAD formats commonly used in Esri ArcGIS, Bentley iModelHub, Bing Maps, Esri Basemaps, Google Maps, Open Street Maps and other services to build stunning holographic images, and the data automatically appears as holograms 3D in your field of vision (Figure 26.35).

Another commonly used application is 'AugView', which is both a mobile GIS, allowing users to view and edit field asset data, and an augmented reality application that allows users to view underground objects that they would not see normally (Augview Limited, 2012). With AugView, anyone, even those less experienced, can visit a completely unknown place and, in a few minutes, without digging a single hole, can see the underground assets, who owns them, customers who are connected to them, and, most importantly, where they are (Figure 26.36).

AugView software on a smartphone or tablet connects to one or more GIS Web servers over a secure connection, and can request map, text or 3D view data. Asset information (cables, pipes, and so on) can be edited directly then by the user in the field, and the data stored on the server are updated immediately and without any additional administrative costs. Augmented reality technology is used to display a 3D model of assets at the top of the video stream, allowing users to 'see' objects on the device that can be hidden in the real world. AugView software offers the new generation of employees a product that meets their demand for innovative and intuitive tools, enabling them to quickly increase productivity and efficiency.

Figure 26.35 AR view with 'vGIS Utilities' application

Underground mapping

Figure 26.36 Viewing underground utilities on the tablet with 'AugView'

Figure 26.37 Trimble 'SiteVision'

A recent solution is Trimble 'SiteVision' (Figure 26.37), which allows users to inspect, plan and report, then quickly extract data for viewing and sharing – without looking for blueprints or drawings. If a 3D model is accessed, the user can simply tap and view the attributes to confirm the integrity of the database, improve security and schedule work on site. You can virtually design ditches or pits on your smartphone and then transmit them to the digging machine equipped with Trimble Earthworks technology, avoiding the destruction of other types of utility networks in the area (Trimble Inc., 1978). During all this time, photos, measurements and textual reports can be taken, in different stages of the intervention, which are sent in real time to the project managers in the office. Trimble SiteVision uses high-precision augmented reality to maximize data value, increase communication, save money, improve security and increase productivity.

Clearly, there are many other applications of this kind and only some have been introduced above. The collaboration of access to databases, online communication systems, precise real-time positioning systems and the contribution of software manufacturers leads to the creation and use of augmented reality products of underground utility maps that are easy to use.

References

Augview Limited (2012) "AugView Augmented Technology" Available at: https://www.augview.net/ (Accessed: 1st November 2020).

Bell, M. (2014) "Strategic Network Level Mapping of Underground Assets Using Ground Penetrating Radar" *FIG Congress, Engaging the Challenges –- Enhancing the Relevance* Kuala Lumpur, Malaysia, 16th–21st June.

Bentley Systems (1984) "Utilities and Communications Networks" Available at: https://www.bentley.com/en/products/product-line/utilities-and-communications-networks-software (Accessed: 1st November 2020).

Esri Romania (1999) "ArcGIS Utility Network" Available at: https://www.esri.com/en-us/arcgis/products/arcgis-utility-network/overview (Accessed: 1st November 2020).

EuroGPR Association (1997) "Introduction to GPR" Available at: http://www.eurogpr.org/vn2/index.php/introduction-to-gpr/ (Accessed: 1st November 2020).

Falorni, P., Capineri, L., Masotti, L. and Pinelli, G. (2004) "3-D Imaging of Buried Utilities by Features Estimation of Hyperbolic Diffraction Patterns in Radar Scans" *GPR 2004 Conference Proceedings Volume 1* Delft, 21st–24th June DOI: 10.1109/ICGPR.2004.180015.

GEOSCANNERS AB Geophysical Survey Solutions (1989) "Post-Processing Software/GeoPointer X" Available at: http://www.geoscanners.es/geopointerx.htm (Accessed: November 2020).

Giannopoulos, A. (2005) "Modelling Ground Penetrating Radar by GprMax" *Construction and Building Materials* 19 pp.755–762 DOI: 10.1016/j.conbuildmat.2005.06.007.

Ground Penetrating Radar Imaging Software (1994) "GPR-SLICE V7.MT" Available at: https://www.gpr-survey.com/ (Accessed: 1st November 2020).

GUIDELINE GEO AB (PUBL) MALÅ & ABEM (2012) "Utility Mapping Software – MALÅ Object Mapper 2018" Available at: https://www.guidelinegeo.com/product/object-mapper-2018/ (Accessed: 1st November 2020).

HEXAGON - IDS GeoRadar (1980) "Utility Detection for Avoidance and Mapping" Available at: https://idsgeoradar.com/applications/utility-detection (Accessed: 1st November 2020).

ImpulseRadar (2015) "ViewPoint App" Available at: https://impulseradargpr.com/viewpoint/ (Accessed: 1st November 2020).

Jaw, S.W. and Hashim M. (2013) "Locational Accuracy of Underground Utility Mapping Using Ground Penetrating Radar" *Tunnelling and Underground Space Technology* 35 pp.20–29.

Manacorda, G., Simi, A. and Miniati, M. (2009) "Mapping Underground Assets with Fully Innovative GPR Hardware and Software Tools" *The North American Society (NASTT) and the International Society for Trenchless Technology (ISTT) International No-Dig Show* Toronto, Canada, 29th March–3rd April, pp.D5041–D5049.

Porsani, J.P., Slob, E., Lima, R.S. and Leite D.N. (2010) "Comparing Detection and Location Performance of Perpendicular and Parallel Broadside GPR Antenna Orientation" *Journal of Applied Geophysics* 70 pp.1–8 DOI: 10.1016/j.jappgeo.2009.12.002.

Saracin, A. (2017) "Using Georadar Systems for Mapping Underground Utility Networks" *Procedia Engineering* 209 pp.216–223.

Sensors & Software Inc. (1988) "EKKO_Project™ GPR Software" Available at: https://www.sensoft.ca/products/ekko-project/overview/ (Accessed: 1st November 2020).

Simi, A., Manacorda, G., Miniati, M., Bracciali, S. and Buonaccorsi, A. (2010) "Underground Asset Mapping with Dualfrequency Dual-polarized GPR Massive Array" *Proceedings of the 13th International Conference on Ground-Penetrating Radar* Lecce, Italy, 21st–25th June, pp.1001–1005.

TOP GEOCART Leica Geosystems (2013) "Leica DX Office Vision" Available at: http://www.top-geocart.ro/software/leica-dx-office-vision_213.html (Accessed: 1st November 2020).

Trimble Inc. (1978) "Trimble SiteVision Software" Available at: https://sitevision.trimble.com/utilities/ (Accessed: 1st November 2020).

USRADAR INC. SUBSURFACE IMAGING SYSTEMS (2019) "Radar Studio – Post Processing GPR Software" Available at: https://usradar.com/gpr-software/radar-studio/ (Accessed: 1st November 2020).

vGIS Inc. (2020) "Locate Infrastructure Faster and Prevent Errors" Available at: https://www.vgis.io/esri-augmented-reality-gis-ar-for-utilities-municipalities-locate-and-municipal-service-companies/ (Accessed: 1st November 2020).

Windsor, C., Capineri, L. and Falorni, P. (2003) "The Classification of Buried Pipes from Radar Scans" *Insight – Non-Destructive Testing and Condition Monitoring* 45 (12) pp.817–821.

27
GEOSPATIAL TECHNOLOGY AND FOOD SECURITY
Forging a four-dimensional partnership

Hillary J. Shaw

Fuzzy four-dimensional food deserts

Our world is four-dimensional and fuzzy. We often represent this world with two-dimensional maps, on paper or computer screen, but topography, the third, vertical, landscape dimension matters, especially to those without access to private motor transport accessing retail facilities. In the long narrow settlements of the South Wales valleys, for example, some houses are 200 metres above the main road where the food stores and bus route are, making access to food problematical for some older residents. Transporting healthier foods home, which have less calories per kilogram than obesogenic ready meals, is difficult for less affluent car-less households, especially if they cannot afford a taxi and lack good Internet access, or are not served by supermarket deliveries. The fourth dimension is time; no landscape is static, and people's access capabilities change over time. Access to shops may improve as broadband is installed, or decline for individuals as they age, or shops close. In less-developed countries, topography contributes to food insecurity as it affects rainfall, transport, military operations, and the path of locust swarms. Meanwhile all these factors can change rapidly over time, as can the economic factors that underlie many famines.

Computer-generated maps are also ultimately four-dimensional. The core elements of GIS mapping are points and lines enclosing polygons; however, these polygons are assigned values; for example, % obese within a census tract or % within a region on incomes below US $1.90 per day, the International Poverty Line. These values change over time, giving the fourth dimension. However, the real world, and people within it, are fuzzy and analogue; attributes that geospatial technology and GIS, inherently binary in nature, have found hard to capture. In 2003, for example, Smith *et al.* asked 'Do Mountains Exist'? The issue for Smith and Mark (2003) was that an observer walking north from Dhaka to Kathmandu would perceive no sudden break where the Ganges Plain give way to the Himalayas, only a gradual increase in hilliness; but GIS mapping demands a binary boundary between 'plains' and 'mountains'. The occurrence of fuzzy fractal landforms such as coastlines has been tackled by Kainz (2007: 4, 20) illustrating how GIS mapping and IT categorization can accommodate non-binary categorizations such as 'tall' or 'short', when many people are somewhat tallish or shortish. The fuzziness problem that Smith and Mark (2003) encountered is also important when considering 'access', as the desert conundrum illustrates. Most people have a concept of 'desert', perhaps as an arid region with sand dunes, maybe some camels and a mirage of an oasis; but now ask, how big does such an area have to be to count as a desert? 1,000 kilometres across, 10 kilometres, 100 metres? At this point most people resort to human-capability oriented answers such as 'it's only a desert if it's too big to be crossed on foot in a day'. However, a fit person with a water bottle could cross a larger tract than a child or elderly person; throw in variables such as do they have a map, compass, survival skills, and the answer grows fuzzier still. The concept of 'desert', as somewhere devoid of the means to support human life, is significant for food security because areas without adequate access (for whom?) have become so-called 'food deserts'.

The use of 'desert' to signify an area where many residents experience difficulty in accessing healthy food (although they will almost certainly have access to some food) originates with a 1996 study of deprivation in Glasgow, Scotland, by the Low Income Project Team (Shaw, 2014: 105); a local resident used the phrase 'food desert' to communicate their perception of the area as lacking in good-quality reasonably priced food retailing. The term was seized upon by the media and academia, and many districts, generally urban and mostly located in the UK and North America, were identified as locales where less-affluent residents might have problems accessing a healthy diet. In 1997 Tessa Jowell, then UK Health Minister, suggested that 500 metres was a reasonable maximum distance for people without private transport to carry food back from shops (Wrigley, 2002: 2034), but 'access to food' has remained a fuzzy concept because it depends on multiple factors unique to each household and individual, and these factors can change over time. Access to food and to healthy food (often different issues) can be classified, concerning food security in both global South and North, into three types of access: 'ability', 'assets', and 'attitude' (Shaw, 2006). In order, these three As relate to:

1. 'Ability' or physical access to food; can the consumer physically get to the food (are they disabled, is there a rural bus service), and carry it home (possibly difficult in hilly areas), or can they access a delivery service. Some disabled cannot physically open food packaging. Physical access can change over time if a new supermarket opens or the village store closes. In some wealthier English villages, residents in their 50s may say they do not want a local shop or bus service, as this keeps the village exclusive, and deters the poor and car-less from moving in; however, some decades later these same villagers may be disabled, unable to drive to a distant supermarket. In the Global South, physical access to food may diminish over a period of months if drought begins to set in, or over days if a locust swarm devours the crops. Peace can become civil war within hours, disrupting travel and food trade.

2. 'Assets' or ability to purchase food. Can the consumer buy food, or afford the delivery fee or the costs of a computer and the Internet? In some deprived UK locales, affording storage space for bulk food, a large freezer, is an issue, without which these poor pay more per food item than the better-off do. In the Global South, conflict will reduce earning abilities just as food prices rise through crop destruction; as the Indian economist Amartya Sen pointed out, this is a 'failure of exchange entitlements' (Sen, 1976). Food still being produced within the famine area may well be exported to districts where wealthier consumers can pay more for it. As with conflict itself, food prices can change within hours, up or down.

3. 'Attitude' or the extent of food knowledge possessed by the consumer. To access food that is physically available and affordable, the consumer still has to know where this

food is located, also how to prepare it to render it edible. In the Global North, states of mind such as fear (of crime for pensioners, or of crossing main roads with children for mothers) may deter access to shops; similar fear may operate in conflict zones. In 2020/2021 fear of COVID-19 has led to many UK consumers ordering food deliveries rather than venturing out to shops. Lack of motivation to eat healthily is another barrier state of mind; many urban consumers know how to prepare healthy food, can afford it, and can physically access it, but are unwilling to make time to cook it. They simply eat less healthy takeaways or microwave ready meals instead.

These three As are all fuzzy attributes, difficult to map. How does one map food knowledge, which may increase over time or even diminish (dementia)? Two adjacent households may possess very different financial capabilities, or one may have a car, the other not. One household may consist of fit adults who can walk for kilometres, next door a single older person who cannot carry food even 100 metres. One should also allow for 'disposable carrying capacity' a concept similar to disposable income. All shoppers on foot have certain large non-food items they must buy such as toilet rolls and pet food if they have a companion animal. Fresh healthy fruit and vegetables are calorie-poor per kilogram compared to obesogenic sugary fatty oven-ready meals (as well as being generally more expensive), and consumers with limited carrying capacity, perhaps using a bus to transport several bags of shopping home, will have limited capability to transport such fresh produce.

In the Global South, it may rain in one valley but not the next. Battlefield front lines can shift within hours and control of port cities can be a crucial tactic to starve the opposing civilian population; in the ongoing Yemen conflict the Saudi Coalition sought to capture the port of Hodeida, to cause famine in the Houthi-held northwest. However, it is important to map food insecurity, both current and potential, in order to ameliorate it, and monitor any progress at famine relief. Otherwise, food aid may not arrive where it is most needed. This may be for political reasons; in 1969 the Nigerian Government blocked food aid to the breakaway Ibo region, whilst 15 years earlier US President Eisenhower signed Public Law 480; the Agricultural Trade Development and Assistance Act of 1954. This debarred US grain exports from Communist-leaning countries, and was used as a threat against President Shastri of India, who was then pursuing a Leftist Nehru-inspired land redistribution programme and had expressed disapproval of the US bombing of Vietnam.

In the Global North, households can fall in and out of being in a 'food desert' situation, for example as local fresh fruit and vegetable food outlets open or close, or as household finances diminish due to recession or government austerity programmes. Again, the COVID-19 pandemic has thrown many people out of work (partly ameliorated by furlough), and this has impacted heavily on poorer households. In the long run, poor diet can stunt children's learning abilities, and raise obesity rates, and for many physiological reasons, obesity is much harder to reverse than to incur. In many Global South countries, obesity is rising and can even co-exist with hunger as it transitions from being a disease of affluence to one of poverty. In Brazil, Egypt, Mexico, and many other low or middle-income states, the wealthy have become less obese as they adopt healthier (but costly) lower meat, higher vegetable diets; meanwhile the poor are more exposed to the penetration of cheap obesogenic meals from fast food outlets such as McDonald's. In Brazil in 1975 low-income women were four times as likely to be underweight as obese, whilst in 1997 low-income women were 40% more likely to be obese than underweight; in 1975 the poorest quartile of Brazilian women were half as likely to be obese as the wealthiest quartile, but in 1997 this poorest quartile were 10% more likely to be obese than the wealthiest quartile (Power and Schulkin, 2009: 129–30). Sometime around 1990 the wealthy-poor orientation of obesity in Brazil had flipped.

The many factors of food security: what has Geospatial Technology ever done for us?

The 1996 World Food Summit (WFS) defined food security as *'when all people at all times have physical and economic access to sufficient, safe and nutritious food to meet their dietary needs and food preferences for an active and healthy life'* (Report of the Committee on World Food Security, 2012: 5, para. 17). The four fuzzy dimensions of food access or food deserts are all captured here: Food access 'at all times' (it can change), physical, economic, and also attitudinal (dietary needs, preferences, cultural norms). More problematical, in terms of measurement of progress towards these goals in the Global North, but as also increasingly in Global South countries such as Brazil, is the *healthy life* stipulation. Once, food security meant surviving, with just a few monarchs being recorded as attaining morbid obesity, and food insecurity meant starvation. Now we have an excess of global calories. However, a 'health premium' remains. The excess cost of a diet rich in fresh healthy fruit and vegetables can be over 50% more than a cheap salty fatty sugary diet in less affluent parts of the UK (Shaw 2014: 119). The response of a Leeds city council member in 2002 to a food survey, in a deprived part of east Leeds, was telling. Within sight of the glittering glass towers of central Leeds, then the UK's second financial centre after London, he said almost matter-of-factly *'Oh yes, we have malnourished kids here'*. Not necessarily undernourished, they were probably eating their way to obesity, but they were deficient in minerals and vitamins.

Geospatial technology and the maps we draft with it can therefore create a varied toolkit, in the Global South and North, to anticipate, monitor, and ameliorate food insecurity. Like the glut of global calories, we have a surfeit of monitoring technology, data sources, real-time mapping, and analysis to pinpoint areas of food insecurity. What is often lacking is the political will and the money. We have satellites that can track and often even anticipate many natural disasters from floods and droughts to locust swarms and crop diseases. Sophisticated models exist for predicting political instability in advance (Goldstone *et al.*, 2010); similarly economic crises can be anticipated. Both food aid in and the flow of information out of a food insecure or famine hit region can be obstructed by national governments, such as in the recent Tigray famine; we have political tools to tackle this problem, but sadly no geospatial technology can facilitate their effective use, nor improve the flow of 'information on the ground' out. The data can be in continuous format (rainfall) or provincial or national data units (economics, civil conflict risk), and in real time (locust swarms, food prices) or static (annual crop production, GDP). Areal data units may change over time. For example, there is little older data for the territory of South Sudan, a state created in 2011, and the states within India have undergone significant and frequent boundary modifications. Combining these formats can be a major challenge, as discussed below.

In the Global North a similarly wide panoply of geospatial tools exists to research the alternative version of food insecurity that exists there; that is, sufficient calories but inadequate nutrition for a healthy life. Physical access to food has been mapped on a data-intensive street-distance methodology from every shop outwards. One should not overlook other outlets besides supermarkets, as corner shops, ethnic-oriented food stores, and street markets also sell fresh produce. One might also include barriers such as hills and main roads, and include access nodes such as bus stops. In a Manhattan-type grid street plan the actual distance A-B would vary as a ratio of the direct line distance between 1.0 (direct street) and 1.4 (diagonal traverse). For most UK urban street patterns, the direct-actual distance is fairly consistent at c. 1.2, meaning that we can divide the map into squares and assume that 400 metres direct equates to a 500 walk (Shaw, 2014: 111–113). For rural areas a different methodology would be necessary, based on villages as nodes, connected or not by buses, and with a grocery store present or not, also allowing for off-village farm shops. The urban 1.2 assumption is prone to two errors; one trivial, the other not. The trivial error is that some cul-de-sac roads and roads on the edge of a town or near a barrier such as a railway line have a distance ratio above 1.2; however, shops are usually located on more connected central street networks and the cul-de-sac issue will only affect the last few metres of consumer travel, so we could ignore this issue if mapping a whole city. The non-trivial problem is that grid squares introduce artificial boundaries and erroneous distance measures can arise if a consumer or shop is near a boundary. This is the MAUP, or Modifiable Areal Unit, Problem issue (Flowerdew, 2009) discussed below.

To investigate financial (asset) and knowledge or cultural (attitudinal) barriers to food security one could use various direct survey tools such as food diaries, questionnaires, and shop price surveys. These are data-intensive methods and assume we already know where the food insecure households are and wish to discover *how* they cope with food issues. Otherwise, we might expect to find food insecurity in, for example, affluent west Sutton Coldfield in Birmingham (UK), where many households are over 1,000 metres from any food retailer. To discover where food insecurity occurs, and what socio-economic factors may contribute to it, we have a wide range of geospatial census tools for many developed countries, where census tracts have been mapped along economic and demographic data. We can use proxies for attitudinal data that is not mapped; for example, cigarette smoking as an inverse measure for health awareness or for cash shortages (Whelan *et al.*, 2002: 2009). Census tract socio-demographic data is readily and cheaply available and large areas can be explored and compared using Excel or SPSS analysis tools. However, there are three main problems to this library-based methodology.

MAUP, spurious correlations, and other geospatial mapping problems

Firstly, census tract data is often only collected every ten years, and then takes a further one to two years to be published. However, households may move every five to seven years on average, and food retail accessibility changes even more rapidly as shops open or close. Mitigating this issue, neighbourhoods often retain their socio-economic status, affluent or deprived, for decades (although gentrification, for example, can cause rapid change). Booth's poverty maps of London from the 1890s show spatial patterns of wealth not unfamiliar to modern eyes (LSE, n.d.). Geospatial mapping is no substitute for 'eyes on the ground' but does enable us to explore 'invisible' socio-economic mechanisms, and then to pinpoint where these mechanisms operate most acutely.

Secondly, there are multiple methodologies for creating census areas, which can complicate efforts to compare food insecurity internationally. Most census tracts have semi-standardized populations, and cover smaller spatial areas in densely populated cities, but France's INSEE, for example, has a special ZUS (*Zone Urbaine Sensibles*), along with ZRUs (*Zones Redynamisation Urbaine*) and ZFUs (*Zones Franches Urbaines*), for deprived urban neighbourhoods. The UK's Office for National Statistics provides local data by MSOA (Middle Layer Super Output Area), which are nested within Local Authority Areas and contain around 7,000 people each. Subject to these constraints a key feature of MSOAs is that (unlike INSEE or US census tracts) they are drawn to be as socially homogenous as possible, resulting in some odd-looking 'gerrymandered' shapes for some tracts. MSOAs may also change over time as urban development changes local populations.

Thirdly there is the MAUP issue. Whenever we map anything by grid or census tract, we are imposing invisible lines; people do not respect borough boundaries when accessing food. It is good practice, when mapping food access for a district within a city, to include outlets within a buffer zone of 500 metres outside the district boundary, a typical maximum walking distance, to cater for consumers close to the border. Census tract boundaries can create spurious correlations as Figure 27.1 illustrates. Each digit 1 represents the location of a low-income household living near the railway with a particular social attribute, e.g., obese children. Each digit 2 represents the location of a high-income household, living near the river; for example, with a child with private music tuition. In 2010 Railbridge City had ten census tracts, and it is likely that a (spurious) positive correlation would appear between the incidence of child obesity and the prevalence of private music tuition. In 2020 the population increased, and there are now 15 census tracts, and it is likely that the positive correlation has persisted or even increased. Often a spurious correlation is exposed over time as it declines towards zero (regression to the mean), but here this regression may not occur. An

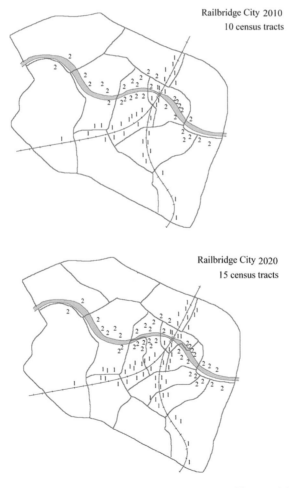

Figure 27.1 Spurious correlation, persistent over time, between variables 1 and 2

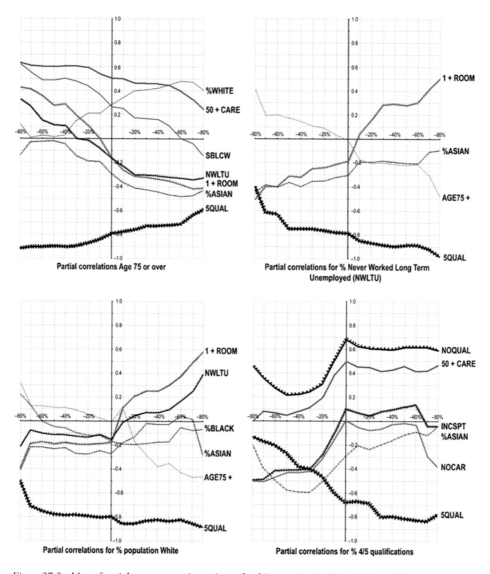

Figure 27.2 Use of social transects to investigate food insecurity in Birmingham, UK

unwary researcher might even conclude that the sedentary nature of music lessons is causing child obesity. The issue is of course the number of census tracts with very few ones or twos; this might not be such an issue with British MSOAs as they are designed to separate out different socio-demographic areas.

As statisticians soon learn, 'correlation is not causation'. If A correlates with B, we still do not know if (i) A has caused B, (ii) B has caused A, (iii) another factor C has caused A and B, or (iv) there is no link between A and B (as with Railbridge City above). Human intuition can help; more obese people in an area is unlikely to reduce the area of parks for exercise, but the reverse could easily be true. In 2020 the COVID-19 pandemic appeared to disproportionately affect BAME people, but was this largely because many BAME workers were in low-pay jobs that could not be done from home, and they often lived in larger more overcrowded households, in inner city areas, using public transport not cars to get to work; all factors that could spread COVID-19. Wilson (2014) has provided amusing examples of totally spurious correlations, but one might reason that his apparent correlation between the declining divorce rate in Maine and the decline in margarine consumption, 2000–2009, could have a factual basis given the region's ageing population, who are more health conscious and tend to be wealthier. Would this result in eating more butter, alongside a calmer less stressed attitude to family life? Apparent correlations can never be ruled in or out without more research and statistical analysis.

Is it a true correlation? The use of social transects in researching obesity in Birmingham, UK

With a dataset of 100+ census tracts, a single correlation figure between variables A and B can be turned into a social transect. Take the tract data and arrange them in value order of a third variable C (C can in fact be A or B). Having calculated the correlation figure for the entire dataset, now remove the 10% of the sample with the highest values of C and recalculate the correlation. Repeat having removed the next 10% highest values of C and so on until one is down to the last 10%; all the time the proportion of tracts with low C values is rising. Now repeat whilst removing the 10% lowest values of C and so on towards a rump selection with very high C values. We now have 19 correlation values for A-B (lowest 10%, 20%... highest 20%, 10%), and have effectively created a transect across our sample area from low to high C values. Plotting the 19 partial sample values may produce a 'random walk' from 10% lowest to 10% highest C, meaning there may be no correspondence between A and B as C varies. However, a consistent shift in correlation value, perhaps even changing sign from −ve to +ve, as C goes from high to low, suggests there is an effect on A and B mediated by C. It also helps to map variables A, B, C across the research area, perhaps using a choropleth map, and to visit the area to see what the characteristics of high/low areas of the variables look like.

This methodology was applied to the city of Birmingham, UK, where mapping obesity levels across 285 MSOA tracts suggested a 'crescent of obesity' in the poorer northwest of the conurbation, in peripheral low-income housing estates away from town centres (Shaw 2012). Figure 27.2 reproduces the social transect results for this city. Correlation values −1 to +1 are plotted on the vertical axis and the 'variable C', whose highest/lowest values were being removed by 10% steps, is labelled below each graph set. The transect line is labelled by a variable B, which is being correlated in all instances with obesity levels (variable A). There is only room for a brief summary of the results here. In the top-right graph, variable %ASIAN changes little in correlation value to obesity as one moves across from areas of high unemployment (left) to low unemployment (to more affluent areas) on the right, suggesting that local unemployment levels make little difference to any relationship between ethnic diversity and obesity levels. Similarly (lower right) having no qualifications (NOQUAL) does not impact much differently on obesity in areas with more degree-level qualified residents compared to fewer of them (variable 5QUAL). However, (top right) living in overcrowded (low income) accommodation in a low unemployment area (RHS) does create a markedly higher risk of obesity; conversely being degree-qualified markedly shifts the obesity rate downwards. On the LHS of this top-right graph set, in poorer areas of high unemployment, being degree-qualified has very little downwards effect on obesity risk (the correlation value approaches zero), yet intriguingly living in overcrowded (poorer) accommodation is actually associated with lower obesity rates than not living in such accommodation, in less affluent areas.

This last observation, along with the obesity choropleth of Birmingham showing the highest rates in peripheral low status housing estates, suggests that unemployed people in poorer central areas have the time to obtain cheap healthy fresh produce from town centre street markets if they live near one, whereas working people in these poorer town centre areas may still be on low wages and get so little financial advantage from being employed to buy fresh produce from regular shops (where it costs more than cheap obesogenic food). These 'working poor' may find their work hours preclude access to cheap street markets. However, on the peripheral housing estates, the unemployed and car-less have no such dietary advantage over the working population. In more affluent areas lower income consumers will find fewer cheap street markets (there may be expensive farmers markets), and there may be more upmarket (costly) supermarkets and fewer cheap discount chains.

The Birmingham data suggests that promoting local street markets, and perhaps encouraging more flexible work hours, could improve access to healthy food for the less affluent. Innovative analytical methods of data analysis as gathered using a range of geospatial technologies can facilitate multi-faceted comparisons of variables and their effect on food security. In Global South nations, we can combine real-time geospatial sensing with socio-demographic data and algorithms to not only monitor progress in relieving food insecurity but to actually predict it before it occurs. Food aid workers in the Global South, and social institutions in the North, can then fine-tune their responses to food insecurity to achieve the maximum improvements using fewest resources. At least, they can suggest the best routes for ameliorating food insecurity. The fact is that no geospatial technology can by itself effect a change in human behaviour; that remains the subject of political persuasion and nudge theory, which is an entirely different subject than this chapter.

References

Committee on World Food Security (2012) "Report of the Committee on World Food Security, 39th Session, Rome" Available at: http://www.fao.org/3/MD776E/MD776E.pdf (Accessed: 3rd March 2021).

Flowerdew, R. (2009) "Understanding the Modifiable Areal Unit Problem" Available at: https://www.ed.ac.uk/files/imports/fileManager/RFlowerdew_Slides.pdf (Accessed: 5th March 2021).

Goldstone J.A., Bates, R.H., Epstein, D.L., Gurr, T.R., Lustik, M.B., Marshall, M.G., Ulfelder, J. and Woodward, M. (2010) "A Global Model for Forecasting Political Instability" *American Journal of Political Science* 54 (1) pp.190–208 DOI: 10.1111/j.1540-5907.2009.00426.x.

Kainz, W. (2007) "Fuzzy Logic and GIS" Available at: https://homepage.univie.ac.at/wolfgang.kainz/Lehrveranstaltungen/ESRI_Fuzzy_Logic/File_2_Kainz_Text.pdf (Accessed: 5th March 2021).

LSE (n.d.) "Charles Booth's London" Available at: https://booth.lse.ac.uk/map/12/-0.0927/51.5053/100/0 (Accessed 5th March 2021).

Power, M.L. and Schulkin, J. (2009) *"The Evolution of Obesity"* Baltimore, MD: John Hopkins University Press.

Sen, A. (1976) "Famines as Failures of Exchange Entitlements" *Economic and Political Weekly* 11(31/33) pp.1273–1280.

Shaw, H.J. (2006) "Food Deserts: Towards the Development of a Classification" *Geografiska Annaler* 88 (B2) pp.231–248 DOI: 10.1111/j.0435-3684.2006.00217.x.

Shaw, H.J. (2012) "Food Access Diet and Health in the UK: An Empirical Study of Birmingham" *British Food Journal* 114 (4) pp.598–616 DOI: 10.1108/00070701211219577.

Shaw, H.J. (2014) *The Consuming Geographies of Food* Abingdon: Routledge.

Smith, B. and Mark, D. (2003) "Do Mountains Exist? Towards an Ontology of Landforms" *Environment and Planning B: Urban Analytics and City Science* 30(3) pp.411–428 DOI: 10.1068/b12821.

Whelan, A., Wrigley, N., Warm, D. and Cannings, E. (2002) "Life in a Food Desert" *Urban Studies* 39 (11) pp.2083–2100 DOI: 10.1080/0042098022000011371.

Wilson, M. (2014) "Spurious Correlations" Available at: http://www.tylervigen.com/spurious-correlations (Accessed: 5th March 2021).

Wrigley, N. (2002) "Food Deserts in British Cities: Policy Context and Research Priorities" *Urban Studies* 39 (11) pp.2029–2040 DOI: 10.1080/0042098022000011344.

28
THE PAST, PRESENT AND FUTURE OF TECHNOLOGIES FOR IMPROVED WATER MANAGEMENT

Leonardo Alfonso

The need for better water management

Water was the first thing ancient civilizations looked for when deciding the place to settle, and it is the first thing we now look for when we explore space. Life depends on the availability of water, which is both finite and vulnerable. It is well known that out of all water on Earth, 97.5% belongs to seas and oceans and that what is available as fresh surface water in rivers, lakes and reservoirs is no more than a 0.007% of the total (Loucks et al., 2005). Every year, on average, this is roughly equivalent to only a 35 km-radius sphere of water!

Unfortunately, humans, a small proportion of the living entities in the planet, are responsible for making water availability challenging for all, triggering global changes such as population growth, deforestation, pollution, urbanization and climate change. It is therefore our responsibility to overcome these challenges and ensure we all can live in a sustainable manner. Arguably, the only way to respond to this situation is by strengthening science, applying and developing technologies that we create, and to keep creating new, innovative tools. One of them is Integrated Water Resources Management (IWRM), which has emerged as a response to this growing pressure on the water resources. It is defined as a process that promotes the coordinated development and management of water, land and related resources in order to maximize the resultant economic and social welfare in an equitable manner without compromising the sustainability of vital ecosystems (GWP, 2000). A base for proper IWRM practices, the development and implementation of ICT technologies is of critical importance.

Technologies to support water management

The water that exists in a natural resource system follows the hydrological cycle, which includes the processes of precipitation, evaporation, infiltration and runoff, among others. Ultimately, the water that can be used is the water that flows in rivers or that is stored in reservoirs, lakes or aquifers. Its amount can only be evaluated after water balance analyses, which can be very complex.

The appearance of computers in the 1960s triggered new methods of analysis (e.g., numerical modelling, based on the numerical solutions of partial differential equations), and

Past, present and future of water management

large engineering and infrastructure projects in the 1970s and 1980s justified vast investments in modelling, measurement devices and control and data acquisition systems. The synergy between numerical modelling and data collection and processing at the end of the 1980s gave birth to Hydroinformatics (Abbott 1991, 2008). Today, Hydroinformatics tools comprise a range of technologies for data collection, numerical and data-driven modelling, optimization, artificial intelligence, machine learning, real-time control and decision support systems (Abrahart *et al.*, 2008), which follows the so-called flow of information Data – Models – Knowledge – Decisions.

Figure 28.1 depicts this logical framework and applies it, in a very generic way, to the case of flood risk management. Data about the presence and flow of water is collected by means of both in-situ sensors and space-borne technologies that are arranged in monitoring networks in a given river basin. These data are then transferred via telecommunication protocols and made available in digital repositories for different users, including water authorities, scientists and practitioners. The available data are then transformed to information by means of data analysis methods, which vary from statistical analysis to numerical and data-driven modelling. Specifically, the role of modelling is to offer ways to understand, predict and forecast the states of a water system by replicating the reality in a simplified, digital way. The outcome of the modelling exercise is a gain in knowledge about the water system under consideration that allows us to understand cause-effect relations among multiple variables and visualize them in ways that can be comprehended by all stakeholders. Finally, this new knowledge and its visualization is used to support decision-making. In the particular case of Figure 28.1, these decisions are about land use planning, operation of reservoirs and the setup of early warning systems of flood events.

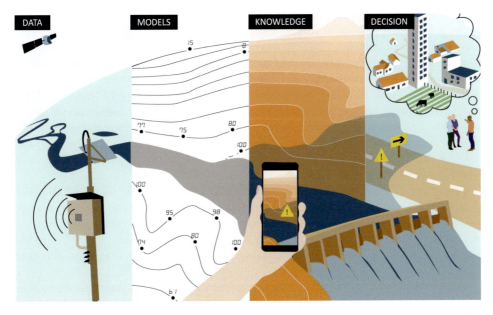

Figure 28.1 Infographics of the information flow in a Hydroinformatics system for the case of flood risk management. Inspired by the presentation of D. Solomatine, available at https://www.slideshare.net/tasstie/dimitri-solomatine-hydroinformatics

In the remainder of this chapter the flow of information will be used to illustrate the kinds of technologies that are used and applied to support water management.

Data

Technologies for data collection include the development of in-situ sensors and the use of Earth observation (EO) systems. Examples of innovative sensor developments include low-cost rain gauges and weather stations (van de Giesen *et al.*, 2014) and citizen science-based rainfall sensors (Alfonso *et al.*, 2015); in-situ sensors are typically used in combination with space-borne technologies, which have had important developments in the last decades. In the last years, citizen science has also been explored as a data collection source for improvement of hydrological and hydraulic modelling (Mazzoleni *et al.*, 2017) and river delta planning (Venturini *et al.*, 2019), among many other applications, most of which take advantage of mobile phone and smartphone technologies (Jonoski *et al.*, 2012).

Some of the available EO systems for the monitoring of the hydrological cycle include the estimation of variables such as rainfall, snowfall, evaporation, snow cover and density, elevation and extent of surface water, surface soil moisture, depth to groundwater, total groundwater storage, water quality, vegetation/land cover/irrigated area, vegetation stress and water vapour, in different spatial and temporal resolutions (McCabe *et al.*, 2017). An interesting example of the use of remote sensing as data source for water availability is presented by Pekel *et al.* (2016), who used more than 4 million scenes from Landsat 5, 7, and 8 acquired between March 1984 and 31st December 2019, and showed the probability of presence of water for the entire time period as well as for the periods 1984–1999 and 2000–2019 to evaluate change detection.

Future data collection technologies that will further improve water management include runoff measurements by the Surface Water and Ocean Topography (SWOT[1]) mission, planned for April 2021, deep soil moisture estimations by the Biomass[2] mission, launched in December 2022, and photosynthesis estimations by the Fluorescence Explorer (FLEX[3]) mission, planned for 2025.

Sensors are crucial, and they have and are being incorporated into every single device we use in our daily life. According to the Financial Times, the amount of data generated every year is estimated to be doubling, with a total expected of 44 zettabytes (10^{21} bytes) by 2020.[4] This advancement in sensors has created the age of Big Data (Butts-Wilmsmeyer *et al.*, 2020) and is leading the development of Internet of Things (IoT), the network of interconnected physical objects with embedded sensors, via the Internet. IoT applications in the water domain are already seen in sustainable water supply (Koo *et al.*, 2015) and early warning systems due to snowmelt (Fang *et al.*, 2015).

Apart from this traditional data, social networks appear as a new source of data with interesting potential to fill data gaps and promote knowledge discovery. Examples include the use of YouTube videos to gauge extreme floods (Le Boursicaud *et al.*, 2016), the use of Twitter to make probabilistic estimations of flood extent (Brouwer *et al.*, 2017) and the use of a combinations of platforms for collaborative flood warning and response (Nespeca *et al.*, 2018).

Digital technologies are needed to process and analyse these data and to produce information to make it available for decision-making. In the water sector, arguably, the most important technology in this regard is modelling, explained in the next section, which has been developing simultaneously with the increase of computational power, capacity of communication protocols via Internet and the evolution of mobile broadbands. However, as data increase, other digital technologies for data analytics are also needed, apart from storage and communication technologies.

Models and data analytics

A model is a simplified description of reality. Although models in the form of physical scale prototypes of a water system are still used, computer-based models are now more popular. Such models are implemented in computer programs that are designed to get input data, perform calculations and produce results in terms of output files, graphs, maps, tables, animations and other visualization techniques.

Models can be classified as physically based and data-driven models. The former uses laws of physics such as the Newton's laws of motion, energy conservation, mass continuity and conservation of momentum, among others, to describe in differential equations the mechanisms of water flow. The computational algorithms used to numerically solve these equations are implemented in software systems that can be used by experts to solve different study cases. The later, data-driven models rely on important volume of observed data and on data analytics methods to establish cause-effect relationships among variables. These methods include machine learning (learning from data using statistical methods that can be as simple as linear regressions or as complex as Artificial Intelligence) and data mining (finding new knowledge by means of classification methods and trend analysis).

Modelling has two primary goals, namely to improve understanding of a water system by reproducing past behaviour and analysing cause and effect, and to make predictions about the water system performance, due to alterations of the physical water system (e.g., river flow changes due to the building of a dam) or due to natural changes in the boundary conditions (e.g., behaviour of river water levels for different rainfall events in the catchment) (Price and Solomatine 2009).

Regarding the primary goal of modelling, some examples of process understanding using models, for very different water systems, are Jung *et al.* (2009), for algae growth; Fleckenstein *et al.* (2010), for surface-groundwater interaction; Mandel *et al.* (2015), for water quality evolution in water distribution networks; and Alonso Vicario *et al.* (2020), for human behaviour during flood evacuation. For the same water systems, examples of model applications to predict the possible effects of changes are Manning *et al.* (2019), for the appearance of algae bloom in lakes; Tran *et al.* (2019), for predicting groundwater recharge, hydraulic heads and river discharges; and Marquez Calvo *et al.* (2019), and Quintiliani *et al.* (2019) for predicting the effect of valves manipulation on water age in water distribution networks.

Artificial Intelligence (AI) is becoming a more and more popular term for the general public, due, among other reasons, to proliferation of news related to profiling of social media or internet browser users. However, the application of AI in the water sector has been explored for two decades. The reader is invited to check the comprehensive reviews of Chau (2006) for AI applications in water quality modelling, Nourani *et al.* (2014) in hydrology and Mosavi *et al.* (2018) for flood prediction. Other AI applications can be found in Solomatine and Ostfeld (2008) for urban water and Bhattacharya and Solomatine (2005) for hydraulics.

The possible combinations of IoT and AI are endless. For example, Figure 28.2 shows a prototype system in which machine learning is used to classify recordings of pipe noises at a household meter and detect anomalies (Seyoum *et al.*, 2017). The principle uses fingerprinting software to analyse the noise recordings and to feed a database. Reports can be then sent either to the users or to the water authorities.

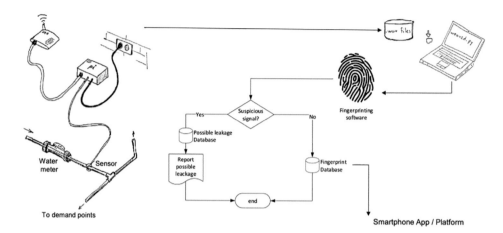

Figure 28.2 System to record pipe noises to be used for machine learning, to recognize anomalies in household water pipes

Case study

An example in which advanced technologies have recently being developed and applied in water management is related to smart water management for critical events, in particular leak reduction in water distribution networks (WDN). In this situation, a drop in pressure in the WDN at a specific location is detected, due to the increased discharge of water, which lost (i.e., not consumed). A direct consequence of this situation is the decay of the WDN capacity to deliver the base demand. This situation can be simulated with water distribution system models, e.g., EPANET (Rossman, 2000), by adding emitters at critical nodes that simulate flow through orifices, using an equivalent burst diameter. These critical nodes can be selected based on the age of the connected pipes, pressures and external stresses such as traffic.

In the framework of the EC H2020 NAIADES project, a methodology to generate critical events and simulate them using modelling and optimization approaches has been developed. The inputs consider historical data of flows and pressures for different statuses of the network, and the outputs consider the same data for a modified network with failures. A well-known drawback of AI methods is that it fully depends on the quality and quantity of historical data produced by sensors. This means it is probable that a critical event such as fire or a big pipe burst has not been observed in the historical records, making the AI unaware of the situation. For this reason, the method aims to complement the observed data with data generated via modelling procedures under a diverse set of critical scenarios (Figure 28.3).

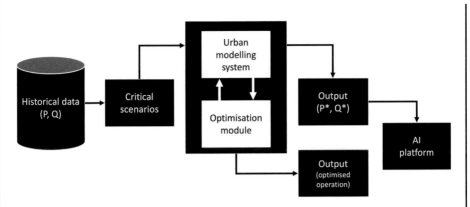

Figure 28.3 Methodology to generate data of critical scenarios to feed the AI platform. Following the methodology, Python scripts were developed to produce input files of pressure decay at different time frames, due to the sudden increase of water demand in all the nodes of the network

Table 28.1 Sample of the seven first records out of 6.2 million for the considered city

Time step (s)	Leak node ID	Leak factor	Reported node	Pressure (mcw)
0	wNode_355	1.5	wNode_355	7.09
3,600	wNode_355	1.5	wNode_355	7.13
7,200	wNode_355	1.5	wNode_355	7.29
10,800	wNode_355	1.5	wNode_355	7.26
14,400	wNode_355	1.5	wNode_355	6.94
18,000	wNode_355	1.5	wNode_355	6.82
21,600	wNode_355	1.5	wNode_355	6.67

As an initial test, the script was set to produce more than 6.2 million records for a district of a European city, for which the first seven are shown in Table 28.1, where time step refers to the elapsed time in seconds, from the 0:00h of simulation until 23:55. Moreover, leak node ID refers to the node in which an artificial increment in demand was assigned, by multiplying its current base flow demand by the leak factor (column 3), considered to be 1.5, 2 and 5. The reported node refers to the node for which the a new pressure is obtained after running the simulation with a change of demand in the leak node ID, by a leak factor. For example, 3,600s after the simulation starts (i.e., at 1:00am), an increment of the demand of node wNode_355 by a factor of 1.5 produces a pressure of 7.13m in node wNode_355 (the same node in this case; however, this value is reported for all nodes in the area).

Decision-making

As discussed above, data are collected and used in models and other data analytics procedures to produce outputs for decision-making. The performance of this process can be studied from the perspective of System Analysis, which includes the optimal functioning of individual components of the system and their interactions to achieve system's goals. This requires the integration of tools such as geographical information systems (GIS), optimization, Multi

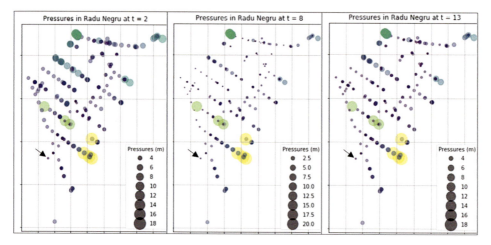

Figure 28.4 Model outputs for different times of the day, showing instant pressures for a critical leak event at node wNode_355 (arrow)

Criteria Analysis and Cost Benefit Analysis, which can form the basis of model-based Decision Support Systems (Jonoski and Seid, 2016).

Optimization, a key field, has also been actively applied in the water sector. Reviews of some applications can be found in Ahmad *et al.* (2014) for the case of reservoir operation and in Mala-Jetmarova *et al.* (2017) for design and operation of water distribution systems.

Due to the fact that uncertainty comes from different sources such as the environment, there are always uncertainties our models, especially in those associated with assumptions in optimization procedures, such as cost and benefit functions. The questions of whether to invest in extra information prior to making a decision (Alfonso *et al.*, 2016), and how to make water decisions under uncertainty (Ma *et al.*, 2020) are active fields of research.

Figure 28.4 shows the model outputs in terms of pressure at 2:00am, 8:00am and 1:00pm, for an increment of the demand due to leak in node wNode_355 (black arrow) by a factor of 1.5, for different times of the day. It can be appreciated that, as expected, pressures drop during the day as water is consumed. This information is currently being used to feed the AI platform using a bigger set of critical events that are being generated.

Future implications for society

Modern digital technology is offering interesting opportunities, not only because of the vast amount of data to be generated, but also because of the large number of available methods to analyse and use it for decision making. This, in line with the development of ICTs, will have implications for society and its relation with water. Continuous development of tools such as the Google Earth Engine, which not only gathers satellite images but also a collection of data analytics methods to be used, will facilitate the assessment of water availability.

Expected advances in the near future include the extended use of robotics for data gathering (already happening with high-resolution drones to complement remote sensing data), operation and control of water systems (improving current control algorithms). Decision Support Systems are likely to be impacted by improved virtual and augmented reality, coupling modelling techniques for design and training purposes. In terms of computational power, quantum computing appears to be promoting the development of new water modelling paradigms.

Finally, increased awareness of citizens in water, energy and health issues (COVID-19 as a palpable recent example) facilitated by data being freely available, easy to access and straightforward to visualize will probably increase the demand of advanced, easy-to-understand 3D animation holograms.

Notes

1 https://swot.jpl.nasa.gov/mission/overview/.
2 https://www.esa.int/Applications/Observing_the_Earth/The_Living_Planet_Programme/Earth_Explorers/Biomass.
3 https://earth.esa.int/web/guest/missions/esa-future-missions/flex.
4 https://www.ft.com/content/9f0a8838-fa25-11e7-9b32-d7d59aace167.

References

Abbott, M.B. (1991) *Hydroinformatics: Information Technology and the Aquatic Environment* Aldershot, Brookfield, WI: Avebury Technical.

Abbott, M.B. (2008) "Some Future Prospects in Hydroinformatics" In Abrahart, R.J., See, L.M. and Solomatine, D.P. (Eds) *Practical Hydroinformatics: Computational Intelligence and Technological Developments in Water Applications* Berlin, Heidelberg: Springer, pp.3–16.

Abrahart, R.J., See, L.M. and Solomatine, D.P. (Eds) (2008) *Practical Hydroinformatics: Computational Intelligence and Technological Developments in Water Applications* Berlin, Heidelberg: Springer.

Ahmad, A., El-Shafie, A., Mohd Razali, S.F. and Mohamad, Z.S. (2014) "Reservoir Optimization in Water Resources: A Review" *Water Resources Management* 28 (11) pp.3391–3405.

Alfonso, L., Chacón, J. and Peña-Castellanos, G. (2015) "Allowing Citizens to Effortlessly Become Rainfall Sensors" *36th IAHR World Congress* The Hague, The Netherlands.

Alfonso, L., Mukolwe, M.M. and Di Baldassarre, G. (2016) "Probabilistic Flood Maps to Support Decision-making: Mapping the Value of Information" *Water Resources Research* 52 (2) pp.1026–1043.

Alonso Vicario, S., Mazzoleni, M., Bhamidipati, S., Gharesifard, M., Ridolfi, E., Pandolfo, C. and Alfonso, L. (2020) "Unravelling the Influence of Human Behaviour on Reducing Casualties During Flood Evacuation" *Hydrological Sciences Journal* 65 (14) pp.2359–2375.

Bhattacharya, B. and Solomatine, D.P. (2005) "Neural Networks and M5 Model Trees in Modelling Water Level–Discharge Relationship" *Neurocomputing* 63 pp.381–396.

Brouwer, T., Eilander, D., van Loenen, A., Booij, M.J., Wijnberg, K.M., Verkade, J.S. and Wagemaker, J. (2017) "Probabilistic Flood Extent Estimates from Social Media Flood Observations" *Natural Hazards and Earth System Sciences* 17 (5) pp.735–747.

Butts-Wilmsmeyer, C. J., Rapp, S. and Guthrie, B. (2020) "The Technological Advancements That Enabled the Age of Big Data in the Environmental Sciences: A History and Future Directions" *Current Opinion in Environmental Science & Health* 18 pp.63–69.

Chau, K.W. (2006) "A Review on Integration of Artificial Intelligence into Water Quality Modelling" *Marine Pollution Bulletin* 52 (7) pp.726–733.

Fang, S., Xu, L., Zhu, Y., Liu, Y., Liu, Z., Pei, H., Yan, J. and Zhang, H. (2015) "An Integrated Information System for Snowmelt Flood Early-warning Based on Internet of Things" *Information Systems Frontiers* 17 (2) pp.321–335.

Fleckenstein, J.H., Krause, S., Hannah, D.M. and Boano, F. (2010) "Groundwater-surface water Interactions: New Methods and Models to Improve Understanding of Processes and Dynamics" *Advances in Water Resources* 33 (11) pp.1291–1295.

GWP (2000) *Water as a Social and Economic Good: How to Put the Principle into Practice* Stockholm: GWP.

Jonoski, A. and Seid, A.H. (2016) "Decision Support in Water Resources Planning and Management: the Nile Basin Decision Support System" In Papathanasiou, J., Ploskas, N. and Linden, I. (Eds) *Real-World Decision Support Systems: Case Studies*. Cham, Switzerland: Springer, pp.199–222.

Jonoski, A., Alfonso, L., Almoradie, A., Popescu, I., van Andel, S.J. and Vojinovic, Z. (2012) "Mobile Phone Applications in the Water Domain" *Environmental Engineering and Management Journal* 11 (5) pp.919–930.

Jung, N.-C., Popescu, I., Kelderman, P. Solomatine, D.P. and Price, R.K. (2009) "Application of Model Trees and Other Machine Learning Techniques for Algal Growth Prediction in Yongdam Reservoir, Republic of Korea" *Journal of Hydroinformatics* 12 (3) pp.262–274.

Koo, D., Piratla, K. and Matthews, C.J. (2015) "Towards Sustainable Water Supply: Schematic Development of Big Data Collection Using Internet of Things (IoT)" *Procedia Engineering* 118 pp.489–497.

Le Boursicaud, R., Pénard, L., Hauet, A., Thollet, F. and Le Coz, J. (2016) "Gauging Extreme Floods on YouTube: Application of LSPIV to Home Movies for the Post-Event Determination of Stream Discharges" *Hydrological Processes* 30 (1) pp.90–105.

Loucks, D.P., van Beek, E. and Stedinger, J.R. (2005) *Water Resources Systems Planning and Management* Paris: UNESCO and WL Delft Hydraulics.

Ma, Y., Li, Y.P., Huang, G.H. and Liu, Y.R. (2020) "Water-energy Nexus Under Uncertainty: Development of a Hierarchical Decision-Making Model" *Journal of Hydrology* 591 pp.125–297.

Mala-Jetmarova, H., Sultanova, N. and Savic, D. (2017) "Lost in Optimisation of Water Distribution Systems? A Literature Review of System Operation" *Environmental Modelling and Software* 93 pp.209–254.

Mandel, P., Maurel, M. and D. Chenu (2015) "Better Understanding of Water Quality Evolution in Water Distribution Networks Using Data Clustering" *Water Research* 87 pp.69–78.

Manning, N.F., Wang, Y-C., Long, C.M., Bertani, I., Sayers, M.J., Bosse, K.R., Shuchman, R.A. and Scavia, D. (2019) "Extending the Forecast Model: Predicting Western Lake Erie Harmful Algal Blooms at Multiple Spatial Scales" *Journal of Great Lakes Research* 45 (3) pp.587–595.

Marquez Calvo, O.O., Quintiliani, C., Alfonso, L., Di Cristo, C., Leopardi, A., Solomatine, D. and de Marinis, G. (2019) "Robust Optimization of Valve Management to Improve Water Quality in WDNs Under Demand Uncertainty" *Urban Water Journal* 15 (10) pp.943–952.

Mazzoleni, M., Verlaan, M., Alfonso, L., Monego, M., Norbiato, D., Ferri, M. and Solomatine, D.P. (2017) "Can Assimilation of Crowdsourced Data in Hydrological Modelling Improve Flood Prediction?" *Hydrology and Earth System Sciences* 21 (2) pp.839–861.

McCabe, M.F., Rodell, M., Alsdorf, D.E., Miralles, D.G., Uijlenhoet, R., Wagner, W., Lucieer, A., Houborg, R., Verhoest, N.E.C., Franz, T.E., Shi, J., Gao, H. and Wood, E.F. (2017) "The Future of Earth Observation in Hydrology" *Hydrology and Earth System Sciences* 21 (7) pp.3879–3914.

Mosavi, A., Ozturk, P. and Chau, K-W. (2018) "Flood Prediction Using Machine Learning Models: Literature Review" *Water* 10 (11) p.1536.

Nespeca, V., Comes, T. and Alfonso, L. (2018) "Information Sharing and Coordination in Collaborative Flood Warning and Response Systems" In Andersson, B., Johansson, B., Carlsson, S., Barry, C., Lang, M., Linger, H. and Schneider, C. (Eds) *Designing Digitalization (ISD2018 Proceedings)* Lund, Sweden: Lund University, pp.1–14 http://aisel.aisnet.org/isd2014/proceedings2018/Transforming/4

Nourani, V., Hosseini Baghanam, A., Adamowski J. and Kisi, O. (2014) "Applications of Hybrid Wavelet-artificial Intelligence Models in Hydrology: A Review" *Journal of Hydrology* 514 pp.358–377.

Pekel, J-F., Cottam, A., Gorelick, N. and Belward, A.S. (2016) "High-resolution Mapping of Global Surface Water and its Long-Term Changes" *Nature* 540 (7633) pp.418–422.

Price, R.K. and Solomatine, D.P. (2009) *A Brief Guide to Hydroinformatics* Delft: UNESCO-IHE.

Quintiliani, C., Marquez-Calvo, O., Alfonso, L., Di Cristo, C., Leopardi, A., Solomatine, D.P. and De Marinis, G. (2019) "Multiobjective Valve Management Optimization Formulations for Water Quality Enhancement in Water Distribution Networks" *Journal of Water Resources Planning and Management* 145 (12) 04019061.

Rossman, L.A. (2000) *EPANET 2 User's Manual* Cincinnati, OH: US Environmental Protection Agency.

Seyoum, S., Alfonso,L., v. Andel, S.J., Koole, W., Groenewegen, A. and van de Giesen, N. (2017) "A Shazam-like Household Water Leakage Detection Method" *Procedia Engineering* 186 pp.452–459.

Solomatine, D.P. and Ostfeld, A. (2008) "Data-Driven Modelling: Some Past Experiences and New Approaches" *Journal of Hydroinformatics* 10 (1) pp.3–22.

Tran, Q.Q., Willems, P. and Huysmans, M. (2019) "Coupling Catchment Runoff Models to Groundwater Flow Models in a Multi-Model Ensemble Approach for Improved Prediction of Groundwater Recharge, Hydraulic Heads and River Discharge" *Hydrogeology Journal* 27 (8) pp.3043–3061.

van de Giesen, N., Hut, R. and Selker, J. (2014) "The Trans-African Hydro-Meteorological Observatory (TAHMO)" *WIREs Water* 1 (4) pp.341–348.

Venturini, A.B., Assumpção, T.H., Popescu, I., Jonoski, A. and Solomatine, D.P. (2019) "Modelling Support to Citizen Observatories for Strategic Danube Delta Planning: Sontea-Fortuna Case Study" *Journal of Environmental Planning and Management* 62 (11) pp.1972–1989.

29
OCEAN MAPPING
Taxonomies of the fluid geospatial

Rupert Allan

The ocean is a site of exhilarating, disruptive, and unreliable narratives for both science and society. As a space that inspires action, its geospatial definition is intangible, yet its biodynamic agency is uncompromising. This dynamism has consistently confounded attempts to map the oceans, with early mapmakers acknowledging its unknowns with the words *Terra Incognita*. Their visualizations, from Ptolemy to Flamsteed, are products of socioeconomics, politics, and the cultural production of space (Paglen, 2009). Charts – like maps – are never 'value-free images' (Harley, 1988: 278), and although Mercator's projection of 1569 was designed for ocean navigation (its 'conformal' angles can be directly steered at a ship's helm), its distortion of area and therefore inaccurate spatial proportions can be misleading (Figure 29.1).

For the purpose of ocean mapping, the United States NOAA (National Oceanic and Atmospheric Administration) uses a polyconic projection for some of its localized charts. For aviation, gnomonic charts show global ('great circle') distances that are analogous to straight (Euclidean) lines between points on the curved surface of the Earth (Figure 29.2).

Designed to serve their civilizations, sea charts are also visually iconic. Their legends constitute knowledge validation and show how humans have had coded marine interests for millennia. The term 'Legend' is uncoincidental, and in technical inscription, these notorious codes have become practiced as Nautical Terms.

Medieval maps frequently represented sections of water with mythical or legendary aquatic creatures that may or may not have existed. Different map projections have diverse practical advantages that variously inform oceanic cartography, and the inability to transfer the surface of the spherical globe to flat paper creates a 'partisan' area. Spatial theorists such as Edward Soja have postulated socially-'interpellated' spaces as simultaneously real and imagined 'Thirdspace' (Soja, 1996; Turner and Davenport, 2005). Contextualized within contemporary digital society, this multiplicity of visualizations is reminiscent of postmodern productions of plural, augmented, or virtual reality/ies (Baudrillard, 1988; Pollock and Bal, 2010). Not unrelated to the cooperative traditions of maritime navigation, the benefits of digital Free Culture (Lessing, 2004) are pervasive in open source nautical software like OpenCPN (https://opencpn.org/) and OpenSeaMap (Calvi, 2015). Interestingly, these digitally enact a time-honoured, decentralized and post-national ethos.

Ocean navigators increasingly leverage satellite Internet technologies such as SpaceX. Despite the 'networked' character of these technologies, their commercially interested

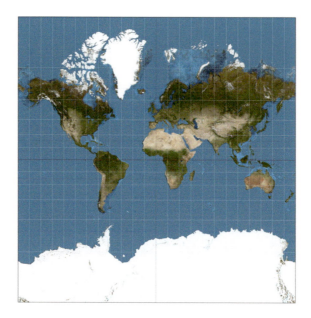

Figure 29.1 Mercator's projection of the world, note comparative continent sizes
Source: Wikipedia.

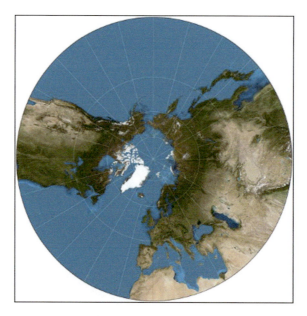

Figure 29.2 Gnomonic projection. Gnomonic projections are said to be the earliest projection type. They show striated curves of spherical shape, and are used in seismic surveying and radio-bearing work (seismic, radio, and other naturally-occurring waves travel along 'great circle' lines)
Source: Wikipedia.

identities imply centralized rather than aggregated architecture. The cultural landscape of mapping and navigating the High Seas is rapidly changing due to this, but the social and environmental dependencies and comparisons presented by such associated systems deserve discussion outside the scope of this chapter.

The ocean remains the world's last uncharted terrain, with less than 10% having been surveyed in any detail (i.e., with sonar) at the time of writing (2023). Although they are technically far from 'all-knowing', modern ocean charts conspicuously omit the term 'Unknown Ocean Area'. Indeed, humans have speculated over the characteristics of the ocean floor and the composition of the water column since maritime exploration began. Today, a range of contemporary methods and technologies are available to complement the mapping of the ocean surface and, to an extent, what lies beneath it. Some of these are outlined below.

Bathymetry

Bathymetry is the measurement of the shape of the seabed. Synonymous with sounding, it is traditionally practised in ocean passage-making from a physical proximity on (or above) the surface of the sea.

Lead-sounding

Tallowed lead lines have been used in navigational 'soundings' of the seafloor since before the Greek and Roman civilizations. These are very long ropes with lead weight sinkers, or 'plummets', at the end that are smeared in sticky wax (tallow). This technique can provide a 'leadsman' – a mental image of the seafloor's profile from many soundings – and is still used today.

Echo-sounding

Conceptually a digital technology, the Echo-Sounder (Sonar) was invented in 1928 by Herbert Grove Dorsey. This measures the time it takes for an underwater sound pulse to travel to – and back from the sea-floor, thereby calculating depth by distance. Today, these are prolific on small craft worldwide and variations are used across the sciences (Figure 29.3).

Operational oceanography

Key techniques of modern operational oceanography include radar altimetry, gravimetry, sea-surface temperature and ocean-colour satellite measurement (Le Traon, 2018: 162). Synthetic aperture radar (SAR), scatterometry, sea-ice and the new sea-surface salinity measurements can all produce impressive ocean modelling simulations, but for rigorous analytics, and instrument calibration in physical operations, Le Traon (p.162) insists that 'in-situ data' are consistently required.

Aerial bathymetry

Aerial bathymetry involves sounding the ocean floor from altitude, usually by aircraft or unmanned aerial vehicle (UAV or drone) and from space via satellite. The technique compares the behaviour of matter frequency waves (through air and water) to analyse water depths/consistencies, inferring vertical relations between air, water, and solid matter (usually geological features).

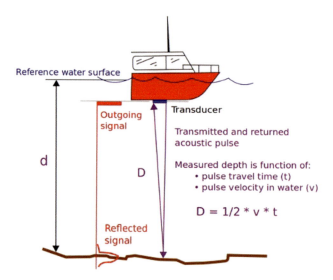

Figure 29.3 Principles of echo sounding
Source: Wikimedia Commons (NRedmayne).

Satellite altimetry

Since the earliest days of mechanized flight, pressure altimetry has used aneroid barometers to express height through air-pressure correlation. Ocean radar altimetry evolved from these ideas and has been used since the 1970s to look at detailed differences in sea levels, mostly using satellite cameras and sensors. Ocean-floor definition can be inferred by calculating waterlevel differences over underwater seamounts (landmasses), where stronger local gravity attracts more water.

Airborne laser bathymetry (ALB)

This technique uses a narrow laser beam to map physical characteristics with very high resolutions. From aircraft, comprehensive imagery can be recorded at a spatial resolution of 30 cm (12 in) or higher by using a reflectometry technique (Parvizi *et al.*, 2017). Corporate satellites are increasingly being utilized for reflectometry, which can also capture images through cloud cover (Gilbert, 2019).

Multispectral remote sensing and Lidar

Multispectral cameras can scan images for a variety of interpretations, ranging from terrestrial moisture saturation to elevation contours. Similarly, ocean colours can be modelled to approximate current and depth. Although Lidar (light detection and ranging) can detect and target a wide variety of materials from aerosols to rocks, the penetration capabilities of this technology for ocean-floor detail is restricted. Therefore, sea-floor investigations using Lidar are often limited to inshore depths, and can combine differential seabed and sea surface laser reflectometry with bathymetric acoustic data.

Corroboration: operational oceanography as physical exploration

The ocean remains mapped globally at relatively low resolutions of approximately 5 km, constrained by remote sensing modelling technologies, as opposed to satellite imagery on land, which renders at least 50 cm resolution (making buildings visually discernible). Because of these resolution limits, the value of deep submarine exploration by autonomous underwater vehicles and remotely operated vehicles remains unrivaled (NOAA, 2021). Submarine surveys can provide higher resolution data, but they have narrow coverage, and are difficult, risky and expensive to implement.

The GEBCO Nippon Project ambitiously attempts to address the difficulty of mapping the world's oceans to a greater resolution by combining drones and distributed data across participating vessels and platforms. Known as *Seabed 2030*, this cooperative project aims to produce 'the definitive, high-resolution bathymetric map of the entire World Ocean' (GEBCO, 2020).

Typical resources for oceanic satellite and aerial data

Copernicus Sentinels (ESA, 2014), Jason-1 (NASA, 2006), GRACE (Dunbar, 2014), Surface Water and Ocean Topography – SWOT (Srinivasan, 2023), and other ocean-oriented missions are able to record land, ice and ocean water light-waves. These use sophisticated analytics and extrapolations from the many hundreds of orbiting satellites hosting scientific projects and resources, and include international collaborations such as EUMETSAT (see https://www.eumetsat.int/).

Long-term global landmass changes, oceanic upwellings, sea-level rise, algae blooms, the water's current behaviours, marine heatwaves, endangered species reproductive potentials and many other oceanographic data trends and events are all covered in the latest Copernicus Marine Service Ocean State Report (Von Schuckmann et al., 2019). For example, the evolution of ice-masses covered in Nansen giving birth to two icebergs is documented by the analytical narratives and environmental news bulletins provided by Sentinel Online (ESA, 2016).

Navigating the oceans

The ancient Greek mathematician Euclid referred to the ocean as a 'self-contained body of knowledge' (Densmore and Heath, 2002). To a ship leaving the safety of land (embarking), the entire landscape of the sea-floor is completely invisible. Without an image of this, the water surface represents a treacherous veil behind which disaster lurks (Figure 29.4). Even with the digital and scientific technologies currently available, the mapping of today's ocean remains subject to time and space anomalies at planetary level On the High Seas,[1] an exact digital location may not coincide with physical conditions. Navigation relies substantially on conjecture to correlate position, direction and speed; and so numbers, theories, charts and screens overlay imperfectly. Notorious historical maritime decisions have successfully relied on anecdotal precedent, figurative history and human adaptation and 'intuition' as much as 'exact science'.

Passage planning

The conditions on the ocean's surface are heavily influenced by the ocean's submarine structure, and ocean navigators can (and must) sense this physicality and correlate its local effects with planetary level causal factors. As a result, time-related transit preparations are 'critical for the safety of lives at sea' (IMO, 1999) (Figure 29.5).

Figure 29.4 'Upwelling' or 'sandbank'? At time of photographic capture, this uncharted island in the territorial waters of Cuba did not appear on any publicly available chart

Source: Allan (2014).

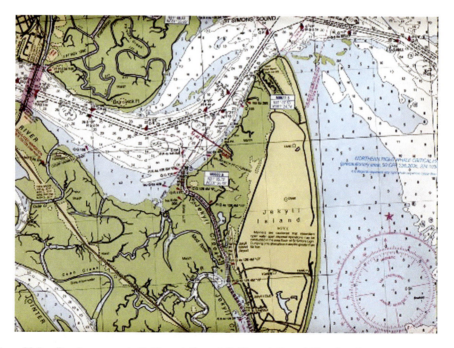

Figure 29.5a Bearing ranges in St Simon's Sound. St Simon's Sound 'Roadway'

Source: NOAA charts, 1998.

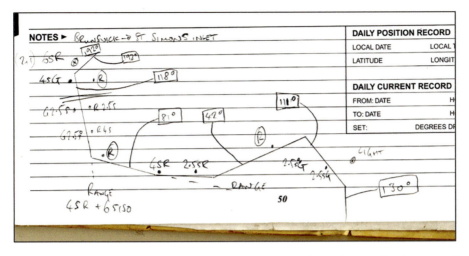

Figure 29.5b   Bearing ranges in St Simon's Sound. St Simon's Sound 'Roadway' night-time navigation vectors, with ranges (aligned lights), channel marker colours and flash intervals (e.g., '4sG' = green flashing every four seconds)

Source: Log of Ketch Sandpiper, October, 2007.

Oceanic theorists have described the sea as a 'perfect and absolute blank' (Anderson and Peters, 2014: 334). Out of sight of fixed landmarks, position is relative to degrees, minutes, and seconds of latitude and longitude (x and y axes), to depths at different tides and to expected angles of movement, but none of this is experienced directly or haptically.

Plane sailing (planar geometry)

As a result, the projection of 'planar geometry' is generally regarded as the safest and most realistic planning and navigational system. Navigators consistently adjust chart 'fixes' (perceived location) by moderating variable relations between their 'local' and 'planetary' position (e.g., latitude adjustment) and re-calibrating indices every 600 miles or so (Figure 29.6). Charts are made accessible for public use by national organizations such as NOAA, and are most useful in cross-reference between digital and analogue versions.

Ocean navigation, like oceanography, is affected by hydrodynamic and meteorological influences. Temperature, topography, visibility and other variables are all altered dramatically with weather changes at sea, and conditions can quickly become life-threatening without urgent, adaptive, and practical engagement.

Weatherfax images (data streams sent by radio as digital files) and downloadable modelling predictions from various methodologies include GFS and ECMWS GRIB files,[2] current flow prediction and wave behaviour images. These can be overlaid up to ten days in advance.

Sailing boats can forecast how the wind will change a passage trajectory, and all mariners can predict hydro-meteorological influence upon passage, and importantly, ('geographic') conditions en route (e.g., flat calm or mountainous).

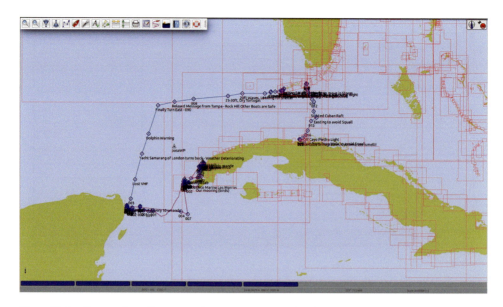

Figure 29.6a Passage from Cancun to Key West, annotated with narrative. Vector Chart
Source: Rupert Allan/OpenCPN, David Register, 2014.

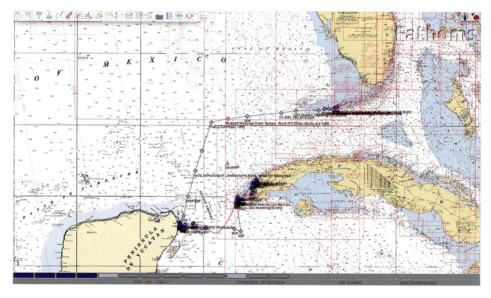

Figure 29.6b Passage from Cancun to Key West, annotated with narrative. Raster Chart whole degrees are indicated by the grid-lines, longitude across, latitude top to bottom
Source: OpenCPN/NOAA Commons, 2014.

Celestial navigation and the sextant

The navigator's job is to frame the ship's exact position on the chart, and how that relates to its physical position. The ability to use at least the North Star (Polaris) to decipher 'north' has been used by many civilizations: the ancient Greeks saw an entire global logic in the way

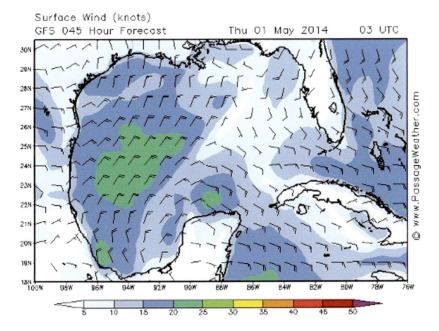

Figure 29.6c Grib File: predictive meteorological modelling: geo-referenced weather forecasting 'Grib' files. Courtesy of www.passageweather.com

Source: National Hurricane Centre: National Oceanic and Atmospheric Administration (NOAA).

the zodiac played out (Rankin, 2016), and Nuer cattle-herders in South Sudan still conceive navigational cattle-herding narratives in the desert constellations (Allan, 2011).

The sextant has played an important role in the pursuit of exact position in post-renaissance western celestial navigation. Exploiting the relatively consistent ocean feature of the water-level horizon, the sextant's split telescopic sight allows an image of this to be seen through one lens, and a view of Sun, Moon or stars through the other (Figure 29.7).

The angle of a star or planet overhead can be assessed by 'bringing-down' the view ('sight') through the lens to coincide with the horizon in the other lens. Exact time/space-related knowledge can be gathered by mathematically understanding how these visible celestial bodies correlate with places on the ground at specific times of day, month, and year (Cunliffe, 2010). This method relies on painstakingly calculated astronomical tables found in almanac publications and can be cosmically extrapolated to within one (nautical) mile of precision (Barrie, 2014) (Figure 29.8).

Celestial navigation is still practised by many ocean sailors, and sextants are important survival equipment when technology (inevitably) fails; few life rafts have electric charging devices on board, and emergency situations continue to demand pre-satellite-navigation skills.

GPS

Celestial navigation references the position of the Earth in relation to other planets and stars. It indexes the constantly shifting eclipses, circuits, and trajectories of satellite movement in space. Digitally, GPS (strictly the US Global Positioning System, but the acronym commonly refers to any iteration of a satellite navigational system) does this too, but then uses distances to artificial celestial bodies, instead of angles.

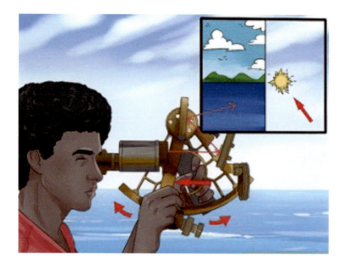

Figure 29.7 Sextant handling
Source: Wikimedia Commons.

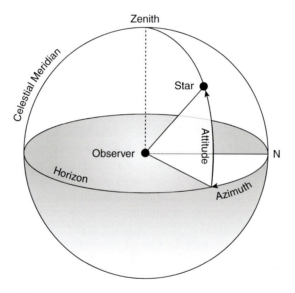

Figure 29.8 Sextant ordnances. Celestial positioning terms
Source: Wikimedia Commons.

Ocean-going GPS systems have become increasingly available over the past four decades and GPS is now available on every ocean-going boat with a smartphone on board. The positioning system analyses the times that signals take to travel from satellites (occupying a known position) as a distance measure.

Pilots of large oil pipeline ships can connect conduits in shifting and rough seas by factoring in wind, wave, and tide conditions and using differential GPS (DGPS). This agile and haptic sensing technology is combined with a pre-digital longwave (analogue) radio system and beacon infrastructure that is only available in US territorial waters. Although bringing highly accurate coordinates (1–3 cm) into the technique, the process is still iterative and re-iterative, and constantly adjusting to multiple planetary variations and *geodetics* (Wölfl et al., 2019).

With the commercial space-race heating up, the US government has stated its plan to more extensively re-deploy this obsolete LORAN system (dubbed 'eLORAN') for security purposes (Glass, 2016), and this acknowledgement of the ongoing usefulness of analogue technologies may imply an important comment on the limits of digital technologies.

Geodetics, distances and the nautical mile

The sea's local surfaces expand and recede with tides, ice-cap evolutions, and geomorphological activity. Obfuscations and variables – such as current, temperature, convection, depth, lack of light, and local magnetic variation – make chart position and GPS position correlation moot. 'Breaking waves, wind, the Coriolis effect, cabbeling and temperature and salinity differences' (Esri, 2015) are compounded by the fact that 'WGS84 rubric (the current GPS coordinate standard) relies on a centred planetary mass which is variable by 2 cm' (Lenart, 2019).

Furthermore, because the Earth is elliptical, rather than a perfect sphere, the length of its circumference varies for each great circle or rhumb line visualized around it. Relative distances are also influenced by this inconsistency (*ibid.*).

The nautical mile

The nautical mile constant is a specific measure used to express distances in geometric terms in maritime and air navigation. Indexing the planetary geometry of a sphere, the circumference of the Earth in a straight line represents 360 degrees around. Differing from a statute mile, a nautical mile has always symbolized one degree-second of this arc, with each degree divided into 60 minutes and then 60 seconds. According to planar geometry, these lines of circumference naturally get shorter as parallels of latitude above or below the equator, so the mile-length becomes different. As a result, latitudinal components must be accommodated in calculations of distance at sea.

The distance of 1,852 m was agreed by the First International Extraordinary Hydrographic Conference in Monaco in 1929 to give an arbitrary value to what became known as the International Nautical Mile (INM). 'The problem of differences between the sea mile and the international nautical mile' has been tackled by developing algorithm software, but 'manual position fixing is still required as a backup for electronic position fixing' (Lenart, 2019: 805).

A sea chart, as a logic system, theory, and method all in one, exists as an aid to transpose the anomaly of a shifting idea of distance. Nautical miles, INMs and statute miles (1,609 m) get scaled together on paper charts, and ocean navigators are used to this dual concept of distance in their habitual practice of chart plotting.

Local compass variation: magnetic and true north

The compass is an instrument used almost without exception on all sea-going vessels to indicate course and position. Standard sea-charts include a 'compass rose', which both orients the chart and depicts the variation between magnetic and true north with declarations of projected yearly deviation corrections. These deviations can be several degrees, and vary locally. Knowledge of deviations is of life-saving impact to small craft navigation. This notification is still extrapolated for paper publications (such as 'Notice to Mariners'), and digital charts automatically adjust GPS position and use software to continuously feedback the adjusted position metrics. It is worth noting that these digital positions do not consistently accommodate various digital imagery offsets.

As a result of so many anomalies, oceanography has traditionally coupled mapping of its 'interior' with more useful – yet 'fuzzy' (Zaitsev et al., 1998) – logics of 'modelling' ocean homeostasis. In the oceanic environment 'the model output allows the calculation and prediction of ocean "parameters" rather than absolutes' (Tsinoremas, 2007).

Dead reckoning – a technique of plane sailing

Digital hardware and connection can be unreliable at sea, and where physically recording absolute distance on a moving fluid remains impossible, multiple navigational methods are considered good practice. Different systems of modelling have been used across maritime cultures to calculate an estimated position (and travel direction). For western navigation, this estimation of position is known as dead-reckoning, and based on the following terms:

Course Over Ground (COG): also known as the 'ground track', this refers to the actual direction of the ship's motion over the seabed. A trick to find ground track inshore, where land or buoys are visible, is to use two points astern of the boat for reference as the boat moves forward, checking for alignments.

Speed Over Ground (SOG): this is the actual speed at which the ship moves, after all currents and water or wind movements have had their influence. If the current is from a position dead-ahead of the ship, the speed of that current (if known) can be subtracted from the speed of the ship through the water (*hull speed*). This should equate to the speed over ground (or actual speed), and should show on a GPS device.

Bearings and Headings: these two terms are often confused, but their poetics imply important differences. A bearing is the angle between a line that joins two points of interest and the line from one of the points to the north, such as a ship's course or a compass reading to a landmark, whereas a heading is an *intended* course, measured in compass points or degrees, and is an expression of action.

Set and Drift: ocean passage relies on mobile influences of tide, wind and momentum. Like weather prediction visualizations, these 'models' are publicly available as digital images in certain areas of the ocean. Although digital chart-plotters make following a course immediate and easy, ocean passages are deceptively difficult to follow without land reference. The value of information on the set (direction) and drift (speed) of water currents is easily overlooked, as is the difference between a course over ground and a course through the water (Figure 29.9).

The limits of instruments

The current trend of substituting human participation for digital 'shortcuts' is often seen by seafarers as a dangerous 'overreliance on these clever machines' by 'a generation of computer-savvy officers who fail to look out of the window at the crucial moment' (George, 2015). In *The Long Way*, Bernard Moitessier (1995: 82) wrote 'concentrating on a magnetized needle prevents one from participating in the real universe, seen and unseen, where a sailboat

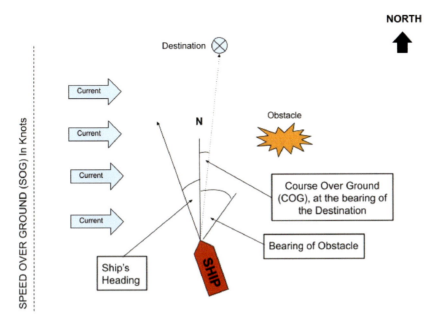

Figure 29.9 To arrive at a destination seen ahead on a bearing of 10 degrees (approximately north-wards), a ship might have to take a heading of 320 degrees, because of the set of a current from 270 degrees (from the West), which will cause drift of the ship eastwards towards an Obstacle found at a bearing of 50 degrees

Source: author.

moves'. He went on to champion instrument-free navigating 'with neither chart nor compass', as part of a growing revival of non–European geospatial 'sensed feedback' practices that are historically found in many seafaring traditions.

Various versions of the *dead-reckoning* technique have been consistently (and safely) used across cultures, by 'late-medieval Arabs, nineteenth-century Micronesians, and early-twentieth-century North Sea trawler skippers' (McGrail, 2006: 53). The recent revival of Marshallese 'wave piloting' (Tingley, 2016) or 'way-finding' (Polynesian Voyaging Society, 2020) as a university-accredited scientific discipline (ethnomathematics) demonstrates an increasing acceptance of diversity in geospatial practice.

Wave piloting places the boat and navigator in a self-contained, multi-systemic anthropomorphic space, which is a microcosm of accumulated cybernetic oceanic balances. Each technical facet applies to external variables beyond the hull perimeter.

Celestial navigation similarly extrapolates local conditions into an inter-planetary cosmology. Cosmic factors (such as the moon) and localized repercussions (such as the tide) remind us that the ocean and seafaring are elementally connected beyond our planetary boundaries. The *nautical mile* conflates planetary distance with planetary time, and would notably be longer or shorter on another planet, depending on that planet's size.

The process of hydraulic 'displacement' – the 'material flow' of a boat (Deleuze and Guattari, 1987: 373) – can be contextualized by the concept of *Cybernetes/Kubernetes* (Weiner, 2019), which is derived from the Greek word for 'steerage'. The 'live-helm' is a physical haptic negotiation between humans, machines, and the environment. It connects physics with metaphysics in a constellation logic that is somehow indicative of a *Technological Sublime* (De Mul, 2011).

Consensus, conventions, cohesion

Humans have attempted to 'govern' a contested ocean space for millennia using technology, treaties and codes. Approximation, adaptation and anomaly are all part of the scientific study of the ocean, and accepted conventions, concessions, and agreed terms of International Maritime Law are as essential as scientific technical practices.

Collision avoidance regulations (ColRegs)

SOLAS (International Convention for the Safety of Life at Sea) regulations and resources are administered by maritime interest groups, institutions and authorities around the world and are observed by navy captains and dinghy sailors alike. Britain's Maritime and Coastal Agency (MCA), and the US National Oceanic and Atmospheric Administration (NOAA) hold online access for much of this documentation.

One example of SOLAS legislation is that all ships, regardless of size, have been required to carry paper charts that cover the intended passage of the vessel at any point of inspection.

In the year 2000, commercial shipping laws changed to accommodate digital chart plotters, and requirements to carry paper charts were eased (See SOLAS Regulations Version V). Some may argue that this was a dangerous change (Cutlip, 2016).

Governance and consensus – rendering of assistance at sea

Treaties and conventions may be the only way to establish consensus within the oceanic space, but these conventions have always been dependent on participation and networks of self-authored collaboration. In *The Governance of the Commons*, Vogler (2000: 11) describes a self-identifying 'relationship of mutual vulnerability between actors'. The ocean is often cited as a 'Commons' paradigm (Latour, 1990), and this social consensus is encoded within human expressions of physical stewardship (as 'hydric' citizenship). Global citizens may now even apply for a passport to prove their World Ocean Nationality (World Ocean Network, 2021), and Article 98 of recently updated UN International Treaty for the Law of the Sea vocalized this cohesion as a 'maritime duty'; a universal principle, incumbent upon humanity created to render assistance:

> Every State shall require the master of a ship flying its flag, in so far as he can do so without danger to the ship, the crew or the passengers: (a) to render assistance to any person found at sea in danger of being lost; (b) to proceed with all possible speed to the rescue of persons in distress.
>
> *(UNCLOS, 1982)*

This imperative to render assistance has been historically central to all those in command of vessels on the High Seas and constitutes the legal precedent for Mediterranean humanitarian rescue boat interventions. This is highlighted by mariners such as Pete Goss (1999) in his decision to abandon competition ranking to rescue a fellow sailor.

However, the governments of the United Kingdom, Australia and the United States have all recently broken these conventions. Although few of these individuals are likely to be mariners, the regrettable decision to prohibit the rendering of aid in such maritime settings (Atger, 2019; Sutherland, 2020) may suggest tacit political agreement to abandon ocean stewardship in general.

Responsible sailors collaborate, allowing distributed observations to raise shared awareness of other sea-craft positions, ocean conditions and 'uncharted hazards' (like semi-submerged – and thus undetectable – shipping containers) (Fretter, 2017; Berti, 2019).

AIS and bridge to bridge networks

Signal language creates a cooperative spatial dynamic between boats and can enhance the sharing of consensual knowledge. Semaphore, Morse code and subsequently Amateur HF ('HAMM') and SSB (Marine Single Side-Band) radio have long been used to form distributed knowledge-sharing networks. Coded flags are still required on all ships' inventories (regardless of size) to indicate their status and security. Despite digital technology, these systems prevail on the High Seas.

Almost every maritime VHF (Very High Frequency) radio now includes the GMDSS (Global Maritime Distress and Safety System). For Search and Rescue (SAR) purposes, its digital signals are encoded with information such as hailing port, size, weight and hull colour.

Most recently, Automatic Identification Systems (AIS) – a life-saving technology that uses radio chatter of these GMDSS identifiers plus GPS position, number of crew and destination port – has evolved to digitally visualize this network. All vessels above 300 tons must have, and deploy, an AIS transponder, according to SOLAS standards (Figure 29.10).

Vessels over 150 gross tons are required to use constantly deployed AIS beacons and radar, and all vessels regardless of size are required to maintain lighting at night. Other stations (usually ships) in the area receive AIS signals in a network-torrenting format and this aggregated data is monitored in turn by stations ashore. Since VHF range works according to 'line of sight', AIS uses satellites and this information can, in turn, be disseminated globally at 406 MHz.

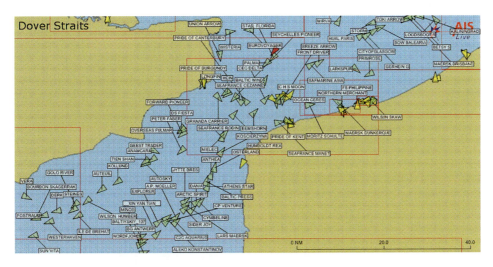

Figure 29.10 AIS in La Manche/The English Channel. AIS – the relational map of hazards (or help). La Manche/English Channel

Source: Global Maritime Distress and Safety System (GMDSS) and Wikimedia Commons.

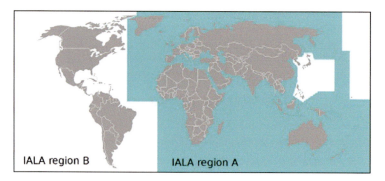

Figure 29.11 IALA Maritime Buoyage System: regions A and B. Rules for mariners. Orthodoxy is promoted by the International Association of Marine Aids to Navigation and Lighthouse Authorities (the IALA). Red (Port) and Green (Starboard) buoys are used to mark the edges of channels and seaways (entrances). IALA buoyage systems A and B are used in different parts of the world, but the IALA has worked to create two buoyage systems from around 30 different systems that were in place

Source: Wikimedia Commons.

Contrary orthodoxies: IALA buoyage

Modern ocean cartography aims to universalize internationally diverse navigation conventions. Visualizations display information such as ordnance, depths, current patterns, shipping channels and direction of travel as clearly as possible. However, historically derived differences in orthodoxies still exist (e.g., red and green channel-marking), and this chart visualizes where the different orthodoxies are embraced (Figure 29.11).

Hydrocentric thinking: the oceanic turn

In their anthology of essays *Atlantic, Science, and Empire in the Atlantic World*, Delbourgo and Dew (2008: 8) reflect how ocean mapping is inextricably bound to language: 'because oceans, too, are socially constructed, with meanings that vary across space and time, any definition of the "Atlantic" will be local and period-bound'. The very term 'Atlantic' has come into use as recently as the mid-eighteenth century, before which different parts had other names – for example, the Southern Atlantic was referred to as the *Aethiopian Sea*. Their comprehensive account of the empire-building tradition of 'charting' recounts how 'would-be empire builders drew lines on maps of water, not just land' (*ibid*.: 7).

Traditionally, masters of ships heading on exploration expeditions paid a visit to these 'empire-builders' – or more accurately, cartographers – who were commissioned to curate and maintain national chart knowledge. Once confirmed by returning sea captains, the cartographers would hand-annotate publications, thereby 'naming' the most recent maritime data.

All maritime states are expected to distribute regular editions of their Notice to Mariners, which include verbal and visual cartography updates on compass variation, new hazards and other maritime 'news'. These are traditionally overlaid onto existing charts as new 'legend' and, as recently as 2013, the Cuban cartography commission was reported to be issuing hand-inked annotations to territorial sea charts to mariners – each of which was individually stamped by official cartographers (Barr, 2013; pers. comm.).

The oceanic gaze and the environment

Ecological theories recognize the inter-dependence of human culture with the sea, often locating anomalous and even anarchic elements of oceanic modelling within a post-humanities context, in the 'repositioning [of] the human among nonhuman actants' (Sanzo, 2018). 'Some ecologists fear that a map of our sea floor will allow extractive industries the chance to profit from these resources, potentially endangering marine habitats and coastal communities in the process' (Cousteau, 1976: 14), and critical work such as *The Manufacturing of Greta Thunberg* (Morningstar, 2020) claims that the environmental story is in danger of being appropriated for dubious economic/personal gain (Klein, 2014). Ecologists also observe proven increases in hurricane activity – an acknowledged result of global climate change (Bailey *et al.*, 2021; Krupnik and Jolly, 2002) – and decipher this as a homeostatic characteristic of the sea revealing oceanic 'agency'. Kingery (2016) describes a 'self-cleaning' planet, quite outside of human influence, which counters the destructive impact of carbon emissions by creating hurricanes to re-balance and potentially stabilize environmental climate change.

Elizabeth DeLoughrey's discussion of 'Sea Ontologies' (2017: 33) traces 'Blue Humanities' back to Cold War geopolitics, claiming that the post-1970s 'spatial turn' went on to produce 'a loosening of nationally-bounded modes of thinking about capital and space'. However, although the ocean subsequently 'became a space for theorizing the materiality of history', DeLoughrey (2017: 34) protests that 'it rarely figured as a material in itself'. This calls for acknowledgement of the sea in terms of 'a new oceanic imaginary'.

The ocean has numerous stakeholders, and their claims often conflict. Audio-visual coverage of 'ocean life', like the BBC's *Blue Planet*, is immensely popular to television-viewing communities, appealing to armchair environmentalists. But this spectatorship also risks 'othering' the ocean, making it palatable on an anthropomorphic level – rather than on its own terms.

'The sea has not been fathomed as a cultural or multispecies ecology' until recently, writes DeLoughrey (2017: 32), and Steinberg (2001: 23) has made a compelling case for viewing the ocean as 'a dynamic force rather than a location'. Elspeth Probyn seeks to disturb the singular and 'othered' ocean to make way for the 'ocean multiple'; a conception of the various forms of 'the oceanic' (Probyn, 2018: 386), whereas DeLoughrey (2017: 33; 2018: 386) sees the sea as a collection of 'decentered ontologies of connection'.

Conclusion: technology as taxonomy

Human interaction with the sea depends, then, on verbalized codes, laws and navigational orthodoxies. Treaties and conventions frequently 'overwrite' the ocean as Nautical Terms and mapping and passage-making have parallel histories, their terms being creative - 'embedded in social contexts of production' (Glasze and Perkins, 2015: 144). Under international maritime protocols, 'articles of passage' (such as logbooks, annotated charts, crew documents, ship's technical certificates) must be legible, as they are scrutinized as a record of 'provenance' by port authorities. Articles of Passage ('Ship's Papers') equate not only to credibility, but arguably to anchor maritime reality itself, and these articles have traditionally been contingent to the granting of 'safe haven'. To be 'undocumented' is to be vulnerable. (A notorious example of this reality failing is the 'invented' logbook of Donald Crowhurst, as noted by Tomalin and Hall, 1995).

Nautical terms are celebrated for descriptive idiom, but this taxonomy also encodes the character of maritime technologies. Nautical terms enshrine unique relationships between abstract and concrete elements. The Beaufort meteorological Scale for Sea State[3] seems unscientific. It embodies the elusive temporal and spatial conditions of the ungovernable ocean but its technical importance exemplifies how taxonomy does, in fact, perform technology. According to Brian Massumi (2002: 69), 'social change is spatially relegated to precarious geographical margins', and, perhaps appropriately, the sea is portrayed in mythologies as an 'anti-establishment' space of liberation and DIY (do-it-yourself) counter-culture (Moran, 2011). Every participant in the mapping of oceans, from cartographer to navigator, citizen scientist (Ebbesmeyer, 2010) to liveaboard (Carling, 2018), does well to remember Jessica Lehmann's (2018: 91) words:

> 'marine cultural heritage might be understood as a frontier in terms not only of spatial territory, but also for knowledge, governance, and politics [and] stakes of marine cultural heritage for different communities, reaching far beyond the sites and objects themselves'.

Notes

1 'High Seas/International Waters' refers to the 64% of the ocean that lies outside the jurisdiction of any nation (i.e., outside its coastal or territorial waters). This definition was expanded in the United Nations Convention on the Law of the Sea (UNCLOS I, 1958).
2 ECMWS Video: Forecasting isn't an Exact Science' Available at: https://www.pennlive.com/news/2016/01/european_vs_american_weather_m.html (Accessed: 20th January 2020).
3 See, for example, https://www.weather.gov/mfl/beaufort (Accessed: 6th May 2023).

References

Allan, R. (2011) "On the Ward: Referendum Aftermath (Part 3)" *Journal of an MSF Logistician in South Sudan* Available at: https://rupertallan.com/2011/02/ (Accessed: 26th January 2023).
Anderson, D. (2006) 'Freedoms of the high seas' In Barnes, R., Freestone, D. and Ong, D.M. (Eds) *The Law of the Sea: Progress and Prospects* Oxford: Oxford University Press, pp.327–346.
Anderson, J. and Peters, K. (2014) *Waterworlds: Human Geographies of the Ocean* Abingdon: Routledge.
Atger, A.F. (2019) "EU Migration Strategy: Compromising Principled Humanitarian Action" *Forced Migration Review: The ETHICS Issue* 61 pp.30–32.
Bailey, H., Hubbard, A., Klein, E.S., Mustonen, K-R., Akers, P.D., Marttila, H. and Welker, J.M. (2021) "Arctic Sea-ice Loss Fuels Extreme European Snowfall" *Nature Geoscience* 14 pp.283–288 DOI: 10.1038/s41561-021-00719-y.
Barnes, T. and Duncan, J. (Eds) (1992) *Writing Worlds* London: Routledge.
Barr, C. (2013) *Cruising Guide to Cuba* Nova Scotia: Yacht Pilot Cruising Guides.
Barrie, D. (2014) *Sextant: A Voyage Guided by the Stars and the Men Who Mapped the World's Oceans* London: William Collins.
Baudrillard, J. (1988) "Simulacra and Simulations" In Poster, M. (Ed.) *Jean Baudrillard: Selected Writings* Stanford, CA: Stanford University Press, pp.166–184.
Berti, A. (2019) "Lost at Sea: How Shipping Container Pollution Affects the Environment" *Ship Technology* (July) Available at: https://www.ship-technology.com/features/containers-lost-at-sea/ (Accessed: 28th February 2020).
Calvi. M. (2015) "Open Source Sails the Seven Seas" Available at https://opensource.com/life/15/7/-opencpn-chartplotter-navigation (Accessed: 26th February 2020).
Carling, J. (2018) "Reflections on Marginal Mobile Lifestyles: New European Nomads, Liveaboards and Shabi Moroccan Men" *Nordic Journal of Migration Research* 4 (1) pp.11–20 DOI: 10.2478/njmr-2014-0002.
Cousteau, J. (1976) "The Perils and Potentials of a Watery Planet" In Menard, W.H. and Scheiber, J. (Eds) *Oceans: Our Continuing Frontier* Del Mar, CA: Publishers Incorporated, pp.13–18.
Cutlip, K. (2016) "AIS and the Challenges of Tracking Vessels at Sea" *Data and Technology in Global Fishing Watch* Available at: https://globalfishingwatch.org/data/ais-and-the-challenges-of-tracking-vessels-at-sea/ (Accessed: 28th February 2020).

Cunliffe, T. (2010) *Celestial Navigation, A Yachtmaster's Guide: What the Yachtmaster Needs to Know* London: John Wiley & Sons.
Delbourgo, J. and Dew, N. (Eds) (2008) *Science and Empire in the Atlantic World* New York: Routledge.
Deleuze, G. and Guattari, F. (1987) *A Thousand Plateaus: Capitalism and Schizophrenia* (trans. Massumi, B.) Minneapolis: University of Minnesota Press.
DeLoughrey, E. (2017) "Submarine Futures of the Anthropocene" *Comparative Literature* 69 (1) pp.32–44 DOI: 10.1215/00104124-3794589.
De Mul, J. (2011) "Wild Systems: The Technological Sublime" *Next Nature Net* Available at https://nextnature.net/2011/07/the-technological-sublime (Accessed: 21st February 2020).
Densmore, D. and Heath, T. (2002) *Euclid's Elements* Santa Fe, CA: Green Lion.
Dunbar, B. (2014) "GRACE - Gravity Recovery and Climate Experiment" Available at https://www.nasa.gov/mission_pages/Grace/index.html (Accessed: 26th January 2023).
ESA (2014, 2015, 2016) "Copernicus Sentinels (1-4, and soon 5 and 6, in 'Copernicus Open Access Hub'" *European Space Agency for Earth Observation* Available at: https://scihub.copernicus.eu/ (Accessed: 26th January, 2023)
Esri (2015) "Ocean Currents in Marine Waters Around Ireland" Available at https://www.arcgis.com/home/item.html?id=f1c9190208f447c1a40a0644d3204d47 (Accessed: 27th January 2023).
Ebbesmeyer, C. and Scigliano, E. (2010) "Flotsametrics and the Floating World: How One Man's Obsession with Runaway Sneakers and Rubber Ducks Revolutionized Ocean Science" New York: Harper.
Fretter, H. (2017) "Could a Floating Shipping Container Sink Your Yacht? How Real is the Danger?" *Yachting World* (17th May) Available at: https://www.yachtingworld.com/news/could-a-floating-shipping-container-sink-your-yacht-is-the-danger-to-sailors-real-or-imagined-107508 (Accessed: 26th February 2020).
George, R. (2015) "Worse Things Still Happen at Sea: The Shipping Disasters We Never Hear About" *The Guardian* (10th January) Available at: https://www.theguardian.com/world/2015/jan/10/-shipping-disasters-we-never-hear-about (Accessed: 24th February 2020).
Gilbert, L. (2019) "Satellites to Reveal Sea State and Much More Than the Eye Can See" *University of New South Wales* Available at: https://phys.org/news/2019-09-satellites-reveal-sea-state-eye.html (Accessed: 26th January 2023).
Glass, D. (2016) "What Happens if GPS Fails?" *The Atlantic* (June) Available at: https://www.theatlantic.com/technology/archive/2016/06/what-happens-if-gps-fails/486824/ (Accessed: 28th February 2020).
Glasze, G. and Perkins, C. (2015) "Social and Political Dimensions of the OpenStreetMap Project: Towards a Critical Geographical Research Agenda" In Arsanjani, J.J., Zipf, A., Mooney, P. and Helbich, M. (Eds) *OpenStreetMap in GIScience: Experiences, Research, Applications* New York: Springer, pp.143-146 DOI:10.1007/978-3-319-14280-7.
Harley, B. (1988) "Maps, Knowledge and Power" Cosgrove, D. and Daniels, S. (Eds) *The Iconography of Landscape; Essays on the Symbolic Representation, Design, and Use of Past Environments* Cambridge: Cambridge University Press, pp.277–312.
GEBCO (2020) "Nippon Foundation-GEBCO Seabed2030 Project" Available at: https://www.gebco.net/about_us/seabed2030_project/ (Accessed 15th March 2023).
Goss, P. (1999) *Close to the Wind* Cambridge, MA: De Capo.
IMO (1999) "Guidelines for Voyage Planning, Section 5" *International Maritime Organization Resolution A.893(21)* Available at: https://wwwcdn.imo.org/localresources/en/KnowledgeCentre/IndexofIMOResolutions/AssemblyDocuments/A.893(21).pdf (Accessed: 27th January 2023).
Kantharia, R. (2019) "MARPOL (The International Convention for Prevention of Marine Pollution For Ships): The Ultimate Guide" *MarineInsight* (4th June) Available at: https://www.marineinsight.com/maritime-law/marpol-convention-shipping/ (Accessed 20th March 2020).
Kingery, K. (2016) "Hurricanes Key to Carbon Uptake by Forests" ScienceDaily (2nd May) Available at: https://www.sciencedaily.com/releases/2016/05/160502161841.htm?fbclid=IwAR2pTbtp9LfGGbtSqOHQhwqiKlnvPqxWHFRjZIUH7tvP3pawbKB30XuhXsE (Accessed: 16th February 2020).
Klein, N. (2014) *This Changes Everything: Capitalism vs. the Climate* New York: Simon and Schuster.
Krupnik, I. and Jolly, D. (2002) *The Earth is Faster Now: Indigenous Observations of Arctic Environmental Change* Fairbanks, AK: Arctic Research Consortium of the United States.

Latour, B. (1990) "On Actor-Network Theory. A Few Clarifications Plus More Than a Few Complications" *Philosophia* 25 (3–4) pp.47–64.

Lehman, J. (2018) "Marine Cultural Heritage: Frontier or Centre?" *International Social Science Journal* 68 (229–230) pp.291–301 DOI:10.1111/ISSJ.12155.

Lenart, A. (2019) "The Sea Mile and Nautical Mile in Marine Navigation" *Journal of Navigation* 72 (3) pp.805–812 DOI: 10.1017/S0373463318000991.

Lessing, L. (2004) *Free Culture* New York: Penguin.

Le Traon, P-Y. (2018) "Satellites and Operational Oceanography" In Chassignet, E., Pascual, A., Tintoré, J. and Verron, J. (Eds) *New Frontiers in Operational Oceanography* Tallahassee, FL: GODAE OceanView, pp.161–190 DOI: 10.17125/gov2018.ch07.

Massumi, B. (2002) *Parables for the Virtual: Movement, Affect, Sensation* Durham, NC: Duke University Press.

McGrail, S. (2006) *Ancient Boats and Ships* (Shire Archaeology) London: Bloomsbury Publishing.

Moitessier, B. (1995) *The Long Way* Chelsea, MI: Sheridan.

Moran, I.P. (2011) "Punk: The Do-it-yourself Subculture" *Social Sciences Journal* 10 (1) 13.

Morningstar, C. (2019) "The Manufacturing of Greta Thunberg" *The Art of Annihilation* Available at: http://www.theartofannihilation.com/the-manufacturing-of-greta-thunberg-for-consent-the-political-economy-of-the-non-profit-industrial-complex/ (Accessed: 16th February 2020).

NASA (2006) "Jason–1: An Ocean Odyssey – Ocean Data from Space" Jason-1 Fact Sheet Available at: https://sealevel.jpl.nasa.gov/system/documents/files/1673_jason-1-fact-sheet- 200610.pdf (Accessed: 15th March 2023).

NOAA (2021) "What is the Difference Between an AUV and an ROV?" *National Ocean Service* Available at: https://oceanservice.noaa.gov/facts/auv-rov.html (Accessed: 26th January 2023).

NOAA (2023) "Global Forecast System – GFS" Available at: https://www.ncei.noaa.gov/products/-weather-climate-models/global-forecast (Accessed: 27th January, 2023).

Paglen, T. (2009) "Experimental Geography: From Cultural Production to the Production of Space" *The Brooklyn Rail* (March) Available at: https://brooklynrail.org/2009/03/express/experimental-geography-from-cultural-production-to-the-production-of-space (Accessed: 23rd February 2020).

Parvizi, R., Henry, J., Honda, N., Donarski, E., Pervan, B.S. and Datta-Barua, S. (2017) "Coordination of GNSS Signals with LiDAR for Reflectometry" *Proceedings of the 30th International Technical Meeting of the Satellite Division of The Institute of Navigation (ION GNSS+ 2017)* Portland, OR, pp.3420–3433 DOI: 10.33012/2017.15347.

Pike, D. (2018) *The History of Navigation* Barnsley: Pen and Sword.

Polynesian Voyaging Society (2020) "The Story of Hōkūle'a" Available at: https://www.hokulea.com/voyages/our-story/ (Accessed: 27th January 2023).

Pollock, G. and Bal, M. (Eds) (2010) *Digital and Other Virtualities: Renegotiating the Image (New Encounters: Arts, Cultures, Concepts)* London: I.B. Tauris & Co.

Probyn, E. (2018) "The Ocean Returns: Mapping a Mercurial Anthropocean" *Social Science Information* 57 (3) pp.386–402 DOI: 10.1177/0539018418792402.

Rankin, W. (2016) *After the Map: Cartography, Navigation, and the Transformation of Territory in the Twentieth Century* Chicago: University of Chicago Press.

Sanzo, K. (2018) "New Materialism(s)" *Genealogy of the Posthuman* Available at: https://criticalposthumanism.net/new-materialisms/ (Accessed: 25th February 2020).

Soja, E. (1996) *Thirdspace: Journeys to Los Angeles and Other Real and Imagined Places* Oxford: Wiley-Blackwell.

Soja, E. (2010) *Seeking Spatial Justice* Minneapolis: University of Minnesota Press.

SOLAS (1974) "International Convention for the Safety of Life at Sea" *Conventions* Available at: https://www.imo.org/en/About/Conventions/Pages/International-Convention-for-the-Safety-of-Life-at-Sea-(SOLAS),-1974.aspx (Accessed: 27th January 2023).

Srinivasan, M. (2023) "Surface Water and Ocean Topography: SWOT" *Jet Propulsion Laboratory* Available at: https://swot.jpl.nasa.gov/ (Accessed: 26th January 2023).

Steinberg, P.E. (2001) *The Social Construction of the Ocean* (Cambridge Studies in International Relations Series) Cambridge: Cambridge University Press.

Sutherland, J. (2020) "EU Turns its Back on Migrants in Distress: New Naval Mission Designed to Avoid Sea Rescues" *Human Rights Watch News* (18th February) https://www.hrw.org/news/2020/02/18/eu-turns-its-back-migrants-distress (Accessed: 26th February 2020).

Tingley, K., (2016) "The Secrets of Wavepilots" *The New York Times Magazine* (17th March) Available at: https://www.nytimes.com/2016/03/20/magazine/the-secrets-of-the-wave-pilots.html (Accessed: 27th January 2023).

Tomalin, N. and Hall, R. (1995) *The Strange Last Voyage of Donald Crowhurst* London: Hodder & Staughton.

Tsinoremas, N. (2007) "Ocean Modelling"; "Research: Atmospheric/Climate/Ecosystem/Geophysical/Ocean Modeling/Archive" Available at: https://ccs.miami.edu/focus-area/climate-and-environmental-hazards/ocean-modeling/ (Accessed: 27th January 2023).

Turner, P. and Davenport, E. (Eds) (2005) *Spaces, Spatiality and Technology* Dordrecht: Springer.

UNCLOS (1958; 1982) "Part VII High Seas Section 1: General Provisions: United Nations Convention on the Law of the Seas" Available at: https://www.un.org/depts/los/convention_agreements/texts/unclos/part7.htm (Accessed: 17th December 2019).

Vogler, J. (2000) *The Global Commons: Environmental and Technological Governance* New York: Wiley & Sons.

Von Schuckmann, K., Le Traon, P., Smith, N., Pascual, A., Djavidnia, S. and Gattuso, J-P. (Eds) (2019) "Copernicus Marine Service Ocean State Report" (3rd Issue) *Journal of Operational Oceanography* 12 (Supplement 1) pp.S1–S123 DOI: 10.1080/1755876X.2019.1633075.

Weiner, N. (2019) *Cybernetics: Or Control and Communication in the Animal and the Machine (Reissue of the 1961 Second Edition)* Cambridge, MA: MIT Press.

Wölfl, A-C., Snaith, H., Amirebrahimi, S., Devey, C.W., Dorschel, B., Ferrini, V., Huvenne, V.A.I., Jakobsson, M., Jencks, J., Johnston, G., Lamarche, G., Mayer, L., Millar, D., Pedersen, T.H., Picard, K., Reitz, A., Schmitt, T., Visbeck, M., Weatherall, P. and Wigley, R. (2019) "Seafloor Mapping – The Challenge of a Truly Global Ocean Bathymetry" *Frontiers in Marine Science* 6 (283) pp.1–16 DOI: 10.3389/fmars.2019.00283.

World Ocean Network (2002) "Who We Are" Available at: https://www.worldoceannetwork.org/en/ (Accessed: 27th January 2023).

Zaitsev, D.A., Sarbei, V.G. and Sleptsov, A.I. (1998) "Synthesis of Continuous-valued Logic Functions Defined in a Tabular Form" *Cybernetics and Systems Analysis* 34 (2) pp.190–195 DOI: 10.1007/BF02742068.

30
GEOSPATIAL TECHNOLOGIES IN TRANSPORT

Shaping and recording everyday lived experiences

Nigel Waters

Transport affects everyone's daily lives in all but a few of the least industrialized countries. Most of us make at least two daily trips from and to our residence: an outward bound and a return journey. These trips may be to work or to school. For those not mobile enough to leave their homes, services must be delivered to them by personal visits from family, friends or businesses. Many will make more trips than for this single purpose of work or education. They will travel to stores, places of entertainment and worship, for recreation or to obtain health services. They will travel on vacation. Their work may involve the delivery of transport services, including, for example, the gig economy of Uber and Lyft. The modeling of the generation and attraction of home-based work and non-work trips is the first step in the 'four-step' model widely used in planning and transportation GIS software such as TransCAD (McNally, 2000; Waters, 1999). Transport affects everyone and it is this complexity that makes it difficult to untie the Gordian knot and describe the vast array of ever-changing geospatial technologies that are used to facilitate and expedite the delivery of these services. The importance of transportation and associated geospatial technologies in our daily lives was re-emphasized to the world by the recent Coronavirus pandemic of 2020.

Transport planners in the industrialized world are usually concerned with the triple bottom line of Sustainable Transportation. The 'triple bottom line' appears to be a universal concept since a search on Google for the four words 'transport triple bottom line' (at the time of writing) revealed a first hit from the Namibia University of Science and Technology (NUST, 2019). The NUST article explains the triple bottom line as providing an emphasis on the Economy, the Social, and the Environmental aspects of transportation. Respectively, NUST defines these three terms as the 'Support [for] economic vitality while developing infrastructure in a cost-efficient manner'; meeting 'social needs by making transportation accessible, safe, and secure; [and the] […] provision of mobility choices for all people (including people with economic disadvantages)'; and, finally, '[creating] solutions that are compatible with […] the natural environment, [and that will] reduce emissions and pollution from the transportation system […]'.

'Transport' is the preferred term in the United Kingdom (UK) and many countries that have a historic connection to the UK (cf. Canada's federal government agency, for example, is 'Transport Canada'). The United States (US) federal agency is the Department of Transportation (US DOT) and all US states also have a Department of Transportation. In the UK, 'transportation' was a punishment originally used for UK convicts and political prisoners who were removed to foreign countries such as the United States and Australia. However, replacing 'transport' with 'transportation' in the Internet search described above yielded the same first hit to NUST.

Geographic information systems for transportation (GIS-T)

Geographic information systems (GIS-T) have been used by Departments of Transportation for many years. For example, in 2013, a survey of the US State Departments of Transportation (plus Washington, DC, Puerto Rico and Alberta, Canada) reported that 88% of the respondents were using GIS technology. Apart from photogrammetry (90% in use) GIS was the most widely used geospatial technology. This was a strong endorsement for GIS technology as an essential tool for Departments of Transportation, but it still implied that six of the 51 responding departments were not using a technology that had been available for decades (Waters, 1999, 2018a). When asked about the top factors holding back the use of new geospatial data tools, the DOTs stated cost, inertia, technical expertise and senior management in that order. A similar survey was sent to sixteen DOT service providers. Thirteen responded and all used GIS technology and provided these to DOTs.

GIS-T use in DOTs in 2019

To understand how the use of GIS has changed in the intervening six years in US DOTs, the best sources of information are the annual GIS-T Conferences sponsored by the American Association of State Highway and Transportation Officials (AASHTO, 2019). At the time of writing the most recent AASHTO Conference, where the presentations are now available, was the 2019 conference. Almost all of the annual presentations (though not the workshops) are eventually made available online.

The 2019 AASHTO survey (43 respondents from various US state DOTs) covered the following topics: organizational issues, data, hardware, software, human resources and funding resources. Organizational issues asked whether GIS services were provided by a single unit (seven respondents), multiple units (20) or as an Enterprise GIS (11) that could be accessed throughout the organization. An Enterprise GIS is generally regarded as the most up-to-date approach for large organizations. Transportation organizations were also asked five questions on the effectiveness of their use of GIS: (1) Had they done a self-assessment of their GIS using a Capability Maturity Model? (2) A Cost/benefit analysis of a GIS application? (3) Return on Investment (ROI) of a GIS application? (4) Had they formed a GIS coordinating group? (5) Had they completed a data governance plan assessing data quality throughout. The responses demonstrated that GIS-T had become one of the standard geospatial technologies now in use by transportation organizations.

Transportation data models

Esri documented 35 data models covering the complete range of geographic data (Esri, 2020a). Three of the most interesting to transportation researchers were the Transportation, Pipeline and Marine Data models. One of the earliest discussions of transportation data models was given by Dueker and Butler (2000). The desirable characteristics of a transportation data model include easy updating and the representation of the entire suite of transportation features and facilities for data sharing. These authors emphasized the importance of using an approach that would include all modes of transportation: roadways, railroads, transit systems, shipping lanes, air routes and pipelines among others. Subsequently, Butler (2008) provided a more detailed set of guidelines and it is these strategies, procedures and principles that are used in the transportation data models currently supported in Esri's transportation geodatabases. An early practical implementation was developed by Esri (2020b) for the New York State Department of Transportation and is widely regarded as a seminal model.

Esri Canada's (2020) overview of how GIS can be used in transportation emphasizes five types of transportation: municipal roads, provincial/state highways, transit, airports and aviation and ports and maritime. To this list rail freight should also be added. Alternatively, a modal breakdown might also be used. Esri Canada also describes eight areas of focus, and each of these transportation types could use GIS and other geospatial technologies to their benefit. These include: (1) Planning. This involves the use of forecasting techniques and spatial analysis to determine how demand for transportation services will change in the future and more specifically where it will change. (2) Infrastructure Asset Maintenance. This includes the construction and maintenance of all the physical assets of urban and regional transportations systems associated with, for example, urban transit systems, airports and maritime facilities. (3) Environmental Impact. Such as the effect of transportation construction and operations on the environment. (4) Operations. For example, urban bus and rail system timetables and stop frequencies, the deployment of snowplows and the determination of speed limits. (5) Safety. Examples include maintaining a record of all traffic accidents and determining the extent to which these are impacted by weather conditions (Medina *et al.*, 2017). Also, analyzing trends in traffic injuries and fatalities and implementing remedial measures. (6) Performance. Evaluating performance can be facilitated by operating urban 'dashboards' that are linked to GIS databases. A comprehensive review of geospatial dashboards for smart cities, together with a timeline, is given by Jing *et al.* (2019). (7) Corridor Management. This includes right-of-way analysis and visualization. (8) Public Engagement. Examples are the provision of timely traffic information during peak travel times and also information concerning damage to infrastructure that may include the collection of volunteered geographic information (VGI) (Hultquist *et al.*, 2018).

Smart geospatial technologies for smart environments

Recently, the *International Journal of Geographical Information Science* published a special issue on the real-time use of geographic information. The introductory article provided a paradigm for real-time GIS in 'smart cities' (Li *et al.*, 2020). Five applications were envisaged including smart transport and mobility, smart disaster management and smart and connected communities. This paradigm argued that smart cities would need to utilize real-time data visualization, simulations and analytics together with the 'ingesting' of real-time geospatial data from sources such as mobile phones, social media, VGI, sensor networks and the Internet-of-Things (IoT). Li *et al.* (2020) state that the real-time geoprocessing technology that is necessary to achieve the required rapid response for many smart city applications would include a NoSQL (non-relational) database, distributed and parallel computing and streaming architecture. They further suggest the use of cloud computing platforms to provide faster distributed computing with processing tasks divided up using software such as MapReduce and Hadoop. The aim would be to reduce the latency of the processing system (i.e., the lag time between input and output of the result). A new complementary technology to supplement cloud-based computing is edge or fog computing where devices such as roadside unit (RSU) sensors or licence plate recognition (LPR) devices collect information from parked and moving vehicles, pre-process it and then forward the reduced dataset to a central processing unit. The bandwidth required is decreased and processing speed is increased. Thus, the system latency is reduced. Ning *et al.* (2019) provide one of the earliest discussions and evaluations of Vehicular Fog Computing (VFC). In 2010, there were 1 billion cars on the road, all potential customers for real-time traffic management in a smart city. Fog computing (FC) extends the cloud computing environment to edge networks.

Autonomous and connected vehicles (AVs and CAVs)

The world is changing with rapid developments in AVs and CAVs. A comprehensive review of this technology and its benefits and challenges has been provided by Litman (2020). Fully autonomous vehicles are unlikely to be introduced before 2030 but the technology has been categorized into a number of levels representing increasing levels of automation that will allow for a gradual evolution of the technology (Letaief *et al.*, 2019). The first body to do so was the US National Highway Traffic Safety Administration (NHTSA). In 2013 this body identified five levels of automation plus a zero level of no automation. These NHTSA categories were deemed to be too broad and a more detailed system was developed by the Society of Automotive Engineers in 2014 (Takács *et al.*, 2018).

This new system included five levels of automation plus a zero level of no automation: Level 1 Driver Assistance includes lane control and cruise control while Level 2, Partial Automation, involves active driver assistance for steering and acceleration. Levels 1 and 2 are Advanced Driver Assistance Systems and are the present state of the art for mass production vehicles. Levels 3, 4 and 5 represent increasing degrees of automation. Level 3 is Conditional Automation where control can pass to the automated system for short periods of time. Level 4 is High Automation where the control of the vehicle is given to the automated system in pre-defined driving modes and control only passes back to a driver when the vehicle comes to a stop (Takács *et al.*, 2018). Finally, Level 5 is a fully automated vehicle that can operate under all conditions and modes without a human driver at all. A primary concern for all automated driving systems (ADS) is the environment for which they are configured. These environments are known as operational driving domains (ODD). In September 2017 the US NHTSA issued guidance on ADS (NHTSA, 2017), which described the minimum characteristics desirable in an ODD. These included roadway types (e.g., interstate, local, among others) where the ADS is to operate; the geographic area (e.g., urban, rural, and so on); speed range; specific environmental conditions (weather types, during daylight or both day and night). This implies that automated vehicles still benefit from connectivity to geospatial technologies such as road databases, environmental conditions and also to vehicle to infrastructure connectivity (V2I) and vehicle-to-vehicle connectivity (V2V) (Tahir *et al.*, 2022).

For over a decade there have been more than a billion cars on the road. Ning *et al.* (2019) estimate that 150 million of these cars will be 'connected' in 2020 with each car generating 30 TB a day. They then suggest that mobile traffic in 2020 will produce 360 exabytes, which they state is about eight times more than what was produced in 2015. In 2020 the average age of vehicles on roads in the USA was almost 12 years. Older vehicles are less likely to be 'connected' and have fewer connection points. New cars are part of the IoT and have 100 or more data points and the processing power of more than 20 computers processing up to 25 GB every hour (Rattigan, 2018). Some of this data is fed back to car manufacturers (especially those working on the development of autonomous vehicles) in real time and since it is location aware it can and is resold to mapping companies and those providing location aware services (Leszczynski, 2019: 15). New cars are likely to be connected to the driver's mobile phone and Bluetooth technology adding to the car's value as a data node, especially if the driver's phone is connected to their social media and related apps. The car becomes not only an information resource consumer but also an information provider. This can then potentially be part of a cloud computing environment of a smart city traffic and transportation management system. The concept is similar to grid-connected home-based renewable energy systems (usually solar) where surplus power is fed back to the regional grid. Vehicular fog computing (VFC) allows information to be harvested from connected vehicles. This approach allows vehicles and RSUs to be part of an environment that is variously referred to as Platform-as-a-Service (PaaS) or Infrastructure-as-a-Service (IaaS). Ning et al. (2020) envisage a three-layer vehicular cloud computing environment: (1) the overarching *cloud layer* composed of a traffic management server (TMS) at the smart city level. The TMS can use the car's location to determine an allocation of resources according to a 'fairness' principle. The nature of this principle is not specified by the authors and remains an active research area; (2) the cloudlet layer receives data from vehicles, RSUs and other sensors, undertakes data reduction processing and then communicates the result to the cloud layer; (3) the fog layer consists of all the primary sensing units including parked and moving vehicles and the sensors mentioned previously. Some of the data remains at the cloudlet or regional level for local traffic monitoring and smart traffic light control and other approaches designed to ease local congestion or prioritize emergency vehicles. The goal is to maximize data collection locally from vehicles and sensors, to process and exploit it regionally at the cloudlet level and finally to minimize real-time data forwarding to the city level TMS. Tactical decisions are made locally in real time and strategic planning is based on periodic uploads (daily, weekly, monthly and yearly). This reduces data movement and sharing between the three interfaces: the fog-vehicle/sensor interface, the fog-fog interface for regional data sharing and the fog-cloud interface to the city-wide transportation data center. Ning et al. (2019) derive equations to determine expected response time for the system. These equations are time and space dependent: the number of regions and time slots becoming decision variables in a mixed integer, non-linear programming problem where the system response time (latency) is to be minimized as an optimization problem. The authors evaluated their proposed system efficacy using taxis in Shanghai. Being obsessive about latency may seem unnecessary but low latency is required by the ever exponentially increasing data volumes and the need for faster decisions in the smart city. Low latency is critical in many applications including the stock market (Lewis, 2014).

If fog computing and the IoT are to work then there must be reliable, fast scalable connectivity. Landaluce et al. (2020) review applications and challenges to what they refer to as the "two fundamental pillars" of the IoT: radio frequency identification (RFID) and wireless sensor networks (WSNs). These two technologies are based on wireless sensing and communication but at present RFID technology has many limitations. Landaluce et al. review these concerns and challenges under the headings and constraints of their limited read/write range, relatively high cost of the readers and problems with present communication protocols that can produce RFID tag 'collisions' when an application has at least one reader and a number of RFID sensors. There are three main areas of RFID sensing in transportation: smart roads; environment-connected vehicles and people-centric integrated transportation including vehicle location in intelligent transportation systems (Zhang et al., 2019). Despite listing six existing technologies for WSNs, Landaluce et al. (2020) note that there are many challenges including reliability, energy consumption, scalability and communication protocols. However, some of these challenges will be overcome with RFID and WSN integration. A convincing application and integration of many of these technologies is provided by the TIMON system (Cankar *et al.*, 2018). TIMON was a research project funded by the European Commission under the Horizon 2020 Programme. It lasted from June 2015 until the end of 2018 and produced 'Enhanced real time services for an optimized multimodal mobility relying on cooperative networks and open data' (TIMON, 2018). In short, it was an intelligent transportation system that provided real-time information to drivers, vulnerable road users (VRUs) and businesses in Ljubljana, Slovenia, a true 'transportation ecosystem in a fog to cloud computing environment'. The data sources were vehicles and VRUs, infrastructure sensors and open data. The technologies involved included hybrid communications using an ITS-5G and an LTE (Long Term Evolution standard) hybrid communication system with the desired low latency for vehicle-to-vehicle (V2V) connectivity. This allowed communications between vehicles and VRUs and also to the cloud. The overarching communication system is V2X (vehicle-to-everything), which includes V2D (vehicle-to-device), V2I (vehicle-to-infrastructure), V2N (vehicle-to-network), V2G (vehicle-to-grid), V2P (vehicle-to-pedestrian) and V2V (vehicle-to-vehicle).

A second set of technologies used in TIMON are artificial intelligence algorithms for traffic congestion prediction in four successive time horizons. Several approaches were tested including a hybrid algorithm that combined a Genetic Algorithm with a Cross Entropy (GACE) approach, a Linear Decreasing Weight-Particle Swarm Optimization and a Steady-State Genetic Algorithm for Extracting Fuzzy Classification Rules. Details are provided in Osaba *et al.* (2016) and Lopez-Garcia *et al.* (2016). The first of these approaches, GACE, produced the best predictions. This research also led to an Application Programming Interface (API) that provided businesses with route planning information that included congestion predictions, updates on road accidents and roadworks and additional data on high traffic density areas and heavy-duty vehicle restrictions among others. Positioning geospatial technologies that were deployed included Global Navigation Satellite System (GNSS) receivers that were linked with Inertial Measurement Units and sensor fusion techniques to allow for accurate positioning in places where satellite signals were unavailable in locations in tunnels, roads surrounded by high buildings and due to various forms of electronic interference. TIMON provided a total of nine services in three groups of beneficiaries: (1) vulnerable road users were given maps of traffic densities and congestion; collision alerts; and dynamic route re-planning for cyclists; (2) drivers were supplied with alerts to increase safety and reduce emissions including emergency vehicle alerts; road hazard warnings; and dynamic route re-planning similar to the service offered to cyclists; (3) general purpose services that were offered to individuals and businesses included a multimodal commuter service; enhanced road traffic information; and, finally, a collaborative ecosystem that facilitates a community of transportation users that can share their experiences in a volunteered geographic information environment (Goodchild, 2008). The TIMON cloudlet environment collects data from open and closed data sources including IoT sensors, vehicle mounted on-board units (OBUs), which communicate with RSUs over dedicated short-range communication channels and from smartphones. The cloud has four logical components: IaaS, automation protocols, monitoring and notifying capabilities and access control management. Cankar *et al.* (2018) also described a proposed fog to cloud design since they argue that the ITS field demands fast and reliable responses that can only be delivered by a Fog and Cloud environment. They considered two possibilities: (1) the Horizon 2020, supported mF2C platform (management Fog-to-Cloud) described in Masip-Bruin *et al.* (2020) and mF2C Project (2017) and (2) the OpenFog Consortium (Minh *et al.*, 2018). The TIMON developers chose the mF2C platform in preference to the architecture specified by the Open Fog Consortium. The latter assumed powerful Fog nodes that were not available in the TIMON ecosystem and represented a more conceptual guidance approach whereas the mF2C platform provided concrete implementations of concepts and components that made mapping more straightforward (Cankar et al., 2018: 269). The Fog layer developed on the mF2C platform offloaded work from the centralized cloud to the distributed fog nodes speeding up processing and lowering latency.

5G and 6G mobile computing environments

Fifth generation technology (5G) for cellular networks began to appear just before 2020. In 2021 the Global mobile Supplier Association (GSA, 2021) issued a report on the state of subscriptions (cellphones) in mid-2020. At that time there were 5.55 billion 4G LTE (Long Term Evolution) subscriptions and just over 137 million 5G subscriptions. In the early 2020s LTE subscriptions appeared to reach their peak and then their growth rate began to decline at the expense of 5G technology. GSA estimates that in 2025 5G will have about 30% of the market and LTE over 50% of the more than 10 billion subscriptions. Ahmad (2015) has estimated that 5G technology will provide many advantages to industries such as connected transportation services including: a 1,000-fold increase in capacity; support for more than a 100 billion connections; speeds of up to 10 Gbit/s speeds and a latency below 1 millisecond. Ahmad states that 5G will offer much greater connectivity for technologies including 2G, 3G, the various versions of LTE and M2M (machine-to-machine). Also, it will provide ultra-high-speed links for high-definition video and low data rates for sensor networks. This will give great flexibility for complex transportation hubs and environments such as airports. For example, Salis and Jensen (2020) and Salis et al. (2019) describe their work with the Horizon 2020 mF2C project building the computer infrastructure for a "smart airport", one of many local transportation environments within the smart city. They describe their fog-to-cloud system as part of the Cloud-to-Things continuum, noting that the global cloud computing market is expected to more than double between 2018 and 2023 to nearly a $625 billion dollar industry much of which will be devoted to transportation applications in the smart city and the smart world of travelers. The main tools provided by a proof-of-concept system developed for Cagliari Airport in Sardinia include benefits to travelers in moving through the airport, proximity marketing to those travelers based on a recommendation system that uses machine learning plus other benefits. Airport managers benefit from heat maps that show travelers' behaviour including bottlenecks in the system allowing tactical and strategic planning and highlighting security concerns.

Impressive as these contributions are based on 4G+ and 5G computing environments, much greater benefits may be expected as the world moves to 6G capabilities. The history of mobile computing technology has been described by (Dang et al., 2020). Their timeline spans the whole range of mobile telecommunications from 1G to the putative 6G. 4G is defined as starting in 2020 and lasting until 2020, at which time 5G begins to takeover while 6G is estimated to start in 2030. The dominance of the various technologies is not absolute as parts of the older technologies can remain important for many years after the newer technology has debuted. These authors provide a qualitative assessment of the expected benefits of 6G technology, which include significant improvements in energy efficiency, affordability and customization and the embedding of artificial intelligence throughout the system. The most dramatic improvements over 5G will be in security secrecy and privacy. Letaief et al. (2019) provided a detailed 'roadmap' of how 6G will be introduced starting with 6G 'visions' in 2020, evaluations in 2025 and products in 2028. Their timeline is more optimistic than some because they argue many products will be introduced in steps reflecting an evolutionary approach to a federated network of products and components. Data rates will be up to 1 TB/second, with high energy efficiency and battery free IoT devices and extremely low end-to-end latency of 1 millisecond. There are many technological problems to solve and each solution requires technological improvements in other components of the system, but there is little doubt that these breakthroughs will provide new opportunities for smart transportation in the smart city. It is a considerable challenge in the IoV environment to establish a highly reliable connectivity in a sustainable, fair manner with low energy consumption among vehicles and from vehicles to road side infrastructure. However, Sodhro *et al.* (2020) provide experimental results for such a system, a self-adaptive, green and reliable approach based on existing 5G technology, which is very much the ITS of the future.

The complexities of mobility sharing technologies, the world of Lyft and Uber

Technology enabled, mobility sharing is a highly complex, multifaceted system that is largely a phenomenon of the twenty-first century and is highly dependent on the geospatial technologies (Shaheen *et al.*, 2017). Mobility sharing can be divided into three groups: sharing of a vehicle, sharing of a passenger ride and sharing of a delivery ride. One of the most popular of the ride sharing services is on-demand ridesourcing. Companies offering these services are variously known as transportation network companies (TNCs), ride hailing, ride booking or simply by their company names such as Uber and Lyft. According to Eidelson (2021) the companies themselves prefer to be known as tech platforms and not as transportation companies so that they are not subject to the labour laws that would apply were they considered to be transportation companies with employees. Presumably the tax benefits of such categorizing are significant as is the case with social media companies that vigorously deny being newspapers even though they deliver the news (Medina and Waters, 2021). Schaller (2021) has examined evidence from Uber and Lyft to determine if their ride sharing activities from 2014 to 2020 have reduced traffic congestion, the need for parking space or vehicle emissions as was hoped. Just prior to the pandemic, levels of carpooling (sharing) rides led to more than a doubling of vehicle miles travelled (VMT) in New York, San Francisco and Boston when comparing patrons' ride hailing trips with their previous mode of travel. Increases in VMT were due primarily from the need for deadheading before pickup and the fact that many patrons switched from public transit, biking or walking. These VMT increases were only partially offset by decreases incurred due to the use of carpooling for first mile/last mile connections to public transit and the lack of a need for cruising for parking when patrons had previously used their private vehicles. Schaller (2021) suggested these observations would apply also to shared autonomous vehicles (SAV) in the future. How these results will endure in a postpandemic world is uncertain. Carpooling and the use of SAVs is likely to be most attractive to former public transit users especially for the 'walk to a stop' model offered by Uber Pool Express and Lyft Shared Rides, but any such movement will add to VMT.

A COVID-19 coda for geospatial technologies in transportation

In 2020, the world changed due to the COVID-19 Pandemic and transportation was arguably the sector of the economy that changed most dramatically. The virus was spread from country to country by international travelers on planes as it had done so for the SARS-1 outbreak in 2002 (Waters, 2018b). The virus that caused the COVID-19 pandemic is occasionally referred to as SARS-2 (Christakis, 2020). Once SARS-2 leaped from one country to the next, community transmission took over. A primary way to curtail the growth in infections was to close a country's borders to air travelers and to lockdown communities. New technologies allowed for online, virtual meetings. AASHTO's 2020 Annual Meeting was for the first time online and this was true for almost every organization within countries and around the world.

Technologies that supported online interactions such as Zoom, Microsoft Teams, Google Meet, Facebook Messenger Rooms, the venerable Skype and a host of others replaced the need for travel. Security and ease of use were issues that affected the popularity of various products. Education at all levels from kindergarten to graduate studies was delivered online dramatically reducing home-based trips to schools and post-secondary educational facilities. People were encouraged to work from home (WFH) and downtown office towers emptied out. WFH became a defining characteristic of the pandemic so much so that the expression and acronym could be found across the globe. For example, an article published in Indonesia (Purwanto, 2020) in the Indonesian language used both the English expression and associated acronym. Purwanto used semi-structured interviews with teachers in an elementary school to determine the advantages and disadvantages of WFH. The former included flexibility in working hours, greatly reduced transportation costs and lower stress levels from the lack of a daily commute. Disadvantages included a lower work motivation, higher home electricity and Internet costs and data security concerns. The invasive nature of WFH may be an issue when more than one adult in a household is involved and when more than one child is also involved in online education. Adequate computer resources and bandwidth may be additional challenges. Social isolation and emotional distancing were found to be a problem in a study in India but may be a concern everywhere. WFH is not an all-or-nothing change brought about by COVID-19 and there is already research that has explored hybrid telecommuting solutions. For example, Mauras *et al.* (2020) explored hybrid telecommuting solutions that include high granularity analysis at the individual level of hybrid telecommuting strategies in the workplace and in primary and high schools. Four telecommuting mitigation strategies were investigated and were then ranked in terms of their ability to reduce the effective reproduction number of the virus (R_e) (see Christakis, 2020: 51 for a discussion of R_e and R_0) to less than 1.0. None of the strategies, only full telecommuting, worked to reduce the spread of the pandemic when the R_0 (R naught) was above 1.7. But once the R_0 had been reduced then the four strategies could be ranked from best to worst: Rotating week-by-week, Rotating day-by-day, On-Off week-by-week and On-Off day-by-day. All of the strategies worked to decrease R_e but the differences between them were slight and so practical and psychological considerations were recommended, with the longer of the rotations the preferred of the four strategies.

The persistence of various forms of telecommuting after the pandemic is over will be determined using the various forms of geospatial transportation approaches discussed above with respect to smart cities. During the COVID-19 pandemic, take-away, kerbside pick-ups and grocery deliveries replaced in-person visits to restaurants, pharmacies and supermarkets. Gyms closed but parks remained open. Travel patterns changed for work, school and recreation and, indeed, for almost every other activity. The world became a quieter place according to Koren (2020). With so many people staying home – and public-transit agencies cutting service as a result – there was significantly less noise from cars, buses, trains and other forms of transportation. Monitoring these changes was an opportunity for geospatial technologies and who better to track these changes but Google. Thus, Google Mobility Reports have been created by those who use Google's technology and turn on their Location History (Google, 2020a). Google created these Reports for almost every country in the world and

Table 30.1 Mobility report for Calgary, Canada and the United States 17th November to 29th December 2020 (percentage increases and decreases in mobility patterns)

	Calgary	Canada	United States
Retail and recreation	−29	−34	−17
Grocery and pharmacy	0	−2	−6
Parks	+19	+20	−8
Transit stations	−68	−65	−39
Workplaces	−60	−59	−45
Residential	+22	+21	+14

were doing this as a service 'so long as public health officials find them useful in their work to stop the spread of COVID-19'. The data are anonymized using artificial noise that is added to the data to prevent identification of individuals but allows useful inferences to be obtained from the travel patterns (Google, 2020b). For confidentiality reasons, the spatial granularity of the data is low. For example, the Canadian province of Alberta (2020 population in 2020 of 4.4 million) is divided into 19 divisions. These divisions vary in size by population and area. Calgary and Strathmore both have their own dataset but in 2020 Calgary's population was 1.5 million while Strathmore's was only 17,000 and thus some data for the latter are suppressed for privacy reasons. There are six datasets recorded and Table 30.1 shows the values for Calgary, Canada and the United States for six areas of mobility: retail and recreation; grocery and pharmacy; parks; transit stations; workplaces and residential. Google's ability to track users of their phones and other software represents a new opportunity for geospatial technologies to track the economic and behavioural impact of the pandemic.

The baseline is the median value, for the corresponding days of the week, during the five-week period from 3rd January to 6th February 2020.

The travel patterns shown for Calgary and Canada are similar: big drops in travel for retail and recreation (discretionary purchases and gyms were deemed unsafe and/or closed) and even larger drops for transit stations and workplaces (people working at home); travel to grocery stores and pharmacies remained more or less unchanged (essential services aided by kerbside pickups); while travel from residences and within neighbourhoods increased as did travel to parks (safe recreation). The differences between Canada and the United States are dramatic especially for travel to parks (possibly due to more unemployed Canadians having government financial support) and workplace and transit stations due to more Canadians working from home. Deng et al. (2020) reported that StatsCan found that over 39% of Canadian workers were teleworking from home in the first week of March 2020, whereas in 2018 only 13% of Canadians were working from home. Dingel and Neiman (2020) have estimated that 37% of Americans could work completely from home. The discrepancies in and between these Canadian and American mobility patterns and changes remain to be explained.

Conclusion

This chapter has described the importance of the triple bottom-line of sustainable transportation that attempts to balance the competing economic, social and environmental demands on modern transportation systems. It has also covered the rise in importance of geographic information systems technology with companies such as Esri managing all aspects of transportation in what has come to be known as GIS T. It has followed Esri Canada's recommendations on how GIS-T may be used to manage all aspects of a modern transportation system. In addition, the importance of the current state-of-the-art and future developments in mobile computing systems including 5G and 6G technology have been discussed. The TIMON system developed as a Horizon 2020 Project of the European Union in the city of Ljubljana, Slovenia, demonstrated to the world what the future of a 'smart city' will be like if vehicle-to-everything (V2X) can be leveraged so the needs of all members of a population, including those of its vulnerable users, are considered. The impact of disruptive technologies such as the ride-hailing services of Lyft and Uber and their interfacing with smart transportation management systems has yet to be determined. Finally, it should be noted that while some of the effects of the COVID-19 pandemic on transportation services has been shown by Google's mobility reports the extent to which these changes will persist remains unknown.

References

AASHTO (2019) "2019 GIS-T State Survey" *Presentation at the 2019 GIS-T Symposium* Kissimmee, FL, 23rd–26th April Available at:https://gis-t.transportation.org/wp-content/uploads/sites/51/2019/04/GIS-T2019StateSummaryPresentationFINAL.pdf (Accessed: 11th February 2020).

Ahmad, M. (2015) "4G And 5G Wireless: How They Are Alike and How They Differ" Available at: http://www.androidauthority.com/4g-and-5g-wireless-how-they-are-alike-and-how-they-differ-615709/.

Butler, J.A. (2008) *Designing Geodatabases for Transportation* Redlands, CA: Esri Press.

Cankar, M., Stanovnik, S. and Landaluce, H. (2018) "Transportation Ecosystem Framework in Fog to Cloud Environment" *IEEE/ACM International Conference on Utility and Cloud Computing Companion (UCC Companion)* pp.266–271 DOI: 10.1109/UCC-Companion.2018.00066.

Christakis, N.A. (2020) *Apollo's Arrow: The Profound and Enduring Impact of Coronavirus on the Way We Live* New York: Little, Brown Spark.

Dang, S., Amin, O., Shihada, B. and Alouini, M.S. (2020) "What Should 6G Be?" *Nature Electronics* 3 (1) pp.20–29.

Deng, Z., Morissette, M. and Messacar, D. (2020) "Running the Economy Remotely: Potential for Working from Home During and After COVID-19" *StatCan Covid-19: Data to Insights for a Better Canada* Available at: https://www150.statcan.gc.ca/n1/pub/45-28-0001/2020001/article/00026-eng.htm (Accessed 2nd January 2021).

Dingel, J.I. and Neiman, B. (2020) "How Many Jobs Can Be Done at Home?" *National Bureau of Economic Research, Working Paper 26948* DOI: 10.3386/w26948.

Dueker, K.J. and Butler, J.A. (2000) "A Geographic Information System Framework for Transportation Data Sharing" *Transportation Research Part C: Emerging Technologies* 8 (1–6) pp.13–36.

Eidelson, J. (2021) "The Gig Economy Is Coming for Millions of American Jobs" *Bloomberg* (21st February) Available at: https://www.bloomberg.com/news/features/2021-02-17/gig-economy-coming-for-millions-of-u-s-jobs-after-california-s-uber-lyft-vote (Accessed: 28th February 2021).

Esri (2020a) "FAQ: Does Esri have industry specific data models?" Available at: https://support.esri.com/en/technical-article/000011644 (Accessed: 11th February 2020).

Esri (2020b) "NY DOT Design Poster ArcGIS Transportation Data Model" Available at: http://downloads2.esri.com/support/TechArticles/Transportation_Data_Model.pdf (Accessed: 16th February 2021).

Esri Canada (2020) "Transportation Overview" Available at: https://www.esri.ca/en-ca/solutions/industries/transportation/overview (Accessed: 17th February 2021).

Global Mobile Supplier Association, (2021) "5G Ecosystem Report, Executive Summary" Available at: https://gsacom.com/reports/ (Accessed: 28th February 2021).

Goodchild, M.F. (2008) "Commentary: Whither VGI?" *GeoJournal* 72 (3–4) pp.239–244.

Google (2020a) "See How Your Community is Moving Around Differently Due to COVID-19" Available at: https://www.google.com/covid19/mobility/.

Google (2020b) "Helping Developers and Organizations Use Differential Privacy" Available at: https://www.youtube.com/watch?v=FfAdemDkLsc&feature=youtu.be&hl=en.

Hultquist, C., Sava, E., Cervone, G. and Waters, N. (2018) "Damage Assessment of the Urban Environment During Disasters Using Volunteered Geographic Information" In Schintler, L.A. and Chen, Z. (Eds) *Big Data for Regional Science* New York: Routledge, pp.214–228.

Jing, C., Du, M., Li, S. and Liu, S. (2019) "Geospatial Dashboards for Monitoring Smart City Performance" *Sustainability* 11 (20) 5648 pp.1–23.

Koren, M. (2020) "The Pandemic Is Turning the Natural World Upside Down" *The Atlantic* (2nd April) Available at: https://www.theatlantic.com/science/archive/2020/04/coronavirus-pandemic-earth-pollution-noise/609316/?utm_source=pocket-newtab (Accessed: 28th February 2021).

Landaluce, H., Arjona, L., Perallos, A., Falcone, F., Angulo, I. and Muralter, F. (2020) "A Review of IOT Sensing Applications and Challenges Using RFID and Wireless Sensor Networks" *Sensors* 20 (9) 2495 pp.1–18.

Leszczynski, A. (2019) "Spatialities" In Ash, J., Kitchin, R. and Leszczynski, A. (Eds) *Digital Geographies* Los Angeles, CA: Sage, pp.13–23.

Letaief, K.B., Chen, W., Shi, Y., Zhang, J. and Zhang, Y.J.A. (2019) "The Roadmap to 6G: AI Empowered Wireless Networks" *IEEE Communications Magazine* 57 (8) pp.84–90.

Lewis, M. (2014) *Flash Boys: A Wall Street Revolt* New York: W.W. Norton & Company.

Li, W., Batty, M. and Goodchild, M.F. (2020) "Real-time GIS for Smart Cities" *International Journal of Geographical Information Science* 34 (2) pp.311–324 DOI: 10.1080/13658816.2019.1673397.

Litman, T. (2020) "Autonomous Vehicle Implementation Predictions Implications for Transport Planning" Victoria, BC, Canada: Victoria Transport Policy Institute Available at: https://www.vtpi.org/avip.pdf (Accessed: 15th February 2021).

Lopez-Garcia, P., Onieva, E., Osaba, E., Masegosa, A.D. and Perallos, A. (2016) "A Hybrid Method for Short-Term Traffic Congestion Forecasting Using Genetic Algorithms and Cross Entropy" *IEEE Transactions on Intelligent Transportation Systems* 17 (2) pp.557–569.

Masip-Bruin, X., Marín-Tordera, E., Ferrer, A.J., Salis, A., Kennedy, J., Jensen, J., Jukan, A., Bartoli, A., Badia, R.M., Cankar, M. and Bégin, M.E. (2020) "mF2C: The Evolution of Cloud Computing Towards an Open and Coordinated Ecosystem of Fogs and Clouds" In Schwardmann, U., Boehme, C., Heras, D.B., Cardellini, V., Jeannot, E., Salis, A., Schifanella, C., Manumachu, R.R., Schwamborn, D., Ricci, L., Sangyoon, O., Gruber, T., Antonelli, L. and Scott, S.L. (Eds) *Euro-Par 2019: Parallel Processing Workshops (Lecture Notes in Computer Science Volume 11997)* Cham, Switzerland: Springer, pp.136–147 DOI: 10.1007/978-3-030-48340-1_11.

Mauras, S., Cohen-Addad, V., Duboc, G., la Tour, M.D., Frasca, P., Mathieu, C., Opatowski, L. and Viennot, L. (2020) "Mitigating COVID-19 Outbreaks in Workplaces and Schools by Hybrid Telecommuting" *medRxiv* DOI: 10.1101/2020.11.09.20228007.

McNally, M.G. (2000) "The Four-step Model" In Hensher, D.A. and Button, K.J. (Eds) *Handbook of Transport Modelling* Oxford: Elsevier, pp.35–52.

Medina, R.M. and Waters, N.M. (2021) "Social Network Analysis" In Fischer, M.M. and Nijkamp, P. (Eds) *Handbook of Regional Science* (2nd ed.) Heidelberg: Springer, pp.725–740 DOI: 10.1007/978-3-642-36203-3_49-1.

Medina, R.M., Cervone, G. and Waters, N.M. (2017) "Characterizing and Predicting Traffic Accidents in Extreme Weather Environments" *The Professional Geographer* 69 (1) pp.126–137.

mF2C project (2017) "Towards an Open, Secure, Decentralized and Coordinated Fog-to-Cloud Management Ecosystem (mF2C)" Available at: https://www.mf2c-project.eu/wp-content/uploads/2018/02/Whitepaper_F2C.pdf (Accessed: 28th February 2021).

Minh, Q. T., Kamioka, E. and Yamada, S. (2018) "CFC-ITS: Context-Aware Fog Computing for Intelligent Transportation Systems" *IT Professional* 20 (6) pp.35–45.

NHTSA (2017) "Automated Driving Systems 2.0: A Vision for Safety" Available at: https://www.nhtsa.gov/sites/nhtsa.dot.gov/files/documents/13069a-ads2.0_090617_v9a_tag.pdf (Accessed: 15th February 2021).

Ning, Z., Huang, J. and Wang, X. (2019) "Vehicular Fog Computing: Enabling Real-Time Traffic Management for Smart Cities" *IEEE Wireless Communications* 26 (1) pp.87–93.

Ning, J., Huang, X., Susilo, W., Liang, K., Liu, X., & Zhang, Y. (2020). Dual access control for cloud-based data storage and sharing. *IEEE Transactions on Dependable and Secure Computing*.

NUST (2020) "Transportation and Sustainability: The Triple Bottom Line" Available at: https://www.nust.na/sites/default/files/documents/TRANSPORTATION%20AND%20SUSTAINABILITY-1.pdf Accessed: 9th February 2020).

Olsen, M.J., Raugust, J.D. and Roe, G.V. (2013) *Use of Advanced Geospatial Data, Tools, Technologies, and Information in Department of Transportation Projects (Synthesis 446)* Washington, DC:

Transportation Research Board Available at: https://nap.nationalacademies.org/catalog/22539/-use-of-advanced-geospatial-data-tools-technologies-and-information-in-department-of-transportation-projects.

Osaba, E., López-García, P., Masegosa, A.D., Onieva, E., Landaluce, H. and Perallos, A. (2016) "TIMON Project: Description and Preliminary Tests for Traffic Prediction Using Evolutionary Techniques" *Proceedings of the 2016 on Genetic and Evolutionary Computation Conference Companion* pp.1471–1472 DOI: 10.1145/2908961.2931635.

Purwanto, A. (2020) "Studi Eksplorasi Dampak Work From Home (WFH) Terhadap Kinerja Guru Selama Pandemi Covid-19" *Journal of Education, Psychology and Counseling* 2 (1) pp.92–100.

Rattigan, K. (2018) "How Much Data Does Your Car Collect?" *Robinson+Cole* Available at: https://www.dataprivacyandsecurityinsider.com/2018/11/how-much-data-does-your-car-collect-heres-a-reminder/?utm_source=feedburner&utm_medium=feed&utm_campaign=Feed%3A+DataPrivacyAndSecurityInsider+%28Data+Privacy+%2B+Security+Insider%29 (Accessed: 20th February 2021).

Salis, A. and Jensen, J. (2020) "A Smart Fog-to-Cloud System in Airport: Challenges and Lessons Learnt" *21st IEEE International Conference on Mobile Data Management (MDM)* Versailles pp.359–364.

Salis, A., Jensen, J., Bulla, R., Mancini, G. and Cocco, P. (2019) "Security and Privacy Management in a Fog-to-Cloud Environment" *European Conference on Parallel Processing* pp.99–111.

Schaller, B. (2021) "Can Sharing a Ride Make for Less Traffic? Evidence from Uber and Lyft and Implications for Cities" *Transport Policy* 102 pp.1–10.

Shaheen, S.A., Bansal, A., Nelson, C. and Adam, A. (2017) *Mobility and the Sharing Economy: Industry Developments and the Early Understanding of Impacts* Berkeley, CA: Transportation Sustainability Research Center Available at: https://escholarship.org/uc/item/96j5r729 (Accessed: 28th February 2021).

Sodhro, A.H., Pirbhulal, S., Sodhro, G.H., Muzammal, M., Zongwei, L., Gurtov, A., de Macêdo, A.R.L., Wang, L., Garcia, N.M. and de Albuquerque, V.H.C. (2020) "Towards 5G-Enabled Self Adaptive Green and Reliable Communication in Intelligent Transportation System" *IEEE Transactions on Intelligent Transportation Systems* 22 (8) pp.5223–5231.

Tahir, M.N., Leviäkangas, P. and Katz, M. (2022) "Connected Vehicles: V2V and V2I Road Weather and Traffic Communication Using Cellular Technologies" *Sensors* 22 (3) 1142.

Takács, Á., Drexler, D.A., Galambos, P., Rudas, I.J. and Haidegger, T. (2018) "Assessment and Standardization of Autonomous Vehicles" *IEEE 22nd International Conference on Intelligent Engineering Systems (INES)* pp.000185–000192.

TIMON (2018) "What is TIMON?" Available at: https://www.timon-project.eu/index.php/about-timom/what-is-timon.html (Accessed: 22nd February 2021).

Waters, N.M. (1999) "Transportation GIS: GIS-T" In Longley, P., Goodchild, M., Maguire, D. and Rhind, D. (Eds) *Geographical Information Systems: Principles, Techniques, Applications and Management* (2nd ed.) New York: Wiley, pp.827–844.

Waters, N.M. (2018a) "History of GIS" In Richardson, D., Castree, N., Goodchild, M.F., Kobayashi, A., Liu, W. and Marston, R. (Eds) *International Encyclopedia of Geography: People, the Earth, Environment, and Technology* New York: Wiley.

Waters, N.M. (2018b) "Tobler's First Law of Geography (Revised)" In Richardson, D., Castree, N., Goodchild, M.F., Kobayashi, A., Liu, W. and Marston, R. (Eds) *International Encyclopedia of Geography: People, the Earth, Environment, and Technology* New York: Wiley.

Zhang, Y., Ma, Y., Liu, K., Wang, J. and Li, S. (2019) "RFID Based Vehicular Localization for Intelligent Transportation Systems" *2019 IEEE International Conference on RFID Technology and Applications (RFID-TA)* Pisa, Italy pp.267–272.

31
GEOSPATIAL TECHNOLOGIES IN ELECTRICAL SYSTEMS

Ivan Bobashev

People the world over depend on electricity for practically all systems of life, from food to education to healthcare. It is not surprising then that demand for electricity in emerging economies has been increasing at staggering rates. In Indonesia, the electrification rate is at 95%, up from 50% in 2000 and in Bangladesh, electricity now reaches more than 80% of the population, up from 20% in 2000 (IEA, 2018). Industrialized countries depend so heavily on electrical systems that the direct and indirect costs of a national blackout are astronomical. A blackout in the United States and Canada in 2003, affecting 55 million people and having only lasted 16–72 hours, resulted in economic losses estimated at $6.4 billion (Abedi et al., 2018).

The electrical grid is arguably the most complicated machine humans have built. Spanning the entire world, it is the single most recognizable indicator of human activity when the Earth is observed from outer space (Figure 31.1).

Understanding the spatial dimension of electrical systems and being familiar with appropriate tools of analysis are essential requisites to being able to maximize the benefits that humanity gains from greater electricity access and reliability. Thus, geographic information systems are vital to the function of the electrical grid. Geographic information systems (GIS) are computer systems that collect, analyse and display data associated with the Earth's surface (National Geographic Society, 2012).

This chapter serves as an introduction to the applications of geospatial technologies in electrical systems. It comprises a brief overview of electrical systems, a discussion of geospatial problems in the industry, and common methods in geospatial technologies for improving electrical systems. It concludes with addressing the limitations of the current state of GIS. Special attention is given to the role of GIS in the process of economic development as well as the rising paradigm of renewable and distributed energy resources. An interdisciplinary approach should allow readers of different backgrounds to have a better understanding of how GIS can support them, and how a geospatial approach can help solve problems in the electricity sector.

Background

A similar principle governs the design of modern electrical grids around the world. A grid consists of four major subsystems: generation, transmission, distribution, and offtake (commonly the customer).

Figure 31.1 Satellite photograph of Earth at night from outer space. Courtesy of NASA Earth Observatory images by Joshua Stevens, using Suomi NPP VIIRS data from Miguel Román, NASA's Goddard Space Flight Center

Generation is a system whereby the primary energy is converted into electrical energy. Generation systems include all coal, atomic, and hydroelectric power stations. Carbon emissions from excessive use of fossil fuels in electricity generation have led to climate change, so countries are rethinking electricity generation and investing in renewable sources such as wind turbines and solar power.

The second component of the grid is transmission. A system of step-up transformers, high voltage lines, and substations is responsible for sending power over great distances. Electricity is collected from the generation station and a step-up transformer brings the electrical current up to a standard high voltage for the transmission lines to carry. Transmission lines carry high voltage electricity to substations in cities, towns, and other locations close to customers. Substations then transform high voltage electricity to medium-voltage feeders (Figure 31.2).

The last subsystem is the distribution network. Distribution is a network of medium voltage lines, step down transformers, and low voltage lines leading to offtakers such as homes, schools, and offices.

The offtaker plays a critical role in the system because it is ultimately the performance of the grid at the level of the offtaker that defines performance as a whole. Offtakers must be prepared to receive power flowing through the distribution network with appropriate infrastructure, usually supplied by the managing utility. The managing utility must also balance the power supplied and the power consumed to avoid grid failures.

Technologies such as smart meters are increasing in popularity in industrialized countries, which automate reporting functions and allow for instantaneous communication between

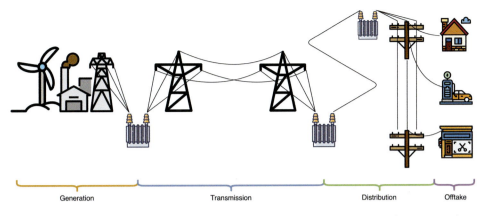

Figure 31.2 Simplified schematic of how modern electrical grids bring electricity from power plants to customers

---

Box 31.1 Dumsor – intermittent power in Ghana

In Ghana, one of the most potent of everyday struggles is intermittent power and frequent outages. These outages are so persistent that the term 'Dumsor' (a combination of the Twi words for on-and-off) is ubiquitous throughout social media, political campaigns, and even popular music all over the country. Those that can afford to run diesel generators during outages spend a disproportionate amount on electricity, while those that cannot afford diesel generators must experience the dangerous consequences of unreliable power. Operating a business efficiently and effectively becomes nearly impossible, while Dumsor at night can make walking the streets an especially harrowing affair.

---

the offtaker and the power supplier. Smart meters report interruptions in service and document all systems data pertaining to which network assets are delivering power to the offtaker.

With the emerging prominence of renewable energy in the United States and Europe, the original model of central power generation is being contested with a distributed energy resource (DER) model. In a DER model, generation happens in many places and at different times, generally closer to the location of the customer or where there is a renewable resource. While presenting a low-carbon alternative, the DER model is not free from the limitations that arise from balancing power throughout the grid. As household power generation, such as rooftop solar power, becomes more popular, the role of the offtaker in the system becomes increasingly complex as the balancing becomes bidirectional.

Utilities in emerging economies face a lack of transparency into operations of the grid at the level of the consumer, which causes significant problems. When considering emerging economies, electrical grids are rapidly expanding to service the basic needs of an increasing population. There is often pressure to rapidly increase electrification, which leads to rushed or illegal interconnection. In countries such as Brazil and South Africa, high numbers of unregistered or illegal offtakers have been responsible for deaths, injuries, and grid failures (Ecer and Mimmi, 2010; Louw, 2019). Proper grid management and grid visibility can mitigate these dangers in emerging economies.

Applications of methods in electrical systems

Transporting electrical energy from the point of generation to the point of use over vast distances is inherently a geospatial challenge. Mapping, planning, and managing grid infrastructure is fundamental to maintaining a steady flow of electrical power to the largest possible amount of people. Geographic information systems are an inextricable part of this process. This section will cover a variety of commonly occurring GIS challenges in the electricity sector, as well as insights into emerging applications and applications specific to developing economies.

Mapping and communication

Maps are familiar visualization tools that can aid stakeholders' understanding of the grid at an intuitive level. GIS visualization can greatly benefit internal discussions between specialized teams within utility companies. While electrical systems are often represented as schematics, they do not constitute a holistic version of the system that includes various contextual factors. Maps, therefore, provide deeper insight into how the system will behave in the real world.

Maps create a shared understanding between project developers, engineers, landowners, policy-makers, or any other potential stakeholder. GIS is commonly used in developing energy systems during negotiations between utilities and landowners over where to build transmission network assets. Optimal locations of transmission lines and substations depend on geographic, environmental, and economic factors, and the most optimal locations will almost certainly pass through privately held land. Maps are opportunities for both developers and landholders to come to a common understanding of where to build assets, who will be affected, the compensation that might be payable, and what other geographic considerations exist. Stakeholders in these negotiations will include – but not be limited to – technicians, lawyers, farmers, and public officials.

Optimizing electrical lines routing

Installing transmission lines is expensive and can have adverse effects on the surrounding environment. A single turn in a transmission line can cost an extra $500,000 in some instances (MISO, 2019). Optimization of transmission routes aims to minimize construction costs and negative environmental impacts while maximizing the safety and reliability of the system.

Geographic least-cost path models can aid in optimizing transmission routing. Solutions presented by Eroğlu and Aydin (2015) used least-cost paths created using layers of raster data, where each cell has an associated traversing cost. Factors influencing costs include elevation changes, vegetation, land rights, and necessary infrastructure. These factors (together with the shortest path, least turns constraints) produce a least-cost path between two points. The model presents the optimal routing depending on different infrastructure classes.

Associating network data with cadastral data

Associating network data with cadastral data can provide insights into landowners and customers without having to manually collect and manage information. Building new infrastructure transmission lines will inevitably involve crossing private land, so understanding relevant landowners becomes imperative. A utility may use an optimized least-cost path to cross-reference with cadastral data and thereby understand who must be approached to come to an agreement. When considering transmission lines that span thousands of miles, such tools can save countless hours of manual surveying.

In areas where the level of technology does not allow real-time monitoring of the grid, associating network data with cadastral data can save grid managers the work of manually maintaining customer information for billing.

Planning for load growth

As communities grow, so will the electrical load on the system. Depending on where there is population and economic activity growth, electrical systems will need to adjust to provide sufficient power to those new customers. Construction of new network assets will therefore depend on the spatial area of the projected growth, and carefully tracking developments can substantially contribute to the foresight that is an essential part of effective planning and investment.

Planning for load growth is particularly relevant to emerging economies, where large swathes of the population are joining the network. In the development sphere, GIS can identify underserved communities. Development organizations that focus on last mile electricity access can more effortlessly identify geographic zones with little or no access to reliable power.

Generally, surveys conducted on the ground by service teams can produce insight into load growth, but these can be very costly. Disparate recording styles and informational systems prevent the most comprehensive visibility and therefore cause inefficient or ineffective decision making. Investment into a centralized information system can mitigate poor visibility, prevent misallocation of resources, and achieve targets more quickly.

Identifying vulnerabilities in grid infrastructure

Electrical systems are subject to vulnerabilities and risks based on the geographic context. Sources of disruptions can be natural hazards, intentional attacks, and random failures (Abedi et al., 2018). Geospatial technologies can be particularly useful in preventing and mitigating natural hazards and random failures.

Natural phenomena can directly disrupt electrical systems, and GIS would be the primary tool for reducing the chances of disruptions. Generally, data on locations of fault lines, flood plains, or any other high-risk areas are publicly available. Some areas may even have compounding hazards. Avoiding these when locating large network assets such as substations is ideal, but in instances where a hazard zone is unavoidable, investment into vulnerability support is prudent. These kinds of analyses are commonplace in early development stages of infrastructure projects.

Random failures can happen at any point in the grid infrastructure, but the further upstream the disruption, the more services and life activities are likely to become negatively affected. Hence, a GIS system can support engineers in developing resilience measures at points where a random failure would cause the most cascading effects to the rest of the grid.

Servicing

GIS enables electricity providers to respond to outages and other servicing needs effectively and efficiently. Power outages may occur in any electrical system and teams of trained electricians are constantly at work servicing network assets. The losses and risks associated with power outage are directly related to the magnitude and duration of the outage. Therefore, implementing GIS to minimize the length of power outages can minimize the damages from them. GIS builds the capacity of utilities to be able to:

Precisely identify the point of failure

Smart-metering technologies allow for real-time status reports at the offtaker level. This way, the utility is notified of the outage and its location instantaneously. Smart meters can also report indicators which give insight into the health of the grid in the area overall. Further innovations in the Internet of things will create a real-time, digitized representation of the status of the grid.

In places where smart-grids are not appropriate or too expensive, more creative solutions are used to identify grid failures. Researchers have developed a method to use mobile phones to map black-outs.

(Klugman et al., 2019)

Direct service team to mend the failure

Coordinating a workforce to mend outages requires the right people with the right equipment to be in the right place. Customized routing systems and network analyses can identify where the most appropriate service team is located and how that team can reach the point of failure the most efficiently. In some cases, servicing machinery may limit the types of roads available to them to reach their destination, and GIS can optimize routing for such constraints.

Optimize routing of teams to address multiple points of failure over time

Resources and servicing machines are not infinite, and therefore resource allocation will certainly play a role in how quickly outages are resolved. Resource-poor areas are often also correlated with higher power outages, making routing GIS disproportionately helpful in emerging economies.

New systems of outage sensors are paving the way to identify outages and clustering them automatically, filtering false positives and pinpointing service interruptions on a map. Thanks to this GIS, dispatch teams need only to refer to the map, which will automatically inform service teams where to go, what kind of service equipment to bring, and so on.

A full stack of these capabilities can significantly enhance livelihoods by minimizing the negative impacts of drawn-out and far-reaching power outages.

Box 31.2 Dumsor, servicing power outages

During Dumsor events in Accra, outages are identified by a chain of call centres. When an outage occurs, the household calls the utility call centre, which gathers complaints until a cluster of power outages emerges, at which point it forwards the cluster to the service dispatch call centre, which then prioritizes its limited-service teams to dispatch to outage locations deemed most critical. The result of such a drawn-out system leaves many citizens of Accra without power for days.

Renewable energy modeling

Unlike fossil fuels, renewable energy sources cannot be stored and transported from their source to a generation station, therefore electricity generating stations need to be located in areas where there is a reliable and consistent source of the energy. Onshore wind farms require large areas with consistently high wind speeds and good service access. Solar energy harvesting, whether it is photovoltaic (PV) or concentrated solar power (CSP), requires clear skies for long periods (Figures 31.3 and 31.4).

Deserts and plains may be obvious candidates for solar power, but areas outside of these biomes may use satellite imagery with high temporal resolution to analyse solar irradiation and cloud cover over the course of the year.

Technical exclusion models can be very useful in modelling renewable energy. Solar and Wind resource data are generally raster surfaces. By collecting and rasterizing technical exclusion areas such as protected areas, urbanized areas, natural features, and terrain features, a technical exclusion filter can rapidly narrow down potential areas for development from large continental areas to specified potential projects (Maclaurin et al., 2019).

Energy storage for renewable energy can also be geographically specific. Pumped hydroelectric storage can drive water up a steep incline to store excess power, then release the water down the incline through turbines to discharge stored power. Pumped hydro is one of the most efficient and cost-effective energy storage technologies but is only feasible where there is access to water with an elevation differential. Digital elevation models (DEM) can help narrow down large areas to appropriate elevation changes. Satellite imagery can identify bodies of water and any other constraints or opportunities.

Geospatial technology methods

This section provides a survey level view of several useful tools and toolsets common in the electrical industry. It is not a comprehensive list but an appropriate starting point for exploration.

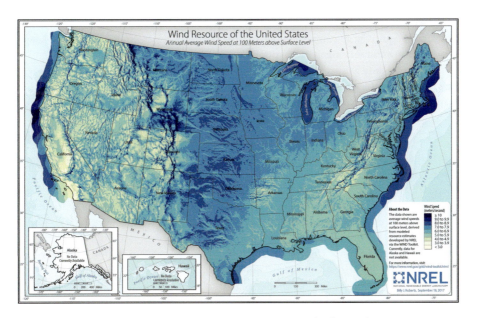

Figure 31.3   United States national average wind map. Courtesy of Billy J Roberts, NREL

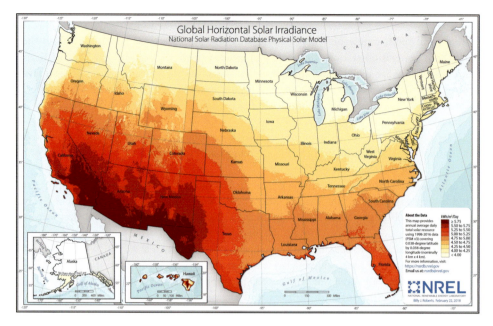

Figure 31.4 United States national global horizontal solar irradiance map. Courtesy of Billy J Roberts, NREL

Visualization and communication

Most people interact with GIS in the form of maps. In fact, for laypeople, GIS and cartography are interchangeable concepts. Maps, indeed an integral subset of GIS, are powerful tools for communicating developments of the grid between stakeholders. Instead of representing electrical systems as schematics, intelligible only to trained professionals, maps can visually represent electrical systems in a way that makes sense to a wide audience. Electrical grids are complex and require large teams of experts with different backgrounds and skills to keep the grid functioning and serving as many people as possible (Figure 31.5).

With more openly available GIS platforms such as QGIS and Google Earth, cartography is now more accessible to professionals without extensive training in GIS. Paid services such as MapBox and Carto can create beautiful designer-grade maps.

Network analysis

The structure of electricity transmission and distribution can best be described as a network. Network analyses are a set of tools which provide insight into interactions between agents connected by the network. A network is a system represented by edges and nodes. In a distribution system, transformers and poles are common nodes, while medium voltage feeders and distribution lines are best represented as edges. This categorization is important in building a digital network for establishing a proper topology. All relevant nodes should be linked to their respective edges. Once this network is built, a set of powerful analyses can be run.

Service area

Network service area is a powerful tool to assess which energy offtakers are serviced by which network assets. A service area analysis takes in point data and network data. The point

Figure 31.5 Visualization of electrical distribution

Box 31.3 Dumsor – measuring impact of installing new transformers

To assist the utility company in Accra, Ghana, researchers were interested in the distribution of power outages relative to transformers. Georeferenced data for feeder lines, distribution lines, and transformer locations was available, but information labelling distribution lines to transformers was limited. A network service area for each transformer was calculated and displayed to show which distribution lines belonged to which transformers. The result was a clearer understanding of which areas of the city were affected by the performance of any given transformer. In addition to this information, areas where there was a gap in service area served as an indication for potential installation of new network assets to increase reliability in those areas.

---

data can represent any facility or point network assets such as substations or transformers, while the network data take in data on distribution line length, voltage, and bare. Constraints of the analysis may include factors such as how far down the line and how far from the line on either side. The result is a polygon which represents the geographic area which is serviced by a particular network asset. One way to consider service area is a similar function to a radial buffer of a point, only bound to a network.

The network service area can be a powerful tool in the case where documentation is scarce and granular data may be unreliable, as it can help establish which network assets serve a particular community (Figure 31.6).

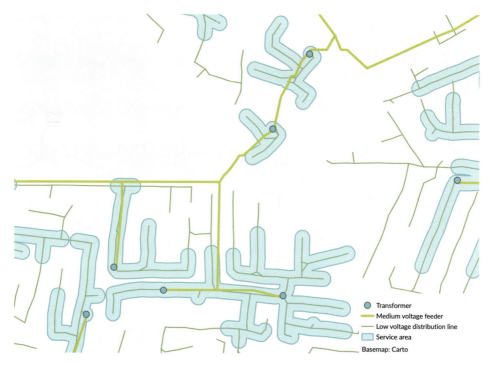

Figure 31.6 Example 150-metre service area of each transformer along the distribution network, extending 25 metres away from the line in either direction

Closest facility

A closest facility analysis is a versatile tool which measures the cost of travelling along a network between facilities and instances. This type of analysis can be a powerful tool for streamlining power outage responses. Consider facilities to be any node network asset (pole, transformer, or substation) and instances to be geographically specified calls reporting power outages. The call locations and network assets are related across the distribution network. The facility with the least-cost path to the outage report becomes a candidate of failure. An automated GIS can then identify a cluster of outage reports associated with a particular network asset i.e., if five calls of power outage come from households connected to the same pole, that pole is a good place to send a team for diagnostics. Five calls for power outage can also come from households connected to different poles, but all be serviced by the same transformer, in which case that transformer becomes a likely candidate for failure. Two outage reports may originate 50 metres from one another but be connected to separate distribution lines below separate transformers. A closest facility analysis would reduce the number of false-positives of clusters created by purely geographic clustering. Clustering using a closest facility analysis can drastically reduce the costs of diagnosing grid failures and reduce the lead time for dispatching service teams with appropriate equipment.

Suitability analysis

When planning installation of any electrical asset -be it new generation stations, substations, or any other electrical infrastructure – geospatial suitability analysis can help identify the most

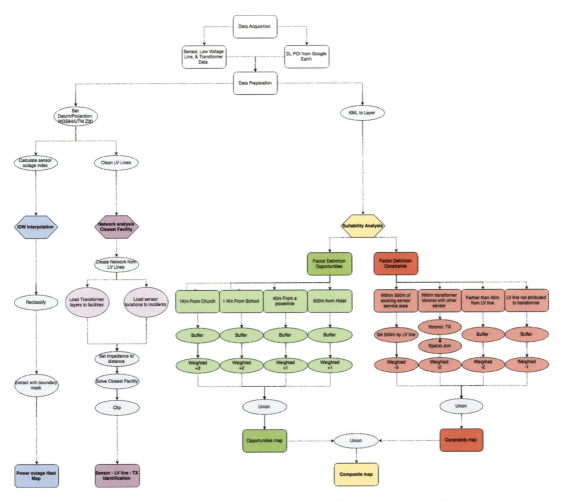

Figure 31.7 Flowchart of GIS workflow, including suitability analysis, network analysis, and interpolation

appropriate location-based a range of relevant criteria. A set of relevant geographic data layers is weighted and combined to synthesize a map indicating least suitable to most suitable locations.

The two principal balancing sides of a suitability analysis are opportunities and constraints. When developing a suitability analysis, it is useful to list all relevant geographic data layers, classify each as either an opportunity or a constraint, and assign a weight to each (Figure 31.7).

Technical exclusion models are also a class of suitability analysis where a number of layers that make development impossible act as filters, leaving only the windows of available areas for development (Maclaurin *et al.*, 2019).

Limitations

The current state of GIS in electrical systems has its limitations related to data quality, data collection, data maintenance, and technical accessibility.

Data quality

Data quality is generally the biggest limitation to GIS solutions in electrical systems. Since GIS is used to digitally model reality, the accuracy, precision, and timeliness of data inputs will translate into the usability of the GIS. Some estimates place the cost of data to be between 60% and 80% of the total cost of a GIS (Meyers, 2005).

Data collection

Collecting data can be a costly endeavour. The costs include the technology and manpower required to survey and physically verify network assets. At the most granular, teams of people collect ground truth geolocation of distribution poles, lines, and transformers, with names, voltages, and customer connections verified regularly. Unmanned aerial vehicles (UAVs) can also collect data, but the hardware, software, and piloting knowhow may be cost-prohibitive. Finally, data collection from satellites can be an option, but good spatial and temporal resolution satellite images will be expensive and processing multiband imagery into useful insights can take technical expertise and time.

Updating changes

Keeping the GIS up-to-date with the most recent state of the grid is a considerable challenge. The distribution grid is constantly changing – new customers are added, overhead lines are damaged by the elements, outdated hardware is replaced. Maintaining a GIS that accurately and precisely mirrors reality requires that documentation be integrated into the system promptly. An open channel of communication between GIS users and service teams would allow for sufficient training in documenting changes in a format that can be easily digestible for the GIS user.

Addressing errors

A limitation of any information system is addressing errors. Effectively addressing false negatives and false positives will result in more reliable, efficient, effective grids. To effectively address errors, a robust system for establishing ground truth is imperative. Many ambitious GIS projects for solving interesting problems fall short because they rely on verification that can only be done manually.

Technical accessibility

While many GIS operations may seem straightforward, technical expertise in the software and theory required to conduct many of these analyses may prevent casual GIS users from being able to gain the most insight from their geospatial data. The level of technical proficiency required is decreasing but is still limiting for many stakeholders. As custom-made Web applications become more commonplace, industry-specific GIS analyses can fall into the hand of larger stakeholder groups, allowing GIS technicians to focus on pushing the boundaries of possibility in the field.

Conclusion

The geospatial question is central to electrical systems. Any utility or other interested stakeholder would be wise to prioritize maintaining a geospatial information system and keeping it functional. To function properly, a GIS needs to be up-to-date and accurate. This may present a bigger challenge than mastering GIS techniques.

GIS can be just as powerful an ally in the context of emerging economies as it is in industrialized countries, providing contextually appropriate solutions. Relatively rudimentary GIS techniques have the ability to save time and resources when collecting, verifying, and processing data. Continued openness and democratization of GIS tools and data will surely contribute to increased access and reliability of the electric grid around the world.

Electrical systems are responsible for alleviating suffering and propelling humans to their highest potential. Maximum benefits that humanity gains from greater electricity access and reliability can only be achieved with a deep understanding of the spatial dimension of electrical systems, the interaction between electrical systems and the physical world, and the tools of analysis. GIS can be a powerful tool for improving electrical systems from the national grid level down to the consumer turning on the lights. This chapter is a starting point for anyone looking to harness the power of GIS to improve electrical systems.

References

Abedi, A., Romerio, F. and Gaudard, L. (2018) "Review of Major Approaches to Analyze Vulnerability in Power System" *Reliability Engineering System Safety* 183 pp.153–172 DOI: 10.1016/j.ress.2018.11.019.

Arderne, C., Zorn, C., Nicolas, C. and Koks, E. E. (2020) "Predictive Mapping of the Global Power System Using Open Data" *Scientific Data* 7 (19) pp.1–12 DOI: 10.1038/s41597-019-0347-4.

Ecer, S. and Mimmi, L. (2010) "An Econometric Study of Illegal Electricity Connections in the Urban Favelas of Belo Horizonte, Brazil" *Energy Policy* 38 pp.5081–5097 DOI: 10.1016/j.enpol.2010.04.037.

Eroğlu, H. and Aydin, M. (2015) "Optimization of Electrical Power Transmission Lines' Routing Using AHP, Fuzzy AHP, and GIS" *Turkish Journal of Electrical Engineering and Computer Sciences* 23 (5) pp.1418–1430.

IEA (2018) *Population Without Access to Electricity Falls Below 1 Billion* Paris: IEA (Available at: https://www.iea.org/commentaries/population-without-access-to-electricity-falls-below-1-billion).

Klugman, N., Adkins, J., Berkouwer, S., Abrokwah, K., Bobashev, I., Pannuto, P., Podolsky, M., Suseno, A., Thatte, R., Wolfram, C., Taneja, J. and Dutta, P. (2019) "Hardware, Apps, and Surveys at Scale: Insights from Measuring Grid Reliability in Accra, Ghana" *Proceedings of the 2nd ACM SIGCAS Conference on Computing and Sustainable Societies* pp.134–144.

Louw, Q. (2019) *Illegal Connections in South African Power Utilities: Is it a Pervasive Problem?* (Master's thesis) University of Johannesburg.

Maclaurin, G.J., Grue, N.W., Lopez, A.J. and Heimiller, D.M. (2019) "The Renewable Energy Potential (reV) Model: A Geospatial Platform for Technical Potential and Supply Curve Modeling" (*Technical Report No. NREL/TP-6A20-73067*) Golden, CO: National Renewable Energy Laboratory (NREL).

Meyers, J. (2005) "GIS in the Utilities" In Longley, P.A. (Ed.) *Geographical Information Systems: Principles, Techniques, Applications and Management* New York: Wiley, pp.801–818).

Midcontinent Independent System Operator (MISO) (2019) *Transmission Cost Estimation Guide, 2019 MISO Transmission Expansion Planning Report (MTEP19)* Carmel, IN: MISO.

National Geographic Society (2012) "GIS (Geographic Information System)" Available at: https://www.nationalgeographic.org/encyclopedia/geographic-information-system-gis/ (Accessed: 10th July 2020).

32

GEOSPATIAL TECHNOLOGIES AND PUBLIC HEALTH

*Fikriyah Winata, Sara McLafferty,
Aída Guhlincozzi and Yiheng Zhou*

Sustaining and improving population health and well-being are critical goals in societies across the globe. Defined by the World Health Organization as 'a state of complete physical, mental, and social wellbeing and not merely the absence of disease', health is not just a biological property but is closely tied to the places and environments people experience in daily life. Characteristics of places including air and water quality, housing quality and density, and access to transportation, food, health care, and social services profoundly affect people's exposure to harmful chemicals and pathogens and their access to health-promoting resources. Thus, enhancing place characteristics can improve population health and wellbeing. Geospatial technologies and geographic information systems (GIS) provide a powerful array of capabilities to help us understand the close interactions between places and health and to plan and implement equitable and effective public health interventions. We argue that for these technologies to be fully effective in improving health, critical and community-based approaches are needed. Our chapter begins with a discussion of the roles of these technologies in understanding health and healthcare inequalities. We then turn to a more detailed discussion of two health topics – infectious diseases, with a focus on COVID-19, and environmental health. The final section highlights the use of geospatial technologies in community settings for improving public health.

Health and healthcare inequalities

Health inequalities refer to the wide disparities in health that exist among places and population groups. An important use of GIS is to map health inequalities at the global, regional, and local scales. There are many ways to measure population health including indicators like infant mortality, life expectancy, perceived health, and subjective wellbeing. At the global scale, GIS have been employed to create cartograms highlighting health disparities among countries (Dorling, 2007). With the sizes of countries changed in proportion to the number of infant deaths, for example, the cartograms show the huge burden of poor infant health in low-income countries. Similar kinds of maps have been created to show health inequalities within countries, states, and cities, and for specific health concerns such as COVID-19 as discussed later in this chapter.

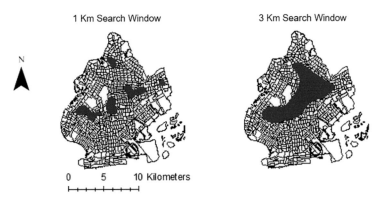

Figure 32.1 Hotspots of infant low birthweight incidence in Brooklyn, NY based on 1 and 2kilometre (radius) search windows

The spatial analysis tools available in GIS also include methods for identifying spatial clusters or *hotspots*, areas in which the risk of disease, or poor (or good) health is elevated. If data are available showing the precise home or work locations of people affected by a disease, point location data, then methods such as kernel density estimation or the spatial scan statistic can be used for hotspot identification (Kirby et al., 2017). When data are available for geographical areas, methods such as local Moran's I or Gi★ may be more appropriate. Researchers and policy analysts have used these methods to pinpoint places in which the risk or incidence of, for example, cancer (Goovaerts and Jacquez, 2005) or infant low birthweight is high (Figure 32.1). Hotspot maps can direct health screening and prevention programs to the communities most in need and suggest environmental or social factors that need to be addressed. Hotspot maps have been employed in planning breast cancer screening services and prenatal care and education programs.

Although valuable for mapping health inequalities and planning interventions, hotspot maps raise significant challenges. Most available hotspot detection methods work by scanning the study area using a localized spatial window (filter) in which health risk is evaluated (Cromley and McLafferty, 2011). In practice, analysts must make important decisions about the size of this window. Using a small window size reveals small, localized hotspots, whereas a larger window size identifies broad regions of elevated risk (Figure 32.1). By changing the window size and other key parameters, the analyst can affect how the map looks and its policy or research impact. Another essential point about hotspot mapping is that for the maps to have an impact, public health resources including screening and prevention programs should be targeted to the hotspot areas. While valuable for spatial targeting of interventions, hotspot maps also raise ethical concerns. The maps can stigmatize certain places as unhealthy or undesirable and thus further marginalize people and places already burdened by poor health. Hotspot mapping should aim to support and fulfill the needs and interests of impacted communities.

Health inequalities can be understood in relation to differences in: (1) *composition* – characteristics of individuals such as age and gender; (2) *context* – characteristics of the local environment including housing quality, access to transportation and services, air and water quality; (3) *collective* – people's social interactions and institutions (Macintyre et al., 2002). These work synergistically and through complex pathways to affect population health and well-being and to shape how health changes dynamically through space and time. Using the overlay capabilities of GIS, analysts can evaluate the associations between the compositional and contextual characteristics of places and population health characteristics. Cebrecos *et al.* (2019) developed a multicomponent model to assess 'heart healthy' urban neighbourhoods in Madrid, Spain based on four domains: the food, physical activity, alcohol, and tobacco environments. The local availability of healthy food stores, density of alcohol outlets, and neighbourhoods walkability were measured using GIS and mapped to show their variability across the city. Statistical results showed that people living in places with more heart-healthy contextual characteristics had lower cardiovascular disease risk.

In addition to analyzing health inequalities, we can map and analyse disparities in access to health services and health-related facilities such as food retail outlets, parks, and stores selling tobacco, alcohol, and other products that influence health. Figure 32.2 maps

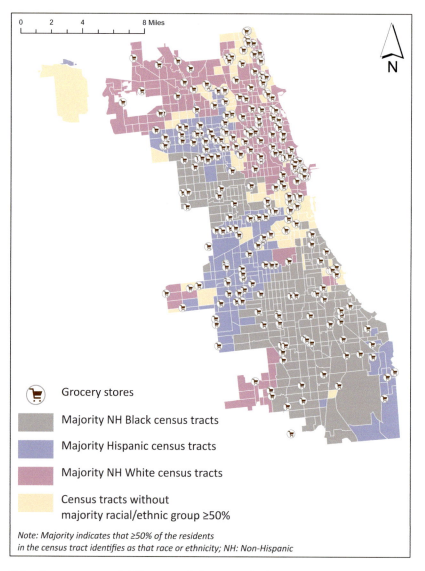

Figure 32.2 Grocery store availability across different racial/ethnic groups in Chicago

Data source: Chicago Data Portal (https://www.chicago.gov/city/en/narr/foia/CityData.html) and US Census Bureau.

grocery store availability across different racial/ethnic groups in Chicago, revealing large areas of the city's south and west side areas that have relatively few grocery stores. Defined as the ability to benefit (or not) from resources and services, access has geographical dimensions that can be assessed via GIS and geospatial technologies. All GIS include tools for measuring distances from population centers to facilities according to diverse metrics such as straight-line distance, network distance, and estimated travel time based on transportation modes such as car, walking, and public transit. Figure 32.3 maps walking areas for grocery stores in Chicago – areas within a ten-minute walk of a grocery store. Overlaying the walking areas with data on the racial and ethnic breakdown of neighbourhoods populations shows disparities among groups in populations who live within walking distance of a store.

Beyond distance, many other approaches are available for analyzing geographical access. A family of methods based on the two-step floating catchment area methodology assess the local supply of services relative to local demand or need (Wang and Luo, 2005). Data on daily travel flows are being incorporated (Widener et al., 2013) to consider access to services within the spatial and temporal constraints of people's everyday lives including travel for work, school, and household responsibilities. These developments make our analyses of healthcare access more sensitive to people's everyday circumstances and challenges. We can also focus on population groups that are marginalized by low income, age, lack of health

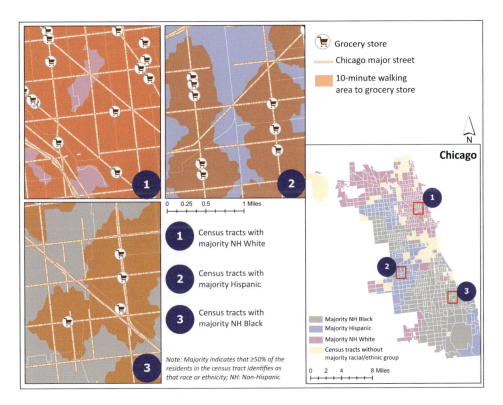

Figure 32.3 Grocery stores within ten-minute walking areas in census tracts with different racial/ethnic groups

Data source: Chicago Data Portal (https://www.chicago.gov/city/en/narr/foia/CityData.html) and United States Census Bureau.

insurance and other barriers to health care to understand if long travel times and limited availability of health services compound these socioeconomic disparities.

Geospatial technologies and infectious disease spread and prevention

Geospatial technologies are widely used for mapping and tracking infectious disease spread and understanding disease occurrence and clusters. Applications range from the traditional mapping in the eighteenth century, when John Snow mapped locations of cholera cases in London, to the current global pandemic of COVID-19. Geospatial technologies play a prominent role in preventing disease spread, understanding *where* and *when* disease transmission occurs, and informing the public of dynamics, patterns, and trends. Geographic information about a disease, analysed in geospatial technologies, can help local, state, and federal authorities make effective decisions in responding to the disease.

Rapid recent developments in information technologies have increased demands to provide reliable and trustworthy information about local disease outbreaks along with global pandemics. The public expects information on how and where diseases are spreading, not only in a matter of days but in minutes or seconds. Real-time information on disease and the mobility of those who may carry it is essential during unprecedented times such as the COVID-19 pandemic. This section outlines pivotal implementations of geospatial technologies in infectious disease control and prevention. We highlight applications on the recent COVID-19 pandemic and emerging infectious diseases (EID) as these two foci are interrelated.

Geospatial technologies and COVID-19 global pandemic

As COVID-19 cases spread rapidly worldwide, various efforts in fighting the disease have been undertaken, including geographic approaches. Geographic applications vary from disease mapping and surveillance, to contact-tracing with the real-time GPS (global positioning system) and Bluetooth connections, to advanced quantitative modeling (McLafferty *et al.*, 2021). In presenting disease distribution, movement, and spread, static maps are still being used. Yet, Web mapping applications in the form of GIS dashboards have become popular in the COVID-19 era. Most countries have built COVID-19 dashboards which include geographic and statistical information of COVID-19 cases, deaths, tests, and other relevant details. COVID-19 dashboards are not only popular at the national level, but also among state and regional-level public health agencies. These COVID-19 dashboards can have societal benefits by allowing the public to access COVID-19-related information anywhere, anytime, and on any device.

Despite their value in presenting and disseminating COVID-19 geographic information, both static mapping and online dashboard approaches are constrained by the geographic scale of data presented. In the United States, the most popular COVID-19 Dashboard is produced by the Center for System Science and Engineering (CSSE) at Johns Hopkins University (https://coronavirus.jhu.edu/). This dashboard presents COVID-19-related information not only within the United States but also globally. Specifically, for the United States, the tab of 'US Map' presents the distribution of COVID-19 cases at a county-level, as this is the most granular data of COVID-19 available country-wide (Figure 32.4).

In contrast, Hong Kong's COVID-19 Dashboard (https://www.coronavirus.gov.hk/) conveys COVID-19 case data based on point representations. The point data include the building name with cases in the past or beyond 14 days. In Singapore's COVID-19 Dashboard

Geospatial technologies and public health

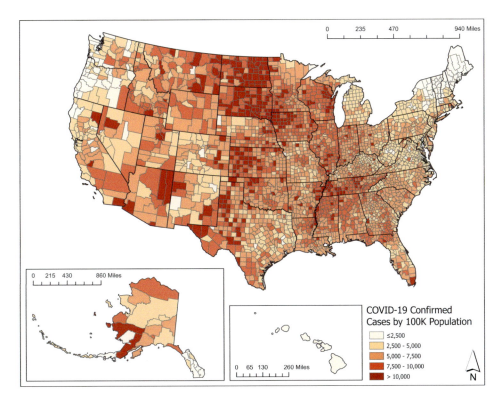

Figure 32.4 County-level distribution of COVID-19 case per 100,000 population in the United States
Data source: The Johns Hopkins Coronavirus Resource Center (CRC), data per 31st December 2020.

(https://co.vid19.sg/singapore/), the public can view clusters of COVID-19 cases, including the names of public facilities where people meet and gather, and residential areas with the number of cases within them. Thus, the spatial scale of data available varies greatly among countries. As there is no global standard in presenting COVID-19 information, each country has its own regulations that reflect legal and ethical concerns about privacy and confidentiality of individual health data. Given this, the geographic representation, analysis, and modeling of COVID-19 depend not only on advanced geospatial technologies but also on privacy concerns that affect data availability.

These interactive COVID-19 GIS dashboards are effective strategies to communicate with the public; however, geospatial technologies are also widely used for research and policy. Countries like China have developed COVID-19 'close contact detector' apps that incorporate people's mobilities and interpersonal contacts. With social distancing and travel restrictions, this app can track users' mobilities within space and time for 14 days, including when a person has had contact with people infected with COVID-19. This collaborative effort between the Chinese government and corporations relies on instant messaging/communication like QQ and WeChat, and online payment service (e.g., Alipay) on mobile phones. When users scan their QR (Quick Response) code, their mobilities, trajectories, and close contacts are retrieved, enabling the government to conduct COVID-19 surveillance. In many other countries, such surveillance efforts would be considered a significant violation of individual privacy and confidentiality.

Geospatial technologies and emerging infectious disease (EID)

The COVID-19 global pandemic is a vivid reminder of the potential for and impacts of EID, newly recognized diseases that spread regionally or globally. Outbreaks of various infectious diseases such as SARS (Severe Acute Respiratory Syndrome), Ebola, *E. coli,* West Nile Virus, dengue fever, Lyme disease, and HIV (Human Immunodeficiency Virus) demand a wide range of implementations of geospatial technologies. Geospatial efforts can inform the public of disease risk and help to better understand the ecology of the disease. Maps of disease incidence rates are one of the most popular applications of geospatial technologies in addressing EID. Yet, geospatial technologies also extend to various domains such as surveillance, modeling, and forecasting.

Most EIDs, including COVID-19, are zoonotic, in which pathogens are transmitted from animals to humans. Understanding the zoonotic host of the pathogen may improve surveillance efforts. For instance, since the emergence of Lyme disease in the late 1970s in the United States, active and passive surveillance of Blacklegged ticks has been conducted not only by government agencies but also researchers and citizen scientists. In understanding Lyme disease risk, updating where ticks are present is crucial. Geospatial technologies have been used to manage and map data on changing tick distributions.

To limit the spread of EIDs, it is important to understand where disease risks are high due to population characteristics and environmental factors. Geospatial technologies can help model the interactions between the disease and environmental factors such as climate (temperature, rainfall, and humidity), land use (forest or agricultural areas), and proximity to water bodies such as rivers, ponds, or lakes, along with socioeconomic and mobility-related characteristics of local population that affect their vulnerability to contact with pathogens. Statistical methods are often used to forecast the occurrence of disease based on past and present data on these contextual and compositional factors.

Geospatial technologies and environmental health

Individuals' health outcomes are not merely determined by their behaviours (e.g., diet, nutrition, physical activity) and genetics. *Where* people live, work, and travel is important for their health. In everyday life, people are exposed to various environmental particles from the air, water, soil, and food, affecting their health and wellbeing. Not all people are afforded the ability to live in healthy environments. Many disadvantaged individuals live, work, and travel to areas contaminated by dangerous chemical components. Thus, compositional and contextual explanations are intertwined in influencing people's environmental health exposures.

Environmental exposures from air, water, soil, food, and sound/noise are ubiquitous in our daily lives. Air pollutants from traffic congestion in metropolitan areas may cause respiratory problems; arsenic and lead contamination in water may be associated with premature birth and low-birth weight. Applications of geospatial technologies extend from data collection and measurement to analysis and intervention. In environmental health assessment, our understanding of geographical methods is paramount, along with the type and quality of data itself. Different spatial data representations, whether points or areal/polygon units such as postal codes or counties, require different methodologies. Health data are often reported for areal units. For example, Figure 32.5 presents the distribution of asthma prevalence (%) in Chicago among adults aged ≥18 in census tracts. In contrast, data on air and water pollution are often presented for point locations such as monitoring sites. Linking these point data to health data for geographic areas involves complex GIS-based spatial methods.

Geospatial technologies and public health

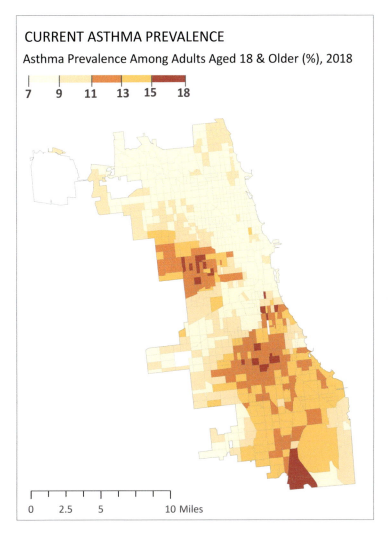

Figure 32.5 Distribution of current asthma prevalence (%) among adults aged >18 in census tracts
Data source: Centers for Disease Control and Prevention (CDC) (2020).

In much environmental health research, geospatial tools have been used to measure and estimate the uneven geographic distribution of environmental hazards (Table 32.1). In Figure 32.6, we apply the method of kriging to create a continuous surface map of ozone pollution in Chicago based on measurements at monitoring sites (point locations). The next step is to analyse whether these varying levels of environmental hazard are associated with health outcomes. Regression analysis is commonly used to assess these associations. Grineski *et al.* (2015) employed geographically weighted regression (GWR) to understand the relationship between environmental and socioeconomic conditions and health inequalities among children in El Paso, TX. Their study examined exposure to $PM_{2.5}$ and the effect of residential pests on children's wheezing severity and found children living in lower-income environs were disproportionately exposed to pests and $PM_{2.5}$ compared to children in higher-income neighbourhood. Time-series environmental data are also important in

Table 32.1 Examples of the use of geospatial tools in environmental health assessment

Exposure	Geospatial tools	Functionality (examples)
Air	Kriging, cokriging	To estimate hourly, daily, weekly, monthly, or a yearly average of air pollution concentrations (e.g., $PM_{2.5}$, Ozone) in any given area.
Water	Cluster analysis (hot spot detection)	To detect clusters of cholera outbreaks.
	Geocoding	To geocode and map the locations of drinking water contaminated by arsenic.
Soil	GPS tracking and survey	To map area with soil contamination by heavy metals.
Noise	GPS tracking with noise sensor	To track the noise level in different places, especially in big cities.

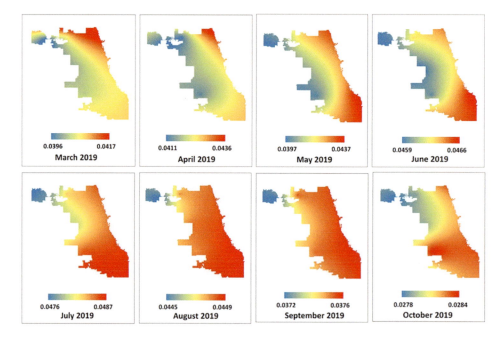

Figure 32.6 Monthly average of Daily Max-8 Hours Ozone concentration (ppm) in Chicago, estimated with Empirical Bayesian kriging method

Data source: United States Environmental Protection Agency (USEPA) (2020).

analyzing environmental influences on health outcomes. Using daily records of asthma cases and $PM_{2.5}$ data (2003–2011) in Jackson, MS, Chang *et al.* (2019) found that even in areas with lower pollutant levels, average daily $PM_{2.5}$ and asthma among Black men were significantly associated.

Exposure to environmental hazards not only reflects the spatial distribution of the hazard but also people's mobilities and behaviours that affect their contact with the hazard. People move from place to place in different timeframes: we often begin the day at home and go to work, school, shopping centers, and other places with different exposure levels.

Thus, the contexts people experience vary over time and space and from person to person, a problem known as the Uncertain Geographic Context Problem (Kwan, 2012). Environmental exposures should be estimated based on people's daily mobilities, not only where they live. Global positioning systems are an increasingly important tool for collecting individual data on mobility so that we can accurately estimate environmental exposures within space and time. A recent study of noise exposure (Ma et al., 2020) incorporated individual's space-time trajectories from GPS trackers to evaluate the effect of noise exposure from different activities and travel modes on mental health. Their assessment found that when noise exposure from outdoor activities and public transportation was higher, individuals' mental health was poorer.

Geospatial technology and community health

In addition to contributing to health research and policy, geospatial technologies have an important role in community efforts to improve health and access to health care. In Chicago, for example, the Little Village Environmental Justice Organization (LVEJO) created the 'EJ and Little Village Industrial Corridor Map' which is an interactive OpenStreetMap Story Map to depict the concentrated environmental justice issues and hazards located in Little Village (Little Village Environmental Justice Organization [LVEJO], 2017). In another example, women concerned about breast cancer in the New York City suburbs conducted door-to-door surveys to identify cancer cases and used both pin maps and GIS to map breast cancer clusters (McLafferty, 2002). These examples of using geospatial technology to address health disparities show the opportunities for community groups to use geospatial tools in confronting local health challenges.

Data

Crucial to any geospatial analysis is data, and data access can be challenging for community groups that want to use GIS and geospatial technologies. Map and geography libraries can be excellent sources of local and community data, especially libraries with a dedicated interest in providing healthcare-related information. Local libraries also may have data and other resources for community mapping of health concerns. However, public libraries are not readily available and accessible in all areas (Figure 32.7), vary greatly in staffing and resources, and increasingly risk closure due to politically motivated cuts in funding.

Other data opportunities exist through direct partnerships with local healthcare clinics and facilities, to use electronic health records (EHR). These can be used for tracking health-related problems, healthcare needs, and utilization. However, EHR data is often difficult to obtain due to privacy concerns, but an innovative approach is to use practice-based research networks or PBRNs which are networks of scholars and physicians working in a participatory research format to conduct community-based research. For example, OCHIN (Oregon Community Health Information Network) is a PBRN, linked with government and academic research organizations. The network developed as a nonprofit corporation with a mix of a volunteer board of directors from mostly member clinics and community health leaders. Using GIS, the organization mapped health clinic service areas and the level of insurance rate amongst patients seeking care (Angier et al., 2014). While the benefits of performing GIS analyses with clinical data may appeal to community groups, the cost of GIS software and need for training make partnering with academic researchers a beneficial strategy.

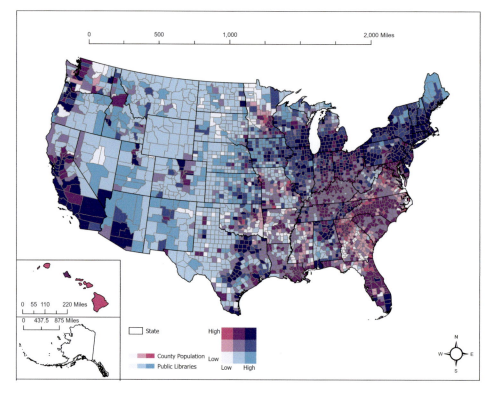

Figure 32.7 Map of density of public libraries and population by county. Alaskan data was too low for mapping. Data provided by Institute of Museum and Library Services and US Census

Partnerships

Community-academic partnerships are valuable in bridging the skills gap between community members and spatial analysis. If a community is interested in using geospatial technology, they can work with academics, including students and faculty experts through partnerships. In Baltimore, Maryland, for example, members of primary care clinics and academic researchers collaborated in mapping healthcare resources and geographic variation in health clinic use from patient address data (Bazemore *et al.*, 2010). The maps revealed areas in which health clinics were in short supply and areas where clinic usage appeared to be low. GIS have also been used to address community health topics in low-income countries. In Malaysia, the Institute for Public Health and partner organizations used handheld GPS units to geocode health facilities and identify gaps in health clinic availability throughout the country (Hazrin *et al.*, 2013).

Establishing and collaborating within community-academic partnerships takes time and mutual adjustment of resource use and expectations, with both success and difficulties. The field of community geography discusses several key goals to effective community-academic partnerships, including reciprocity, collaborative knowledge production, and shared power between partners (Robinson *et al.*, 2017). For successful partnerships with community organizations, collaboration requires establishing common ground through mutual trust and shared initiatives and creating an accessible environment for all participants (Chang *et al.*, 2020).

Examples of health-related partnerships that strongly engage with community residents are only beginning to appear in the scholarly literature. For example, to address concerns about the limited availability of healthy food stores in some communities, geographer Jerry Shannon (Shannon, 2015) fostered a community-based project on food deserts. The researcher provided smartphones to community members to capture photos, complete food diaries, and record GPS data from food paths. This data was then mapped and analysed by the researcher alone and with community participants to identify constraints and concerns about food access (Shannon, 2015). This project is an important example of the value of incorporating qualitative data from community residents as community groups embrace GIS, both on their own and through partnerships, in efforts to improve health and healthcare access.

Conclusion

Health and access to health services are closely bound up with the places and environments people experience in daily life. One need only look to the COVID-19 pandemic, with its rapid worldwide spread and shifting geographies of disease incidence and mortality, to see how people's mobilities and place-based socioeconomic vulnerabilities affect their exposure to disease agents and their resulting health outcomes. Our chapter argues that the close ties between health and place create diverse opportunities to use geospatial technologies to better understand health and wellbeing and to create more effective and equitable public health interventions.

Our brief review of applications of geospatial technologies in the areas of health disparities, infectious disease, environmental health, and community health reveals several key ongoing trends. First is the emerging effort to make geospatial health data more widely accessible for diverse stakeholders through efforts such as hotspot mapping and creation of geospatial data dashboards for COVID-19. Members of the public and community groups want to know about the health issues they face and health-related resources available, and efforts are underway to disseminate this information via legible maps and interactive dashboards. Another ongoing trend is the use of GPS-based devices to document spatiotemporal changes in environmental contaminants and people's dynamic exposures to those contaminants, as well as to pinpoint how diseases such as COVID-19 spread via shifting patterns of human mobility and contact. Although the explosion of real-time health data holds great promise for research and policy, we have noted that it also requires novel methods of spatial and spatiotemporal analysis to identify meaningful trends and associations. A final theme is the growing effort to use GIS and geospatial technologies to address community health concerns. Use of these technologies is widespread among health researchers and governmental health agencies, but these efforts may be detached from the needs of community residents and organizations for access to healthy spaces and environments and quality healthcare services. Ongoing efforts to create academic-community partnerships around important local health issues and to incorporate qualitative information on community members' health-related experiences and perceptions present exciting opportunities to mobilize geospatial technologies to reduce the wide inequalities in health and healthcare access among places and populations.

References

Angier, H. Likumahuwa, S., Finnegan, S., Vakarcs, T., Nelson, C., Bazemore, A., Carrozza, M. and DeVoe, J.E. (2014) "Using Geographic Information Systems (GIS) to Identify Communities in Need of Health Insurance Outreach: An OCHIN Practice-Based Research Network (PBRN) Report" *Journal of the American Board of Family Medicine* 27 (6) pp.804–810 DOI: 10.3122/jabfm.2014.06.140029.

Bazemore, A., Phillips, R.L. and Miyoshi, T. (2010) "Harnessing Geographic Information Systems (GIS) to Enable Community-Oriented Primary Care" *Journal of the American Board of Family Medicine* 23 (1) pp.22–31 DOI: 10.3122/jabfm.2010.01.090097.

Cebrecos, A., Escobar, F., Borrell, L.N., Díez, J., Gullón, P., Sureda, X., Klein, O. and Franco, M. (2019) "A Multicomponent Method Assessing Healthy Cardiovascular Urban Environments: The Heart Healthy Hoods Index" *Health & Place* 55 pp.111–119 DOI: 10.1016/j.healthplace.2018.11.010.

Centers for Disease Control and Prevention (CDC) (2020) "Most Recent National Asthma Data" Available at: https://www.cdc.gov/asthma/most_recent_national_asthma_data.htm.

Chang, H.H., Pan, A., Lary, D.J., Waller, L.A., Zhang, L., Brackin, B.T., Finley, R.W. and Faruque, F.S. (2019) "Time-Series Analysis of Satellite-Derived Fine Particulate Matter Pollution and Asthma Morbidity in Jackson, MS" *Environment Monitoring Assessment* 191 (Supplement 2) 280 DOI: 10.1007/s10661-019-7421-4.

Chang, H., Granek, E.F., Ervin, D., Yeakley, A., Dujon, V. and Shandas, V. (2020) "A Community-Engaged Approach to Transdisciplinary Doctoral Training in Urban Ecosystem Services" *Sustainability Science* 15 (3) pp.699–715 DOI: 10.1007/s11625-020-00785-y

Cromley, E. and McLafferty, S. (2011) *GIS and Public Health* (2nd Edition) New York: Guilford Press.

Dorling, D. (2007) "Worldmapper: The Human Anatomy of a Small Planet" *PLoS Medicine* 4 (1) pp. 13–18 DOI: 10.1371/journal.pmed.0040001.

Goovaerts, P. and Jacquez, J. (2005) "Detection of Temporal Changes in the Spatial Distribution of Cancer Rates using Local Moran's I and Geostatistically Simulated Spatial Neutral Models" *Journal of Geographical Systems* 7 pp.137–151 DOI: 10.1007/s10109-005-0154-7.

Grineski, S.E., Collins, T.W. and Olvera, H.A. (2015) "Local Variability in the Impacts of Residential Particulate Matter and Pest Exposure on Children's Wheezing Severity: A Geographically Weighted Regression Analysis of Environmental Health Justice" *Population Environment* 37 pp.22–43 DOI: 10.1007/s11111-015-0230-y.

Hazrin, H., Fadhli, Y., Tahir, A., Safurah, J., Kamaliah, M.N. and Noraini, M.Y. (2013) "Spatial Patterns of Health Clinic in Malaysia" *Health* 05 (12) pp.2104–2109 DOI: 10.4236/health.2013.512287.

Kirby, R.S., Delmelle, E. and Eberth, J.M. (2017) "Advances in Spatial Epidemiology and Geographic Information Systems" *Annals of Epidemiology* 27 (1) pp.1–9 DOI: 10.1016/j.annepidem.2016.12.001.

Kwan, M-P. (2012) "The Uncertain Geographic Context Problem" *Annals of the Association of American Geographers* 102 (5) pp.958–968.

Little Village Environmental Justice Organization (LVEJO) (2017) "EJ and Little Village Industrial Corridor Map" Available at: http://www.lvejo.org/our-community/map/ej-and-little-village-industrial-corridor-map/ (Accessed: 1st August 2021).

Macintyre, S., Ellaway, A. and Cummins, S. (2002) "Place Effects on Health: How Can We Conceptualise, Operationalise, and Measure Them?" *Social Science & Medicine* 55 (1) pp.125–139 DOI: 10.1016/S0277-9536(01)00214-3.

Ma, X., Huang, H., Wang, Y., Romano, S., Erfani, S. and Bailey, J. (2020) "Normalized Loss Functions for Deep Learning with Noisy Labels" *International Conference on Machine Learning* pp. 6543–6553.

McLafferty, S.L. (2002) "Mapping Women's Worlds: Knowledge, Power and the Bounds of GIS" *Gender, Place & Culture* 9 (3) pp.263–269 DOI: 10.1080/0966369022000003879.

McLafferty, S.L., Guhlincozzi, A. and Winata, F. (2021) "Counting COVID: Quantitative Geographic Approaches to COVID-19" In Andrews, G.J., Crooks, V., Pearce, J. and Messina, J. (Eds) *COVID-19 and Similar Futures: Geographical Perspectives, Issues and Agendas* Cham: Springer, pp.409–416.

Robinson, J.A., Block, D. and Rees, A. (2017) "Community Geography: Addressing Barriers in Public Participation GIS" *The Cartographic Journal* 54 (1) pp.5–13 DOI: 10.1080/00087041.2016.1244322.

Shannon, J. (2015) "Rethinking Food Deserts Using Mixed Methods GIS" *CityScape* 17 (1) pp.85–96.

United States Environmental Protection Agency (USEPA) (2020) "CMAQ: The Community Multiscale Air Quality Modeling System" Available at https://www.epa.gov/cmaq (Accessed: 24th December 2020).

Wang, F. and Luo, W. (2005) "Assessing Spatial and Nonspatial Factors for Healthcare Access: Towards an Integrated Approach to Defining Health Professional Shortage Areas" *Health & Place* 11 (2) pp. 131–146 DOI: 10.1016/j.healthplace.2004.02.003.

Widener, M., Farber, S., Neutens, T. and Horner, M. (2013) "Using Urban Commuting Data to Calculate a Spatiotemporal Accessibility Measure for Food-Environment Studies" *Health & Place* 21 pp.1–9 DOI: 10.1016/j.healthplace.2013.01.004.

33
APPLICATIONS OF GIScience TO DISEASE MAPPING
A COVID-19 case study

Leah Rosenkrantz and Nadine Schuurman

Infectious diseases are intrinsically linked to our geography. Their transmission is dependent on both the spatial and temporal dimensions of human activity. Globalization has highlighted these relationships by connecting once isolated communities and countries. Trade liberalization, labour mobility and modern transportation networks are moving human populations, and subsequently the infectious diseases they carry (Ruiz Estrada and Khan, 2020). Consequently, epidemics and pandemics have the potential to become more common and intense. The tools we use to track their spatial and temporal diffusion are a means to fight the spread of infection and its myriad effects (Joi, 2020; Kamel Boulos and Geraghty, 2020). Geographic information systems (GIS) are at the forefront of these efforts to control infectious diseases.

While GIS itself was only pioneered in the 1960s, maps in their own right have long been used for epidemiological purposes. The most well-known example is John Snow's 1854 map of cholera in London, which helped him determine that a water pump on Broad Street was making people ill. However, the use of maps to visually represent aspects related to disease dates back even further to the time of the Black Plague in the fourteenth century (Koch, 2017), and since then maps have served as effective visuals for communicating geolocated data for numerous other disease outbreaks. With the rise of computerized GIS in the 1960s, the possibilities for analyzing, visualizing and detecting patterns of disease expanded drastically, and have allowed us to both ask and answer much more complex questions than was historically possible. Maps and other geographic visualizations are also valuable tools for communication with the ability to rapidly communicate in a single glance what words and tables cannot. These capabilities have proven critical in our ability to manage infectious diseases, as illustrated by the COVID-19 pandemic.

COVID-19 emerged in late 2019 in Wuhan, China caused by the SARS-CoV-2 virus. In a matter of months, the disease spread to nearly every country in the world. Maps and online dashboards began to track active cases and deaths from the virus in near real time, coupling Big Data with GIS technologies to understand the scale and spread of the virus. This chapter uses COVID-19 as an example of how both geospatial and non-geospatial technologies can be harnessed to prevent and manage the spread of disease outbreaks around the world. We begin by examining the data that supports the maps and geospatial analyses developed to model the spread of COVID-19 and its numerous socio-economic impacts. We then explore the numerous applications of GIScience at each stage of the pandemic using this data. Finally, we consider the lessons learned from COVID-19 and suggest the ways in which we can use geospatial technologies to prepare for similar futures.

Dealing with the data

Behind any geographic visualization are the data that power it. These data can be classified broadly into two groups: spatial and non-spatial. During the pandemic, there were unique challenges to obtaining and analysing both data types. Overcoming these challenges took time as there was a general lack of understanding of the importance of both data quality and resolution. Consequently, the initial maps and geospatial analyses that emerged at the start of COVID-19 were affected by poor quality and insufficiently granular data. These factors contributed to the proliferation of low-resolution choropleth and graduated circle maps at the start of the pandemic (Everts, 2020; Rosenkrantz et al., 2020). Below, the challenges of both spatial and non-spatial data are examined.

Non-spatial data

The beginning of the COVID-19 pandemic was marked by the rapid spread of the virus around the world, with high transmission rates overwhelming under-prepared health systems. As a result, public health records suffered extensive irregularities in data collection and reporting of COVID-19 cases (Platt, 2020; Smart, 2020). COVID-19 case counts were highly dependent on the testing capacity of a region, and as a result, cases were likely underestimated anywhere testing capacity was limited (Brunsdon, 2020; Rinner, 2021). Moreover, since testing capacity and reporting strategy varied over time and place, it was extremely difficult to analyse case counts longitudinally or across regions, making this indicator challenging to work with. Inconsistencies in data also arose when labs at the beginning of the pandemic were still adjusting to the flood of testing; in many instances, daily spikes in cases often meant a backlog of tests being cleared rather than true daily counts.

While hospitalization and death data were (and generally are for any disease) more reliable, they still resulted in some major limitations for geospatial analysis and visualization. First, due to the virus' incubation period, both indicators are representative of the recent past and not the present. There was also evidence that mortality-related events were not being systematically tested and coded, which led to substantial undercounts of death for the pandemic (*The Economist*, 2020). Lastly, these indicators were often disproportionately affected by outbreaks in long-term care homes, as people over 65 were especially susceptible to COVID-19 and at an increased risk for hospitalization and death (Walsh and Semeniuk, 2020; Yourish et al., 2020).

Taken together, such data limitations were a major roadblock to geospatially tracking COVID-19. Accordingly, the best maps created to visualize the COVID-19 pandemic accounted for these data irregularities. Figure 33.1 shows case counts of COVID-19 using a five-day rolling average. Rolling averages help to prevent major events (like the clearing of a testing backlog or a change to how cases get reported) from skewing the data (Johns Hopkins Coronavirus Resource Center, 2020a). Figure 33.2 shows the prevalence of confirmed cases of COVID-19 per 100,000 people in Ottawa (Ottawa Public Health, 2021). Importantly, their data was filtered to *exclude* long-term care homes and retirement residences as outbreaks in these communities would have excessively skewed the visualization of COVID-19 data. To be clear, it is not that the cases in long-term care homes are any less important or

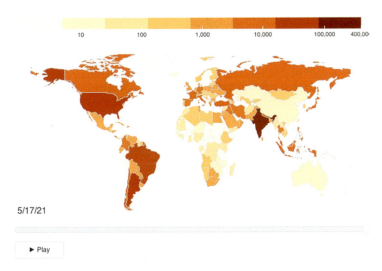

Figure 33.1 Johns Hopkins animated global map of daily confirmed cases of COVID-19 using a five-day moving average (Johns Hopkins Coronavirus Resource Center, 2020a)

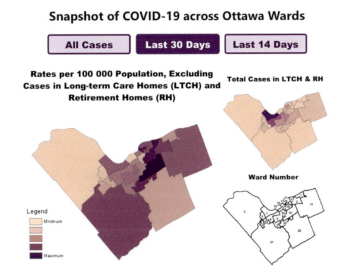

Figure 33.2 Confirmed cases of COVID-19 per 100,000 people in Ottawa by Ward – excluding cases in long-term care homes and retirement homes, which are shown separately (Ottawa Public Health, 2021)

concerning, rather they are generally well isolated and pose less of a transmission risk to the community at-large and it is therefore helpful to separate them out to get a better understanding of communities facing the highest risk.

Spatial data

The lack of high-resolution spatial data on COVID-19 was another major limitation to the types of geographic analyses and visualizations that were possible, especially in North

America where COVID-19 data was predominantly reported at the county, city, or state level to maintain patient privacy (LA County Department of Public Health, 2020; NYC Department of Health and Mental Hygiene, 2020). Though this level of data provided a general overview of the situation, it limited more comprehensive analyses that could have better directed resources and tailored preventative measures to specific communities. Moreover, this kind of data assumed a sort of stasis – misrepresenting the fact that individuals, and subsequently the virus, readily moved from one community to the next.

To resolve these initial issues, GIScientists tapped into geospatial and non-geospatial technologies alike. These included technologies like global positioning systems (GPS), Wi-Fi positioning systems (WFPS), cell phone tower triangulation, Bluetooth tracking, closed-circuit television (CCTV) and credit card transactions, to name but a few. While some are more effective than others, each of these technologies in some way or another allowed researchers to tap into the trajectory data of both infected and non-infected individuals as they moved about their daily lives (Rosenkrantz et al., 2020), revealing higher-resolution data of where and to whom disease was transmitted, and acting as a pivotal transition away from the slow and often laborious efforts of manual contact tracing.

The use of some of these technologies (especially those connected to mobile phones) has made contact tracing near instantaneous – when a newly diagnosed individual is identified, algorithmic contact tracing can spread this information rapidly back through a past contact network to let individuals know they may be at risk and that they should take appropriate measures to social distance themselves and promptly get tested (Ferretti et al., 2020).

Some countries like China and South Korea implemented a top-down, centralized approach of these technologies to track their citizens movements. The Chinese government collaborated with both Alibaba and Tencent – two of China's Internet giants – to develop a QR (quick response) code system to prevent COVID-19 spread and control the movement of high-risk individuals (Gan and Culver, 2020; Ienca and Vayena, 2020; Whitelaw et al., 2020). The 'Health Code' was generated from data collected from location data about a person's recent potential exposures to the virus, and functions as a COVID-19 health status certificate for residents. The Health Code was used to grant access to public places, like transit systems, hotels, and restaurants by assigning individuals to one of three colours: green, yellow or red (Gan and Culver, 2020). Green represented low exposure risk, yellow represented medium exposure risk, and red represented high exposure risk, with each colour accordingly designating the user's freedom of movement. Although authorities did not make the health codes compulsory, citizens without the app or a smartphone were not able to leave their residential compounds or enter most public places without one (Whitelaw et al., 2020).

South Korea also implemented tools for aggressive contact tracing of the virus, with technologies like GPS from vehicles and mobile phones, bank card records, and security camera footage used to get around people being unwilling to disclose or unable to recall their close contacts in the weeks leading up to their infection (Lewis, 2020). Laws passed since the 2015 Middle East respiratory syndrome (MERS) outbreak in the country specifically allow authorities to collect and make public information on the infected person's demographics and location history – sometimes down to the minute – through widely broadcasted alerts (Zastrow, 2020).

Top-down enforcement of such technologies seemingly proved to be an effective strategy for controlling the spread of the virus in both China and South Korea. However, issues of privacy with regard to these technologies must be acknowledged. Many people around the world vocalized concern, viewing these measures as an encroachment on individual privacy and speculating about the stigmatization that may occur if an infected person's identity is revealed (Calvo, Deterding and Ryan, 2020; Ienca and Vayena, 2020). In fact, numerous instances have been recorded where the identity of infected individuals was exposed due to inadequate de-identification of their data (Kitchin, 2020; Singer and Sang-Hun, 2020). For reasons both legal and ethical, numerous countries chose not to pursue this route. Instead, efforts to obtain higher resolution data focused on what is known as volunteered geographic information (VGI). VGI is a term coined by the geographer Goodchild (2007) to describe the increasingly popular phenomena of citizens engaged in the creation of geographic information. With regard to COVID-19, governments were encouraging citizens to volunteer their health and location data to actively track themselves.

A number of local-scale VGI Web and mobile apps like *Outbreaks Near Me* (2020), *COVID Symptom Tracker* (2020), and *Safepaths* (2020) were developed early on in the pandemic and required users to either download an app or visit a webpage to collect information on users' location and symptomology. Companies such as Kinsa Health, who were already in the VGI space before the pandemic hit were able to use existing technology in their smart home thermometers to collect volunteered health and location data on feverish illness in the United States (Kinsa Health, 2020). Though their data does not distinguish COVID-19 from other feverish illnesses, it can detect and map *abnormal* spikes in fevers, serving as an important early detector of COVID-19 hot spots (McNeil Jr., 2020; Kinsa Health, 2020).

Initiatives by Kinsa Health and those previously mentioned were either research or enterprise-driven and were not integrated into policy or practice by either the US or Canada's administrations, where they originated (Whitelaw et al., 2020). However, there are examples where this is the case. On a national scale, Singapore developed 'TraceTogether', a mobile app that used Bluetooth technology to send anonymized IDs between cell phones of people in close contact with each other. If a person had downloaded the app to their phone and later tested positive for COVID-19, the anonymized IDs were used to quickly identify and contact people who recently came in close proximity to the individual and may be at high risk of infection (Flatten, 2020). While the app was slow to be adopted by users, with only 17% adoption rate after the first month (Vaughan, 2020), it eventually gained wider popularity and became a key tool in helping Singapore maintain one of the lowest per-capita COVID-19 mortality rates in the world throughout the pandemic (Lee, 2020).

High-resolution case or symptom data is not the only means researchers have to curb the spread of the pandemic. Social media has also proven an important data source in the geographer's toolkit for monitoring sentiment towards the pandemic, and consequently predicting outbreaks. Forati and Ghose (2020) examined COVID-19 related Twitter activity in May of 2020 and found that tweets containing hashtags that were categorized as either against containment measures, as downplaying the virus, or as disseminating disinformation (e.g., '#Plandemic', '#CovidPropaganda', '#Scamdemic') largely originated from Orlando, FL; Dallas, TX; Palm Beach, FL; Houston, TX; Los Angeles, CA; and Watchung, NJ (Forati and Ghose, 2020). Results of their analysis showed that cities with the highest usage of these hashtags one month were the same cities which experienced substantial surges in COVID-19 cases only two months later (Forati and Ghose, 2020). They also ran a multiscale geographically weighted regression model to investigate the relationship between the total number of these kinds of tweets per county and the number of confirmed COVID-19 cases per county. The model obtained a high adjusted R^2, explaining 96.7% of the total variance in COVID-19 incidence rates. Research such as this provides a strong argument in the event of future pandemics to explore geographically tagged social media data as a means of understanding where and why diseases are spreading at multiple spatial scales.

The many applications of GIS

Despite early data challenges, geographers were quick to apply their knowledge and expertise to the most pressing problems of the pandemic as they evolved. In this section, we break down the pandemic into distinct 'phases'. We use these phases as a framework for discussing how geographers used geospatial technologies to meet the varying needs of society across numerous sectors at that time. It is important to acknowledge however that what we present here is just a small sampling of the different applications of GIS during the pandemic and that to discuss them all could be a book in itself. Instead, we present what we feel are some of the most noteworthy uses of GIS during the major events of the COVID-19 crisis. We would also like to note that while we lay these phases out sequentially for the purpose of showcasing different geospatial approaches, much of what was happening was occurring simultaneously.

Outbreaks emerge around the world

At the beginning of the pandemic, the primary role of geographers was tracking the spread of the disease as SARS-CoV-2 quickly spread from one country to the next. Likely the most well-known example of this is the John Hopkins University (JHU) COVID-19 dashboard developed by Lauren Gardner and Ensheng Dong, which illustrates the number and location of confirmed COVID-19 cases, deaths, and recoveries for all affected countries on a world map (Dong *et al.*, 2020; Johns Hopkins Coronavirus Resource Center, 2020a). The dashboard was first shared publicly on 22nd January 22, nearly two months before it was designated a pandemic with the aim of providing researchers, public health officials and the general public an accessible and interactive tool to track the outbreak on a world-wide scale as it unfolded (Dong *et al.*, 2020).

Soon after the JHU dashboard gained popularity, numerous other GIS-based dashboards began to pop up, with many focused on communicating more granular case data about specific countries, states, provinces, or counties (Everts, 2020; Kamel *et al.*, 2020). While these dashboards have been largely beneficial for both public health authorities and the public alike for illustrating a broad sense of where the disease is progressing across space and time, they have also garnered criticism for obscuring risk groups and small-scale patterns of the disease by 'glossing over' the uneven geographies of infections and mortality (Everts, 2020). Dashboards have also been criticized for promoting a territorial approach to managing the pandemic via border closures and lockdowns, when what is really needed, critics argue, is a more localized approach targeting the most severe outbreaks where and when they happen (Everts, 2020).

Going beyond mere visualization are the geospatial statistics that help us elucidate patterns in disease spread. In many ways, these help us to move beyond the arbitrary nature of borders and territoriality as their analyses are not limited by boundaries. Kernel density analysis calculates a magnitude per unit area from point or polyline data by comparing the spatial distribution of cases relative to the underlying population. The outcome is a continuous estimate of spatially varying risk, without the need to aggregate data – that is, it is 'unfettered by arbitrary geographical boundaries' (Elson *et al.*, 2021). This approach was employed by Elson *et al.* (2021) in analysing the spatial distribution of COVID-19 in England at discrete times between January and June of 2020. They found that widespread transmission was already underway when certain lockdown measures were first implemented, and that the distribution of cases was greatest in many, but not all, large urban areas. Ultimately, they were able to use kernel density estimation to determine where risk of COVID-19 infection differs significantly in England, allowing for more accurate identification of areas within the country that required better public health messaging or tailored intervention measures. An advantage of using the kernel density method is enhanced privacy as individual homes or buildings are never identified.

Another example of a type of geospatial analysis that can help elucidate the spread of disease during a pandemic is the prospective space-time scan statistic (Kulldorff, 2001). Like the kernel density analysis, the space-time scan statistic can also detect spatial clusters irrespective of predefined geographical boundaries. Unlike kernel density, the space-time scan statistic allows for detection of clusters both spatially *and* temporally. The space-time scan statistic uses a cylindrical window in three dimensions, with the base of the cylinder representing space and the length of the cylinder representing time. The cylinder window is flexible in both its size and time, meaning that for each possible circle location and size, each possible starting date is considered. In total, this method considers a large number of distinct cylindrical windows, each a possible candidate for identifying a cluster of events over time. A maximum likelihood ratio is calculated for all possible cylinders to identify the cylinder that constitutes the most likely clusters and is still 'alive' at the end of the study period (i.e., ignoring those that existed historically but are no longer present) (Kulldorff, 2001).

Prospective space-time analysis is highly useful for monitoring disease outbreaks, as it allows for detection of active, emerging clusters and the relative risk of these clusters at each site. Numerous researchers have used this type of analysis to study the spatial dispersion of COVID-19 in their communities (Andrade *et al.*, 2020; Desjardins, Hohl and Delmelle, 2020; Ferreira *et al.*, 2020; Hohl *et al.*, 2020; Kim and Castro, 2020; Masrur *et al.*, 2020; Kan *et al.*, 2021).

The spatial scan statistic imposes a circular window on the map and lets the centre of the circle move over the area so that at different positions the window includes different sets of neighbouring census areas. If the window contains the centroid of a census area, then that whole area is included in the window. For each circle centroid, the radius of the circular window is varied continuously from 0 up to a maximum radius so that the window never includes more than 50% of the total population at risk. In this way, the circular window is flexible both in location and in size. In total the method creates a very large number of distinct circular windows, each with a different set of neighbouring counties within it, and each a possible candidate for containing a cluster of events.

Understanding the disease's spread

Tracking disease outbreaks is only one aspect of spatial monitoring. The other is understanding why it is spreading. One way to do this is to build an explanatory model using variables that best account for the spread of COVID-19. While there are several models that can be used to do this, such as an Ordinary Least Squares (OLS) Regression model, a Spatial Lag Model, or a Spatial Error Model, these global models all assume a spatial stationarity in the relationship between the explanatory and dependent variables (in the case of OLS, it does not assume a spatial dependence at all). This is limiting, as we know infectious diseases like COVID-19 are intricately linked to our geography. The solution? Using a local model, like geographically weighted regression (GWR) or multiscale GWR which allows parameters to be derived for each location separately, and as such, incorporates geographic context into the models. Multiscale GWR takes this a step further, by also allowing analysis of relationships at different scales.

A notable study by Mollalo *et al.* (2020) makes the case for using GWR and MGWR in their study modelling COVID-19 incidence at the county level in the contiguous United States. They identified four place-based sociodemographic and economic risk factors (narrowed down from a candidate list of 35) that best explained significant variation in COVID-19 incidence by county. These were: income inequality, median household income, percentage of nurse practitioners and percentage of Black females. They then used these four variables to compare the performance of both global and local models alike, and ultimately showed GWR and MGWR to out-perform any of the global models tested. Using GWR and MGWR, the researchers were also able to determine where in the US these variables were most influential in explaining COVID-19 spread. For example, Mollalo *et al.* (2020) found that income inequality was especially influential in explaining COVID-19 incidence rates for counties in New York, New Jersey, and Connecticut, but was less so for counties in Arizona or Texas. Numerous other studies also made use of GWR or MGWR to a similar effect on different populations and study areas (Liu *et al.*, 2020; Sannigrahi *et al.*, 2020; Mansour *et al.*, 2021). Ultimately these studies helped to further our understanding of the place-based nature of explanatory factors and related spatial patterns that were driving COVID-19 spread.

A racial reckoning

When stories of suspected health inequities began to emerge in the early summer of 2020, little evidence was available to corroborate them since almost no public health organization had originally collected this data (Alliance for Healthier Communities, 2020; Bowden, 2020; CDC, 2020; Johns Hopkins Coronavirus Resource Center, 2020b). Despite these data shortfalls, geographers were in a unique position to respond. They quickly harnessed the power of geospatial data linkage and analysis properties, comparing COVID-19 case data with the most recent census or community survey data to expose the uneven and unjust geographies of the pandemic (Everts, 2020; The COVID Tracking Project, 2020). These maps allowed researchers to determine the racial breakdown of areas hit especially hard by the disease, as well as monitor changes in overall death to better pinpoint inconsistent patterns of mortality attributable to racial disparities in risk and access to healthcare during an outbreak (Gaglioti *et al.*, 2020; Millett *et al.*, 2020; Mollalo, Vahedi and Rivera, 2020; Anaele, Doran and McIntire, 2021; Andersen *et al.*, 2021; Iyanda *et al.*, 2021).

Researchers were also able to examine COVID-19 testing accessibility in predominantly non-white areas. When they found significant inequities in testing between predominantly Black and Hispanic communities versus predominantly white communities, they were able to call out for more equitable testing (Credit, 2020; Mody *et al.*, 2020). Even as countries and counties began the imperative work of collecting racial data on COVID-19, GIS continued to play an important role in this area of research, with spatial statistics like Local Moran's I explaining better the association between dependent variables such as race and ethnicity and COVID-19 than non-spatial statistics such as global regression (Iyanda *et al.*, 2021). This was critical work done at a time of renewed recognition of racial injustices, particularly in the United States and Canada, but also around the world.

Vaccine roll-out

When vaccines for the virus were developed, geographers once again responded. In some places, geographers worked to determine which neighbourhoods should be prioritized for vaccine rollout based on the overall vulnerability of the population residing there and current outbreaks of the disease. Both Ontario and British Columbia, Canada opted for this approach in their vaccine distribution plans, allowing them to maximize the prevention of death and long-term morbidity, slow the spread of COVID-19, and best manage health care system capacity (Brown *et al.*, 2021; Province of British Columbia, 2021). Health officials from both provinces argued that such an approach is superior to prioritization of vaccines based on age alone, with Ontario suggesting that 17.8% of COVID-19 hospitalizations, 19.4% of ICU admissions, and 10.0% of deaths were prevented using this strategy (Brown *et al.*, 2021). Given the misinformation and hesitancy around vaccines, geographers were also able to apply GIS to monitor the spatial distribution of vaccine hesitancy. A Web-based dashboard developed by the US Department of Health and Human Services provides an excellent illustration of this application, showing federal survey data that identifies the percentage of the population within a county estimated to be 'vaccine hesitant' or 'strongly vaccine hesitant' (APSE, 2021). Understanding the spatial distribution of vaccine hesitancy proved critical for the United States in supporting state and local communication and outreach efforts in their vaccine campaigns.

Diminished case counts

As case counts diminished with more people vaccinated, geographers once again moved onto other important applications of GIScience, helping our society move back to a new 'normal'. Some continued to focus on health and health care, examining the secondary health impacts of the pandemic (Griffin, 2021), or trying to understand which communities were hit particularly hard by the virus and why (Cheung, 2021; Yang *et al.*, 2021). Others have turned their attention to reviving the economy. Unacast, a location data company, has built a COVID-19 retail scoreboard to help businesses understand the commercial and economic impact of the pandemic across different industries and geographies (Valentine, 2021). Not only is this an important tool for understanding the state of the industry where business owners operate, but it can also help identify early signals of a potential recovery in different regions.

Preparing for similar futures

Since the 1990s, the world has seen a rapid succession of disease outbreaks and near misses, with some of the top researchers around the world warning of the next 'big one'. Despite these warnings, governments around the world were largely unprepared to deal with the COVID-19 pandemic. They ignored warning signs, underestimated the magnitude of the problem, and prioritized other spending over pandemic preparedness, costing countless people their lives and livelihoods when COVID-19 hit.

COVID-19 will certainly not be the last pandemic we face as a society. And while we do not know which disease will come next, we can be better prepared and learn from our mistakes.

What does this mean with regard to GIScience? It means we continue developing geospatial and non-geospatial technologies that can be used to collect higher resolution spatial trajectory data. It means we organize maps of businesses and supply chains that can be 'at-the-ready' and respond dynamically to help meet new resource demands when outbreaks emerge. It means we continue to use GIS to address the gross inequities in our health care systems – that is, ensuring both equitable access to health care and equitable treatment no matter your age, gender, race, or country of residence. Lastly, it means that we continue to study COVID-19 retrospectively. There is much that can be learned from both a geospatial and non-geospatial perspective in how this pandemic was handled that can help prepare us for similar futures. We should be sure not to let these important lessons go to waste.

Conclusion

GIS took on an enormous role throughout many aspects of the COVID-19 pandemic. This chapter summarized the key means through which geospatial data and analyses shaped our responses at each stage of the pandemic. As we have shifted away from the 'panic' mode that accompanies crises like this one, we must now work to break the cycle that so often results in forgetting and moving on. We know another pandemic looms in our future, and GIScience will most certainly play a critical role like it has done for COVID-19. It is time now for us to learn from, and improve upon, the ways GIS can be mobilized in pandemics so we can be as prepared as possible for similar futures that lie ahead.

References

Alliance for Healthier Communities (2020) "Statement from Black Health Leaders on COVID-19's Impact on Black Communities in Ontario" Available at: https://www.allianceon.org/news/Statement-Black-Health-Leaders-COVID-19s-impact-Black-Communities-Ontario (Accessed: 7th July 2020).

Anaele, B. I., Doran, C. and McIntire, R. (2021) "Visualizing COVID-19 Mortality Rates and African-American Populations in the USA and Pennsylvania" *Journal of Racial and Ethnic Health Disparities* 8 (6) pp.1356–1363 DOI: 10.1007/s40615-020-00897-2.

Andersen, L.M., Harden, S.R., Sugg, M.M., Runkle, J.D. and Lundquist, T.E. (2021) "Analyzing the Spatial Determinants of Local Covid-19 Transmission in the United States" *Science of the Total Environment* 754 (142396) pp.1–10 DOI: 10.1016/j.scitotenv.2020.142396.

Andrade, L.A., Gomes, D.S., de Oliveira Góes, M.A., de Souza, M.S.F., Teixeira, D.C.P., Ribeiro, C.J.N., Alves, J.A.B., de Araújo, K.C.G.M. and dos Santos, A.D. (2020) "Surveillance of the First Cases of COVID-19 in Sergipe Using a Prospective Spatiotemporal Analysis: The Spatial Dispersion and its Public Health Implications" *Revista da Sociedade Brasileira de Medicina Tropical* 53 (e20200287) pp.1–5 DOI: 10.1590/0037–8682–0287–2020.

ASPE (2021) "Vaccine Hesitancy for COVID-19: State, County, and Local Estimates" Available at: https://aspe.hhs.gov/pdf-report/vaccine-hesitancy (Accessed: 23rd April 2021).

Bowden, O. (2020) "Canada's Lack of Race-Based COVID-19 Data Hurting Black Canadians: Experts" *Global News* (2nd May) Available at: https://globalnews.ca/news/6892178/black-canadians-coronavirus-risk/ (Accessed: 7th July 2020).

Brown, K.A., Stall, N.M., Joh, E., Allen, U., Bogoch, I.I., Buchan, S.A., Daneman, N., Evans, G.A., Fisman, D.N., Gibson, J.L., Hopkins, J., Van Ingen, T., Maltsev, A., McGeer, A., Mishra, S., Razak, F., Sander, B., Schwartz, B., Schwartz, K., Siddiqi, A., Smylie, J., Jüni, P. (2021) "COVID-19 Vaccination Strategy for Ontario Using Age and Neighbourhood-Based Prioritization" *Ontario COVID-19 Science Advisory Table* (26th February) DOI: 10.47326/ocsat.2021.02.10.1.0.

Brunsdon, C. (2020) "Modelling Epidemics: Technical and Critical Issues in the Context of COVID-19" *Dialogues in Human Geography* 10 (2) pp.250–254 DOI: 10.1177/2043820620934328.

Calvo, R. A., Deterding, S. and Ryan, R. M. (2020) "Health Surveillance During Covid-19 Pandemic" *British Medical Journal* 369 (m1373) DOI: 10.1136/bmj.m1373.

CDC (2020) "COVID-19 in Racial and Ethnic Minority Groups, Centers for Disease Control and Prevention" Available at: https://stacks.cdc.gov/view/cdc/89820/cdc_89820_DS1.pdf (Accessed: 21st February 2023).

Cheung, C. (2021) "Here's Why Communities Become COVID Hotspots" *The Tyee* Available at: https://thetyee.ca/News/2021/05/20/Vancouver-Fraser-Valley-COVID-Hotspots/ (Accessed: 20th May 2021).

COVID Symptom Tracker (2020) "COVID Symptom Tracker – Help Slow the Spread of COVID-19" Available at: https://covid.joinzoe.com/ (Accessed: 10th April 2020).

Credit, K. (2020) "Neighbourhood Inequity: Exploring the Factors Underlying Racial and Ethnic Disparities in COVID-19 Testing and Infection Rates Using ZIP Code Data in Chicago and New York" *Regional Science Policy & Practice* 12 (6) pp.1249–1271 DOI: 10.1111/rsp3.12321.

Desjardins, M.R., Hohl, A. and Delmelle, E.M. (2020) "Rapid Surveillance of COVID-19 in the United States Using a Prospective Space-time Scan Statistic: Detecting and Evaluating Emerging Clusters" *Applied Geography* 118 (102202) pp.1–7 DOI: 10.1016/j.apgeog.2020.102202.

Dong, E., Du, H. and Gardner, L. (2020) "An Interactive Web-based Dashboard to Track COVID-19 in Real Time" *The Lancet Infectious Diseases* 20 (5) pp.533–534 DOI: 10.1016/S1473-3099(20)30120-1.

Elson, R., Davies, T.M., Lake, I.R., Vivancos, R., Blomquist, P.B., Charlett, A. and Dabrera, G. (2021) "The Spatio-temporal Distribution of COVID-19 Infection in England Between January and June 2020" *Epidemiology & Infection* 149 (e73) pp.1–6 DOI: 10.1017/S0950268821000534.

Everts, J. (2020) "The Dashboard Pandemic" *Dialogues in Human Geography* 10 (2) pp.260–264 DOI: 10.1177/2043820620935355.

Ferreira, R.V., Martines, M.R., Toppa, R.H., Assunção, L.M., Desjardins, M.R. and Delmelle, E.M. (2020) "Applying a Prospective Space-time Scan Statistic to Examine the Evolution of COVID-19 Clusters in the State of São Paulo, Brazil" *medRxiv* DOI: 10.1101/2020.06.04.20122770.

Ferretti, L., Wymant, C., Kendall, M., Zhao, L., Nurtay, A., Abeler-Dörner, A., Parker, M., Bonsall, D. and Fraser, C. (2020) "Quantifying SARS-CoV-2 Transmission Suggests Epidemic Control with Digital Contact Tracing" *Science* 368 (6491) pp.1–7 DOI: 10.1126/science.abb6936.

Flatten (2020) Available at: https://flatten.ca/#home (Accessed: 10th April 2020).

Forati, A.M. and Ghose, R. (2020) "Geospatial Analysis of Misinformation in COVID-19 Related Tweets" *Applied Geography* 133 (102473) pp.1–10 DOI: 10.1016/j.apgeog.2021.102473.

Gaglioti, A., Douglas, M., Li, C., Baltrus, P., Blount, M. and Mack, D. (2020) "County-Level Proportion of Non-Hispanic Black Population is Associated with Increased County Confirmed COVID-19 Case Rates After Accounting for Poverty, Insurance Status, and Population Density" (*White Paper*) *Morehouse School of Medicine* Available at: https://www.msm.edu/RSSFeedArticles/2020/May/documents/County-Level-Proportion-of-AA-Case-Rate-of-COVID19.pdf (Accessed: 10th April 2021).

Gan, N. and Culver, D. (2020) "China is Fighting the Coronavirus with a Digital QR Code: Here's How it Works" *CNN Business* (16th April) Available at: https://www.cnn.com/2020/04/15/asia/china-coronavirus-qr-code-intl-hnk/index.html (Accessed: 26th April 2021).

Goodchild, M.F. (2007) "Citizens as Sensors: The World of Volunteered Geography" *GeoJournal* 69 (4) pp.211–221 DOI: 10.1007/s10708-007-9111-y.

Griffin, K. (2021) "COVID-19: SFU Researchers Map Secondary Health Effects of COVID-19" *Vancouver Sun* (19th May) Available at: https://vancouversun.com/news/local-news/covid-19-sfu-researchers-map-secondary-health-effects-of-covid-19 (Accessed: 20th May 2021).

Hohl, A., Delmelle, E.M., Desjardins, M.R. and Lan, Y. (2020) "Daily Surveillance of COVID-19 Using the Prospective Space-time Scan Statistic in the United States" *Spatial and Spatio-Temporal Epidemiology* 34 (100354) pp.1–8 DOI: 10.1016/j.sste.2020.100354.

Ienca, M. and Vayena, E. (2020) "On the Responsible Use of Digital Data to Tackle the COVID-19 Pandemic" *Nature Medicine* 26 (4) pp.463–464 DOI: 10.1038/s41591-020-0832-5.

Iyanda, A.E., Boakye, K.A., Lu, Y. and Oppong, J.R. (2021) "Racial/Ethnic Heterogeneity and Rural-Urban Disparity of COVID-19 Case Fatality Ratio in the USA: A Negative Binomial and GIS-Based Analysis" *Journal of Racial and Ethnic Health Disparities* 9 (2) pp.708–721 DOI: 10.1007/s40615-021-01006-7.

Johns Hopkins Coronavirus Resource Center (2020a) Available at: https://coronavirus.jhu.edu/ (Accessed: 17th May 2021).

Johns Hopkins Coronavirus Resource Center (2020b) "COVID-19 Racial Data Transparency" Available at: https://coronavirus.jhu.edu/data/racial-data-transparency (Accessed: 27th April 2021).

Joi, P. (2020) "5 Reasons Why Pandemics Like COVID-19 Are Becoming More Likely" *Gavi* Available at: https://www.gavi.org/vaccineswork/5-reasons-why-pandemics-like-covid-19-are-becoming-more-likely (Accessed: 20th April 2021).

Kamel Boulos, M.N. and Geraghty, E.M. (2020) "Geographical Tracking and Mapping of Coronavirus Disease COVID-19/Severe Acute Respiratory Syndrome Coronavirus 2 (SARS-Cov-2) Epidemic and Associated Events Around the World: How 21st Century GIS Technologies are Supporting the Global Fight Against Outbreaks and Epidemics" *International Journal of Health Geographics* 19 (1) pp.1–12 DOI: 10.1186/s12942-020-00202-8.

Kan, Z., Kwan, M-P., Wong, M.S., Huang, J. and Liu, D. (2021) "Identifying the Space-time Patterns of COVID-19 Risk and Their Associations with Different Built Environment Features in Hong Kong" *Science of the Total Environment* 772 (145379) pp.1–12 DOI: 10.1016/j.scitotenv.2021.145379.

Kim, S. and Castro, M. C. (2020) "Spatiotemporal Pattern of COVID-19 and Government Response in South Korea (as of May 31, 2020)" *International Journal of Infectious Diseases* 98 pp.328–333 DOI: 10.1016/j.ijid.2020.07.004.

Kinsa Health (2020) "US Health Weather Map" Available at: https://healthweather.us (Accessed: 20th April 2020).

Kitchin, R. (2020) "Civil Liberties *or* Public Health, or Civil Liberties *and* Public Health? Using Surveillance Technologies to Tackle the Spread of COVID-19" *Space and Polity* 24 (3) pp.362–381 DOI: 10.1080/13562576.2020.1770587.

Koch, T. (2017) *Cartographies of Disease* Redlands, CA: Esri Press.

Kulldorff, M. (2001) "Prospective Time Periodic Geographical Disease Surveillance Using a Scan Statistic" *Journal of the Royal Statistical Society: Series A (Statistics in Society)* 164 (1) pp.61–72 DOI: 10.1111/1467-985X.00186.

LA County Department of Public Health (2020) Available at: http://publichealth.lacounty.gov/media/Coronavirus/ (Accessed: 9th April 2020).

Lee, Y. (2020) "Singapore App Halves Contact-Tracing Time, Top Engineer Says" *Bloomberg* (8th December) Available at: https://www.bloomberg.com/news/articles/2020-12-08/singapore-app-halves-contact-tracing-time-leading-engineer-says (Accessed: 1st June 2021).

Lewis, D. (2020) 'Why Many Countries Failed at COVID Contact-Tracing – But Some Got it Right" *Nature* 588 (7838) pp.384–387 DOI: 10.1038/d41586-020-03518-4.

Liu, F., Wang, J., Liu, J., Li, Y., Liu, D., Tong, J., Li, Z., Yu, D., Fan, Y., Bi, X., Zhang, X. and Mo, S. (2020) "Predicting and Analyzing the COVID-19 Epidemic in China: Based on SEIRD, LSTM and GWR Models" *PLoS One* 15 (8) e0238280 pp.1–22 DOI: 10.1371/journal.pone.0238280.

Mansour, S., Al Kindi, A., Al-Said, A., Al-Said, A., Atkinson, P. (2021) "Sociodemographic Determinants of COVID-19 Incidence Rates in Oman: Geospatial Modelling Using Multiscale Geographically Weighted Regression (MGWR)" *Sustainable Cities and Society* 65 (102627) pp.1–13 DOI: 10.1016/j.scs.2020.102627.

Masrur, A., Yu, M., Luo, W. and Dewan, A. (2020) "Space-time Patterns, Change, and Propagation of COVID-19 Risk Relative to the Intervention Scenarios in Bangladesh" *International Journal of Environmental Research and Public Health* 17 (16) 5911 DOI: 10.3390/ijerph17165911.

McNeil Jr., D.G. (2020) "Restrictions Are Slowing Coronavirus Infections, New Data Suggest" *The New York Times* (30th March) Available at: https://www.nytimes.com/2020/03/30/health/-coronavirus-restrictions-fevers.html (Accessed: 20th April 2020).

Millett, G.A., Jones, A.T., Benkeser, D., Baral, S., Mercer, L., Beyrer, C., Honermann, B., Lankiewicz, E., Mena, L., Crowley, J.S., Sherwood, J. and Sullivan, P.S. (2020) "Assessing Differential Impacts of COVID-19 on Black Communities" *Annals of Epidemiology* 47 pp.37–44 DOI: 10.1016/j.annepidem.2020.05.003.

Mody, A., Pfeifauf, K., Bradley, C., Fox, B., Hlatshwayo, M.G., Ross, W., Sanders-Thompson, V., Maddox, K.J., Reidhead, M., Schootman, M., Powderly, W.G. and Geng, E.H. (2020) "Understanding Drivers of Coronavirus Disease 2019 (COVID-19) Racial Disparities: A Population-Level Analysis of COVID-19 Testing Among Black and White Populations" *Clinical Infectious Diseases* 73 (9) pp. e2921–e2931 DOI: 10.1093/cid/ciaa1848.

Mollalo, A., Vahedi, B. and Rivera, K. M. (2020) "GIS-Based Spatial Modeling of COVID-19 Incidence Rate in the Continental United States" *Science of the Total Environment* 728 p. (138884) pp.1–8 DOI: 10.1016/j.scitotenv.2020.138884.

NYC Department of Health and Mental Hygiene (2020) "New York City Health Coronavirus Data" Available at: https://github.com/nychealth/coronavirus-data (Accessed: 9th April 2020).

Ottawa Public Health (2021) "Mapping of Confirmed COVID-19 in Ottawa" Available at: https://www.ottawapublichealth.ca/en/reports-research-and-statistics/mapping-products.aspx (Accessed: 18th May 2021).

Outbreaks Near Me (2020) Available at: https://outbreaksnearme.org/ca/en-CA/ (Accessed: 27th April 2021).

Platt, B. (2020) "Canada's Public Data on COVID-19 Is (Mostly) a Mess. Here's How to Find the Useful Info" *The Guardian* (2nd April) Available at: http://www.theguardian.pe.ca/news/canada/-canadas-public-data-on-covid-19-is-mostly-a-mess-heres-how-to-find-the-useful-info-433201/ (Accessed: 7th July 2020).

Province of British Columbia (2021) "Vaccines in High-transmission Neighbourhoods" Available at: https://www2.gov.bc.ca/gov/content/covid-19/vaccine/neighbourhood (Accessed: 23rd April 2021).

Rinner, C. (2021) "Mapping COVID-19 in Context: Promoting a Proportionate Perspective on the Pandemic" *Cartographica* 56 (1) pp.14–26 DOI: 10.3138/cart-2020-0020.

Rosenkrantz, L., Schuurman, N., Bell, N. and Amram, O. (2020) "The Need for Giscience in Mapping COVID-19" *Health & Place* 67 (102389) pp.1–4 DOI: 10.1016/j.healthplace.2020.102389.

Ruiz Estrada, M.A. and Khan, A. (2020) *Globalization and Pandemics: The Case of COVID-19.* (*SSRN Scholarly Paper ID 3560681*) Rochester, NY: Social Science Research Network DOI: 10.2139/ssrn.3560681.

Safepaths (2020) "Private Kit: Safe Paths; Privacy-by-Design Contact Tracing using GPS+Bluetooth" Available at: http://safepaths.mit.edu/ (Accessed: 10th April 2020).

Sannigrahi, S., Pilla, F., Basu, B., Basu, A.S. and Molter, A. (2020) "Examining the Association Between Socio-Demographic Composition and COVID-19 Fatalities in the European Region Using Spatial Regression Approach" *Sustainable Cities and Society* 62 (102418) pp.1–14 DOI: 10.1016/j.scs.2020.102418.

Singer, N. and Sang-Hun, C. (2020) "As Coronavirus Surveillance Escalates, Personal Privacy Plummets" *The New York Times* (23rd March) Available at: https://www.nytimes.com/2020/03/23/technology/coronavirus-surveillance-tracking-privacy.html (Accessed: 1st June 2021).

Smart, A. (2020) "Lack of Data Makes Coronavirus Curve Hard to Predict, B.C. Researcher Says, British Columbia" *CTV News* (24th March) Available at: https://bc.ctvnews.ca/lack-of-data-makes-coronavirus-curve-hard-to-predict-b-c-researcher-says-1.4865714 (Accessed: 7th July 2020).

The COVID Tracking Project (2020) "The COVID Racial Data Tracker" Available at: https://covidtracking.com/race (Accessed: 27th April 2021).

The Economist (2020) "Fatal Flaws: Covid-19's Death Toll Appears Higher Than Official Figures Suggest" (4th April) Available at: https://www.economist.com/graphic-detail/2020/04/03/covid-19s-death-toll-appears-higher-than-official-figures-suggest (Accessed 21st February 2023).

Valentine, S. (2021) "Covid-19 Retail Impact Scoreboard" *Unacast* (20th May) Available at: https://www.unacast.com/covid19/covid-19-retail-impact-scoreboard (Accessed: 20th May 2021).

Vaughan, A. (2020) "There Are Many Reasons Why Covid-19 Contact-tracing Apps May Not Work" *New Scientist* (17th April) Available at: https://www.newscientist.com/article/2241041-there-are-many-reasons-why-covid-19-contact-tracing-apps-may-not-work/ (Accessed: 1st June 2021).

Walsh, M. and Semeniuk, I. (2020) "Long-term Care Connected to 79 Per Cent of COVID-19 Deaths in Canada" Available at: https://www.theglobeandmail.com/politics/article-long-term-care-connected-to-79-per-cent-of-covid-19-deaths-in-canada/ (Accessed: 7th July 2020).

Whitelaw, S., Mamas, M.A., Topol, E. and Van Spall, H.G.C. (2020) "Applications of Digital Technology in COVID-19 Pandemic Planning and Response" *The Lancet Digital Health* 2 (8) pp.e435–e440 DOI: 10.1016/S2589–7500(20)30142-4.

Yang, T-C., Kim, S., Zhao, Y. and Choi, S-W.E. (2021) "Examining Spatial Inequality in COVID-19 Positivity Rates Across New York City ZIP Codes" *Health & Place* 69 (102574) pp.1–8 DOI: 10.1016/j.healthplace.2021.102574.

Yourish, K., Lai, K.K.R., Ivory, D. and Smith, M. (2020) "One-Third of All U.S. Coronavirus Deaths Are Nursing Home Residents or Workers" *The New York Times* (9th May) Available at: https://www.nytimes.com/interactive/2020/05/09/us/coronavirus-cases-nursing-homes-us.html (Accessed: 7th July 2020).

Zastrow, M. (2020) "South Korea is Reporting Intimate Details of COVID-19 Cases: Has it Helped?" *Nature News* (18th March) DOI: 10.1038/d41586-020-00740-y.

34
GEOSURVEILLANCE AND SOCIETY

Rob Kitchin

Surveillance concerns the systematic monitoring of people, places and systems. It has long been a feature of societies, used to observe the lives and activities of citizens in order to effect law and order, secure the loyalty of subjects, monitor the efficiency and productivity of workers and provide useful information for public bodies and companies (Lyon, 2007). The nature and depth of surveillance has varied over time and space, shaped by the political ideology and economy of state governance and the development of new techniques and technologies. In the twenty-first century, the drive towards ubiquitous computing (network connections available everywhere) and pervasive computing (computation embedded into everything) has radically extended the scope and extent of surveillance. Networked digital technologies typically produce big data; that is, large volumes of real-time, exhaustive (within a system), fine-grained, uniquely indexical, relational data (Kitchin, 2014), creating a step-change in the breadth (more aspects of everyday life) and depth (fine-grained spatially and temporally) of surveillance.

Geospatial technologies are a key component of this new surveillance regime, playing a fundamental role in two key respects. First, they enable georeferenced data produced through other means (e.g., surveys, administrative systems, sensors) to be visualized, made sense of, and acted upon. Second, they perform geosurveillance, producing a rich stream of detailed, georeferenced location and movement data, most of which are big data. In other words, geospatial technologies enable the fine-grained, exhaustive monitoring and tracking of places and spatial behaviour for large populations, which was previously impossible to accomplish. For example, it is possible to track location and movement of millions of people simultaneously through a variety of networked technologies:

- controllable digital high-definition CCTV (closed circuit television) cameras (increasingly used with facial recognition software);
- smartphones and associated apps that track phone location via cell masts, GPS (global positioning system), or Wi-Fi connections;
- smart devices such as GPS-enabled fitness trackers and smart watches;
- sensor networks that capture passing phone identifiers such as MAC (media access control) addresses, enabling the tracking of movement along streets or through malls;
- public wireless networks that record actual and attempted connections by Internet-enabled digital devices (e.g., laptops, tablets, smartphones, cars);

- smart card tracking that capture the scanning of barcodes/RFID chips of cards used to enter buildings or use public transport;
- vehicle tracking using Automatic Number Plate Recognition (ANPR) cameras, unique ID transponders for automated road tolls and car parking, and on-board GPS;
- other staging points, such as the use of ATMs (automatic teller machines), credit card use, metadata tagging of photos uploaded to the Internet, geotagging of social media posts;
- electronic tagging of children and parolees with GPS tracking devices;
- shared calendars that provide date, time, and location of meetings.

(Kitchin, 2015)

In addition, satellites and drones monitor large portions of the planet at highly granular resolutions, taking up fixed orbits to provide a continuous stream of data about a location. For example, the ARGUS-IS project, unveiled by DARPA (Defense Research Projects Agency) and the US Army in 2013, is a 1.8-gigapixel video surveillance platform operated from a drone with a resolution of six inches from an altitude of 20,000 feet. Capturing 12 frames per second the system can track in real-time up to 65 moving objects (Anthony 2013). The FBI (Federal Bureau of Investigation) and Department of Homeland Security have an active programme of Cessna aircraft circling many US cities daily using high-resolution video cameras, MAC address sensors, and 'augmented reality' software to superimpose other geospatial information such as property ownership onto the video feed (Aldhous and Seife, 2016).

For those that control these systems, individuals are no longer lost in the crowd; the monitoring of location and movement is pervasive, continuous, automatic, and relatively cheap. These forms of geosurveillance then are producing vast data shadows; that is, data generated by third parties about people and places. They are complemented by data footprints; that is, data produced knowingly by people about themselves, commonly known as sousveillance (Mann et al., 2003). For example, millions of people track their personal health by capturing their performance (e.g., miles walked/run/cycled, hours slept and types of sleep), consumption (e.g., food/calorie intake), physical states (e.g., blood pressure, pulse), and emotional states (e.g., mood, arousal) using smartphone apps and dedicated technologies (Lupton, 2016). They share their personal and family stories (via Facebook and blogs), personal thoughts (via Twitter, chat rooms and online reviews), and family photographs and videos (via Instagram and YouTube). Many of these utilize GPS to track the place of measurement and posting so that the data are georeferenced.

Geosurveillance and sousveillance are having pronounced transformative effects on three key aspects of society. They are actively reshaping the practices of governance and governmentality. They are producing new expressions of capitalism, creating new markets and means to accumulate profit. They are eroding the right to privacy and challenging civil liberties. The next three sections detail the changes occurring in three respects, followed by an illustrative discussion of their intersection in response to the COVID-19 pandemic.

Geosurveillance and governance

Surveillance technologies have long been a part of managing and governing societies. Geospatial technologies are central to contemporary governance regimes, both with respect to public space and workplaces. As well as providing a finer, more systematic grid of surveillance, these technologies are changing the nature of governance and governmentality in meaningful ways. The practices of governance are becoming more technocratic, algorithmic, and automated. Here, governance is increasingly performed through technical, codified, data-driven systems, which have become essential architectures for managing populations, controlling, and regulating infrastructure, monitoring and directing government work, and communicating with the general public. The notion of smart cities is predicated on the use of such systems, which aim to more efficiently and effectively manage various aspects of urban life (Townsend, 2013). Geospatial technologies are a key component of such real-time governance systems, perhaps best exemplified through control rooms and associated spatial media, such as dashboards. For example, the Centro De Operacoes Prefeitura Do Rio in Rio de Janeiro, Brazil (COR) is a data-driven city operations centre that continuously monitors and manages the city and also acts as a coordinated emergency management centre. COR pulls together into a single control room real-time data streams from thirty-two agencies and twelve private concessions (e.g., bus and electricity companies), with the data used to manage and coordinate service and infrastructure delivery, maintenance, and performance (Luque-Ayala and Marvin, 2016). The centre is complemented by a virtual operations platform that enables city officials to log in from the field to access real-time information. For example, police at an accident scene can use the platform to see how many ambulances have been dispatched and their expected arrival time, and to upload additional information.

These technocratic, algorithmic systems often operate a form of 'automated management' (Dodge and Kitchin, 2007); that is, they work in automated, autonomous, and automatic ways, with systems directly regulating service delivery and citizen behaviour. In such automated systems, human involvement in their operation is limited to three levels of participation: human-in- (humans make key decisions), human-on- (algorithms make key decisions with human oversight and intervention), and human-off-the-loop (the algorithms make the key decisions) (Docherty, 2012). These systems are often concerned with regulating spatial behaviour and travel, and can rely on geospatial data. An intelligent transport system generally operates as a human-on-the-loop system, with the system automatically phasing traffic lights across the road network based on real-time data feeds, though a human controller can intervene if needed. Such automation can be contentious; for example, there is an active debate as to whether drone strikes should be administered by human-in, -on, or -off systems, with the decision to launch based on geospatial intelligence related to mobile phone location (Docherty, 2012).

In addition, governance is also becoming more predictive and anticipatory. Data have long been used to profile and segment populations, and to predict how people will behave in different scenarios and how best to manage them, but these processes have become much more sophisticated, fine-grained, widespread, and routine. The state is increasingly embracing predictive profiling and using data and algorithmic governance to make decisions about how to treat people and manage populations, allocate funding and resourcing, and deliver frontline services (Eubanks, 2017). This includes geodemographic techniques that segment and socially/spatially sort communities and places and are used to determine decisions and spatially target resources by public bodies. Many of the systems deployed use predictive modelling to assess the likelihood of particular conditions or outcomes in order to direct further attention and pre-empt situations arising. Such anticipatory governance has been a feature of air travel for a number of years, with passengers profiled regarding their security risk (Amoore, 2006). Predictive analytics has been extended into policing in general, with a number of police forces using them to anticipate the location of future crimes and to direct police officers to increase patrols in those areas, and also to identify potential suspects (Jefferson, 2018). In such predictive systems a person's data shadow does more than follow them; it precedes them. People and communities are held accountable and treated in relation to predictions rather than actual actions.

In turn, the underlying governmentality of societies is being transformed. Governmentality is the logics, rationalities, and techniques that render societies governable and enable government to enact governance (Foucault, 1991). Big data has widened, deepened, and intensified the surveillance gaze and how governance is enacted, shifting governmentality from disciplinary regimes, in which people self-regulate behaviour based on the fear of surveillance and sanction, to control regimes in which people are corralled and compelled to act in certain ways, their behaviour explicitly or implicitly steered or nudged (Deleuze, 1992). The nature of observation and management is moved from a model where an external observer is needed, to one where observation is an integral aspect of performance, with behaviour no longer adjusted in case of observation but actively reshaped (Kitchin *et al.*, 2020). For example, the work of a retail checkout employee used to be monitored periodically by a supervisor on the shop floor, then via a CCTV system, and is now continually modulated by the till as they perform their work; the act of scanning is the act of surveillance. Relatedly, a transport network controlled by an intelligent transport system explicitly or implicitly nudges travel behaviour. Here, driving is modulated by traffic light sequencing and the act of driving itself becomes a site of administration (Dodge and Kitchin, 2007). In such systems, surveillance and its associated data shadow become continuous, pervasive, distributed, persistent, reactive to the subject's behaviour, but outside of the subject's control (Cohen, 2013). In more authoritarian states, geospatial technologies and geosurveillance play an active role in closely and oppressively policing and managing populations, through both disciplinary and control logics. Across all jurisdictions then, geospatial technologies are directly implicated in shifting the logics of governance in ways that have profound implications for civil liberties and citizenship.

Geosurveillance and data capitalism

Data has become a vital ingredient for leveraging value and the production and accumulation of capital, and gives rise to what Sadowski (2019) terms 'data capitalism'. That is, a form of capitalism wherein data are themselves a form of capital and not simply a commodity that can be converted into monetary value (data intrinsically have value and they produce value) and where value and profit are driven in the main or in large part by extracting value from data. Data capitalism encompasses the diverse ways data are used to accumulate profit, including optimizing systems, managing and controlling systems, modelling probabilities and planning future activities, designing and creating new products and markets, and growing the value of assets or slowing depreciation (Sadowski, 2019). A specific form of data capitalism is surveillance capitalism, which actively includes the use of geospatial technologies and their data.

Surveillance capitalism derives value and profit through the capture and monetization of data about people and places (Zuboff, 2019). In the case of geosurveillance, this involves extracting value from georeferenced data through the creation of spatial products, such as location-based targeted advertising and geodemographics (the spatial profiling and sorting of customers and places). Indeed, there is now a multi-billion-dollar market for georeferenced data globally, with a suite of specialist data companies producing, purchasing, consolidating, and linking datasets to create data products. The holdings of data brokers can be vast. In 2019, Acxiom was thought to hold data on '2.5 billion addressable consumers' in 'more than 62 countries' across more than '10,000 attributes' (Melendez and Pasternack, 2019). Equifax, a consumer credit reporting agency, has collected information on over 800 million individual consumers and over 88 million businesses worldwide (Pinchot *et al.*, 2018). Experian holds financial and purchase records for 918 million people in America, Europe, and Asia (Christl, 2017). IRI is reported to have pulled together data from more than 85,000 US retail outlets, with Nielsen collating data from 900,000 stores in more than 100 countries, and Oracle having data on billions of transactions from 1,500 chain retailers (Christl, 2017). A number of brokers specialize in location and movement data, and in spatial profiling. For example, Groundtruth, a location-focused data broker, enables companies to install their proprietary Software Development Kit into smartphone apps to deliver location-sensitive adverts, in the process collecting user location data. Their reach is about 120 million smartphone users that generate 30 billion location points a month (Smith, 2019).

In part then, extensive geosurveillance is being driven by a desire to extract value from geospatial data. As with other forms of capitalism, profit is accumulated through uneven and unequal processes of exploitation that seek to extract maximum profit at the expense of others (Sadowski, 2019). Indeed, for some surveillance capitalism is a form of modern-day colonialism in which accumulation occurs through data dispossession, with the labour of producing data rendered cheap or free, communal resources are enclosed and personal resources ensnared, and control of these exploitative relationships reside with the data extractors (Thatcher et al., 2016). In other words, the users of geospatial media provide labour (e.g., clicking, swiping, typing, uploading) and data (the product of those labours) for free to those that control the means of production (Sadowski, 2019). Through this colonizing process previously non-commodified aspects of daily life are privatized and converted into commodities and a new terrain for capital investment and exchange is realized (Thatcher et al., 2016).

For those providing labour and data, accumulation through data dispossession can have a profound effect on the services and opportunities extended to them, such as job offers, credit lines, insurance policy issued, tenancy approved, or what price goods and services cost, or whether a place receives targeted interventions or how it is policed (Angwin, 2014). Such profiling and sorting can work in discriminatory ways, overly misrepresenting and unfairly targeting certain groups, and reproducing and deepening inequalities by limiting life chances (Eubanks, 2017). And some companies use profiles in ways that are highly exploitative, for example offering poor deals with 'abusive terms (balloon payments, hidden fees, brutish penalty clauses) to the most vulnerable populations', or direct marketing commercial and political ads to susceptible communities (Roderick, 2014). Moreover, it can be difficult to opt out of data being captured or from data products and services. Of 212 data brokers operating in the United States examined by Angwin (2014), only 92 allowed opt-outs, and only 11 of 58 mobile location tracking businesses offered opt-outs. Moreover, opting out is not synonymous with deletion, but rather might mean omitting an individual's data from products while keeping them in the database, or only using their data in anonymous, aggregated forms (Kuempel, 2016). Whether one wants it or not then, geospatial data are being used by companies in ways that are to their advantage, but not necessarily to whom or where the data relate.

Geosurveillance, privacy, and civil liberties

Privacy – to reveal selectively oneself to the world – is a condition that many people value and expect. At a fundamental level, privacy provides a means for individuals to be able to define and present themselves; to manage what others know about them (Solove, 2006). Privacy ensures that other civil liberties, related to how individuals are treated based on what others know about them, are maintained. Privacy is understood to be a basic human right and entitlement in most jurisdictions enshrined in national and supra-national laws. The use of geosurveillance in new regimes of governance and capitalism is having a profound effect on privacy and associated civil liberties, enabling a range of privacy and predictive privacy harms. This is particularly the case with respect to identity privacy (to protect personal and confidential data), territorial privacy (to protect personal space, objects, and property), and locational and movement privacy (to protect against the tracking of spatial behaviour) (Kitchin, 2016).

The consensus of academics, lawyers, and civil liberties groups concerned about privacy is that extensive surveillance through networked, big data technologies has placed enormous strain on present legal frameworks in Europe and North America. In both geographic spheres, privacy legal provisions draw extensively on the OECD's (Organization for Economic Co-operation and Development) fair information practice principles (FIPPs). These prioritize personal rights regarding the generation, use, and disclosure of personal data, and place obligations on data controllers and processors (Solove, 2013). Other jurisdictions have their own approaches (see DLA Piper, 2019) or have limited legal protections; for example, only eight out of 55 Sub-Saharan African countries had data protection legislation in place in 2013 (Greenleaf, 2013).

In the big data age, FIPPs are being challenged and undermined in several ways. A key premise of big data is that they are repurposed so that additional value can be leveraged. Repurposing runs counter to data minimization, one of the key FIPPs. Data minimization stipulates that data controllers and processors should only generate data necessary to perform a particular task, that the data are only retained for as long as they are required to perform that task, and that the data generated should only be used for this task (Tene and Polonetsky, 2012). The solution has been to repackage personally identifiable information through de-identifying them or creating derived data that is exempt from the provisions. However, unless carefully undertaken it is possible to re-identify data through various techniques. One example relating to geospatial data concerns a dataset released by the New York City Taxi and Limousine Commission in 2013 relating to 173 million individual cab rides. The taxi drivers' medallion numbers were anonymized but were quickly de-identified, enabling information related to pickup and drop-off times, locations, fare and tip amounts to be tied to specific drivers, and to infer their home address, income, and religion (by if they took breaks to pray at set times). By combining this dataset with other public information, like celebrity blogs, it was possible to determine home addresses and where and who was visited (Metcalf and Crawford, 2016).

FIPPs do not relate to predictive privacy harms or group privacy harms. Predictive modelling can generate inferences about an individual and places that reinforce or create stigma and harm (Crawford and Schultz, 2014). For example, tracking data that reveals a person frequents gay bars, leading to the inference that the person is likely to gay, could be harmful if shared inappropriately. Yet, as no data about sexuality has been directly collected it is exempt from FIPPs provisions. Similarly, co-proximity and co-movement with others can be used to infer political, social, and/or religious affiliation, potentially revealing membership of particular groups (Leszczynski, 2017). Moreover, at present, approaches to privacy focus almost exclusively on individual interests and personal harm (Taylor et al., 2017). However, this individual focus fails to recognize that aggregated data relating to groups can lead to group privacy harms by enabling group members to be targeted and treated with minimal protections (Rainie et al., 2019). This is particularly problematic with respect to marginalized groups who are already collectively victimized through actions that indiscriminately target members.

Conclusion

Geospatial technologies undoubtedly have many productive uses and make positive contributions to society and economy. They are also key agents of geosurveillance and have troubling effects with respect to governance, capitalism, and civil liberties. The tension between productive uses and troubling effects has been well illustrated through the use of geospatial technologies and georeferenced data to try to limit the spread of COVID-19 (Kitchin, 2020; Taylor et al., 2020). A number of new and existing technologies designed to restrict movement (smartphone apps, facial recognition and thermal cameras, biometric wearables, smart helmets, drones, and predictive analytics) were rapidly developed or re-orientated for contact tracing, quarantine enforcement, travel permission, social distancing/movement monitoring, and symptom tracking.

In South Korea, the government utilized surveillance camera footage, smartphone location data, and credit card purchase records to track positive cases and their contacts (Singer and Sang-Hun, 2020). Singapore quickly launched TraceTogether, a Bluetooth-enabled app that detects and stores the details of nearby phones to enable contact tracing, with dozens of other countries launching similar apps shortly after (Taylor et al., 2020). Moscow authorities rolled out an app system to pre-approve journeys and routes (Ilyushina, 2020). Taiwan deployed a mandatory phone-location tracking system to enforce quarantines (Timberg and Harwell, 2020). In some parts of China citizens were required to scan QR codes when accessing public spaces and transit systems to verify their infection status and gain permission (Goh, 2020). A number of companies offered, or actively undertook, repurposing of their platforms and data as a means to help tackle the virus. In Germany, Deutsche Telekom provided aggregated, anonymized information to the government on peoples' movements; likewise, Telecom Italia, Vodafone and WindTre did the same in Italy (Pollina and Busvine, 2020). Palantir monitored and modelled the spread of the disease to predict the required health service response for the Center for Disease Control in the US and the National Health Service in the UK (Hatmaker, 2020), and a number of cyber-intelligence companies such as NSO Group, Cellebrite, Intellexa, Verint Systems, and Rayzone Group offered their people tracking services to governments (Schectman et al., 2020).

For many politicians, policy makers, and citizens the use of these surveillance technologies was legitimated by the need to contain the virus and save lives, regardless of their effects on civil liberties. For others, their deployment and the extensive geosurveillance enacted was highly problematic, raising a number of civil liberties and political economy concerns. Much of the public debate focused on privacy, since the technologies demand fine-grained knowledge about location, movement, social networks, and health status, and what else might be done with these data (Taylor et al., 2020). However, there were also concerns relating to governance given the technologies socially and spatially sorted people, redlining who could and could not mix, move and access spaces and services, and the extent to which these technocratic measures would creep into other domains and persist after the pandemic (in the same way heightened security measures persisted after 9/11). The Chinese government indicated that some of its intervention systems will remain in place post-pandemic, and although the Singapore government assured citizens that TraceTogether would only be used for contact tracing, a few months later it changed the terms to include movement data being available to the police for criminal matters (Mohan, 2021). In addition, there were worries about the lack of due process, oversight, and the right to redress and to opt out from systems (McDonald, 2020). Others were troubled that pursuing surveillance-based solutions in collaboration with industry legitimated and normalized the methods and logics of surveillance capitalism, while at the same time opening up sensitive public health data to private interests (Taylor et al., 2020). Moreover, the use of privately generated geosurveillance by states enabled the 'covidwashing' surveillance practices while simultaneously opening up new markets (Kitchin, 2020).

Despite arguments that public health trumped civil liberty concerns, when pressured by civil liberties organizations governments were able to ensure that appropriate safeguards were put in place to protect civil liberties (e.g., anonymization, encryption, not sharing data and deleting after two weeks, discontinuation at end of pandemic, publishing code and data protection assessments) while still being able to use the technologies for the purpose intended. In other words, it was proven that public health interventions could be enacted while preserving civil liberties when sufficient public pressure is applied, and the same is undoubtedly the case with respect to other uses of geospatial technologies. What this concluding discussion highlights is that there are a number of ethical and data justice issues relating to the use of geospatial technologies and the vast volumes of data they produce. These issues pose numerous moral dilemmas and inconvenient truths, which are often avoided or glossed over by the geospatial community. While some, notably within the Critical GIS community, do try to explore such issues and moderate their practices, it is clear that wider normative debates about geosurveillance are still required.

Acknowledgements

This chapter draws heavily on previously published work, in particular Kitchin (2015), Kitchin (2016), Kitchin (2020), and Kitchin et al. (2020). The research was supported by the European Research Council (grant ERC-2012-AdG-323636) and Science Foundation Ireland (grant 15/IA/3090).

References

Aldhous, P. and Seife, C. (2016) "Spies in the Sky: See Maps Showing Where FBI Planes Are Watching From Above" *BuzzFeed* (6th April) Available at: https://www.buzzfeednews.com/article/peteraldhous/spies-in-the-skies.
Amoore, L. (2006) "Biometric Borders: Governing Mobilities in the War on Terror" *Political Geography* 25 pp.336–351.
Angwin, J. (2014) *Dragnet Nation* New York: St Martin's Press.
Anthony, S. (2013) "DARPA Shows Off 1.8-Gigapixel Surveillance Drone, Can Spot a Terrorist from 20,000 Feet" *ExtremeTech* (28 January) Available at: https://www.extremetech.com/extreme/-146909-darpa-shows-off-1-8-gigapixel-surveillance-drone-can-spot-a-terrorist-from-20000-feet.
Christl, W. (2017) *Corporate Surveillance in Everyday Life* Vienna: Cracked Labs.
Cohen, J.E. (2013) "What is Privacy For?" *Harvard Law Review* 126 pp.1904–1933.
Crawford, K. and Schultz, J. (2014) "Big Data and Due Process: Toward a Framework to Redress Predictive Privacy Harms" *Boston College Law Review*, 55 pp.93–128.
Deleuze, G. (1992) "Postscript on the Societies of Control" In Szeman, I. and Kaposy, T. (Eds) (2010) *Cultural Theory: An Anthology* John Wiley & Sons, pp.139–142.
DLA Piper (2019) "Data Protection Laws of the World" Available at:https://www.dlapiperdataprotection.com/#handbook/world-map-section.
Docherty, B. (2012) *Losing Humanity: The Case Against Killer Robots* New York: Human Right Watch, New York.
Dodge, M. and Kitchin, R. (2007) "The Automatic Management of Drivers and Driving Spaces" *Geoforum* 38 (2) pp.264–275.
Eubanks, V. (2017) *Automating Inequality: How High-Tech Tools Profile, Police, and Punish the Poor* New York: St Martin's Press.

Foucault, M. (1991) "Governmentality" In Burchell, G., Gordon, C. and Miller, P. (Eds) *The Foucault Effect: Studies in Governmentality* Chicago: University of Chicago Press, pp.87–104.

Goh, B. (2020) "China Rolls Out Fresh Data Collection Campaign to Combat Coronavirus" *Reuters* (26th February) Available at: https://www.reuters.com/article/us-china-health-data-collection/china-rolls-outfresh-data-collection-campaign-to-combat-coronavirus-idUSKCN20K0LW.

Greenleaf, G. (2013) "Scheherazade and the 101 Data Privacy Laws: Origins, Significance and Global Trajectories" *Journal of Law, Information & Science* 23 (1) pp.4–49.

Hatmaker, T. (2020) "Palantir Provides COVID-19 Tracking Software to CDC and NHS, Pitches European Health Agencies" *TechCrunch* (1st April) Available at: https://techcrunch.com/2020/04/01/palantircoronavirus-cdc-nhs-gotham-foundry/.

Ilyushina, M. (2020) "Moscow Rolls Out Digital Tracking to Enforce Lockdown: Critics Dub it a 'Cyber Gulag'" *CNN* (14th April) Available at: https://edition.cnn.com/2020/04/14/world/-moscow-cyber-tracking-qrcode-intl/index.html.

Jefferson, B.J. (2018) "Predictable Policing: Predictive Crime Mapping and Geographies of Policing and Race" *Annals of the American Association of Geographers* 108 (1) pp.1–16.

Kitchin, R. (2014) *The Data Revolution: Big Data, Open Data, Data Infrastructures and Their Consequences* London: Sage.

Kitchin, R. (2015) "Spatial Big Data and the Era of Continuous Geosurveillance" *DIS Magazine* Available at: http://dismagazine.com/issues/73066/rob-kitchin-spatial-big-data-and-geosurveillance/.

Kitchin, R. (2016) *Getting Smarter About Smart Cities: Improving Data Privacy and Data Security* Dublin:, Data Protection Unit, Department of the Taoiseach.

Kitchin, R. (2020) "Civil Liberties or Public Health, or Civil Liberties and Public Health? Using Surveillance Technologies to Tackle the Spread of COVID-19" *Space and Polity* 24 (3) pp.362–381 DOI: 10.1080/13562576.2020.1770587.

Kitchin, R., Coletta, C. and McArdle, G. (2020) Governmentality and urban control. In Willis, K. and Aurigi, A. (eds) *The Companion to Smart Cities*. Routledge, London, pp. 109–122.

Kuempel, A. (2016) "The Invisible Middlemen: Critique and Call for Reform of the Data Broker Industry" *Northwestern Journal of International Law & Business* 36 (1) pp.207–234.

Leszczynski, A. (2017) "Geoprivacy" In Kitchin, R., Lauriault, T. and Wilson, M. (Eds) *Understanding Spatial Media* London: Sage, pp.239–248.

Lupton, D. (2016) *The Quantified Self* Cambridge: Polity.

Luque-Ayala, A. and Marvin, S. (2016) "The Maintenance of Urban Circulation: An Operational Logic of Infrastructural Control" *Environment and Planning D: Society and Space* 34 (2) pp.191–208.

Lyon, D. (2007) *Surveillance Studies: An Overview* Cambridge: Polity.

Mann, S., Nolan, J. and Wellman, B. (2003) "Sousveillance: Inventing and Using Wearable Computing Devices for Data Collection in Surveillance Environments" *Surveillance and Society* 1 (3) pp.331–355.

McDonald, S. (2020) "The Digital Response to the Outbreak of COVID-19" *Centre for International Governance Innovation* (30th March) Available at: https://www.cigionline.org/articles/digital-response-outbreak-covid-19.

Melendez, S. and Pasternack, A. (2019) "Here Are the Data Brokers Quietly Buying and Selling Your Personal Information" *The Fast Company* (2nd March) Available at: https://www.fastcompany.com/90310803/here-are-the-data-brokers-quietly-buying-and-selling-your-personal-information.

Metcalf, J. and Crawford, K. (2016) "Where Are Human Subjects in Big Data Research? The Emerging Ethics Divide" *Big Data & Society* 3 (1) pp.1–14.

Mohan, M. (2021) "Singapore Police Force Can Obtain Tracetogether Data for Criminal Investigations" *CNA* (4th January) Available at: https://www.channelnewsasia.com/news/singapore/singapore-police-force-can-obtain-tracetogether-data-covid-19-13889914.

Pinchot, J., Chawdhry, A.A. and Paullet, K. (2018) "Data Privacy Issues in the Age of Data Brokerage: An Exploratory Literature Review" *Issues in Information Systems* 19 (3) pp.92–100.

Pollina, E. and Busvine, D. (2020) "European Mobile Operators Share Data for Coronavirus Fight" *Reuters* (18th March) Available at: https://www.reuters.com/article/us-health-coronavirus-europe-telecoms/europeanmobile-operators-share-data-for-coronavirus-fight-idUSKBN2152C2.

Rainie, S.C., Kukutai, T., Walter, M., Figueroa-Rodríguez, O.L., Walker, J. and Axelsson, P. (2019) "Issues in Open Data: Indigenous Data Sovereignty" In Davies, T., Walker, S., Rubinstein, M. and Perini, F. (Eds) *The State of Open Data: Histories and Horizons* Cape Town and Ottawa: African Minds and International Development Research Centre, pp.300–319.

Roderick, L. (2014) "Discipline and Power in the Digital Age: The Case of the US Consumer Data Broker Industry" *Critical Sociology* 40 (5) pp.729–746.

Sadowski, J. (2019) "When Data is Capital: Datafication, Accumulation, and Extraction" *Big Data & Society* 5 (1) pp.1–12.

Schectman, J., Bing, C. and Stubbs, J. (2020) "Cyber-intel Firms Pitch Governments on Spy Tools to Trace Coronavirus" *Reuters* (28th April) Available at: https://www.reuters.com/article/us-health-coronavirusspy-specialreport/special-report-cyber-intel-firms-pitch-governments-on-spy-tools-to-tracecoronavirus-idUSKCN22A2G1.

Singer, N. and Sang-Hun, C. (2020) "As Coronavirus Surveillance Escalates, Personal Privacy Plummets" *New York Times* (23rd March) Available at: https://www.nytimes.com/2020/03/23/technology/coronavirus-surveillance-tracking-privacy.html.

Smith, H. (2019) "Monetizing Movement: Groundtruth" In Graham, M., Kitchin, R., Mattern, S. and Shaw, J. (Eds) *How to Run a City Like Amazon, and Other Fables* Oxford: Meatspace Press, pp.570–605.

Solove, D.J. (2006) "A Taxonomy of Privacy" *University of Pennsylvania Law Review* 154 (3) pp.477–560.

Solove, D. (2013) "Privacy Management and the Consent Dilemma" *Harvard Law Review* 126 pp.1880–1903.

Taylor, L., Floridi, L. and van der Sloot, B. (2017) "Introduction: A new perspective on privacy" In Taylor, L., Floridi, L. and van der Sloot, B. (Eds) *Group Privacy: New Challenges of Data Technologies* Cham, Switzerland: Springer, pp.1–12.

Taylor, L., Sharma, G., Martin, A. and Jameson, S. (2020) *Data Justice and Covid-19: Global Perspectives* London: Meatspace Press.

Tene, O. and Polonetsky, J. (2013) "Big Data for All: Privacy and User Control in the Age of Analytics" *Northwestern Journal of Technology and Intellectual Property* 11 (5) pp.240–273.

Thatcher, J., O'Sullivan, D. and Mahmoudi, D. (2016) "Data Colonialism Through Accumulation by Dispossession: New Metaphors for Daily Data" *Environment and Planning D: Society and Space* 34 (6) pp.990–1006.

Timberg, C. and Harwell, D. (2020) "Government Efforts to Track Virus Through Phone Location Data Complicated by Privacy Concerns" *Washington Post* (19th March) Available at: https://www.washingtonpost.com/technology/2020/03/19/privacy-coronavirus-phone-data/.

Townsend, A. (2013) *Smart Cities: Big Data, Civic Hackers, and the Quest for a New Utopia.* New York: W.W. Norton & Co.

Zuboff, S. (2019) *The Age of Surveillance Capitalism: The Fight for the Future at the New Frontier of Power* New York: Profile Books.

35
GEOSPATIAL TECHNOLOGY AND JOURNALISM IN A POST-TRUTH WORLD

Amy Schmitz Weiss

Introduction

Geospatial technology has provided a wealth of new forms and avenues of how we can understand the world around us. As alluded to in previous chapters of this book, geospatial technology has expanded into the application of sensors, machine learning and drones. The application of geospatial technology in society crosses many industries from archaeology to journalism, and efforts that range from humanitarian causes to tracking the spread of diseases. This chapter in particular focuses on the implications geospatial technology has on the journalism field.

It is probably no surprise to see that journalism is in a moment of change and crisis worldwide. Economic, political, social and cultural influences are each playing a significant role in how the press is able to operate in different regions of the world. As newsrooms deal with smaller staffs to produce more content in an era of efficiency, challenges arise in terms of being able to cover all the news in a given area or community. In addition, pressures (e.g., the market, ownership, audience) from outside the newsroom can hinder journalists and their work. Furthermore, in some extreme cases journalists are being threatened, censored or killed for doing their job, which can result in less news coverage of such communities.

This can be compounded by the fact that the tenets of truth and accuracy are also in upheaval worldwide. News organizations face a number of challenges because of institutions and/or individuals who aim to spread propaganda, fake news and misinformation to confuse and dissuade the public.

A recent report by the Knight Foundation found that trust in the news media is at a very low point,

> Most US adults, including more than nine in 10 Republicans, say they personally have lost trust in the news media in recent years. At the same time, 69% of those who have lost trust say that trust can be restored.
>
> *(Knight Foundation, 2019)*

In this 'post-truth' world as some have called it (McIntyre, 2018), journalism faces some steep challenges like trust and accountability (Schudson, 2019). However, the field can also see this moment as an opportunity. Geospatial technology will not be the sole panacea for the news industry, but it can provide another lens by which news organizations can tell important, factual and accurate stories that impact the communities they cover. And in turn, it can provide a different kind of spatial news experience to the consumer.

A spatial news experience is one where places and spaces are at the center of the story. Spatial journalism helps to expand on this concept by examining how space, place and location whether physical, augmented or virtual can be incorporated into the process and practice of journalism (Schmitz Weiss, 2015, 2018a, 2018b, 2020). For example, research has identified that news consumers are seeking out locative information and news that is near to where they live, work and play (Schmitz Weiss, 2018a). Furthermore, there are news organizations that are employing this spatial journalism approach in an effort to help dispel stereotypes in the community, as a meaning maker for the community, and serve as another layer of local news to the consumer (Schmitz Weiss, 2020).

This notion of seeking locative news and information nearby has been explored from a wider spectrum than journalism. Research has shown that locative information can help the public have better context of a specific area (Humphreys, 2007; Humphreys and Liao, 2011; Frith, 2015), experience satisfaction with a specific location (Frith, 2015; Oppegaard and Rabby, 2015), and learn about a particular area (Gordon and de Souza e Silva, 2011; Humphreys and Liao, 2011; Frith, 2015).

How does locative information and news connect to geospatial technology? Well, geospatial technology can be the conduit that helps connect locative and spatial information to newsworthy information. From a news perspective, this connection can be found in the ways that mapping and sensors are being used in the journalism industry.

This chapter will explore how geospatial technology has been used in journalism and where it can take the news industry next in this post-truth world. Before delving into the uses of geospatial technology in journalism today, it is important to know the history of how the news industry has used mapping technologies to explain spaces and places in the storytelling process.

History of mapping in journalism

Maps have been used in journalism for several centuries. The initial maps featured in the press would not be exactly the kind of maps we would expect today in the newspaper. According to Monmonier (1989), the first newspaper map might be considered Benjamin Franklin's political cartoon snake showing a broken snake as broken up colonies in 1754 (1989). Over time, maps have played a greater role in the storytelling process. In the century to follow, maps would help in the storytelling process to show shipping and trade routes as well as the places where a crisis or conflict broke out (Monmonier, 1989). These maps would show the actual places and moments with specific placemarkers. By the twentieth century, the application of maps in the news process would help compliment stories to show socioeconomic information like crime, poverty levels and health indicators in a given area. Furthermore, maps would also show political information – how a political campaign evolved in a region and which candidates were ahead in a poll or election in a given county (Monmonier, 1989).

By the 1970s and 1980s, maps were commonplace and widely accepted in the news process. However, the ability to layer multiple pieces of information to a given location on a map would become the next frontier for mapping in the news industry.

History of GIS in journalism

The use of geographic information systems (GIS) would enter the market in the 1980s and 1990s as alluded to in previous chapters. For the news industry, its application in newsrooms began in the 1990s with enterprise news stories by several news organizations. For example, the *Miami Herald* used GIS to map the path of Hurricane Andrew to show how shoddy construction and faulty inspections resulted in the large amount of damage to homes in the area from the hurricane. This mapping was the result of not just getting one dataset but compiling multiple datasets including wind contour information and creating their own GIS map file from scratch as well as spending significant time doing analysis of the area. Their reporting and GIS analysis allowed them to identify specific areas and neighbourhoods they could visit to do more research, interview sources and take photos to show the damage. Their big investigative story launched in December 1993. Months after the story came out, the county changed building codes and inspection practices. The newspaper won the Pulitzer Prize for their story, one of the highest honours in the news industry (Herzog, 2003).

Another example is the enterprise story by the *Providence Journal* in 1998 to map lead poisoning in children in Rhode Island (Herzog, 2003). This was done by collecting health department data of lead tests and connecting those with census tract information. With this data mapped, the GIS analysis showed that the census tracts with the worst rates were in the state's largest cities. Furthermore, the GIS analysis showed that in those cities, the neighborhoods that had the highest poisoning rates had high concentrations of immigrants and minorities. The newspaper published a massive investigation and then followed up a few years later to do another story on the location of residential lead hazards. They were able to identify 'hot spots' or clusters of lead hazards and published a series of stories in May 2001. After the investigation came out, the state department changed the rules on lead home inspections and the US EPA announced it would conduct its own inspections in the state for federal compliance. The newspaper won multiple awards at the state and national level for their news coverage (Herzog, 2003). These two investigations are just a small sample of the GIS stories produced in the 1990s by several journalists.

Following the 1990s into the 2000s, the use of GIS in newsrooms would grow as more journalists would learn the techniques of GIS and acquire datasets from local, national and international entities to layer them into powerful infographics for a variety of news stories (Herzog, 2003; Cairo, 2016). Reporters would use GIS to show the results of local and national elections, the impacts of a natural disaster like a hurricane or tornado in a community or the socioeconomic differences with access to education by census tract in a city or region. These are just a few examples of numerous infographics created over the past three decades.

Today, GIS remains an important element in many newsrooms and its adoption has spread further and wider with the development of open-source tools and technologies, educational resources and specific journalist communities that have helped to strengthen and support this approach.

Sensors and journalism

As GIS has become a bigger part of the news industry, another geospatial technology that is evolving in the news industry is sensors. The use of sensors in the news industry has evolved over the past three decades (Pitt, 2014). Journalists used to use manual or analogue sensors to get information like air or water quality. However, these processes were arduous and expensive. Nowadays, digital or electronic sensors have created a new pathway to tell stories with collecting digital data often instantaneously and continuously in a given area that has transformed storytelling for the journalist and how the news consumer experiences a particular sensor-based story.

For example, Ghost Factories was an enterprise investigative project by *USA Today* in 2012 that explored the amount of lead in the soil from 400 sites across the country ('Ghost Factories', 2012). The reporters used soil sensors to gather the data and map the locations of these sites.

Another example is The Harlem Heat Project by WNYC in New York in 2016 that explored temperature rise in apartments in the Harlem area. Reporters used heat and humidity sensors to find that indoor temperatures were much higher inside than outside that put sensitive populations like the elderly and children at risk ('Harlem Heat Project'" 2016). As part of this project, the reporters went into the community to collect the data from individuals who had the sensors in their homes and to also check in on them. This approach allowed them to gather the sensor data but also to build a rapport with the community in telling their stories. The project resulted in multiple stories which also included a map that showed the heat index indoors and outdoors. The map also featured an auditory component. The reporters discovered that the participants were experiencing high temperatures in their homes, 'We found that the heat index inside our participants' homes regularly stay higher than 80 degrees, for days and nights on end, not even cooling off the way it does outside when the sun goes down' (Gonzalez and Palazzolo, 2016). The project ended with a community workshop on ways to help mitigate the issue in the community.

In the past year, two other projects of similar scope (Code Red and Bitter Cold) launched in the Baltimore area to explore temperature in Baltimore homes in the summer and winter using temperature sensors to explore the health implications of rising or freezing temperatures on residents.

The Code Red project utilized the same approach of sensors from the Harlem Heat Project in which they deployed eighty sensors into the community. They interviewed a myriad of sources, delved into multiple reports and datasets for the story,

> We spent a year gathering and analyzing data and talking with climate and health experts. We built temperature and humidity sensors to help understand what it was like to live in rowhouses in Baltimore's hottest neighborhoods. Our data analysis identified McElderry Park as the city's hottest neighborhood, and we spent a summer reporting on the community and the people who live there.
>
> *(Code Red, 2019)*

These examples demonstrate that sensor data can be mapped to specific locations and show how much is happening in the environment in that particular area whether that is one's home or a neighbourhood. However, this is just the beginning of the power of sensors and mapping in the news industry. Remote and mobile sensing provide another powerful layer in the storytelling process.

Remote sensing is the ability to collect data from the energy that is bounced off the Earth's surface in a given area and then collected from satellites above the Earth (National Oceanic and Atmospheric Administration [NOAA], n.d.). The potential for this approach can show a variety of impacts environmentally when one examines data collected by NOAA and other entities that show the changes in the ocean temperature, coastal erosion, wetland management, etc. In journalism, remote sensing has mainly been used for weather forecasts, but is now being used for other purposes, 'For decades, satellite imagery helped to forecast the weather. Now journalists use the images to break stories, to verify information, and to cover war zones and the effects of climate change' (Shrestha and Vora, 2019). Shrestha and Vora (2019) identify that satellite imagery can help reporters access places that are difficult to reach and see possible trends or patterns in a given area over time. However, satellite imagery does require a certain amount of expertise for use and analysis, otherwise journalists could misinterpret what they see leading to misleading or inaccurate information. There are a growing number of companies that are now providing satellite images for free to news organizations than before (Shrestha and Vora, 2019), but there is still more to be explored in the news industry as to how remote sensing can be applied in a variety of news contexts.

The other area of sensors is mobile sensing. Mobile sensing is where the data is captured as one is in movement – often sensors capture data in a static place over time. However, digital technology has advanced that many sensors are now mobile devices that can capture data as one moves around or the sensor moves around.

For example, *The Financial Times* conducted a mobile sensing project in 2019 that explored air quality in five major cities around the world: Beijing, Lagos, London, New York and São Paulo. The reporters wore personal air pollution monitors for a few weeks that captured the air quality in real-time as they moved around the city. The reporters were able to map their data and show the extent of poor air quality during a typical day when in traffic or taking the subway (Hook *et al.,* 2019). This intimate way of obtaining air quality data daily directly in their own personal life changes the idea of air quality data from the traditional forms of accessing large governmental datasets and making the data from these sensors more relatable and personal. Thus, creating an opportunity for the reader to have a different understanding of the air that they breathe around them from this unique context.

For example, one could acquire the collection of sound or motion for noise pollution in a given neighbourhood or the collection of changes in nature around one's home. Each of these could be captured through the existing applications on the mobile device while in movement. This form of crowdsourced data collection can be seen as a form of citizen science. Citizen scientists are individuals in society who are not specialists but are interested and engaged in some way of contributing to science and its data collection (Hicks *et al.*, 2019). As mobile devices continue to proliferate around the world and become more widely adopted, it will be easier for the news public to understand mobile sensing and also become possible participants like citizen scientists in the collection of data from their mobile device in the future.

When one considers the potential for using geospatial technology in the news process with remote and mobile sensing approaches, more stories can be explored with invisible issues like poor air quality or soil contamination that can be made visible in geographic and spatial form in a news story.

The post-truth world and geospatial technology

As alluded to in previous sections of this chapter, geospatial technology has helped the news industry be able to communicate important information through maps and location. For example, the public is better able to know about air quality in a specific region or understand the amount of contaminants in the soil in their community through the power of storytelling and mapping the information or data in an infographic or data visualization. Furthermore, recent technology like sensors provide another layer to the storytelling process that is just now beginning to evolve.

The role of geospatial technology in the post-truth world lies in its ability to provide truth, accuracy and facts for a given area or location. When one can obtain truthful and accurate information about what is happening in their community, it can help assuage an already skeptical news audience. However, if the facts or data about a given location are manipulated or fake, geospatial technology can lose its ability to be trusted or credible. As Monmonier (2018) states in *How to Lie with Maps*, it is quite easy for one to plot data on a map but there are so many subjective decisions made with the type of technology used to make the map, the projection of how the map is presented, the ways symbols are created on the map and the overall design that can result in information in the map that is misleading, inaccurate or fake. One can also have the outright purpose to mislead or misrepresent data in a map that can result in inaccurate or fake information.

However, it takes a reporter to verify and corroborate the locative information and data to make sure it is factual and accurate beforehand. Geospatial technology cannot exist on its own in the news industry – it requires the reporter's know-how to make sure the locative information is legitimate and accurate.

As alluded to earlier, trust in the media has hit an all-time low, but there is opportunity to regain it back according to a recent Knight Foundation report,

> While the loss of trust in the news media is concerning in a democratic society, a more encouraging sign is that most of those who have lost trust believe it can be regained. Specifically, 69% of U.S. adults who say they have lost trust in the news media over the past decade say their trust can be restored.
>
> *(Knight Foundation, 2019)*

Geospatial technology can provide truth in location and place if used correctly and accurately in the appropriate hands of an ethical reporter. It can help to show directly and clearly how a community is shaped by the information layers that exist in specific locations and the stories that are contained within. Geospatial technology will not be the sole answer to the overall news industry's credibility and truth troubles, but with the proper measures of how it is applied in the storytelling process, may help to move the barometer in a positive direction with the public. If one is able to provide truth in a location or area with the richness of a narrative that only a journalist can provide, the public can have a more certain sense of what is happening around them and make better decisions as they go about their daily life.

Conclusion

We are only at the beginning of the potential of what geospatial technology can do for society and specifically its application in journalism. The next frontier for geospatial technology in journalism is the realm of spatial data science. Spatial data science (Anselin, 2016) explores the depths of geospatial data and provides an exciting time for journalists to build their analytical skills to delve deeper into the why and how behind geographical data for a story. As Luc Anselin of the University of Chicago notes, spatial data science provides geographers (and we can also say journalists) insights into the aspects of spatial mismatch and spatial disparities (Anselin, 2016) in a specific area that may not have been known before. For a journalist, being able to dig into this kind of data to find these potential issues can only enlighten and strengthen the news coverage in a specific community when this analytical approach is taken.

Spatial data science is just one way that geospatial technology can impact the future of the news industry. The news industry has the opportunity to see how they can best adopt and adapt current and future geospatial technology to their news process and how it can impact the communities they serve. In this post-truth era, spatial and locative information will become more important and crucial.

As noted at the beginning of the chapter, geospatial technology has the ability to give the news consumer a spatial news experience that gives them the opportunity to explore the places and spaces around them in a different way. It allows for the idea of location to be opened up beyond just a street block or neighbourhood, to open one's mind to the possibilities of what information and news lays in the many layers of the spaces we move in for work, life and play. As news consumers continually seek out locative information that is near to them, it will be those news organizations that understand this that will succeed. They will see how geospatial technology with solid and ethical news reporting can provide a spatial news experience that is being sought in the world today.

References

Anselin, L. (2016) "Spatial Data, Spatial Analysis and Spatial Data Science" Available at: https://www.youtube.com/watch?v=lawWM6jQYEE (Accessed: 15th January 2020).

Cairo, A. (2016) *The Truthful Art* San Francisco, CA: New Riders.

Code Red (2019) "Heat & Inequality" Available at: https://cnsmaryland.org/interactives/summer-2019/code-red/neighborhood-heat-inequality.html (Accessed: 15th January 2020).

Frith, J. (2015) *Smartphones as Locative Media* Cambridge: Polity Press.

Ghost Factories (2012) *USA Today* (13th July) Available at: *http://usatoday30.usatoday.com/news/nation/-lead-poisoning* (Accessed: 15th January 2020).

Gonzalez, S. and Palazzolo, A. (2016) "Hear the Heat: Our Song Demonstrates What It Felt Like Inside Harlem Homes This Summer" *WNYC* Available at: *https://www.wnyc.org/story/harlem-heat-song* (Accessed: 17th June 2020).

Gordon, E. and de Souza e Silva, A. (2011) *Net Locality* Malden, MA: Wiley-Blackwell.

Harlem Heat Project (2016) *WNYC* Available at: https://www.wnyc.org/series/harlem-heat-project (Accessed: 15th January 2020).

Herzog, D. (2003) *Mapping the News, Case studies in GIS and Journalism* Redlands: Esri Press.

Hicks, A., Barclay, J., Chilvers, J. Armijos, M.T., Oven, K., Simmons, P. and Haklay, M. (2019) "Global Mapping of Citizen Science Projects for Disaster Risk Reduction" *Frontiers in Earth Science* 7 (226) pp.1–18 DOI: 10.3389/feart.2019.00226.

Hook, L., Munshi, N., Hornby, L., Kuchler, H., Schipani, A., Bernard, S. and Harlow, M. (2019) "How Safe is the Air We Breathe?" *Financial Times* (4th September) Available at: https://www.ft.com/content/7d54cfb8-cea5-11e9-b018-ca4456540ea6 (Accessed: 17th June 2020).

Humphreys, L. (2007) "Mobile Social Networks and Social Practice: A Case Study of Dodgeball" *Journal of Computer-Mediated Communication* 13 (1) pp.341–360 DOI: 10.1111/j.1083-6101.2007.00399.x.

Humphreys, L. and Liao, T. (2011) "Mobile Geotagging: Reexamining Our Interactions with Urban Space" *Journal of Computer-Mediated Communication* 16 (3) pp.40.

Knight Foundation (2019) "State of Public Trust in Local News" Available at: https://knightfoundation.org/reports/state-of-public-trust-in-local-news/ (Accessed: 15th January 2020).

McIntyre, L. (2018) *Post-Truth* Cambridge: MIT Press.

Monmonier, M. (1989) *Maps with the News* Chicago: University of Chicago Press.

Monmonier, M. (2018) *How to Lie with Maps* Chicago: University of Chicago Press.

National Oceanic and Atmospheric Administration (NOAA) (n.d.) "What Is Remote Sensing?" Available at: https://oceanservice.noaa.gov/facts/remotesensing.html (Accessed: 19th February 2020).

Oppegaard, B. and Rabby, M. (2015) "Proximity: Revealing New Mobile Meanings of a Traditional News Concept" *Digital Journalism* 4(5) pp.621–638 DOI: 10.10 80/21670811.2015.1063075.

Pitt, F. (2014) *Sensors and Journalism*. New York: Tow Center for Digital Journalism, A Tow/Knight Report.

Schmitz Weiss, A. (2015) "Place-Based Knowledge in the Twenty-First Century" *Digital Journalism* 3 (1) pp.116–131 DOI: 10.1080/21670811.2014.928107.

Schmitz Weiss, A. (2018a) "Journalism Conundrum: Perceiving Location and Geographic Space Norms and Values" *Westminster Papers in Communication and Culture* 13 (2) pp.46–60 DOI: 10.16997/wpcc.285.

Schmitz Weiss, A. (2018b) *Geolocated News: How Place, Space and Context Matters for Mobile News Users* Washington, DC: Association for Education in Journalism & Mass Communication Convention.

Schmitz Weiss, A. (2020) "Journalists and Their Perceptions of Location: Making Meaning in the Community" *Journalism Studies* 21 (3) pp.352–369 DOI: 10.1080/1461670X.2019.1664315.

Schudson, M. (2019) "The Fall, Rise, and Fall of Media Trust" *Columbia Journalism Review* (Winter) Available at: https://www.cjr.org/special_report/the-fall-rise-and-fall-of-media-trust.php (Accessed: 19th February 2020).

Shrestha, S. and Vora, P. (2019) "Satellites Give Newsrooms Eyes in the Sky" Available at: https://nyujournalismprojects.org/newsliteracy2019/topics/satellite-imagery-for-news/ (Accessed: 17th June 2020).

36
ADVANCING SUSTAINABILITY RESEARCH THROUGH GEOSPATIAL TECHNOLOGY AND SOCIAL MEDIA

Yaella Depietri, Johannes Langemeyer, Derek Van Berkel and Andrea Ghermandi

Social media (SM) are Internet-based applications that enable people to communicate and share resources and information through Web 2.0 sites (i.e., sites supporting user-generated content). Originally based on existing social relationships among individuals, primarily at universities in the United States (Castells, 2011), SM use has now extended across social groups, societies, and continents. Advances in mobile and smartphone technologies, like low-cost GPS (global positioning system) receivers and photo cameras, are of critical importance in such applications (Heipke, 2010). Considering that more than half of the world's population is estimated to use SM networks, user-generated content, like photographs, videos, text, or audio, has great potential for democratizing data generation and sharing globally. Platforms such as Flickr, Twitter, Weibo, Wikimapia, OpenStreetMap, and to some extent Instagram, allow their users to upload geolocated content and researchers to retrieve such information. Thanks to the technological advancements, the quality of the resulting geospatial data is generally good, despite being produced by inexperienced users (Zielstra and Hochmair, 2013).

This wide usage of SM presents new opportunities for public involvement in monitoring and managing the environment, using passive crowdsourcing of data for purposes other than originally intended. SM content is indeed becoming an important source of data in geographical and environmental studies, especially in the form of passive crowdsourcing (Connors *et al.*, 2012). Numerous applications have already leveraged this continuous and direct flow of data on human-nature interactions and perceptions in various related contexts (Ghermandi and Sinclair, 2019; Ghermandi *et al.*, 2023).

In this chapter, we present some brief information on how to collect and process SM data. We then review the main fields of applications of SM data in sustainability research, such as the assessment of human-nature interactions, landscape value, and urban planning. Finally, we discuss potential limitations of SM-based approaches in the context of the mentioned applications.

Collecting and processing social media data

Different options are available to researchers for the mining of georeferenced SM data. These include (Toivonen *et al.*, 2019): manual data download, which requires only limited technical skills but is very time consuming, thus greatly limiting the volume of data that can be retrieved; Web scraping and Web crawling, which use automated scripts to access and retrieve data from the public Web, but often risks violating the SM platforms' terms of service; reliance on third parties' software, such as InVEST (Wood et al., 2013), TAGS (Hawksey, 2010), or COSMOS (Burnap et al., 2015), which however may be discontinued or not kept up-to-date; or through Application Programming Interfaces (APIs), which is the most frequent approach used by researchers (Ghermandi and Sinclair, 2019). An API is 'an interface of a computer program that allows the software to 'speak' with other software' (Lomborg and Bechmann, 2014). Several SM companies, indeed, make their databases (or part of them) accessible through dedicated APIs, via which researchers can collect data, relying on moderate programming skills and purposely written code (e.g., Sinclair et al., 2020). Another option to retrieve data through APIs is to rely on third-party software, such as the 'photosearcher' and 'academictwitteR' packages in R (Fox et al., 2020; Barrie and Ho, 2021) or the 'twitterhistory' module in Python (Fink, 2021).

Through the APIs, geotagged photographs or tweets can be retrieved together with the associated metadata, which may include user ID, photo ID, timestamp, geotags in the form of GPS coordinates, and other text in the form of user-provided tags or photo titles. APIs can also be used to retrieve or ascertain certain socio-demographic information on the users themselves (e.g., nationality, location of residence, gender), based on the public information contained in their SM profile (Ghermandi et al., 2020a; Sinclair et al., 2020; Väisänen et al., 2021). One drawback of research based on APIs is that access might be limited to only a part of the data and metadata (Lomborg and Bechmann, 2014). Moreover, changes in the platforms' terms of service can abruptly limit or remove accessibility to the API, as the case of Twitter recently demonstrated (Ghermandi et al., 2022).

Geolocated SM data is associated with spatial coordinates and can easily be processed and represented by uploading the data as vector data through a GIS software (such as ArcGIS or QGIS). Photographs and text content can be analysed and coded manually (Calcagni et al., 2022) or by applying artificial intelligence and machine learning, such as computer vision for automated image content analysis (Ghermandi et al., 2020b, 2022). Among the best-known commercial providers that offer image recognition capabilities, one may count Google Cloud Vision, Clarifai, Microsoft Azure Computer Vision, IBM's Watson Visual Recognition, and Amazon Rekognition. Reliance on a specific provider of computer vision services may have implications insofar as the characterization of outdoor recreational activities, biophysical environment (e.g., wildlife and vegetation), and feelings associated with the photographs are concerned (Ghermandi et al., 2022).

Supporting landscape and urban planning

One of the main applications of crowdsourced data in sustainability research is the identification of landscape features of interest and the assessment of their cultural values. Different methods have traditionally been used to capture these values, such as (public) participatory GIS (PPGIS or PGIS) (Brown and Fagerholm, 2015; Depietri and Orenstein, 2020), in-depth, face-to-face interviews (Norton et al., 2012), and focus group discussions (Orenstein et al., 2015). The rise of SM has opened new pathways for monitoring and value assessment (Tieskens et al., 2018; Ghermandi et al., 2023), often providing opportunities to complement these approaches (Depietri et al., 2021). Information about preferences, public interests, and perceptions regarding a landscape can be assessed through manual content analysis (Van Berkel et al., 2018) or through automated image content analysis (Ghermandi et al., 2020b, 2022), as detailed above. Additional information can be collected through the analysis of tags and the text associated with the picture (Ghermandi et al., 2020a; Calcagni et al., 2022).

Assessments based on passively crowdsourced data offer numerous advantages compared to survey-based methods, principally due to the large amount of relatively easily accessible and publicly shared data (Ghermandi and Sinclair, 2019), as well as to the broad spatial extents, from country to global scales, including traditionally data-scarce and under-researched regions (Sinclair et al., 2019). The detailed spatial resolution, and the availability of in-situ and site-specific data are additional strengths of this method (Figueroa-Alfaro and Tang, 2017). SM-based data also allow for obtaining quasi real-time information, analyse temporal patterns of activities and, in some cases, cover relatively long-time spans (Bubalo et al., 2019). Finally, the lack of response biases, such as respondents answering in a socially desirable manner, which are common for survey and interview approaches, are absent in this method.

Monitoring and assessing tourism and recreation

From a methodological standpoint, the cultural value of a landscape is often challenging to assess due to its immaterial and intangible nature (Langemeyer et al., 2018). Methods to estimate the interest in and the value attributed to a landscape are, for instance, the measurement of visitors flows, their behaviour, and the spatial distribution of visits. These are well captured by SM data (Wood et al., 2020). Broad temporal patterns and trends in visitation (Sessions et al., 2016) as well as fine-scale surveys of user's movement can be provided at low costs with this method (Gosal et al., 2018). Furthermore, such techniques uniquely capture behaviours, like off-trail use (i.e., areas that are outside marked routes) and users' preferences in areas that are remote or may be hard to reach for researchers (Rossi et al., 2019). Research on SM has also refined methods to identify changes in landscape cultural preferences and recreational patterns depending on the type of visitor group (e.g., locals, domestic or international tourists) (Ghermandi et al., 2020a; Sinclair et al., 2020).

From the analysis of the subject of the pictures, and/or of the tags and text, it is also possible to infer preferences and motives behind visitation. It is for instance possible to link, with high degrees of precision, visitation intensity to specific types and features of the landscape (e.g., type of forest, water bodies, particular species of trees and animals, geological formations, cultural elements of the landscape) and derive conclusions regarding the relationships between elements of the landscape and their cultural value (Bernetti et al., 2019; Van Berkel et al., 2018). By superimposing maps of spatial intensity of visitation with land cover and land use maps, the interactions between built and natural capitals in the creation of value can also be explored (Langemeyer et al., 2018; Depietri et al., 2021; Lingua et al., 2022).

A growing number of studies also explore the possibility to integrate SM data with economic valuation techniques to assess the monetary touristic or recreational value of a site. Techniques based on revealed preferences such as the travel-cost-method have found wide application in the economic analysis of the (recreational) benefits derived from landscapes (e.g., Langemeyer et al., 2015); their integration with SM data is a promising field of development (Ghermandi, 2018; Sinclair et al., 2020).

One drawback of this data is the challenge of classifying user-specific activities across sites from photographic and text content (Teles da Mota and Pickering, 2020). Often, recreational activities (e.g., hiking, swimming, biking) are not the explicit subject of the photograph and user-generated tags. While individuals might engage in multiple activities, photographic capture is likely curated and may miss specific spatial behaviour. Moreover, text and photograph may not capture actual affective responses and values of users. This requires degrees of subjective assessments by researchers that possibly introduces biases (Angradi et al., 2018). One option to alleviate these potential biases is coupling SM analysis with in-situ surveys of behaviour and affective responses (Depietri et al., 2021).

Landscape aesthetics and other cultural benefits

Amongst the different cultural benefits provided by a landscape that can be captured with a SM-based assessment is the aesthetic value (Figueroa-Alfaro and Tang, 2017; Langemeyer et al., 2018; Tieskens et al., 2018). Scenic views of the landscape from lookout points and specific elements of interest are often the principal subject of the images uploaded (Depietri et al., 2021; Pickering et al., 2020). As many of the nature and landscape photographs are likely to be taken for aesthetic reasons (Richards and Friess, 2015), the suitability of the SM-based method to capture aesthetic values is well acknowledged (e.g., Bernetti et al., 2019).

Other cultural benefits that can be derived from a landscape, such as heritage, sense of place, relational and spiritual values (Ginzarly et al., 2019; Nummi, 2018), are however challenging to capture through the content of crowdsourced photographs only, but may be revealed in the analysis of associated user-generated text and other metadata (Calcagni et al., 2022). This more in-depth analysis can reveal the relational interactions people have with nature or specific landscapes, which refer to the dynamic and context-specific relationships that people have with components of nature and, through nature, among people (Chan et al., 2018; Schröter et al., 2020).

Consequently, SM platforms may serve as valuable data sources for revealing the 'digital co-construction of values' assigned to the environment (Calcagni et al., 2019; Langemeyer and Calcagni, 2022). Promising techniques that combine the analysis of the content of the uploaded photographs with the associated tags and texts, such as topic models and sentiment analysis, may provide more information about these subjective values overall, and about the complete cultural experience at a site (Langemeyer et al., 2023).

Urban diversity and vitality

In urban contexts it is often difficult to collect empirical data on life in cities and on urban vitality without the use of large scale, labour-intensive surveys (De Nadai et al., 2016). Social media and mobile technologies can help collect this crucial information, and foster citizen engagement and participation for improved urban planning (Kleinhans et al., 2015). For instance, De Nadai et al. (2016) used mobile phone records and Web data to develop proxies for urban vitality and diversity in four Italian cities, offering insights on how urban dwellers experience cities as a whole.

Social media also offer unprecedented possibilities for codifying local knowledge about places (Graham, 2010; Shelton et al., 2015). Crowdsourced big data can help to understand contemporary urban processes, such as how people engage with the public urban space, social needs (such as security, play, isolation, and encounter) (Cerrone et al., 2018), can support urban planning in data source contexts of the global south (Zapata et al., 2022), and may help explain phenomena such as gentrification (Amorim Maia et al., 2020). Overall, this information has been used to rethink how we conceptualize, define, and design neighbourhoods or other urban (Hermes et al., 2018; Shelton et al., 2015).

Ecology and conservation

The available ecological data to inform decision making on conservation are often limited or fragmented (Schmidt *et al.*, 2010). User-generated content produced via various SM platforms provide new and additional opportunities to collect ecological data and information, for instance, on biodiversity, species niches, distribution, and abundance (Di Minin *et al.*, 2015; Jiménez-Valverde *et al.*, 2019). Crowdsourced based methods have proven to contribute to forest monitoring by capturing the dynamics of ecosystems in response to biotic and abiotic changes or disturbances (Daume *et al.*, 2014). The term "iEcology" has recently been introduced to characterize a new field of research that relies on harnessing online resources, including, but not limited to, SM data, for the analysis of ecological patterns and processes (Jarić *et al.*, 2020).

SM data in support to environmental conservation is also particularly valuable for providing data on the social context (i.e., on stakeholders, events or perceptions and demands), related to forest degradation, supplying evidence-based information on factors and obstacles to forest conservation and management (Daume *et al.*, 2014). Thanks to the strengths of the method in assessing visitation, it is possible to identify conservation sites under pressure from visitors and to improve the sustainability of ecosystem management and monitoring, adapting human activities to the specificities of each site (Hausmann *et al.*, 2019; Mancini *et al.*, 2018). Social media are also used to track illegal wildlife trade (Di Minin *et al.*, 2018). Dealers often traffic online, and this disturbingly large amount of data can be retrieved and analysed through machine-learning algorithms, investigating human behaviour with respect to these matters and support the implementation of corrective actions. The field of research in conservation biology dedicated to the exploration of new forms of digital data, including SM data, in application to the cultural dimension of the conservation of biodiversity is known as 'conservation culturomics' (Correia *et al.*, 2021).

Complementing Earth observations

Earth observation increasingly engages with innovative techniques, including non-traditional sources of data such as mobile phone and SM data to address current societal challenges (Anderson *et al.*, 2017). Citizen science is indeed newly evolving based on advanced technology and smartphones, allowing wide segments of the society to contribute in the collection of empirical data (Mazumdar *et al.*, 2017). While Earth observations are generally based on satellite imagery, in-situ observations, including SM data streams, are now used to complement this information in various ways (Salcedo-Sanz *et al.*, 2020). For instance, SM data are used to supplement remotely sensed data (Sudmanns *et al.*, 2020) or to validate land cover data (e.g., through the Geo-Wiki project). Some variability is to the quality and distribution of the data, temporally, geographically, and depending on the land cover type (Fonte *et al.*, 2015). In the urban context, OpenStreetMap data have been used to map land cover types with good agreement with the Europe Environment Agency (EEA) Urban Atlas maps (Arsanjani *et al.*, 2013). Moreover, SM data may be applied for the determination of different types of land uses (as distinct from land cover), including the utilization of land for recreational, residential, or work purposes (Zhou and Zhang, 2016).

Informing disaster risk reduction

Passively crowdsourced data can serve a vital role in disaster risk reduction, contributing new insights to: (1) public debates surrounding disasters; (2) monitoring the evolving situation before, during, and after these events; (3) contributing to emergency response and management; (4) collaborative development of methods and good practices; (5) creating social cohesion, which strengthens the response and coping capacity of a community; (6) promoting action, such as charitable donations; and (7) enhancing research (Alexander, 2014).

Despite the quantity and wide availability of remote sensing data, information gaps often make these data inappropriate in a hazard context where real-time information is necessary for coping with the event. Limitations of the remote sensing instruments, their carrier platforms, or because of atmospheric interference (Cervone et al., 2017) and repeat rate may cause spatial and temporal gaps in data flows during critical moments (Rossi et al., 2015). SM present a huge potential to overcome these information gaps given the large amounts of data generated in real time that can provide or support a more flexible and scalable emergency service (Cervone et al., 2017; Rossi et al., 2015). During hazard events, early emergency response can be implemented based on the monitoring of the areas at higher risk as well as the occurrence of impacts through ground information from SM (Cervone et al., 2017). Probabilistic flood maps from posts on Twitter mentioning locations of flooding have, for instance, demonstrated good potential to gain near-real time insight into the situation of flooding (Brouwer et al., 2017). A drawback of the use of SM during an emergency relates, however, to the potential dissemination of rumours, undermining public authority (Alexander, 2014).

Limitations of methods based on SM data

The respect of privacy and ethical use of the data are critical aspects when dealing with crowdsourced data from SM (Ghermandi et al., 2023). Sensitive personal information, for which the user has a reasonable expectation of privacy, needs to be protected, anonymized, and securely stored. For the time being, policies and protocols still need to be developed to establish consistent ethical standards and protocols for collecting, treating, and storing crowdsourced data (Mazumdar et al., 2017; Di Minin et al., 2021). An important step to increase transparency and legitimacy would be to differentiate between the purposes for which SM data are used (e.g., facilitating scientific research for the public good compared to commercial data uses) (Ghermandi et al., 2023).

At the same time, preserving the maximum amount of metadata and provenance information related to crowdsourced observations is critical for assessing data quality and user context (Mazumdar et al., 2017). For instance, the lack of detailed information on user demographics is a well-known limitation of the method, specifically when applied to assess cultural values of a landscape. This lack of information can potentially lead to biases in the outcomes of the study in terms of representativeness of the groups of users (Ghermandi and Sinclair, 2019; Ilieva and McPhearson, 2018). Digital divides, related to age, gender, and social power relations (Huang et al., 2013; Muñoz et al., 2019), cause uncertainty regarding the reliability and validity of the information obtained, and must be specifically taken into account for different SM data sources and geographical contexts (each SM network has a particular user community). A particularly critical aspect for SM-based research is also due to the fluctuations in popularity of different SM platforms, as well as to the accessibility to data (Teles da Mota and Pickering, 2020). The latter is also primarily related to many SM companies' restrictive data sharing policies when it comes to scientific research.

Some other technical limitations include: noise in the data, such as images uploaded more than once or at the wrong location (Huang et al., 2013); generally unstructured heterogeneous nature of the data, which requires work to clean and prepare the data for analysis (Cervone et al., 2017; Ilieva and McPhearson, 2018); variable spatial accuracy caused by the change in intensity of cellular signal or variable precision of GPS (global positioning system) receivers (Figueroa-Alfaro and Tang, 2017); and *a posteriori* mapping, with consequent uncertainty regarding the precision of the geolocation (Muñoz et al., 2019). Due to these potential biases, an increasingly critical aspect for future research will be the accurate interpretation and verification of SM data through artificial intelligence.

Conclusions

Despite some important limitations, social media data provide key opportunities for sustainability research, which are unprecedented in time and spatial scales. Such passively crowdsourced data often provide additional or complementary information to traditional survey-based or remote sensing approaches and have found many applications in landscape and urban planning, as well as in environmental monitoring and management. The growing research in the field is opening new venues to provide real-time data and increase the capacity to respond and cope with natural hazards. Perhaps the main concerns related to the use of this data relate to the examination of potential biases and ethical and privacy issues. However, the development of standards and protocols for the handling of the data is expanding and is likely to act in response and temper these concerns.

References

Alexander, D.E. (2014) "Social Media in Disaster Risk Reduction and Crisis Management" *Science and Engineering Ethics* 20 pp.717–733 DOI: 10.1007/s11948-013-9502-z.

Amorim Maia, A.T., Calcagni, F., Connolly, J.J.T., Anguelovski, I. and Langemeyer, J. (2020) "Hidden Drivers of Social Injustice: Uncovering Unequal Cultural Ecosystem Services Behind Green Gentrification" *Environmental Science & Policy* 112 pp.254–263 DOI: 10.1016/j.envsci.2020.05.021.

Anderson, K., Ryan, B., Sonntag, W., Kavvada, A. and Friedl, L. (2017) "Earth Observation in Service of the 2030 Agenda for Sustainable Development" *Geo-Spatial Information Science* 20 pp.77–96 DOI: 10.1080/10095020.2017.1333230.

Angradi, T.R., Launspach, J.J. and Debbout, R. (2018) "Determining Preferences for Ecosystem Benefits in Great Lakes Areas of Concern from Photographs Posted to Social Media" *Journal of Great Lakes Research* 44 (2) pp.340–351 DOI: 10.1016/j.jglr.2017.12.007.

Arsanjani, J.J., Helbich, M., Bakillah, M., Hagenauer, J. and Zipf, A. (2013) "Toward Mapping Land-use Patterns from Volunteered Geographic Information" *International Journal of Geographical Information Science* 27 (12) pp.2264–2278 DOI: 10.1080/13658816.2013.800871.

Barrie, C. and Ho, J. (2021) "academictwitteR: An R Package to Access the Twitter Academic Research Product Track v2 API Endpoint" *Journal of Open Source Software* 6 (62) 3272 pp.1–2 DOI: 10.21105/joss.03272.

Bernetti, I., Chirici, G. and Sacchelli, S. (2019) "Big Data and Evaluation of Cultural Ecosystem Services: An Analysis Based on Geotagged Photographs from Social Media in Tuscan Forest (Italy)" *iForest - Biogeosciences and Forestry* 12 (1) pp.98–105 DOI: 10.3832/ifor2821-011.

Brouwer, T., Eilander, D., van Loenen, A., Booij, M.J., Wijnberg, K.M., Verkade, J.S. and Wagemaker, J. (2017) "Probabilistic Flood Extent Estimates from Social Media Flood Observations" *Natural Hazards and Earth System Sciences* 17 (5) pp.735–747 DOI: 10.5194/nhess-17-735-2017.

Brown, G. and Fagerholm, N. (2015) "Empirical PPGIS/PGIS Mapping of Ecosystem Services: A Review and Evaluation" *Ecosystem Services, Best Practices for Mapping Ecosystem Services* 13 pp.119–133 DOI: 10.1016/j.ecoser.2014.10.007.

Bubalo, M., van Zanten, B.T. and Verburg, P.H. (2019) "Crowdsourcing Geo-information on Landscape Perceptions and Preferences: A Review" *Landscape and Urban Planning* 184 pp.101–111 DOI: 10.1016/j.landurbplan.2019.01.001.

Burnap, P., Rana, O., Williams, M., Housley, W., Edwards, A., Morgan, J., Sloan, L. and Conejero, J. (2015) "COSMOS: Towards an Integrated and Scalable Service for Analysing Social Media on Demand" *International Journal of Parallel, Emergent and Distributed Systems* 30 pp.80–100 DOI: 10.1080/17445760.2014.902057.

Calcagni, F., Amorim Maia, A.T., Connolly, J.J.T. and Langemeyer, J. (2019) "Digital Co-construction of Relational Values: Understanding the Role of Social Media for Sustainability" *Sustainability Science* 14 pp.1309–1321 DOI: 10.1007/s11625-019-00672-1.

Calcagni, F., Nogue' Batalle', J., Baró, F., Connolly, J.J.T. and Langemeyer, J. (2022) "A Tag is Worth a Thousand Pictures: Social Media Data and Metadata Analysis to Uncover Cultural Ecosystem Services Values Distribution" *Ecosystem Services* 58 pp.1–16 DOI: 10.1016/j.ecoser.2022.101495.

Castells, M. (2011) *The Rise of the Network Society* New York: John Wiley & Sons.

Cerrone, D., López Baeza, J. and Lehtovuori, P. (2018) "Integrative Urbanism: Using Social Media to Map Activity Patterns for Decision-making Assessment" *Proceedings of IFKAD 2018* Delft, The Netherlands, 4th–6th July, pp.1094–1107.

Cervone, G., Schnebele, E., Waters, N., Moccaldi, M. and Sicignano, R. (2017) "Using Social Media and Satellite Data for Damage Assessment in Urban Areas During Emergencies" In Thakuriah, P. Tilahun, N. and Zellner, M. (Eds) *Seeing Cities Through Big Data: Research, Methods and Applications in Urban Informatics* Cham, Switzerland: Springer, pp.443–457 DOI: DOI: 10.1007/978-3-319-40902-3_24.

Chan, K.M., Gould, R.K. and Pascual, U. (2018) "Editorial Overview: Relational Values: What Are They, and What's the Fuss About?" *Current Opinion in Environmental Sustainability, Sustainability Challenges: Relational Values* 35 pp.A1–A7 DOI: 10.1016/j.cosust.2018.11.003.

Connors, J.P., Lei, S. and Kelly, M. (2012) "Citizen Science in the Age of Neogeography: Utilizing Volunteered Geographic Information for Environmental Monitoring" *Annals of the Association of American Geographers* 102 (6) pp.1267–1289 DOI: 10.1080/00045608.2011.627058.

Correia, R.A., Ladle, R. and Roll, U. (2021) "Introduction" *Conservation Biology* 35 (2) pp.395–397 DOI: DOI: 10.1111/cobi.13700.

Daume, S., Albert, M. and von Gadow, K. (2014) "Forest Monitoring and Social Media – Complementary Data Sources for Ecosystem Surveillance?" *Forest Ecology and Management, Forest Observational Studies: "Data Sources for Analysing Forest Structure and Dynamics"* 316 pp.9–20 DOI: 10.1016/j.foreco.2013.09.004.

De Nadai, M., Staiano, J., Larcher, R., Sebe, N., Quercia, D. and Lepri, B. (2016) "The Death and Life of Great Italian Cities: A Mobile Phone Data Perspective" *Proceedings of the 25th International Conference on World Wide Web* Montréal, Québec, pp.413–423 DOI: 10.1145/2872427.2883084.

Depietri, Y., Ghermandi, A., Campisi Pinto, S., Orenstein, D.E. (2021) "Public Participation GIS Versus Geolocated Social Media Data to Assess Urban Cultural Ecosystem Services: Instances of Complementarity" *Ecosystem Services* 50 pp.101–277.

Depietri, Y. and Orenstein, D.E. (2020) "Managing Fire Risk at the Wildland-Urban Interface Requires Reconciliation of Tradeoffs Between Regulating and Cultural Ecosystem Services" *Ecosystem Services* 44 (101108) pp.1–13 DOI: 10.1016/j.ecoser.2020.101108.

Di Minin, E., Fink, C., Hausmann, A., Kremer, J. and Kulkarni, R. (2021) "How to Address Data Privacy Concerns when Using Social Media Data in Conservation Science" *Conservation Biology* 35 (2) pp.437–446.

Di Minin, E., Fink, C., Tenkanen, H. and Hiippala, T. (2018) "Machine Learning for Tracking Illegal Wildlife Trade on Social Media" *Nature Ecology & Evolution* 2 pp.406–407 DOI: 10.1038/s41559-018-0466-x.

Di Minin, E., Tenkanen, H. and Toivonen, T. (2015) "Prospects and Challenges for Social Media Data in Conservation Science" *Frontiers in Environmental Science* 3 (63) pp.1–6 DOI: 10.3389/fenvs.2015.00063.

Figueroa-Alfaro, R.W. and Tang, Z. (2017) "Evaluating the Aesthetic Value of Cultural Ecosystem Services by Mapping Geo-Tagged Photographs from Social Media Data on Panoramio and Flickr" *Journal of Environmental Planning and Management* 60 (2) pp.266–281 DOI: 10.1080/09640568.2016.1151772.

Fink, C. (2021) "twitterhistory: A Python Tool to Download Historical Twitter Data" *Zenodo* (26th March) pp.1–6 DOI: 10.5281/ZENODO.4471195.

Fonte, C.C., Bastin, L., See, L., Foody, G. and Lupia, F. (2015) "Usability of VGI for Validation of Land Cover Maps" *International Journal of Geographical Information Science* 29 (7) pp.1269–1291 DOI: 10.1080/13658816.2015.1018266.

Fox, N., August, T., Mancini, F., Parks, K.E., Eigenbrod, F., Bullock, J.M., Sutter, L. and Graham, L.J. (2020) "'photosearcher' Package in R: An Accessible and Reproducible Method for Harvesting Large Datasets from Flickr" *SoftwareX* 12 (100624) pp.1–6 DOI: 10.1016/j.softx.2020.100624.

Ghermandi, A. (2018) "Integrating Social Media Analysis and Revealed Preference Methods to Value the Recreation Services of Ecologically Engineered Wetlands" *Ecosystem Services, Assessment and Valuation of Recreational Ecosystem Services* 31 pp.351–357 DOI: 10.1016/j.ecoser.2017.12.012.

Ghermandi, A., Camacho-Valdez, V. and Trejo-Espinosa, H. (2020a) "Social Media-based Analysis of Cultural Ecosystem Services and Heritage Tourism in a Coastal Region of Mexico" *Tourism Management* 77 (104002) pp.1–9 DOI: 10.1016/j.tourman.2019.104002.

Ghermandi, A., Depietri, Y. and Sinclair, M. (2022) "In the AI of the Beholder: A Comparative Analysis of Computer Vision-assisted Characterizations of Human-nature Interactions in Urban Green Spaces" *Landscape and Urban Planning* 217 (104261) pp.1–10 DOI: 10.1016/j.landurbplan.2021.104261. /j.landurbplan.2021.104261.

Ghermandi, A. and Sinclair, M. (2019) "Passive Crowdsourcing of Social Media in Environmental Research: A Systematic Map" *Global Environmental Change* 55 pp.36–47 DOI: 10.1016/j.gloenvcha.2019.02.003.

Ghermandi, A., Langemeyer, J., Van Berkel, D., Calcagni, F., Depietri, Y., Vigl, L.E., Fox, N., Havinga, I., Jäger, H., Kaiser, N. and Karasov, O. (2023) "Social Media Data for Environmental Sustainability: A Critical Review of Opportunities, Threats, and Ethical Use" *One Earth* 6 (3) pp.236–250.

Ghermandi, A., Sinclair, M., Fichtman, E. and Gish, M. (2020b) "Novel Insights on Intensity and Typology of Direct Human-Nature Interactions in Protected Areas Through Passive Crowdsourcing" *Global Environmental Change* 65 (102189) pp.1–10 DOI: 10.1016/j.gloenvcha.2020.102189.

Ginzarly, M., Pereira Roders, A. and Teller, J. (2019) "Mapping Historic Urban Landscape Values Through Social Media" *Journal of Cultural Heritage* 36 pp.1–11 DOI: 10.1016/j.culher.2018.10.002.

Gosal, A.S., Newton, A.C. and Gillingham, P.K. (2018) "Comparison of Methods for a Landscape-Scale Assessment of the Cultural Ecosystem Services Associated with Different Habitats" *International Journal of Biodiversity Science, Ecosystem Services & Management* 14 pp.91–104 DOI: 10.1080/21513732.2018.1447016.

Graham, S. (2010) *Disrupted Cities: When Infrastructure Fails* Abingdon: Routledge.

Hausmann, A., Toivonen, T., Fink, C., Heikinheimo, V., Tenkanen, H., Butchart, S.H.M., Brooks, T.M. and Di Minin, E. (2019) "Assessing Global Popularity and Threats to Important Bird and Biodiversity Areas Using Social Media Data" *Science of the Total Environment* pp.617–623 DOI: 10.1016/j.scitotenv.2019.05.268.

Hawksey, M. (2010) "Using Google Spreadsheet to Automatically Monitor Twitter Event Hashtags and More" *MASHe* Available at: https://hawksey.info/blog/2010/06/using-google-spreadsheet-to-automatically-monitor-twitter/ (Accessed: 24th March 2021).

Heipke, C. (2010) "Crowdsourcing Geospatial Data" *ISPRS Journal of Photogrammetry and Remote Sensing (ISPRS Centenary Celebration Issue)* 65 pp.550–557 DOI: 10.1016/j.isprsjprs.2010.06.005.

Hermes, J., Van Berkel, D., Burkhard, B., Plieninger, T., Fagerholm, N., von Haaren, C. and Albert, C. (2018) "Assessment and Valuation of Recreational Ecosystem Services of Landscapes" *Ecosystem Services, Assessment and Valuation of Recreational Ecosystem Services* 31 pp.289–295 DOI: 10.1016/j.ecoser.2018.04.011.

Huang, H., Gartner, G. and Turdean, T. (2013) *Social Media Data as a Source for Studying People's Perception and Knowledge of Environments* Vienna: Verlag der Österreichischen Akademie der Wissenschaften.

Ilieva, R.T. and McPhearson, T. (2018) Social-Media Data for Urban Sustainability" *Nature Sustainability* 1 pp.553–565 DOI: 10.1038/s41893-018-0153-6.

Jarić, I., Correia, R.A., Brook, B.W., Buettel, J.C., Courchamp, F., Di Minin, E., Firth, J.A., Gaston, K.J., Jepson, P., Kalinkat, G., Ladle, R., Soriano-Redondo, A., Souza, A.T. and Roll, U. (2020) "iEcology: Harnessing Large Online Resources to Generate Ecological Insights" *Trends in Ecology & Evolution* 35 pp.630–639 DOI: 10.1016/j.tree.2020.03.003.

Jiménez-Valverde, A., Peña-Aguilera, P., Barve, V. and Burguillo-Madrid, L. (2019) "Photo-Sharing Platforms Key for Characterising Niche and Distribution in Poorly Studied Taxa" *Insect Conservation and Diversity* 12 pp.389–403 DOI: 10.1111/icad.12351.

Kleinhans, R., Van Ham, M. and Evans-Cowley, J. (2015) "Using Social Media and Mobile Technologies to Foster Engagement and Self-Organization in Participatory Urban Planning and Neighbourhood Governance" *Planning Practice and Research* 30 pp.237–247 DOI: 10.1080/02697459.2015.1051320.

Langemeyer, J., Baró, F., Roebeling, P. and Gómez-Baggethun, E. (2015) "Contrasting Values of Cultural Ecosystem Services in Urban Areas: The Case of Park Montjuïc in Barcelona" *Ecosystem Services* 12 pp.178–186 DOI: 10.1016/j.ecoser.2014.11.016.

Langemeyer, J. and Calcagni F. (2022) "Virtual Spill-over Effects: What Social Media Has to Do with Relational Values and Global Environmental Stewardship" *Ecosystem Services*, 53 (101400) pp.1–2 DOI: 10.1016/j.ecoser.2021.101400.

Langemeyer, J., Ghermandi A., Keeler B. and Van Berkel D. (2023) "The Future of Crowd-sourced Cultural Ecosystem Services Assessments" *Ecosystem Services* 60 (101518) pp.1–4 DOI: 10.1016/j.ecoser.2023.101518.

Langemeyer, J., Calcagni, F. and Baró, F. (2018) "Mapping the Intangible: Using Geolocated Social Media Data to Examine Landscape Aesthetics" *Land Use Policy* 77 pp.542–552 DOI: 10.1016/j.landusepol.2018.05.049.

Lingua, F., Coops, N.C. and Griess, V.C. (2022) "Valuing Cultural Ecosystem Services Combining Deep Learning and Benefit Transfer Approach" *Ecosystem Services* 58 101487 pp.1–16 DOI: 10.1016/j.ecoser.2022.101487.

Lomborg, S. and Bechmann, A. (2014) "Using APIs for Data Collection on Social Media" *The Information Society* 30 pp.256–265 DOI: 10.1080/01972243.2014.915276.

Mancini, F., Coghill, G.M. and Lusseau, D. (2018) "Using Social Media to Quantify Spatial and Temporal Dynamics of Nature-Based Recreational Activities" *PLoS One* 13 (e0200565) pp.1–19 DOI: 10.1371/journal.pone.0200565.

Mazumdar, S., Wrigley, S. and Ciravegna, F. (2017) "Citizen Science and Crowdsourcing for Earth Observations: An Analysis of Stakeholder Opinions on the Present and Future" *Remote Sensing* 9 pp.1–21 DOI: 10.3390/rs9010087.

Muñoz, L., Hausner, V.H. and Monz, C.A. (2019) "Advantages and Limitations of Using Mobile Apps for Protected Area Monitoring and Management" *Society & Natural Resources* 32 pp.473–488 DOI: 10.1080/08941920.2018.1544680.

Norton, L.R., Inwood, H., Crowe, A. and Baker, A. (2012) "Trialling a Method to Quantify the 'Cultural Services' of the English Landscape Using Countryside Survey Data" *Land Use Policy* 29 pp.449–455 DOI: 10.1016/j.landusepol.2011.09.002.

Nummi, P. (2018) "Crowdsourcing Local Knowledge with PPGIS and Social Media for Urban Planning to Reveal Intangible Cultural Heritage" *Urban Planning* 3 pp.100–115 DOI: 10.17645/up.v3i1.1266.

Orenstein, D.E., Zimroni, H. and Eizenberg, E. (2015) "The Immersive Visualization Theater: a New Tool for Ecosystem Assessment and Landscape Planning" *Computers, Environment and Urban Systems* 54 pp.347–355 DOI: 10.1016/j.compenvurbsys.2015.10.004.

Pickering, C., Walden-Schreiner, C., Barros, A. and Rossi, S.D. (2020) "Using Social Media Images and Text to Examine How Tourists View and Value the Highest Mountain in Australia" *Journal of Outdoor Recreation and Tourism* 29 (100252) pp.1–12 DOI: 10.1016/j.jort.2019.100252.

Richards, D.R. and Friess, D.A. (2015) "A Rapid Indicator of Cultural Ecosystem Service Usage at a Fine Spatial Scale: Content Analysis of Social Media Photographs" *Ecological Indicators* 53 pp.187–195 DOI: 10.1016/j.ecolind.2015.01.034.

Rossi, C., Stemberger, W., Bielski, C., Zeug, G., Costa, N., Poletto, D., Spaltro, E. and Dominici, F. (2015) "Coupling Crowdsourcing, Earth Observations, and E-GNSS in a Novel Flood Emergency Service in the Cloud" *Proceedings of the 2015 IEEE International Geoscience and Remote Sensing Symposium (IGARSS)* pp.2703–2706 DOI: 10.1109/IGARSS.2015.7326371.

Rossi, S.D., Barros, A., Walden-Schreiner, C. and Pickering, C. (2019) "Using Social Media Images to Assess Ecosystem Services in a Remote Protected Area in the Argentinean Andes" *Ambio* 49 pp.1146–1160.

Salcedo-Sanz, S., Ghamisi, P., Piles, M., Werner, M., Cuadra, L., Moreno-Martínez, A., Izquierdo-Verdiguier, E., Muñoz-Marí, J., Mosavi, A. and Camps-Valls, G. (2020) "Machine Learning Information Fusion in Earth Observation: A Comprehensive Review of Methods, Applications and Data Sources" *Information Fusion* 63 pp.256–272 DOI: 10.1016/j.inffus.2020.07.004.

Schmidt, K.A., Dall, S.R.X. and Gils, J.A.V. (2010) "The Ecology of Information: An Overview on the Ecological Significance of Making Informed Decisions" *Oikos* 119 pp.304–316 DOI: 10.1111/j.1600-0706.2009.17573.x.

Schröter, M., Başak, E., Christie, M., Church, A., Keune, H., Osipova, E., Oteros-Rozas, E., Sievers-Glotzbach, S., Oudenhoven, A.P.E. van, Balvanera, P., González, D., Jacobs, S., Molnár, Z.,

Pascual, U. and Martín-López, B. (2020) "Indicators for Relational Values of Nature's Contributions to Good Quality of Life: The IPBES Approach for Europe and Central Asia" *Ecosystems and People* 16 pp.50–69 DOI: 10.1080/26395916.2019.1703039.

Sessions, C., Wood, S.A., Rabotyagov, S. and Fisher, D.M. (2016) "Measuring Recreational Visitation at U.S. National Parks with Crowd-sourced Photographs" *Journal of Environmental Management* 183 pp.703–711 DOI: 10.1016/j.jenvman.2016.09.018.

Shelton, T., Poorthuis, A. and Zook, M. (2015) "Social Media and the City: Rethinking Urban Socio-Spatial Inequality Using User-generated Geographic Information" *Landscape and Urban Planning, Special Issue: Critical Approaches to Landscape Visualization* 142 pp.198–211 DOI: 10.1016/j.landurbplan.2015.02.020.

Sinclair, M., Ghermandi, A., Moses, S.A. and Joseph, S. (2019) "Recreation and Environmental Quality of Tropical Wetlands: A Social Media Based Spatial Analysis" *Tourism Management* 71 pp. 179–186 DOI: 10.1016/j.tourman.2018.10.018.

Sinclair, M., Mayer, M., Woltering, M. and Ghermandi, A. (2020) "Using Social Media to Estimate Visitor Provenance and Patterns of Recreation in Germany's National Parks" *Journal of Environmental Management* 263 (110418) pp.1–12 DOI: 10.1016/j.jenvman.2020.110418.

Sudmanns, M., Tiede, D., Lang, S., Bergstedt, H., Trost, G., Augustin, H., Baraldi, A., Blaschke, T. (2020) "Big Earth Data: Disruptive Changes in Earth Observation Data Management and Analysis?" *International Journal of Digital Earth* 13 pp.832–850 DOI: 10.1080/17538947.2019.1585976.

Teles da Mota, V. and Pickering, C. (2020) "Using Social Media to Assess Nature-Based Tourism: Current Research and Future Trends" *Journal of Outdoor Recreation and Tourism* 30 (100295) pp.1–11 DOI: 10.1016/j.jort.2020.100295.

Tieskens, K.F., Van Zanten, B.T., Schulp, C.J.E. and Verburg, P.H. (2018) "Aesthetic Appreciation of the Cultural Landscape Through Social Media: An Analysis of Revealed Preference in the Dutch River Landscape" *Landscape and Urban Planning* 177 pp.128–137 DOI: 10.1016/j.landurbplan.2018.05.002.

Toivonen, T., Heikinheimo, V., Fink, C., Hausmann, A., Hiippala, T., Jarv, O., Tenkanen, H. and Di Minin, E. (2019) "Social Media Data for Conservation Science: A Methodological Overview" *Biological Conserversation* 233 pp.298–315 DOI: 10.1016/j.biocon.2019.01.023.

Väisänen, T., Heikinheimo, V., Hiippala, T. and Toivonen, T. (2021) "Exploring Human–Nature Interactions in National Parks with Social Media Photographs and Computer Vision" *Conservation Biology* 35 pp.424–436 DOI: 10.1111/cobi.13704.

Van Berkel, D.B., Tabrizian, P., Dorning, M.A., Smart, L., Newcomb, D., Mehaffey, M., Neale, A. and Meentemeyer, R.K. (2018) "Quantifying the Visual-sensory Landscape Qualities that Contribute to Cultural Ecosystem Services Using Social Media and LiDAR" *Ecosystem Services, Assessment and Valuation of Recreational Ecosystem Services* 31 pp.326–335 DOI: 10.1016/j.ecoser.2018.03.022.

Wood, S.A., Guerry, A.D. and Silver, J.M., Lacayo, M. (2013) "Using Social Media to Quantify Nature-based Tourism and Recreation" *Scientific Reports* 3 (2976) pp.1–7 DOI: 10.1038/srep02976.

Wood, S.A., Winder, S.G., Lia, E.H., White, E.M., Crowley, C.S.L., Milnor, A.A. (2020) "Next-generation Visitation Models Using Social Media to Estimate Recreation on Public Lands" *Scientific Reports* 10 (15419) pp.1–12 DOI: 10.1038/s41598-020-70829-x.

Zapata, E., Calcagni F., Baró F. and Langemeyer J. (2022) "Using Crowdsourced Imagery to Assess Cultural Ecosystem Services in Data-scarce Urban Regions: The Case of the Metropolitan Area of Cali, Colombia" *Ecosystem Services* 56 (101445) pp.1–13 DOI: 10.1016/j.ecoser.2022.101445.

Zhou, X. and Zhang, L. (2016) "Crowdsourcing Functions of the Living City from Twitter and Foursquare Data" *Cartography and Geographic Information Science* 43 (5) pp.393–404 DOI: 10.1080/15230406.2015.1128852.

Zielstra, D. and Hochmair, H.H. (2013) "Positional Accuracy Analysis of Flickr and Panoramio Images for Selected World Regions" *Journal of Spatial Science* 58 (2) pp.251–273 DOI: 10.1080/14498596.2013.801331.

37
CRISIS AND HAZARD MAPPING

Amelia Hunt

Humanitarian objectives set out to 'alleviate humanitarian suffering without discrimination. However, people must first be visible' (IFRC, 2018: 29). The UN Data Revolution Group describes data as 'the lifeblood of decision-making' (IEAG, 2014: 4), without which key humanitarian issues cannot be addressed, yet unmapped areas are often in low- and middle-income countries, the ones most exposed to crises (IFRC, 2018). Humanitarian mapping resolves to define 'ground objects, such as buildings and roads, to know where the vulnerable people are located during a humanitarian crisis' (Chen et al., 2018: 1713), yet in many ways, they serve a purpose beyond the crux of a crisis alone. Humanitarian maps can be utilized across all areas of the disaster response cycle – prevention, preparedness, response, and recovery – straddling emergency response and international development. Crisis maps are a subset of humanitarian mapping, focusing on the specific 'crisis' segment of the cycle.

To understand the application of crisis maps, it is important to consider the nature of a 'crisis'. Traditionally, crisis mapping has been associated with rapid onset crises such as earthquakes and natural disasters, but in reality, it covers mapping within all humanitarian contexts, including slow-onset crises such as migration or protracted conflict (Ziemke, 2012). These maps pull together information from social media, SMS messages, aerial imagery, and surveys to give 'voice to the distributed voiceless' (Meier, 2011b). Produced accurately and utilized effectively, maps have the opportunity to empower at-risk and affected populations to define the way they receive humanitarian assistance (Harvard Humanitarian Initiative [HHI], 2011).

History and main players

Humanitarian maps are the product of coordinated contributions from digital humanitarians, local community mappers, and field responders, yet this has not always been the case. Historically, needs assessment maps were produced by information management specialists deployed from formal humanitarian entities (Hunt and Specht, 2019), yet as access to the Internet and technology has expanded, crisis maps are no longer, 'the purview of an elite' (Bolletino in Meier, 2015: xx).

In the humanitarian sphere, the first widely recognized humanitarian map is John Snow's 1854 cholera map (Brody et al., 2003), which, for the first time on record, plotted epidemiological data geographically, aiding efforts to identify and contain the outbreak. As humanitarian disasters have become more frequent, the need for geo-tagged field data has increased; however, an absence of baseline maps has often meant there was little to plot onto (Adams, 2006). Humanitarian mapping as we see it today is an innovation born from the digital revolution and Open Data movements (Stauffacher *et al.*, 2012). Open access modern-day mapping, using platforms like OpenStreetMap (OSM Wiki, 2019), a free map of the world, has meant that almost anyone, anywhere in the world can contribute cartographic data. Users can trace satellite imagery at a 'mapathon', generate survey data from their village, or validate map contributions from their living room. The multitude of actors contributing information is defined as 'crowdsourced'. Crowdsourced crisis mapping relies on the collective action of global citizens to map spatial data for a common humanitarian goal (Anderson-Tarver, 2015). Map contributors are often coordinated through Volunteer and Technical Communities (V&TCs) (Capelo *et al.*, 2012). One such V&TC is the Humanitarian OpenStreetMap Team (HOT), who establish tasks of unmapped, at-risk areas for a network of global mappers (HOT, 2019). In times of crisis, these groups are called to action to produce maps.

The Haiti earthquake in 2010 was a pivotal point in the history of crisis mapping. Ushahidi (2020a), 'testimony' in Swahili, began in 2008 helping local people map their concerns, and together with The Harvard Humanitarian Initiative (HHI), they had begun investigating hypothetical applications of humanitarian maps. Crisis maps had been loosely tested during the 2009 Philippines Typhoon (Phillips and Verity, 2016), yet Haiti was different, with over 800 features updated on OpenStreetMap within the first 48 hours (Parr, 2015). The quantity and speed at which data was contributed meant that crisis maps quickly became an invaluable resource for frontline responders. It was the first time that online volunteers had collectively built functioning crisis maps (GDPC, 2017), and it proved that digital humanitarians had a crucial role to play in collating and analysing data (Weinandy, 2016).

> On the timeline of the Internet's evolution, the 2010 Haiti earthquake response will be remembered as the moment when…thousands of citizens around the world collaborated [...] to help make sense of a largescale calamity and give voice to an affected population.
>
> *(HHI, 2011: 11)*

Despite the success of mapping in Haiti, the event also highlighted fundamental issues. Within the chaos of a disaster, emergency responders had limited capacity to evaluate data accuracy or respond to every mapped need (Hunt & Specht, 2019). With no established communication channels between formal and online humanitarians, the utilization of maps stemmed largely from individual risk takers and a desperation for data (HHI, 2011). Information flows between relief groups was informal and based on personal relationships (Nelson *et al.*, 2010). The use of social media data and opensource software raised concerns around data accuracy and protection (Hunt and Specht, 2019).

Over the years, formal humanitarians, affected populations, and V&TCs have worked together to attempt to overcome these challenges. In 2012, the Digital Humanitarian Network (DHN, 2019) was founded to bridge the gap between V&TCs and traditional humanitarians. Through DHN and other partnerships, slowly, awareness and trust in crowdsourced data have increased. A key success is The Missing Maps project, which aims to map information gaps in geographic areas most vulnerable to disasters (MissingMaps, 2020). The initiative was founded by actors from both the traditional and digital humanitarian spheres – British

Red Cross, American Red Cross, HOT and Médecins Sans Frontières (MSF). Utilizing OSM, remote volunteers add key map features, whilst local mappers add micro-level details, such as place names, water points, hospitals, and historical flood data.

As open source technology has expanded, networks have become inherently more inclusive, with contributions from local and global citizens, grassroot and international non-governmental organizations (NGOs), public and corporate entities (Givoni, 2016). Crucially, the role of affected communities has expanded – local people act as data users as well as data providers, with the power to use open access maps for citizen-led initiatives. Crowdsourced maps have democratized humanitarian response (Hunt and Specht, 2019).

Case studies

The Humanitarian Coalition (2020) describes humanitarian crises as, 'an event or series of events that represents a critical threat to the health, safety, security or wellbeing of a community or other large group of people'. This incorporates natural disasters such as earthquakes, typhoons, floods, and epidemics, as well as man-made disasters such as armed conflict and political warfare. The following case studies showcase maps developed during various crises; however, map production does not always equate to utilization. This is sometimes due to the complexity of a disaster or open data restrictions, yet policies and cultural attitudes towards new technology also play a role (Hunt and Specht, 2019).

Natural disasters

2015 Nepal earthquake

In 2015, a magnitude 7.8 earthquake hit Nepal. Over 8,790 people lost their lives, and a third of the population were affected, both in the initial earthquake and the 300 aftershocks (NPC, 2015). The scale of the damage and the remote locations of many affected communities made it difficult for responders to understand where to prioritize response efforts (NPC, 2015).

Kathmandu Living Labs (KLL, 2020a), a non-for-profit committed to creating digital solutions to improve urban resilience, were key actors in the response. 'Those who were seeking help did not know where to find it and those who were trying to help did not necessarily know where exactly their help was needed' (KLL, 2020b). Within 24 hours KLL established an Ushahidi map, 'QuakeMap.org', to map the needs of affected people against relief supplies (Ushahidi, 2020b). Replicating methods from Haiti, data was collated from social media and SMS messages to understand needs on the ground (KLL, 2020b). Over 350 Quakemap reports were acted on by the Nepalese army and NGOs to monitor where relief was needed most urgently and access issues affecting delivery (Ushahidi, 2020b). Quakemap reports and OSM basemaps helped inform strategic operational maps such as the one in Figure 37.1.

Requests for support submitted to Quakemap were tagged by location and urgency to connect relief agencies to specific people in need (KLL, 2020b). Another V&TC, Standby Task Force (SBTF, 2015), created infrastructural damage maps based on Twitter photographs using a tagging app, MicroMappers. Volunteers were asked to rate Twitter photos based on the level of damage observed (MicroMappers, 2015). In both instances, crisis maps informed humanitarian agencies of which locations were most in need of emergency support (Figure 37.2).

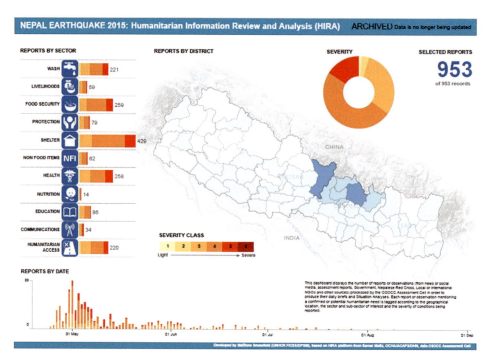

Figure 37.1 Smawfield, M. (2015) Nepal Earthquake 2015: Humanitarian Information Review and Analysis (HIRA), Nepal. Available at: http://data.unhcr.org/hira/ (Accessed: 6th April 2020)

Figure 37.2 Qatar Center for Artificial Intelligence & Standby Task Force (2015) MicroMap of Nepal Earthquake – April 2015, Kathmandu. Available at: http://maps.micromappers.org/2015/nepal/images/#close (Accessed: 6th January 2020)

Compared to Haiti, four times the number of online mappers responded to Nepal within 48 hours of the disaster (Ganesh, 2015). Parr's (2015) analysis of mapping activities from 25th April–6th May 2015 shows that 9,989 new mappers joined OpenStreetMap, many of whom contributed to the creation, modification, or deletion of over a million features in Nepal. The increase in OSM members can largely be attributed to the media attention surrounding the earthquake, yet despite the best intentions of new mappers, an increase in first-time volunteers raised concerns around map accuracy (Parr, 2015). In an After Action Review Report from SBTF (2015), one longstanding mapper commented, 'the flood of new volunteers made it really hectic [...] especially on an activation of this scale'. Emergency responders were concerned about the accuracy of maps produced by first-time mappers, as well as the potential for human bias in rating damage seen on social media images. Additional online capacity resulted in more crisis map contributions, yet not necessarily effective use of the maps produced.

The under-utilization of crisis maps during the Nepal earthquake is in part attributable to data validity concerns, but the crisis highlights a more human issue. Interviews conducted with field personnel and digital humanitarians suggest that a failure to recognize the significance of crisis maps in Nepal came down to individuals (Hunt and Specht, 2019). Traditional humanitarians had requested the support of V&TCs at head-quarter level, yet on the frontline, crowdsourced maps were often dismissed by response leads. Attitudes towards crowdsourced, open source maps do vary across established humanitarian organizations, yet, in an emergency, it is often personal beliefs that define their success or dismissal.

Ebola epidemic

The 2013–2016 West Africa Ebola outbreak resulted in an estimated 11,310 deaths (WHO, 2016). The response was heavily criticized, in particular for the lack of usable, timely information. Affected areas were either unmapped, or lacked comprehensible maps, which made tracking the outbreak next to impossible (Fast and Waugaman, 2016). The complexity of the contexts in Guinea, Libya, and Sierra Leone, cannot be overlooked – these countries were recovering from years of conflict, with fragile health services and fluid borders (USAID, 2017). Data collection across the region also relied on paper systems that could not be incorporated into online maps (USAID, 2017) (Figure 37.3).

The outbreak began in Guinea in December 2013, yet was not officially declared by the World Health Organization (WHO) until March 2014, and not raised as a Public Health Emergency of International Concern until August 2014 (CDC, 2019). Aware of the escalating situation, MSF deployed a GIS specialist to Guéckédou, Guinea, in March to begin plotting route maps over the course of two months (Koch, 2015). Almost 250 digital volunteers supported remotely by tracing granular level basemaps onto OpenStreetMap (Lüge, 2014).

The lack of international attention to meant that DHN were not formally activated by humanitarian organizations until August – eight months after patient zero (SBTF, 2014). The limited capacity and resources dedicated to mapping at the start of the crisis slowed efforts at contact tracing. Koch (2015) notes that, 'transmission mapping at the beginning of the outbreak might have alerted field medical personnel that the outbreak that they thought was contained in May was in fact expanding'.

Learning from the 2014 outbreak, in May 2018, when an Ebola outbreak was declared in the Democratic Republic of the Congo (WHO, 2018), mappers were ready to respond. To aid medical personnel on the ground, digital volunteers produced OSM basemaps that were overlaid with local-level health surveys. Whilst online mappers plotted cartographic features such as buildings and roads, mappers on-the-ground tagged key features such as hospitals

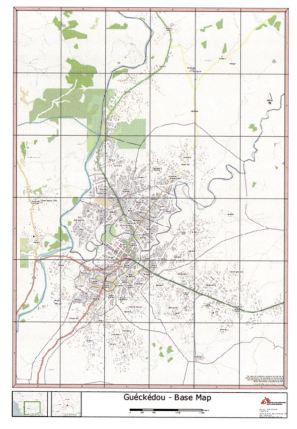

Figure 37.3 Médecins Sans Frontières (2014) Guéckédou – Base Map, Guinea. Available at: https://reliefweb.int/sites/reliefweb.int/files/resources/GIN_Gueckedou_OSM_A1.pdf (Accessed: 20th April 2020)

and schools to provide health officials with baseline data to track and trace patients. As the response continued, neighbouring country Uganda began preventative measures to map informal and formal entry points along the DRC-Uganda border (HOT, 2018). Pinpointing available health facilities, key services, and infrastructure, such as water points or marketplaces, proved vital in understanding the spread of disease and containing the outbreak.

Humanitarians continue to research new methods to ensure future crisis maps are timely and accurate. Motorcycle mapping is one such study that has demonstrated the use of bikes for local people to travel vast areas mapping municipal boundaries, access routes, and health facilities (Allan *et al.*, 2019). Empowering local people to map their surroundings before a crisis is crucial in ensuring adequate treatment and tracing of future epidemics (Figure 37.4).

Man-made

Libya

In February 2011, anti-government protests began in Libya. What started as peaceful demonstrations soon escalated, with armed forces killing civilians, and retaining and torturing activists

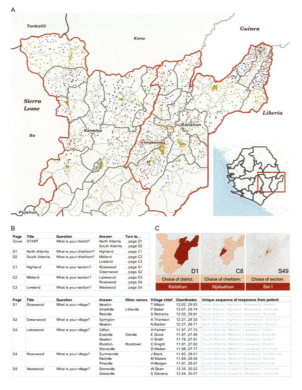

Figure 37.4 Allan, R., Gayton, I., Monk, E.J.M. Yee, K-P. (2019) Surveyed Villages Within the Catchment Aarea of Nixon Memorial Methodist Hospital Based in Segbwema, Sierra Leone" Available at: https://doi.org/10.1093/trstmh/trz063 (Accessed: 5th May 2020)

(INSCT, 2012). For the first time, UNOCHA formally requested the activation of V&TCs to support the humanitarian response (Meier, 2011a). Led by SBTF, a network of volunteers mapped conflict areas tagged with witness reports (Zak, 2011). Within two days, over 220 incident reports had been mapped (Meier, 2011a), and the layers of social media data provided an impression of the true scale of the crisis (Stottlemyre and Stottlemyre, 2012). Digital volunteers also mapped 3W data – Who, What, Where – which helped organizations plan cluster-specific operations.

The Libya Crisis Map provided a platform for the most vulnerable, but the open access nature meant that these were not the only voices heard. Amongst the humanitarian reports contributed, military information, such as the location of tanks, was also uploaded to influence international interventions (Stottlemyre and Stottlemyre, 2012). Conflicting information demands exasperated the mass of data (Stottlemyre and Stottlemyre, 2012), lessening the ability to monitor human rights violations and warn local citizens, for which the map had been designed. The map also presented security concerns noted in other politically volatile contexts. The same maps that can empower activists and protect civilians in areas of conflict can also be used by opposing forces to trace, repress, and persecute (Meier, 2011b) (Figure 37.5).

Complex

Complex disasters are the consequence of multiple natural and human factors resulting in a humanitarian crisis (Humanitarian Coalition, 2020). One of the most significant complex

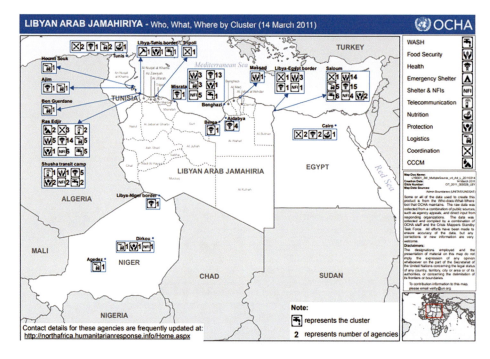

Figure 37.5 MapAction and UNOCHA (2011) LIBYAN ARAB JAMAHIRIYA – Who, What, Where by Cluster (14th March 2011). Available at: https://reliefweb.int/map/libya/-libyan-arab-jamahiriya-who-what-where-cluster-14-march-2011 (Accessed: 16th January 2020)

emergencies of our times is migration, characterized by 'people moving due to fear of persecution, conflict and violence, human rights violations, poverty and lack of economic prospects, or natural disasters' (IFRC, 2017). Figures from 2019 estimate that there are 271.6 million migrants worldwide (GMDP, 2019), and 70.8 million forcibly displaced individuals, of which 25.9 million are refugees seeking protection outside of their national countries (UNHCR, 2019a). As climate change increases, populations find themselves in unstable contexts exasperated by climate shocks – natural disasters, such as droughts and floods, are heightened, and fragile contexts become more volatile as food insecurity rises (UNFCC, 2017).

Syria crisis

In 2007, a three-year drought in Eastern-Syria, worsened by climate change, caused drops in food production resulting in food price spikes and rural-urban migration (Benko, 2017). Economic instability, food insecurity, political tensions, and unsustainable agricultural and environmental policies were all contributing factors leading to civil war (Cane et al., 2015). Since the war in Syria began in 2011, thousands of civilians have fled the country as refugees.

Lebanon, which has the highest population of refugees per capita, has banned permanent refugee camps (Medair, 2019). Although it officially hosts 1 million Syrian refugees, around 550,000 additional unregistered people seek refuge in unstructured, informal settlements where complex decentralized structures create blackholes of data (Sewell, 2020). Humanitarian organizations such as Medair have been surveying and mapping informal refugee

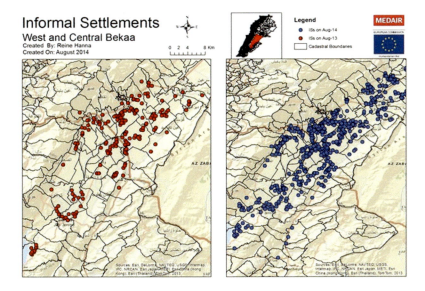

Figure 37.6 Reine Hanna Medair (2014) Informal Refugee Settlements, West and Central Bekaa, Lebanon (August 2014). Available at: https://reliefweb.int/report/lebanon/syrian-refugee-crisis-medair-identifies-unofficial-settlements-refugee-numbers-soar-3 (Accessed: 16th May 2020)

settlements using mobile technologies like OpenDataKit (Medair, 2020). To date, Medair have records of 302,209 Syrian refugees across 6,000 informal camps (Sewell, 2020), who they are now able to support through health, shelter, water, and sanitation programmes (Medair, 2019) (Figure 37.6).

Refugees in Uganda

Almost 1.4 million refugees reside in Uganda, 62% of whom have fled from conflict in South Sudan, and a large proportion from the Democratic Republic of the Congo (UNHCR, 2020). Globally, it is estimated that 80% of refugees live below the international poverty line and are dependent on humanitarian assistance (UNHCR, 2019b). Refugee settlements such as Bidibidi, the second largest refugee settlement in the world (Strochlic, 2019) with over 280,000 inhabitants (UNHCR, 2018), are more populated than most Ugandan cities yet lack appropriate maps. Without sufficient data, host countries and humanitarian agencies have difficulty implementing refugee services. In 2015, HOT Uganda began working with local partners in thirty-three refugee zones (HOT, 2019) to create a true impression of community assets and needs as mapped by local leaders and refugees themselves (HOT, 2019). The facilities map in Figure 37.7 highlights gaps in services to help refugee agencies target their programmes.

A crisis map gives visibility to displaced people and can open up access to humanitarian relief and protection against human rights violations. However, as with any other data visualization, maps are open to interpretation. Refugee agencies can use map demographics to plan food distribution, schools, maternity clinics, and trauma centres. Yet amidst the controversy of the current refugee crisis, maps designed to support the most vulnerable can also fuel prejudices against them. Migration is a complex humanitarian crisis, yet in many instances,

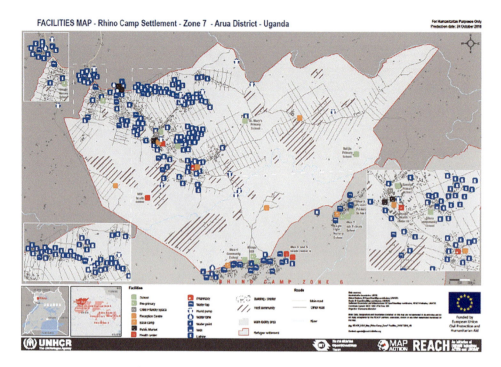

Figure 37.7 HOT, MapAction, REACH & UNHCR (2018) Facilities Map – Rhino Camp Settlement – Zone 7 – Arua District – Uganda

data on refugee numbers deduced from settlement maps has been framed as a 'crisis' for host countries rather than for displaced persons (Georgiou and Zaborowski, 2017).

Challenges and the future

Whilst formal humanitarian organizations are largely based on hierarchical, centralized structures, digital entities tend to be decentralized and linear (Weinandy, 2016). Structural differences influence organizational practices – traditional entities rely on established processes and targets, whilst informal organizations sway towards experimental and innovative methods (Weinandy, 2016). The effectiveness of cross-sectoral collaboration therefore relies on a willingness to exchange and combine traditional and crowdsourced practices. It also demands a bottom-up approach, with local communities leading data collection, collation, and analysis – data that defines their lives (Hunt and Specht, 2019). Humanitarian mapping brings together a multitude of actors from different spheres – trust and collaboration is crucial.

Aside from the human challenges of crisis mapping, the technology itself raises issues. The influx of unverifiable, crowdsourced big data during the chaos of a response has not always proved useful or usable (Hunt and Specht, 2019). As Meier (2014) writes, '[the] overflow of information generated during disasters can be as paralyzing to disaster response as the absence of information'.

The ability to mine Big data into precise, useful information is always challenging, but within a crisis is next to impossible (Verity and Whipkey, 2015). Poor data can not only result in the redundancy of crisis maps, but, worse still, hold negative implications for affected communities. In the humanitarian sphere, wrong information can mean unnecessary loss of life, so data accuracy is essential, and mapped information must come with an expiration date. Following the 2015 Nepal earthquake, initial damage assessment maps were used by government to prioritize areas most in need of financial initiatives – 'the assessment ultimately scripted the kinds of earthquake recovery that took place in post-earthquake Nepal: enabling some, and rendering others unthinkable' (Lord and Sodden, 2018). Effectively, affected populations mapped outside of high-damage areas became invisible and were excluded from recovery support. Crisis maps are useful for rapid relief assessments, but there is a risk that they inscribe long-standing impressions of community needs.

The timeliness and accuracy of data inputted into crisis maps is important, but so is the data calculated from cartographic visualizations. Crisis maps visualize affected lives, with layers of geotagged humanitarian relief providing a clear indication of where needs are met. To the donor community, this represents where funds were effectively invested. It is common practice to calculate the number of affected people from the number of mapped buildings multiplied by the average residents per household. This is useful during a crisis to estimate the scale of lives affected, yet is this data entirely accurate in stating humanitarian impact? Is the success of a response actually quantifiable or is it influenced by donor demands? In a study of 2013 Typhoon Haiyan, Cornelio *et al.* (2016) observed that mapped data was presented to donors as impact reports, yet not shared back to the affected population. Maps provide an insight into lives affected during a crisis, yet responsible mapping demands feedback from those affected, with qualitative studies supplementing mapped impressions.

Engaging communities in citizen-led mapping parallels a broader shift from classical humanitarianism to 'resilience humanitarianism'. Traditional humanitarianism focused on the provision of aid within a crisis, whilst resilience methods insist that at-risk populations are included in the mitigation, preparedness, response, and recovery stages of a disaster (Hilhorst, 2018). The rise of bottom-up information systems has helped to catalyse change within humanitarian operations from aid-driven methods to citizen-led relief (HHI, 2011). Not only is crowdsourced information important to ensure that the needs of vulnerable populations are represented, but Goodchild and Glennon (2010) ascertain that data generated from multiple observers is inherently more accurate. To further allow local people to contribute to the mapped world, communications infrastructure needs to be established before crises hit.

The frequency and magnitude of natural disasters will continue to increase (IPCC, 2012), and mapping has to shift from 'crisis mapping' in an emergency, to 'humanitarian mapping' for resilience. From the first examples of crisis mapping, Milner and Verity (2013: 47) note, 'those V&TCs that successfully collaborated with the formal humanitarian system were those that developed relationships prior to crisis'. Acting from across the globe, digital volunteers are often detached from the complexity of the crisis they are supporting. Both parties should be trained in the application of humanitarian maps before disaster strikes. It is this human element that will be the greatest test of crisis mapping. Technology such as artificial intelligence, drones, and Big data scraping will continue to improve (Wright and Verity, 2020), yet as IEAG (2014: 24) write, 'Strengthening national capacities will be the essential test of any data revolution' (IEAG, 2014: 24). Humanitarian mapping of the future should empower citizens to identify data gaps within their own communities so that policies and programmes can address visualized needs (DataShift, 2015).

When data in presented in mapped form, it is easy to forget the human narrative behind the visualizations. Data points on humanitarian maps represent more than geophysical features, they represent crisis-affected communities. On the surface, a map is a representation of a geographic region with physical features, but for humanitarian mapping, the true value of a map lies in the layers of data underneath, the people who generate it and those that it serves.

References

Adams, M. (2006) "Mapping as a Tool for Planning and Coordination in Humanitarian Operations" *Humanitarian Exchange* 36 pp.29–32 Available at: https://odihpn.org/wp-content/uploads/2007/01/humanitarianexchange036.pdf (Accessed: 23rd February 2020).

Allan, R., Gayton, I., Monk, E.J.M. and Yee, K-P. (2019) "Determination of True Patient Origin Through Motorcycle Mapping: Design and Implementation of a Community-defined Geographic Infrastructure Surveillance Tool in Rural Sierra Leone" *Transactions of the Royal Society of Tropical Medicine and Hygiene* 113 (9) pp.572–575 DOI: 10.1093/trstmh/trz063.

Anderson-Tarver, C. (2015) *Crisis Mapping the 2010 Earthquake in OpenStreetMap Haiti* Boulder, CO: ProQuest Dissertations Publishing.

Benko, J. (2017) "How a Warming Planet Drives Human Migration" *The New York Times* (19th April) Available at: https://www.nytimes.com/2017/04/19/magazine/how-a-warming-planet-drives-human-migration.html.

Brody, H., Paneth, N., Rachman, S., Rip, M. and Vinten-Johansen, P. (2003) *Cholera, Chloroform and the Science of Medicine: A Life of John Snow* Oxford: Oxford University Press.

Cane, S., Kelley, C.P., Kushnir, Y. Mohtadi, M.C. and Seager, R. (2015) "Climate Change in the Fertile Crescent and Implications of the Recent Syrian Drought" *Proceedings of the National Academy of Sciences* 112 (11) pp.3241–3246 DOI: 10.1073/pnas.1421533112

Capelo, L., Chang, N. & Verity, A. (2012) *Guidance for Collaborating with Volunteer and Technical Communities* Geneva: United Nations Office for the Coordination of Humanitarian Affairs.

Centers for Disease Control and Prevention (2019) "2014-2016 Ebola Outbreak in West Africa" Available at: https://www.cdc.gov/vhf/ebola/history/2014-2016-outbreak/index.html (Accessed: 15th April 2020).

Chen, J., Fan, H., Zhou, Y. and Zipf, A. (2018) "Deep Learning From Multiple Crowds: A Case Study of Humanitarian Mapping" *IEEE Transactions on Geoscience and Remote Sensing* 57 (3) DOI: 10.1109/TGRS.2018.2868748.

Cornelio, J.S., Longboan, L., Madianou, M. and Ong, J.C. (2016) "The Appearance of Accountability: Communication Technologies and Power Asymmetries in Humanitarian Aid and Disaster Recovery" *Journal of Communication* 66 (6) pp.960–81 https://doi.org/10.1111/jcom.12258.

DataShift (2015) "Theory of Change" Available at: https://civicus.org/thedatashift/about/theory-of-change (Accessed: 23rd February 2020).

Digital Humanitarian Network (DHN) (2019) "History and today" *Digital Humanitarian Network* Available at: https://www.digitalhumanitarians.com/ (Accessed: 2nd January 2020).

Fast, L. and Waugaman, A. (2016) *Fighting Ebola with Information: Learning from Data and Information Flows in the West Africa Ebola Response.* Washington: USAID.

Ganesh, A. (2015) "OpenStreetMap Rallies for Nepal" Mapbox (April 29) Available at: https://www.mapbox.com/blog/mapping-nepal/ (Accessed: 15th April 2020).

GDPC (2017) "Crisis Mapping" *Global Disaster Preparedness Centre* Available at: https://www.preparecenter.org/topics/crisis-mapping (Accessed: 20th February 2020).

Georgiou, M. and Zaborowski, R. (2017) "Media Coverage of the "Refugee Crisis": A Cross-European Perspective" (*Report DG1(2017)03*) Council of Europe Available at: https://rm.coe.int/1680706b00.

Givoni, M. (2016) "Between Micro Mappers and Missing Maps: Digital Humanitarianism and the Politics of Material Participation in Disaster Response" *Environment and Planning D: Society and Space* 34 (6) pp.1025–1043 DOI:10.1177/ 0263775816652899.

Global Migration Data Portal (GMDP) (2019) "World: Key Migration Statistics" *GMDP* Available at: https://migrationdataportal.org/?i=stock_abs_&t=2019 (Accessed: 29th January 2020).

Goodchild, M.F. Glennon, J.A. (2010) "Crowdsourcing Geographic Information for Disaster Response: A Research Frontier" *International Journal of Digital Earth* 3 pp.231–241 DOI: 10.1080/17538941003759255.

Harvard Humanitarian Initiative (HHI) (2011) *Disaster Relief 2.0: The Future of Information Sharing in Humanitarian Emergencies* Washington, DC & and London: UN Foundation and Vodafone Foundation Technology Partnership.

Hilhorst, D. (2018) "Classical Humanitarianism and Resilience Humanitarianism: Making sense of two brands of humanitarian action" *Journal of International Humanitarian Action* 3 (15) DOI: https://doi.org/10.1186/s41018-018-0043-6

Humanitarian Coalition (2020) "What is a Humanitarian Emergency?" Available at: https://www.humanitariancoalition.ca/info-portal/factsheets/what-is-a-humanitarian-crisis (Accessed: 20th April 2020).

Humanitarian OpenStreetMap Team (HOT) (2018) "Digital Information for Ebola Preparedness: Point of Entry Training and Data Collection in Uganda" Humanitarian OpenStreetMap Team" Available at: https://www.hotosm.org/projects/building-digital-information-for-ebola-preparedness-uganda-point-of-entry-poe-training-and-data-collection-exercise/ (Accessed: 15th November 2019).

Humanitarian OpenStreetMap Team (HOT) (2019) "Bridging Data Gaps: Mapping Refugee Contexts in East Africa" *Humanitarian OpenStreetMap Team* Available at: https://www.hotosm.org/projects/bridging-data-gaps-mapping-refugee-contexts-in-east-africa/ (Accessed: 5th January 2020).

Hunt, A and Specht, D. (2019) "Crowdsourced Mapping in Crisis Zones: Collaboration, Organization and Impact" *Journal of International Humanitarian Action* 4 (1) pp.1–11 DOI: 10.1186/s41018-018-0048-1.

IEAG (2014) "A World That Counts: Mobilising the Data Revolution for Sustainable Development" (*Report*) Available at: https://www.undatarevolution.org/wp-content/uploads/2014/11/A-World-That-Counts.pdf (Accessed: 21st February 2023).

IFRC (2017) *IFRC Global Strategy on Migration 2018–2022* Geneva: International Federation of Red Cross and Red Crescent Societies.

IFRC (2018) *World Disasters Report 2018: Leaving No One Behind* Geneva: International Federation of Red Cross and Red Crescent Societies.

INSCT (2012) *Mapping the Libya Conflict* New York: Institute for National Security and Counterterrorism, Syracuse University.

IPCC (2012) Allen, S.K., Barros, V., Dokken, D.J., Ebi, K.L., Field, C.B., Mach, K.J., Mastrandrea, Midgley, P.M., M.D., Plattner, G.-K., Qin, D., Stocker, T.F. and Tignor, M. (Eds) *Managing the Risks of Extreme Events and Disasters to Advance Climate Change Adaptation Field, The Intergovernmental Panel on Climate Change*, Cambridge: Cambridge University Press.

Kathmandu Living Labs (KLL) (2020a) "About" Available at: http://www.kathmandulivinglabs.org/about (Accessed: 5th March 2020).

Kathmandu Living Labs (KLL) (2020b) "QuakeMap.org" Available at: http://www.kathmandulivinglabs.org/projects/quakemaporg (Accessed: 5th March 2020).

Koch (2015) *Disaster Medicine and Public Health Preparedness* Koch T. (2015) "Mapping Medical Disasters: Ebola Makes Old Lessons, New" *Disaster Medicine and Public Health Preparedness* 9 (1) pp.66-73 DOI: 10.1017/dmp.2015.14.

Lord, A. and Sodden, R. (2018) "Mapping Silences, Reconfiguring Loss: Practices of Damage Assessment & Repair in Post-earthquake Nepal" *ACM* 2 (161) pp.1–21 DOI: 10.1145/3274430.

Lüge, T. (2014) *GIS Support for the MSF Ebola Response in Guinea in 2014* Geneva: Médecins Sans Frontières.

Medair (2019) "Medair Health Project in Bekaa Valley – Lebanon, Health and Nutrition Knowledge, Practices and Coverage Household Survey Analysis 2019, Part 1 – Analysis Report" *Medair and Knowledge for Development* Available at: https://www.medair.org/wp-content/uploads/2019/07/Medair-KPC-2019-Report-Part-1-Final-1.pdf.

Medair (2020) "Innovating Through Technology: Improving the Way We Work to Serve People Better" *Medair* Available at: https://www.medair.org/medair/innovating-through-technology/ (Accessed: 2nd January 2020).

Meier, P. (2011a) "Crisis Mapping Libya: This is No Haiti" *iRevolutions* (4th March) Available at: https://irevolutions.org/2011/03/04/crisis-mapping-libya/ (Accessed: 10th May 2020).

Meier, P. (2011b) "Using a Map to Bear Witness in Egypt" *iRevolutions* (3rd February) Available at: http://irevolutions.org/2011/02/03/egypt-jan25-ushahidi/ (Accessed: 10th May 2020).

Meier, P. (2014) "Establishing Social Media Hashtag Standards for Disaster Response" *iRevolutions* (5th November) Available at: https://irevolutions.org/2014/11/05/social-media-hashtag-standards-disaster-response/ (Accessed: 17th February 2020).

Meier, P. (2015) *Digital Humanitarians: How BIG DATA Changes the Face of Humanitarian Response* Boca Raton, FL: CRC Press.

MicroMappers (2015) "Maps" Available at: https://micromappers.wordpress.com/maps/ (Accessed: 29th January 2020).

Milner, M. and Verity, A. (2013) "Collaborative Innovation in Humanitarian Affairs: Organization and Governance in the Era of Digital Humanitarianism" (*Report*) Available at: http://blog.verityth-ink.com (Accessed: 21st February 2023).

MissingMaps (2020) "About" Available at: https://www.missingmaps.org/about/ (Accessed: 23rd January 2020).

Nelson, A., Sigal, I. and Zambrano, D. (2010) *Media, Information Systems Communities: Lessons from Haiti* Unknown: Internews & CDAC Foundation.

NPC (2015) *Post Disaster Needs Assessment: Executive Summary* Kathmandu: Government of Nepal National Planning Commission.

OpenStreetMap Wiki (2019) "About OpenStreetMap" Available at: https://wiki.openstreetmap.org/wiki/About_OpenStreetMap (Accessed: 16th January 2020).

Parr, D.A. (2015) "Crisis Mapping and the Nepal Earthquake: The Impact of New Contributors" *Kartographische Nachrichten* 65 pp.151–155 DOI: 10.1007/BF03545120.

Phillips, J. and Verity, A. (2016) *Guidance for Developing a Local Digital Response Network* Geneva: DHNetwork

SBTF (2014) "SBTF Ebola Response" Available at: http://standbytaskforce.maps.arcgis.com/home/webmap/viewer.html?webmap=e7436b33968e49c4a8831a91ba6e9227 (Accessed: 5th May 2020).

SBTF (2015) "Nepal Earthquake Deployment-May 2015: After Action Review" Available at: https://www.standbytaskforce.org/wp-content/uploads/2015/06/newsletternepalearthquakeafteractionreport.pdf

Sewell, A. (2020) "How Aid Groups Map Refugee Camps that Officially Don't Exist" *Wired* (20th January) Available at: https://www.wired.com/story/aid-groups-map-refugee-camps-officially-dont-exist/ (Accessed: 29th May 2020).

Stottlemyre, S. and Stottlemyre, S. (2012) "Crisis Mapping Intelligence Information During the Libyan Civil War: An Exploratory Case Study" *Policy & Internet* 4 (3–4) pp.24–39.

Stauffacher, D., Hattotuwa, S. and Weekes, B. (2012) "The Potential and Challenges of Open Data for Crisis Information Management and Aid Efficiency: A Preliminary Assessment" *ICT4Peace Foundation* Available at: http://ict4peace.org/wp-content/uploads/2012/03/The-potential-and-challenges-of-open-data-for-crisis-information-management-and-aid-efficiency.pdf (Accessed: 31st August 2016).

Strochlic, N. (2019) "In Uganda, a Unique Urban Experiment is Under Way" *National Geographic* (April) Available at: https://www.nationalgeographic.com/magazine/2019/04/how-bidibidi-uganda-refugee-camp-became-city/ (Accessed: 21st February 2023).

UNFCC (2017) "Climate Change Is A Key Driver of Migration and Food Insecurity" *United Nations Framework Convention on Climate Change* (16th October) Available at: https://unfccc.int/news/climate-change-is-a-key-driver-of-migration-and-food-insecurity (Accessed: 29th April 2020).

UNHCR (2018) "Uganda Refugee Response Monitoring, Settlement Fact Sheet: Bidi Bidi" *United Nations High Commissioner for Refugees* (June) Available at: https://reliefweb.int/sites/reliefweb.int/files/resources/66780.pdf (Accessed: 21st February 2023).

UNHCR (2019a) "Figures at a Glance" *United Nations High Commissioner for Refugees* Available at: https://www.unhcr.org/en-au/figures-at-a-glance.html (Accessed: 10th April 2020).

UNHCR (2019b) "Uganda Country Refugee Response Plan Nairobi" *United Nations High Commissioner for Refugees Regional Refugee Coordination Office (RRC)* Available at: https://data2.unhcr.org/en/documents/download/69674 (Accessed: 21st February 2023).

UNHCR (2020) "Uganda Comprehensive Refugee Portal" *United Nations High Commissioner for Refugees & the Office of the Prime Minister Government of Uganda* Available at: https://data2.unhcr.org/en/country/uga (Accessed: 16th February 2020).

USAID (2017) *Fighting Ebola with Information: Learning from the Use of Data, Information, and Digital Technologies in the West Africa Ebola Outbreak Response* Washington, DC: US Agency for International Development.

Ushahidi (2020a) "About Ushahidi" Available at: https://www.ushahidi.com/about (Accessed: 17th March 2020).

Ushahidi (2020b) "QuakeMap" Available at: https://www.ushahidi.com/case-studies/quakemap (Accessed: 15th March 2020).

Verity, A. and Whipkey, K. (2015) *Guidance for Incorporating Big Data into Humanitarian Operations* Geneva: Digital Humanitarian Network.

Weinandy, T. (2016) "Volunteer and Technical Communities in Humanitarian Response" *UN Chronicle* 8 (1) Available at: https://www.un.org/en/chronicle/article/volunteer-and-technical-communities-humanitarian-response (Accessed: 21st February 2023).

WHO (2016) "Situation Report: Ebola Virus Disease" *World Health Organization* (10th June) Available at: https://apps.who.int/iris/bitstream/handle/10665/208883/ebolasitrep_10Jun2016_eng.pdf;jsessionid=925A4C4361088A3A2DFE5C14616D8C9B?sequence=1 (Accessed: 21st February 2023).

WHO (2018) "Ebola Virus Disease: Democratic Republic of the Congo" *World Health Organization (External Situation Report 01)* Available at: https://apps.who.int/iris/bitstream/handle/10665/273640/SITREP_EVD_DRC_20180807-eng.pdf?ua=1 (Accessed: 21st February 2023).

Wright, J. and Verity, A. (2020) "Artificial Intelligence Principles: for Vulnerable Populations in Humanitarian Contexts" (*Report*) *Digital Humanitarian Network* Available at: https://digitalhumanitarians.com/artificial-intelligence-principles-for-vulnerable-populations-in-humanitarian-contexts/ (Accessed: 21st February 2023)

Zak, D. (2011) "Mapping the Egypt Protests and Libya Crisis" *Ushahidi* (7th April) Available at: https://www.ushahidi.com/blog/2011/04/07/mapping-the-egypt-protests-and-libya-crisis (Accessed: 10th April 2020).

Ziemke, J. (2012) "Crisis Mapping: The Construction of a New Interdisciplinary Field?" *Journal of Map and Geography Libraries: Advances in Geospatial Information* 8 (2) pp.101–17.

38
HUMANITARIAN RELIEF AND GEOSPATIAL TECHNOLOGIES

John C. Kostelnick

Humanitarian relief and geospatial technologies

Each year, increasing numbers of people around the globe are at risk of natural disasters, human conflicts, and other types of humanitarian crises. In 2020, an estimated 168 million people were expected to require some form of humanitarian aid or assistance, with that number expected to rise to 212 million people by the year 2022 (UNOCHA, 2019). Although precise definitions vary, a humanitarian crisis generally may be defined as an event or events that significantly threaten human health, safety, and wellbeing (Humanitarian Coalition, 2015). Two types of general events may contribute to a humanitarian crisis: natural hazards (e.g., earthquake, flood, drought, typhoon) and human-made hazards (e.g., armed conflicts, transportation, or industrial accidents) (IFRC, 2020). A humanitarian crisis typically becomes more catastrophic as the number of vulnerable people impacted by the crisis increases and as the crisis spans a wider geographical extent. In addition to injury and loss of life, humanitarian crises may displace significant numbers of people from their homes as refugees or internally displaced persons (IDPs). Many humanitarian crises are consequences of sudden catastrophic events (such as an earthquake), but others develop more gradually until a 'tipping point' is reached (such as the gradual spread of a disease epidemic) (InterNews, 2014). Humanitarian response is defined as efforts to alleviate impacts of the crisis through coordination between humanitarian relief organizations, government agencies, and donors (InterNews, 2014).

As the number of people at risk to a humanitarian crisis has increased in recent years, geospatial technologies have emerged on the front lines to protect human lives. Reliable and accurate information, including locational information, is critical for staging an effective humanitarian response in the immediate aftermath of a crisis, and for adverting a crisis before it occurs. Geospatial technologies provide critical capabilities for capture, storage, analysis, and visualization of locational information throughout the life-cycle of a crisis. Before a crisis event, geospatial technologies such as geographical information systems (GIS) may be used to assess a region's vulnerability to a specific type of crisis, such as a flood or drought, as a way to plan for or mitigate potential impacts before disaster hits. After a crisis occurs, imagery captured from satellites or unmanned aerial vehicles (UAVs) may be used to assess damage caused by an earthquake that strikes a community. Although it is impossible to measure the overall positive impact, geospatial technologies have a critical, indispensable role for confronting humanitarian crises head on. In addition to safeguarding civilian lives impacted by a crisis, geospatial technologies also protect humanitarian relief workers, improve critical response times, and increase efficiency of cost, time, and labour to a crisis response.

This chapter provides an overview of commonly used geospatial technologies in humanitarian relief activities and examples of their use during times of crisis, including web/cloud mapping technologies, global navigation satellite systems (GNSS) and locational technologies, geographical information systems (GIS), remote sensing, social media, artificial intelligence, and crowdsourced mapping.

Web/cloud mapping technologies

Maps are indispensable for answering important geographical questions during a humanitarian crisis, such as 'where is damage from the typhoon most severe?' and 'where should relief supplies be routed to reach people in need?'. Maps may be used for a range of purposes and audiences in humanitarian relief activities. For example, relief workers use maps in the field to assess damage, to coordinate the delivery of relief supplies, to locate victims in need of assistance, and to monitor critical facilities and resources (e.g., hospitals, water, and food supplies). Maps may be produced for civilian audiences to alert them of active hazards in their communities. Finally, maps may also be produced for donors or government agencies to display the status or overall progress of humanitarian relief operations.

In the past, maps used by humanitarian relief workers were often printed and updated by hand in the field, but today these maps are stored 'in the cloud' commonly and streamed on demand as interactive maps over the Web to computers and mobile devices. Individual map layers stored in the cloud as Web mapping services (WMS) may be combined to create interactive maps. Web maps called mashups merge operational data, often collected by government or humanitarian relief organizations, with re-usable basemaps, such as a terrain maps, roads maps, or satellite images served on demand to map users. A specific genre of Web maps commonly called crisis maps are characterized by real-time damage reports and requests for assistance contributed by both crisis victims and volunteers displayed on a basemap (Liu and Palen, 2010).

Web maps have at least four notable advantages for humanitarian relief applications over their static counterparts. First, Web maps are easily distributed to wide audiences, including both humanitarian relief workers and civilian populations, which provides a common operational picture to the many organizations that often coordinate in response to a crisis. Second, Web maps may be updated frequently, thus providing 'real time' situational awareness as a crisis unfolds. Third, Web maps may seamlessly integrate a range of digital data types into a common display, including data collected in the field as well as social media content and crowdsourced information from volunteer mappers. Fourth, Web maps offer interactivity features that allow the map user to customize the content displayed on the map as well as the zoom level or map scale, which may be beneficial for providing both the macro and micro perspectives on the crisis. A potential limitation of Web maps is that they require a reliable Web connection, which may be problematic in a crisis area if Internet or cellular service is disrupted or otherwise unavailable.

Figure 38.1 is an example of a Web map dashboard by Johns Hopkins University that provides situational awareness of confirmed novel coronavirus (COVID-19) cases reported by 20th March 2020 as the virus spread globally following the first reported cases in China

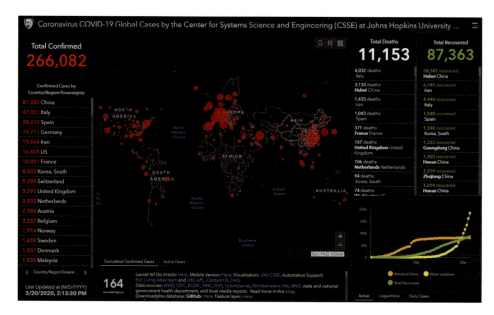

Figure 38.1 A Web map displaying real-time status of confirmed cases of the novel coronavirus (COVID-19) around the world as of 20th March 2020

Source: Johns Hopkins Center for Systems Science and Engineering, Copyright 2020, all rights reserved https://www.arcgis.com/apps/opsdashboard/index.html#/bda7594740fd40299423467b48e9ecf6.

in December 2019. The geographical distribution of the outbreak is displayed on the map, along with the total number of deaths and recoveries in the accompanying tables adjacent to the map. Charts in the lower right of the dashboard track the spread of the disease over time. The map is updated daily, as frequently as every 15 minutes in some parts of the world, by automated and manual methods as new cases are confirmed (Dong et al., 2020). The map may be used by the general public to track the overall spread of coronavirus over space and time, and may provide critical information to public health officials and government leaders in their efforts to contain the pandemic.

Another example of a Web map is the US Geological Survey's Earthquake Hazards map, which provides real-time status of earthquake activity around the world (Figure 38.2). Earthquakes displayed on the map may be filtered by magnitude, location, or time of occurrence. Seismic measurements are sent to the map automatically from sensors positioned at fixed stations around the world. When earthquake tremors are detected at a station, data is transmitted within minutes for display on the map through GeoJSON, a Web service format that is commonly used to update mashups in real time. Such geospatial technologies are an example of how networks of sensors may record measurements at locations and then feed automatically into Web maps for early warning detection systems that are critical for minimizing casualties in the immediate aftermath of a natural disaster.

Global navigation satellite systems (GNSS) and locational technologies

Location is critical during a humanitarian crisis: for victims, for relief supplies, and for the hazard itself. Locational technologies, such as global navigation satellite systems (GNSS)

Humanitarian geospatial technologies

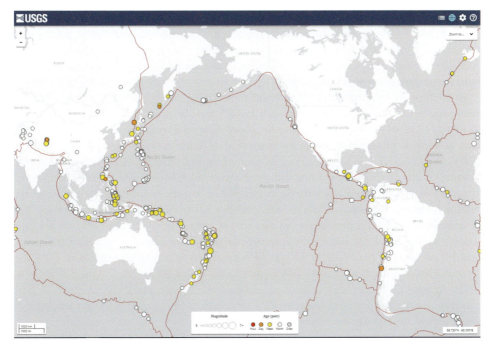

Figure 38.2 Earthquake Web map, centered on the Pacific tectonic plate, by the US Geological Survey (USGS). The map displayed was customized to display earthquakes of 4.5 magnitude or greater for the 30-day time period of 17th February to 17th March 2020

Source: US Geological Survey (USGS), https://earthquake.usgs.gov/earthquakes/map/.

(more commonly referred to as GPS, from the Global Positioning System devised and operated by the United States) that include GLONASS, developed by Russia, and Galileo, developed by the European Union, provide accurate ways to pinpoint precise locational positions on a map. Through automated computation of geographical coordinates using the principles of trilateration and triangulation, locational technologies allow humanitarian relief workers to improve the accuracy of their maps, provide a viable way to track the location of relief supplies in transit to a crisis scene, and allow victims to report their location as they seek assistance. Laser rangefinders may be utilized with satellite-based locational technologies to calculate positional offsets from a known location by calculating the azimuth and distance to a new location of interest that may be inaccessible due to safety concerns.

Humanitarian relief supplies and other assets moved to the scene of a crisis may be tracked and monitored by locational technologies such as GPS that are commonly used for navigation by airplanes, ships, and land vehicles. Coordinates from locational technologies may be displayed on Web maps to display real-time movement of assets during travel towards a destination in a crisis zone, and to estimate time and distance to destination.

Many mobile phones and other devices have built-in GPS and location-based services, which may be used to map and collect geospatial data at a crisis scene and then transmit the data to the cloud for storage or display on a Web map. As a practical matter, satellite-based locational technology has advantages over cellular location-based services which may be disrupted due to damaged or destroyed cellular towers during times of disaster. Photographs taken by phones commonly are geotagged by the internal GPS of the phone, where latitude/longitude coordinates are embedded in the image, and then may be mapped by location. Mobile phone users in peril during a crisis may have their location determined automatically through the location-based services of the phone, or may share their locations manually through SMS text message, which in both cases allows their location to be available to emergency responders.

In addition to locating crisis victims and monitoring the supply train of humanitarian relief supplies, locational technologies such as GPS may be used to improve the positional accuracy of hazardous features that are commonly mapped during humanitarian crises. One such example is minefield mapping by humanitarian mine action personnel as part of the global effort to clear land contaminated by landmines and other explosive remnants of war. In order to ensure civilians and demining personnel are safe, the perimeter of a suspected minefield must be mapped as accurately as possible during the technical survey phase of the landmine clearance process. A traditional approach has been to use a measuring tape and compass and to record azimuth and distance measurements by hand, but GPS mapping has become standard and can potentially improve the overall positional accuracy of the minefield perimeter on a map (GICHD, 2011). Positional error is possible with both GPS and non-GPS approaches, but accuracy may be improved with GPS especially when differential GPS is utilized and as the size of the minefield increases (Berger and Dunbar, 2007). A laser rangefinder may be employed with the GPS to calculate offsets that allows for mapping of the minefield at a safe distance away from its perimeter.

Geographical information systems (GIS)

Geographical information systems (GIS) provide a versatile role for storing, analysing, and visualizing geospatial data that are essential for preparing for and responding to a humanitarian crisis. A benefit of GIS is the capability to integrate several types of data sources commonly used during a crisis (basemaps, operational layers, field data, imagery, statistical data) into a single software platform. LandScan, a global population dataset for estimating populations at risk, is an example of a GIS dataset that is well-suited for humanitarian relief applications (Figure 38.3) (ORNL, 2019). LandScan is a compilation of best available population census data from every country that is disaggregated through a spatial model that incorporates variables such as elevation, slope, land cover, nighttime lights, roads, and urban areas to estimate where population is most likely to live within a country (Dobson et al., 2000).

Spatial analysis in GIS may uncover patterns in datasets and answer important questions during a crisis. For example, distance operations such as buffering may be used to define the perimeter of an active hazard zone or to specify an evacuation zone based on a specified distance from the hazard. Overlay analysis may be used to calculate the number of critical facilities (e.g., hospitals) inside of a hazard zone to assess overall impact. Network analysis may be utilized to determine the optimal route for transporting relief supplies along a road network in the quickest manner, while considering obstacles along the road (e.g., fallen trees) that may hinder travel. Multi-criteria analysis may be used to combine variables into an overall risk assessment, such as an analysis of land cover, slope, and wind direction/speed to determine risk for wildfire. Maps, charts, and other visualizations are common products derived from GIS analyses, and may be used to inform decision-makers as well as to communicate to general audiences.

To demonstrate the use of GIS during a humanitarian crisis situation, imagine a hypothetical earthquake event that strikes a densely populated urban area within a country. First,

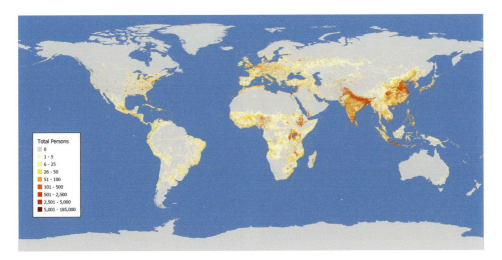

Figure 38.3 Landscan population dataset for estimating populations at risk during humanitarian crises. Population estimates are displayed by thirty arc-second raster cells, approximately 1-km resolution

Source: LandScan 2018™, ORNL, UT-Battelle, LLC.

a buffer analysis may be performed to define a potential impact zone of the earthquake, centered on the epicentre of the earthquake and extended to a distance in proportion to the earthquake's magnitude. Next, an overlay analysis may be performed with the buffered area and a population dataset (such as LandScan) to estimate population residing inside the impact zone. The population inside of the impact zone could be summarized by the GIS to derive a total estimate of the population that may be in need of assistance. Such an analysis may be performed as part of a hypothetical crisis simulation to inform disaster planning, or once an actual crisis has occurred as an initial estimate of impact. GIS analyses for 'what if' scenarios may be used to assess vulnerability and impact for many types of hazards and risks, such as flooding, drought, or the spread of a contaminant.

In some cases, humanitarian relief organizations may develop customized GIS platforms for specialized tasks. The Information Management System for Mine Action (IMSMA) developed by the Geneva International Centre for Humanitarian Demining (GICHD) is one such example (GICHD, 2019).[1] IMSMA offers versatile functionality, such as mobile data collection of reported landmine incidents in a community, hazard mapping during technical surveys, site suitability analyses to determine clearance priorities, and visualization of hazard reduction over time as minefields are cleared and land is released back to civilians.

Remote sensing

Remotely sensed imagery that is collected from sensors fixed to satellites, airplanes, UAVs, and other platforms is another important geospatial technology commonly deployed in humanitarian applications. Satellite or aerial imagery is particularly beneficial for assessing the overall extent of a disaster's impact, particularly areas that are inaccessible due to damage. Satellite and aerial imagery provide an important source of information for damage assessment in the immediate aftermath of a disaster, and allows for change detection by comparison of pre- and post-disaster images (Figure 38.4). Satellite imagery with a spatial resolution

Figure 38.4 Satellite images of San Juan, Puerto Rico before (left) and after (right) Hurricane Maria in September 2017. Flooding and damage to building rooftops and trees are evident in the post-disaster image

Source: Digital Globe Open Data, https://www.digitalglobe.com/ecosystem/open-data/hurricane-maria, Creative Commons Attribution Non-Commercial 4.0, https://creativecommons.org/licenses/by-nc/4.0/legalcode.

as fine as 30 cm is available from commercial satellites today, which provides ample detail to detect damage to infrastructure following a disaster, or to monitor ethnic cleansing and human rights violations (Crampton, 2009). Many commercial satellites also are configured with sensors that may be tasked over a specific geographical area for on-demand image acquisition in the immediate aftermath of a disaster. Sensors on many satellites are able to image in both visible and non-visible bands of the electromagnetic spectrum. Thermal imagery can detect heat from wildfires, and near-infrared images may monitor vegetation condition during times of drought. In addition to post-disaster assessment, remotely sensed imagery may be used to mitigate an impending crisis, such as the use of imagery for drought monitoring to predict future food or freshwater shortages.

Unmanned aerial vehicles provide additional flexibility for remotely sensed image acquisition. Given their small size and portability, UAVs affixed with cameras and sensors may be launched virtually anywhere for quick, on-demand acquisition of high-resolution imagery in the aftermath of a disaster. UAVs are relatively inexpensive options compared to other sources of commercially available satellite and aerial imagery, and also offer flexibility for acquiring imagery as frequently as needed. UAVs often are preprogrammed with specific route plans and then navigate automatically to their destinations and return by GPS guidance. UAVs have been operationalized in many areas of humanitarian response, including humanitarian mine action where acquired imagery has been used for diverse purposes such as identifying patterns of landmines in the ground, monitoring progress of demolition operations, and showing donors the impacts of landmine clearance to a community (Cruz *et al.*, 2018).

Light detection and ranging (Lidar) remote sensing is another type of geospatial technology for humanitarian applications that is beneficial for creating high-resolution 3D models of the Earth's surface. Lidar is an active remote sensing system with sensors that generate pulses of infrared light that reflect off objects on the Earth's surface and then return to the sensor for measurement. 'First returns' are the reflected pulses that interact with the tallest objects on the Earth's surface (e.g., buildings, trees) and 'last returns' are pulses that penetrate to the 'bare' Earth and are useful for generating surface digital elevation models (DEMs) that may be used in GIS-based models to assess risk for hazards such as floods or landslides. Lidar data collected in the aftermath of a disaster may be used to create 3D maps of collapsed buildings for damage assessment, as well as area and volume estimates of damage to individual buildings by comparing pre- and post-disaster Lidar datasets.

Social media

Social media (e.g., Twitter, Instagram, Facebook) has become a powerful platform for the creation and dissemination of massive amounts of humanitarian-related information that may reach wide audiences instantaneously. In early 2023, there were 556 million active Twitter users, 2 billion Instagram users, and over almost 3 billion Facebook users (Statista, 2023). Many social media platforms allow a user to geotag a location with their post, which transforms the content into a coordinate that may be mapped as a point to provide locational context to content contained in the post. For example, Twitter users have the option to geotag tweets, which adds a location to the content of the tweet. Twitter users generate an estimated 500 million total tweets each day, although only a fraction (~3%) of tweets are geoenabled or otherwise not publicly available for mapping due to user privacy settings (Meier, 2015). These geotweets may be imported into mashups and other custom Web applications using Application Programming Interfaces (API), where content may be filtered and displayed by geographic location, time of the tweet, or content included in the tweet, such as a keyword (e.g., 'tsunami') or hashtag (e.g., #coronavirus). Many Web mapping dashboards for crisis management include automated feeds from geo-enabled social media content, which allows emergency managers to monitor general trends as well as to respond to alerts for specific incidents. Social media content may be helpful for contextualizing a crisis to emergency responders and government officials, especially in the early stages when information may be scarce or incomplete. Important challenges remain, however, for the use of social media as a source of information in crisis response. First, social media content has varying levels of accuracy and completeness, and evaluating data quality to filter out misinformation from quality information is an important challenge (Goodchild and Glennon, 2010). Second, the mere volume of information available from social media creates challenges for sorting through content to extract important trends and patterns during times of crisis.

Artificial intelligence

Big data, as a result of technological innovations such as automated sensors, mobile phones, and social media, are increasingly available today for humanitarian response. A key challenge is to develop methods for analysing big data in order to extract and visualize meaningful insights and patterns into humanitarian workflows (Burns, 2018). Artificial intelligence (AI) and machine learning algorithms are sophisticated methods for uncovering hidden patterns in data that may be translated into important information that is beneficial during a humanitarian crisis. For example, artificial intelligence has been applied to humanitarian relief activities such as analysing social media content, and also for training computers to extract geographical features (e.g., damaged buildings, roads, and other infrastructure in the aftermath of an earthquake) automatically from aerial or satellite imagery. Key advantages of artificial intelligence applications in crisis response are the dramatic decrease in human labour requirements due to automation as well as improved mapping accuracy afforded by sophisticated machine learning algorithms. Artificial intelligence continues to emerge as a promising area for analysis of big data in humanitarian relief.

Crowdsourcing

The rise of Web 2.0 in the early 2000s paved the way for crowdsourcing, a phenomenon that has had a significant impact on humanitarian response. Web 2.0 developments ushered in a new era for the Internet, which shifted from a top-down, one-way stream of information for user consumption to a two-way, bottom-up platform where users can generate and share their own content (e.g., pictures, text, videos) with other users through social media and other Web platforms. Goodchild (2007) coined the term Volunteered Geographic Information (VGI) to refer to crowdsourced information that includes a locational component. Volunteers, often working remotely in coordination with other volunteers, are essential to the creation of VGI maps and geospatial data such as Humanitarian OpenStreetMap. The 2010 earthquake that devasted Haiti was a pivotal moment that demonstrated the life-saving capabilities of crowdsourced information through the dedicated efforts of hundreds of volunteer mappers working remotely (Harvard Humanitarian Initiative, 2011). Locations of victims in need of assistance were plotted by volunteers using the Ushahidi platform, while other volunteers collaborated to create an updated map of Haiti for OpenStreetMap to support the relief workers on the ground. The term 'digital humanitarianism' (Meier, 2015) has been coined in the aftermath of the Haiti earthquake to describe the rise in volunteers who have embraced crowdsourcing and other digital technologies for humanitarian response.

Since the Haiti earthquake, organized groups of digital humanitarian mappers have deployed to assist during other times of crisis. MapGive, sponsored by the US Department of State, is one such initiative. MapGive organizes high priority areas in need of better maps and geographical data to support humanitarian relief missions. From the MapGive website (https://mapgive.state.gov/), volunteers first select a project, then check out a mapping tile for a region from the Humanitarian OpenStreetMap Team Tasking Manager website, and finally digitize or trace features of interest such as buildings or roads using satellite imagery as a backdrop before checking in the tile upon completion (Figure 38.5). The newly mapped tile is validated by an experienced mapper and finally added to the Humanitarian OpenStreetMap basemap for use by crisis responders. The Standby Task Force is another organization of volunteer mappers that is mobilized during times of crisis to extract information from social media platforms such as Twitter, and to assist with various crowdsourced mapping and satellite image interpretation tasks. TomNod, a crowdsourcing initiative that was recently retired, brought together volunteers who mapped features from high-resolution satellite imagery during humanitarian crises such as the search for missing Malaysian Airlines Flight 370 in 2014. MapGive, Standby Task Force, and TomNod have all made it possible to engage masses of volunteers, many working remotely far away from the scene of a crisis, as digital humanitarians by providing opportunities for them to create geospatial data through a 'divide and conquer' group effort.

Ethical issues

Although geospatial technologies clearly have a significant impact on protecting lives during a humanitarian crisis, it is important to consider several ethical issues related to their implementation to prevent misuse, or even outright abuse. Concerns over privacy related to geospatial technologies such as GPS or UAVs, for example, may be extended to extreme

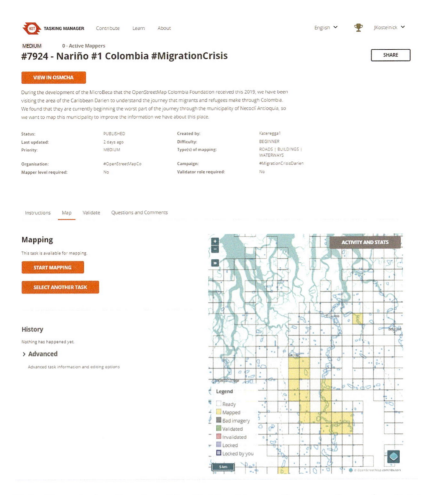

Figure 38.5 Humanitarian OpenStreetMap Tasking Manager for a crowdsourced project to map roads, buildings, and waterways in response to the Colombian refugee crisis
Source: Humanitarian OpenStreetMap Team, https://tasks.hotosm.org/project/7924?task=225.

examples, such as the use of locational technologies to track or even coerce individuals through 'geoslavery' (Dobson and Fisher, 2003). As such, a paradox exists: the same technologies used to alleviate a humanitarian crisis also may be abused as the catalyst for a crisis in the first place. Although extreme examples such as these may be easy to identify, other ethical issues are more nuanced, and may be overlooked by even the most well-intentioned volunteers or professionals. For example, volunteer mappers with limited experience in geospatial data creation or unfamiliarity with a foreign location on the other side of the world may unknowingly contribute data of questionable accuracy during a crowdsourced mapping activity in response to a humanitarian crisis.

Likewise, the type of visualization methods selected by a cartographer or geospatial analyst may evoke unintended reactions by an audience. For example, a visualization with a high degree of realism (e.g., 3D, virtual reality) or the use of icons in pictorial map symbology may be beneficial for conveying the dangers associated with an actual hazard or crisis to an audience, but also has the potential to sensationalize maps that forecast hazard risk especially in the absence of any portrayal of uncertainties in a simulation such as a flood prediction model (Kostelnick et al., 2013). More generally, Web maps, mashups, and social media may communicate very realistic and evocative images of human suffering in places both near and far away from the observer, which may create dilemmas as to the intended response that is evoked from the use of such visualizations. In one way, interactive Web maps of a crisis may generate feelings of empathy by those unaffected, and perhaps prompt action through financial or volunteer support for an impacted area. As such, the map interface may invoke transparency in the 'knowledge politics' to recruit donors for crisis relief efforts (Elwood and Leszczynski, 2013). Alternatively, data-driven Web maps may leave users with an overly abstract and sanitized view of a hazard (Kostelnick and Kostelnick, 2016), while mashups that incorporate images of a crisis may fall short of their intended purpose by desensitizing viewers and limiting communication of the broader complexities of the crisis (Parks, 2009). Ethical issues such as these require careful consideration given the high stakes involved with humanitarian response.

Summary

The integration of geospatial technologies into many areas of humanitarian response has expanded greatly in recent years, and promises to continue into the future as new technologies are developed. Geospatial technologies have impacted humanitarian response by providing new sources of data, expanding mobile data collection in disaster-affected areas, involving more people in data creation, developing new methods for data analysis, and providing new ways for visualizing geographical data for responders, decision-makers, and the public alike. Important challenges remain, though, including privacy issues as technologies such as GPS and UAVs become more pervasive in society as well as other ethical issues. In addition, there remains a 'digital divide' between those disaster-impacted people who have access to such technologies, and those who do not due to social inequalities (Madianou, 2015). Geospatial technologies will continue to provide an important role for adverting crisis before it happens, and for protecting lives for when disaster occurs.

Note

1 See https://www.gichd.org/en/imsma-core/#!g=1&slide=0 for more detailed information and examples of how IMSMA is used to support humanitarian demining.

References

Berger, S. and Dunbar, M. (2007) "The Accuracy of Measuring Perimeter Points: Use of GPS vs. Bearing and Distance" *Geneva International Centre for Humanitarian Demining (GICHD)* Available at: https://www.gichd.org/fileadmin/GICHD-resources/rec-documents/Accuracy_of_Measuring_Perimeter_Points_Berger_2006.pdf (Accessed: 16th March 2020).

Burns, R. (2018) "Synthesizing Geoweb and Digital Humanitarian Research" In Thatcher, R., Eckert, J. and Shears, A. (Eds) *Thinking Big Data in Geography: New Regimes, New Research* Lincoln: University of Nebraska, pp.214–228.

Crampton, J. (2009) "Cartography: Maps 2.0" *Progress in Human Geography* 33 (1) pp.91–100 DOI: 10.1177/0309132508094074.

Cruz, I., Jaupi, L., Sequesseque, S.K.N. and Cottray, O. (2018) "Enhancing Mine Action in Angola with High-resolution UAS IM" *The Journal of Conventional Weapons Destruction* 22 (3) 5 Available at: https://commons.lib.jmu.edu/cisr-journal/vol22/iss3/5 (Accessed: 18th March 2020).

Dobson, J.E., Bright, E.A., Coleman, P.R., Durfee, R.C. and Worley, B.A. (2000) "LandScan: A Global Population Database for Estimating Populations at Risk" *Photogrammetric Engineering and Remote Sensing* 66 pp.849–857.

Dobson, J.E. and Fisher, P.F. (2003) "Geoslavery" *IEEE Technology and Society Magazine* 22 pp.47–52.

Dong, E., Hongru, D. and Gardner, L. (2020) "An Interactive Web-based Dashboard to Track COVID-19 in Real Time" *The Lancet Infectious Diseases* 20 pp.533–534 DOI: 10.1016/S1473-3099(20)30120-1.

Elwood, S. and A. Leszczynski. (2013) "New Spatial Media, New Knowledge Politics" *Transactions of the Institute of British Geographers* 38 (4) pp.544–549 DOI: 10.1111/j.1475–5661.2012.00543.x.

Geneva International Centre for Humanitarian Demining (GICHD) (2011) "A Guide to Land Release: Technical Methods" Available at: https://www.gichd.org/fileadmin/GICHD-resources/rec-documents/Land-Release-Tech-Methods-Apr2011.pdf (Accessed: 20th March 2020).

Geneva International Centre for Humanitarian Demining (GICHD) (2019) "Information Management System for Mine Action (IMSMA)" Available at: https://www.gichd.org/en/imsma-core/ (Accessed: 19th March 2020).

Goodchild, M.F. (2007) "Citizens as Sensors: The World of Volunteered Geography" *GeoJournal* 69 pp.211–221.

Goodchild, M.F. and Glennon, J.A. (2010) "Crowdsourcing Geographic Information for Disaster Response: A Research Frontier" *International Journal of Digital Earth* 3 (3) pp.231–241 DOI: 10.1080/17538941003759255.

Harvard Humanitarian Initiative (2011) *Disaster Relief 2.0: The Future of Information Sharing in Humanitarian Emergencies* Washington, DC and London: UN Foundation & Vodafone Foundation Technology Partnership.

Humanitarian Coalition (2015) "What is a Humanitarian Emergency" Available at: https://www.humanitariancoalition.ca/info-portal/factsheets/what-is-a-humanitarian-crisis (Accessed: 9th March 2020).

International Federation of Red Cross and Red Crescent Societies (IFRC) (2020) "Types of Disasters: Definition of a Hazard" Available at: https://www.ifrc.org/en/what-we-do/disaster-management/about-disasters/definition-of-hazard/ (Accessed: 9th March 2020).

Internews (2014) "A Manual for Trainers & Journalists and an Introduction for Humanitarian Workers" Available at: https://reliefweb.int/sites/reliefweb.int/files/resources/IN140220_HumanitarianReportingHANDOUTS_WEB.pdf (Accessed: 9th March 2020).

Kostelnick, C.J. and Kostelnick, J.C. (2016) "Online Visualizations of Natural Disasters and Nazards: The Rhetorical Dynamics of Charting Risk" In Gross, A.G. and Buehl, J. (Eds) *Science and the Internet: Communicating Knowledge in a Digital Age* Amityville, NY: Baywood Press, pp.157–190.

Kostelnick, J.C., McDermott, D., Rowley, R.J. and Bunnyfield, N. (2013) "A Cartographic Framework for Visualizing Risk" *Cartographica* 48 (3) pp.200–224 DOI: 10.3138/carto.48.3.1531.

Liu, S.B. and Palen, L. (2010) "The New Cartographers: Crisis Map Mashups and the Emergence of Neogeographic Practice" *Cartography and Geographic Information Science* 37 (1) pp.69–90.

Madianou, M. (2015) "Digital Inequality and Second-order Disasters: Social Media in the Typhoon Haiyan Recovery" *Social Media + Society* (July–December) pp.1–11 DOI: 10.1177/2056305115603386.

Meier, P. (2015) *Digital Humanitarians: How Big Data is Changing the Face of Humanitarian Response* Boca Raton, FL: CRC Press.

Oak Ridge National Laboratory (ORNL) (2019) "LandScan 2018" Available at: https://landscan.ornl.gov/ (Accessed: 19th March 2020).

Parks, L. (2009) "Digging into Google Earth: An Analysis of 'Crisis in Darfur'" *GeoForum* 40 pp.535–545 DOI: 10.1016/j.geoforum.2009.04.004.

Statista (2023) "Most Popular Social Networks Worldwide as of January 2023, Ranked by Number of Active Users" Available at: https://www.statista.com/statistics/272014/global-social-networks-ranked-by-number-of-users/ (Accessed: 21st February 2023).

United Nations Office for the Coordination of Humanitarian Affairs (UNOCHA) (2019) "Global Humanitarian Overview 2020" Available at: https://www.unocha.org/sites/unocha/files/GHO2020_v9.1.pdf (Accessed: 9th March 2020).

39
GEOSPATIAL TECHNOLOGY AND THE SUSTAINABLE DEVELOPMENT GOALS (SDGs)

Doug Specht

The Sustainable Development Goals, widely known as the SDGs, were formally adopted by the United Nations (UN) in September 2015. Recognizing the failures of their predecessors, the Millennium Development Goals (MDGs), the SDGs, through 17 development objectives aim to transform the lives of everyone on the planet by 2030 (Holloway *et al.*, 2018; Solís *et al.*, 2018; Xiao *et al.*, 2018). The notion of sustainability can be traced back to the definition given by the then Norwegian Prime Minster, Gro Harlem Brundtland, who, in a 1987 report entitled 'Our Common Future', expressed concern around the limitations of the planet's resources and their unequal distribution. However, over time, the concept of sustainable development came to mean many different things to different groups of people leading to disjointed efforts to achieve sustainable development and end poverty. The 2030 agenda, then, is designed to unite countries under a single set of guiding principles that will lead to transformations in economic, social and environmental well-being (Scott and Rajabifard, 2017).

Recognizing that previous attempts have failed due to a lack of monitoring and tracking processes, the 17 SDGs are further broken down into 169 smaller targets within a global indicator framework (*ibid.*). This framework means that the SDGs are expected to do better than the MDGs, an outcome that is seen as ever more crucial – some 400 million people remain trapped in extreme poverty (less than $1.90 a day) and 1.2 billion in low-income status (less than $3.20 a day) (ESCAP, 2018); the global climate is changing rapidly, mainly due to increasing atmospheric concentrations of carbon dioxide (CO_2) and methane (CH_4), and is creating one the greatest global challenges faced by society, putting billions of people at risk if the SDGs are not met (Giuliani *et al.*, 2017; Jia *et al.*, 2020). Additionally, by 2050 over 60% of the world's population is expected to be living in urban areas (ESCAP, 2018), which adds further concerns around meeting the SDGs.

Aims of the SDGs

The overarching aims of the SDGs are captured in the title of the 2014 plan that brought them to life; 'Transforming our World'. The inclusion of sustainability in their name also alludes to a shift towards the qualitative and quantitative continuity of resources, and a state of equilibrium between humans and the planet and an interconnectivity between each of the goals, something that was missing from the MDGs (Acharya and Lee, 2019; Choi et al., 2016). A further change in the establishment of the SDGs was the call for a globally connected, and data driven, effort to achieve these aims, combining local monitoring with international data collection (Moomen et al., 2019; Sarvajayakesavalu, 2015; Solís et al., 2018). The SDGs then emphasize the importance of technological support and digital infrastructures, especially in developing countries, with digital tools and data being seen as crucial in the delivering of their success (Yanga and Rajabifard, 2019).

It has been widely acknowledged that success will only be achieved through national statistical information within countries, requiring vast improvements in data collection, modelling and linking (Choi et al., 2016; Scott and Rajabifard, 2017). This will need to include the integration of different datasets on physical, chemical, biological and socio-economic systems (Giuliani et al., 2017). While these aims present a great many challenges, it has become increasingly clear that geographic information (GI) and monitoring tools do, and will, play a significant role in achieving these aims; indeed, many of the goals and targets of the SDGs can be analysed, mapped, discussed and/or modelled within a geographic context (UN-GGIM, 2012). This notion had been put forward as early as 1998, when Al Gore presented his 'Digital Earth' model, and at the 2002 World Summit on Sustainable Development in Johannesburg, South Africa, which also called for the wider use of Earth observation and GIS technologies in development (Scott and Rajabifard, 2017). Yet in 2012, the UN Secretary General noted that the integration of geospatial information was still being underutilized in better decision making (United Nations, 2012).

Following mapping exercises of geospatial data sources and the SDGs performed by UN-GGIM (see Figure 39.1), it is now abundantly clear just how pivotal the role of GI and geospatial technologies will be in supporting sustainable development. GI has been shown to be able to provide data on water cycles, air quality, forests, land use patterns, crop yields (ESCAP, 2018); predicting and measuring poverty on a global scale (Acharya and Lee, 2019); trace outbreaks of illnesses such as malaria (Solís et al., 2018); track forced labour and modern slavery (Acharya and Lee, 2019); monitor and protect cultural heritage (Xiao et al., 2018); and much more besides.

Using geospatial technology to support the SDGs

Before examining the specifics of the technology used in the monitoring and promoting of the SDGs, it is worth first looking more closely at what aspects of the SDGs can be supported by geospatial interventions. It is estimated that 20% of indicators can be directly measured with Geographic Information and many more indirectly (Acharya and Lee, 2019; Arnold et al., 2019), UN-GGIM have also defined fourteen Global Fundamental Geospatial Data themes[1] that support SDG indicator monitoring and this may continue to grow as data becomes better integrated (Scott and Rajabifard, 2017).

At the most basic level, topographical maps, digital elevation models and satellite imagery can offer important data on land use and migration, both key information in understanding what is needed to achieve the SDGs (Arnold et al., 2019). With 514 Earth observation satellites launched by 2011, and 200 more planned before 2030, huge amounts of data can now be collected about the Earth's surface (Jia et al., 2020). Coupled with machine learning methods, data from these satellites have become essential in understanding the ecology, hydrology, geological and physical characteristics of the Earth in a multitemporal manner that is key to the monitoring of the SDGs and the conceptualizing of complex problems involved in reaching them (Acharya and Lee, 2019; Arnold et al, 2019; Jia et al., 2020; Solís et al., 2018).

Doug Specht

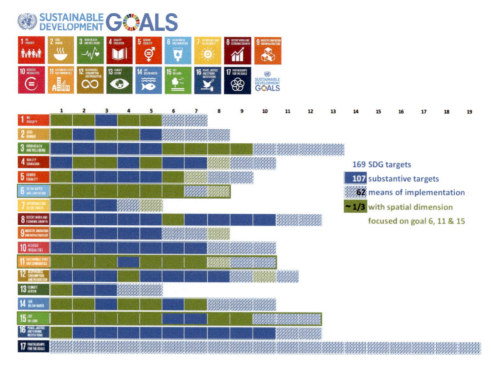

Figure 39.1 Sustainable Development Goals measurable through geographic information. Based upon the work of GEO, UN-GGIM and Ordnance Survey

Earth observation is perhaps the most widely used geospatial technology in relation to the SDGs. The UN Task Team on Satellite Imagery and Geospatial Data called for further use of these tools in their 2017 report (United Nations, 2017), and the Group on Earth Observation (GEO) have identified numerous indicators that can be measured this way (GEO, 2016). For example, knowledge about Land Cover is essential to several SDG indicators, including indicator 11.3.1,[2] 6.6.1[3] and 15.3.1[4] (Arnold et al., 2019). Observations of water turbidity can help identify water quality, supporting SDG6[5] and indicator 6.3.2[6] (Holloway et al., 2018). Space applications can also support SDG7[7] through the mapping of potential renewable energy sources and the tracking of biomass stocks or untapped energy resources (ESCAP, 2018). In addition to satellites, unmanned aerial vehicles (UAVs) and drones can add additional layers of data through remote sensing and on-demand high resolution imagery, further adding to the potential of Earth observation data in support of the SDGs (ESCAP, 2018).

Not all the technology being used is high-end though, citizen science – scientific work carried out by members of the public – and volunteers are also contributing large amounts of geographic information to help in the fulfilment of the SDGs. Vast amounts of data are now being generated and shared through online platforms by volunteers and humanitarian or crisis mappers around the world (Solís et al., 2018; Hunt and Specht, 2019). These data, collected through participatory GIS projects or through Volunteered Geographic Information (VGI), have great potential to inform the SDGs, as well as to increase the visibility of the goals to a wider population (Arnold et al., 2019; Solís et al., 2018).

Bringing all these data, as well as from other sources, together requires a herculean effort, with numerous frameworks and organizations being established to aid in the linking of geospatial Data and the SDGs. UN-GGIM, the Global Geospatial Information Management Group of the UN Statistics Division, is key in the development of these frameworks, serving as a body for global policy making (ESCAP, 2018). The European Union also developed the INSPIRE Directive to enhance the EUs Spatial Data Infrastructure (European Commission, 2007). The Open Geospatial Consortium (OGC) and the International Organization for Standardization (ISO)/Technical Committee (TC) 211 Geographic Information are also leading on developing open standards allowing for the better integration of data and improved interoperability (Giuliani et al., 2017). These all come together to help establish a Spatial Data Infrastructure, an idea first posited in 1993 by the US National Research Council, but which is now being realized in service of the SDGs (Arnold et al., 2019; Giuliani et al., 2017).

Geospatial data, then, can be used in a wide variety of ways to support the SDGs as well as local development and well-being. Governments around the world are already, or planning to, increase their ability to collect and analyse geographic data (ESCAP, 2018; Giuliani et al., 2017). The future will also see Artificial Intelligence, Digital Twins, Virtual Reality, the Internet of Things and humans as sensors, playing an even greater role in the collection and analysis of these data, allowing for further advances in sustainable development and regional decision-making support systems (Acharya and Lee, 2019). The pace at which geographic technology is changing is huge, and there is now a wide range of both new and traditional technologies being used in support of the SDGs.

Applying geospatial technology to implement and monitor the SDGs

Geospatial technology is any technology that incorporates spatial information (Acharya and Lee, 2019), and geospatial data describes the location and relationship of the features or phenomena on or above the Earth's surface (Arnold et al., 2019). One of the most widely used technology groups in supporting the SDGs is satellites and the numerous types of imaging they produce, including Land Cover Classification (Holloway et al., 2018; Arnold et al., 2019); Global surface temperature data from the AVHRR thermal infrared band (4,5) (Jia et al., 2020); crop monitoring via GEOGLAM10 (Giuliani et al., 2017); weather and climate via JAXA (ESAP, 2018); settlements, via the Global Urban Footprint (GUF) project of the German Earth Observation Centre (Arnold et al., 2019); and many more. This part of this chapter will examine some of these technologies, and others, in more detail, before exploring some of the advantages and concerns around these data-driven solutions.

Land cover classification mapping is a hugely important part of the monitoring the SDGs, as discussed above there are numerous Goals and Indicators that can be supported through this work. Observations of the Earth's land use have been collected since 1972 following the launch of the first Landsat satellites. Starting with the collection of 60–80 m spatial resolution and four spectral bands through Landsat-MSS, Landsat-TM later brought 30 m resolution and seven spectral bands in the 1980s (Acharya and Lee, 2019).

As the importance of this data has grown, it has led to many projects and companies working to collect such data. These range from the broader use of things such as the Google Earth Engine, which has developed a number of tools, including Trends Earth[8] to the more specialist high-spatial-resolution satellites such as Sentinel, SPOT, RapidEye, ALOS, Worldview, GeoEye, KompSat, SkySat, TripleSat and Pléiades (Acharya and Lee, 2019). Some of the more recent projects include the Copernicus Global Land Service – LandCover10010 – which was established by the European Commission and has been collecting detailed land use data since 2015 (Arnold *et al.*, 2019). China's GlobcLand30, developed by the National Center of China, now also provides open-access 30 m resolution global Land Cover data. There are also specialist sensors and arrays used for picking out particular characteristics of land use. The P-band Synthetic Aperture Radar (SAR), for example, is able to penetrate cloud cover and vegetation to examine landcover below the treeline, which is particularly useful in tropical and northern forested regions (Jia *et al.*, 2020). Additions to this technology, including the Advanced-SAR and the phased array L-Band SAR (PALSAR), are used for the specific examination of agriculture, forestry and hydrology (*ibid.*).

Earth observation techniques also allow us to better monitor climate and weather patterns. The European funded PRUDENCE and ENSEMBLES projects are designed to provide high-resolution regional climate predictions, feeding into the European Climate Adaptation Platform (CLIMATEADAPT) run by the European Commission and the European Environment Agency, with the aim of helping to develop strategies for mitigating and adapting to climate change (Giuliani *et al.*, 2017). The Japan Aerospace Exploration Agency (JAXA) have similarly launched projects to monitor climate and weather for adaptation and mitigation purposes. In 2015 they launched Himawari-8, a weather satellite covering the Asia-Pacific region, which feeds into their National Meteorological and Hydrological Services providing real-time rainfall data within the region. In 2018 the project was extended and now covers the world (ESCAP, 2018). Married with projects such as the Global Precipitation Measurement (GPM) project, which consists of six satellites that can observe precipitation to a radius of 5 km, and which observes 90% of the globes surface every three hours, it is now possible to have extremely accurate information (Jia *et al.*, 2020). There are of course further projects. The Suomi-NPP Visible Infrared Imaging Radiometer Suite (VIIRS) provides a range of measurements for climate studies (Acharya and Lee, 2019). AVHRR, MODIS and MERIS all also provide important data on rainfall, clouds and pollutants leading to a clearer understanding of climate (Jia *et al.*, 2020).

Oceans and other water bodies too can be monitored through geospatial technologies, and primarily Earth observation data. In 2018, China, in partnership with France, launched CFOSAT (China France Oceanographic Satellite). Fitted with two radars, the SWIM Spectometer and the SCAT Scattermeter can for the first time simultaneously measure wind, waves and other ocean data, providing oceanographers with unprecedented data allowing for the monitoring of the marine environment, disaster management and risk reduction (ESCAP, 2018). Similar advances have been made in clean water and sanitation management, in which remote sensing techniques have been applied to surface water mapping to provide high quality, and long-term monitoring of water quality parameters. Special projects such as the Sentinal-1 Program for Water Management in Low-income Countries have proved to be especially effective in supporting SDG-6 (Acharya and Lee, 2019).

The urban environment is one of the central concerns of the SDGs, and this too can be monitored and managed through geographical information. The smart cities agenda is one such way, and is dealt with separately in this book, but other projects also feed data into the sustainable development of cities. The Global Human Settlement Layer (GHSL) – developed by the Earth Observation Center at the German Aerospace Center – maps settlement patterns worldwide at a scale of ~12 m (Arnold *et al.*, 2019). The project has built up urban development reference points from 1975, 1990, 2000 and 2014, allowing for a better understanding of urban development. Combining satellites with ground sensors, including human sensors, and RS imagery allows for the complex world of cities to be carefully monitored for their sustainability (Acharya and Lee, 2019).

It is clear then that a huge range of geospatial technologies are being used in relation to the SDGs. While many of these are Earth-observation-based – a reflection of the global ambition of the SDGs – more traditional methods are also used such as TIGER (Topographical Integrated Geographic Encoding and Referencing), which brings together advanced topographical maps in the United States, including geology, buildings, water and administrative boundaries (Acharya and Lee, 2019). Since the 1970s, GIS and remote sensing have allowed us to collect a large amount of data about the Earth, but now as (global positioning systems) become ubiquitous to everyday life, individuals are also becoming sensors and producing vast amounts of data about our planet that can feed into the SDGs (Sarvajayakesavalu, 2015; Van Genderen, 2017; Xiao et al., 2018). As more people around the world gain access to smartphones, the idea of citizens as sensors will ultimately increase.

It is now clear that these technologies have already had a profound effect on our ability to achieve, and measure the impact of, the Sustainable Development Goals. As noted in the introduction, the precursors to the SDGs, the Millennium Development Goals, often failed to recognize the need for data and specifically geographic information, and UN-GGIM have suggested that this is a reason for their failure (Choi et al., 2016).

The geospatial community is now then in a unique position to integrate data in a more holistic and sustainable way, bringing interoperability along the entire data value chain. This will allow for better storing, visualizing and assessing of the data that is vitally important for the fulfilment of the SDGs (Giuliani et al., 2017; Van Genderen, 2017). The increase in the number of satellite launches will further increase the amount of data and monitoring available, especially in relation to climate change, natural disasters and sustainable agriculture (ESCAP, 2018). Furthermore, as more of this imagery is made 'open licence' and of higher spatial and temporal resolutions, there becomes an increased impetus for countries to implement and integrate the necessary structures to improve the quality of their data in order to feed into the wider data system (Arnold et al., 2019).

The introduction of NSDIs (National Spatial Data Infrastructures) has also enabled the development of more central points of data collection, and linkages to other national databases, and it is clear these will become ever more important (Scott and Rajabifard, 2017). Technological advancements in Web services and Internet infrastructures have also increased our ability to share and distribute data (Solís et al., 2018). Adding VGI into this mix also allows for a shift away from full state dependence and towards allowing citizens themselves to engage in the creation and monitoring of data related to the SDGs that most affect their lives on a local scale (ibid.).

Geospatial technologies can then provide timely information and data for supporting the SDGs. Including the monitoring of natural disasters and their clear-up (ESCAP, 2018); examining landcover use (Arnold et al., 2019); using RS to protect, restore and promote the sustainable use of terrestrial ecosystems, combat deforestation and halt biodiversity loss [SDG-15] (Acharya and Lee, 2019); to monitor the growth of cities (Choi et al., 2016); and to use remote sensing to provide insights into urban and environmental conditions on a scale never seen before (ESCAP, 2018). This data are now also cheaper, more available and more standardized than ever before (Holloway et al., 2018; Scott and Rajabifard, 2017). It would indeed seem wise for the United Nations, and other actors, to continue to push for the integration of geospatial technology and information as part of the Sustainable Development agenda. However, for all these benefits, this remains a challenging task for many countries, and stark differences in data, analysis and implementation persist and have the potential to create greater divides between those countries able to access geospatial technology and information, and those who are left behind (Arnold et al., 2019).

One of the significant issues of using Geographic Data in the implementation and monitoring of the SDGs is the lack of data supply, as opposed to the increasing demand. One of the principal issues remains the cost of producing such data on the ground. While this has led to an explosion of satellite and remote monitoring, it runs the risk of returning us to single points of information that are incomplete or unrepresentative (Holloway et al., 2018; Specht and Feigenbaum, 2018). The processing of data to ensure it can be used effectively for evidence-based decision making is also very expensive and time consuming, and many developing, or smaller nations, do not have the resources to process or utilize this information (ESCAP, 2018). Where gaps are being plugged by citizen scientists and VGI there are also, justifiably, concerns around the quality of the data produced; projects by OpenStreetMap, for example, have had very mixed results (Currion, 2010; Quinn and Bull, 2019).

Where data exists, the problem of integration persists. Despite NSDIs, ISOs and the work of organizations such as the OGC and UN-GGIM, there remains very little connection between some datasets and the desires and practical applications related to the SDGs (Scott and Rajabifard, 2017). A lack of information sharing by some countries (ESCAP, 2018); low demand for some types of data that causes it to be produced in too small a quantity to be effective (Giuliani et al., 2017); low rates of updating; shortages of information about accuracy (Arnold et al., 2019); and a lack of timely data (UN-GGIM, 2015) all lead to issues of interoperability and reduce the effectiveness of geospatial information in pursuit of the SDGs.

These issues, while predominantly affecting the developing world, where costs and infrastructure are especially problematic, are not limited to these parts countries (Sarvajayakesavalu, 2015; Scott and Rajabifard, 2017). The most developed parts of the world are also struggling to create the necessary linkages and infrastructures to allow them to grapple with the vast amounts of data available and to turn this into viable datasets for use in achieving the SDGs (Scott and Rajabifard, 2017). This problem illustrates well the myth that having large datasets can, in and of themselves, provide insights that solve complex issues (Solís et al., 2018). It is clear then that challenges persist, including infrastructure, data sharing, data quality, patchy data collection and a lack of interoperability. Even where interoperability exists, it presents new issues of data volume, time and cost of analysis, and how these feed into policy making (Giuliani et al., 2017). The issue of ensuring decision makers use the data available to them to enact policies that are supportive of achieving the SDGs is perhaps the biggest hurdle to be faced by the geospatial community (Sarvajayakesavalu, 2015).

Conclusions

Considerable progress is being made in relation to both the achievement of the Sustainable Development Goals and the use of geospatial technology in supporting and monitoring them. Building on the long history of using GI to monitor and map the environment, great advances have been made in terms of collection, sorting and sharing of data (Scott and Rajabifard, 2017). However, there is much still to be done. Some of the SDG goals and indicators are hugely ambitious and will require data and data processing that is not yet defined (Arnold et al., 2019). Many governments still face significant impediments to providing sustainable information infrastructures and turning information into decisions (UN-GGIM, 2012). A lack of complimentary data also hinders the ability of data processes to validate data collected through remote sensing and satellite processes (Holloway et al., 2018). Increases in data require increases in data processing and the hardware to do this is expensive and requires highly skilled operatives, leading to data pooling in areas of wealth, which can produce biases and problems in analysis and interpretation that is taking place far from the source of collection (Acharya and Lee, 2019).

While these issues remain, there are still a great many things to celebrate in relation to geospatial technology and the Sustainable Development Goals. Earth observation, Volunteered Geographic Information, International Data Standards and NSDIs are all helping move society closer to achieving the ambitious aims of the UN. Only through the constant tracking, monitoring and comparisons of the SDGs will we be able to avoid the failures of the MDGs. Data must be made interoperable, open and usable as quickly as possible. And we must enact a shift that means these data are then used in a way that is supportive to sustainable development (Sarvajayakesavalu, 2015; Solís et al., 2018).

As Scott and Rajabifard (2017) have noted, geography and the geospatial community is underpinned by the philosophy that the field provides an integrative framework that allows it to support the requirements of projects that need multiple information communities to come together in a timely and effective manner. It is this philosophy, underpinned by advances in technology and a global effort led by UN-GGIM, OGC, and TC211 to push for global standards for data, that will ensure that geospatial technology and geographers are at the core of the efforts to achieve the SDGs and improve the health of the planet and the lives of people around the globe and usher in the fourth wave of environmentalism, in which technological innovations allow us to scale up solutions and supercharge the approaches that came before (Laresen, 2020).

Notes

1. See https://undesa.maps.arcgis.com/apps/Cascade/index.html?appid=4741ad51ff7a463d833d18cbcec29fff.
2. SDG-11 3.1: Ratio of land consumption to population growth.
3. SDG-6 6.1: Change in the extent of water related ecosystems.
4. SDG-15 3.1: Proportion of land that is degraded over total land area.
5. SDG-6: Clean water and sanitation.
6. SDG-6: 3.2: Percentage of bodies of water with good ambient water quality.
7. SDG-7: Ensure access to affordable, reliable, sustainable and modern energy for all.
8. See http://trends.earth/docs/en/.

References

Acharya, T.D. and Lee, D.H. (2019) "Remote Sensing and Geospatial Technologies for Sustainable Development: A Review of Applications" *Sensors and Materials* 31 (11) pp.3931–3945.

Arnold, S., Chen, J. and Eggers, O. (2019) "Global and Complementary (Non-Authoritative) Geospatial Data for SDGs: Role and Utilisation" *(Report)* Available at: https://ggim.un.org/documents/Report_Global_and_Complementary_Geospatial_Data_for_SDGs.pdf.

Choi, J., Hwang, M., Kim, G., Seong, J. and Ahn, J. (2016) "Supporting the Measurement of the United Nations' Sustainable Development Goal 11 Through the Use of National Urban Information Systems and Open Geospatial Technologies: A Case Study of South Korea" *Open Geospatial Data, Software and Standards* 1 (4) pp.1–9.

Currion, P. (2010) "If All You Have is a Hammer, How Useful is Humanitarian Crowdsourcing?" *Medium* Available at: https://medium.com/@paulcurrion/if-all-you-have-is-a-hammer-how-useful-is-humanitarian-crowdsourcing-fed4ef33f8c8.

ESCAP (2018) "Good Practices and Emerging Trends on Geospatial Technology and Information Applications for the Sustainable Development Goals in Asia and the Pacific" (Staff Working Paper: Information and Communications Technology and Disaster Risk Reduction Division) Available at: https://www.unescap.org/resources/policy-and-technical-paper-good-practices-and-emerging-trends-geospatial-technology-and.

European Commission (2007) "Directive 2007/2/EC of the European Parliament and the Council of 14 March 2007 Establishing an Infrastructure for Spatial Information in the European Community (INSPIRE)" Available at: https://eur-lex.europa.eu/legal-content/EN/ALL/?uri=CELEX%3A32007L0002.

GEO (2016) "Earth Observations and Geospatial Information: Supporting Official Statistics in Monitoring the SDGs" Available at: http://www.un.org/ga/search/view_doc.asp?symbol=A/RES/70/1andLang=E.

Giuliani, G., Nativi, S., Obregon, A., Beniston, M. and Lehmann, A. (2017) "Spatially Enabling the Global Framework for Climate Services: Reviewing Geospatial Solutions to Efficiently Share and Integrate Climate Data & Information" *Climate Services* 8 pp.44–58.

Holloway, J., Mengersen, K. and Helmstedt, K. (2018) "Spatial and Machine Learning Methods of Satellite Imagery Analysis for Sustainable Development Goals" *Proceedings of the 16th Conference of International Association for Official Statistics (IAOS)* International Association for Official Statistics (IAOS), pp.1–14.

Hunt, A. and Specht, D. (2019) "Crowdsourced Mapping in Crisis Zones: Collaboration, Organisation and Impact" *Journal of International Humanitarian Action* 4 (1) pp.1–11.

Jia, G., Zhang, L., Zhu, L., Xu, R., Liang, D., Xu, X. and Bao, T. (2020) "Digital Earth for Climate Change Research" In Guo, H., Goodchild, M. F. and Annoni, A. (Eds) (2020) *Manual of Digital Earth* New York: Springer, pp.473–494.

Laresen, R. (2020) "SDG and the Fourth Wave of Environmentalism – A Walk in the Park" *Medium* Available at: https://medium.com/@ragnvald/sdg-and-the-fourth-wave-of-environmentalism-a-walk-in-the-park-8fc73cfafd58.

Moomen, A., Bertolotto, M., Lacroix, P. and Jensen, D. (2019) "Inadequate Adaptation of Geospatial Information for Sustainable Mining Towards Agenda 2030 Sustainable Development Goals" *Journal of Cleaner Production* 238 (117954) pp.1–8.

Quinn, S. and Bull, F. (2019) "Understanding Threats to Crowdsourced Geographic Data Quality Through a Study of OpenStreetMap Contributor Bans" In Valcik, N. and Dean, D. (Eds) *Geospatial Information System Use in Public Organizations* Abingdon: Routledge, pp.80–96.

Sarvajayakesavalu, S. (2015) "Addressing Challenges of Developing Countries in Implementing Five Priorities for Sustainable Development Goals" *Ecosystem Health and Sustainability* 1 (7) pp.1–4.

Scott, G. and Rajabifard, A. (2017) "Sustainable Development and Geospatial Information: A Strategic Framework for Integrating a Global Policy Agenda into National Geospatial Capabilities" *Geo-Spatial Information Science* 20 (2) pp.59–76.

Solís, P., McCusker, B., Menkiti, N., Cowan, N. and Blevins, C. (2018) "Engaging Global Youth in Participatory Spatial Data Creation for the UN Sustainable Development Goals: The Case of Open Mapping for Malaria Prevention" *Applied Geography* 98 pp.143–155.

Specht, D. and Feigenbaum, A. (2018) "From the Cartographic Gaze to Contestatory Cartographies" In Bargués-Pedreny, P., Chandler, D. and Simon, E. (Eds) *Mapping and Politics in the Digital Age* Abingdon: Routledge, pp.39–55.

UN-GGIM (2012) *Monitoring Sustainable Development: Contribution of Geospatial Information to the Rio+20 Processes* New York: United Nations Available at: http://ggim.un.org/2nd%20Session/GGIM%20paper%20for%20Rio_Background%20paper_18May%20 2012.pdf.

UN-GGIM (2015) *Future Trends in Geospatial Information Management: The Five to Ten-Year Vision* (2nd ed.) New York: United Nations Committee of Experts on Global Geospatial Information Management Available at: http://ggim.un.org/docs/UN-GGIM-Futuretrends_Second%20edition.pdf.

United Nations (2012) *Global Geospatial Information Management* New York: United Nations.

United Nations (2017) "Earth Observations for Official Statistics: Satellite Imagery and Geospatial Data Task" (*Team Report*) Available at: https://unstats.un.org/bigdata/taskteams/satellite/UNGWG_Satellite_Task_Team_Report_WhiteCover.pdf.

Van Genderen, J. (2017) "Perspectives on the Nature of Geospatial Information" *Geo-Spatial Information Science* 20 (2) pp.57–58.

Xiao, W., Mills, J., Guidi, G., Rodríguez-Gonzálvez, P., Barsanti, S.G. and González-Aguilera, D. (2018) "Geoinformatics for the Conservation and Promotion of Cultural Heritage in Support of the UN Sustainable Development Goals" *ISPRS Journal of Photogrammetry and Remote Sensing* 142 pp.389–406.

Yanga, Z. and Rajabifard, A. (2019) "SDGs, Digital Tools and Smart Cities" *Coordinates* (March) Available at: https://mycoordinates.org/sdgs-digital-tools-and-smart-cities.

40
MAPS OF TIME
Menno-Jan Kraak

The temporal perspective

Maps tell stories. Maps do this well because they present us with an abstract and selected view of geographic reality. It can be argued that each map has, next to a geospatial and attribute component, a temporal component. This temporal component defines the when, the time stamp of the map. Examples are weather maps representing the rain fall and temperature Wednesday morning at nine o'clock, or a map showing the average temperature in 2020. These time stamps can refer to a single moment in time, or to a time interval for which the data has be accumulated and processed. Both on a different temporal scale are considered as snapshots. Figure 40.1 provides some other examples. The map of Iceland (Figure 40.1a) was drawn by Ortelius in 1587 based on information known at that time. Users of this historical map will experience it as old because of its design and somewhat unfamiliar incorrect shape of Iceland. The OpenStreetMap of Iceland (Figure 40.1b) on the right is based on data available at the beginning of 2021. Both maps of Iceland are snapshots and do not display events. Events are about change, about movements of goods, of flooding, of urban sprawl, of changing temperatures, and so on. For maps displaying events we can argue these to be *maps of time*. These maps have been designed to tell the story of change, to help understand changing patterns and trends over time, and help us monitor processes.

Old and new maps of time

One of the first maps of time, and probably one of the most famous maps of all times, is shown in Figure 40.2a. It is the 'Carte Figurative des pertes successives en hommes de l'Armée Française dans la campagne de Russie 1812–1813' by Charles Joseph Minard. It depicts Napoleon's Russian campaign and was designed in 1869 as a pamphlet against war. The flow map shows how Napoleon's army started to march for Moscow with around half a million soldiers but only came back with just over 10,000. This map and other works by Minard (see also Figure 40.5) have been described in our cartographic domain by Funkhouser (1937) and Robinson (1967). Its fame really spread after Tufte's discussion of the map in his book *The Visual Display of Quantitative Information* (1983). Tufte is an authoritative expert on information design, and his description of the map states: '[…] how multivariate complexity can

Figure 40.1 Temporal perspective: (a) Ortelius map of Iceland from 1587, (b) OpenStreetMap's Iceland (2021)

Source: (a) http://mappingiceland.com/wp-content/uploads/2017/09/177.jpg. (b) OpenStreetMap-authors / licence: CC BY-SA.

be subtly integrated into graphical architecture, integrated so gently and unobtrusively that the viewers are hardly aware that they are looking into a world of four or five dimensions'. His quote 'It may well be the best statistical graphic ever drawn' is echoed by many others, who have been inspired by Minard's map and used it as a benchmark to evaluate their own software or mappings. He described the character of Minard's maps well with the sentence 'Graphic elegance is often found in simplicity of design and complexity of data'. An overview of the many variants and applications of this map is found in the book *Mapping Time Illustrated by Minard's Map of Napoleon's Russian Campaign of 1812* (Kraak, 2014). The complete works of Minard have recently been described in *The Minard System: The Complete Statistical Graphics of Charles-Joseph Minard* (Rendgen, 2018).

Minard's map visually expresses the narrative of an event and shows what happened during a time interval of six months. The map does answer a generic temporal question like: 'What happened during the given period?'. However, can the map answer basic temporal questions like 'When?', 'How often?', or 'In what order?' This often can only be answered at a general level based on the title, since flow maps lack a direct time reference. Of course, one can reason about what happens and deduct a few answers, but there are other more efficient graphic representations to support temporal reasoning as will be explained in later sections.

The map in Figure 40.2b is of a completely different nature. It is not a static map like Minard's map. It is a dynamic interactive map showing the air traffic in the European and Western Asian airspace. Symbols represent airports and the current location of airplanes. The map is updated in near real-time based on streaming location data received by ground stations and satellites. Clicking a plane symbol provides detailed information on a particular flight such as the airplane type, airline, departure time, estimated arrival, the flight path, etc. The airport symbols give access to the airport's timetable of arrivals and departures. All kinds of additional map options related to weather or traffic control can be overlayed depending on one's subscription level.

Types of time

Time can be considered as absolute or relative. Absolute time is described based on calendars and clocks, such as the event happened on 15th August between 15:15 and 17:35. For relative time events are compared to each other; for example, 'the event happened before

Maps of time

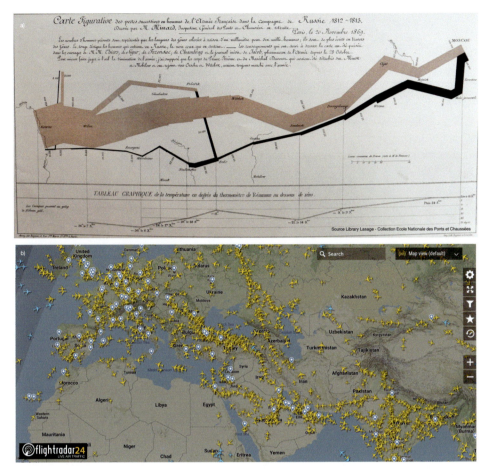

Figure 40.2 Old and modern classics: (a) Minard's "Carte Figurative des pertes successives en hommes de l'Armée Française dans la campagne de Russie 1812–1813"; (b) real time flight information with situation at 03-02-2021 UCT 08:00

Source: (a) Library Lasage - Collection Ecole Nationale des Ponts et Chaussées. (b) https://www.flightradar24.com.

the outbreak of the COVID-19 pandemic'. Absolute time can be linear or cyclic. Linear events happen without repetition, like human life from birth to death. Cyclic events repeat themselves forever, such as the seasons. Linear time is typically plotted against a timeline, and cyclic time on a circular clock.

Figure 40.3a shows an example of changing municipal boundaries over time. This is a linear event with discrete time stamps referring to the exact moments of change. At the beginning of 1974 municipalities merged and since that moment a new configuration came into existence. The timeline shows the duration of the different municipal configurations. The application of the visual variable value from dark (old) to new (light) is used to strengthen notion of the flow of time. Figure 40.3b displays a continuous event, the changing wind flow around the Faroe Islands. The map series show the daily situation at 22.00 for five consecutive days. The temporal information is indicated as text at the lower bottom of each image.

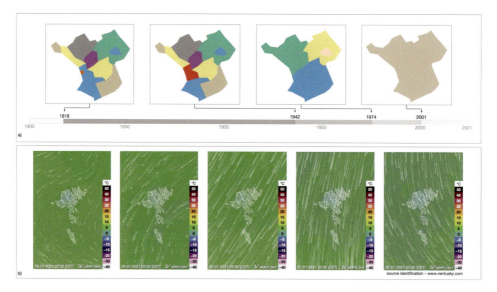

Figure 40.3 Nature of time: (a) linear – changing municipal boundaries Steenwijkerland; (b) continuous – wind patterns above the Faroe Islands

Source: www.ventusky.com.

Change

Mapping time is about mapping change. It enables one to understand spatial temporal patterns and trends. Typically, it is possible to distinguish three kinds of change based on the components of spatial data: change in the existence of a phenomena, change in its location, or change in attributes (respectively Figure 40.4a–c). The time series of topographic maps in Figure 40.4 demonstrates these types of change. Existential change is about the appearance or disappearance of an phenomena. Comparing 1937 and 1965 reveals that new land appears in the 1965 map. Attribute change can be quantitative or qualitative. By comparing the 1909 and 1937 maps it can be seen that number of inhabitants of the town of Blokzijl has decreased. Comparing the maps of 1965 and 2005 exemplifies qualitative a change in land use. The highlighted parcel switched from pasture to arable land. Locational change is about the movement of a phenomena or may relate to an expanding/shrinking of a space. Comparing 1909 with 1937 shows that the lighthouse has been moved up the pier extension. An example of expansion can be seen by comparing 1937 with 1965 and looking at the built-up area of the town of Blokzijl.

Representing change: a single map or a series of maps?

Bertin (1967) was one of the first to systematically describe how to map change. He suggested to design a single map using the visual variables to express the changes or use a series of maps each expressing part of the change. This series of maps is what Tufte (1983) calls small multiples, a set of images in a logical (temporal) order telling the story of change. Actually, Minard presented such a series of maps in 1866 when he published three adjacent maps showing the import of raw cotton in Europe for the years 1858, 1864, and 1865.

To illustrate the use of a single map and a series of map a simple dataset on the Faroe Islands' tunnels is used. The maps and table in Figure 40.5a I and II, respectively, show their

Maps of time

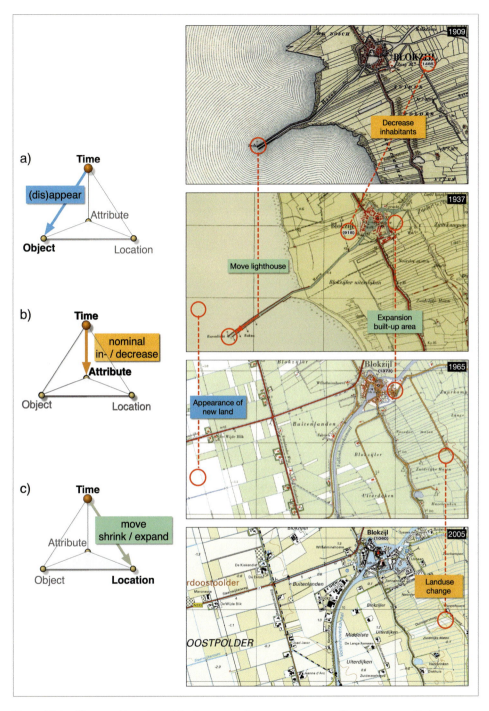

Figure 40.4 Change: (a) existential change – the appearance or disappearance of phenomena; (b) attribute change – qualitative or quantitative change; (c) locational change – move or shrink/expand of phenomena

Source topographic maps: Kadaster Geo Information.

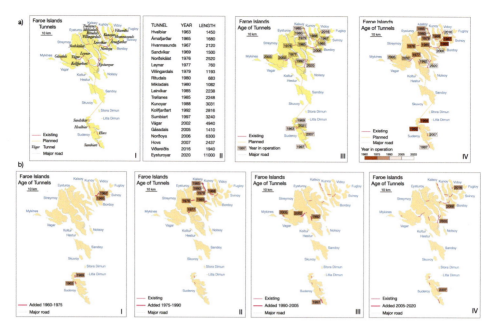

Figure 40.5 Display of change: tunnel built on the Faroe Islands 1960–2020: (a) single map − I. no temporal hierarchy/II. temporal hierarchy; (b) series of maps − I. Period 1960–1975, II. Period 1975–1990, III. Period 1990–2005, IV. Period 2005–2020

location, the year when the tunnels came into service, as well as their length. The objective of the map is to give an overview of when these tunnels became operational. A straightforward solution is to add the labels with the opening year to the map (Figure 40.5a III). Such a map indeed contains the information but does not offer any sense of time. Answering the question 'Where is the oldest tunnel?' requires scanning the whole map. To avoid this the temporal hierarchy of the data has to be visualized. Therefore, time should be considered as an attribute, and an appropriate visual variable has to be applied. Value is a good choice since it offers a sense of (temporal) order. In the example in Figure 40.5a IV the years have been grouped in four time periods for simplification. It is common to use the darkest value tint for the older categories. The map can now more easily answer questions about the temporal distribution over space. In well-designed single maps, the notion of time is expressed by visual variables. An interactive map environment would make answering these questions even easier as will be explained later in the chapter.

Using small multiples, or adjacent maps, will split the event into smaller time intervals. In Figure 40.5b, the four time intervals from Figure 40.5a IV each get their own map. To compare these maps, one has to be sure the scale and design are the same for all maps. The notion of time is derived by the order of the individual maps, and often the map with the oldest information is put at the start of the reading direction. One could argue that reading a map is like reading a comic book. In the case of the use of many maps in a series, like one for each tunnel, the maps will have to be simplified and reduced in scale to be able to compare. The use of more maps than would fit on a single page is not advised. In the case of a map series like the one shown in Figure 40.5b it is good to realize that the choice of the time intervals used to group the tunnels has an influence on how the time series will be interpreted.

Maps of time

Change maps

The single-map solution to represent events does often require an innovative design approach and data wrangling to be able to compile the map. However, several well-known thematic map types are quite suitable for this. The famous Minard map, a flow map, shown in Figures 40.2a and 40.6dII, is such an example. Flow maps summarize events, but time might not always be explicit. The flow lines emphasize the character and (schematic) route of the movements of goods or people for a certain time interval. Often time has to be deducted from the title or legend of the map. Figure 40.6a shows an example of a flow map

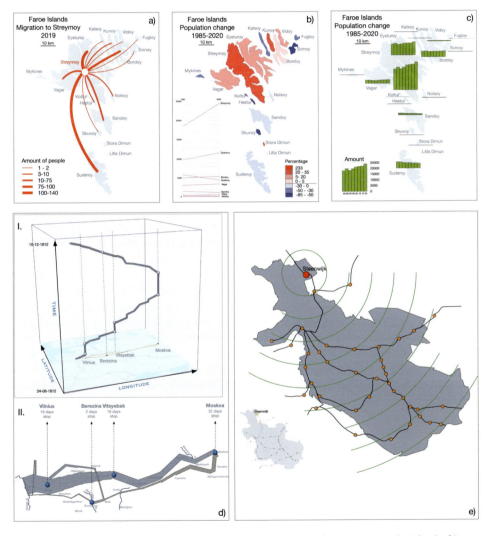

Figure 40.6 Map types to represent change: (a) flow map of Faroe's migration to the island of Streymoy in 2019; (b) choropleth map of Faroe's relative population change between 1985 and 2020 (inset slope chart with absolute change); (c) diagram maps of Faroe's relative population change between 1985 and 2020; (d) Space-Time-Cube of Minard's Carte Figurative (compare with Figure 40.2a); (e) cartogram of travel times from Steenwijk to other railway station in the province of Overijssel

representing the migration of people from all other islands of the Faroe to the island of Streymoy in the year 2019.

The choropleth in Figure 40.6b shows the population change on the Faroe Islands between the year 1985 and 2020. It is an example of showing increase and decrease in a single map using a diverging colour scheme centering around zero percent change, with red for increase and blue for decrease. The problem of this approach is that percentual change only shows the relative data perspective. That is why next to the choropleth a slope chart based on absolute numbers is given. From the diagram it becomes clear that the highest absolute change was on Streymoy. The highest increase on the choropleth is 233% for the island of Stora Dimun. It is not even visible in the slope chart because of the low values that show an increase from three inhabitants in 1985 to ten in 2020. Alternatively, the absolute numbers could have been plotted on top of the choropleth map with proportional symbols. It also depicts one of the main cartographic challenges to deal with a wide range of values in a single map. The choropleth does not reveal any local spatial patterns or temporal trends for the years in between and as such simplifies the time series.

The diagram map (Figure 40.6c) is another option to show changes. In this example the population change on the Faroe Islands is represented by a bar graph with five-year intervals for each island. The diagram allows more nuance for the in-between period of the total time interval. The trend on island with a larger population such as Eysturoy or Suderoy can be followed, but because of a wide range between the minimum and maximum values, no trends are visible for the islands with a small population. An interactive map that allows mouse-over techniques could be helpful. If one would like to display individual years to not miss patterns or trends in the time series the so-called spark-lines could be a solution. According to Tufte (2006), sparklines are 'data-intense, design-simple, word-sized graphics'. However, diagram maps are not the most suitable solution to provide a quick overview as choropleths do, but diagrams do work for individual enumeration districts. The questions one would like to ask should drive the choice of map type, and the option to interactively vary and compare map types, depending on the questions, would be ideal.

The Space-Time Cube (STC) as seen in Figure 40.6dI was originally developed by Hägerstrand (1970) as a major element of his theory of time-geography and visualizes space and time in a cube. The cube's horizontal plane is used to display space, and along the cube's vertical axis time is shown (Kveladze et al., 2018). A map with the simplified path of Napoleon's campaign is shown in the horizontal plane. The path is also shown along the vertical (time) axis as a three-dimensional line. This so-called space-time path is one of the main components of the STC concept. Vertical segments in the path imply no movement. The location does not change, but time is passing. These stops like in Moscow are not obvious in Minard's original map (Figure 40.6dII). However, in combination with the map the STC is able to answer most temporal questions discussed before. It is obvious that questions emphasizing 'Where?' are best answered by the map, and those related to time by the STC. Minard's well-praised map is indeed very effective, but for many temporal questions it is unable to give answers.

The time cartogram in Figure 40.6e does not necessarily visualize an event or change, but in the map 'geography' is replaced by time. In a cartogram, enumeration districts are not scaled by their geography but by the value of an attribute. This could be number of inhabitants, income or oil production. In a time cartogram, the area is scaled to travel time, and the shape of an area changes accordingly. In the example the surface represents travel time by train from the city of Steenwijk to all other railroad stations in the province of Overijssel in the Netherlands. The small inset shows the province of Overijssel in its true geographic

Maps of time

shape. The circles in the map represent thirty-minute intervals. With respect to the city of Steenwijk the other stations are push out or pulled in along a virtual line representing the true angle of direction between the stations. All other points in the map are then mapped accordingly resulting in the cartogram's shape (Ullah and Kraak, 2014).

Interaction

Today we are not limited to static single maps or series of maps. (On-line) interactivity offers options to play with the design, to create different perspectives, or to access map content and the data behind the map (Roth, 2013). Having the tunnel map in mind imagine how one could include extra layers, such as the built-up area or relief to put the tunnels in context. Alternatively, selecting a three-dimensional map would change the perspective on the need for tunnels, while search/query and filter functions would support asking different (spatio)temporal questions.

Maps like those in Figure 40.7a can come alive in an interactive environment. Moving the mouse over existing tunnels can reveal their names and the years they opened for traffic. Queries such as 'Show me the oldest tunnel' could indeed highlight that particular tunnel. Having the proper functionality available offers users the ability to temporally identify, locate, and compare map objects that represent events. Additionally, it is possible to add all kinds of annotations such as photos, drawings, videos, text blocks, and links (see section on story maps).

Figure 40.7a is an example of a static map comparing two situations: today's tunnels and tomorrow's tunnels. These before and after maps are more impressive in an interactive environment. Figure 40.7b shows such an example. With a kind of slider, the user can drag a map from one period over the map to another period. In Figure 40.7b the dramatic effect of COVID-19 on air traffic is shown. During March 2020 the airspace above Europe was very busy, but a month later air traffic had decreased dramatically. This type of functionality is used frequently by newspapers, for example, in their online editions to show the effect of flooding or other natural disasters using imagery.

Animation is yet another popular method to show change (Harrower and Fabrikant, 2008). Animations are composed of an ordered series of maps (frames) that are displayed in

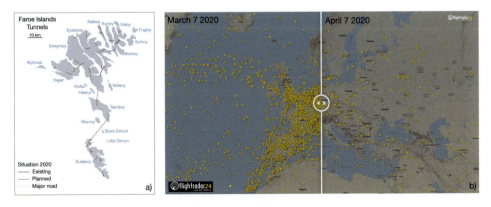

Figure 40.7 Comparing situations before and after: (a) Faroe's tunnels built in the past and those planned for the future; (b) the situation in the European airspace on 7th March 2020 and on 7th April 2020 after the COVID outbreak

Source: https://www.flightradar24.com.

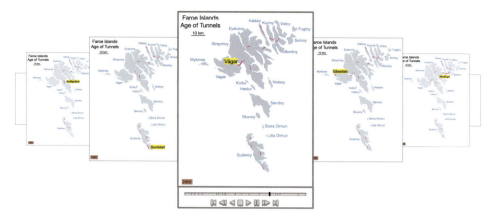

Figure 40.8 Animation of tunnel building in the Faroe Islands

a (temporal) sequence giving the user an experience of movement and change. Each map frame can be pre-computed as shown in Figure 40.3a or computed (near) real time as in Figure 40.3b. The individual frames can each represent an equal time interval, such as in Figure 40.5b, or could represent irregular events such as in Figure 40.8. In the first case the time interval for each frame is fifteen years. This results in a regular relation between display time (e.g., three seconds per frame) and world time (e.g., frame represents fifteen years). The animation would have a temporal scale of one second for five years. In the animation represented by Figure 40.8 the temporal scale is not that clearly defined. Here each frame represents the year in which a particular tunnel became operational. The relation between display time, with three seconds per frame, and world time is irregular because the time between opening tunnels varies between a single year to more than a decade. A solution would be to vary display time for each frame, but in the case of no activity the user will likely become impatient. Alternatively, one can indicate on the time slider when new tunnels become operational, as is done in the figure. The current frame is highlighted on the slider and a year label is given in the frame. However, if there are too many actions to observe besides showing where the new tunnel is located the effect of the animation in telling the story might be limited. Observing all these frames is more complex than studying a single map, and the user is likely to miss many smaller changes. This effect is known as change blindness. However, we know animation is helpful in presenting overall trends. This type of animations does require a good temporal legend and this legend can also function as a user interaction tool (e.g., to allow one to pause, reverse, or fast forward the animation, jump to a selected moment in time, etc.), to be able to study local spatial and temporal trends.

Exploration

In more complex situations one might not only be interested in understanding particular changes but may want to see potential causalities or relations with other phenomena. This process of getting insights into previously unknown relation and trends is called exploration (Dykes *et al.*, 2005). During exploration one 'plays' with all kind of maps and diagrams to stimulate visual thinking, allowing one to reason about what the newly created maps and diagrams show. In exploration Shneiderman's visual information seeking mantra is the strategy often followed (Shneiderman, 1996). It is roughly based on a three-step process,

Maps of time

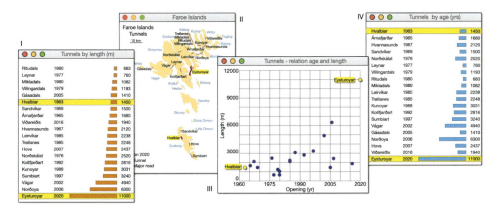

Figure 40.9 Temporal exploration: comparing the age of tunnels and their length in an interactive linked view environment highlighting he oldest and newest tunnel: I. Tunnels by length; II. Tunnel map; III. Scatterplot length versus year; IV. Tunnels by age

starting with the overview first, followed by zoom and filter, and finally providing details on demand.

This is illustrated with Figure 40.6d, which shows Napoleon's Russian campaign in the space-time-cube and Minard's map. In an exploratory environment the maps would be linked together, such that an action in one of the maps views would highlight the same features in the other map views. The figure as such presents the overview mode. If one would be interested in when and for how long Napoleon stayed in Moscow, the next step would be to zoom in on the area and period of interest and establish in detail his arrival and departure from the city. The map view would be used to zoom in on the area and the space-time-cube to narrow down the time interval. The details can be obtained from the space-time-cube by clicking the vertical segment in Moscow, which will reveal arrival and departure.

Figure 40.9 is another example of temporal exploration. If one wonders about a possible relation between the age and length of the tunnels one could again 'play' with the available information. The two tables show sorted diagrams of the tunnels, in view I by length in view I and in view IV by age. View II shows the map and view III a scatterplot with length along the vertical axis and age along the horizontal axis. In all views the oldest and newest tunnels are highlighted. It can be visually deducted that over time tunnels do become longer. This trend would even be more explicit if future tunnels were included because over 10 km tunnels are currently being built (see Figure 40.7a). The examples discussed here are relatively simple cases with small datasets, but they do explain the potential of visual exploration. The exploration process can be time driven but will not be successful without location and attribute.

An integrated approach

The previous section discussed the search for potential relations between the observed changes and other phenomena in a highly interactive mapping environment. The story of exploration or any sequence of events can also be told in so-called story maps. Today storytelling with maps is also popular in journalism and science communication but is not necessarily new. The combination of maps, texts, images, videos, etc., has been seen in earlier cartographic multimedia solutions at the end of the last century (Cartwright *et al.*, 1999). However, with today's tools it is much easier to create these interactive visual narratives.

Story maps can be created according different visual storytelling genres, all following a kind of linear narrative. The visual storytelling genres proposed by Segel and Heer (2010) were translated into the cartographic domain by Roth (2020). The genres are static map (linearity guided by map layout (e.g., Minard's maps could contain numbered annotations to chronologically follow the campaign); longform infographics (e.g., linearity guided by browser scroll functionality with oldest at the top; see Figure 40.10); dynamic slideshows (e.g., linearity guided by swiping individually slides with oldest on the left); narrated animations (e.g., linearity guided by animation designer with oldest first); personalized story maps (e.g., linearity enforced by the user with oldest first); and multimedia visual experiences (e.g., linearity guided by layout and hyperlinking with oldest first, but hyperlinks allowing for time travel).

Figure 40.10 shows a simplified variation of the longform infographics. For this chapter all information is placed in a single image fitting the page, but in the true application one will scroll down via the browser to follow the temporal narrative along the vertical line, and relevant information will pop up when time-wise appropriate. The example shows when and where the most recent tunnel, the Eysturoyar tunnel, was built. The map background shows the tunnel location in detail. The timeline starts with an act of parliament that made the decision to build the tunnel. Moving along the timeline the deal with the contractors and the actual start of building is indicated illustrated by two maps, showing how travel time and

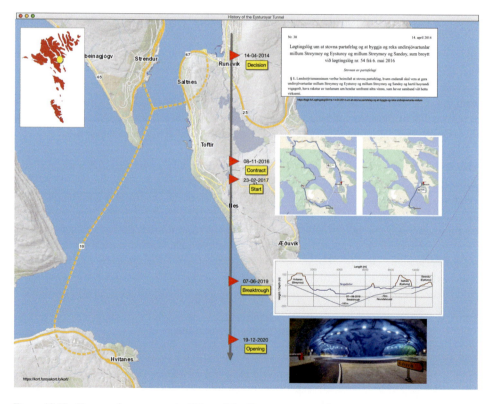

Figure 40.10 Temporal story map: building of the Eysturoyar tunnel

Source: ©Umhvørvisstovan: kort.foroyakort.fo; https://logir.fo/Logtingslog/30-fra-14-04-2014-um-at-stovna-partafelag-og-at-byggja-og-reka-undirsjovartunlar-millum; 2021 Google LLC, used with permission; https://en.wikipedia.org/wiki/Eysturoyartunnilin CC BY-SA 4.0.

distance are reduced. The moment of breakthrough when the builders from the north and south met each other underwater is illustrated by a profile of the tunnel. The final moment along the timeline is the opening at the end of 2020, illustrated by one of the unique features of the tunnel, the underwater round-about.

Conclusion

In this chapter different options to map events were discussed. Usually, the map design process is influenced by a set of factors, such as the nature of the (temporal) data, the message to bring across, and the audience and their use environment. Maps of time are no different. However, most maps are only good at answering particular temporal questions. Even the highly praised Minard map cannot answer the simple 'When?' question. What is needed are options for the user to choose different visual perspectives on the temporal data using a variety of maps and diagrams, each with its own strength combined in an environment that visually links all visualizations. This will allow for the understanding of temporal patterns and trends in a spatial context.

References

Bertin, J. (1967) *Semiology Graphique* Den Haag: Mouton.
Cartwright, W., Peterson, M. and Gartner, G. (Eds) (1999) *Multimedia Cartography* Berlin: Springer.
Dykes, J., MacEachren, A.M. and Kraak, M-J. (Eds) (2005) *Exploring Geovisualization* Amsterdam: Elsevier.
Funkhouser, H.G. (1937) "Historical Development of the Graphical Representation of Statistical Data" *Osiris* 3 pp.269–404.
Hägerstrand, T. (1970) "What About People in Regional Science?" *Papers in Regional Science* 24 (1)pp.7–24.
Harrower, M. and Fabrikant, S. (2008) "The Role of Map Animation in Geographic Visualization" In Dodge, M., McDerby, M. and Turner, M. (Eds) *Geographic Visualization: Concepts, Tools and Applications* New York: Wiley.
Kraak, M-J. (2014) *Mapping Time: Illustrated by Minard's map of Napoleon's Russian Campaign of 1812* San Diego, CA: Esri Press.
Kveladze, I., Kraak, M-J. and van Elzakker, C.P.J.M. (2018) "Cartographic Design and the Space-Time Cube" *The Cartographic Journal* 55 (4) pp.73–90 DOI: 10.1080/00087041.2018.1495898.
Rendgen, S. (2018) *The Minard System: The Complete Statistical Graphics of Charles-Joseph Minard* Princeton, NJ: Princeton Architectural Press..
Robinson, A.H. (1967) "The Thematic Maps of Charles Joseph Minard" *Imago Mundi* 21 pp.95–108.
Roth, R.E. (2013) "An Empirically-Derived Taxonomy of Interaction Primitives for Interactive Cartography and Geovisualization" *IEEE Transactions in Visualization and Computer Graphics* 19 (12) pp.2356–2365.
Roth, R.E. (2020) "Cartographic Design as Visual Storytelling: Synthesis and Review of Map-Based Narratives, Genres, and Tropes" *The Cartographic Journal* 58 (1) pp.83–114 DOI: 10.1080/00087041.2019.1633103.
Segel, E. and Heer, J. (2010) "Narrative Visualization: Telling Stories with Data" *IEEE Transactions on Visualization and Computer Graphics* 16 (6) pp.1139–1148.
Shneiderman, B. (1996) "The Eyes Have It: A Task by Data Type Taxonomy for Information Visualization" *Proceedings of the IEEE Symposium on Visual Languages* Boulder, CO.
Tufte, E.R. (1983) *The Visual Display of Quantitative Information* Cheshire, CT: Graphics Press.
Tufte, E.R. (2006) *Beautiful Evidence* Cheshire, CT: Graphics Press.
Ullah, R. and Kraak, M-J. (2014) "An Alternative Method to Constructing Time Cartograms for the Visual Representation of Scheduled Movement Data" *Journal of Maps* 11 (4) pp.674–687 DOI: 10.1080/17445647.2014.935502.

41
GEOSPATIAL TECHNOLOGIES IN ARCHAEOLOGY

Alexander J. Kent and Doug Specht

As a discipline that aims to understand past human activity through the retrieval and collection of material culture, archaeology relies upon the analysis and interpretation of geospatial data. From the very beginning of archaeological practice, maps (and plans) have been amongst the discipline's most fundamental tools (Gillings et al., 2019). Archaeologists continue to utilize a wide range of geospatial technologies, whose techniques have arguably become as important (if not more so) than others in the field, such as radiocarbon dating (McCoy, 2021). Relevant geospatial technologies include geographic information systems (GIS), global positioning systems (GPS), aerial and satellite imagery and light detection and ranging (Lidar). Archaeology has also been quick to embrace these technologies in constructing and manipulating 3D models, detecting previously unknown sites, characterizing environments using multispectral and hyperspectral sensors and improving the visualization and analysis of artefacts with or without their retrieval from the ground. This range of monitoring and cataloguing tools provides archaeologists with greater capacity to develop local and global databases of human history that, collectively, cover vast areas of the Earth's surface (Rayne et al., 2017). Coupled with the digitalization of legacy data, these datasets allow archaeologists to establish not only the provenance and context of artefacts, but also to monitor damage and disturbance to archaeological sites (Harrower and Comer, 2013; Rayne et al., 2017). This chapter will explore these applications in greater detail on the ground, from the air and in three dimensions.

Geographical information systems

Since archaeology is an inherently spatial practice, geographic information systems (GIS) – which allow the storage, organization, analysis and visualization of spatial data – are ideally suited to enhancing archaeological fieldwork, and were already being used extensively in the 1980s and early 1990s. Building on the capacity of GIS to expand the functionality of paper maps through the storage, organization and visualization of data in layers, early applications in archaeology were focused primarily on the management of archaeological data. The potential to facilitate the identification of spatial patterns increasingly led to the use of GIS for the exploration of potential correlations between sites and human behaviours (Richards-Rissetto, 2017a, 2017b).

Figure 41.1 A typical GIS interface for an archaeological project, with layers of data arranged in a hierarchical structure (left) and visualized on screen as a map (right). Courtesy of Christian Gugl

As the functional capabilities of GIS advanced, they allowed for the creation of more than just 'pretty pictures' of historic sites, and became valuable for the analysis, management and representation of archaeological data in cartographic formats (Ebert, 2004) (Figure 41.1). Initially, the essential mapmaking functionality within GIS was often downplayed as an output-and-display or read-only tool, with visualization considered merely as an activity for communicating with a wider audience. Data analysis, however, offered ways to explore and generate new knowledge. This tripartite view, representative of how GIS is perceived in other fields, precluded the recognition of information visualization as a process for generating knowledge – a key role now recognized in recent geovisualization and geovisual analytics studies (Dykes *et al.*, 2005; Lloyd and Dykes, 2011). The shift in focus to exploit the benefits of geographic visualization for creating knowledge is still gaining traction in the archaeological community and can offer further opportunities to analyse the complex spatial and temporal data inherent to the discipline (Gupta and Devillers, 2017).

A significant advantage of GIS for archaeology is its capacity to integrate a range of datasets, such as soils, geology, vegetation, hydrology, and terrain, with archaeological data. Accurate georeferencing (preserving the real-world locations) of these layers allows the spatial relationships between natural and human made features to be made apparent. However, some archaeological data are also cultural and qualitative in nature, and the process of digital conversion their use in a GIS is complex and can lead to significant issues of representation (Leszczynski, 2009). Some approaches to GIS may be environmentally deterministic and overly quantitative, due to practical issues involving data and software availability, and the underlying theoretical assumptions driving research objectives. Some of these limitations can be overcome and cultural information can be integrated into GIS analyses if they are explicitly grounded in archaeological and/or social theory and interpreted within a society's particular historical, socio-political and ideological circumstances (Llobera, 1996; Lock and Pouncett, 2017). To overcome these shortcomings, archaeologists cannot regard GIS as an unbiased tool. Instead, they should consider GIS as a form of practice that must be situated within archaeological theory and use theory-inspired cultural variables that see places as socially created, as well as linked to both space and time (Tschan et al., 2000).

Aerial and satellite imagery

Archaeologists have long understood the importance of being able to view sites from the air. This is useful for establishing their scale and for revealing spatial patterns in landscapes and buildings to provide historical and geographical context. Archaeologists first used photography from hot-air balloons for these purposes since at least the early 1900s (Figure 41.2) and with the advent of the aeroplane, archaeology was amongst the first non-military applications of aerial imagery. Techniques for the latter were pioneered by Crawford and Poideboard in the 1930s for discovering new sites and placing them in their wider geographical context (Ur, 2013). The availability of declassified military aerial photography, such as that taken by the Royal Air Force (RAF) during the Second World War over Great Britain, provided a hitherto unparalleled resource for archaeologists to interpret historical landscapes from the air.

Operating at higher altitudes, satellite remote sensing has been invaluable in allowing archaeological research to take place in areas that are either excessively remote or otherwise inaccessible, for example, due to conflict. As this imagery has become more readily available, archaeologists embraced the possibilities this offered to advance the field (Ur, 2013). The combination of the availability of low-cost and high-resolution satellite imagery has empowered archaeologists to overcome the practical limitations of working in inaccessible

Figure 41.2 Aerial photograph of the pyramids at Giza taken by Eduard Spelterini in 1904 from a hot air balloon about 600 metres above ground (Spelterini, 1928)

regions. The declassification of Cold-War-era US intelligence imagery from the CORONA and GAMBIT satellite programs in 1998 (Harrower and Comer, 2013) and the commercial availability of 1-metre resolution IKONOS imagery from 1999, for example, presented significant steps forward. Archaeologists in the Near East recognized the potential of this new dataset, with the examination of sites located in the now-inundated floodplain of the Turkish Euphrates (Kennedy, 1998) a prime example. The free availability of natural-colour aerial and satellite imagery via Google Earth has (since 2004) revolutionized accessibility to this resource for archaeologists and for interpreting landscapes.

Remote sensing technologies encompass the ability to detect energy that is reflected at wavelengths beyond our visible spectrum. Landsat, a program of Earth observation satellites first launched in 1974, is equipped with sensors that can detect infrared light, and has been used to identify regions of arable soil around Petra in Jordan (Comer, 2013). Similarly, Savage *et al.* (2013) used hyperspectral imagery (which records narrow bands of wavelengths) of Jordan from NASA's Hyperion sensor platform to identify ancient copper mining and processing – an important activity that began in the area during the third millennium BC and physically transformed the surrounding landscapes.

Today, a broader range of remote sensing apparatus is available, which includes active systems that are not dependent on reflected energy from the Sun. Synthetic Aperture Radar (SAR) offers a versatile method of detecting natural and historic landscape features, including topography, vegetative structure and superficial roughness in studies of ice or geological formations. If SAR data are collected and analysed interferometrically, they can also be used to develop digital surface models of the ground or vegetative canopies, depending on the wavelength used. The SAR returns are influenced by the capacity of materials encountered to conduct electricity, they are generally unaffected by cloud cover, and under certain conditions they can penetrate vegetation and soil to reveal what lies beneath (Harrower and Comer, 2013).

A more recent method is light detection and ranging, or Lidar. Another active system, this combines laser scanning with onboard altimetry and GPS (global positioning system), and has been used to re-examine well-researched sites and reveal new discoveries. For example, Megarry and Davis (2013) combined Lidar with Worldview-2 imagery to study the vicinity of 'Bru na Bóinne (Ireland), demonstrating the enormous utility of high-resolution sensors even in landscapes that have already been studied intensively. The capacity for Lidar to 'see' through gaps in tree cover has led to new archaeological discoveries. Chase *et al.* (2013) obtained Lidar data for a 200 km^2 area that included the site of Caracol, in Belize. The high-resolution Lidar coverage, at 20 points per m^2, was sufficient to produce a 'bald-earth' digital surface model of the landscape that recorded the surface beneath the dense, tropical vegetation. Since the central monuments had been made physically inaccessible due to the dense tropical vegetation surrounding them, the Lidar survey presented the first opportunity for an entire Mayan city to be mapped. Similarly, other Mayan sites have been mapped using Lidar to create an effective 3D digital model (Figure 41.3).

Applications of 3D archaeological data

The addition of the third dimension introduces the possibility of appreciating the volume and depth of spatial relationships (indexicality) of elements such as buildings to each other and to the landscape (Figure 41.4). These models can also be linked to descriptive and/or metadata to inform users about not only what they are looking at, but also how modelling decisions were made, and the data sources used to make those decisions (Richards-Rissetto,

Figure 41.3 Lidar image of the grand plaza of the Maya city of Tikal in present-day Guatemala (Juan Carlos Fernandez Diaz/National Center for Airborne Laser Mapping)

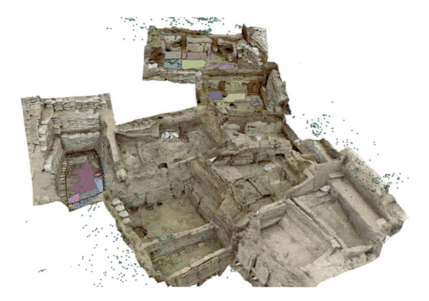

Figure 41.4 A 3D visualization and digital model of the south shelter at Catalhöyük, Turkey (Berggren et al., 2015)

2017a). When combined with technologies such as virtual reality (VR) it is possible to turn archaeological sites and data into immersive learning spaces, leading to new ways of looking at these data, potentially leading to new insights (Richards-Rissetto, 2017b).

However, 3D archaeological reconstructions often forgo the natural environment. Hills, streams and plants, when present, often serve as a backdrop rather than active agents in cultural transformation. The application of 3D technologies to digitally remove artefacts from their surroundings represents a difficult issue and some archaeological projects have tried to counter this, such as the MayaCityBuilder project. This involved the creation of an interactive 3D visualization of Copan's ancient buildings that are situated within georeferenced terrain. The 3D environment enables students and scholars to explore Copan from a ground-based perspective and incorporates paleoenvironmental and ethnobotanical data to create a database of images and 3D models that are linked to the descriptive attributes of plant habitats, such as elevation range, cultural affiliation and time period. The project aims to allow scholars to interchange not only georeferenced architectural models but also environmental models within a 3D environment. Georeferenced 3D architectural, vegetation, hydrological and topographic data combine the locational precision of GIS with the human perception of 3D, allowing researchers to explore archaeological and other data from the ground to the sky and back again. In this way, GIS becomes less 'environmentally-deterministic' and continues to move towards the development of new GIS methods in landscape archaeology (Richards-Rissetto, 2017b).

Limitations

Archaeologists are now able to collect and analyse data at an unprecedented scale. This has led to significant findings, improved documentation and the essential management of landscapes at risk of disappearing (Rayne *et al.*, 2017). Geospatial technologies such as Lidar are allowing for the creation of inordinate amounts of georeferenced 3D data (Richards-Rissetto, 2017a). However, these tools and datasets are not without their issues. As a starting point, remote sensing datasets are not equally helpful across the globe. Whereas high-resolution imagery is useful for analysing alluvial landscapes, such as the Tigris-Euphrates and Nile river valleys, it is less useful for mountainous regions that typically introduce topographic and atmospheric effects.

In addition, despite the promise that advancing geospatial techniques provide successive new insights, the creation of digital geospatial models or simulations is based on several assumptions. Firstly, we cannot use these to recreate the past, but offer biased interpretations of how we believe selected past human dynamics may have functioned (Whitley, 2017). Secondly, we may put some effort into developing the parameters and details that go into a model, but choices over the selection of elements are influenced by the research questions pursued, and the ways in which they are articulated within the programming framework. Thirdly, models are still works of human interpretation. They should not be taken to be as inherently more accurate than other archaeological ideas, regardless of how much detail or quantification they contain. Computational modelling and geospatial analyses (like all statistics) are not independently objective, nor should we treat them that way (Whitley, 2017).

To overcome, or move beyond, such criticisms and those that suggest that the quantitative nature of GIS divorces archaeology from human experience, Llobera (1996) contends that we need some middle ground/bridging concepts to: (1) situate models and methods within context-rich narratives; (2) explore how processes play out within particular contexts rather than seek universal norms of behaviour; (3) shorten the gap between empirical information and narratives; and (4) generate multiple scenarios as feedback to results. These recognize and acknowledge the diversity of approaches to archaeological modelling and interpretation, and, fundamentally, that archaeological artefacts themselves can only present selective – and therefore inherently biased – perspectives on the past.

Conclusion

Archaeologists have long recognized that mapping lies at the core of their discipline. The tools and techniques for the gathering, organization, presentation and interpretation of geospatial data have evolved to offer new insights more efficiently. Their application will only continue to grow as the costs of using them are reduced to broaden accessibility for researchers. The culmination of these tools and data on a global scale has allowed for considerable advancements in our understanding of human history, as well as allowing for better conservation and management of archaeological sites. The introduction of 3D technologies has also improved the accessibility of sites and artefacts, while 'digital twins' will help to ensure that future generations continue to have access to our past. Of course, as these datasets grow, so too does apprehension about data quality, privacy, and how to best manage these geospatial datasets through geographic data standards.

References

Berggren, Å., Dell'Unto, N., Forte, M., Haddow, S., Hodder, I., Issavi, J., Lercari, N., Mazzucato, C., Mickel, A. and Taylor, J.S. (2015) "Revisiting Reflexive Archaeology at Çatalhöyük: Integrating Digital and 3D Technologies at the Trowel's Edge" *Antiquity* 89 pp.433–448.

Chase, A.F., Chase, D.Z. and Weishampel, J.F. (2013) "The Use of LiDAR at the Maya Site of Caracol, Belize" In Comer, D.C. and Harrower, M.J. (Eds) *Mapping Archaeological Landscapes from Space* New York: Springer, pp.187–197.

Comer, D.C. (2013) "Petra and the Paradox of a Great City Built by Nomads: An Explanation Suggested by Satellite Imagery" In Comer, D.C. and Harrower, M.J. (Eds) *Mapping Archaeological Landscapes from Space* New York: Springer, pp.73–83.

Dykes, J., MacEachren, A.M. and Kraak, M.J. (2005) *Exploring Geovisualization* Amsterdam: Elsevier.

Ebert, D. (2004) "Applications of Archaeological GIS" *Canadian Journal of Archaeology/Journal Canadien d'Archéologie* 28 (2) pp.319–341.

Gillings, M., Hacıgüzeller, P. and Lock, G. (Eds) (2019) *Re-mapping Archaeology: Critical Perspectives, Alternative Mappings* Abingdon: Routledge.

Gupta, N. and Devillers, R. (2017) "Geographic Visualization in Archaeology" *Journal of Archaeological Method and Theory* 24 (3) pp.852–885.

Harrower, M.J. and Comer, D.C. (2013) "Introduction: The History and Future of Geospatial and Space Technologies in Archaeology" In Harrower, M.J. and Comer, D.C. (Eds) *Mapping Archaeological Landscapes from Space* New York: Springer, pp.1–8.

Kennedy, D. (1998) "Declassified Satellite Photographs and Archaeology in the Middle East: Case Studies from Turkey" *Antiquity* 72 (277) pp.553–561.

Leszczynski, A. (2009) "Quantitative Limits to Qualitative Engagements: GIS, its Critics, and the Philosophical Divide" *The Professional Geographer* 61 (3) pp.350–365.

Llobera, M. (1996) "Exploring the Topography of Mind: GIS, Social Space and Archaeology" *Antiquity* 70 (269) pp.612–622.

Lloyd, D. and Dykes, J. (2011) "Human-centered Approaches in Geovisualization Design: Investigating Multiple Methods through a Long-term Case Study" *IEEE Transactions on Visualization and Computer Graphics* 17 (12) pp.2498–2507.

Lock, G. and Pouncett, J. (2017) "Spatial Thinking in Archaeology: Is GIS the Answer?" *Journal of Archaeological Science* 84 pp.129–135.

McCoy, M.D. (2017) "Geospatial Big Data and Archaeology: Prospects and Problems Too Great to Ignore" *Journal of Archaeological Science* 84 pp.74–94.

Megarry, W. and Davis, S. (2013) "Beyond the Bend: Remotely Sensed data and Archaeological Site Prospection in the Boyne Valley, Ireland" In Comer, D.C. and Harrower, M.J. (Eds) *Mapping Archaeological Landscapes from Space* New York: Springer, pp.85–95.

Rayne, L., Bradbury, J., Mattingly, D., Philip, G., Bewley, R. and Wilson, A. (2017) "From above and on the Ground: Geospatial Methods for Recording Endangered Archaeology in the Middle East and North Africa" *Geosciences* 7 (4) pp.1–31.

Richards-Rissetto, H. (2017a) "What can GIS+ 3D Mean for Landscape Archaeology? *Journal of Archaeological Science* 84 pp.10–21.

Richards-Rissetto, H. (2017b) "An Iterative 3D GIS Analysis of the Role of Visibility in Ancient Maya Landscapes: A Case Study from Copan, Honduras" *Digital Scholarship in the Humanities* 32 (Supplement 2) pp.195–212.

Savage, S.H., Levy, T.E. and Jones, I.W. (2013) "Archaeological Remote Sensing in Jordan's Faynan Copper Mining District with Hyperspectral Imagery" In Comer, D.C. and Harrower, M.J. (Eds) *Mapping Archaeological Landscapes from Space* New York: Springer, pp.97–110.

Spelterini, E. (1928) *Über den Wolken/Par dessus les nuages* Zürich: Brunner & Co.

Tschan, A.P., Raczkowski, W. and Latałowa, M. (2000) "Perception and Viewsheds: Are They Mutually Inclusive? In Lock, G. (Ed.) *Beyond the Map: Archaeology and Spatial Technologies* Amsterdam: IOS Press, pp.28–48.

Ur, J.A. (2013) "CORONA Satellite Imagery and Ancient Near Eastern Landscapes" In Comer, D.C. and Harrower, M.J. (Eds) *Mapping Archaeological Landscapes from Space* New York: Springer, pp.21–31.

Whitley, T.G. (2017) "Geospatial Analysis as Experimental Archaeology" *Journal of Archaeological Science* 84 pp.103–114.

42
MAPPING PLANETARY BODIES
Trent Michael Hare

Introduction

Planetary mapping, also referred to as planetary cartography, is a practice that enables planetary exploration and scientific research. To piece together the story of a planetary body, whether a planet, moon, asteroid, or comet, a scientist must be able to rely on the locations of the observations made on the surface of that body. Standards such as uniformity of coordinate systems, accurate horizontal and vertical positioning, and methods used to spatially register these data into a reliable and trusted science data product are critical.

Besides sketches using the naked eye, mapping of Earth's moon began in the early 1600s with the development of the telescope. Earth-based telescopes were used until spacecraft were launched in the mid-1900s. In 1964, the US Ranger VII spacecraft transmitted 4,316 high-resolution telescopic vidicon television camera photos before impacting the Moon. These images represent the first significant catalog of remotely sensed planetary images as taken by a satellite. Using the Ranger images as well as images subsequently acquired by NASA's five Lunar Orbiter flights (1966–1967) and the Apollo missions (1961–1972), the *United States' Defense Mapping Agency (DMA)* was tasked with producing the first controlled lunar photomosaic and topographic maps (Figure 42.1; Schimerman, 1973). Prior to modern digital image processing, these original images were manually tied together and then artistically blended using airbrush painting techniques to make a consistent base map without seams or differences in solar illumination.

To help support and coordinate consistency of created extraterrestrial maps, the International Astronomical Union (IAU) established the Working Group on the Cartographic Coordinates and Rotational Elements of Planets and Satellites in 1976. This working group strives to report triennially on the preferred rotation rate, spin axis, prime meridian, and reference surface for planets and satellites (Archinal *et al.*, 2018). These standardized parameters are the foundation for building critical data products and to ensuring that feature locations are dependable from one map to another. Despite these efforts, all planetary maps are not perfectly co-registered. As knowledge of the shape of the body evolves and data accuracy improves with subsequent missions, surface features are expected to shift from map to map due to the original uncertainties in shape and the gathered data. Methods that address this uncertainty and the importance of continually improving the spatial locations for planetary maps will be discussed in the next section.

Mapping planetary bodies

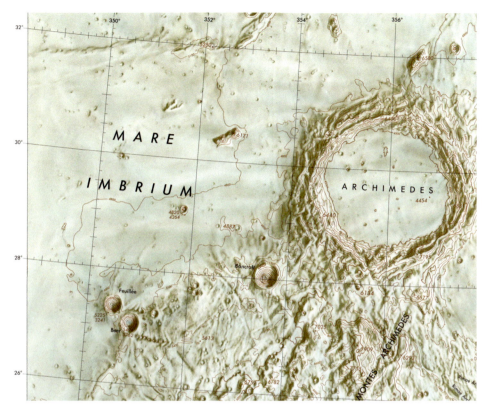

Figure 42.1 Lunar topographic maps of the Mare Imbrium region extracted from the Lunar Map Series (LM # 41), scale 1:1,000,000, published by the Defense Mapping Agency, Aerospace Center, in 1976

Source: NASA, LPI.

Lastly, similar to the production of terrestrial maps, extraterrestrial maps require a large number of choices beyond the IAU-recommended details of the coordinate system, including the map projection, the scale (often limited by the available resolution of the data), the extent of an individual digital product, the defined attributes and cartographic symbology, and the occurrence and extent of geographic place names. Thus, the section 'Base map construction: photogrammetric processing' of this chapter discusses typical processing for imaging instruments. Sections 'Map projections' and 'Nomenclature' introduce map projections and standardized nomenclature. Section 'Planetary GIS' provides a summary for geographic information system (GIS) mapping, highlighting geologic mapping as one of many use cases. Lastly, sections 'Planetary spatial data infrastructures' and 'PSDI use case: MarsGIS' speak to standardization efforts in the planetary community focused on spatial data infrastructures (SDIs) and highlight one use case to help select the best regions on Mars for eventual human missions.

Base map construction: photogrammetric processing

Digital image processing was not technically practical until the late 1980s. Although much of the process can now be automated, the generation of image mosaics is still a complicated

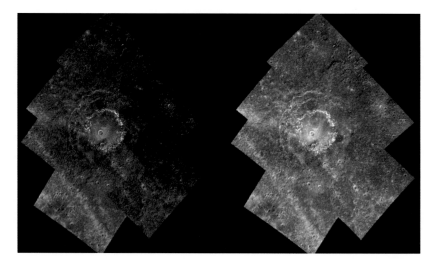

Figure 42.2 Example image mosaic of Raditladi crater (258 km diameter) from NASA's Mercury Dual Imaging System (MDIS) on the MESSENGER spacecraft showing no photometric correction (left), then the same image mosaic after applying the photometric corrections to the images prior to the mosaic creation (right)

Source: NASA/USGS (K. Becker).

process. With the growing volume of collected datasets for bodies like the Moon or Mars, the generation and updating of global base maps continue to be difficult.

As employed for terrestrial data sets, planetary image processing includes correcting for typical camera distortions, location geometry, and surface topography (Laura *et al.*, 2020). Planetary software commonly updates radiometry parameters and performs photometric corrections. This is because the collection of planetary data often follows a 'greedy' strategy, wherein missions cannot always collect images at optimal times of the day or consistent season. Radiometric calibration recalculates the image values based on exposure time, flat-field observations, dark current observations, and other factors that describe the characteristics of the physical imaging system. Photometric corrections help to adjust the brightness and contrast of images acquired under different solar illuminations and viewing geometries, the goal being to adjust the resulting images to appear as if obtained under similar conditions (Figure 42.2).

Although now significantly better with modern missions, spatial errors on the order of kilometers were common for planetary datasets and their derived base maps. As we send more landers, rovers, and eventually humans to the Moon, Mars, and other bodies, minimizing spatial inaccuracies is becoming increasingly more important. As more data are gathered across existing and forthcoming missions, and techniques improve, it is critical to improve and regenerate products to continuously improve their positional accuracies. This not only supports the ability to safely land robotic spacecrafts and humans on a planetary surface but also allows for better cross-instrument and mission data fusion.

For planetary data, positional accuracy primarily depends on the ability to measure or reconstruct the position of the spacecraft and the pointing of the instrument. An informational system to capture these positional data is called SPICE (spacecraft, planet, instrument, C-matrix (pointing), and events). The use of SPICE has become an essential component in the development and operation of the majority of NASA's and other international planetary

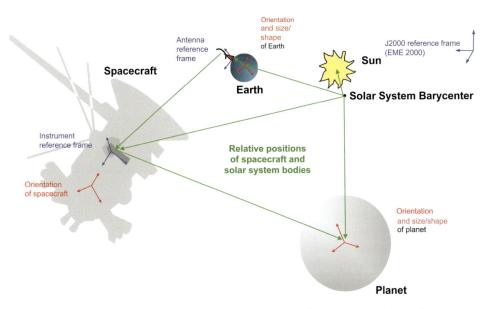

Figure 42.3 Spice framework. During the capture of an image from a spacecraft, to gather accurate SPICE, the vehicle's ephemeris (trajectory), the body's ephemeris, orientation and size, the instrument's field of view, shape, orientation, and lastly the internal timing is critical to understand an image's approximate location on that body

Source: Adapted from NASA's Navigation and Ancillary Information Facility. Here J2000 (also known as EME2000) defines a frame based on the Earth's equator and equinox, determined from observations of planetary motions, plus other data.

missions (Figure 42.3). Spatial location is also dependent on our knowledge of the shape and rotation of the targeted body. Even with excellent SPICE (positions) and knowledge of the shape of the body, decreasing positional errors further requires photogrammetric (or related) processing techniques.

The photogrammetric process of adjusting images as a group, called a bundle or block adjustment, helps to spatially locate individual images relative to each other (Edmundson et al., 2012). This process also establishes a control network that can be used for follow-on data production. A control network consists of a set of well-defined topographic or image points whose latitudes and longitudes have been precisely computed. In general, the construction of planetary control networks has been either (1) derived photogrammetrically or (2) constructed from laser altimeter instrument (Lidar) as flown from an orbiting spacecraft.

Two examples for photogrammetric image processing systems currently used in the planetary community are ISIS (Integrated Software for Imagers and Spectrometers) and VICAR (Video Image Communication and Retrieval). Since the 1970s, the Astrogeology Science Center at the US Geological Survey (USGS) has supported ISIS and its predecessors, Planetary Image Cartography System (PICS) and Flagstaff Image Processing System (FIPS). To list a few examples, ISIS is currently used to process images from NASA's Mars Odyssey, Mars Global Surveyor, Mars Reconnaissance Orbiter, Lunar Reconnaissance Orbiter, and the Mercury MESSENGER mission. The VICAR system began in 1966 at the Jet Propulsion Laboratory (JPL). This system processed data from NASA missions like Ranger, Surveyor, Mariner, Viking, and Voyager. It has now been upgraded and ported to nearly all modern operating systems and is still used by many facilities to process data for Earth, Mars, and other bodies. VICAR is also currently being used by several NASA rover missions and by ESA's High Resolution Stereo Camera (HRSC) aboard the Mars Express spacecraft.

The photogrammetric applications described above necessitate both the SPICE and the collected images to be returned to Earth for processing. However, efforts are being instigated to shift the processing closer to the planetary body. While we have returned petabytes of data from various planetary missions, the farther we venture into the Solar System, the more dramatically reduced the amount of data we can send back to Earth due to bandwidth limitations (Deutsch, 2020). A recent study pointed to the potential benefits of expanding the available storage and relocating the initial photogrammetric processing to the spacecraft (Vander Hook et al., 2020). Once these data have been processed and prioritized, only then would a subset of data be sent back to Earth for analysis. Because the concept of never returning all data from a mission seems disconcerting, Vander Hook et al. (2020) also proposed the concept of a small *data cycler*, whose job is to simply fly to the planet and back, locally gather the data from the orbiting satellite(s), and then bring the data back to Earth. In this scenario, not only are high priority data still being returned in a timely manner, but the full catalog of collected data can be returned by using one or more data cyclers.

Map projections

Map projections are mathematical equations for transforming a three-dimensional (3D) body onto a two-dimensional (2D) plane or Cartesian coordinate system. While seemingly a mundane task, the definition and standardization of coordinate systems greatly influence the usability of any one product. Although dozens of defined map projections are available, a much smaller subset are more commonly used in both terrestrial and extraterrestrial mapping efforts. Unfortunately, users will encounter some quirks in the use of map projections on planetary bodies, including positive west longitudes systems and geocentric (planetocentric) latitude systems (Hargitai et al., 2019). Understanding which system to use for which body is defined by the IAU (Archinal et al., 2018). Although many planetary bodies can be defined as triaxial ellipsoids (using three radius values: sub-planetary equatorial, along-orbit equatorial, and polar axes), the IAU also commonly defines a best-fit mean radius, simplifying the definition for the map projection to use a single spherical radius. Though using a perfect sphere rather than an ellipsoid can increase distortions in the data, it also greatly increases useability across software applications.

As stated by Snyder (1987), no one projection is best for mapping, and care must be taken to choose a projection that is suitable for the area of study given that every projection incurs some type of spatial distortion. Most printed maps of the planets and their satellites have been based on conformal projections (angle preserving): Mercator for low latitudes, polar stereographic for high latitudes, Lambert conformal conic for intermediate latitudes at small scales, and transverse Mercator for large-scale maps. However, for modern digital products, the usability of the data across various mapping applications, referred to a product's interoperability, is just as important (Hare et al., 2017). Important qualities for digital interoperability are the use of a simple map projection; thus, global planetary data are commonly downloaded using the meters-based simple cylindrical map projection (also known as plate carrée, equidistant cylindrical, or equirectangular). This projection (whether in meters or similarly rectangular-shaped version in degrees, usually referred to as geographic) is also often used for streaming global datasets to Web mapping applications (i.e., Cesium, NASA Treks) or desktop GIS applications like Esri's ArcGIS Pro or OSGeo's QGIS. Because polar regions are grossly distorted in a simple cylindrical projection, high-latitude products are commonly released in a meters-based polar stereographic map projection (latitude ranges from $\pm\sim 60°$ to $90°$).

Mapping planetary bodies

As humanity explores more irregularly shaped (non-spherical) objects like asteroids, comets, and small moons, common map projections are often overly simplistic and cannot always capture the shape of the body. While a best-fit sphere still may be defined, the distortions are often too significant to use for science and engineering purposes. Thus, the planetary community is currently researching new or modified map projections. For example, the morphographic projection (Stooke, 2015) essentially warps the body to the output 2D map by using a unique radius at every transformed point. In addition, fully 3D-enabled mapping applications are being developed to address the limitations of using standard map projections on irregularly shaped bodies. But until these new map projections and 3D applications are more widely accepted and available, the community will often still resort to highly distorted map projections to support interoperability. For example, NASA's Near Earth Asteroid Rendezvous (NEAR) team members define a latitude- and longitude-based (rectangular) system (Figure 42.4) and reproject the body into the simple cylindrical map projection for digital distribution (Zuber et at., 2000).

3D software applications and accepted 3D formats are becoming more common. For example, the interactive Small Body Mapping Tool (SBMT), created at Johns Hopkins University Applied Physics Laboratory, allow data acquired of the asteroid Eros to be displayed within a 3D interactive environment, directly addressing the limitations of typical 2D map projections that are defined using latitude, longitude, and radius. To support such irregular

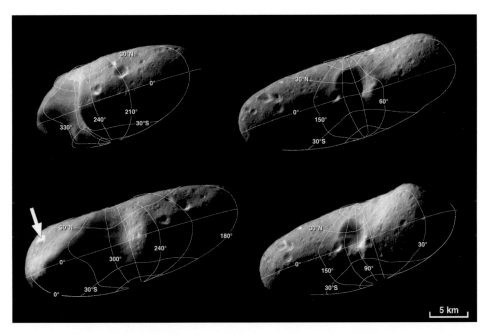

Figure 42.4 The potato-like shape of the asteroid Eros demonstrates how more typical map projections, meant to be used with ellipsoid or spherical bodies, will quickly break down. Four image mosaics produced by the NEAR mission team and the latitude- and longitude-based reference system the team defined to locate points on the asteroid's surface are shown here. As is typical for many planetary bodies, positive west longitude directions are also used here. The prime meridian defined by the NEAR team is drawn though a large, bright crater at one end of Eros (white arrow). Once projected to simple cylindrical or geocentric, the now wavy grid-lines will snap straight, often grossly distorting the image data underneath

Source: NASA/JPL/JHUAPL. url: https://photojournal.jpl.nasa.gov/catalog/PIA03112.

bodies, these applications generally use triangulated (or tessellated) 3D points in a metric *X, Y, Z* geocentric system, often called a shapemodel. *SBMT*, like any typical desktop GIS application, allows users to overlay various layers and accurately measure and collect catalogs of surface features (e.g., craters) using these defined 3D shapemodels (Ernst *et al.*, 2018).

Nomenclature

Terrestrial maps commonly show placenames like towns or labeled rivers and roads. Planetary nomenclature is used to uniquely identify a feature on the surface of a planet or satellite so that the feature can be easily located and described (Hunter *et al.*, 2019). The IAU has been the authority of planetary and satellite nomenclature since its first organizational meeting in Brussels in 1919. The first goals were to normalize various, often discrepant nomenclature systems for the Moon and Mars used across different countries. The names defined come from cultures all over the world and generally follow a theme for a feature type and body. For example, on Mars, larger craters are named after deceased scientists, authors, or artists who have contributed significantly to the study or lore of Mars. For Mercury, large valleys are named after abandoned cities (or towns or settlements) of antiquity. The current nomenclature database is managed by the Astrogeology Science Center on behalf of the IAU (https://planetarynames.wr.usgs.gov/). The USGS and IAU recommend that maps created for digital or hard-copy distribution should completely and accurately display internationally approved nomenclature, based on the map's scale (Figure 42.5).

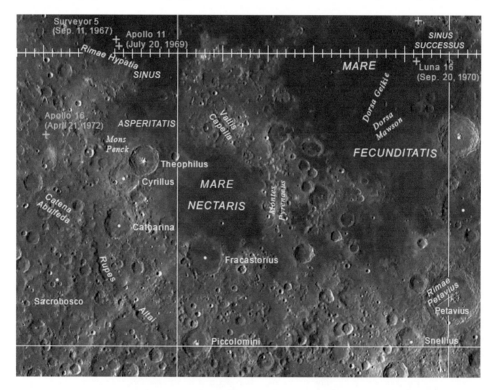

Figure 42.5 Extracted portion of the original map (Hare *et al.*, 2015) centered on Mare Nectaris. This map at 1:10,000,000 scale is only showing names for the larger features on the Moon and is overlain on Lunar Reconnaissance Orbiter Wide Angle Camera (WAC) global mosaic

Source: NASA/ASU/USGS.

Planetary GIS

GIS applications have been used for planetary projects for many years (Frigeri *et al.*, 2011; Hare *et al.*, 2003; Hargitai, 2019). Planetary researchers rely on GIS technologies because they can efficiently overlay multiple datasets, collect feature catalogs (e.g., craters, structural ridges, or channels), and provide enhanced geospatial analytical techniques. Direct planetary support is available across several commercial and open-source, Web-based and desktop GIS software applications. This has resulted from the planetary community directly working with the software developers, writing extensions or plugins, and working with standards-based organizations like the Open Geospatial Consortium (OGC, Hare *et al.*, 2018). Though these companies and working groups recognize planetary mapping as a niche field, their support of planetary science through specialized file formats and analytical techniques continues to grow. Also, extremely capable GIS applications are largely written exclusively for planetary analysis and mapping, for example, Arizona State University's JMARS and the already mentioned NASA Treks and the Small Body Mapping Tool (SBMT).

A popular practice of GIS in planetary research is its use to characterize a planet's geologic and structural history. Geologic maps provide a local, regional, or global contextual framework for summarizing thematic research. They are also frequently used for specialized investigations such as targeting regions of interest for data collection and for characterizing sites for landed missions (Rossi and van Gasselt, 2018). Within the planetary community, but especially for NASA-funded researchers, geologic maps that are intended for publication by USGS are a well-defined and deliberately controlled process. For decades, modern planetary geologic mapping published by the USGS are standardized, peer reviewed, and edited using a wide range of cartographic software, technical procedures, and publication requirements (Skinner *et al.*, 2019). Consistent adherence to these community standards regarding mapping methods and representation is critical and must be thoroughly considered and described. These methods apply with respect to the defined base map, data collection, attribution, symbology (Nass *et al.*, 2011), documentation, and distribution formats (Hare *et al.*, 2017). The goal in using these strict standardized methods is to ensure the map details are readily understood, cross comparable, and used across the community by terrestrial and planetary researchers (Figure 42.6).

Planetary spatial data infrastructures

Since 2014, the planetary community has worked to define SDIs. An SDI is a robust framework for (1) data and data products, (2) metadata and data access mechanisms, (3) standards, (4) policies, and (5) a user community that helps to define and standardize data and data access necessary to meet the specified goals (Laura and Beyer, 2021; Laura *et al.*, 2017). For planetary spatial data infrastructure (PSDI), Laura and Beyer (2021) include the following as foundational geospatial data products: (1) geodetic control networks, (2) topography, and (3) rigorously controlled and orthorectified images tied to a standardized reference frame. They also define more derived or subsequent products like geologic or compositional maps and feature catalogs as *framework* products. Framework products are no less important but are enhanced by being tightly coupled or tied to the foundational datasets.

One of the major tenants of any SDI is a well-established control network that ties all data products. Previously, we noted that the IAU helps define the body size and reference meridian, and we also described the process for using control networks to build photogrammetrically

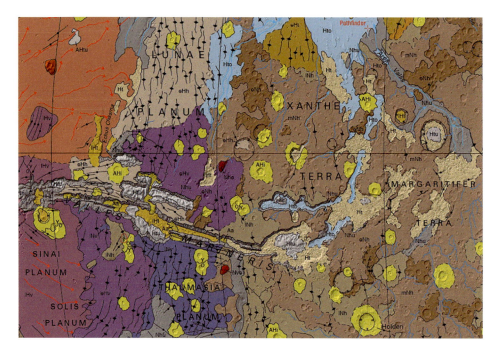

Figure 42.6 Extracted portion of the original global geologic map of Mars (Tanaka *et al.*, 2014) showing Valles Marineris (canyon system more than 4,000 km long). For scale-based mapping constraints, vertex spacing of drafted line work was set at 5 km (four vertices per millimeter at 1:20,000,000 scale) and the minimum feature length accepted was 100 km

Source: USGS.

bundled image mosaics. When an image mosaic or base map is created in this fashion, it is called a *controlled* mosaic. An example for a PSDI *foundational* controlled mosaic includes the Mars Thermal Emission Imaging System (THEMIS) base maps (Fergason and Weller, 2018). When derived from a Lidar instrument, a dataset of topographic control points are adjusted as a group and generally converted to a digital elevation model (DEM). DEM-derived control networks have the added benefit of deriving orthographically rectified images, a process that removes geometric distortions in the images due to changes in topography. We note that stereo imagery (Fergason *et al.*, 2020; Kirk *et al.*, 2017), stereo-photoclinometry (Gaskell *et al.*, 2008), or dense reconstructed image-based control networks (Bland *et al.*, 2020) can also be used to generate local or global DEMs.

Controlled mosaics and DEMs should be emphasized because they form the foundation for any terrestrial or non-terrestrial SDI and can help other datasets be co-aligned. There is a 'lesser' tier of provisional products that are also commonly created and released by the planetary community but should not be considered foundational. These are often called semi-controlled or uncontrolled mosaics. Semi-controlled mosaics are generally made from orthographically rectified, spatially filtered images and might be spatially tied to one another, but they are not tied to a wider control network. Uncontrolled mosaics are generally 'quick-look' mosaics, which also are geometrically corrected using only initial (SPICE) tracking data only. No attempt is made to adjust the images to each other or a control network. Uncontrolled mosaics will commonly contain geometric and radiometric distortions and overlapping images and may have large errors and discontinuities.

Greeley and Batson (1990: 70) state that 'Except for unusual circumstances, only controlled photomosaics and ortho-photomosaics are published in the form of maps. Uncontrolled and semi-controlled mosaics are used for data cataloging and interim scientific reporting, but they are not formal cartographic products'. We now consider their described maps above to include digital mosaics and not just printed maps. Though the creation of controlled mosaics is ideal, uncontrolled and semi-controlled mosaics are often released due to the time required to properly generate a controlled mosaic. For some missions, the ability to more precisely track the spacecraft's location has allowed for better registered and tonally blended semi-controlled mosaics. However, the community should always strive to support controlled mosaics, as foundation products, because they facilitate science results and interoperability of data gathered by different missions. Similarly, the scientific community should recognize and make efforts to document the inherent spatial limitations and accuracies associated with observations made on semi-controlled or uncontrolled mosaics.

PSDI use case: MarsGIS

MarsGIS is a NASA and community-based initiative being developed to support both landing site analysis and eventual human operations on Mars. The MarsGIS working group (WG) strives to integrate efforts for NASA in support of the Mars Human Landing Site Study (HLS2). To help understand the MarsGIS initiative, mapping their goals to a PSDI is useful.

PSDI theme – foundation data products

Both existing Mars foundational datasets and framework data products (e.g., geology, mineralogy, feature catalogs) will be used within the MarsGIS PSDI. Next, we enumerate the currently available Mars foundational datasets in order of increasing spatial resolution (Laura and Beyer, 2021). Note that all these datasets are tied to Mars Orbiter Laser Altimeter (MOLA) spot observations, which have an absolute horizontal uncertainty of 100 m and 3 m vertical uncertainty. All subsequent datasets will, in the best case, share these absolute accuracies and in the worst case, cumulatively contribute additional uncertainty.

- MOLA topography, near-global coverage, 463 m/pixel, created by NASA Goddard and released from the Planetary Data System (PDS).
- Mars digital image mosaic (MDIM 2.1), global coverage, 231 m/pixel, created by the USGS and released from the USGS Annex.
- THEMIS controlled day/night infrared mosaics, near-global coverage, 100 m/pixel, created by USGS and released from the USGS Annex.
- High Resolution Stereo Camera (HRSC) topography and derived orthorectified images, regional coverage, ~12 m/pixel images and ~50 m/pixel topography, created by the HRSC Team and released by the Planetary Science Archive (PSA) and PDS.
- Controlled Context Camera (CTX) near-global coverage, 5–6 m/pixel, versions created by the California Institute of Technology and the Southwest Research Institute (Robbins *et al.*, 2023).
- High Resolution Imaging Science Experiment (HiRISE) topography and derived orthorectified images, very sparse coverage, ~0.25 to 1 m/pixel, created by the University of Arizona, USGS, and Arizona State University, released by PDS.

If additional datasets are defined, the MarsGIS WG will recommend, when possible, that the data be geospatially controlled to MOLA, the current IAU-recommended control network and reference frame for Mars. For a list of data products for Mars and references, please see Laura and Beyer (2021).

PSDI theme – data access and metadata

A priority goal for the MarsGIS WG is to design and maintain a strategic investment plan for Mars data and infrastructural services that transcends individual missions. This includes goals for deploying a MarsGIS data catalog/registry for the discovery and access of existing data products and the development of standards and best practices on how to characterize, capture, and present uncertainty and distortion in metadata. These infrastructural services and data catalogs will help to separate the data storage and access from the tools that will consume it (Figure 42.7).

In addition to supporting infrastructural services, the MarsGIS WG will support well-documented guidelines and best practices for community-available processing workflows (on-line services and tools). Lastly, the MarsGIS WG will also research and determine processes to geospatially link features (e.g., Gale crater, Valles Marineris, mountains) to relevant published research to improve ease of access and searchability.

PSDI theme – standards

The MarsGIS WG has stated several goals related to supporting standards. These include the metadata and data access standards mentioned above as well as the following:

- The promotion of common data formats for interoperability between different applications and facilities;
- The establishment of cartographic standards (e.g., symbologies) for engineering elements required for Exploration Zones, leveraging existing cartographic standards when possible (e.g., from the Federal Geographic Data Consortium [FGDC]); and
- Defining standards and best practices for conversion, distributing, visualizing, and archiving temporal datasets (e.g., 3D + time).

PSDI theme – policies

The MarsGIS WG established responsibilities and policies within the working group's charter as summarized next.

- The MarsGIS WG will be chartered through the Mars Exploration Program to integrate and manage MarsGIS efforts for NASA.
- The MarsGIS WG will maintain and coordinate, with appropriate NASA management, a set of MarsGIS goals.
- The MarsGIS WG shall operate by consensus management. Decisions and recommendations will be communicated to the Mars Exploration Program.

PSDI theme – user community

The MarsGIS WG has prioritized fostering a community of practice to support MarsGIS coordination, technical task execution, and sharing of knowledge and capabilities. This includes providing guidance to self-moderated citizen science initiatives and public outreach

Mapping planetary bodies

Figure 42.7 Showing JPL's MarsTrek (above) and ASU's JMARS (below) as initial consumers and GIS interfaces for the MarsGIS planetary spatial data infrastructure (PSDI) services. Both applications showing foundational datasets, the MDIM 2.1 mosaic (above) and MOLA topography (below) see https://marstrek.jpl.nasa.gov and https://jmars.asu.edu/

efforts to aid them in processing data and publishing their GIS-ready datasets in standardized formats.

In summary, the MarsGIS WG, in concert with a small representative group of the planetary community (e.g., mission planners and engineers, scientists, data providers, outreach specialists), have developed initial goals to support landing site analysis and eventual human operations on Mars. As described, incorporating these goals into a PSDI will highlight the areas for further study. As defined in a PSDI, one of the next steps for the MarsGIS WG will be to perform a knowledge inventory to help identify strategic gaps in foundational data products (Laura and Beyer, 2021), the current state of data interoperability in off-the-shelf geospatial tools, and available data access methods, as well as to engage the user community.

Conclusion

Staying abreast of the volume of data and maps generated is already difficult for planetary researchers and will only become more challenging with the forthcoming missions to the Moon, Mars, and beyond. The efforts to streamline, standardize, and simplify the complexity of the gathered datasets into usable derived maps will continue to be a priority. As outlined above, most planetary researchers will need to continue to be data experts or lean heavily on data experts. Mapping planetary bodies involves corrections to distortion and calibration; and updates to the spatial location generally require modern photogrammetric techniques. Once processed, distributing these datasets enables researchers to work with common forms and formats.

Several efforts are underway to lessen the complexity resulting from interoperability of data formats to supporting direct access to the data in custom and commercial GIS mapping applications. Efforts as defined under planetary spatial data infrastructures can help to even further lessen the barrier of entry for researchers and the public alike. Understanding when a dataset can be considered foundational will help to ensure consistency, and ultimately trust in a dataset for use in general research applications and mission planning, as well as robotic and human mission support tasks.

Acknowledgments

I am grateful to those in the planetary mapping community who continue to support advances in the field. Many of the topics discussed above continue to be deliberated and often promoted by members of the Mapping and Planetary Spatial Infrastructure Team (MAPSIT). Those discussions continue to impact our mapping community for the better. Lastly, thanks to James Skinner, Jr., Janet Slate, Marc Hunter, Janet Richie, and Tenielle Gather for their valuable edits. Any use of trade, firm, or product names is for descriptive purposes only and does not imply endorsement by the US Government.

References

Archinal, B.A., Acton, C.H., A'Hearn, M.F., Conrad, A., Consolmagno, G.J., Duxbury, T., Hestroffer, D., Hilton, J.L., Kirk, R.L., Klioner, S.A., McCarthy, D., Meech, K., Oberst, J., Ping, J., Seidelmann, P.K., Tholen, D.J., Thomas, P.C. and Williams, I.P. (2018) "Report of the IAU Working Group on Cartographic Coordinates and Rotational Elements: 2015" *Celestial Mechanics and Dynamical Astronomy* 130 (22) DOI: 10.1007/s10569-017-9805-5.

Bland, M.T., Weller, L.A., Mayer, D.P. and Archinal, B.A. (2020) "A Global Shape Model for Saturn's Moon Enceladus from a Dense Photogrammetric Control Network" *ISPRS Annals*

Photogrammetry, Remote Sensing and Spatial Information Sciences V-3-2020 pp.579–586 DOI: 10.5194/isprs-annals-V-3-2020-579-2020.

Deutsch, L.J. (2020) "Towards Deep Space Optical Communications" *Nature Astronomy* 4 (907) DOI: 10.1038/s41550-020-1193-1.

Edmundson, K.L., Cook, D.A., Thomas, O.H., Archinal, B.A. and Kirk, R.L. (2012) "Jigsaw: The ISIS3 Bundle Adjustment for Extraterrestrial Photogrammetry" *ISPRS Annals of Photogrammetry, Remote Sensing and Spatial Information Sciences*, I-4-2012 pp.203–208.

Ernst, C.M., Barnouin, O.S. and Daly, R.T. (2018) "The Small Body Mapping Tool (SBMT) for Accessing, Visualizing, and Analyzing Spacecraft Data in Three Dimensions" *49th Lunar and Planetary Science Conference* The Woodlands, TX (Abstract #1043) Available at: https://www.hou.usra.edu/meetings/lpsc2018/pdf/1043.pdf.

Fergason, R.L. and Weller, L. (2018) "The Importance of Geodetically Controlled Data Sets: THEMIS Controlled Mosaics of Mars, a Case Study" *Planetary Science Informatics and Data Analytics Conference* St. Louis, MO (Abstract #2081) Available at: https://www.hou.usra.edu/meetings/informatics2018/pdf/6030.pdf.

Fergason, R.L., Hare, T.M., Mayer, D.P., Galuszka, D.M., Redding, B.L., Smith, E.D., Shinaman, J.R., Cheng, Y. and Otero, R.E. (2020) "Mars 2020 Terrain Relative Navigation Flight Product Generation: Digital Terrain Model and Orthorectified Image Mosaics" *51st Lunar and Planetary Science Conference* DOI: 10.5066/P906QQT8.

Frigeri, A., Hare, T.M., Neteler, M., Coradini, A., Federico, C. and Orosei, R. (2011) "A Working Environment for Digital Planetary Data Processing and Mapping Using ISIS and GRASS GIS" *Planet Space Science (PSS)* 59 pp.1265–1272 DOI: 10.1016/j.pss.2010.12.008.

Gaskell, R.W., Barnouin-Jha, O.S., Scheeres, D.J., Konopliv, A.S., Mukai, T. Abe, S., Saito, J., Ishiguro, M., Kubota, T., Hashimoto, T., Kawaguchi, J., Yoshikawa, M., Shirakawa, K., Kominato, T., Hirata, N. and Demura, H. (2008) "Characterizing and Navigating Small Bodies with Imaging Data" *Meteoritics and Planetary Science* 43 (6) pp.1049–1061.

Greeley, R. and Batson, R.M. (1990) *Planetary Mapping (Cambridge Planetary Science Old Books Series)* Cambridge: Cambridge University Press.

Hare, T.M., Hayward, R.K., Blue, J.S., Archinal, B.A., Robinson, M.S., Speyerer, E.J., Wagner, R.V., Smith, D.E., Zuber, M.T., Neumann, G.A. and Mazarico, E. (2015) "Image Mosaic and Topographic Map of the Moon: US Geological Survey Scientific Investigations Map 3316" (*2 sheets*) (Available at: https://dx.doi.org/10.3133/sim3316).

Hargitai, H. (Ed.) (2019) *Planetary Cartography and GIS (Lecture Notes in Geoinformation and Cartography)* Cham, Switzerland: Springer.

Hargitai, H., Willner, K. and Hare, T. (2019) "Fundamental Frameworks in Planetary Mapping: A Review" In Hargitai H. (Ed.) *Planetary Cartography and GIS (Lecture Notes in Geoinformation and Cartography)* Cham, Switzerland: Springer, pp.75–101.

Hunter, M., Hayward, R. and Hare, T. (2019) "Planetary Nomenclature" In Hargitai, H. (Ed.) *Planetary Cartography and GIS (Lecture Notes in Geoinformation and Cartography)* Cham, Switzerland: Springer, pp.65–74.

Kirk, R.L., Howington-Kraus, E., Edmundson, K., Redding, B., Galuszka, D., Hare, T. and Gwinner, K. (2017) "Community Tools for Cartographic and Photogrammetric Processing of Mars Express HRSC Images" *International Society for Photogrammetry and Remote Sensing* XLII-3/W1 pp. 69–76 DOI: 10.5194/isprs-archives-XLII-3-W1-69-2017.

Laura, J.R. and Beyer, R. (2021) "Knowledge Inventory of Foundational Data Products in Planetary Science" *Planetary Science Journal* 2 (18) pp.1–28 DOI: 10.1002/essoar.10501479.1.

Laura, J.R., Hare, T.M., Gaddis, L.R., Fergason, R.L., Skinner, J.A., Jr., Hagerty, J.J. and Archinal, B.A. (2017) "Towards a Planetary Spatial Data Infrastructure" *International Society for Photogrammetry and Remote Sensing* 6 (6) 181 DOI: 10.3390/ijgi6060181.

Laura, J.R., Mapel, J. and Hare, T. (2020) "Planetary Sensor Models Interoperability Using the Community Sensor Model Specification" *Earth and Space Science* 7 (6) pp.1–17 DOI: 10.1029/2019EA000713.

Nass, A., van Gasselt, S., Jaumann, R. and Asche, S. (2011) "Implementation of Cartographic Symbols for Planetary Mapping in Geographic Information Systems" *Planetary and Space Science (PSS)* 59 (Special Issue: Planetary Mapping) pp.1255–1264 DOI: 10.1016/j.pss.2010.08.022.

Robbins, S.J., Kirchoff, M.R. and Hoover, R.H. (2023) "Fully Controlled 6 Meters per Pixel Equatorial Mosaic of Mars from Mars Reconnaissance Orbiter Context Camera images" Version 1 *Earth and Space Science* 10 (e2022EA002443) pp.1–12 DOI: 10.1029/2022EA002443.

Rossi, A.P. and van Gasselt, S. (Eds) (2018) *Planetary Geology (Astronomy and Planetary Sciences Series)* Cham, Switzerland: Springer.

Schimerman, L.A. (Ed.) (1973) *Lunar Cartographic Dossier (Volume I)* St Louis, MO: NASA and the Defense Mapping Agency (Available at: http://www.lpi.usra.edu/lunar_resources/lc_dossier.pdf).

Skinner, J.A., Jr., Hare, T.M., Fortezzo, C.M. and Hunter, M.A. (2019) *Planetary Geologic Mapping Handbook* Flagstaff, AZ: US Geologic Survey (Available at: http://planetarymapping.wr.usgs.gov/Page/view/Guidelines).

Snyder, J.P. (1987) "Map Projections: A Working Manual" *US Geological Survey Professional Paper 1395* DOI: 10.3133/pp1395.

Stooke, P. (2015) "Stooke Small Bodies Maps" (V3.0. MULTI-SA-MULTI-6-STOOKEMAPS-V3.0) NASA Planetary Data System.

Tanaka, K.L., Skinner, J.A., Jr., Dohm, J.M., Irwin, R.P., III, Kolb, E.J., Fortezzo, C.M., Platz, T., Michael, G.G. and Hare, T.M. (2014) "Geologic Map of Mars" (US Geological Survey Scientific Investigations Map 3292, 1:20,000,000) (*Pamphlet to Accompany Map*) DOI: 10.3133/sim3292.

Vander Hook, J., Castillo-Rogez, J., Doyle, R., Stegun-Vaquero, T., Hare, T.M., Kirk, R.L., Bekker, D., Cocoros, A. and Fox, V. (2020) "Nebulae: A Proposed Concept of Operation for Deep Space Computing Clouds" *IEEE Aerospace Conference* Big Sky, MT DOI: 10.1109/AERO47225.2020.9172264.

Zuber, M.T., Smith, D.E., Cheng, A.F., Garvin, J.B., Aharonson, O., Cole, T.D., Dunn, P.J., Guo, Y., Lemoine, F.G., Neumann, G.A., Rowlands, D.D. and Torrence, M.H. (2000) "The Shape of 433 Eros from the NEAR-Shoemaker Laser Rangefinder" *Science* 289 (5487) pp.2097–2101 DOI: 10.1126/science.289.5487.2097.

PART IV

New ontologies and strategies for geospatial technologies

43
TOWARD THE DEMOCRATIZATION OF GEOSPATIAL DATA
Evaluating data decisioning practices

Victoria Fast, Nikki Rogers and Ryan Burns

A geospatial data revolution

With the proliferation of more affordable data production devices like smart phones, social media technologies, and consumer-grade unmanned aerial vehicles, there has been a surge in the collection, storage, and processing of digital geospatial data, leading to what some have termed the fourth paradigm of sciences: data-intensive scientific discovery (Hey, 2012; Kitchin, 2013). Miller and Goodchild (2015) theorize data-driven geography as developing in response to this; rather than mapping the world using large-scale censuses and topological surveys, we are increasingly turning to datasets created from sensors and quotidian instruments to gain new insights. Many of these data sources can be understood as big data, not only referring to datasets of large volume, but also to the velocity at which data are produced, the variety of data types, the resolution at which data represent some phenomenon, exhaustivity of scope, and flexibility of extensionality and scalability (Graham and Shelton, 2013; Kitchin, 2013, 2014a; Miller and Goodchild, 2015).

The sources of these new geospatial datasets are vast. Notably, three novel data concepts—open data, urban data, and volunteered data—are changing how data are created, shared, and used. Open data means being able to download machine-readable data for free, at the spatial and temporal level desired, and with an unrestricted licence (Lauriault, 2013). The open data movement is occurring alongside increases in available data, fostered by more engaged citizenry, and a desire for increased public transparency through open government (Gurstein, 2011; Sieber and Johnson, 2015). Second, in a rush to create smart cities, where city services are digitally mediated and improved through data, we see data-driven urbanism emerge (Kitchin, 2014b; Kitchin et al., 2015; Ashton et al., 2017). Cities across the globe are saturated with data-producing monitors and public-facing data access platforms, recording and reporting millions of interactions and transactions that take place within cities; generating vast volumes of data of movements of people, vehicles, waste, social media interactions, and financial transactions (Ojo et al., 2015; Welch et al., 2016). Lastly, new technology has also led to increases in publicly produced data, or volunteered geographic information (VGI), where citizens both actively and passively contribute information with a location component (Fast and Rinner, 2014; See *et al.*, 2016). Collectively, these new data streams are changing the data landscape, allowing unprecedented geospatial data production and access.

However, dissemination of geospatial data is shockingly unequal. Data ecosystems are not neutral but are influenced by social, political, economic, and cultural values, and are designed to produce certain effects (Kitchin *et al.*, 2018). Data democratization – defined as a process of enabling community actors to access data and to use it to effect social change (Treuhaft, 2006) – exists within particular social and technical contexts. It is influenced by the availability of financial resources and data literacy, the local political context and organizational culture, and other interrelated components (Kitchin and Lauriault, 2014; Craglia and Shanley, 2015; Tenney and Sieber, 2016). For instance, municipal open data platforms retain the privilege of the city to disseminate their own data rather than the data holdings of various civic stakeholders; and while purportedly 'anyone' can access the datasets, they are often in complex technical formats and specifications that require a great deal of technical and quantitative analysis expertise (Burns and Wark, 2019). These influences can amplify power differentials between various groups (Chou *et al.*, 2014; Wiig, 2016; Ash *et al.*, 2018).

Data democratization, and the sociotechnical systems in which data are created, used, and shared, is infinitely complex. In order to better understand the democratization of geospatial data, we need to look closer at data decisioning practices. We define data decisioning as the conditions, and decisions that enable or restrict the availability, dissemination, access, and use of spatial data to support mapping, analysis, or further study. In this chapter, we critically evaluate how four different data decisioning practices enable and restrict data sharing. We have identified four broad levels of data decisioning: open, controlled access, closed, and blind spot. Of course, these categories are fluid, non-sequential, and not entirely mutually exclusive. For example, a data blind spot (data that is not currently collected) may start being collected and become openly available without first being closed or controlled. Within each category, we explore data decisioning practices to better understand the democratization of geospatial data (Figure 43.1).

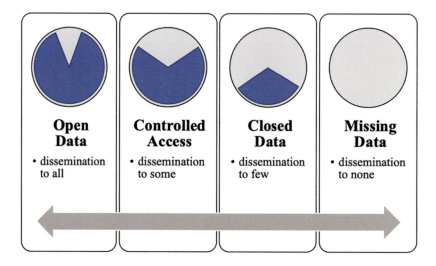

Figure 43.1 Data decisioning practices that enable to restrict (left to right) data dissemination

Open data

'Open data' is defined by Janssen *et al.* (2012: 258) as 'non-privacy-restricted and non-confidential data which is produced with public money and is made available without any restrictions on its usage or distribution'. To be open, data must be stored in machine-readable formats with appropriate metadata (data about the data) with no or minimal restrictions on the use of this data (Gurstein, 2011; Hu and Li, 2017). The widespread best-practice is that governments and non-for-profit organizations make public and other data available in electronic form over the Web (Gurstein, 2011). Many believe that citizen-government interactions are dependent on such opening of data and government acceptance of citizen feedback (Sieber and Johnson, 2015).

The geographic community has been a pioneer of open data long before the open data movement, as seen through the Canadian data initiatives Geogratis, GeoConnections, and other open geospatial data portals (Lauriault and Kitchin, 2014). These initiatives, along with civic groups for open data such as civicaccess.ca, datalibre.ca, and visiblegovernment.ca, helped to pave the way for the open data movement in Canada (Lauriault and Kitchin, 2014). Then, in 2010, the Canadian federal government partnered with municipal, provincial, and territorial governments on open data initiatives resulting in established online interactive portals where raw data are provided to the public (Baculi *et al.*, 2017). As of 2016, there were 98 open data portals across these various levels of government in Canada (Baculi, Fast and Rinner, 2017). In 2015, 83% of these datasets contained geographic content, and 55% were GIS-ready, meaning that they could be opened in GIS software immediately without further handling (Baculi *et al.*, 2017). We suspect the number of open data portals, and data within them, to have grown exponentially since then.

SDIs of open data

Spatial data infrastructures (SDIs) are foundational to open data strategies. An SDI encompasses 'the technology, policies, standards, and human resources necessary to acquire, process, store, distribute, and improve utilization of geospatial data, services, and other digital resources' (Hu and Li, 2017: 1). Sharing and centralizing datasets from different government agencies is challenging, and data redundancies can result when the same dataset is accessible through different platforms (Hu and Li, 2017). SDIs help alleviate these inefficiencies by providing basic infrastructure; organizations and departments can more effectively share and collaborate with consistent spatial data collection, storage, and analysis processes (Rajabifard and Williamson, 2001). Also essential are the open data portals, commonly Socrata, that act as the interface for people accessing openly available geospatial datasets (Hu and Li, 2017). Specific examples include Data.gov in the United States and Open.Canada.ca in Canada. These portals provide a user friendly, one-way transfer of data, via direct download or through an application programming interface (API), and may offer various tools, such as the ability to visualize data with charts and maps (Sieber and Johnson, 2015).

Uneven geographies of open data

Paradoxically, with the vast amounts of data being produced, access to such information is still quite limited to certain geographies and groups of people (Kitchin, 2013). Given that open data are mostly provided by individual levels of government, there is uneven investment in, and quality of, open data across jurisdictions (Johnson *et al.*, 2017). Social scientists have called this uneven access to data 'data poverty' or the 'data divide'. Some municipalities have no open data infrastructure, while others adopt the 'open by default' tactic, whereby the assumption is that their governments must make all data open unless there is justification to keep it closed (Chatwin and Landry, 2018). The 'data divide' is especially pronounced when comparing the open data of rural and urban municipalities. Rural municipalities may have to rely on data from different levels of jurisdiction, each having its own licensing arrangements, leading to issues of interoperability among datasets and additional costs (Johnson et al., 2017).

A similar impediment to effective open data use has been termed the 'digital divide'. Simply opening data can be distinguished from providing meaningful access to open data to all citizens (Gurstein, 2011). As Gilbert (2010) has shown, effectively engaging technology often requires significant social and cultural capital, rather than mere 'access'. Despite intentions to make data more accessible, openly available data are often provided in a specialist format, using government-specific language, and requires specialized software to download and use–limiting who can meaningfully use the data. As Burns and Wark (2019) explain, the database itself acts as a space for the production of social meaning. It is comprised of various data models, organizational schemes, and file formats that have been shaped by societal norms. Thus, there remains uneven access to open data among different groups of people. Moreover, data has systemic biases; the way it is collected and analysed is influenced by assumptions of society that ultimately limit the kinds of data that are collected, and the type of analyses performed (Graham and Shelton, 2013).

Controlled access

Despite the open data movement's many successes, most data are still not considered common property (Welch et al., 2016). Before the proliferation of open data portals, nearly all government data was kept internally and only shared publicly in heavily manicured forms (Sieber and Johnson, 2015). For example, Canada's national census data, administered by Statistics Canada, were not always openly accessible. Public use microdata files (PUMFs) were only available via payment. As access to microdata was cumbersome and out of reach, researchers instead turned to US or international data, potentially inhibiting Canadian research (Boyko and Watkins, 2011). This did not change until 1996 with the launch of the Data Liberation Initiative, which advocated for affordable access to Statistics Canada data resources (Boyko and Watkins, 2011). Now, despite widespread adoption of open data portals, governments still hold a considerable amount of data that is not open (except in rare cases of open-by-default adoption, described above). In these instances, the data are subject to what we call *controlled access*, which we characterize as 'access for some people, but not for all'. This category assumes that the data are sharable, which requires it to be stored in machine-readable formats with metadata (Gurstein, 2011). Here, we explore data decisioning related to controlled data: individual/institutional data sharing agreements, and related privacy concerns.

Individual and institutional data sharing agreements

Data sharing can be defined as the transfer of data between two or more people or agencies, and is often accompanied by a data sharing arrangement (Welch et al., 2016). Data sharing agreements may involve fees and restrictions on what and how the data can be published. In fact, data sharing agreements can stipulate uneven dissemination, for instance restricting the purposes to which it can be used by each of the parties. As data sharing arrangements are created between people, they are subject to the complexities of social and political relationships between individuals and agencies (Tulloch and Harvey, 2008). Welch et al. (2016) used a 2012 National Survey of US municipal government managers to predict sharing behaviour of municipal government bodies. Their findings suggest that rather than regulatory mechanisms, persuasive mechanisms are more likely to play a role in data sharing behaviour. Tulloch and Harvey (2008) also identify both the context and attitude of those involved in data sharing as contributing to the outcome of data negotiations. By context they mean that the location and institution influence data sharing decisions, and by attitude they mean that some individuals view data sharing differently. Some may view data as power and seek to hold onto it, whereas others view power in terms of being able to share data.

Institutional data sharing is also influenced by relationships. In interviews with government employees, it was noted that several institutions were more likely to share data with those whom they know and trust (Ghose and Appel, 2016). Conversely, they were likely to make it difficult for people to gain access to data if there was no prior relationship. Likelihood of data sharing decreased when one of the parties lacks trust that the requested data will be used properly (Welch et al., 2016). Unless access can be negotiated, certain researchers—including geographers—may become cut off from access (Kitchin, 2013). Media have also criticized governments for a lack of data sharing across agencies (e.g., police and child welfare), arguing that uneven access leads to negative consequences such as less effective service delivery (Welch et al., Feeney and Park, 2016). There are many benefits of opening data, including increasing the use of the data beyond reasons it was originally collected for (Tulloch and Harvey, 2008). For instance, municipal staff can benefit from the research and analysis produced by university researchers, if the latter have secure access to temporally and geographically relevant data holdings. Overall, we advocate that where possible, government bodies should provide data openly, because controlled access relies heavily on data decisioning practices that are fraught with inequity.

Privacy concerns

Data may be withheld for many reasons, such as government wanting to retain control over data usage or the monetary value of the data, but most frequently privacy concerns is cited as the reason for restricting the open sharing of data (Welch et al., 2016). Data providers may avoid opening datasets that pose privacy risks and/or require an expensive anonymization process (Johnson et al., 2017). At the government level, the value of anonymization costs is often seen as being a trade-off with privacy concerns and ethical adherence, meaning that they seek a balance between expenditures and protections (Burns and Shanley, 2013). The data provider may even choose to omit certain categories of data, especially in cases of sexual assault or domestic violence crimes (Johnson et al., 2017). Geospatial attributes of data enhance this complication because it has been argued that releasing an individual's location could reveal their identity (Johnson et al., 2017).

Geographic information systems (GIS) are capable of integrating vast amounts of georeferenced data, and have the potential to be more privacy invasive than many other information technologies, thus causing alarm for privacy advocates (Richardson et al., 2015). Access to high-resolution imagery from remote sensing platforms is now commonplace, and imagery can be combined with census data, land use, and social surveys, further increasing risk of data misuse (Richardson et al., 2015).

Protected geoprivacy would comprise a critical element of a democratized data context, and we here highlight several ways to approach such a need. One approach is through more strict government regulation (Judge and Scassa, 2010). However, this can hinder beneficial research uses of spatial data. Geographic masking – aggregation, affine transformations, random perturbation, and the donut method – can be used to protect privacy (Richardson et al., 2015). Although geographic masking can be used to obscure the exact location of a data point, this process can be costly. It is important for current and future GIS professionals to be guided by legal and ethical practices that can help prevent the misuse of personally identifiable data. The Urban and Regional Information Systems Association (URISA) has created a 'GIS code of ethics' that promotes best practices and principles to guide users conducting research in a GIS environment (Richardson et al., 2015). Moreover, ethical components and implications of working with georeferenced data can be embedded within undergraduate and graduate courses.

Closed data

A key component of controlled access, just explored, assumes that the data are sharable and machine-readable with adequate metadata imbedded in an SDI. We define closed data as data that exists, but most people cannot see or use them. There are many reasons data remain closed. For example, the original dataset is lost, broken, or in a digital data format that is obsolete or hard to access (floppy disk). A common example of closed data is a dataset that contains no or inadequate metadata, obscuring its origins, the attributes' meanings, the contexts of its collection, and so forth. This messy bedroom of data management effectively precludes the data from being used beyond its original purpose. In situations where a machine-readable dataset exists and holds adequate metadata, data remains closed due to absence of a mechanism for sharing the data. As described above, SDIs facilitate data collection, standardization, visualization, and dissemination infrastructure; without it, data promulgation is unlikely. While lots of government data are closed, academics, businesses, and NGOs/communities perpetually under-share their data. For legal, practical, and technical reasons, government open data platforms cannot accept and publish non-government data, even high-quality data from a reliable data source; meaning, these non-government actors have no access to existing SDI to open or provided controlled access of their data.

From the not-for-profit and community perspective, Gray et al. (2016) argue that data generated by community organizations can richly contribute to a city's open data ecosystem. They capture important ways of knowing and interacting with a community's urban environment. For example, a neighbourhood association may collect information related to hotspots of citizens' fear, usage of community resources, or constituents' demographics. For Gray et al. (2016), these grassroots data production practices can provide useful material for city administrators, and for cultivating insights across a range of non-government stakeholders. These forms and topics of grassroots data may also challenge state-sanctioned open data claims to transparency and accountability.

Increasingly (but not nearly enough), local and institutional jurisdictions are developing the capacity to share what was previously considered closed data. Robinson and Mather (2017) identify the important role of libraries as civic infomediaries. Libraries can go beyond simple data provision by creating conditions in which data are more accessible, understandable, and usable by civic society. An example of the critical role of libraries is the Scholars GeoPortal Dataverse – enabled by the Ontario Council of University Libraries. They provide the hardware and software infrastructure, plus detailed guides, to students, researchers, faculty, and staff of participating Canadian universities to access large geospatial datasets. The platform makes otherwise closed data accessible to a wider audience, although it still remains in the category of controlled access.

Data blind spot

Lastly, some data simply do not exist, for various reasons. We refer to this phenomenon as a data blind spot. Critical data studies recognize that data are more than an arrangement of objective facts – they exist within particular social and technical contexts (Kitchin and Lauriault, 2014; Burns, 2018). On a fundamental level, deciding to collect and maintain a dataset, or not, is subject to socio-political influence. Blind spots are not forgotten data; they are structurally created through a complex system of power. Insufficient attention has been directed to the forms of data, knowledge, people, and places that might be left out of data models. Open data research has typically focused on characterizing and typologizing platforms (Mirri et al., 2014; Sieber and Johnson, 2015), but a small and growing body of work is exploring open data with an eye toward its attendant social and political shifts (Gray, 2014; Barns, 2016). Data scientists and researchers are now recognizing that data and its supporting technologies are embodied within past and present social, political, and economic rules (Burns and Wark, 2019).

Citizen-generated data are starting to fill in some of the data blind spots. Broadly termed volunteered geographic information (VGI), the literature describes the phenomenon of citizen-generated data with many different terms, including crowdsourcing, citizen science, and user-generated geographic content (See et al., 2016). VGI refers to the creation of digital spatial data by citizens using technology such as cell phones, digital cameras, GPS (global positioning system), and Web mapping platforms such as Ushahidi, Maptionnaire, and Survey123 (Lamoureux and Fast, 2019). OpenStreetMap is arguably the most studied source of VGI (Haklay, 2010), but there are thousands of initiatives globally that are filling immediate data needs, such as during a crisis response and emergency management (Haworth, 2018).

The plethora of shiny new big data sources warrants a more critical examination of the data we have, and more importantly the data we still do not have. Data blind spots can be eclipsed by the 'wow factor' of seemingly comprehensive and exhaustive datasets. For example, Strava, the app that tracks the movement of millions of everyday athletes all over the world, provides government and researchers with billions of data points on cycling and fitness behaviour in their city. These data come at a cost (controlled access), but they far surpass the volume of data collected from traditional, analogue cycling counts, and provide valuable data that supports cycle planning (Sun et al., 2017; McArthur and Hong, 2019). However, often, more data is just a smokescreen for phenomena excluded from the outset, an allusion of completeness. Researchers need to stop and consider: who's represented? Who's left out? If we forget about the seniors who cycle daily but don't own a smart phone, or young children who cycle for fun, then the data-enabled city is only a mechanism for amplifying inequalities and exclusions.

Conclusion

This chapter has interrogated data decisioning practices along a continuum from open, controlled, closed, and not collected. Data decisioning practices are dynamic and complex, and we hope that by illuminating them here, individuals and organizations begin to examine their data decisioning practices. The democratization of data for mapping requires government and other institutions – academic researchers, businesses, and community groups – to work toward open data, where appropriate, as it reduces some of the socio-political, and highly bias, data decisioning practices inherent in controlled and closed data. Opening data can be transformative since it expands the range of actors that can use geospatial information, providing opportunities for more direct participation in science, politics, and social action.

References

Ash, J., Kitchin, R. and Leszczynski, A. (2018) "Digital Turn, Digital Geographies?" *Progress in Human Geography* 42 (1) pp.25–43 DOI: 10.1177/0309132516664800.

Ashton, P., Weber, R. and Zook, M. (2017) "The Cloud, the Crowd, and the City: How New Data Practices Reconfigure Urban Governance?" *Big Data & Society* (June) pp.1–5 DOI: 10.1177/2053951717706718.

Baculi, E., Fast, V. and Rinner, C. (2017) "The Geospatial Contents of Municipal and Regional Open Data Catalogs in Canada" *Journal of the Urban and Regional Information Systems Association"*, 28 pp.39–48.

Barns, S. (2016) "Mine Your Data: Open Data, Digital Strategies and Entrepreneurial Governance by Code" *Urban Geography* 37 (4) pp.554–571 DOI: 10.1080/02723638.2016.1139876.

Boyko, E. and Watkins, W. (2011) "The Canadian Data Liberation Initiative An Idea Worth Considering" *International Household Survey Network* Working Paper No.006 Available at: https://www.federationhss.ca/sites/default/files/2021-08/IHSN-WP006.pdf.

Burns, R. (2018) "Datafying Disaster: Institutional Framings of Data Production Following Superstorm Sandy" *Annals of the American Association of Geographers* 108 (2) pp.569–578 DOI: 10.1080/24694452.2017.1402673.

Burns, R. and Shanley, L. (2013) "Connecting Grassroots to Government for Disaster Management: Workshop Summary" *The Woodrow Wilson International Center for Scholars* Available at: http://www.scribd.com/doc/165813847/Connecting-Grassroots-to-Government-for-Disaster-Management-Workshop-Summary.

Burns, R. and Wark, G. (2019) "Where's the Database in Digital Ethnography? Exploring Database Ethnography for Open Data Research" *Qualitative Research*, 20 (5) pp.598–616 DOI: 10.1177/1468794119885040.

Chatwin, M. and Landry, J-N. (2018) "Making Cities Open by Default: Lessons from Open Data Pioneers" Available at: https://drive.google.com/file/d/1I27YcdEPDa5V3ttKvPun7oIuVoQ-ioRz/view.

Chou, S., Li, W. and Sridharan, R. (2014) "Democratizing Data Science: Effecting Positive Social Change with Data Science" *Conference on Knowledge Discovery and Data Mining* New York, pp.24–27 DOI:10.1.1.478.3295.

Craglia, M. and Shanley, L. (2015) "Data Democracy – Increased Supply of Geospatial Information and Expanded Participatory Processes in the Production of Data" *International Journal of Digital Earth* 8 (9) pp.679–693 DOI: 10.1080/17538947.2015.1008214.

Fast, V. and Rinner, C. (2014) "A Systems Perspective on Volunteered Geographic Information" *ISPRS International Journal of Geo-Information* 3 (4) pp.1278–1292 DOI: 10.3390/ijgi3041278.

Ghose, R. and Appel, S. (2016) "Facilitating PPGIS Through University Libraries" *The Cartographic Journal* 53 (4) pp.341–347 DOI: 10.1080/00087041.2016.1227567.

Gilbert, M. (2010) "Theorizing Digital and Urban Inequalities: Critical Geographies of 'Race', Gender and Technological Capital" *Information Communication and Society* 13 (7) pp.1000–1018 DOI: 10.1080/1369118X.2010.499954.

Graham, M. and Shelton, T. (2013) "Geography and the Future of Big Data, Big Data and the Future of Geography" *Dialogues in Human Geography* 3 (3) pp.255–261 DOI: 10.1177/2043820613513121.

Gray, J. (2014) "Towards a Genealogy of Open Data" *SSRN Electronic Journal* DOI: 10.2139/ssrn.2605828.

Gray, J., Lämmerhirt, D. and Bounegru, L. (2016) "Changing What Counts: How Can Citizen-Generated and Civil Society Data Be Used as an Advocacy Tool to Change Official Data Collection?" *SSRN Electronic Journal* DOI: 10.2139/ssrn.2742871.

Gurstein, M. (2011) "Open Data: Empowering the Empowered or Effective Data Use for Everyone?" *First Monday* 16 (2–7) pp.1–23.

Haklay, M. (2010) "How Good is Volunteered Geographical Information? A Comparative Study of OpenStreetMap and Ordnance Survey Datasets" *Environment and Planning B: Planning and Design* 37 (4) pp.682–703 DOI:10.1068/b35097.

Haworth, B. T. (2018) "Implications of Volunteered Geographic Information for Disaster Management and GIScience: A More Complex World of Volunteered Geography" *Annals of the American Association of Geographers* 108 (1) pp.226–240 DOI: 10.1080/24694452.2017.1321979.

Hey, T. (2012) "The Fourth Paradigm – Data-Intensive Scientific Discovery" In Kurbanoğlu, S., Al, U., Erdoğan, P.L., Tonta, Y. and Uçak, N. (Eds) *E-Science and Information Management (IMCW 2012 Communications in Computer and Information Science, Volume 317)* Berlin and Heidelberg: Springer, p.1 DOI: 10.1007/978-3-642-33299-9_1.

Hu, Y. and Li, W. (2017) "Spatial Data Infrastructures" *Geographic Information Science & Technology Body of Knowledge* DOI:10.22224/gistbok/2017.2.1.

Janssen, M., Charalabidis, Y. and Zuiderwijk, A. (2012) "Benefits, Adoption Barriers and Myths of Open Data and Open Government" *Information Systems Management* 29 (4) pp.258–268 DOIDOI: 10.1080/10580530.2012.716740.

Johnson, P.A., Sieber, R., Scassa, T., Stephens, M. and Robinson, P. (2017) "The Cost(s) of Geospatial Open Data" *Transactions in GIS* 21 (3) pp.434–445 DOI: 10.1111/tgis.12283.

Judge, E.F. and Scassa, T. (2010) "Intellectual Property and the Licensing of Canadian Government Geospatial Data: An Examination of GeoConnections' Recommendations for Best Practices and Template Licences" *The Canadian Geographer* 54 (3) pp.366–374 DOI: 10.1111/j.1541–0064.2010.00308.x.

Kitchin, R. (2013) "Big Data and Human Geography: Opportunities, Challenges and Risks" *Dialogues in Human Geography* 3 (3) pp.262–267 DOI: 10.1177/2043820613513388.

Kitchin, R. (2014a) *The Data Revolution: Big Data, Open Data, Data Infrastructures & Their Consequences* London: Sage.

Kitchin, R. (2014b) "The Real-time City? Big Data and Smart Urbanism" *GeoJournal* 79 (1) pp.1–14 DOI: 10.1007/s10708-013-9516-8.

Kitchin, R. and Lauriault, T.P. (2014) "Towards Critical Data Studies: Charting and Unpacking Data Assemblages and Their Work" In Thatcher, J., Eckert, J. and Shears, A. (Eds) *Thinking Big Data in Geography: New Regimes, New Research* Lincoln, NE: University of Nebraska Press, pp.3–20.

Kitchin, R., Lauriault, T.P. and McArdle, G. (2015) "Knowing and Governing Cities Through Urban Indicators, City Benchmarking and Real-time Dashboards" *Regional Studies, Regional Science* 2 (1) pp.6–28 DOI: 10.1080/21681376.2014.983149.

Kitchin, R., Lauriault, T.P. and McArdle, G. (2018) "Data and the City" In Kitchin, R., Lauriault, T.P. and McArdle, G. (Eds) New York: Routledge, pp.1–13.

Lamoureux, Z. and Fast, V. (2019) "The Tools of Citizen Science: An Evaluation of Map-based Crowdsourcing Platforms" *Spatial Knowledge and Information Canada* 7 (4) pp.1–7.

Lauriault, T.P. (2013) "Geospatial Data Preservation Primer" (*Canadian Geospatial Data Infrastructure Information Product 36e*) GeoConnections/Hickling Arthurs Low Corporation DOI: 10.2139/ssrn.2639851.

Lauriault, T.P. and Kitchin, R. (2014) "A Genealogy of Data Assemblages: Tracing the Geospatial Open Access and Open Data Movements in Canada" Presented at *American Association of Geographers Annual Meeting* Tampa Bay, FL.

McArthur, D.P. and Hong, J. (2019) 'Visualising Where Commuting Cyclists Travel Using Crowdsourced Data' *Journal of Transport Geography* 74 pp.233–241 DOI: 10.1016/j.jtrangeo.2018.11.018.

Miller, H.J. and Goodchild, M.F. (2015) "Data-driven Geography" *GeoJournal* 80 (4) pp. 449–461 DOI: 10.1007/s10708-014-9602-6.

Mirri, S., Prandi, C., Salomoni, P., Callegati, F. and Campi, A. (2014) "On Combining Crowdsourcing, Sensing and Open Data for an Accessible Smart City" *of the 8th International Conference on Next Generation Mobile Applications, Services and Technologies, NGMAST 2014* pp.294–299 DOI: 10.1109/NGMAST.2014.59.

Ojo, A., Curry, E. and Zeleti, F.A. (2015) "A Tale of Open Data Innovations in Five Smart Cities" *Proceedings of the Annual Hawaii International Conference on System Sciences* IEEE Computer Society, pp.2326–2335 DOI: 10.1109/HICSS.2015.280.

Rajabifard, A. and Williamson, I.P. (2001) "Spatial Data Infrastructures: Concept, SDI Hierarchy and Future Directions" *GEOMATICS'80 Conference* Tehran.

Richardson, D.B., Kwan, M-P., Alter, G. and McKendry, J.E. (2015) "Replication of Scientific Research: Addressing Geoprivacy, Confidentiality, and Data Sharing Challenges in Geospatial Research" *Annals of GIS* 21 (2) pp.101–110 DOI: 10.1080/19475683.2015.1027792.

Robinson, P. and Mather, L. W. (2017) "Open Data Community Maturity: Libraries as Civic Infomediaries" *Journal of the Urban and Regional Information Systems Association* 28 pp.31–38.

See, L., Mooney, P., Foody, G., Bastin, L., Comber, A., Estima, J., Fritz, S., Kerle, N., Jiang, B., Laakso, M., Liu, H-Y., Milčinski, G., Nikšič, M., Painho, M., Pődör, A., Olteanu-Raimond, A-M. and Rutzinger, M. (2016) "Crowdsourcing, Citizen Science or Volunteered Geographic Information? The Current State of Crowdsourced Geographic Information" *ISPRS International Journal of Geo-Information* 5 (5) pp.1–23 DOI: 10.3390/ijgi5050055.

Sieber, R. and Johnson, P. (2015) "Civic Open Data at a Crossroads: Dominant Models and Current Challenges" *Government Information Quarterly* 32 pp.308–315 DOI: 10.1016/j.giq.2015.05.003.

Sun, Y., Du, Y., Wang, Y. and Zhuang, L. (2017) "Examining Associations of Environmental Characteristics with Recreational Cycling Behaviour by Street-Level Strava Data" DOI: 10.3390/ijerph14060644.

Tenney, M. and Sieber, R. (2016) "Data-Driven Participation: Algorithms, Cities, Citizens, and Corporate Control" *Urban Planning* 1 (2) p.101–113 DOI: 10.17645/up.v1i2.645.

Treuhaft, S. (2006) "The Democratization of Data: How the Internet is Shaping the Work of Data Intermediaries" (*University of California Berkeley Working Paper*) Oakland, CA.

Tulloch, D.L. and Harvey, F. (2008) "When Data Sharing Becomes Institutionalized: Best Practices in Local Government Geographic Information Relationships" *Urban and Regional Information Systems Association* 19 pp.51–59.

Welch, E.W., Feeney, M.K. and Park, C.H. (2016) "Determinants of Data Sharing in U.S. City Governments" *Government Information Quarterly* 33 (3) pp.393–403 DOI: 10.1016/j.giq.2016.07.002.

Wiig, A. (2016) "The Empty Rhetoric of the Smart City: From Digital Inclusion to Economic Promotion in Philadelphia" *Urban Geography* 37 (4) pp.535–553 DOI: 10.1080/02723638.2015.1065686.

44
DEVELOPING GEOSPATIAL STRATEGIES

Mark Iliffe

The geospatial need

The American polymath James Baldwin, in his unfinished memoir 'Remember this House', detailed his extensive experiences with the civil rights leaders' Martin Luther King, Malcolm X and Medgar Evers opined the universal and transcendental belief that 'Not everything that is faced can be changed. But nothing can be changed until it is faced' (Peck and Strauss, 2017). Set against our current societal issues of vast social and economic inequality, exposure to risk and compounded by the currently ongoing COVID-19 pandemic and the coming troubles of climate change, our global stage is being bleakly set.

However, within each of these challenges, we have the opportunity to transform our societies, economies, and the environment to not just meet these challenges but go beyond them to enable a secure, prosperous, and safe future for all of humanity. Many domains will opine that they have the answers to solving these seemingly intractable and cascading set of problems, yet all approaches to resolving societal injustice, strengthening resilience to risk, or effecting actions to combat climate change all begin with one fundamental basis: Geography.

Everything humanity has done, and will do in future, is inherently related to a geographic location, yet the analysis and interrogation of geospatial information is not yet, but should be, a universal commodity. Take the efforts of cities to reintegrate societies and reach across societal divides, or identifying outbreaks of disease. Geography is the common denominator, yet is often considered so fundamental that the base geographical skills, strategies, and methods are not as commonly deployed to combat the challenges, yet, this is beginning to change, and change so that Geography is widely being used as a force for good across our world, this is primarily being driven by geospatial information, as the 'digital currency' of Geography.

At its core, geospatial information is the data that links data with a specific place. These data could be statistical data, with a defined geographic extent, an environmental area, or an orthoimage from a satellite, drone, or plane. The integration of different forms of geospatial information can lead to identifying trends, such as sea-level rise, urban renewal, climate change, and response to pandemics. Simply without an understanding of Geography, any decision-maker, be it the leader of a country deciding what action(s) they take to fight a pandemic, an urban planner to determine how best to improve urban connectivity through public transport, or a family planning their journey home, these decisions are interwoven and grounded by geospatial information, whether we see it or not.[1]

Accordingly, this chapter discusses the space around developing geospatial strategies, hopefully, to inform and strengthen the use and awareness of geospatial information.

A brief, non-definitive history of how we got to here

Arguably, the key to developing a geospatial strategy is understanding the historical depth and breadth that has provided us with the foundation from which we can build. Several works have aimed to definitively cover the history and craft of Cartography, from several directions, including 'The History of Cartography' series developed by the University of Wisconsin-Maddison; 'Cartography' (Field, 2018); a historical analysis of Chinese cartography (Cheng-Siang, 1978); a longitudinal examination of globalization and the intersectionality of geography, technology, and institutions (Sachs, 2020); to the other seminal and well-known publications or hidden gems that extensively chart the beginning of how we have understood, and then influenced, the world around us.

Cartography as a tool that supported decision-making presciently came into its own in the late 1790s through to the mid-1800s, with the formation of the world's first national mapping agency, the Ordnance Survey in 1791. This was followed by Jon Snow's famous mapping of cholera around the Broad Street water pump in 1854 (and incidentally the founding of what is now the science of epidemiology). Significant though these changes were, decisions based from Geography remained primarily decisions of a higher-order, made by a privileged few at the apex of their professions, generals, kings, emperors, and presidents[2] identifying their domains or scientists at the very forefront of innovative thinking.

In the post-Second World War geopolitical landscape, global actors like the United Nations, complementing individual efforts of nation-states (such as the United States providing access to GPS and Landsat 'free at the point of delivery'), have demonstrated an immense need for geographic information. The first movers in taking advantage of this need and riding the technological wave were/are companies like Esri and Hexagon (formerly Intergraph) in developing geographic information system (GIS) software, enabling the digitization of Cartography and then a basis for deriving meaningful analytical insights from such geospatial information. These grew the need for interoperability and the standardized transmission of data and analysis through the development of standards,[3] enabling geospatial information, as with other forms of modern data, to be *Findable, Accessible, Interoperable,* and *Reusable* (Wilkinson *et al.*, 2016).

Yet perversely, while the need for comprehensive geospatial strategic thinking has consistently been recognized,[4] especially commercially and within local and national governments, the decisive global steps to strengthen the coordination and coherence of geospatial information have only really started to gain traction since the 2000s. Here, country-led/global initiatives like the Group on Earth Observations[5] (GEO) and the United Nations Committee of Experts on Global Geospatial Information Management[6] (UN-GGIM) complement globally relevant initiatives from commercial collectives like the World Geospatial Industry Council; humanitarian interests such as Missing Maps; or volunteer efforts such as OpenStreetMap, there is a global geospatial ecosystem that provides some of the fundamental components with which to build off a geospatial strategy, irrespective of whether you are the National Surveyor responsible for the cadastral management and lands of a country, or an innovator with a potentially groundbreaking idea.[7]

The landscape of geospatial information is further enhanced by a new wave of strategic thinking, advanced in no small part by the efforts of the global geospatial information management community, which simultaneously enables the convening of events that lead to the Moganshan Declaration (UNWGIC, 2018), the production of reports like the Future Trends on Geospatial Information Management – the five to ten-year vision (UN-GGIM, 2020), and generally providing a platform for the development of effective frameworks and strategies that enable countries to build and strengthen their national capacity.

Commensurately, the innovations and impact achieved by the commercial geospatial information industry have simultaneously benefited from, and empowered by, these efforts. To be clear, the geospatial community is more encompassing than can be discussed here, but highlights could include Esri leveraging concepts originally pioneered for understanding, cataloguing, and managing land use, which is, as of the publication of this book, currently enabling the exploration of Mars[8]; Google making the breakthroughs that balance innovation and usability in a complex environment, distilling complex topics of Geodesy, Global Navigation Satellite Systems (GNSS), Cartography, and Telecommunications, into a 'blue dot' that tells you that you need to leave the motorway at Junction 21, not 22 due to an upcoming accident; or OpenStreetMap starting in 2004 as a hobby project to map London, which evolved to be a crucial tool for humanitarians and first responders (Zook, 2010). Academia also plays its part in advancing the geospatial agenda, pushing the frontier of knowledge through research or educating the next generation of geographers.

Towards developing a geospatial information strategy: *why, who, when, what, and how*

Presumably, as a reader of this text, you are a geographer, or are working in a Geography-associated subject where geospatial information is considered a helpful component. However, many who use geospatial information for their decision-making often do not realize that they are drawing on an extensive and established domain, and so invent terminology, practices, and methods, decoupled from the already existing prevailing geospatial ecosystem. Many disciplines fall into this trap, chiefly environmentalists and statisticians, but also several others. As geospatial information often finds itself as a fundamental interlocutor that enables the entry, analysis, and dissemination of data for all forms of decision making, it is vital to bring together the appropriate elements to develop a geospatial strategy, and then promote said strategy to those that would most benefit from it.

So, what is a strategy? Primarily, strategies are anchored by a vision that enables those following the strategy to commit to goals, identify resources, and enable collective planning to realize and benefit from the stated strategic vision. Accordingly, when developing a geospatial strategy, several elements need to be considered and some that can be considered. While your strategy should be a long-term plan designed to help achieve your objectives, crucially, there is 'no one way' to succeed with many potential paths to nirvana.

Perhaps it is also best to consider what a strategy is not. Strategic thinking goes beyond an immediate problem, operation, or task. Furthermore, it is not, but informs, tactical thinking – while more thought needs to be brought into problems and tasks, but a level of critical thinking, consultation, and shared-thinking that would be necessitated by a strategy is not required to accomplish the task at hand. A practical and so simple it is often overlooked rubric[9] are the 'five Ws and one H':

> **Who** are the primary beneficiaries of your strategy?
> *Are they users, consumers, citizens?*

What is your vision?
What's your guiding philosophy?
When do you need to deliver the strategy?
Always have a timeline… and stick to it, even if you're late!
Where does this strategy affect?
A global strategy will naturally different to a local one.
Why are you developing this strategy?
What is the main 'driver' or drivers that necessitate this development?
How will you deliver the strategy?
Can you ensure that you have the means and resources to deliver your strategy? How will you communicate, promote, consult and ensure that it is practically used?

From these fundamental questions, the ability to develop a coherent geospatial strategy becomes attainable. Yet, they are just a basis for any future strategy and share many overlaps with many other domains. Still, if the geospatial strategy is not 'right', the impact will cascade through leaving a trail of unintended consequences behind it – in part, this is due to the very nature of geospatial information as it is often the basis for integrating data to enable informed decision-making.

Arguably 'Seeing Like a State: How Certain Schemes to Improve the Human Condition Have Failed' (Scott, 1998) is a close to perfect treatise on what happens when the geographic dimension is not considered – which through documenting various historical case studies, Scott reviews 'what might have been' had certain factors, primarily anchored to the understanding of the local geography by which a strategy is formulated, executed, and deemed a failure, been considered. Today, the ubiquity of geospatial information allows the ability to incorporate various perspectives and experiences; accordingly, strategic failures should be rarer; but, the stakes are often higher. However, if a strategy is executed that is responsive and 'fit-for-purpose' within its intended environment, multiple benefits that are much greater than the sum of their parts will emanate, with potential issues of outdated or insufficient data being able to be ameliorated by a variety of innovative approaches.

Ramani Huria and the Zanzibar Mapping Initiative, two geospatial activities within East Africa, developed their geospatial strategies to focus on achieving tangible, practical outcomes that impacted the 'local' environment. Still, they were formed from 'first' strategic principles, without an overarching geospatial framework due to the lack of developed infrastructure. But each activity achieved outcomes due to the establishment of a clear vision, resources, means, and method (Figure 44.1).

Towards establishing a (national) geospatial strategy

The start of collective thinking regarding the development of modern[10] geospatial strategies is perhaps the National Spatial Data Infrastructure or 'NSDI' (Williamson *et al.*, 2003), which can broadly be defined as a 'framework of policies, standards, technology and institutional arrangements that facilitate data providers to publish and users to access and integrate, distributed heterogeneous geospatial information' (Scott, 2019). Yet, there are often limitations with the development and uptake of an NSDI within the national context, in part due to the growing complexity of available geospatial information, compounded by a poor data infrastructure and the need to think beyond silos and outside of 'traditional' structures to take an integrated approach to decision-making.[11]

Developing geospatial strategies

Figure 44.1 Community mapping and identification of areas prone to flooding in Dar Es Salaam, Tanzania

Towards taking an integrated approach: grounded by strategic frameworks

At the national level, a country's geospatial information strategy is primarily executed by the National Mapping Agency/National Geospatial Information Agency; but there are often other institutions and agencies within the national context that contribute and complement each other, but often due to the often-fragmented nature within a national context a truly integrated environment for national geospatial information is for many still out of reach.

Climate change, COVID-19, and other challenges (alongside the more 'traditional' activities of cadastral and street mapping) have illuminated the extant fragility, need, and urgency to go beyond the NSDI with a more integrated approach, through a dedicated push to develop a framework that can easily be deployed and fit with prevailing national contexts. The current evolution and culmination of the work of the global geospatial information management community, even within this relatively short twenty-year timeframe, is the Integrated Geospatial Information Framework (IGIF).[12,13] Part 1: Overarching Strategic Framework communicates the IGIF's vision and mission statements through seven (7) underpinning principles, eight (8) goals, and nine (9) strategic pathways, all of which are aligned to strategic national and global drivers.

One of the critical factors of an integrated approach is that there is a focus on incorporating existing capabilities, subsequently strengthening or building capacity where needed. This enables progress to made equally, irrespective of national circumstances, in a responsive and agile manner. Figure 44.2 shows the United Kingdom's use of the IGIF, which is followed by a brief discussion of the implementation of the IGIF within the UK's national strategy.

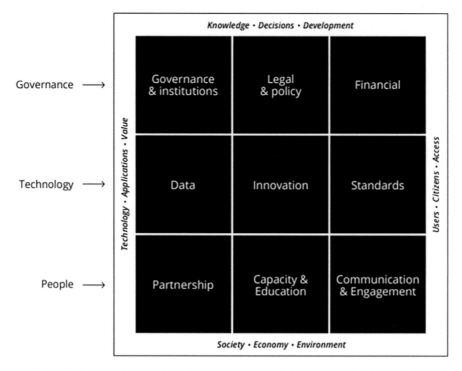

Figure 44.2 The integrated geosaptial information framework (as referenced by the United Kingdom)

Unlocking the power of location: the United Kingdom's geospatial strategy, 2020–2025

The statement 'you can't use an old map to explore a new world' looms large over the UK's Geospatial Strategy 2020–2025. The Strategy sets a forward-looking vision for the coordination of geospatial activity at the national level. 'A national location data framework describes all the elements – including data assets, standards and technologies, policies and guidance, people and organizations – that are required to unlock the power of location […]. The UK has chosen to adopt the IGIF by implementing a national location data framework that is consistent with the IGIF'. Notably, the strategy highlights the economic benefits that geospatial information brings to the national economy, such as

> the opening of Ordnance Survey's highly detailed MasterMap[14] will be made completely open, with the remaining data being made freely available up to a threshold of transactions […] this will release an estimated £130 million of economic value each year" and "Initial research carried out in 2018 suggested that location data has a potential economic benefit to the UK of up to £11 billion per year.
>
> *(Cabinet Office, 1998)*

While each of these three case studies is unique, whether being an encompassing, high-level geospatial strategy by one of the most technologically advanced countries in the world, or directed and straightforward geospatial interventions, each embraces the opportunities afforded by geospatial information, subsequently developing a suitable strategy to achieve success. Notably, given the constraints of this chapter and the intent of this book, examining strategies for developing a geospatial strategy within a commercial environment would be out of scope; however, SparkGeo's 'Strategic Geospatial: Observations for Technology and GIS Leaders' (Cadell, 2021) provides an apt entry point into this domain, with other resources like Geospatial World[15] complementing the oeuvre – but following this fundamental rule: *have a business model.*

Going beyond: situating geospatial information for our future

There are many phrases on the futility of predicting the future; however, it is inevitable that the advancement of geospatial information and the consummate innovation and technological transformation that it will bring should significantly change how we approach strategic planning. National mapping agencies primarily had the same form and function since their first establishment from the late 1790s until the 1970s when GIS fundamentally transformed their business practices.

Primarily, nationally focused frameworks like the IGIF anchor to the prevailing national legal and policy arrangements. However, sometimes these environments could be developed further, or miss challenges that are unknown to us. We may be better prepared than ever to respond to emergent challenges and ensure that the overarching tenet of the 2030 Agenda for Sustainable Development, 'leave no-one behind and reach those furthest behind first' is met. Perhaps this is best summarized by the 2018 Moganshan Declaration, observing that 'geospatial technologies and innovations have been unequally adopted, and that there is an urgent need to effectively bridge the geospatial digital divide to achieve "digital transformation" [that we need]'. Empowered by technology, geospatial information has the capacity to ascertain economic progress through analysing the prevalence of electricity and light during the night or warning citizens of an impending tsunami – all through a device that is as essential as clothes for many in our global society. The opportunity for us to ensure an equitable and prosperous future for all on this planet but a better, safer, and more prosperous future is within our grasp, but we cannot and must not leave anyone behind. We have one world, and while we are 'all in this together', there are a few clouds on that horizon.

Although it is now possible to integrate and apply cutting edge mathematical techniques to identify motifs and features within our datasets, there is also a fundamental tension on whether we should, and if such work is undertaken, to what extent are the sacrosanct boundaries of the individual privacy questioned? An unfortunate example is illustrated by Target, an American retailer, identifying, and then targeting, one of its customers with maternity-specific marketing based on an adjustment to their 'usual' spending, and much to her and her family's surprise (Corrigan *et al.*, 2014). Consumer platforms such as Google, Amazon, Facebook, and Apple are ubiquitous in the devices, tools, and practices that generate a mass of personal information, almost all it accompanied by a geographic location; complemented the second tier of telephony companies that collect masses of metadata (but not the content), with others from Palantir (international development) or Cambridge Analytica (political 'insights') integrating data for their own purposes, which may be moving in areas where national law has no current basis.

Here, ethical considerations, privacy, and confidentiality come to the fore, made more complex by how policy and legal arrangements are often considered by many as a static instrument. Yet, these constantly evolve to incorporate prevailing societal demands or reflect new government directions; for instance, COVID-19 brought a plethora of specialized approaches into the public consciousness – like track and trace or contact tracing – that would have been an anathema to their consciousness.

Artefacts like the Locus Charter, developed by a broader coalition within the geospatial community (as discussed in Chapter 48 of this book), highlight the work still to be done. Practical innovations within the data domain, such as Linked Data or technologies like Blockchain, also offer opportunities that are yet to be fully realized.

Summary

Geospatial information and its enabling digital technologies now pervade our lives, whether we see or not – and it is more often than not. We are collectively empowered by this geospatial opportunity, enabling us to analyse the digital footprints, irrespective of their nationality, of almost all humans on our planet or observe the impact of these footprints in our global, national, and local environment. Geospatial information simultaneously underpins and records our interactions with the world around us, cataloguing our behaviour to ever-increasing levels of detail or informing how policies and decisions are made for the safety and security of all. Our future can only be understood through leveraging geospatial information appropriately, underpinned by existing frameworks and strategies yet to be devised.

So how do we, as individuals or strategists in our respective environments, benefit from strategic thinking? In effect, the better we manage geospatial information, the more significant benefit that can be reaped from it. Geospatial information is at the forefront of our human thirst for knowledge, and its innovative basis will be the bedrock of our adventurous reach beyond humanity's humble origins. The integration of statistics and other forms of data with geospatial information has proved decisive in our ongoing fight against COVID-19. Further, strategies and policies yet to be developed will have geospatial information at their very foundation.

In developing a coherent geospatial strategy, we can now leverage prevailing frameworks, concepts, innovations, experiences, theories, and technologies, all to help further the commodification of geospatial information that empowers our daily lives. The past 20 years have been further enshrined by the fact that geospatial information goes beyond map making and indeed underpins decision-making for all in our society. The start of this decade, diplomatically speaking, wasn't the most auspicious of beginnings and is already being overshadowed by challenges that seemingly have the potential to cascade and amplify in their severity; but through the appropriate development and application of geospatial information, we can not only break away from these shadows but reach those that have been left for too long on the fringes of our global society. We may not know what the future will truly bring, but one thing is sure: Our future will be empowered by geospatial information, and we can all benefit, but we must develop suitable strategies to fully benefit and face the challenges that are coming, while vanquishing those that are already here.

Geospatial strategies in the wild

The following is a non-exhaustive list of prevailing geospatial strategies in use by countries or other adjacent institutions:

National strategies

- Australia – Geoscience Australia Strategy 2028 https://www.ga.gov.au/strategy-2028
- Namibia – The National Spatial Data Infrastructure 2015–2020: Strategy and Action Plan: https://cms.my.na/assets/documents/p1a4vbovsv4qjsi5310ij88sg1.pdf

- United Kingdom – Unlocking the Power of Location: The UK's Geospatial Strategy, 2020–2025: https://assets.publishing.service.gov.uk/government/uploads/system/uploads/attachment_data/file/894755/Geospatial_Strategy.pdf
- United States of America – The Federal Geographic Data Committee's National Spatial Data Infrastructure (NSDI) Strategic Plan: https://www.fgdc.gov/nsdi-plan/nsdi-strategic-plan-2021-2024.pdf

Intergovernmental Organizations, United Nations, European Union

- The Integrated Geospatial Information Framework – http://igif.un.org
- European Union – INSPIRE (Infrastructure for Spatial Information in Europe) – Directive 2007/2/EC http://data.europa.eu/eli/dir/2007/2/oj
- The Locus Charter: https://ethicalgeo.org/wp-content/uploads/2021/03/Locus_Charter_March21.pdf
- Ramani Huria – The Atlas of Flood Resilience in Dar es Salaam: http://documents1.worldbank.org/curated/en/200421524092301920/pdf/Ramani-Huria-the-atlas-of-flood-resilience-in-Dar-es-Salaam.pdf
- United Nations Geospatial Network– The Geospatial Blueprint
- http://ggim.un.org/meetings/GGIM-committee/10th-Session/documents/2020_UN-Geospatial-Network-Blueprint.pdf
- UN-GGIM – The Compendium on Licensing Geospatial Information: http://ggim.un.org/documents/E-C20-2018-9-Add_2-Compendium-on-Licensing-of-Geospatial-Information.pdf
- UN-GGIM – UN-GGIM Strategic Framework 2018–2022 E/C.20/2018/4/Add.1

Private sector

- Accenture – Navigating OpenStreetMap – https://www.accenture.com/_acnmedia/PDF-126/Accenture-Navigating-OpenStreetMap.pdf
- Esri – Defining and Executing a Geospatial Strategy: https://proceedings.esri.com/library/userconf/fed19/papers/fed_24.pdf
- Google – Unlocking value with location intelligence – https://mapsplatform.withgoogle.com/location-intelligence-report/home.html
- SparkGeo – Strategic Geospatial: Observations for Technology and GIS Leaders: https://sparkgeo.com/wp-content/uploads/2020/02/strategic-geospatial-0-6.pdf
- WGIC – Resilient Infrastructure – Geospatial and Building Information Modelling: https://wgicouncil.org/publication/reports/report-resilient-infrastructure-geospatial-bim/?utm_source=referral&utm_medium=press_release&utm_campaign=wfeo-hlf-white-paper

Notes

1 However, this chapter will not cover the strategic applications of geospatial information within the military domain. There is a vast amount of literature on how geospatial information empowers and enables this domain and interlinks with other domains. Key words would include Human Factors; 'Command, Control, Communications, Computers, Intelligence, Surveillance and Reconnaissance (C4ISR)'; Sense-Making; and Situational Awareness; among many others.
2 *An Atlas of Empire* (Horrabin, 1937) provides a pre-Second World War (c. 1937) perspective of global colonial possessions, followed up with his 1943 *Atlas of Post War Problems* that imagines a post-war world that could have been.
3 Primarily, the Open Geospatial Consortium, Technical Committee 211 of the International Organization for Standardization and the International Hydrographic Organization undertake

the brunt of the development of international standards, however, organizations like the World Wide Web Consortium (W3C) and Organization for the Advancement of Structured Information Standards (OASIS) and others within the standards development ecosystem play a role in ensuring this.

4 Starting with the Economic and Social Council (ECOSOC) resolution 131 (VI) of 19th February 1948 on 'Co-ordination of Cartographic Services of Specialized Agencies and International Organizations' which recognized the importance of maps to global activities and the benefits of coordinating cartographic services of the United Nations and its member states. Following a process which focused on the development of regional capacity, enabling some regions to develop their own geospatial strategies, perhaps at the unintended detriment of others. The road to ameliorating this situation was started by the creation of UN-GGIM in 2011, yet when compared with other globally relevant institutional structures that have shaped and guided their domains, geospatial information is comparatively underdeveloped.

5 GEO is a partnership of more than 100 national governments and in excess of 100 Participating Organizations that envisions a future where decisions and actions for the benefit of humankind are informed by coordinated, comprehensive and sustained Earth observations (see https://www.earthobservations.org/geo_community.php).

6 UN-GGIM is the apex intergovernmental mechanism for making joint decisions and setting directions with regard to the production, availability and use of geospatial information within national, regional and global policy frameworks. Led by United Nations Member States, UN-GGIM aims to address global challenges regarding the use of geospatial information, including in the development agendas, and to serve as a body for global policymaking in the field of geospatial information management (see http://ggim.un.org/).

7 PlanetLabs started in 2013 in a 'nondescript brick building near the intersection of 2nd and Bryant Streets in San Francisco' and as of 2021 has several constellations enabling close to near-real time observation of the Earth (see https://gigaom.com/2013/08/02/at-planet-labs-the-space-industry-is-back/).

8 https://www.gim-international.com/content/news/how-geospatial-technology-is-vital-for-exploring-mars.

9 The Wikipedia entry on the topic (https://en.wikipedia.org/wiki/Five_Ws) provides a useful diversion from Aristotle to Kipling in examining the fundamental nature of problem solving and strategic thinking.

10 Cadastres, the management of land use, and other domains requiring geospatial information have had their own mechanisms – yet were broadly superseded by the NSDI.

11 A comprehensive discussion that synthesizes the interconnectivity and mutual importunate of sustainable development and geospatial information is found in (Scott and Rajabifard, 2017).

12 The IGIF comprises of three separate, but connected, documents: Part 1 is an Overarching Strategic Framework; Part 2 is an Implementation Guide; and Part 3 is a Country-level Action Plan. The three parts comprise a comprehensive Integrated Geospatial Information Framework that serve a country's needs in addressing economic, social and environmental factors; which depend on location information in a continually changing world. The Implementation Guide communicates to the user what is needed to establish, implement, strengthen, improve, and/or maintain a national geospatial information management system and capability (see http://igif.un.org).

13 The author stresses that the work of the IGIF is much larger than can be covered in this chapter; after three years since its inception, several countries have developed their national geospatial information strategy to be in alignment with it, with many more working to benefit from it. Arguably, part of the IGIF's success is the vast global consultation, with representatives from over 135 countries providing their guidance, along with a formal 'adoption' as an overarching framework. The history and journey of the IGIF makes this truly 'country-owned and country-led', yet some of the historical papers published before its formal development, such as Scott (2019), Scott and Rajabifard (2017), and practically every published report of UN-GGIM in each of its ten annual sessions, testify to its truly comprehensive, participatory and reflective nature. On a personal note, the gravity to ensure that it is treated appropriately weighs heavy on the author's shoulders.

14 The Ordnance Survey is the UK's national mapping agency.

15 www.geospatialworld.net.

References

Cabinet Office (1998) *An Initial Analysis of the Potential Geospatial Economic Opportunity* London: HM Government Available at: https://assets.publishing.service.gov.uk/government/uploads/system/uploads/attachment_data/file/733864/Initial_Analysis_of_the_Potential_Geospatial_Economic_Opportunity.pdf.

Cadell, W. (2021) "SparkGeo" *Strategic Geospatial: Observations for Technology and GIS Leaders* (18th May) Available at: https://sparkgeo.com/wp-content/uploads/2020/02/strategic-geospatial-0-6.pdf.

Cheng-Siang, C. (1978) "The Historical Development of Cartography in China" *Progress in Human Geography* 2 (1) pp.101–120 DOI: 10.1177%2F030913257800200106

Corrigan, H.B., Craciun, G. and Powell, A.M. (2014) "How Does Target Know So Much About Its Customers? Utilising Customer Analytics to Make Marketing Decisions" *Marketing Education Review* 24 (2) pp.159–166 DOI: 10.2753/MER1052-8008240206.

Field, K. (2018) *Cartography. A Compendium of Design Thinking for Mapmakers* Redlands, CA: Esri Press.

Horrabin, J.F. (1937) *An Atlas of Empire* New York: Alfred A. Knopf.

Peck, R. and Strauss, A. (2017) *I Am Not Your Negro* New York: Vintage International.

Sachs, J.D. (2020) *The Ages of Globalisation* New York: Colombia University Press.

Scott, J.C. (1998). *Seeing Like A State* New Haven, CT and London: Yale University Press.

Scott, G. (2019) "The IGIF: Improving and Strengthening NSDIs and Geospatial Information Management Capacities" In Scott, G. (Ed.) *First International Workshop on Operationalising the Integrated Geospatial Information Framework* Santiago: United Nations Economic Commission for Latin America and the Caribbean, pp.1–8. Available at: https://www.cepal.org/sites/default/files/presentations/-igif-improving-strengthening-nsdis-greg-scott-un-ggim.pdf.

Scott, G. and Rajabifard, A. (2017) "Sustainable Development and Geospatial Information: A Strategic Framework for Integrating a Global Policy Agenda into National Geospatial Capabilities" *Geo-Spatial Information Science* 20 (2) pp.59–76 DOI: 10.1080/10095020.2017.1325594.

UN-GGIM (2020) *Future Trends in Geospatial Information Management: The Five to Ten Year Vision* New York: UN-GGIM. Available at: https://ggim.un.org/meetings/GGIM-committee/10th-Session/documents/Future_Trends_Report_THIRD_EDITION_digital_accessible.pdf.

UNWGIC (2018) *Moganshan Declaration – The Geospatial Way to a Better World* Deqing, China: United Nations World Geospatial Information Congress. Available at: http://ggim.un.org/unwgic/documents/moganshan_declaration_draft_final.pdf.

Wilkinson, M.D., Dumontier, M., Aalbersberg, I.J., Appleton, G., Axton, M., Baak, A., Blomberg, N., Boiten, J.W., da Silva Santos, L.B., Bourne, P.E. and Bouwman, J. (2016) "The FAIR Guiding Principles for Scientific Data Management and Stewardship" *Scientific Data* 3 (1) pp.1–9.

Williamson, I.P., Rajabifard, A. and Feeney, M-E. F. (Eds) (2003) *Developing Spatial Data Infrastructures: Developing Spatial Data Infrastructures* Melbourne: CRC Press.

Zook, M.G. (2010) "Volunteered Geographic Information and Crowdsourcing Disaster Relief: A Case Study of the Haitian Earthquake" *World Medical & Health Policy* 2 (2) pp.7–33 DOI: 10.2202/1948-4682.1069

45
MAP THINKING ACROSS THE LIFE SCIENCES

Rasmus Grønfeldt Winther

> The encyclopedic arrangement of our knowledge […] is a kind of world map which is to show the principal countries, their position and their mutual dependence, the road that leads directly from one to the other.
>
> *Jean le Rond d'Alembert, 'Preliminary Discourse' (1751)*

This chapter supplements existing geospatial ontologies from the perspective of mapping practices of other disciplines. I draw on what I elsewhere call *map thinking*, namely,

> philosophical reflection concerning what standard geographic maps are and how they are made and used. The purpose of such contemplation is to explore the promises and limits of representations – cartographic and beyond. […] Map thinking massages the imagination; excavates hidden assumptions; challenges and synthesizes dualisms; and invites us to reflect on space and time – including the future.
>
> *(Winther, 2020: 4–5)*

Map thinking is rich in rewards. The power and pervasiveness of traditional maps – of cartographic and GIS representations – will be familiar to readers of this volume. But map thinking in the natural sciences may not be, although it is highly prevalent across the life sciences.

The sciences are suffused with the rich traditions and practices of mapping and cartography (Robinson and Petchenik, 1976; Wood, 1992; Harley, 2001; Jacob, 2006; Brotton, 2018; Edney, 2019). Let us consider one form of map thinking: map analogizing. Maps and mapping have served as significant analogies for knowledge and representation at least since the eighteenth century in the Western tradition (see epigraph). The basic form of the map analogy is 'a scientific theory is a map of the world' (Winther, 2020: 35). This thesis is revelatory: imagining modeling and theorizing across the sciences as mapping projects illuminates some obscured but key features of scientific representation – its plurality, context-dependence, and purposiveness.

In order to fully capture the power of map analogizing, we must be more concrete. The charge of this chapter will be to first identify three map types, which I call *literal*, *causal*, and *extreme-scale*. While actual cartographic objects – that is, maps in the familiar sense – serve as the core inspiration for map thinking and the basic map analogy, I expand the meaning of maps and mapping considerably. Mapping in science also includes the making of various representations, sometimes dynamic, of processual networks or causal connections identified via statistics or experiments, and of the world after scaling it down or up in an extreme manner.

It is this panorama of map types, and the spaces that they chart, that I turn to in this chapter. For me, 'mapping is a communal and personal representational effort to imagine and control the different kinds of space of distinct map types' (Winther, 2020: 40). The three map types help shed light on representational processes in three life science projects that I have not previously explored: the evolution of Darwin's finches according to Peter and Rosemary Grant, Kurt Kohn's biochemical causal maps, and the extreme-scale gene expression maps of the Allen Human Brain Atlas. Map thinking these influential scientific research programs provides insight into their methods and purposes. It also helps us look differently at the cartographic object – the traditional map – itself.

Literal maps of Darwin's finches

Darwin's finches are an iconic case of evolution in action. The fifteen or so species differ in beak size and shape, and in feeding habits.[1] Islands are the ecological, evolutionary, and geospatial context. The cactus finch (*Geospiza scandens*), for instance, is a ground finch typically living in areas with the Galápagos 'prickly pear' cactus *Opuntia*, from which it feeds (flowers, fruits, and insects living on the cactus). In contrast, the woodpecker finch (*Camarhynchus pallidus*) amazingly uses a stick or cactus spine to pry out grubs and worms from trees. These species are adapted to unique lifestyles (cf. Lack, 1947: 146; Grant, 1981, 1986; Grant and Grant, 1989, 2002, 2008).

These birds take centre stage in David Lack's classic of evolutionary ecology, *Darwin's Finches*. This work notably established the birds' adaptive radiation, including a model of allopatric speciation (geographical separation). Of 27 figures in Lack's book, 13 (48%) are, or contain, what I will call *literal maps*. A literal map is a visual rendition with geospatial objects at geographic scales. Literal maps become scientific literal maps when they are used as a scientific representation and thereby assist in the scientific work of explanation, prediction, understanding, data organization, and so on. For example, a topographic map becomes a scientific literal map when deployed in a scientific project, such as an ecological or geological one (Winther, 2020: 38). In the biological sciences, literal scientific maps can include maps of ecosystems or of the distribution of particular species.

Lack's book starts with two maps of the Galápagos. The first situates the archipelago in the Pacific Ocean, labels Ecuador and Panama on the South American mainland, and outlines the 1,000-fathom line. The second, larger-scale map, names 13 of the islands both in English and Spanish. These maps ground Lack's discussions of the distribution, variability, and phylogeny of Darwin's finches, and of the evolutionary processes of their specialization, speciation, and adaptive radiation.[2] These scientific literal maps provide a pragmatic context for Lack's evolutionary theorizing. They are the means with which narratives of evolutionary speciation are told.

Literal maps themselves can also serve as abstract theoretical models. Here again Darwin's finches are a useful case study. For instance, literal maps play an especially central role in Peter Grant's efforts to articulate a model of allopatric speciation.[3] Grant's theory or model of allopatric speciation consists of the following five steps (Grant, 1986: 264–265; cf. a similar four-step model in Grant, 1981: 654–655):

1 *Founding.* Speciation starts with a founder population from the mainland colonizing an island (e.g., San Cristóbal);

2   *Cross-Island Migration.* New populations migrate repeatedly across various islands; populations change due to natural selection, and individuals become adapted to local conditions on each island;
3   *Sympatric Reproductive Isolation and Full Speciation.* Eventually a transmuted population meets the original founder population, with which it does not generally interbreed; speciation occurs fully as reproductive isolation and ecological differentiation of the transmuted and original populations are fine-tuned in sympatry;
4   *Further Migration.* Populations of the new species undergo analogous specialization and adaptation through repeated dispersal (analogous to 2);
5   *Further Speciation.* New species that had evolved in (2)−(3) meet after (4), and the same process of reproductive isolation and ecological differentiation promotes further speciation and multi-species adaptive radiation (analogous to 3).

A map of the Galápagos (Figure 45.1) represents the circular, cross-island movement starting with founding (step 1), by a small immigrant population, on the island of San Cristóbal. Subsequently, there were three cross-island migrations (step 2) from San Cristóbal to, successively, Española, Floreana, and Santa Cruz. A small number of individuals from the derived

Figure 45.1   Galápagos map embedding the first three steps of the five-step theory or model of allopatric speciation, printed in different permutations in Grant (1981, 1986, 2008), Grant and Grant (2002) (e.g., speciating finches get smaller in 2002 and bigger in 2008). Redrawn for clarity and geospatial precision by Mats Wedin. (Republished with permission of Princeton University Press, from *Ecology and Evolution of Darwin's Finches* (Princeton Science Library Edition), Grant, PR. (p. 264), figure 74 (2017; originally published in 1986); permission conveyed through Copyright Clearance Center Inc.)

populations on Santa Cruz then migrated back to San Cristóbal in step 3. Regarding the last two stages, Grant contends: 'The cycle of events was repeated many times, each involving an allopatric phase (step 4) and a secondary contact phase (step 5), and resulting in the formation of 13 species, possibly more' (Grant, 1986/2017: 264). Grant's map-based model of adaptive radiation via allopatric speciation depicts an evolutionary machine for churning out new finch species.

The literal map of Figure 45.1 and its attendant model also comprise a *causal* map, which illustrates the evolutionary and ecological processes of migration, local adaptation, and competitive exclusion among nearly related varieties or species. The basic five-step model shown for four islands could even be extended via, for instance, the branching and iterative colonization of distinct sets of islands and the associated multiplication of speciation cycles.

In their impressive book from 2008, Peter and Rosemary Grant reprinted a more visually compelling version of the literal and causal map from 1986 (Grant and Grant, 2002: 134, figure 4), representing the theory of allopatric speciation 'as a model, which is an abstraction designed to capture the essence of speciation from a mass of particulars' (Grant and Grant, 2008: 28). The map model plays an epistemic role in the three 'stages' of adaptive radiation that they classify. Here Darwin's finches are only at the first stage, and the third stage is radiation leading to differences among major branches of the tree of life (Grant and Grant, 2008: 153–160). For instance, the evolution of genetic incompatibilities occurs 'in sympatry at step 3 of the map model' (Grant, 2015: pers. comm). Although only five of 89 figures and colour plates in Grant and Grant's (2008) book are maps (i.e., figures 1.1, 2.2, 3.1, 5.3, and 11.2), geospatial information is contained in many of the non-map figures.

Ecology texts are still replete with species maps serving as literal maps, sometimes doubling as causal maps, even if other representations such as mathematical graphs and data charts are more common today than in the mid-twentieth century. Just a cursory glance at one standard ecology textbook, (i.e., Begon *et al.*, 2006), indicates the ongoing liberal use of maps. In this example, of the 21 figures in Chapter 1, nine (43%) are geographic (or near-geographic) literal maps, or contain such maps in them. Future qualitative and quantitative study could track the variety of purposes to which the cartographic object is put in ecology by exploring its varied uses in ecological textbooks, professional books, and articles. A historical perspective would track changes in the relative frequency of species or ecosystem maps over time, compared to other visual representations.

Geospatial imagination and visualization practices suffuse ecological and evolutionary theories and visualizations. Considering how and why this occurs could clarify the purposes of cartographic practices themselves, including the explicit representation of dynamic, causal processes.

Causal maps of biochemistry

Biochemistry is important to emergent interdisciplinary fields such as biomedicine, systems biology, and synthetic biology. Causal maps of biochemical reactions are typically 'topologically accurate in the same sense as the London Underground map is' (Winther, 2020: 39). Space is highly abstracted, with key objects (molecular agents) represented via symbols and spatial organization. According to Kurt Kohn of the National Cancer Institute in Bethesda, Maryland, USA, the plane of graphical space is a convenient organizer of reaction sets into functional classes (e.g., replication, transcription, or cell cycle control).[4]

Consider also the 'biochemical pathway maps' adorning the walls of many labs, which were first produced by the Swiss pharmaceutical company Roche in 1965.[5] These graphics render key biochemical reactions among molecular agents on a large causal map. Reactants such as the sugar glycogen or the lipid cholesterol are represented with nodes. Arrows denote various types of reaction, including covalent modification, non-covalent binding, and enzymatic stimulation.

Although the graphical plane of such biochemical causal maps often contains some elements of extreme-scale maps (e.g., the cell membrane), it is much more concerned with the topology and temporality of causal relations – representing which reactants, catalysts, and so forth give rise to which products, under which conditions. Such graphical depictions of causal networks are valuable for experimenting and modeling in the molecular life sciences.

Specialist biochemical research relies heavily upon map thinking for understanding biological processes. In characterizing gene regulatory networks, for instance, Douglas Erwin and Eric Davidson examine the complexities involved in genes taking input from and regulating one another, concluding that 'the total map of their interactions has the form of a network' (Erwin and Davidson, 2009: 142). Or consider some relevant articles titles: 'toward a protein-protein interaction map of the budding yeast' (Ito *et al.*, 2000); 'detailed map of a cis-regulatory input function' (Setty *et al.*, 2003); 'a map of the interactome network of the metazoan *C. elegans*' (Li *et al.*, 2004); and others (Collins and Barker, 2007; Cui *et al.*, 2007; Zhao *et al.*, 2020).

Biochemistry maps are more concerned with representing causal influence than physical structure. Proximity on these maps thus tends to represent causation, at least when an arrow or some other causal indicator is present. As with all other abstractions, *the (causal) map is not the reality*. The threat of pernicious reification looms in concretizing abstractions (Winther, 2020: 90–94).

Molecular interaction maps (MIMs)

Consider molecular interaction map (MIM) methodology, a visualization tool for representing molecular interactions among proteins developed by Kurt Kohn and collaborators (e.g., Kohn, 1998, 1999, 2001; Kohn *et al.*, 2004, 2006). MIMs have achieved some market penetration in systems and synthetic biology.[6] Let us consider three broad classes of pragmatic features of Kohn's MIM visualization tool: (i) *desiderata*, (ii) *purposes*, and (iii) *conventions*.

Desiderata (or desired features) of MIMs include (i) a unique, singular, and unambiguous depiction of each molecular kind (Kohn, 1998: 1065–1066, 1999: 2704; Kohn *et al.*, 2006: 11); (ii) a clear network topology (Kohn, 1998: 1066); (iii) extensible notation for multimolecular complexes (Kohn, 1999: 2704, 2001: 86); (iv) reliable map coordinates (visual MIM) or interaction number (electronic MIM) for location and identification purposes (Kohn, 2001: 84; Kohn *et al.*, 2006: 10–11); and (v) general and abstract single diagrams capturing cellular and molecular types and states (Kohn, 2001: 84; Kohn *et al.*, 2006: 10–11). MIMs should be as explicit as possible.

Purposes or outcomes of MIMs include (i) translating MIM diagrams 'into an input file for computer simulation' (Kohn *et al.*, 2006: 10, citing Kohn, 1998, 2001; Kohn *et al.*, 2004); (ii) suggesting novel experimental questions or empirical interpretations (Kohn, 1998: 1066, 1999: 2703, 2001: 84, 88); (iii) 'impos[ing] a discipline of logic and critique' (Kohn, 1999: 2703); and (iv) understanding how 'biological effects' emerge from 'molecular interactions' (Kohn, 1999: 2707).

MIMs require visual conventions. Basic *objects* such as 'elementary molecular species' – including proteins, protein domains, or DNA promoter sites – are depicted in call boxes (Kohn *et al.*, 2006: 3, figure 2). Basic *processes* such as covalent modification or inhibition are represented with various kinds of arrows (Kohn, 2001: 85, figure 1; Kohn *et al.*, 2006: 3–4, figures 3 and 4). Bertin's map, discussed below, also differentiates objects (e.g., French departments) from processes (e.g., migration).

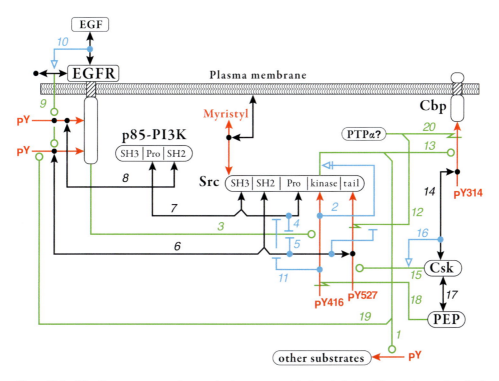

Figure 45.2 The Src enzyme regulates and triggers many biochemical signalling pathways involved in cell division, survival, motility, and adhesion (Sen and Johnson, 2011; Cirotti *et al.*, 2020). In an unregulated, activated state, Src is often implicated in cancer progression, making the *SRC* gene an oncogene (Stehelin *et al.*, 1976); J. Michael Bishop and Harold E. Varmus won a Nobel Prize in 1989 for discovering the gene. Kohn's MIM here shows different domains of Src, and details both intra- and intermolecular interactions. It thus helps us tell a story about what causes Src to open up (its active state) or close (its inactive state). Roughly put: Src amino acid location Y416 (a tyrosine) must be phosphorylated for Src protein activation (#2; 'a bar behind the arrowhead signifies necessity', 'the [blue] node represents the phosphorylated species', Kohn, 2001: 85). The binding of EGF to the transmembrane protein EGFR (#10) triggers Src activation (#3; the open circle indicates 'enzymatic stimulation of a reaction' Kohn, 2001: 85). An active Src can phosphorylate other protein substrates (#1). But the Y416 phosphorylation is inhibited by intramolecular binding which closes Src (#4 and #5). (See Kohn, 2001: 85, 89–91 for further explication.). Partial map key: 'black for binding interactions […]; red for covalent modifications and gene transcription; green for enzyme actions; blue for stimulation and inhibition' (Kohn, 2001: 90). (Reproduced from Kohn, 2001: 90, figure 8, with the permission of AIP Publishing; permission conveyed through Copyright Clearance Center Inc.)

In Figure 45.2, note how Src and EGFR function as objects, while the different types of arrows denote various kinds of temporal processes. Conventions, together with desiderata and purposes, permit explicit and useful graphical renditions of complex biochemical pathways, whereby basic objects and processes are identified in a manner that can be automated and understood logically.

MIMs grant a synoptic view of the biochemical landscape, focusing on causal, topological relations rather than (tiny; inverse scale) geospatial features. And they are only one visualization technique deploying the map analogy in making causal maps.[7] Even so, Kohn writes, 'a molecular interaction map can be used in much the same way as a road map or electronic circuit diagram' (Kohn, 1999: 2703; cf. Kohn, 1999: 2704–2707, 2001: 84; Kohn *et al.*, 2006: 10–11). Furthermore, 'a coordinate grid and an alphabetical list of molecules' permits finding single molecules in a manner 'analogous to the way towns are found on a roadmap' (Kohn *et al.*, 2006: 10–11).

Maps of French interdepartmental migration (1954)

Deep similarities between causal and literal maps can be gleaned by turning to migration maps (cf. Winther, 2020: 180–187). Following the spatial map analogy, we can connect biochemical causal maps, as represented in Figure 45.2, to Figure 45.3 (cf. Winther, 2020: 36, figure 2.2). Recall the three pragmatic assumption kinds identified for MIMs: desiderata, purposes, and conventions. Overlapping desiderata between Kohn's methodology and Bertin's map include unique depiction of each object kind (e.g., molecule; department capital), a clear network topology, and reliable map coordinates. Second, both abstractions share purposes such as summarizing data, suggesting novel interpretations, clarifying patterns and processes, and understanding how general features can emerge from lower-level processes (e.g., biochemical; basic migration). Finally, both maps contain basic objects and processes. Rich empirical, causal, and temporal information is summarized in both representations via resonant diagrammatic conventions.

The maps are of course not the same. For instance, arrows imply reactions in MIMs, whereas they capture geographic movement in Bertin's map. Interestingly, Bertin's map takes very large patterns and puts them in a smaller, digestible graphic form while Kohn's map does the inverse. Analogies are always partial.

Extreme-scale maps of gene expression in brains

Some recent Big Science projects aim to produce atlases or simulations of the brains of humans, rats, and other animals (see http://www.brain-map.org/; https://blogs.cuit.columbia.edu/rmy5/bam/; https://braininitiative.nih.gov/; http://bluebrain.epfl.ch/; Markram *et al.*, 2015). I shall focus on the gene expression maps surveyed, abstracted, and visualized by the Allen Human Brain Atlas (Hawrylycz *et al.*, 2012; Shen *et al.*, 2012).

Knowing the chromosomal location of genes does not tell us their function. Focusing on differential gene activation among tissue types, or organ regions, is essential to understanding a fundamental biological question: how do sameness and homogeneity become difference and heterogeneity, at the genetic, cellular, and tissue levels? Gene expression must be understood – and mapped – in the context of localized intra- and intercellular space. Resulting gene expression maps may either reduce space (such maps are at the very large or high end of standard geospatial scale) or *amplify* space (i.e., inverse scale, Winther, 2020: 71–73), depending on whether they represent, respectively, an object larger than the map, such as a brain, or microscopic objects and processes such as genes and cellular location.

The Allen Human Brain Atlas aims to construct a 'comprehensive map of transcript usage across the entire adult brain' (Hawrylycz *et al.*, 2012: 391). Surveying approximately 20,000 genes across roughly 170 brain structures, the atlas maps the brain's transcriptome architecture (Hawrylycz *et al.*, 2012: 392–394),[8] with an eye towards future studies of the function and dynamics of distinct brain regions, down to the neuronal level.[9]

Deploying high-throughput experimental practices and significant computational power, three kinds of maps were constructed. Recall first that genes together with biochemical machinery produce corresponding messenger RNA during transcription. The presence of

Map thinking across the life sciences

Figure 45.3 This is figure 3 of Bertin's composite map of 'Interdepartmental migrations in France' (1954), which grants a synoptic view of the migration landscape in mid-1950s metropolitan France. Two maps are superimposed here: one showing all migration among all French metropolitan departments (administrative regions), except for Paris (Bertin, 1983: 350, figure 1); and another depicting all migration between Paris and all other departments (Bertin, 1983: 350, figure 2). Each target or source 'empty space' represents the capital, or rough centre, of one of approximately ninety departments. Migration quantities of more than 2% of original department population/year are represented with black arrows or triangles. White arrows or triangles indicate migration of less than 2% per year of the department's population, but more than 10,000 migrants. For both black and white symbols, area is proportional to absolute migration quantity. In figure 1, regular black arrows or a few non-Paris-pointing black triangles, capture all significant interdepartmental migration in all pairwise combinations (e.g., note multiple thick arrows pointing to Lyon). All migration vis-à-vis Paris is represented in figure 2 with triangles, whether black (to Paris) or white (from Paris). This composite map was created by Serge Bonin, laboratoire de Cartographie, École Pratique des Hautes Études, Bertin (1983: 350), figure caption: 351. Redrawn for clarity by Mats Wedin. (Republished with permission from Jacques Bertin, *Semiologie graphique. Les diagrammes, les reseaux, les cartes*, 2005: 350, (c) Ed. de l'EHESS, Paris.)

a particular transcriptional RNA product in a cell or tissue area thus signifies the presence of an active gene. These extreme-scale maps embody gene expression spatial information:

i   *Global microarray maps* are produced via an 'all genes, all structures' strategy relying on microarray technology to produce approximately 10 million microarray expression

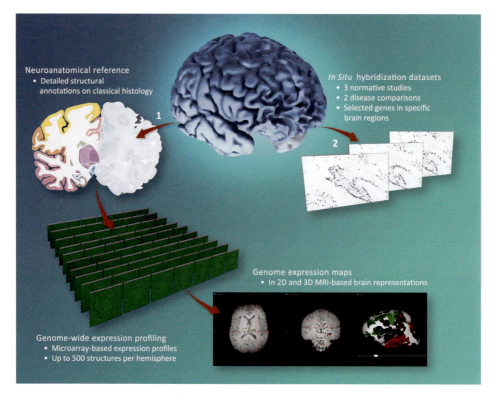

Figure 45.4 Two basic experimental strategies for producing gene expression maps of human brains. (Reproduced from Shen *et al.*, 2012: 712, figure 1, Copyright (2012), with permission from Elsevier.)

datapoints per brain (11,414 differentially expressed genes × 900 neuroanatomical sites) (Hawrylycz *et al.*, 2012: 391, 394; Shen *et al.*, 2012: 711). These datapoints can be represented in zoomable 2D and 3D brain maps (Figure 45.4, bottom).

ii *Heat maps* display in tabular form gene set expression overlap among pair-wise brain structures. These are also generated from microarray data (see, e.g., 'the genetic geography of the brain': http://casestudies.brain-map.org/ggb).

iii *Histological maps* employ in situ hybridization maps to permit local visualization, in particular tissues or cells, of the transcript products of genes (Figure 45.4, top).[10]

Global microarray, heat microarray, and histological gene expression maps permit data mining and visual inference about potential causal mechanisms, helping identify the causal structure of development, and furthering the exploration and discovery of the genetic geography of human brains.[11] For instance, Allen Atlas maps may be used 'to identify molecular networks that underlie brain structure and function and which are often targets of disease; and to characterize cell types and circuitry that drive behavior and thought' (Shen *et al.*, 2012: 714).

Rethinking geospatial ontologies through the life sciences

The life sciences explore processes at many scales, from the vast continents and epochs of ecology and evolution to the miniscule magnitudes of biochemistry and genetics. Literal maps are particularly instructive for examining Darwin's finches, but we must also amplify space with extreme-scale maps if we wish to represent microscopic objects and processes. Bringing multiple levels into focus helps provide a fuller picture of life.

There is a strong drive in the philosophy of science to analyse explanation in terms of causation (e.g., Cartwright, 1989, 2007; Craver, 2007; Glennan, 2017): to explain a phenomenon is to model the processes causing it, thereby allowing for understanding and prediction. Thus, to understand or predict the evolution of beak size or shape, the production of certain molecular objects or agents in biochemical reactions, or the path from genotype to phenotype, we must produce causal models of these phenomena.[12] Literal or extreme-scale maps can be understood as causal maps when they serve as heuristics, inspiration, or even actual models in scientific causal projects. Analogously, geospatial maps can serve as causal maps insofar as they help us explain, understand, or predict phenomena in the social or behavioural sciences, such as human migration or voting patterns.

Finally, time has received increasing attention in the geospatial sciences in recent years (e.g., Yuan, 2008; Andrews, 2021; Kraak and Ormeling, 2021: chapter 8). Here geographers, cartographers, and GIS specialists might benefit from a detour through philosophy of science, especially discussions of temporal dimensions and dynamics in different sciences (e.g., van Fraassen, 1989, 2008; Winther, 2006, 2020: chapters 6–8). Causes occur in time, making temporal ontologies worth investigating.

Conclusion

Examining map thinking in the life sciences can help us rethink geospatial ontologies and practices. Map thinking is highly general. Map analogizing may yet end up working in both directions: not only does science rely on the cartographic map, but the cartographic map may yet be floodlit, transformed, and fractured by attention to map thinking in science.

Acknowledgments

Peter Grant, Michael Hawrylycz, Anthony Hunt, Peter Mashman, Lucas McGranahan, and Marie Raffn provided critical feedback. Mats Wedin assisted with the figures and Laura Laine secured figure permissions. Gratitude to Alexander J. Kent for guidance and follow-through.

Notes

1. Grant and Grant (2008: 3) write '14 or 15'; Lamichhaney et al. (2015) have closer to 18 or so species in their genetic, phylogenetic, and morphological analyses.
2. Lack (1947) offers a succinct three-page summary of his book's 16 chapters on pages 160–163; for state-of-the-art phylogenetic inferences on Darwin's finches, see Lamichhaney et al. (2015, 2018).
3. In his first extended book on Darwin's finches, one of Peter Grant's main interlocutors was David Lack. Grant and Grant (2008: xvii) also dedicate a book to Lack's memory.
4. See, e.g., the large figure 6 of Kohn (1999: 2708–2711), with map key on page 2705, figure 1.
5. See, e.g., http://biochemical-pathways.com/#/map/1.
6. See, e.g., http://discover.nci.nih.gov/mim/index.jsp, and citations listed in Kohn et al. (2006: 11).
7. See, e.g., iPath http://pathways.embl.de/; KEGG Pathway Maps: http://www.genome.jp/kegg/kegg3a.html.
8. Hawrylycz and colleagues surveyed the total genome (roughly 30,000 genes, according to this study: Hawrylycz et al., 2012: 393). Since differences in phenotypic structure and function must be correlated with differences in gene expression patterns, they focused on genes not expressed uniformly across the brain.

9 Importantly, gene expression patterns across individuals seem to be highly – but not perfectly – correlated, with over 90% overlap (Hawrylycz et al., 2012: Supplementary Figure 2).
10 A seminal article notes: 'The hybridization of [experimentally produced] RNA to the [cell's own] DNA in a cytological preparation should exhibit a high degree of spatial localization, since each RNA species hybridizes only with sequences to which it is complementary' (Gall and Pardue, 1969: 378). Gilbert (2007: 362) reviews in-situ hybridization.
11 Further kinds of gene expression maps were also produced, and the principal components correlated highly with spatial brain structure (Hawrylycz et al., 2012: 398, figure 6). Moreover, human embryo brain studies complement this one (Miller et al. 2014); on mouse brains, see Ko et al. (2013).
12 Another tradition in the philosophy of science, influenced by theoretical physics, views explanation, understanding, and prediction more in terms of formal mathematics (e.g., Friedman, 1983; van Fraassen, 1989, 2008). Indeed, in physics – and even in the biological and social sciences – formal *state-space maps* showing the abstract topography of, and voyages in, mathematical phase spaces are crucial (e.g., Nolte, 2010; Winther, 2020, chapters 6 and 8; see Grant and Grant, 2002: 137–139 for state-space maps of the evolution and ecology of Darwin's finches).

References

d'Alembert, J-B. le R. (2009) "Preliminary Discourse." *The Encyclopedia of Diderot & d'Alembert Collaborative Translation Project* (trans. Schwab, R.N and Rex, W.E.) Ann Arbor, MI: Michigan Publishing, University of Michigan Library Available at: http://hdl.handle.net/2027/spo.did2222.0001.083 (Accessed: 8th May 2023). Originally published in 1751 as "Discours Préliminaire" *Encyclopédie ou Dictionnaire raisonné des sciences, des arts et des métiers* Paris: David, Briasson, Durand and Le Breton, Vol.1 pp.i-xlv.

Andrews, G.J. (2021) "Bios and Arrows: On Time in Health Geographies" *Geography Compass* 15 (e12559) pp.1–19 DOI: 10.1111/gec3.12559.

Begon, M., Townsend, C.R. and Harper, J.L. (2006) *Ecology: From Individuals to Ecosystems* (4th ed.) Malden, MA: Blackwell.

Bertin, J. (1983) *Semiology of Graphics: Diagrams, Networks, Maps* (trans. Berg, W.J.) Madison, WI: University of Wisconsin Press. Originally published as (1967) *Sémiologie graphique: Les diagrammes, les réseaux, les cartes* Paris: La Haye Mouton and Gauthier-Villars. Reprinted in 2005 by Les Éditions de l'École des hautes études en sciences sociales (l'EHESS), Paris.

Brotton, J. (2018) *Trading Territories: Mapping the Early Modern World* (2nd ed.) Chicago: University of Chicago Press.

Cartwright, N. (1989) *Nature's Capacities and Their Measurement* New York: Oxford University Press.

Cartwright, N. (2007) *Hunting Causes and Using Them: Approaches in Philosophy and Economics* Cambridge: Cambridge University Press.

Cirotti, C., Contadini, C. and Barilà, D. (2020) "SRC Kinase in Glioblastoma: News from an Old Acquaintance" *Cancers* 12 (6) pp.1–21 DOI: 10.3390/cancers12061558.

Collins, F.S. and Barker, A.D. (2007) "Mapping the Cancer Genome" *Scientific American* 296 (3) pp.50–57.

Craver, C. (2007) *Explaining the Brain: Mechanisms and the Mosaic Unity of Neuroscience* New York: Oxford University Press.

Cui, Q., Ma, Y., Jaramillo., M, Bari, H., Awan, A., Yang, S., Zhang, S., Liu, L., Lu, M., O'Connor-McCourt, M., Purisima, E.O. and Wang, E. (2007) "A Map of Human Cancer Signaling" *Molecular Systems Biology* 3 (152) pp.1–13 DOI: 10.1038/msb4100200.

Edney, M.H. (2019) *Cartography: The Ideal and Its History* Chicago: University of Chicago Press.

Erwin, D.H. and Davidson, E.H. (2009) "The Evolution of Hierarchical Gene Regulatory Networks" *Nature Reviews Genetics* 10 pp.141–148 DOI: 10.1038/nrg2499.

Friedman, M. (1983) *Foundations of Space-Time Theory: Relativistic Physics and Philosophy of Science* Princeton, NJ: Princeton University Press.

Gall, J.G. and Pardue, M.L. (1969) "Formation and Detection of RNA-DNA Hybrid Molecules in Cytological Preparations" *Proceedings of the National Academy of Sciences (USA)* 63 (2) pp.378–383 DOI: 10.1073/pnas.63.2.378.

Gilbert, S.F. (2007) "Fate Maps, Gene Expression Maps, and the Evidentiary Structure of Evolutionary Developmental Biology" In Laubichler, M.D. and Maienschein, J. (Eds) *From Embryology to Evo-Devo: A History of Developmental Evolution* Cambridge, MA: MIT Press, pp.358–374.

Glennan, S. (2017) *The New Mechanical Philosophy*. New York: Oxford University Press.
Grant, P.R. (1981) "Speciation and the Adaptive Radiation of Darwin's Finches" *American Scientist* 69 (6) pp.653–663.
Grant, P.R. (1986) *Ecology and Evolution of Darwin's Finches* (reprinted 2017) Princeton, NJ: Princeton University Press.
Grant, B.R and Grant, P.R. (1989) *Evolutionary Dynamics of a Natural Population: The Large Cactus Finch of the Galapagos* Chicago: University of Chicago Press.
Grant, P.R. and Grant, B.R. (2002) "Adaptive Radiation of Darwin's Finches" *American Scientist* 90 (2) pp.130–139 DOI: 10.1511/2002.2.130.
Grant, P.R. and Grant, BR. (2008) *How and Why Species Multiply: The Radiation of Darwin's Finches* Princeton, NJ: Princeton University Press.
Harley, J.B. (2001) *The New Nature of Maps: Essays in the History of Cartography* (ed. Paul Laxton) Baltimore, MD: Johns Hopkins University Press.
Hawrylycz, M.J., Lein, E.S., Guillozet-Bongaarts, A.L., Shen, E.H., Ng, L.,Miller, J.A., van de Lagemaat, L.N., Smith, K.A., Ebbert, A., Riley, Z.L., Abajian, C., Beckmann, C.F., Bernard, A., Bertagnolli, D., Boe, A.F., Cartagena, P.M., Chakravarty, M.M., Chapin, M., Chong, J., Dalley, R.A., Daly, B.D., Dang, C., Datta, S., Dee, N., Dolbeare, T.A., Faber, V., Feng, D., Fowler, D.R., Goldy, J., Gregor, B.W., Haradon, Z., Haynor, D.R., Hohmann, J.G., Horvath, S., Howard, R.E., Jeromin, A., Jochim, J.M., Kinnunen, M., Lau, C., Lazarz, E.T., Lee, C., Lemon, T.A., Li, L., Li, Y., Morris, J.A., Overly, C.C., Parker, P.D., Parry, S.E., Reding, M., Royall, J.J., Schulkin, J., Sequeira, P.A., Slaughterbeck, C.R., Smith, S.C., Sodt, A.J., Sunkin, S.M., Swanson, B.E., Vawter, M.P., Williams, D., Wohnoutka, P., Zielke, H.R., Geschwind, D.H., Hof, P.R., Smith, S.M., Koch, C., Grant, S.G.N. and Jones, A.R. (2012) "An Anatomically Comprehensive Atlas of the Adult Human Brain Transcriptome" *Nature* 489 (7416) pp.391–399 DOI: 10.1038/nature11405.
Ito, T., Tashiro, K., Muta, S., Ozawa, R., Chiba, T. Nishizawa, M., Yamamoto, K., Kuhara, S. and Sakaki, Y. (2000) "Toward a Protein–Protein Interaction Map of the Budding Yeast: A Comprehensive System to Examine Two-Hybrid Interactions in All Possible Combinations Between the Yeast Proteins" *Proceedings of the National Academy of Sciences (USA)* 97 (3) pp.1143–1147 DOI: 10.1073/pnas.97.3.1143.
Jacob, C. (2006) *The Sovereign Map: Theoretical Approaches in Cartography Throughout History* (trans. Conley, T.) Chicago: University of Chicago Press. Originally published as (1992) *L'empire des cartes: Approche théorique de la Cartographie à travers l'histoire* Paris: Albin Michel.
Ko, Y., Ament, S.A., Eddy, J.A., Caballero, J., Earls, J.C., Hood, L. and Price, N.D. (2013) "Cell Type-Specific Genes Show Striking and Distinct Patterns of Spatial Expression in the Mouse Brain" *Proceedings of the National Academy of Sciences (USA)* 110 (8) pp.3095–3100 DOI: 10.1073/pnas.1222897110.
Kohn, K.W. (1998) "Functional Capabilities of Molecular Network Components Controlling the Mammalian G1/S Cell Cycle Phase Transition" *Oncogene* 16 pp.1065–1075 DOI: 10.1038/sj.onc.1201608.
Kohn, K.W. (1999) "Molecular Interaction Map of the Mammalian Cell Cycle Control and DNA Repair Systems" *Molecular Biology of the Cell* 10 (8) pp.2703–2734 DOI: 10.1091/mbc.10.8.2703.
Kohn, K.W. (2001) "Molecular Interaction Maps as Information Organizers and Simulation Guides" *Chaos* 11 (1) pp.84–97 DOI: 10.1063/1.1338126.
Kohn, K.W., Aladjem, M.I., Weinstein, J.N. and Pommier, Y. (2006) "Molecular Interaction Maps of Bioregulatory Networks: A General Rubric for Systems Biology" *Molecular Biology of the Cell* 17 (1) pp.1–13 DOI: 10.1091/mbc.E05–09–0824.
Kohn, K.W., Riss, J., Aprelikova, O., Weinstein, J.N., Pommier, Y. and Barrett, J.C. (2004) "Properties of Switch-Like Bioregulatory Networks Studied by Simulation of the Hypoxia Response Control System" *Molecular Biology of the Cell* 15 (7) pp.3042–3052 DOI: 10.1091/mbc.e03–12–0897.
Kraak, M.J. and Ormeling, F. (2021) *Cartography: Visualization of Geospatial Data*. Boca Raton, FL: CRC Press.
Lack, D. (1947/1983) *Darwin's Finches* Cambridge: Cambridge University Press.
Lamichhaney, S., Berglund, J., Almén, M., Maqbool, K., Grabherr, M., Martínez-Barrio, A., Grabherr, M., Martinez-Barrio, A., Promerová, M., Rubin, C-J., Wang, C., Zamani, N., Grant, B.R., Grant, P.R., Webster, M.T. and Andersson, L. (2015) "Evolution of Darwin's Finches and Their Beaks Revealed by Genome Sequencing" *Nature* 518 pp.371–375 DOI: 10.1038/nature14181.
Lamichhaney, S., Han, F., Webster, M.T., Andersson, L., Grant, B.R. and Grant, P.R. (2018) "Rapid Hybrid Speciation in Darwin's Finches" *Science* 359 (6372) pp.224–228 DOI: 10.1126/science.aao4593.

Li, S., Armstrong, C.M., Bertin, N., Ge, H., Milstein, S., Boxem, M., Vidalain, P-O., Han, J-D.J., Chesneau, A., Hao, T., Goldberg, D.S., Li, N., Martinez, M., Rual, J-F., Lamesch, P., Xu, L., Tewari, M., Wong, S.L., Zhang, L.V., Berriz, G.F., Jacotot, L., Vaglio, P., Reboul, J., Hirozane-Kishikawa, T., Li, Q., Gabel, H.W., Elewa, A., Baumgartner, B., Rose, D.J., Yu, H., Bosak, S., Sequerra, R., Fraser, A., Mango, S.E., Saxton, W.M., Strome, S., Van Den Heuvel, S., Piano, F., Vandenhaute, J., Sardet, C., Gerstein, M., Doucette-Stamm, L., Gunsalus, K.C., Harper, J.W., Cusick, M.E., Roth, F.P., Hill, D.E. and Vidal, M. (2004) "A Map of the Interactome Network of the Metazoan *C. elegans*" *Science* 303 (5657) pp.540–543 DOI: 10.1126/science.1091403.

Markram, H., Muller, E., Ramaswamy, S., Reimann, M.W. Abdellah, M., Sanchez, C.A., Ailamaki, A., Alonso-Nanclares, L., Antille, N., Arsever, S., Kahou, G.A.A., Berger, T.A., Bilgili, A., Buncic, N., Chalimourda, A., Chindemi, G., Courcol, J-D., Delalondre, F., Delattre, V., Druckmann, S., Dumusc, R., Dynes, J., Eilemann, S., Gal, E., Gevaert, M.E., Ghobril, J-P., Gidon, A., Graham, J.W., Gupta, A., Haenel, V., Hay, E., Heinis, T., Hernando, J.B., Hines, M., Kanari, L., Keller, D., Kenyon, J., Khazen, G., Kim, Y., King, J.G., Kisvarday, Z., Kumbhar, P., Lasserre, S., Le Bé, J-V., Magalhães, B.R.C., Merchán-Pérez, A., Meystre, J., Morrice, B.R., Muller, J., Muñoz-Céspedes, A., Muralidhar, S., Muthurasa, K., Nachbaur, D., Newton, T.H., Nolte, M., Ovcharenko, A., Palacios, J., Pastor, L., Perin, R., Ranjan, R., Riachi, I., Rodríguez, J-R., Riquelme, J.L., Rössert, C., Sfyrakis, K., Shi, Y., Shillcock, J.C., Silberberg, G., Silva, R., Tauheed, F., Telefont, M., Toledo-Rodriguez, M., Tränkler, T., Van Geit, W., Díaz, J.V., Walker, R., Wang, Y., Zaninetta, S.M., DeFelipe, J., Hill, S.L., Segev, I. and Schürmann, F. (2015) "Reconstruction and Simulation of Neocortical Microcircuitry" *Cell* 163 (2) pp.456–492 DOI: 10.1016/j.cell.2015.09.029.

Miller, J.A., Ding, S.L., Sunkin, S.M., Smith, K.A., Ng, L., Szafer, A., Ebbert, A., Riley, Z.L., Royall, J.J., Aiona, K., Arnold, J.M., Bennet, C., Bertagnolli, D., Brouner, K., Butler, S., Caldejon, S., Carey, A., Cuhaciyan, C., Dalley, R.A., Dee, N., Dolbeare, T.A., Facer, B.A.C., Feng, D., Fliss, T.P., Gee, G., Goldy, J., Gourley, L., Gregor, B.W., Gu, G., Howard, R.E., Jochim, J.M., Kuan, C.L., Lau, C., Lee, C-K., Lee, F., Lemon, T.A., Lesnar, P., McMurray, B., Mastan, N., Mosqueda, N., Naluai-Cecchini, T., Ngo, N-K., Nyhus, J., Oldre, A., Olson, E., Parente, J., Parker, P.D., Parry, S.E., Stevens, A., Pletikos, M., Reding, M., Roll, K., Sandman, D., Sarreal, M., Shapouri, S., Shapovalova, N.V., Shen, E.H., Sjoquist, N., Slaughterbeck, C.R., Smith, M., Sodt, A.J., Williams, D., Zöllei, L., Fischl, B., Gerstein, M.B., Geschwind, D.H., Glass, I.A., Hawrylycz, M.J., Hevner, R.F., Huang, H., Jones, A.R., Knowles, J.A., Levitt, P., Phillips, J.W., Sestan, N., Wohnoutka, P., Dang, C., Bernard, A., Hohmann, J.G. and Lein, E.S. (2014) "Transcriptional Landscape of the Prenatal Human Brain" *Nature* 508 pp.199–206 DOI: 10.1038/nature13185.

Nolte, D.D. (2010) "The Tangled Tale of Phase Space" *Physics Today* 63 (4) pp.33–38.

Robinson, A.H. and Petchenik, B.B. (1976) *The Nature of Maps: Essays Toward Understanding Maps and Mapping* Chicago: University of Chicago Press.

Sen, B. and Johnson, FM. (2011) "Regulation of SRC Family Kinases in Human Cancers" *Journal of Signal Transduction* (865819) pp.1–14 DOI: 10.1155/2011/865819.

Setty, Y., Mayo, A.E., Surette, M.G. and Alon, U. (2003) "Detailed Map of a Cis-regulatory Input Function" *Proceedings of the National Academy of Sciences (USA)* 100 (13) pp.7702–7707 DOI: 10.1073/pnas.1230759100.

Shen, E.H., Overly, C.O., Jones, A.R. (2012) "The Allen Human Brain Atlas: Comprehensive Gene Expression Mapping of the Human Brain" *Trends in Neurosciences* 35 (12) pp.711–714 DOI: 10.1016/j.tins.2012.09.005.

Stehelin, D., Varmus, H.E., Bishop, J.M. and Vogt, P.K. (1976) "DNA Related to the Transforming Gene(s) of Avian Sarcoma Viruses is Present in Normal Avian DNA" *Nature* 260 (5547) pp.170–173 DOI: 10.1038/260170a0.

Van Fraassen, B. (1989) *Laws and Symmetry* New York: Oxford University Press.

Van Fraassen, B. (2008) *Scientific Representation: Paradoxes of Perspective* New York: Oxford University Press.

Winther, R.G. (2006) "Parts and Theories in Compositional Biology" *Biology and Philosophy* 21 pp. 471–499 DOI: 10.1007/s10539-005-9002-x.

Winther, R.G. (2020) *When Maps Become the World* Chicago: University of Chicago Press.

Wood, D. (1992) *The Power of Maps* New York: Guilford Press.

Yuan, M. (2008) "Temporal GIS and Applications" In Shekhar, S. and Xiong, H. (Eds) *Encyclopedia of GIS* Boston, MA: Springer, pp.1147–1150 DOI: 10.1007/978-0-387-35973-1_1373.

Zhao, LY., Song, J., Liu, Y., Song, C-X. and Yi, C. (2020) "Mapping the Epigenetic Modifications of DNA and RNA" *Protein Cell* 11 pp.792–808 DOI: 10.1007/s13238-020-00733-7.

46
SPATIAL ANTHROPOLOGY
Understanding deep mapping as a form of visual ethnography

Les Roberts

Introduction: geospatial humanities and spatial anthropology

In the 2010s, the term 'spatial humanities' steadily nudged its way into the lexicon of an interdisciplinary field of scholarship that in some way or another engages with questions of culture, space and place. Its rather broad-brushed credentials in this respect are such as to underscore the rich, if rather fuzzy intersections between arts, culture and humanities disciplines and those that can claim more solid affiliations with the rigours of spatial thinking and practice. As a general term of reference that seeks to foreground the spatial dynamics of everyday social and cultural practices, *spatial humanities* functions admirably in the task of delineating a very particular set of approaches and orientations that have the capacity to bring into interdisciplinary alignment otherwise discreet areas of arts and humanities research. From studies in film, photography and visual culture, to literature, popular music, performance, or to those that map the proliferating landscapes of digital culture, spatiality offers a common point of engagement and exchange. However, it is also the case that the very generality and capaciousness of 'spatial humanities' can – and, indeed, often does – work against it inasmuch as it is simply too catch-all, lacking, in particular, specificity in terms of method.

If we review much of the literature that has expressly promoted the *idea* of spatial humanities (e.g., Bodenhamer *et al.*, 2010, 2015; Gregory and Geddes, 2014), it is the geospatial rather than the more diffusive 'spatial' around which many of the interventions in this area appear to be positioning themselves. From this, we may deduce that spatial humanities represents a consolidation of humanities scholarship that has ostensibly turned towards geospatial computing methods and the application of geographical information systems (GIS) technologies to forge new research directions in subject areas where such tools have hitherto made little, if any, inroads. In this respect, the take-up of the term '*geo*spatial humanities' in a special issue of the *International Journal of Geographical Information Science* (Murrieta-Flores and Martins, 2019), seems a more appropriate way of framing research that occupies a specific corner of a field that consists of an otherwise wide-ranging and eclectic array of approaches, methods and disciplinary dispositions. Bringing together everything that can be compressed into the term 'spatial' with everything to which the label 'humanities' in some way speaks does little to steer attention towards the concrete practical concerns that are prompted by the *doing* of research on culture, space and place. The application of geospatial methods (which are themselves not necessarily uniform) does not seamlessly translate to everything that spatial humanities research represents or extends to *in practice*.

For these and other reasons, as a researcher whose work has routinely combed through the cultures of space and the spaces of culture, it is the term 'spatial anthropology', not spatial humanities, that I have increasingly found myself reaching for (Roberts, 2018). The aim of this chapter, then, is, firstly, to provide an introduction to spatial anthropology understood as a critical reflection on, and immersion in, everyday spatial practices. It does this by casting the focus of attention towards a specific iteration of spatial anthropology in practice, namely *deep mapping* (another label that has gained popularity in recent years and which, as I go on to discuss below, also struggles to corral a uniformity of meaning and practice, or marshal a clear epistemological consensus). Deep mapping, it is argued, can be made better sense of through the prism of spatial anthropology and by considering to what extent it may qualify as a form of visual ethnography. Another, secondary, aim of the chapter is to offer some further reflections on questions of interdisciplinarity, and the ways in which space and spatiality dialectically unravel when mapped across the boundaries that otherwise hold epistemological understandings of space 'in place'. The implications for arts and humanities disciplines, especially against the backdrop of an ongoing neoliberalization and marketization of the academy in which humanities scholarship is all too often under assault and humanities scholars are all too often called upon to justify their existence (Mayall, 2016; Davies, 2020), are such as to warrant a degree of vigilance lest the drift towards an overly positivistic and instrumental discourse of culture, space and place crowd out what makes the arts and humanities explicitly *human* in their orientation towards the world.

Understanding deep mapping

The first thing to observe when introducing the term 'deep mapping' is that it does not necessarily have to resemble anything that might be understood as a 'mapping' in the more conventional sense. Nor does it presuppose the production of an artefact or resource that fits the routine definition of a 'map'. It is undoubtedly an approach that falls very comfortably under the spatial humanities banner, but in many other respects exactly what deep mapping is can be quite hard to pin down with any degree of precision. This plurality of meaning was the driving impetus behind the publication *Deep Mapping* (Roberts, 2015–2016), the aim of which was to draw together scholarship that in some way staked a claim on the concept of deep mapping but which, taken as a whole, did not necessarily add up to a coherent field of practice. There is clearly a significant gulf between approaches to deep mapping that explore its capacity to *model* the world as a big-data driven resource in which 'a near limitless range and quantity of sources can be included, interrogated, manipulated, archived, analysed, and read' (Ridge *et al.*, 2013: 184), and deep mapping conceived of, rather more simply, as literary thick description that provides a finely detailed spatial compendium that is rooted in the cultures, histories and everyday stories that give shape to very localized particularities of place. In the latter case (literary thick description), it is as a written 'map' (or deep map, to be more precise) that the idea of deep mapping was first introduced. This was the book *PrairyEarth (a Deep Map)* by the writer William Least Heat-Moon, published in 1991. Whether or not the book itself qualifies as a map is a consideration that seems less worthy of attention than that which seeks to understand the ways the process invested in the writing and researching of the book qualifies as *mapping*. In other words, it is not the representational medium (in this case a book) that we need to consider but the method employed by, and the disposition of Heat-Moon as a writer engaged in the *practice* of deep mapping. As a dense, 'deeply' layered and richly textured literary survey of Chase County in the US state of Kansas, *PrairyErth* is a deep map only in rather loose metaphorical terms. But as a spatial and creative intervention in the world, the mapping of *PrairyErth* is revealed and given substance to in the detail ascribed to its practice. As a self-styled 'secretary of under-life' (1991: 367), Heat-Moon's inclinations are to burrow down from the surface in order to excavate that which is hidden or buried beneath thinly layered deposits of topsoil or asphalt. Deep mapping in this sense is as much a process of archaeology as it is cartography. With this comes an emphasis on verticality (Schiavini, 2004–2005): the 'plumbing of a place's depth' (Gregory-Guider, 2005: 5). Horizontality, by corollary, is a property of what Trevor Harris (2015: 30) refers to as the 'thin map' (rather than 'shallow map', the antonym of deep map, which, for Harris, carries misleading connotations of superficiality and lack of substance). The concept of a thin map accords with Michel De Certeau's description of the city as viewed from above (as in Google Earth imaging, for example), in that it represents a 'rational organization [of space that represses] all the physical, mental and political pollutions that would compromise it [...] flattening out of all the data in a plane projection' (1984: 94). A deep map, by contrast, can be measured by the extent to which it provides the opportunity for the map-reader/user to 'dive within', as artist, filmmaker and transcendental meditation advocate David Lynch might put it (2006). Heat-Moon, as Wydeven (1993: 134) notes, 'encourages us to fit ourselves in the creases [of maps]', a nice turn of phrase that neatly captures the materiality and performativity of placing oneself within the multi-scalar locative dimensions that are opened up through the act of deep mapping. What we understand as deep mapping cannot, therefore, be reduced to the otherwise a-spatial and a-temporal domain of the map. It denotes an anthropology of practice. People are doing things when they engage in deep mapping; what it is they are doing becomes the focus of a spatial anthropology: a culture of mapping practice (Roberts, 2012a, 2018; Wood, 2012).

With this in mind, at this point we need to turn our attention more directly towards consideration of *performance* but also push the discussion beyond examples of literary deep mapping (such as *PrairyErth*), to get a better understanding of deep mapping as an expressly *visual* mode of spatiocultural intervention. In the canon of scholarship oriented around deep mapping, the important emphasis placed on performance is most notably explored by the archaeologists Mike Pearson and Michael Shanks, whose book *Theatre/Archaeology* distils a reoriented and quintessentially interdisciplinary view of landscape, one that pays heed to 'the grain and patina of place [...] the interpenetrations of the historical and the contemporary, the political and the poetic, the factual and the fictional, the discursive and the sensual' (2001: 64–65). For Pearson and Shanks, deep mapping extends to 'everything you might ever want to say about a place' (*ibid.*: 65). By resolutely stepping into the domain of arts and creativity, these advocates of deep mapping put the *depth* criterion to work in ways that are designed to allow the layered temporalities of place and space to flow – like a wellspring of spatial stories – into a landscape that is as much embodied and affectively configured as it is material. As a performance, deep mapping can be looked upon as an *invocation* of place rather than merely a cartographic rendering of that place. In this important respect, it is an endeavour that finds resonance with the sensibilities and creative praxis of the artist. Given this, it is instructive to think about deep mapping as a practice that sits outside, or at least goes against the grain of an academic habitus in which, in the words of anthropologist Tim Ingold, 'the default setting for research is science' (2019: 664). 'Art, and not science,' Ingold contends,

> is exemplary in the practice of research [...]. Rather than seeking to hold the world to account, or to extract its secrets through force or deception, research would then mean

going along with it, entering into its relations and processes and following their evolution from the inside.

(ibid.)

There is, then, a fundamental creativity at work in the practice of deep mapping. It is cartography as art rather than science, if, by the latter, we mean 'a specialized mode of inquiry dedicated to testing hypotheses through the collection and analysis of data under controlled conditions' and an understanding of research driven by an inclination 'to conclude with everything rather than to begin from it' (*ibid.*: 664, 667). Accordingly, it is not all that surprising to discover that, alongside the proponents of a literary deep mapping (a la Heat-Moon), much of the activity that has explicitly traded under the deep mapping banner has been initiated by visual and performance artists. Clifford McLucas (2000), for example, argues that deep maps need to be 'slow' and 'sumptuous'; their form 'unstable, fragile and temporary'. Requiring the engagement of 'both the insider and outsider', for McLucas, deep maps 'will not seek the authority and objectivity of conventional cartography. They will be politicized, passionate and partisan. They will involve negotiation and contestation over who and what is represented and how'. Writing in a similar vein, the artist Iain Biggs stresses the importance of maintaining a necessarily loose and 'open' approach to deep mapping and to thereby avoid 'becoming complicit in its "disciplining"' (2010: 21). This openness of practice is predicated on resistance to the formalizing of a discourse or method of deep mapping that in some way reins it in as an otherwise 'knowledgeable, passionate, polyvocal engagement with the world' (*ibid.*: 8). Accordingly, in their ethnographically informed case study based in rural North Cornwall, Biggs and fellow artist Jane Bailey describe a deep mapping process that consists of 'observing, listening, walking, conversing, writing and exchanging [...] of selecting, reflecting, naming, and generating [and] of digitizing, interweaving, offering and inviting' (2012: 326). As an *undisciplined* practice of cartography, what this description of open methods usefully points to is the way that very little of what deep mappers are doing is in fact oriented towards the production of maps so much as immersing themselves in the warp and weft of a lived and fundamentally intersubjective spatiality. Whether or not we wish to call what emerges from this process a 'map' (or the process itself 'mapping') seems to me less important than the fact that it is taking place at all. What counts is the method or methods by which the researcher-cum-artist (after Ingold) or researcher-as-bricoleur (Denzin and Lincoln, 2011; Roberts, 2017–2018) is able to *dive within* and re-emerge with knowledges of spaces and places that are not hemmed in or disciplined by the representational frames that are imposed upon them.

Reconciling qualitative enquiry and geospatial methods

From the starting point of geospatial computing, – which, as noted earlier, is where much that is categorized as spatial or *geospatial* humanities research typically sets out from – the question of method is one that offers reduced capacity to go 'off grid', so to speak. The performative, creative and embodied approaches to deep mapping that Biggs, McLucas or Pearson and Shanks subscribe to are difficult to reconcile with those for whom 'Deep maps should [in the first instance] be able to model aggregate quantitative data and help reveal spatial patterns, and also represent rich qualitative data while conveying its uniqueness, nuance, ambiguity and contingency' (Ridge *et al.*, 2013: 181). Although it didn't represent a starting point as such, my own segue into geospatial humanities came about in 2007–2008 when exploring the potential of GIS as a research tool for better understanding the spatialities of film. What had started as an archival project – *City in Film: Liverpool's Urban Landscape and the Moving Image* – had developed into an online spatial database of archive films shot in and/or of the city of Liverpool in the northwest of England. Working as part of a small interdisciplinary team comprising architects and film studies scholars (my own contribution reflecting a background in film/cultural studies and anthropology), the aim of the *City in Film* project (2006–2008) was to compile information about films that depicted Liverpool's historical urban landscape, from the very earliest (1897) through to the 1980s. The films were almost exclusively non-fiction, ranging from documentaries, municipal films of the city (films produced by the Corporation of Liverpool/Liverpool City Council), newsreels, actualities, to amateur films and collections amassed or produced by local cine-clubs, a particularly rich and largely neglected corpus of archival research material. A key distinctive feature of the database that developed from this research is that film titles can be searched by location (building, street, district, etc.), by architectural characterization or function (industrial and commercial, public buildings and spaces, leisure, maritime and so on), and by spatial usage (reflecting the social and anthropological practices that inscribed particular landscapes with meaning as lived spaces: everyday life, festivals and parades, contested and political, transit and mobility, for example). On completion, the *City in Film* database had information on over 1,700 films, along with the spatial and geographical information (some titles more detailed than others) that tied the films to points on a map. The logical next step, therefore, was to set about projecting the film data onto a map of Liverpool, a significant shift in the direction of the research which led to serious engagement with GIS, and a follow-on project called *Mapping the City in Film: a Geo-Historical Analysis* (see Hallam, 2012; Roberts, 2012b; www.liverpool.ac.uk/communication-and-media/research/groups/cityfilm/).

To rehearse even a small part of what could be said about the spatialities of the moving image would take us well beyond what can or needs to be unpacked here. But as a spatial humanist who has always sought to bring an anthropological disposition to scholarship on film, space and place, my abiding interest in the *Mapping the City in Film* research has been to understand in what ways the representational spaces of film can be thought of as maps, or can be mapped, and what this might mean when considered through the particular lens of spatial anthropology. It has also been to throw a spotlight on the question of how, and to what extent, the audio-visual medium of film can be re-envisioned as a constellation of *spatial practices*; and of how an exercise in cinematic cartography can productively inform interdisciplinary and multimedia forms of engagement with cultural mapping and deep mapping. Suffice it to say, the interdisciplinarity that is a key component of initiatives such as *Mapping the City in Film* brings with it certain challenges and opportunities. As Julia Hallam and I discuss in the introduction to the book *Locating the Moving Image* (Roberts and Hallam, 2014), often these expose the kind of fault lines delineated by Ridge et al. (see previous citation), where the modelling of aggregate data and analyses of spatial patterns is weighed against the need for qualitative richness that is able to tap into and convey 'uniqueness, nuance, ambiguity and contingency': qualities more associated with the impressionistic and affective cartographies of place that artists and filmmakers are drawn to than the altogether more rigorous, geometric and 'objective' mediations of geospatial science. Mindful of the potential contradictions, as well as – importantly – benefits that GIS mapping and visualization technologies can bring to research on film, space and place, *Mapping the City in Film* was designed to accommodate both cartographic and ethnographic/qualitative modes of analysis. The latter took the form of ethnographic work and semi-structured interviews conducted with amateur filmmakers and cine-club members in and around Merseyside. What this strand of research brought to the overall objective of *mapping* a city in film was to flesh out crucial layers of contextual meaning surrounding the production and consumption of moving images of Liverpool's urban landscape. The expressly *visual* mediations of the city – whether cartographic or cinematographic – by which, as researchers, we glean and interpret geographical and historical information, are thus made open to and resonant with wider spaces of representation that are not fixed or delimited in terms of the specificities of any given medium. By bringing the geospatial abstractions of digital mapping into inchoate alliance with the concrete spaces (Lefebvre, 1991) of lived experience and embodied memory, the city *in* film is pushed beyond the representational frames of the medium (and beyond the aesthetics of architectural form) to service a more holistically conceived project of spatial anthropology (Figures 46.1 and 46.2).

When embarking on the *Mapping the City in Film* project in 2007–2008, the value and place of GIS tools and methods in film research (and humanities research more generally) had been little explored, and in this respect it found productive points of connection with other fields of humanities research, such as literature and popular music studies (Cohen, 2012; Cooper *et al.*, 2016), which were making similarly tentative forays into the world of geospatial computing. Although steps towards the development of qualitative GIS were gathering pace at the time (Cope and Elwood, 2009), initiatives which, as we have seen, have more latterly coalesced around the interdisciplinary field of spatial humanities, the added value of GIS is perhaps best measured not in terms of what it has been able to offer as a resource and stand-alone research tool in its own right (if, in the advent of locative media and ever

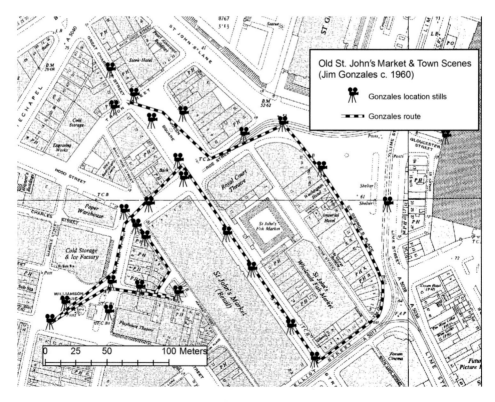

Figure 46.1 Map showing locations showing in the amateur film Old St John Market and Town Scenes (Jim Gonzales/Liver Cine Group, c. 1960; Ordnance Survey map © Crown Copyright and Landmark Information Group Limited (2010). All rights reserved (1955))

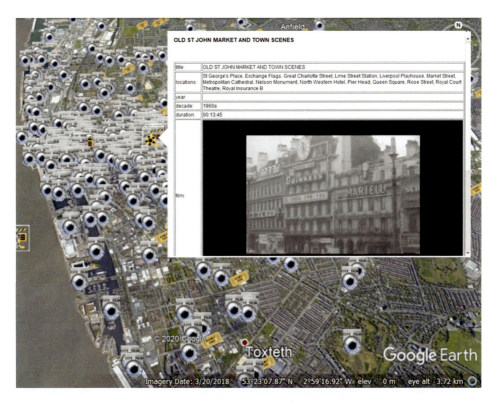

Figure 46.2 Screenshot of Google Earth map of Liverpool film locations featuring embedded video of the amateur film Old St John Market and Town Scenes (Google Earth, 20th March 2018, 53°23′07.87″N and 2°59′16.92″W, accessed 4th May 2020)

proliferating digital methods, I was embarking on the *Mapping the City in Film* project today, it is unlikely I would be making use of GIS resources in the same way, if at all). The value of geospatial tools and methods, I am suggesting, can be attributed to the part they have played in helping instill a richer spatiocultural consciousness across cultural and humanities disciplines. In my own work, and reflecting on some of the directions my research has taken in the wake of the *City in Film* projects, it is less the contribution that GIS has directly made that is apparent (indeed, if the measure is its ability to deliver qualitative nuance, ambiguity and contingency, then it has been found severely wanting). It is rather the more diffuse role geospatial technologies have played in helping shape a *spatial praxis* that has allowed the differential qualities of space (Lefebvre, 1991) to be played out and given voice to in ways that do not seek to fix or territorialize the spaces of culture or the cultures of space, but to put them to work anthropologically: to confront and engage with them as *lived* spaces.

Why spatial anthropology matters

If some iterations of deep mapping veer towards the totalizing and systematic, and show an epistemological predilection for *data* over what might otherwise be thought of as *culture* (narrative, texts, images, video, audio, practice, performance, thick description), or the mechanics of *visualization* over the hermeneutics of *visuality*, then a spatial humanities that is conversant with the objectives and methods of visual ethnography clearly occupies a different position on the deep mapping spectrum. David Bodenhamer (2015: 23), remarks that, at its best, GIS-based deep mapping is an 'ideal storyboard for humanists', offering a conceptual, technological, and spatial framework adapted to the need to tell spatial stories that are harvested from 'experiential as well as objective space' and that are replete with the 'rich contradictions and complexities' that ordinarily, as abstract representations of space, maps fall short of conveying. While geospatial humanities may offer a frame within which to project, albeit in rather limited form, the experiential attributes of space and place, without the porosity of spatial practices that allow for meaningful immersion in experiential spaces (that is, practices that extend out to the performative and embodied imbrications of lived space, or what De Certeau [1984] and Merleau-Ponty [2014] referred to as 'anthropological space') there is only so far GIS can take us beyond the representational abstraction of conventional (i.e., *thin*) maps. As part of the toolkit of the visual ethnographer, geospatial methods are of value only when deployed as means rather than ends. In my own case, reflecting on the role played by GIS in the *Mapping the City in Film* project and how it has gone on to inform, or not, my subsequent interventions in the field of spatial anthropology, then its value can be gauged in terms of what geospatial methods were able to facilitate rather than determine in any absolutist sense (Figure 46.3).

'Cestrian Book of the Dead' (liminoids.com/projects/sandsofdee) started as a project that sought to geo-reference, using ArcGIS, historical sites of death by drowning on the Dee

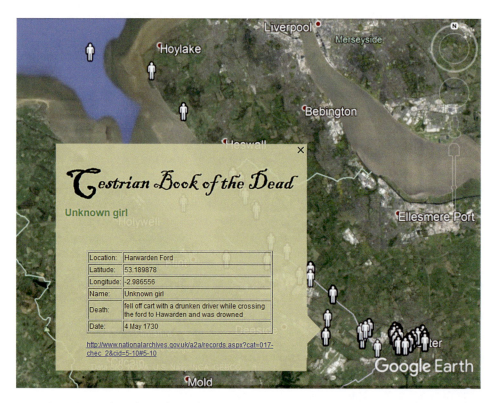

Figure 46.3   Screenshot of Google Earth map showing the location of the Dee Estuary. The example shown from *The Cestrian Book of the Dead* marks the site of drowning of an unknown girl who was crossing the ford to Hawarden in 1730 (Google Earth, 20th September 2019, 53°11′34.29″N and 2°58′54.10″W, accessed 4th May 2020)

in North West England and North Wales. Information was drawn from the City of Chester Coroner Records, dating back to the 1500s. The mapping was the starting not the end point. It helped to set in motion a research process that is very much engaged in the social and cultural landscapes of the estuary as they are experienced, imagined and put into practice today (Roberts, 2018: 187–236). Similarly, the project that became 'Heterotopolis' began with the mapping of former sites of cinemas across Liverpool and Merseyside (177 in total). This provided a performative stage upon which to engineer the production of local spatial stories and photographically 'map' the locations as they presented themselves to me as part of what, by then, had become site-specific fieldwork (*ibid.*: 119–156). Another project, 'Concrete Island' (liminoids.com/projects/concreteisland), used geospatial tools (GIS and GPS tracking data), alongside digital video, photography, sound recording and autoethnographic narration to capture the experience of being marooned, for a day and a night, on a traffic island located between the north and southbound carriageways of the M53 Mid-Wirral Motorway (*ibid.*: 63–90). What these, and other, geospatial interventions help illustrate is the way that digital tools (including, but by no means limited to, geospatial technologies such as GIS) can be productively utilized as part of mixed or bricolage-style (Roberts, 2017–2018) qualitative research methods aimed at mining the deeply layered textures, affects and experiential fabric of lived spaces. In this respect, spatial anthropology matters because what people do with, make of, and understand about the everyday spaces they live, move through and dwell within matters. Understanding deep mapping as visual ethnography is thereby simply a means to re-orientate ourselves towards the world not with the intent of capturing or visualizing that world, but to assist in the task of living it.

References

Bailey, J. and Biggs, I. (2012) "'Either Side of Delphy Bridge': A Deep Mapping Project Evoking and Engaging the Lives of Older Adults in Rural North Cornwall" *Journal of Rural Studies* 28 (4) pp. 318–328 DOI: 10.1016/j.jrurstud.2012.01.001.

Biggs, I. (2010) "The Spaces of 'Deep Mapping': A Partial Account" *Journal of Arts and Communities* 2 pp.5–25 DOI: 10.1386/jaac.2.1.5_1.

Bodenhamer, D.J. (2015) "Narrating Space and Place" In Bodenhamer, D.J., Corrigan, J. and Harris, T.M. (Eds) *Deep Maps and Spatial Narratives* Bloomington: Indiana University Press, pp.7–27.

Bodenhamer, D.J., Corrigan, J. and Harris, T.M. (Eds) (2010) *The Spatial Humanities: GIS and the Future of Humanities Scholarship* Bloomington: Indiana University Press.

Bodenhamer, D.J., Corrigan, J. and Harris, T.M. (Eds) (2015) *Deep Maps and Spatial Narratives* Bloomington: Indiana University Press.

Cohen, S. (2012) "Urban Musicscapes: Mapping Music-Making in Liverpool" In Roberts, L. (Ed.) *Mapping Cultures: Place, Practice, Performance* Basingstoke: Palgrave, pp.123–143.

Cooper, D., Donaldson, C. and Murrieta-Flores, P. (Eds) (2016) *Literary Mapping in the Digital Age* London: Routledge.

Cope, M. and Elwood, S. (2009) *Qualitative GIS: A Mixed Methods Approach* London: Sage.

Davies, W. (2020) "How the Humanities Became the New Enemy Within" *The Guardian* (28th February) Available at: http://www.theguardian.com/commentisfree/2020/feb/28/humanities-british-government-culture (Accessed: 2nd March 2020).

De Certeau, M. (1984) *The Practice of Everyday Life* (trans. Rendall, S.) London: University of California Press.

Denzin, N.K. and Lincoln, Y.S. (2011) "The Future of Qualitative Research" In Denzin, N.K. and Lincoln, Y.S. (Eds) *The Sage Handbook of Qualitative Research* (4th ed.) London: SAGE, pp. 681–684.

Gregory, I. and Geddes, A. (Eds) (2014) *Towards Spatial Humanities: Historical GIS and Spatial History* Bloomington: Indiana University Press.

Gregory-Guider, C.C. (2005) "'Deep Maps': William Least Heat-Moon's Psychogeographic Cartographies" *eSharp* 4 Available at: http://www.gla.ac.uk/research/az/esharp/issues/4/ (Accessed: 23rd December 2015).

Hallam, J. (2012) "Civic Visions: Mapping the City Film 1900–1960" *Culture, Theory and Critique* 53 (1) pp.37–58 DOI: 10.1080/14735784.2012.657912.

Harris, T. (2015) "Deep Geography – Deep Mapping: Spatial Storytelling and a Sense of Place" In Bodenhamer, D.J., Corrigan, J. and Harris, T.M. (Eds) *Deep Maps and Spatial Narratives* Bloomington: Indiana University Press, pp.28–53.

Heat-Moon, W.L. (1991) *PrairyErth (A Deep Map)* Boston, MA: Houghton Mifflin.

Ingold, T. (2019) "Art and Anthropology for a Sustainable World" *Journal of the Royal Anthropological Institute* (N.S.) 25 (4) pp.659–675 DOI: 10.1111/1467–9655.13125.

Lefebvre, H. (1991) *The Production of Space* (trans. Nicholson-Smith, D.) Oxford: Blackwell.

Lynch, D. (2006) *Catching the Big Fish: Meditation, Consciousness and Creativity* London: Penguin.

Mayall, D. (2016) "Why Are the Humanities Always Under Fire? We Need Them More Than Ever" *Times Higher Education* (8th July) Available at: https://www.timeshighereducation.com/blog/why-are-the-humanities-always-under-fire-we-need-them-more-than-ever (Accessed: 13th February 2020).

McLucas, C. (2000) "Deep Mapping" Available at: http://cliffordmclucas.info/deep-mapping.html (Accessed: 23rd December 2015).

Merleau-Ponty, M. (2014) *Phenomenology of Perception* (trans. Landes, D.A.) Abingdon: Routledge.

Murrieta-Flores, P. and Martins, B. (2019) "The Geospatial Humanities: Past, Present and Future" *International Journal of Geographical Information Science* 33 (12) pp.2424–2429 DOI: 10.1080/13658816.2019.1645336.

Pearson, M. and Shanks, M. (2001) *Theatre/Archaeology* London: Routledge.

Ridge, M., Lafreniere, D. and Nesbit, S. (2013) "Creating Deep Maps and Spatial Narratives Through Design" *International Journal of Humanities and Arts Computing* 7 (1–2) pp.176–189 DOI: 10.3366/ijhac.2013.0088.

Roberts, L. (2012a) "Mapping Cultures – A Spatial Anthropology" In Roberts, L. (Ed.) *Mapping Cultures: Place, Practice, Performance* Basingstoke: Palgrave, pp.1–25.

Roberts, L. (2012b) *Film, Mobility and Urban Space: A Cinematic Geography of Liverpool* Liverpool: Liverpool University Press.

Roberts, L. (Ed.) (2015–2016) "Deep Mapping" (Special Issue) *Humanities* 4–5 Available at: http://www.mdpi.com/journal/humanities/special_issues/DeepMapping.

Roberts, L. (Ed.) (2017–2018) "Spatial Bricolage: Methodological Eclecticism and the Poetics of 'Making Do'" (Special Issue) *Humanities* 6–7 Available at: http://www.mdpi.com/journal/humanities/special_issues/spatial_bricolage.

Roberts, L. (2018) *Spatial Anthropology: Excursions in Liminal Space* London: Rowman and Littlefield.

Roberts, L. and Hallam, J. (2014) "Locating the Moving Image: Outline of a New Empiricism" In Hallam, J. and Roberts, L. (Eds) *Locating the Moving Image: New Approaches to Film and Place* Bloomington: Indiana University Press, pp.1–30.

Schiavini, C. (2004–2005) "Writing the Land: Horizontality, Verticality and Deep Travel in William Least Heat-Moon's *PrairyErth*" *Revista di Studi Americani* 15–16 pp.93–113.

Wood, D. (2012) "The Anthropology of Cartography" In Roberts, L. (Ed.) *Mapping Cultures: Place, Practice, Performance* Basingstoke: Palgrave, pp.280–303.

Wydeven, J.J. (1993) "Review of PariryErth (a deep map)" *Great Plains Quarterly* 13 (2) pp.133–134.

47
THE QUANTUM TURN FOR GEOSPATIAL TECHNOLOGIES AND SOCIETY

Daniel Sui

The end of history, as proclaimed by Francis Fukuyama (1992) at the end of the Cold War in the early 1990s, has, in retrospect, accelerated the arrival of the spatial century. With the increasing use of geospatial information and technologies in almost every aspect of society, the first spatial century in human history continues to unfold in front of us as we enter the third decade of the twenty-first century (Sui, 2010). The ongoing global pandemic triggered by COVID-19 serves as a potent reminder of the complex roles geospatial technologies play in a global crisis. On the one hand, geospatial technologies have been used by a plethora of both professional and amateur users to map and to understand the spatial and temporal dynamics of COVID-19 (Shaw and Sui, 2021). On the other hand, geospatial technologies have also been used to spread mis/disinformation from local to global scale, intentionally or unintentionally, thus contributing to a global infodemic that parallels the global pandemic (Kent, 2020; Stephens, 2020; Zhao et al., 2021). The twin roles of geospatial technologies played during this crisis are particularly perplexing, especially in the context of growing political polarization, uninformed populism, and rising nationalism. More than ever, we need global solidarity and cooperation if we are to tackle global challenges effectively, ranging from climate change and the global pandemic, to cybersecurity and economic disparity.

The development of geospatial technologies is at a crossroads, and this volume has identified many productive pathways to address the challenging, sometimes paradoxical, impacts of geospatial technologies on society. This chapter proposes an alternative direction for geospatial technologies and society – it is about time for the GIS community to make a quantum turn, which potentially could serve both as an extension and synthesis of the materials covered in this book and the literature on GIS and society in the previous three decades. The primary goal of this chapter is to continue to catalyse an informed discussion on the potential utilizations of quantum science and technology for geospatial information science and technologies and their multifaceted applications in society from local to global levels.

The rest of this chapter is organized as follows. The core concepts developed in quantum mechanics are presented, which are essential for understanding the basic elements of the quantum information science and technology. The next section discusses the major implications of the quantum turn for GIS research and practice, followed by a summary and conclusion.

Core quantum concepts in context

At their very core foundation, the current generation of geospatial technologies are developed according to a Newtonian world view, which is based upon the following five basic assumptions:

1. Objectivism: that objects exist independent of the subjects who observe them;
2. Materialism: that the elementary units of reality are physical objects;
3. Reductionism: that larger objects can be reduced to smaller ones;
4. Determinism: that objects behave in law-like ways; and
5. Mechanism: that causation is mechanical and local.

Quantum theory, initially developed in the early twentieth century, challenges all five assumptions at the sub-atomic level (for an overview, see Zettili, 2009 and Pratt, 2021). Systems are not independent of observers; physical objects dissolve into ghost-like processes; and wholes cannot be reduced to parts. Quantum mechanics recognizes that elements exist as both waves and particles, and where an object's state is a wave function that collapses upon measurement. Furthermore, the world does not behave deterministically and causation is non-local – a phenomenon known as quantum entanglement.

Importantly, these findings do not necessarily invalidate the classical worldview at the macro level, since quantum states normally 'decohere' into classical ones above the molecular level, which is why the everyday world appears to us as conforming to the classical worldview. Decoherence has been a barrier to developing a unified quantum theory encompassing both micro and macro levels, and is a fundamental obstacle to the quantum consciousness hypothesis in particular (Wendt, 2015). Nevertheless, at the nano-level the quantum revolution has decisively overturned the claim of the classical worldview to provide a complete description of reality.

More specifically, espousing a quantum turn for geospatial technologies and society, understanding the following the core quantum concepts is essential – each of which defies our common-sense Newtonian understanding of the world but has been proven true by decades of experimentation.

Complementarity: holism

This refers the tenet that a complete knowledge of phenomena on atomic dimensions requires a description of both wave and particle properties, the 'wave-particle duality'. The principle was announced in 1928 by the Danish physicist Niels Bohr (1885–1962). For example, light comprises particles (photons) traveling at light speed in wave formation. A photon's position in space at any moment is a combination of forward velocity and position in the wave cycle. However, it is impossible to observe both the wave and particle aspects simultaneously.

Uncertainty: non-determinism

The uncertainty principle asserts that our ability to completely measure of quantum states is fundamentally limited. Knowledge about two dimensions is complementary, and the more accurately we measure one property (e.g., particle speed) and less likely we can measure accurately another (e.g., particle position). This fundamental uncertainty holds for multiple attributes: particle spin in different directions, momentum, and position. It is not a limitation in measurement technology or experimental sophistication, but a fundamental attribute of the mathematical formalism that underlies quantum theory and is held by a century of experimentation.

Entanglement: non-locality

Some multi-particle quantum systems are linked, so that measuring one particle immediately alters the properties of the other, even when they are physically separated. Once described by skeptical Einstein and his colleagues (Einstein *et al.*, 1935) as 'spooky action at a distance', these phenomena certainly negate the idea of local realism, in which every event has an immediate cause.

Superposition: potentiality

Another challenging concept about the nature or the behaviour of matter at the sub-atomic scale claims that, while we do not know what the state of any object is, it is actually in all possible states simultaneously, as long as we do not look to check. The act of measurement itself causes the object to be limited to single possibility – Schrodinger's cat – simultaneously alive and dead.

Decoherence refers to the loss of information and change in state of a quantum system when it interacts with the broader world. It is why quantum phenomena are not seen in our macro environment – a quantum particle is rarely completely isolated from its environment. Further, this decoherence occurs when the system is observed: the particle and the environment are bound together as one system, including any observer as part of the environment. Since the observer and observed cannot be completely separated, such interaction with the environment will lead to an ebbing away of quantum states, thus leading to decoherence.

The idea that quantum phenomena, characterized by non-linearity, non-locality, non-determinism, and potentiality, are relevant to macroscale systems has traditionally been dismissed, if not ridiculed. This is because quantum coherence apparently cannot be maintained in these larger contexts that favour random scattering, vibration, and motion. Nonetheless, efforts to translate fundamental quantum concepts into fields beyond physics have sought with varying degrees of success of acceptance.

Admittedly, there are diverse opinions on quantum theory's relevance and scalability beyond the subatomic level, ranging from whole-hearted embrace, considerable skepticism, and all the way to wholesale rejection. However, these diverse opinions on quantum theory's scalability and applicability do not seem to slow down the broader scientific and scholarly community's continued engagement with the quantum paradigm, ranging from physical sciences (Lloyd, 2006; Prencipe, 2019), social sciences (Haven and Khrennikov, 2017), to humanities (Barad, 2007), and the arts (Denti, 2020). Recent research in quantum biology (Al-Khalili and McFadden, 2014) suggests that some processes, such as photosynthesis, bird navigation, and sense of smell, are quantum rather than classical, and that relatively small numbers of highly ordered particles can make a difference through processes such as quantum tunnelling. In other words, it is increasingly argued that the division between the quantum and classical worlds appears not to be fundamental. Recent results and advances in these efforts and developments to apply quantum concepts beyond the subatomic world have been very encouraging and exciting.

Quantum implications for geospatial technologies and society

Like scholars in many other fields, both physical and human geographers have tried to link quantum theory to their research (Peterman, 1994; Harrison and Dunham, 1998), but, unlike other fields, these early efforts have not generated any followers (Smith, 2016). Indeed, the geographical quantum path blazed by these pioneers was ahead of its time for geography, but technological and social-political environments have undergone profound changes during the past ten years. We would miss a major opportunity for both technological and theoretical advances if we do not resume our efforts to re-examine the implications of quantum theory for geospatial technologies and society.

With the recent advances in quantum computing and growing complexity of the global challenges facing humanity today, more than ever, it is an imperative for the geospatial community to engage the quantum paradigm for two primary reasons. First, geospatial technologies are essentially computer-based technologies for spatial data handling at their core. Recent advances in quantum computing and communication could potentially be a game changer that may disrupt everything we do since the invention of computers in the late 1940s after the Second World War. Second, humanity is ill served if we can continue to operate and organize our social activities according to siloed, objective, reductionist, and deterministic world views while the world is increasingly connected, subjective, holistic, and indeterministic, and probabilistic.

The quantum turn entails that the geospatial community should make concerted efforts to embrace the theories, algorithms, techniques, and new ethics and practices inspired by quantum information science and technology (QIST). Needless to say, it is beyond the scope of this chapter to cover the technical details due to the breadth and complexity of QIST, but it is possible to offer an outline of a broad-brush road map for further exploration.

Ontological and epistemological implications

Ontology is concerned about what exists and what is real, whereas epistemology tries to answer the question of how we know what we know. Conceptually, ontology and epistemology are intertwined and mutually implicated. Critiques of GIS have recognized the ontological and epistemological deficiencies early on (Sui, 1994), but the alternatives offered, including feminist approaches, post-structuralism, critical social theory, and actor network theory, seem to present a formidable critique of positivist tradition of GIS practice, yet have had little impact on GIS applications and their vast array of users. A quantum perspective potentially can offer a more holistic ontological foundation and more inclusive epistemology. As Barad (2007: ix) succinctly articulated 'To be entangled is not simply to be intertwined with another, as in the joining of separate entities, but to lack an independent, self-contained existence. Existence is not an individual affair'.

Practices of being and knowing are not isolable and are mutually implicated. The separation of epistemology from ontology is a reverberation of a metaphysics that assumes an inherent difference between human and nonhuman, subject and object, mind and body, matter and discourse. The quantum turn aims to end the separation epistemology from ontology. Barad (2007) proposed 'agential realism' as a quantum-inspired ontological-epistemological framework that aims to synthesize and create an integrated understanding of the role of human vs. nonhuman, material vs. discursive, and natural vs. cultural factors in scientific and technological practices, thereby moving such considerations beyond the well-worn debates that pit constructivism against realism, agency against structure, and idealism against materialism. To Barad (2010: 261), '[...] phenomena are not located in space and time; rather, phenomena are material entanglements enfolded and threaded through the spacetimemattering of the universe'.

Although not explicitly framed as a quantum framework per se, the splatial (space-place) framework Shaw and Sui (2019) developed is a significant step forward towards a more holistic framework that links absolute space, relative space, relational space, and mental space, which is conceptually consistently with Barad's agential realism. The splatial framework is one of the very few comprehensive theoretical framework in GIScience developed so far that aims to incorporate both the observer and the observed in a unified framework mediated by both relative and relational space.

Drawing from the concept of superposition, Bittner (2017) proposed a quantum theory of geographic fields that allows for the possibility of representing multiple incompatible states simultaneously at a given point of geographic field. This quantum theory of geographic fields provides a new level of synthesis and understanding of indeterminacy and ontological vagueness in the geographic world. Bittner's (2019a, 2019b) subsequent contributions provide fresh new thinking on representing geographic objects with indeterminate boundaries that have plagued the GIS community for decades (Burrough and Frank, 1995).

A quantum-inspired ontology (based upon entanglement and superposition) and epistemology (based upon the inseparability of the observer and the observed) would greatly facilitate the consilience of the positivist practices and post-positivist critiques of GIS applications in society. Quantum social theory equips us with a more robust framework deal with misinformation, political polarization, and conspiracy theories – defining characteristics of quantum politics in a world turned upside turn (Becker, 1991; Thornhill, 2018).

Methodological and technological implications

The methodological and technological implications of the quantum turn for geospatial technologies and society may be the most obvious. Methodologically, the use of 'quantum-like' mathematical and statistical models to study probabilistic dynamical systems has increasingly become popular. Recognizing that the observer is an integral part of what is being observed, Q method is one of the very few available quantitative techniques for eliciting, evaluating, and comparing human subjectivity. The quantum turn will revitalize its use in geographical studies according to its early successes (Robbins and Krueger, 2000).

In the emerging quantum social science, Haven and Khrennikov (2013) describe its goal as the investigation of economics, finance, psychology, sociology, and other domains of inquiry with the help of formal models and concepts used in quantum physics. Quantum decision-making recognizes that judgments and decisions are influenced by context, and that entangled systems cannot, in theory, be modelled as separate systems. Importantly, this statistical approach does not assume that quantum physical effects are really part of the social world. Although it applies quantum mathematics to social and cognitive phenomena, it considers that these phenomena are based on classical information processing consistent with the neuronal paradigm of neurophysiology and cognitive science. It draws attention to quantum theory as a statistical theory, recognizing that the interference of probabilities is a basic statistical feature of quantum theory. Quantum formalisms are merely considered a more effective way of processing incomplete information and accounting for the interference of probabilities in macroscopic quantum systems.

Related to the generative visions of quantum mechanics, the emerging quantum-Bayesian (QBism) approach considers how beliefs and experiences guide agents in their interactions with the world. Taking an agent's actions and experiences as the central concerns of the theory, QBism deals with common questions in the interpretation of quantum theory about the nature of wavefunction superposition, quantum measurement, and entanglement.

Beyond these specific quantum-inspired methodologies for possible geospatial research, we also need to keep in mind Google's declaration on quantum supremacy (aka advantage) in quantum computing (Connover, 2019). By exploiting collective properties of quantum states, such as superposition and entanglement, to perform computation, quantum computers have been proven to be able to solve certain computational problems substantially faster than classical computers. The study of quantum computing is a subfield of quantum information science. The goal of quantum supremacy or quantum advantage is to demonstrate that a programmable quantum device can solve a problem that no classical computer can solve in any feasible amount of time (irrespective of the usefulness of the problem). As of now, there are generally four ways to build quantum computers, but fundamental to all is the quantum bit (qubit or qbit), which, entangled in all possible states of superpositions, provides an exponential advantage over classical computers that are based upon digital bits in binary states of either 0 or 1. Preliminary results have shown many well-known problems, such as the Prisoner's dilemma, the travelling salesman problem, and the four-colour map problem, can be solved or resolved using quantum computers where classic computers have hit a brick wall.

Goodchild (2018) argues that it is possible to imagine a very different birth of GIS based on globes rather than maps, and with positional uncertainty and spatial resolution addressed at the outset using hierarchical data structures for the globe. These structures also have the advantages of a congruent geography operating at multiple scales. GIS today still reflects in part the constraints of computing in the mid-1960s using von Neumann computer. With the latest advances in quantum computing, it will be natural to reimagine the next generation of GIS according to digital globes instead of two-dimensional maps. Indeed, the potential impacts of quantum supremacy for geospatial technologies cannot be overestimated.

Ethical and practical implications

Throughout human history, our worldviews have often been shaped by the dominant machines of the time (Siegfried, 2008). Like the clock in the seventeenth century and steam engine in the nineteenth century, the computer became the dominant machine in the late twentieth century and we are all, like it or not, 'Turing's man' according to Bolter (1984). Now, as quantum computers are on the cusp to become dominant in the coming years, it is not a stretch of the imagination that quantum computers, along with all the concepts and theories behind them, could potentially become new cultural forces to shape and reshape our ethics and practices in the coming decade.

With its non-local, non-deterministic, and participatory approach to social change, the emerging quantum social theory may appear strange and counterintuitive in relation to classical understandings of society (Zohar, 1994). Yet, as more quantum-inspired social scientists have demonstrated (Busemeyer and Bruza, 2012; Wendt, 2015), by taking non-local, hidden, and subjective factors seriously and explicitly, a social science based on quantum physics arguably conforms better to how we experience the world relative to a social science based on classical physics. O'Brien (2016) further argues for a shift from an Enlightenment paradigm to an Enlivenment paradigm, or a cultural worldview that emphasizes the importance of lived experience, embodied meaning, material exchange, and subjectivity. An Enlivenment perspective may be key to effectively addressing complex issues facing humanity today, such as increasing global polarization, the growing economic disparity, and worsening global environmental change. Indeed, uninformed populism and nationalist approaches go against an entangled worldview as espoused by the quantum turn.

The encouraging news is that emerging GIS practices are deeply connected to the six senses of the whole new mind (Sui, 2015). Geodesign, location-based story-telling, synthesis, empathy, play, and meaning are conceptually consistent with quantum ethics because all these new practices tend to emphasize: the role of observer; aspects of reality not captured by the database; the whole is more than the sum of the parts; nothing is an island by itself; and everything is inter-connected; and so on.

As already demonstrated by ground-breaking work in international politics and international relationships (Portugali, 1993; Der Derian, 2011, 2013; Zanotti, 2018), the quantum turn has potentially become a disruptive paradigm changer to fundamentally change our worldview and our practices across the social sciences (Wendt, 2015). O'Brien (2016) lays out a road map of how the quantum perspective could help us better cope with the challenges posed by the global climate change through meaningful social transformation. For future GIS practices, the quantum turn may have even bigger impacts when considering the widespread application and implementation of GIS in society. However, more concerted efforts in education and training are needed.

Summary and concluding remarks

Nature uses only the longest threads to weave her patterns, so each small piece of her fabric reveals the organization of the entire tapestry.

Richard Feynman

We are caught in an inescapable network of mutuality, tied in a single garment of destiny. Whatever affects one directly, affects all indirectly.

Martin Luther King

With the growing application of geospatial data and technologies in our individual daily lives, as well as in our collective efforts to address various challenging problems facing society, we have come to an inflection point in the evolution of geospatial technologies and society as we enter the third decade of the twenty-first century. As a predominant tool for spatial data handling, GIS technology is deeply rooted in the cartographic tradition. Throughout its history, GIS researchers, developers, and users have been keenly aware of its methodological and conceptual deficiencies. During the past four decades, we have witnessed the analytical, computational, and critical turns in the evolution of GIS to tackle these deficiencies. All these efforts have greatly enriched the theoretical foundation and methodological effectiveness of geospatial technologies.

Closely connected to the post-positivist turn but distinctly different from them, the quantum turn has the following three characteristics that deserve special attention. Firstly, the quantum turn will fundamentally change our worldview from a Newtonian fragmented/reductionist perspective to a more holistic one. Consequently, this new world will compel us to seek new ontologies as well as epistemologies. Secondly, the quantum turn will soon be supported by a set of new technologies based upon quantum computing and quantum communication – perhaps the most disruptive development of computing technologies according to Alan Turing and Von Neuman; thus enabling us to ask and to resolve questions that were not possible before. Thirdly, the quantum turn restores humans – the observer – into the environmental, social, economics, and political issues we are trying to address using geospatial technologies. The human-centred quantum turn could potentially move us greatly towards a more efficient, equitable, and sustainable society. Recognizing a fundamentally indeterministic future could empower us to be more open, innovative, and creative.

Quantum concepts such as complementarity, superposition, entanglement, and uncertainty provide a strong basis for recognizing and positioning people (as opposed to technology) at the core as both problems and solutions facing humanity today. As new scientific discoveries create global entanglements, ubiquitous media produce profound observer effects, advances in quantum science and technology offer geospatial technologies and society a new alternative at the theoretical/conceptual, methodological/technical, and social/ethical guidelines. The quantum turn is not abandoning what the GIScience community accomplished from the positivist and critical traditions, it is in fact of new level of synthesis of the two, but more versatile and robust, positioning ourselves to better deal with the challenges outlined above.

Perhaps more importantly, the quantum turn will propel GIS to be more than a computational science in search of new algorithms, but also a humanistic science in search of purpose and meanings, especially those phenomena that lies beyond the limits of computation – our new *terrae incognita*.

References

Al-Khalili, J. and McFadden, J. (2014) *Life on the Edge: The Coming of Age of Quantum Biology* London: Bantam.
Barad, K. (2007) *Meeting the Universe Halfway: Quantum Physics and the Entanglement of Matter and Meaning* Durham, NC: Duke University Press.
Barad, K. (2010) "Quantum Entanglements and Hauntological Relations of Inheritance: Dis/continuities, Spacetime Enfoldings, and Justice-to-come" *Derrida Today* 3 (2) pp.240–268.
Becker, T. (Ed.) (1991) *Quantum Politics: Applying Quantum Theory to Political Phenomena* New York: Praeger.
Bittner, T. (2017) "Towards a Quantum Theory of Geographic Fields" In Clementini, E., Donnelly, M., Yuan, M., Kray, C., Fogliaroni, P. and Ballatore, A. (Eds) *13th International Conference on Spatial Information Theory (COSIT 2017), Schloss Dagstuhl–Leibniz-Zentrum fuer Informatik, Dagstuhl, Germany, Leibniz International Proceedings in Informatics (LIPIcs)* 86 pp.5:1–5:14 DOI: 10.4230/LIPIcs.COSIT.2017.5.
Bittner, T. (2019a) "Is There a Quantum Geography?" In Tambassi, T. (Ed.) *The Philosophy of GIS* Cham, Switzerland: Springer Geography, pp.209–239.
Bittner, T. (2019b) "Why Classificatory Information of Geographic Regions is Quantum Information" *Spatial Information Theory, International Conference, COSIT 2019* Available at: https://drops.dagstuhl.de/opus/volltexte/2019/11108/pdf/LIPIcs-COSIT-2019-16.pdf (Accessed: 8th May 2023).
Bolter, J.D. (1984) *Turing's Man: Western Culture in the Computer Age* Chapel Hill, NC: University of North Carolina.
Burrough, P. and Frank, A.U. (Eds) (1995) *Geographic Objects with Indeterminate Boundaries* (GISDATA Series II) London: Taylor and Francis.
Busemeyer, J. and Bruza, P. (2012) *Quantum Models of Cognition and Decision* Cambridge: Cambridge University Press.
Connover, E. (2019) "Google Claimed Quantum Supremacy in 2019 — and Sparked Controversy" *ScienceNews* (16th December) (Available at: https://www.sciencenews.org/article/google-quantum-supremacy-claim-controversy-top-science-stories-2019-yir).
Denti, R. (2020) "What is Quantum Art" *International Academy MAQ* (Available at: https://www.iamaq.org/what-is-quantum-art).
Der Derian, J. (2011) "Quantum Diplomacy, German-US Relations and the Psychogeography of Berlin" *The Hague Journal of Diplomacy* 6 (3–4) pp.373–392.

Der Derian, J. (2013) "From War 2.0 to Quantum War: The Superpositionality of Global Violence" *Australian Journal of International Affairs* 6 (5) pp.570–585.

Einstein, A., Podolsky, B. and Rosen, N. (1935) "Can Quantum-mechanical Description of Physical Reality be Considered Complete?" *Physical Review* 47 pp.777–780.

Fukuyama, F. (1992) *The End of History and the Last Man* New York: Free Press.

Goodchild, M.F. (2018) "Reimagining the History of GIS" *Annals of GIS* 24 (1) pp.1–8 DOI: 10.1080/19475683.2018.1424737.

Harrison, S. and Dunham, P. (1998) "Decoherence, Quantum Theory and Their Implications for the Philosophy of Geomorphology" *Transactions of the Institute of British Geographers* 23 pp.501–514 DOI: 10.1111/j.0020-2754.1998.00501.x.

Haven, E. and Khrennikov, A. (Eds) (2017) *The Palgrave Handbook of Quantum Models in Social Science* London: Palgrave Macmillan.

Haven, E. and Khrennikov, A. (2013) *Quantum Social Science* Cambridge: Cambridge University Press.

Kent, A.J. (2020) "Mapping and Counter-Mapping COVID-19: From Crisis to Cartocracy" *The Cartographic Journal* 57 (3) pp.187–195 DOI: 10.1080/00087041.2020.1855001.

Lloyd, S. (2006) *Programming the Universe: A Quantum Computer Scientist Takes on the Cosmos* New York: Vintage.

O'Brien, K.L. (2016) "Climate Change and Social Transformations: Is it Time for a Quantum Leap?" *WIREs Climate Change* 7 pp.618–626.

Peterman, W. (1994) "Quantum Theory and Geography: What Can Dr. Bertlmann Teach Us? *The Professional Geographer* 46 pp.1–9 DOI: 10.1111/j.0033-.

Portugali, J. (1993) *Implicate Relations: Society and Space in the Israeli-Palestinian Conflict* Berlin: Springer.

Pratt, C.J. (2021) *Quantum Physics for Beginners: From Wave Theory to Quantum Computing* London: Stefano Solimito.

Prencipe, M. (2019) "Quantum Mechanics in Earth Sciences: A One-century-old Story" *Rendiconti Lincei. Scienze Fisiche e Naturali* 30 pp.239–259 DOI: 10.1007/s12210-018-0744-1.

Robbins, P. and Krueger, R. (2000) "Beyond Bias? The Promise and Limits of Q Method in Human Geography" *The Professional Geographer* 52 (4) pp.636–648.

Shaw, S.L and Sui, D.Z. (Eds) (2021) *Mapping COVID-19 in Space and Time: Understanding the Spatial and Temporal Dynamics of a Global Pandemic* Berlin: Springer.

Shaw, S.L. and Sui, D.Z. (2019) "Understanding the New Human Dynamics in Smart Spaces and Places: Towards a Splatial Framework" *Annals of the AAG* 110 (2) pp.339–348 DOI: 10.1080/24694452.2019.1631145.

Siegfried, T. (2008) *The Bit and the Pendulum: From Quantum Computing to M Theory – the New Physics of Information* New York: John Wiley & Sons.

Smith, T.S.J. (2016) "What Ever Happened to Quantum Geography? Toward a New Qualified Naturalism" *Geoforum* 71 (1) pp.5–8.

Stephens, M. (2020) "A Geospatial Infodemic: Mapping Twitter Conspiracy Theories of COVID-19" *Dialogues in Human Geography* 10 (2) pp.276–281 DOI: 10.1177/2043820620935683.

Sui, D.Z. (2015) ""Emerging GIS Themes and the Six Senses of the New Mind: Is GIS Becoming a Liberation Technology?" *Annals of GIS* 21 (1) pp.1–13.

Sui, D.Z. (2010) "The Spatial Century and the Renaissance of Geography" *GeoWorld* (March) pp.17–19.

Sui, D.Z. (1994) "GIS and Urban Studies: Positivism, Post-positivism, and Beyond" *Urban Geography* 14 (3) pp.258–278.

Thornhill, J. (2018) "Quantum Politics and a World Turned Upside Down" *The Financial Times* (6th October) (Available at: https://www.ft.com/content/e4eabea2-c2f9-11e8-8d55-54197280d3f7).

Wendt, A. (2015) *Quantum Mind and Social Science* Cambridge: Cambridge University Press.

Zanotti, L. (2018) *Ontological Entanglements, Agency and Ethics in International Relations: Exploring the Crossroads* Abingdon: Routledge.

Zettili, Q. (2009) *Quantum Mechanics: Concepts and Applications* (2nd ed.) New York: John Wiley and Sons.

Zhao, B., Zhang, S., Xu, C., Sun, Y. and Deng, C. (2021) "Deep Fake Geography? When Geospatial Data Encounter Artificial Intelligence" *Cartography and Geographic Information Science* DOI: 10.1080/15230406.2021.1910075.

Zohar, D. (1994) *The Quantum Society: Mind, Physics and a New Social Vision* New York: Morrow.

48

THE LOCUS CHARTER

Towards ethical principles and practice for location data services

Denise McKenzie and Ben Hawes

The Locus Charter is a set of proposed common international ethical principles to help users of location data to make informed and responsible decisions. The Charter was launched in 2021, building on work by two programs, the Benchmark Initiative in the UK and EthicalGEO in the United States, and on initial development in international roundtables.

This chapter explores some of the ethical issues related to the development and operation of location data services, the role of codes of data ethics, and the background and rationale for the Locus Charter and its foundational ethics principles. We include the text of the Locus Charter. We then summarise some complementary programs that focus on data ethics challenges in specific areas of geospatial applications, and include the Gather principles for sanitation data, as an example. Finally, we consider how the Charter and other activity focused on ethical practice with location data might develop in the future.

Ethics and geospatial data

In the past three decades, as explored elsewhere in this volume, geospatial technologies have continued to evolve in capability, precision, and usability. This has in part been driven by the expansion of the Internet and the rise in location services embedded in mobile, wearable, and other connected devices. More places and more things are mapped digitally by more devices than would have seemed possible only a decade ago, even to many geospatial data experts. Most significantly, deploying the power of geospatial technology is no longer just the domain of GIS experts, and is available in many sectors and to many users with no previous training in geospatial techniques.

Over the same period, increasing digitization across industry sectors and functions has shown repeatedly that some distinct kinds of ethical risk tend to result from using data to manage a process or environment. The most commonly highlighted risks are unfair treatment resulting from bias in datasets; disproportionate intrusions into the privacy of individuals, groups, or organizations; and changes in the distribution of power in markets and other relationships.

These effects are not new in themselves. Sectors (notably healthcare) with a history of working with sensitive data have established systems for identifying and addressing the characteristic challenges that can result. However, the rapid spread of datafication has dramatically increased the potential scale, range, and incidence of related risks. This has prompted greatly increased activity in research into, and practice of, data ethics. Until recently, that activity did not include much attention to ethics in the use of geospatial data. Our work has aimed at that gap.

Principles to promote ethical practice

Ethical codes set out norms of behaviour for communities, defined by organizational membership, profession, sector, or other affiliation or context (Illinois Institute of Technology, 2021). Codes offer ways of identifying, exploring, and resolving value-laden questions to support decisions on what action to take in behaviour to comply with agreed good practice within that group. Ethical principles operate within law and tend to set out socially and morally positive aspirations rather than only minimal requirements.

Most professions have developed more or less formal ethical guidelines over time. In recent years, many professional groups have had to consider whether their existing ethical code adequately governs the digital realm and their use of data, and in many cases, they have developed specific data ethics codes as well. Data ethics guidelines have also been developed by and for data analysts and data scientists, within and across digital technology domains, including (increasingly) artificial intelligence (Accenture, 2016).

There has been work across domains to transfer lessons about data ethics. A 2017 joint report by the Royal Society and the British Academy set out high-level principles for data governance and use, and provided an overarching framework to support more domain-specific principles (The Royal Society, 2017). The Open Data Institute has published a Data Ethics Canvas to support practical exploration of values and challenges (ODI, 2019). Omidyar Network has produced the Ethics Explorer to support technology organizations' exploration of responsible technology development (Omidyar Network, 2020).

Geospatial technology has not by any means been an ethics-free area of activity. There has been extensive exploration of mapping functions as potentially democratizing and liberating, and conversely as potentially oppressive, particularly from post-colonial perspectives. The politics of mapping continues to generate new themes, concepts, and critique (Bargues-Pedreny *et al.*, 2018). However, until recently codes of geospatial ethics have been generally confined to specific national organizations and associations such as the URISA GIS Code of Ethics (URISA, 2003) with little sign of international activity on codes, charters, or agreed principles specifically relating to how data is used.

Background to the Locus Charter

The Benchmark Initiative (Benchmark), based in the United Kingdom, was launched by Omidyar Network and Ordnance Survey in 2019 to raise awareness of potential risks and harms from using location data and identify solutions that realize benefits while minimizing negative impacts. Benchmark has delivered a series of public thought leadership events and an entrepreneur program. In parallel with Benchmark, the American Geographical Society (also in partnership with Omidyar Network) launched the EthicalGEO Initiative, USA, which has supported research fellows investigating issues at the intersection of ethics and geospatial sciences.

These collaborating initiatives share these propositions:

- that geospatial data applications are powerful and applicable to many contexts, and with that power necessarily comes some potential for risk and harm, as has been the case with other powerful data technologies;

- that attention to risks and challenges is essential to realizing the social and economic benefits of geospatial technology and to assuring public trust, and should not be seen as limiting the usefulness of applications;
- that users of geospatial data can learn from practical data ethics in other fields and adapt lessons to their own needs and circumstances.

The Benchmark dialogues have investigated: bias, transparency, privacy, imbalances of power, anonymity, contact tracing, managing social distancing in public areas and workspaces, smart cities, designing for accessibility, data colonialism, tracking waste, and data relating to human migration and humanitarian crises. Four entrepreneur teams were funded to develop practical solutions in transport, mobility, development & mobile technology (Benchmark Initiative, 2021).

EthicalGEO has supported new research into emerging challenges and opportunities, including high-tech survey tools to empower informal land rights activists, providing low-tech training for communities experiencing environmental injustice, evaluating individual concepts of privacy, and providing educators with resources to teach students about geoprivacy. EthicalGEO has also assembled an online knowledge repository building resources for researchers working to understand ethics and geospatial technology (EthicalGEO, 2021).

At the start of both initiatives, ethical use of location data was not a live topic in public policy and media discourse, but that changed. Early in 2020, both Apple and Android moved to restrict default location tracking by apps, and location tracking is increasingly a contested issue among platforms and app providers (Nellis and Paresh, 2020). In January 2020, the Irish data protection regulator announced an investigation into Google's use of location data in the EU.

The COVID-19 pandemic significantly increased global interest in the use, effectiveness, and fairness of different applications for contact tracing, for monitoring activity in public spaces and workplaces, and for enforcing social distancing. This served to highlight the importance and relevance of the Benchmark and EthicalGEO work, and supported our joint mission to bring location data into the global data ethics debates and data ethics into geospatial practice.

Throughout our work, there has been a great deal of interest among users of geospatial data to address and find solutions to challenges. However, it became clear that users of location data had no shared principles, guidelines, or frameworks specific to geospatial data, to work with. Put simply; there was a lack of clarity on what questions to ask and how to discover where risks and harms could occur from uses of location data. Existing data ethics guidance tends to be silent about location data or consider it as only relating to privacy, as another type of personal data can be used to identify individuals.

After many discussions with geospatial practitioners and organizations around the world, we concluded that shared ethical principles could improve clarity, trustworthiness, and trust, and help to realize greater overall benefits from the use of geospatial data technologies.

There will be different priorities in ethics for geospatial in different regions, countries, groups, and contexts. As the community builds, users of location data are encouraged to point to critical issues that have been missed and alternative priorities and approaches. The Charter is proposed as an initial focus for improving mutual appreciation of perspectives, experiences, and expertise. International collaboration and knowledge sharing are seen as key to success. The rationale for developing the Locus Charter is explained in more detail in the preamble in Box 48.1.

Box 48.1 Locus Charter (full text)

Preamble: The Locus Charter proposes that wider, shared understanding of risks and solutions relating to uses of location data can improve standards of practice, and help protect individuals and the public interest. We hope the Charter can improve understanding of risk, so those can be managed, and the many benefits of geospatial technologies can be realized for individuals and societies.

'Locus' is the Latin word for 'place'.

Our Vision: A world where location data is utilized for the betterment of the world and all species that live in it.

WHO WE ARE: An international collaboration of governments, organizations, and individual practitioners seeking to ensure the ethical & responsible use of location data throughout the world.

AUDIENCE: The Charter is written for individuals and organizations who use location data or have responsibility for activities that create, collect, analyse, and store location data.

Why now?

Data about places and people in places has an ever-increasing range of uses. One person can be at the same time the user of a mobile map, and the object of tracking in digital mapping by many organizations they are not aware of. The proliferation of mapping serves many purposes, with diverse costs and benefits, from street, up to city, national, and international scale. How the costs and benefits from mass and multiple use of location data are distributed is a new dimension of power relations. Ubiquitous digital mapping can serve individual freedoms, but can also intensify imbalances of power.

There is already widespread concern – among the public, the media, governments, and professionals in many fields – about potential negative effects from the data ecosystems that now operate around us. It is now clear that data-driven applications can come with specific kinds of risks, including of undue manipulation, discrimination, opacity, and undermining personal privacy.

The field of data ethics is concerned with identifying the risks and developing mitigations, to support responsible use of data. Targeted data ethics principles and guidelines can improve understanding of rights, obligations, and responsibilities, and support better decisions. Ethical guidelines are evolving for different contexts (e.g., healthcare; online platforms), professions (e.g., data scientist; researcher; manager), or categories of data (personal medical; online consumer; financial). At present, there is no common set of guidelines for responsible use of location data. This is a serious gap. Ethical principles are not new in mapping, but the number, variety, and accessibility of digital mapping has created risks and opportunities that are new in kind and scale. Users of location data should have help to understand potential harms from their activity, to manage risks, and communicate how they make sensitive decisions.

Geospatial technologies are developing alongside Artificial Intelligence, the Internet of Things, and Robotics. Users should be as well informed about risks as users of those technologies are increasingly expected to be. The Geospatial Revolution is a major factor in the latest wave of digital transformation that is extending digital capabilities much further into the physical world, and weaving together on- and offline into new forms. Both in terms of new capabilities and of new risks, considerations about X, Y, Z & T may now require the same attention in data ethics, as those around ones and zeroes online have in recent years. We can learn from data ethics development in other fields, and adapt and develop those to support use of location data. The development of this Charter has been informed by the extensive work already done on data ethics in many fields

Many of the sources of potential harm identified in other fields can occur in relation to location data, including bias in datasets, privacy intrusion, and misuse of power imbalances in markets. But while the examples are very useful, it is important to recognize that risks relating to use of location data can have impacts that are specific to those uses, to people and to places. To be followed around where you go in the world has parallels with being tracked around the internet, but it is not the same thing. Also, while a lot of data regulation is framed around the rights of individuals, use of location data can affect groups of people, including some already under pressures relating to where they are, like refugees and other migrants. Location data lends specific powers, which can imply specific responsibilities.

Location Data and Ethics in Practice

We hope the Locus Charter will stimulate discussion, debate, and a community of commitment to ethics around location.

We believe that common understanding of risks and potential solutions will help geospatial data professionals make better, more informed decisions. Discussions with major sector groups and organizations and our own work all suggest that geospatial data professionals – like people in other data sectors – believe in making applications work better for society, but most have had relatively little training on data ethics in practice. In any case, the potential impacts of uses of data change as more data and more granular data becomes available. Our experiences throughout the Benchmark Initiative and EthicalGEO have been that these professionals will value practical tools and terms for understanding ethical questions, resolving any conflicts and risks, and informing their own decisions on what responsible practice is in particular situations.

Digitization and datafication are advancing at speed, changing impacts and benefits in new areas, and changing what is possible and what is desirable. Converging technologies are particularly hard to predict. Principles and guidelines sometimes need to evolve too. We hope that the Locus Charter can deliver a useful stage in the evolution of geospatial applications, and help to bring together the parallel work being done on good practice in different geospatial application areas, and in different fora.

We believe that we can all make better decisions if many perspectives from all around the world are brought together in a shared expression of what characterizes responsible and informed professional practice with location data.

Founding Principles

1. Realize opportunities: Location data offers many social and economic benefits, and these opportunities should be realized responsibly.
2. Understand impacts: Users of location data have responsibility to understand the potential effects of their uses of data, including knowing who (individuals and groups) and what could be affected, and how. That understanding should be used to make informed and proportionate decisions, and to minimize negative impacts.
3. Do no harm: Physical proximity amplifies the potential harms that can befall people, flora, and fauna. Data users should ensure that the individual or collective location data

pertaining to all species should not be used to discriminate, exploit, or harm. Rights established in the physical world must be protected in digital contexts and interactions.
4. Protect the vulnerable: Vulnerable people and places can be disproportionately harmed by the misuses of location data, and may lack the capacity to protect themselves. In these contexts, data users should take additional care, act proportionately, and positively avoid causing harm.
5. Address bias: Bias in the collection, use, and combination of location datasets can either remove affected groups from mapping that conveys rights or services, or amplify negative impacts of inclusion in a dataset. Therefore, care should be taken to understand bias in the datasets and avoid discriminatory outcomes.
6. Minimize intrusion: Given the intimate and personal nature of location data, users should avoid unnecessary and intrusive examination of people's lives and the places they live in, that would undermine human dignity.
7. Minimize data: Most business and mission applications do not require the most invasive scale of location tracking available in order to provide the intended level of service. Users should comply with practices that adhere to the data minimization principle of using only the necessary personal data that is adequate, relevant, and limited to the objective, including abstracting location data to the least invasive scale feasible for the application.
8. Protect privacy: Tracking the movement of individuals through space and time gives insights into the most intimate aspects of their lives. In the rare cases when aggregated and anonymized location data will not meet the specific business or mission need, location data that identifies individuals should be respected, protected, and used with informed consent where possible and proportionate.
9. Prevent identification of individuals: As an individual's mobile location data is situated within more and more geospatial context data, its anonymity erodes, measures should be put in place to prevent subsequent use of the data resulting in identification of individuals or their location.
10. Provide accountability: People who are represented in location data collected, combined, and used by organizations should be able to interrogate how it is collected and used in relation to them and their interests, and appeal those uses proportionate to levels of detail and potential for harms.

Global developments and activity in geospatial data ethics

As above, we are not aware of another initiative to propose common international principles with this focus, but there is strong interest in ethical practice in several geospatial organizations. We have engaged actively in these fora, and we encourage the broadest range of contributors to try the principles and use them as the basis for their discussions and explorations. The Locus Charter will be of more value as more experts and organizations bring their perspectives to it. The Locus Charter Community invites organizations and individuals to join the community and share their activity and experiences, including how they might use, build, on or adapt the Charter. The following are references to some key initiatives promoting and investigating geospatial data ethics.

Geospatial Commission, United Kingdom

In 2020 in its first national Geospatial Strategy, the UK Government committed 'to develop and maintain guidance on how to unlock value from location data while mitigating ethical and privacy risks, ensuring compliance with legal principles and retaining the trust of citizens' (Geospatial Commission, 2020).

Gather principles for sanitation data, United Kingdom

Gather works internationally with municipal sanitation organizations to help collect, share, and analyse data so that they can get toilets to people in their communities who need them most. They use the data to create city-wide maps that help decision-makers know where and how to improve sanitation. Their goal is to help improve access to safely managed sanitation for 5 million people in four emerging cities by 2025. Gather developed these data principles based on their experience and support from the Entrepreneur in Residence program of the Benchmark Initiative (see Box 49.2). Learning from our successes, challenges, and failures: we will share the lessons from our successes, challenges, and failures in working to improve the integrity of location data in line with The Nakuru Accord: Failing better in the WASH Sector.

Geonovum ethical framework, the Netherlands

Geonovum is a foundation supported by the Government of the Netherlands dedicated to supporting more effective use of geospatial information. In 2019, Geonovum developed a draft a framework to guide ethical use of location data, inspired by existing ethical codes, enriched by input from workshops. See more: https://www.geonovum.nl/themas/geo4covid/ethical-framework.

Real Estate Data (RED) Foundation data ethics principles, United Kingdom

The RED Foundation is an initiative set up to ensure the real estate sector benefits from an increased use of data, avoids some of the risks that this presents and is better placed to serve society. The foundation has established a steering group of leading experts and six principles that RED asks companies to sign up to demonstrate that they are working towards (see more: https://www.theredfoundation.org/dataethics).

Worldwide Web Consortium (W3C), Global

In January 2021, the W3C released an Interest Group Note from the Spatial Data on the Web Interest Group titled the Responsible use of Spatial Data. From the introduction, the purpose of the document is

> to raise awareness of the ethical responsibilities of both providers and users of spatial data on the web. While there is considerable discussion of data ethics in general, this document illustrates the issues specifically associated with the nature of spatial data and both the benefits and risks of sharing this information implicitly and explicitly on the web.
>
> *(see https://www.w3.org/TR/responsible-use-spatial/)*

> **Box 48.2 Gather principles**
>
> We commit to:
>
> - Work for the social good: we commit to using data for the social good and will actively seek to prevent unintended, harmful consequences that could occur when collecting, sharing, and analysing the data. We believe that sanitation data should only be collected if it is to be used to improve the infrastructure and services for vulnerable communities.
> - Work in collaboration with others: we will work with organizations and individuals within the sanitation sector to improve the integrity of location data for sanitation infrastructure and services.
> - Be transparent: we will make sure that we:
>   - publicly communicate our reason for collecting, storing, sharing, and analysing the data;
>   - communicate our reason for collecting, storing, sharing, and analysing the data with those we collect data about;
>   - welcome critique and feedback from all stakeholders, particularly those affected by the data;
>   - ensure data diversity: we commit to collecting data that is inclusive and well-representative so that we surface under-represented groups who have different needs to the majority of the population or geography;
>   - protect people's privacy: we will protect the privacy of individuals when we collect data about them. We will only collect, store, share, and analyse data with their express permission and will delete it when the permission expires. Data should only be held for the minimal amount of time needed to make good decisions;
>   - Be open to sharing data without compromising privacy: we will work together to improve data sharing between organizations that work in the same country or city. We believe that this openness will reduce the repetition and overlap of data collections and allow for geospatial analysis at the country or city level;
>   - Work to make the data equable: we will make sure that the data we collect, store, share, and analyse does not require skills or technology that excludes key stakeholders from being able to analyse it.
>
> See more at: https://benchmark.gatherhub.org/principles

Mobile data for development handbook, Global

Mobile for Development (M4D) is a global team within GSMA (the international organization of mobile telephony operators), connecting GSMA's operator members, tech innovators, the development community, and governments, to harness the power of mobile in emerging markets. M4D guides incorporating ethical decision-making into governance models and making informed and proportionate choices that recognize the imperatives of protecting the privacy and vulnerable groups and realizing collective benefit from data for development (see https://www.gsma.com/publicpolicy/mobilepolicyhandbook/mobile-for-development).

Global open data for agriculture and nutrition (GODAN)

Godan is a network with more than a thousand members across national governments, non-governmental organizations, and international and private sector organizations. GODAN supports global efforts to make agricultural and nutritionally relevant data available, accessible, and usable for unrestricted use worldwide, building high-level policy and public and private institutional support for open data. As part of its mission, GODAN facilitates the use of codes of conduct, voluntary guidelines and sets of principles around using farm data, and has published an online Code of Conduct Toolkit to help organizations and partnerships develop data practice codes to fit their needs, 'providing the conceptual basis for general, scalable guidelines for everyone dealing with the production, ownership, sharing and use of data in agriculture' (see https://www.godan.info/codes).

These are initiatives with which Benchmark and EthicalGEO have engaged. Other initiatives are exploring themes including machine learning with location data, Earth observations, the bias of algorithms, and privacy. With global interest and investment continuing to increase, initiatives will emerge to explore issues of geospatial data ethics, and there is hope that collaboration will continue and grow. This is just the beginning.

The future of geospatial data ethics

From the work to date, there are several areas where geospatial data ethics may develop.

The Charter's first Principle is about the importance of realizing benefits from geospatial data. While addressing risks and challenges does require focusing on negative impacts and harms, many users of geospatial data have rightly highlighted that such a focus needs to be balanced with broad recognition of the significant benefits that geospatial data offers. There may well be scope to develop a more detailed shared understanding of what positive ethical data use looks like in practice. In particular, geospatial data is likely to play a significant role in future measures to address climate change by better targeting and conservation of resources and managing the impacts of climate change in throughout the world. Those applications could be supported by a more developed and more widely shared view of ethically positive data use for global collective interests and the planet. Geospatial data ethics development may connect productively with other activity to support ethically positive practices, particularly work towards the Sustainable Development Goals.

There is likely to be more exploration of how to understand and advance collective interests in geospatial data in other respects too. The European General Data Protection Regulation (GDPR) is focused on data privacy for individuals. However, geospatial data will often necessarily include information about groups or communities and potentially have uses and impacts better understood at the collective rather than individual level. Location-enabled data will often be collected without meaningful (or any) permission, and accumulating it may have implications that are better considered in terms of collective rights.

For instance, a city government negotiating with a ride-sharing company for access to the company's data about demand and traffic movements seeks cumulative rather than personal insights and will act for collective interests. It seems likely that this collective aspect of geospatial data ethics – both in terms of risks and potential benefits – is insufficiently developed or expressed.

This may be most important in civic spaces and contexts. Individuals have limited power to represent their interests around how data is used in local environments and by national governments. Furthermore, inequalities may be implicit between the technically enabled data collection community and the communities who are subject to that collection and analysis, as examined by 'Projects by IF' as part of the Benchmark entrepreneurs' programme. Work is emerging in local data and civic trusts and ethics for guiding local use of emerging data technologies. In 2020, the Mayor of London announced work to develop a set of criteria to 'guide emerging technology in London'. Geospatial data ethics could and should be a key part of developments of this nature.

Increasingly, location-enabled data is also being used for localized economic analysis and planning of infrastructure, transport, and economic interventions. Exactly how that data is used, and in whose interests, is likely to attract increasing analytical and political attention.

Digital tracking of the location of individual people may have a long way yet to develop, in the ways it can be done, what the data can be used for, what the ethical implications are, and what possible counter-measures and protections may develop. Most of the dialogue and debate about digital tracking of individual members of the public is still about tracking online behaviour only. Even there, while a great deal of concern is expressed, public attitudes and behaviour have not reached any very settled norms. More tracking of individuals in physical places is likely to generate similar commercial opportunities and similar concerns, but also some different ones. Being followed around the Internet and being followed around town are not quite the same. Control over location privacy may well come to be a valued feature of consumer app services. In her analysis of 'surveillance capitalism' working through the accumulation and use of personal data by major digital companies, Shoshana Zuboff makes a case for protecting autonomy and a sense of private space. Private space in the literally physical sense may become an increasingly contested area of data ethics and practice (Zuboff 2019).

Conclusion

Location data ethics is at the beginning of its journey and will continue to evolve as more voices join the international dialogue. What is clear is that knowledge sharing and collaboration will be key to our shared understanding of what ethical and responsible practice is for the development and operation of location data services. It is critical that all practitioners join this journey and work together to develop our understanding in order to realize and share the great benefits that ethical and responsible use of location data can bring to the world.

References

Accenture (2016) *Universal Principles of Data Ethics: 12 Guidelines for Developing Ethics Codes* Dublin: Accenture.
Bargues-Pedreny, P., Chandler, D. and Simon, E. (2018) *Mapping and Politics in the Digital Age* Abingdon: Routledge.
Benchmark Initiative (2021) "Benchmark" (*Blog*) Available at: https://benchmarkinitiative.com/blog (Accessed: 14th May 2021).
EthicalGEO (2021) "Knowledge Repository – EthicalGEO" Available at: https://ethicalgeo.org/-knowledge-repository/ (Accessed: 14th May 2021).
Geospatial Commission (2020) "Unlocking the Power of Location: The UK's Geospatial Strategy 2020 to 2025" Available at: https://www.gov.uk/government/publications/unlocking-the-power-of-locationthe-uks-geospatial-strategy/unlocking-the-power-of-location-the-uks-geospatial-strategy-2020-to-2025 (Accessed: 14th May 2021).
Illinois Institute of Technology (2021) "Ethics Code Collection" *Center for the Study of Ethics in the Professions* Available at: http://ethicscodescollection.org/ (Accessed: 14th May 2021).

Nellis, S. and Paresh, D. (2020) "Apple, Google Ban Use of Location Tracking in Contact Tracing Apps" *Reuters* Available at: https://www.reuters.com/article/us-health-coronavirus-usa-apps-idUSKBN22G28W (Accessed: 14th May 2021).

ODI (2019) "Data Ethics Canvas" *ODI* Available at: https://www.theodi.org/article/the-data-ethics-canvas-2021/#:~:text=What%20is%20the%20Data%20Ethics,and%20reflect%20on%20the%20responses (Accessed: 12th January 2023).

Omidyar Network (2020) "Ethical Explorer" Available at: https://ethicalexplorer.org/ (Accessed: 14th May 2021).

The Royal Society (2017) "Data Management and Use: Governance in the 21st Century – A British Academy and Royal Society Project" *The Royal Society* Available at: https://royalsociety.org/-topics-policy/projects/data-governance/ (Accessed: 14th May 2021).

URISA (2003) "GIS Code of Ethics" *URISA* Available at: https://www.urisa.org/about-us/gis-code-of-ethics/ (Accessed: 14th May 2021).

Zuboff, S. (2019) *The Age of Surveillance Capitalism: The Fight for a Human Future at the New Frontier of Power* London: Hachette.

INDEX

Note: **Bold** page numbers refer to tables; *italic* page numbers refer to figures and page numbers followed by "n" denote endnotes.

AASHTO survey 425
Advanced Land Observing Satellite (ALOS) 229
advanced-SAR 536
Advanced Spaceborne Thermal Emission and Reflection Radiometer (ASTER) 229
aerial-based data sources 347
aerial bathymetry 405
aerial imagery *556,* 556–557
aerial photogrammetry (manned and unmanned systems) 231
aesthetics: definition of 79; experience 80–81; of GIS 81–82; software applications 81; story mapping 81
affordances: as design 82; design features 83; of qualitative GIS research practice 83; technology or practice 82–83
African GEOSS (AfriGEOSS) 61
Agricultural Production Systems Simulator (APSIM) product 65, *66*; WPS 65, *66*
agriXchange 61
agro-ecological data 68
AHAB 220
AHAH (Access to Healthy Assets & Hazards) 293
airborne digital cameras 206; large-format cameras 206–208, *207*; laser scanning technologies 213–220; medium-format cameras 208–209, *209*; multiple frame cameras 210–212; small-format cameras 209–210, *210*
airborne film cameras 203, *204*
airborne laser bathymetry (ALB) 406
airborne laser scanning (ALS) *214*; airborne Lidar 177, *178*; bathymetric laser scanners 217–220, *219, 220*; 3D coordinates 178; full waveform digitization 178, *179*; and MLS, comparing *185,* 185–186; principle of *214*; technologies 213–220; topographic laser scanners 214–217, *216*; unmanned airborne systems 178, *179,* 180; use of ALS 178
airborne non-digital mapping technologies 203; airborne film cameras 203, *204*
airborne photogrammetric mapping: digital photogrammetric workstation (DPW) 221–222; digital terrain models 222–223; laser scanning 223; manned airborne platforms 202; orthophotographs 222–223; orthophotomosaics 222; 'point cloud' 221; software developments 223–224; spaceborne remote sensing techniques 202; unmanned aerial vehicles (UAVs) 202–203
airbrush-produced elevation hillshading 39, *39*
Alexandria Digital Library 47
ALOS Phased Array *230*
Amazon Rekognition 495
analogue stereo-plotting instruments 203, *205*
analytical maps 56
analytical plotters 203–204, *205*
ancestral domain delimitation 118n1
Anti-Eviction Mapping Project 89
Apple Maps 162, 164–165, *165*
application programming interfaces (APIs): ecosystems 256; kinds of **259**; managed 257–258; Map Tile 257; open frameworks 258; server programming 258; social media 495; spatial servers 258; static map image 257; toolkits 259; wrappers 257
APSIM 66

Index

ArcGIS 261–262
archaeology: aerial and satellite imagery *556, 556*–557; applications of 3D archaeological data 557–559, *558*; geographic information systems 554–556, *555*; limitations 559
ARGUS-IS project 477
artificial intelligence (AI) 243; application of 397; deep learning 319–320, *320*; defined 313; ethical questions and implications 325–327; future of 324–325; Future Today Institute 326–327; Geospatial (*see* Geospatial AI); ML (*see* machine learning (ML)); neural networks 316–318, *317*; Strong AI (True Intelligence or Artificial General Intelligence) 315, *316*; in Sustainable Development Goals *325*; systems 313–314, *314*; technologies 315–316; Weak AI (narrow AI) 314–315, *316*
artificial neural networks (ANNs) 302, 316
augmented reality (AR) 1, 236, *238*, 239, *239*, 241, 334, 339, 379, 380, 382, 400, 477
'AugView' 380, *381*
Australian Institute of Cartographers 23
auto-demarcation 118n1
autoencoder neural networks 318
automated driving systems (ADS) 427
Automatic Identification Systems (AIS) 417
autonomous vehicles (AVs) 427–430

base maps 56
bathymetry: aerial 405
beam divergence (BD) 176, *176*
Bertin, Jacques 544, 604, 606
big data, geovisualization of: cartograms 294, *294*; defined 291; ethical considerations 295–296; General Data Protection Regulation (GDPR) 295; geodemographic 292; impacting on society 295; secondary data 291; secure data infrastructures 296–297; summarizing and simplifying data 292–293; visualizing big data 293–294; working with 292
Bing Maps 134, 162, 347
Black Lives Matter (BLM) protests 89
bluetooth tracking 466
body mapping 91
Boolean logic 50
border thinking, concept of 89
Building Information Modelling (BIM) data 343, 344
Business Process Modelling and Notation (BPMN) 66; FME environment 67; Genetic Agro-ecological Sustainability Proposal model *68*; and WPS 67

cadastral/topographic databases 347
Canada Geographic Information System (CGIS) 43–46
Cartesian coordinates 50, 182–183, *183*
cartographic generalization 302–304
case-based reasoning 316
Cassini, Giovanni Domenico 13, 14
cave automatic virtual environment (CAVE) 242
celestial measurement 8
celestial navigation 415
cell phone tower triangulation 466
cellular automata approach 281
CFOSAT (China France Oceanographic Satellite) 536
change detection 281
chronometers: adoption of 18–19; to manage deep-water course corrections 18, *18*; 'time balls' 19, *19*
circle growth 303
citizen science 69–71; CLIMAS 70; COBWEB 70; concept of RRI 70; FOODIE VGI 70, *71, 72*; Geo Wiki 70; OpenStreetMap 70; SDI4 Apps project 71, *72*; SWE4CS 70
City Geography Markup Language (CityGML) 47, 345
City Information Modelling (CIM) 344
CityLab 102
'CityMapper' 217
Clarifai 495
climate forecasting data 68
closed-circuit television (CCTV) 466
clustering methods 277
CODATA 72, 73
cognitive maps *(whakapapas)* 123
collision avoidance regulations (ColRegs) 416
community health: data access 459, *460*; partnerships 460–461; spatial clusters or hotspots 451, *451*
Community In-Power and Development Association (CIDA) 91
community mapping 118n1
complex disasters 511–514; refugees in Uganda 513–514, *514*; Syria crisis 512–513, *513*
Confederated Villages of Lisjan 92
connected vehicles (CAVs) 427–430
construction-industry standard BIM 47
Consultative Group on International Agricultural Research (CGIAR) 61
convolutional neural networks (CNNs) 318
Copernicus Sentinels 407
CORONA satellite program 557
COSMOS 495
counter-mapping: air pollution and minority population *117*; autodemarcation in Brazilian Amazon 116; community land rights 113, 125; Cree maps of harvest locations 111; critiques of 114; Dene maps of traplines 111; empowerment from 89–90, 109; history of Indigenous counter-mapping 92–93, 110–113; Inuit Land Use and Occupancy Project

111, *111*; map biographies 110; mapping of Indigenous territories *112,* 112–115, 119n5; mapping of mobilities 116; material acts of transforming place names and geographical markers 116; power and critical GIS 54; proliferation of 112; to spatial (racialized) inequalities 116

Course Over Ground (COG) (ground track) 414

COVID-19 global pandemic: coda for geospatial technologies 431–433; geospatial and non-geospatial technologies 463–464, *465*; global impact of 2; lockdowns 102; mapping 3; and public health 454–455, *455*; rates 89; web map displaying confirmed cases of 521, *522*; work from home (WFH) 432

COVID Symptom Tracker (2020) 467

credit card transactions 466

crisis and hazard mapping: case studies 507; challenges and future 514–516; complex disasters 511–514; history and main players 505–507; man made disasters 510–511; natural disasters 507–510

critical cartography 133

critical GIS: emergence of 50–52; and power 50, 53–55; in Web 2.0 age, new opportunities and challenges 55–56

Cromalin proofing process 39

crowdsource or commercial data 347

crowdsourcing 69

culture lettering 37, *38*

Cybernetes/Kubernetes, concept of 415

CycloMedia 162

DARPA (Defense Research Projects Agency) 477

Database Management System (DBMS) PostgreSQL 248

DataBio 61

data capture 3, 62, 185, 207, 223

data classification schemes 104

data interoperability 60; semantic 60; structural 60, 63; syntactic 60, 63

data modelling, spatial analysis and modelling 265; abstraction principles 264; attributes 265; coordinate system (CS) perspective 267; digital elevation model (DEM) 267; discrete global grid systems (DGGS) 268; discrete objects 264; feature abstraction, principles of 265; level of detail (LoD) 265; matrix sampling 267; multi-geometries 266; nearest neighbour interpolation 267; network data structures 268; on raster nodes 267; simple features (SF) 265; spatial fields 266–267; triangulated irregular network (TIN) 267–268; Voronoy diagram 268

data model transformations *273*; contouring 274; spatial density of objects 274; spatial interpolation 274; vector-raster and raster-vector conversions 273

data processing 3, 161, 164, 209, 223, 225, 538, 575

DBSCAN 278

3D city models 167–168, *168,* 343, 348

dead reckoning 414, 415

decision tree (DT) 301, 306, *308, 309*

decision tree supported with genetic algorithms (DT-GA) 304, 306

decolonial mapping 90

'deductive reckoning' (or 'dead reckoning') 7

deep learning (DL) 302, *320*; artificial intelligence 319–320, *320*; models 304, 306

deep mapping 134, 137–139; application of GIS 137–138; conventional maps 138; digital deep maps 138–139; greedy deep map 139; psychogeographic cartography 138; spatial anthropology 614–616

Demeter 61

democratization: closed data 584; controlled access 582–584; data blind spot 585; geospatial data revolution 579–580, *580*; individual and institutional data sharing agreements 582–583; open data 581–582; privacy concerns 583–584

Derrida, Jacques 100

3D geometric models 344

diazo process *34*

differential GPS (DGPS) 154, 161, 163, 412

digital elevation model (DEM) 221, 276, 527; aerial photogrammetry (manned and unmanned systems) 231; Agisoft photoscan 231, *232*; airborne or aerial data sources 229; ALS 230–231; applications and quality of 231–233; brief history of 227; computed or secondary 226; data sources and generation methods 227–228; definitions of 226–227; digital surface model (DSM) 227; digital terrain model (DTM) 227; ground-based data sources 228; mean sea level (MSL) 226; measured or primary 226; raster 226; satellite-based data sources 228–229; Triangular Irregular Network (TIN) 226

digital photogrammetric workstation (DPW) 204, 206, 221–222

digital spatial technologies 56

digital terrain model (DTM) 221, 222–223, 347

Dimensionally Extended 9-Intersection Model (DE-9IM) 269–270

distance measuring indicator/instruments (DMI) 154, 366

distributed energy resource (DER) model 439

dot map simplification algorithm 303

DPIA (Data Protection Impact Assessment) 296

drawn map: manuscript map image 29; negative map artwork 30, 35–37, *36–39*;

photo-mechanical processes 29–30, 37, 39–40, *40*; positive map artwork 30, *31–36, 32, 35*
drones: brief history of 190–193; civilian drone industry 190, 193; consumer drones 193; debates 196; defining 188–189; delivery drones 194; democratization of aerial view 198; descent of *191*; GPS receiver 190; imagery 193, 194; international humanitarian law and differentiation 197–198; light detection and ranging technology (Lidar) 193; military drones 189, 196–197; modern drones 189; privacy and surveillance 196; regulating and managing 195–196; specialists use 194–195
3D Visualization 373, *374*

East India Company 18
Ebola epidemic 509–510, *510, 511*
echo-sounding 405, *406*
Economic and Social Council (ECOSOC) 598n4
EKKO_Project 368–369, *370*
electrical systems: addressing errors 448; associating network data with cadastral data 440–441; background 437–439; data collection 448; data quality 448; Dumsor, intermittent power in Ghana 439, 442, 445; Earth observed from outer space 437, *438*; geospatial technology methods 443; identifying vulnerabilities in grid infrastructure 441; limitations 447; mapping and communication 440; modern electrical grids 438, *439*; network analysis 444–446; optimizing electrical lines routing 440; planning for load growth 441; renewable energy modeling 443; servicing 441–442; suitability analysis 446–447, *447*; technical accessibility 448; United States national average wind map *443*; United States national global horizontal solar irradiance map *444*; updating changes 448; visualization and communication 444, *445*
electric optical (EO) sensors 193
electromagnetic (EM) energy 171; beam divergence 176, *176*; cross-matrix 175, **176**; 2.5D and 3D point clouds 174–176, *175*; exploiting lasers 172–173; geodesic morphological ground filtering method 175, *175*; interaction with objects 172; measurement principles 172–174; NIR light 172, *173*; phase-shift 172, 174; pulse measurements or time-of-flight (ToF) 174; triangulation-based 174
emerging infectious disease (EID) 456, *457*
Enfolding 102
ENSEMBLES project 536

Environmental Justice (EJ) organizations 91
ethics: and geospatial data 632–633
ethnocartography 118n1
EU-funded projects 61
EUMETSAT 407
European Federation for Information Technologies in Agriculture (EFITA) 61
European Open Science Cloud (EOSC) 68, 72
evolutionary learning 301
exploratory maps 56
extended reality (XR) technologies: applications and impact on society 243–244; augmented reality (AR) technologies 236, 237, *238*, 239, *239*, 241, *242*; 'augmented virtuality' (AV) 236–237; definitions 236–240; immersion and interactivity 240–241; mixed reality (MR) technologies 236, 237, 239, *239*; user's sense of illusion 240, *241*; virtuality continuum 236, *237*; virtual reality (VR) technologies 236–237, *238,* 239, 242–243
Extensible Markup Language (XML) 65

Facebook 77, 162
face recognition 316
FAIR (Findable, Accessible, Interoperable, and Reusable) 62–63, 72–73
Farm Oriented Open Data in Europe (FOODIE) 70
feedforward neural networks 318
feminist GIS research 54
feminist mapping: binaries and hierarchies 104–105; *Cartography and Geographic Information Science* (CaGIS) 99–100; context and pluralism 102–103; emotion and embodiment 103–104; 'feminist GIS' procures, Google Trends *99,* 99–100; futures or gallery of possibilities 100–105; geographers and mapmakers 98; intersectional feminism 98–99; labour 105; past and present 98–100; power 101; principles for data feminism 100, **100**
The Financial Times 490
Flamsteed, John 16, 403
Flickr 134, 494
Fluorescence Explorer (FLEX) mission 396
Fog Computing (FC) 427
Food and Agricultural Organization (FAO) 61, 63–64
FOODIE 61
food security: census tracts 388; COVID-19 pandemic 390–391; food insecurity 389, *390*; fuzzy four-dimensional food deserts 384–387; geospatial mapping problems 388; geospatial technology 387–388; MAUP issue 389; spurious correlation 389, *389*; use of social transects in obesity, Birmingham, UK 391–392
Foucault, Michel 100

Foursquare 77
free and open-source software (FOSS) 2, 246; *see also* free and open source software for geospatial applications (FOSS4G)
free and open source software for geospatial applications (FOSS4G): brief history of 247–248; current state of 248–250; GeoForAll 247; importance and relevance of 251–252; learning and adopting 249; Open Data 247; openness 246–247; Open Standards 247; OSGeo Education Initiative 247; rapid growth of 248; to society and current challenges 251–252; software selection and adoption 250–251; use of 250
'Free-Culture' 403
freehand technique 37, *38*
French curve drafting set 32, *33*
full waveform digitization (FWD) 178, *179*

Galileo 153
GAMBIT satellite program 557
Gather principles for sanitation data, United Kingdom 638
GEBCO Nippon Project 407
General Data Protection Regulation (GDPR) 196, 295
General Geographic Object Database (GGOD) 305, *305*
generalization: cartographic 299–300, 304
genetic algorithm (GA) 301–302, 316
genetic data 68
'geo-coding' practice 55
geodemographics 293
geographically weighted regression (GWR) 457
Geographic Resources Analysis Support System (GRASS) 62
GeoJSON 522
geometric transformation model (GTM) 182
geomorphometry 227, 276
Geonovum ethical framework, the Netherlands 638
geo-phenotypic information 68
GeoPointer X 373, *375*
georeferencing 177, *177*
Geospatial AI (GeoAI) 1; advantages and limits 324; geospatial big data 321–322; and GIS *323*, 323–324; supercomputing 322–323
Geospatial Commission, United Kingdom 638
geospatial ethics 637–641
geospatial humanities 613–614
Geospatial One-Stop 47
geospatial standards: citizen science 69–71; connection with national standards 69; definition of 60; emergence of cross-disciplinary enablers 64–68; evolution of geospatial data standardization 60; future research for 72–73; GIS 61–62; historical background 61–62; levels of interoperability 60, 63; OGC standards development (*see* Open Geospatial Consortium (OGC)); soil data exchange 63
geospatial strategies: community mapping and identification *593*; developing 591–592; establishing (national) geospatial strategy 592; geospatial information for our future 595–596; geospatial need 589–590; history of 590–591; integrated approach 593, *594*; United Kingdom's geospatial strategy, 2020–2025 594–595; in the wild 596–597
geospatial technologies and society: complementarity, holism 624; core quantum concepts 624; development of 623; entanglement: non-locality 625; ethical and practical implications 628–629; methodological and technological implications 627–628; ontological and epistemological implications 626–627; quantum implications 626; superposition: potentiality 625; uncertainty: non-determinism 624–625
GEOSS (Group on Earth Observation System of Systems) 68, 72
geostatistics 277
geosurveillance and society: and data capitalism 479–480; and governance 477–479; networked technologies 476–477; privacy and civil liberties 480–481
geoweb 55, 79
'ggplot2' package 260
Ghost Factories 489
GIScience to disease mapping: diminished case counts 471; futures 471–472; non-spatial data 464–465; outbreaks 468–469; racial reckoning 470; rise of computerized GIS 463; spatial data 465–467; understanding disease's spread 469–470; vaccine roll-out 470–471
GIS (geographic information system): 1, 61–62, 122, 437; accuracy and uncertainty 45–46; archaeology 554–556, *555*; *ArcMap* or *QGIS* 77, 80; assembling 78–79; assumptions 44–45; CGIS 43–46; consumerization of 48; critical GIS 100; data models 47; DGGS 46; evolving visions 47–48; exploiting Internet 47–48; 'feminist GIS' 99–100; flattening Earth 46; flowchart of workflow *447*; Google Maps 48; interfaces 80; mapping buildings and cities 343; maps 81; maps and computers 42–44; multiple 'digital *despositif*' 79; neogeography 48; object-oriented models 47; participatory GIS (PGIS) projects 79; social constructivism 133–134; SYMAP package 43, 44; technology 77–78; utility networks and 376–378, *376–378*
Global Human Settlement Layer (GHSL) 536

'global intimate maps' 54
global microarray maps 607–608
global navigation satellite system (GNSS) 1, *148*, 151, 153–154, 171, *173*, 182–185, 214, 228, 334, 354–355, 362, 363, 366, 429, 521–522, 591
Global open data for agriculture and nutrition (GODAN) 61, 251, 640
global positioning system (GPS): APIs 259; archaeological applications 554, 557; big data 136; counter-mapping 116, 117; COVID-19 pandemic 454, **458**, 466; deep mapping 621; democratization of data 585; differential GPS (DGPS) 154, 161, 163, 412; drones 190, 193; ethics of 528, 530; geospatial strategies 590; geosurveillance 476–477; geovisualization of GPS tracks *294*; humanitarian applications 523–524, 526; location-based services 336; mapping mobility and public health 459–461; mapping oceans 411–414, 417; mapping sacred sites 93; mapping underground 366, *366*, 368–369, 372, 373; mobile mapping technologies 147–148, *148*, *149*, 151, 153–163, *154*, *155*, *156*, *157*, *158*, *160*, *161*, *167*, 169; participatory rural appraisal 111; photogrammetry 206, 208–210, *209*, *210*, 212–214, *212*, *214*, 216–218, 221, 228; rural Indigenous knowledges 122–123, 125, 127; social media 494–495, 500; spatial modelling 270, 279; Sustainable Development Goals (SDGs) 537
Global Soil Map Markup Language (GSMML) 63
GLONASS 153, 523
GO FAIR 73
Google Cloud Vision 495
Google Earth 56, 134, 150, 347, *620*
Google Maps 48, 134, 163–164
GPR-SLICE 370, *371,* 372, *372*
GPS *see* global positioning system (GPS)
GPS/IMU sub-system: Applanix POS/LV unit 156, *156*; fibre-optic gyros 155; MEMS (Micro Electro-Mechanical Systems) technology 155; odometer or distance measuring instrument 156; ring laser gyros 155
GRACE 407
graphic lettering pen set 30, *31*
GRASP project 67
Greenwich meridian 20
Greenwich Observatory 16
grid operations *272,* 272–273; focal analysis 273; global analysis 273; local analysis 272; zonal analysis 273
ground-based data sources 347
ground-based laser scanning 180–185; mobile laser scanning (MLS) 183–185; multi-beam Lidar systems 181, *181*; point density 181; scan mechanism 180; terrestrial laser scanning (TLS) 180, *180, 181,* 182–183, *183*
ground control points (GCP) 171
ground penetrating radar (GPR) 165; antenna and frequency **361**; data acquisition software 367–372; data georeferencing 365–367, *366*; operating principles of 360–363; software for post-processing data 372–376; systems with antenna arrays 362, *362*; systems with single antenna 361–362, *362*; system with two antennas 364, *364*
Group on Earth Observation System of Systems (GEOSS) 61

The Harlem Heat Project 489
Harley, J. Brian 100, 133
'Health Code' 466
heat maps 608
histological maps 608, *608*
Humanitarian OpenStreetMap Team (HOT) 251
humanitarian relief: artificial intelligence 527–528; crowdsourcing 528; ethical issues 528–530; geographical information systems (GIS) 524–525; and geospatial technologies 520–521; GNSS and locational technologies 522–524; remote sensing 525–527, *526*; social media 527; web/cloud mapping technologies 521–522
hydraulic displacement 415
hydrography 7
Hyperbolus Automatic Detection Algorithm (HADA) 365

IBM's Watson Visual Recognition 495
IKONOS 351
image recognition 319–320
ImpulseRadar ViewPoint 368
Indian People Organizing for Change (IPOC) 92
Indigenous cartography 122
inertial measurement unit (IMU) 148, 154, 228, 355, 366
Infrastructure-as-a-Service (IaaS) 428
infrastructure, mobile mapping for: railway and tramway infrastructure surveys 166, *167*; road infrastructure applications 165, *166*; underwater multi-beam sonar systems 167, *167*
Infrastructure of Spatial Information in Europe (INSPIRE) 61, 63, 69
Instagram 494
instrumentation 7
Integrated Water Resources Management (IWRM) 394
interactive maps 56

interferometry SAR (InSAR) 228
internally displaced persons (IDPs) 520
International Cartographic Association (ICA) 335
International Civil Aviation Organization (ICAO) 195
International Nautical Mile (INM) 413
International Organization for Standardization (ISO) 60, 64
International Society for Precision Agriculture (ICPA) 61
International Union of Soil Sciences Working Group 64
Internet mapping services 122
Internet of Things (IoT) 396, 398
InVEST 495
IQMaps 375, *375*
ISODATA 278
Israeli Defense Forces (IDF) 196

Japan Aerospace Exploration Agency (JAXA) 229, 536
JavaScript Topology Suite (JSTS) 260
Joint Authorities for Rule-making on Unmanned Systems (JARUS) 195
journalism: history of GIS 487–488; history of mapping in 487; news media 486; 'post-truth' world 486–487, 490–491; sensors and journalism 488–490
Jupiter's satellites, eclipses of 13–15, 20

knowledge-based systems 316
kriging 277

#LandBack movement 89–90
large-format cameras 206–208, *207*
laser ranging, profiling and scanning devices 147, 148, 151–153; 2D laser profilers or scanners 151; 'full-circle' scanning/profiling capability 152, *152, 153*; multi-scanner units 152; purpose-built TOF 2D laser scanners 152; ranging measurements 151; series of 2D profiles 152; three-dimensional model 151;
laser scanning, use of 171, 223, 351
latitude, location at sea: back staff 9, 10–11, *12, 13*; cross staff 10, *11*; 'double altitude and elapsed time' technique 13; early navigational instruments 9; *kamal* 8, 10; mariner's astrolabe 8, *9, 10*; mariner's quadrant 8, *9*; mirrors, use of 11; observations of Polaris 13; octants 10, 11–12, *14, 15*; 'rule of the Sun' 8; stellar and solar techniques 8–9; terrestrial latitude 8; *Ursa minor* 8
learned by indicators of spatial association (LISA) 285
Leica DX Office Vision 375, *376*
lethal surveillance 196

lettering colour separation 37, *38*
Level of Detail (LOD) 240, *240*, 345–346, *346*, *348*, 348–349, *349, 350, 352, 353*
license plate recognition (LPR) devices 426
Lidar (light detection and ranging): 3D modelling with *175–176*, 230–231; applications of AI in 320; archaeological applications of 554, 557–559, *558*; basic principles and techniques of 171–187, *173*, 228; bathymetry with 405; data processing and photogrammetry with 221, *221*, 223–224; deriving heights of buildings from 351–353; humanitarian applications of 526–527; instruments 152, *153*; limitations of 176; mapping buildings and structures with 343, 347; mobile laser scanning 183; multi-beam systems 181; planetary mapping with 565, 570; scanning mechanisms of 178, *178–179*; unmanned airborne systems 178–180, *179*, 193
life mapping techniques 91
location-based services (LBS): advances of 333, 339–340; analysis of LBS-generated data 338; cartography and GIScience 332; communication aspects 337; contribution 333; 'data-driven era' 335; definition 332; development of 332–338; engineering discipline 334; evaluation methodology questions 337; historical information *337*; human-centred concepts 335; as interface *333*; modelling aspects 336; personalization 332; personalized services 336–337; popularity 333; potential of location 331–334; research agenda *338*; research challenges *336*, 337; selection 332; societal and behavioural implications 338; success of 334–338; ubiquitous availability 332; ubiquitous positioning 335–336; users of 337
location intelligence 343
Locus Charter: background to 633–637; ethics and geospatial data 632–633; future of geospatial data ethics 640–641; gather principles 639; Gather principles for sanitation data, United Kingdom 638; Geonovum ethical framework, the Netherlands 638; Geospatial Commission, United Kingdom 638; global developments and activity in geospatial data ethics 637–640; Global open data for agriculture and nutrition (GODAN) 640; Mobile for Development (M4D) 639; principles to promote ethical practice 633; Real Estate Data (RED) Foundation data ethics principles, United Kingdom 638; Worldwide Web Consortium (W3C), Global 638
longitude: celestial clock 13–14; chronometers *18*, 18–19, *19*; determination at sea 7, 15–19;

eclipses of Jupiter's satellites 13–15; field observer, equipment 14–15; location on land 13–15; lunar eclipses 13; lunar distances 16–17, *17*
Look Around 164–165
LORAN system 413
lunar distances 16–17, *17*; Maritime nations 17; measuring method 16–17; sextant 16, *17*; tracking Moon's motion 16

machine learning (ML): *320*; application of 300; cartographic generalization 299–300, 304; decision tree 301, *308, 309*; district group examples 306, *307, 308*; DL (*see* deep learning (DL)); evolutionary learning 301; General Geographic Object Database (GGOD) 305, *305*; genetic algorithm (GA) 301–302; models in settlement selection for small-scale maps 304–305; parallel selection processes 306; performance of 306–310, **307**; random forest models 302; reinforcement learning 301; stages 306; supervised learning (predictive learning) 300; types of 300–301; unsupervised learning (descriptive learning) 300–301; *see also* artificial intelligence (AI)
MALA Object Mapper 368
man-made disasters 510–511, *512*
manned airborne platforms 202
mapping 56; analytical 56; base maps 56; countermapping (*see* counter-mapping); exploratory 56; interactive 56; marine 20; mental 91; story 91; terrestrial 20
Mapping Black Futures 103–104
mapping buildings and cities: BIM data 343; building facade extraction 353–354; building footprint extraction 350–351; challenges and constraints 344, 344–345; 3D geometric models 344; GIS data 343; indoor measurements 354–355; level of details and standards 345–347; Lidar 343; mapping plays 343; methodology for modelling cities 347–349; mobile mapping, or location intelligence 343; roof extraction 351–353; UAS/UAV 343
mapping planetary bodies 562–563; map projections 566–568, *567*; nomenclature 568, *568*; photogrammetric processing 563–566, *564, 565*; planetary GIS 569; planetary spatial data infrastructures 569–571; PSDI use case: MarsGIS 571–574
mapping subaltern: alternative approaches 115–118; counter-mapping 109, 110; debates and ambivalences 113–115; dominant representational field 110; social groups 109–110
Maps 2.0 55
maps of time: change 544, *545, 547*, 547–549; exploration 550–551, *551*; integrated approach 551–553; interaction *549,* 549–550; old and modern classics *543*; old and new maps 541–542; single map or series of maps 544–546, *546*; story maps 551–553; temporal perspective 541, *542*; types of time 542–543, *544*

map thinking: causal maps of biochemistry 603–604; extreme-scale maps of gene expression in brains 606, *607*, 608; literal, causal, and extreme-scale 600; literal maps of Darwin's finches 601–603, *602*; maps of French interdepartmental migration (1954) 606, *607*; molecular interaction map (MIM) methodology 604–606, *605*; through the life sciences 608–609
Map Tile APIs 257
marine mapping 20
MayaCityBuilder project 558
medium-format cameras 208–209, *209*
mental mapping 91
Mercator, Gerhardus 20, 403
Microsoft Azure Computer Vision 495
Middle East respiratory syndrome (MERS) outbreak 466
Millennium Development Goals 537
'Million Dollar Hoods' 101
Minard, Charles Joseph 541, 542, 544, 548
minimum bounding rectangle (MBR) matrix approach 269
missing and murdered Indigenous women and girls (MMIWG) 89
Missing and Murdered Indigenous Women Database (MMIWD2) 101
Mobile for Development (M4D) 639
mobile laser scanning (MLS) 183–185; and ALS, comparing *185,* 185–186; IMU and GNSS 183–184, *184*
mobile mapping 343; Apple Maps and Look Around 164–165; Commercial Off-the-Shelf (COTS) 156–157; compact mobile mapping systems 157, *157*; Google Maps and Street View 163–164; hand-portable mobile systems 158; Here Technologies 161–162; imaging devices 148–150; for imaging services 162–165; inertial measuring unit (IMU) 148; for infrastructure 165–167; laser ranging, profiling and scanning devices 147, 148, 151–153; more compact systems 157; multiple camera units 149, *150*; network of cables and interface cards 148, *149*; operations, applications and service providers 159; positioning (geo-referencing) devices 153–156; for road navigation and cartography 159–162; systems 156–158; technologies 148; for three-dimensional city modelling 167–168; TomTom 159–161; vehicle-mounted mobile mapping system 148, *148*; wearable

or portable backpack systems 158; Zeb Discovery 158, *158*
Mobile Mapping Systems (MMS) 351
modifiable areal unit problem (MAUP) 285
molecular interaction map (MIM) methodology 604–606, *605*
monochrome topographic map 32, *34*
Morrill Act of 1862 101
multi-beam Lidar systems 181, *181*
multicriteria decision analysis (MCA) 279
multiple camera units 149, *150*
multiple cartography 122
multiple frame cameras 210–212; generating multi-spectral imagery 212–213; generating oblique images 210–212
'multiple-pulses-in-the-air' (MPiA) 216
multi-spectral imagery, multiple frame cameras: in United States 212–213

Namibia University of Science and Technology (NUST) 424–425
National Aeronautics and Space Administration (NASA): MERRA-2 mission of 65, 66
National Community Reinvestment Coalition 89
natural disasters: Ebola epidemic 509–510, *510, 511*; 2015 Nepal earthquake 507, *508,* 510
natural language processing 320
Naval Observatory 20
navigation 7
Navteq 161, *161, 162*
near-infrared (NIR) part 172
negative map artwork 30, 35–37, *36–39*; airbrush-produced elevation hillshading 39, *39*; Cromalin proofing process 39; culture lettering 37, *38*; freehand technique 37, *38*; lettering colour separation 37, *38*; linework 35, *36*, 37; photo-lithographers 39–40, *40*; scribing technique 37, *37*; topographic map 35, *36,* 37
neogeography 48, 55, 134–136
2015 Nepal earthquake 507, *508,* 510
Netflix 320
network analysis: closest facility analysis 446; service area 444–445, *446*
neural networks: advantages and disadvantages 318; aim of 316–317; principle of 317, *317*
New Not Normal 103
Newton's laws of motion 397
non-spatial operations *272*; aggregation 271–272; (re)calculation 271; classification 272; interpolation 272; join 272
non-stationarity 284
NSDIs (National Spatial Data Infrastructures) 537
Nui Dat, South Vietnam *33*

Object Mapper software *369*
object-oriented models 47
oblique images, multiple frame cameras: aerial photography 211–212; block configuration 211; concentric (or star-type) configuration 211; fan configuration 211; IGI Dual-DigiCAM system 212, *212*; image data collection 210; 'Maltese Cross' configuration 211, 212; MIDAS (Multi-Image Digital Acquisition System) system 212
ocean mapping: aerial bathymetry 405; airborne laser bathymetry (ALB) 406; AIS and bridge to bridge networks 417, *417, 418*; bathymetry 405; celestial navigation and sextant 410–411, *412*; collision avoidance regulations (ColRegs) 416; consensus, conventions, cohesion 416; corroboration, operational oceanography 407; dead reckoning 414; echo-sounding 405, *406*; geodetics, distances and the nautical mile 413; gnomonic projections 403, *404*; governance and consensus 416–417; GPS 411–413, *412*; hydrocentric thinking, oceanic turn 418; IALA buoyage 418; lead-sounding 405; limits of instruments 414–415; local compass variation 413–414; Mercator's projection of world 403, *404*; multispectral remote sensing and Lidar 406; nautical mile 413; navigating oceans 407; oceanic gaze and environment 419; operational oceanography 405; passage planning 407, *408,* 409, *409*; plane sailing (planar geometry) 409, *410*; satellite altimetry 406; technology as taxonomy 419–420; typical resources for oceanic satellite and aerial data 407
odometers 154
online digital technologies 56
onshore triangulations 20
Open Climate Workbench 252
OpenCPN and OpenSeaMap 403
open data: defined 581; SDI of 581; uneven geographies 581–582
Open Geospatial Consortium (OGC) 60; Agriculture Domain Working Groups (DWGs) 61, 62, 63; APIs 73; Catalogue Services for Web (CSW) 65; Geography Markup Language (GML) 62, 63, 65; GeoPackage standard 66; Geoscience Markup Language (GeoSciML) 63, 64; Observations & Measurements (O& M) 64; Sensor Model Language (SensorML) 64; Sensor Observation Service (SOS) 64; Sensor Planning Service (SPS) 65; SensorThings API (STA) 64, 73; Soil Data Interoperability Experiment 63; standards development 61; Standards Working Groups (SWGs) 62; Testbed-15 project 72; WaterML 65; Web Coverage Service

(WCS) 65, 66; Web Feature Service (WFS) 62, 65; Web Map Service (WMS) 65; Web Processing Service (WPS) 65–68, *66*
Open Science Framework (OSF) 68
OpenStreetMap (OSM) 48, 104, 134, 248, 494, *529*
operational driving domains (ODD) 427
Oracle 162
Ordinary Least Squares (OLS) Regression model 469
Organization for Economic Cooperation and Development (OECD) 313
Organization for the Advancement of Structured Information Standards (OASIS) 73
orthophotographs 222–223
orthophotomosaics 222
OSF (Open Science Framework) 72–73
Outbreaks Near Me (2020) 467
Ovi Maps 162

Paris Observatory 14–15
participatory land use mapping 118n1
participatory mapping: mapping allotments, Mountain Maidu homelands 92–93; mapping sacred sites, Winnemem Wintu homelands 93; PPGIS 495; principles for Indigenous methodologies 90–91; visions for community revitalization, West Port Arthur, Texas 91; visualizing alternative land futures 92
participatory resource mapping 118n1
phased array L- Band SAR (PALSAR) 536
photogrammetry 1, 193
photo-lithographers 39–40, *40*
photo-mechanical era of cartography: computer-assisted cartography 23; copyboard 29; drawing map 29–35; manual map production 25, *26–28*, 29; (*Journal of the Australian Institute of Cartographers*) map-production focused papers, 1954–1978 24–25; 'negative' map artwork 35–37; photogrammetric Block Adjustment 24; photo-mechanical production 37–40; postscript 40; pre-computer map production 23; process camera 25, *29, 30*
Pickles: *Ground Truth: The Social Implications of Geographic Information Systems* 51
plane sailing (planar geometry) 409, *410*
Plate Carree projection 46
Platform-as-a-Service (PaaS) 428
'point cloud' 221; point clouds 2.5D and 3D 174–176
point clustering thematic feature simplification 303
point pattern analysis 277
positioning (geo-referencing) devices 153–156; CORS (Continuously Operating Reference Stations) 154; differential GPS (DGPS) method 154; dual-frequency type 154; Global Navigation Satellite System (GNSS) 153; GPS receiver 153, *154,* 155; IMU 154, 155, *156*; multi-path effects 154; 'urban canyons' 154
positive map artwork: Cartographic Branch, Survey Office *31*; diazo process *34*; French curve drafting set 32, *33*; graphic lettering pen set 30, *31*; monochrome topographic map 32, *34*; Nui Dat, South Vietnam *33*; Rotring isograph pen 30, 32, *32*; Wrico lettering guides 32, *34*
positivism: and GIS, debates on 52–53; strategic 52–53
programmatic integration with GIS software: ArcGIS 261–262; QGIS 261
PRUDENCE project 536
PSDI use case: MarsGIS: data access and metadata 572; foundation data products 571–572; policies 572; standards 572; user community 572, *573,* 574
psychogeographic cartography 138
Ptolemy, Claudius 7, 403; *Geography* 7
public health: and community health 459–461; and COVID-19 global pandemic 454–455; defined 450; and emerging infectious disease (EID) 456; and environmental health 456–459, **458,** *458*; health and healthcare inequalities 450–454, *452, 453*; and infectious disease spread and prevention 454
pushbroom line scanner 207–208, *208*
Python 260

QGIS 261
Queering the Map 102
QUICKBIRD 351

R 260
race and mapping: anti-racist decolonial praxis 94; countermapping as power-building 89–90; dislocation and racialized exclusion 87–88; indigenous decolonial mapping 90; participatory mapping 90–93; resistant ways of being and knowing 88–89; and settler-colonialism 86–87
Races of Europe (Ripley) 87
'Radical Law' 303
radio frequency identification (RFID) 428
random forest (RF) 304, 306
RCD30 medium-format (80 megapixel) camera 217
Real Estate Data (RED) Foundation data ethics principles, United Kingdom 638
recommendation systems 320
reconciliation 86
recurrent neural networks (RNNs) 318
refugees in Uganda 513–514, *514*
reinforcement learning 301
'Relief Maps' 104

remote sensing: aesthetics of 81; agricultural applications 69, 72; AI 322, 325; Airborne Laser Scanning 177; archaeological applications 556–557, 559; discipline 99; drones 191; emergence of GIS 43–44; FOSS4G 248; humanitarian applications 521, 525–526; Indigenous mapping 111, 122, 126; journalism 489–490; modelling the world 46; oceanography 407; OGC standards 61; photogrammetry 202, 229, 231; privacy 583; social construction of space 134; social media 499–500; spatial analysis 281; Sustainable Development Goals (SDGs) 534, 536–538; water management 396, 400
Research Data Alliance (RDA) 61, 72
Responsible Research and Innovation (RRI) 70
RIEGL 220
road-side unit (RSU) sensors 426
Robinson, Arthur H. 541
robotics 316
Rotring isograph pen 30, 32, *32*
rural and Indigenous knowledges (RIK): Catequilla hill 124, *124*; cognitive maps *(whakapapas)* 123; criticism 126–127; 3D modelling 128; geospatial literacy programs 128–129; geospatial technologies and efforts 127–129; inclusion in geospatial technologies 125–126; Indigenous cartography 122; multiple cartography 122; process cartography 122–124, 127, 129

Safepaths (2020) 467
SARS-1 outbreak 431
satellite altimetry 406
satellite-based data sources 347
satellite imagery *556*, 556–557
scribing technique 37, *37*
sensor-driven mapping 2
sextant 16, *17,* 19–20
shared autonomous vehicles (SAV) 431
SHARE Program Library Agency 247
Shuttle Radar Topography Mission (SRTM) 229
SIEUSOIL 61
Silicon Valley-based Zipline company 194
simultaneous localisation and mapping (SLAM) algorithms 355
small-format cameras 209–210, *210*
small-format digital frame cameras 148
smart farming 68
smart geospatial technologies 426–427
'smart' technology 55
social constructivism: arrival of 139–140; big data, rise of 136–137; critical cartography 133; deep mapping 134, 137–139; GIS 133–134, 140; neogeography 134–136; VGI (*see* volunteered geographic information (VGI))

social media (SM): collecting and processing 494–495; disaster risk reduction 498–499; Earth observations 498; ecology and conservation 498; landscape aesthetics and other cultural benefits 497; limitations of methods 499–500; monitoring and assessing tourism and recreation 496–497; supporting landscape and urban planning 495–496; urban diversity and vitality 497
software development kits (SDKs) 242
Sogorea Te Land Trust (STLT) in Oakland, California 92
Soil and Terrain (SOTER) Databases 63
solar energy harvesting 443
SOTER Markup Language (SoTerML) 63
Sovereign Bodies Institute 89
Space-Time Cube (STC) 548
spatial analysis and modelling: application of *278, 279, 280, 281, 284, 285*; carbon dioxide (CO_2) concentrations 263; colocation and correlation 283; combinations and optimum solutions 279–280, *280*; conceptualization 264; data modelling 264–268; data model transformations 273–274; dependency and autocorrelation 283–284; distribution and structure 277; dynamics and evolution 280–281, *281*; grid operations *272,* 272–273; grouping and polygons 277–279, *278*; heterogeneity and non-stationarity *284,* 284–285, *285*; hierarchy and multiscaling 285; influence and dominance 279, *279*; morphometry and shapes 276–277; movement and trajectories 282; networks and accessibility 282–283; non-spatial operations 271–272; recognition and enrichment 276, *276*; searching and joining 275–276; spatial operations 270–271; spatial relations 268–270; visualization and mapping 274, *275*
spatial anthropology: deep mapping 614–616; geospatial humanities 613–614; Google Earth map *620*; qualitative enquiry and geospatial methods 616–619, *618, 619*; visualization 619–621
spatial autocorrelation 283
Spatial Data Infrastructure (SDI) 65
spatial error model 284, 469
spatial heterogeneity 284, 285
spatial index 275
spatial join 275
Spatial Lag Model 284, 469
spatial libraries for popular programming languages: base libraries 261; JavaScript/Node.js 260; Python 260; R 260
spatial operations *271*; bounding geometries 270; buffering 271; construction 270; overlay and dissolve 271; skeletons and medial axes 270; subsetting and type conversion 270

spatial relations 268–270; directional relations 269, *269*; metric relations 268–269, *269*; spatial queries 268; topological relations *269*, 269–270
spatial statistics 277
speech recognition 319
Speed Over Ground (SOG) 414
stellar altitudes 8
stereo SAR 228
story mapping 91
strategic positivism 52–53
STREAM (Subsurface Tomographic Radar Equipment for Assets Mapping) 362
Street View *163*, 163–164, *164*
supervised learning (predictive learning) 300
Surface Water and Ocean Topography (SWOT) mission 396, 407
sustainable development goals (SDGs) 69; aims of 532–533; monitoring and tracking 532; using geospatial technology 533–538, *534*
SYMAP package 43, 44
synthetic aperture radar (SAR) 228; interferometry SAR (InSAR) 228; stereo SAR 228; TanDEM-X (TDX) 229; TerraSAR-X (TSX) 228
synthetic aperture radar (SAR) interferometry 227
Syria crisis 512–513, *513*

TAGS 495
TanDEM-X (TDX) 229
TeleAtlas company 159, *160*
Teledyne Optech 220
telegraph 20
Tencent 466
'TerrainMapper' 217
TerraSAR-X (TSX) 228
terrestrial laser scanning (TLS) 1, 177, 180, *180*, *181*, 182–183, *183*, 227, 228, 354
terrestrial mapping 20
TIGER (Topographical Integrated Geographic Encoding and Referencing) 537
"The Time Ball, Royal Observatory, Greenwich" *19*
'time balls' 19, *19*
time signals 19, *19*, 22
TIMON system 429, 434
Tobler, Waldo R. 283
Tomlinson Roger F. 43, 44, 48, 50
TomTom 159–161, *160*
'TraceTogether' 467
traditional desk-based mapping technologies 56
traffic management server (TMS) 428
TransCAD 424
transportation: AVs and CAVs 427–430; Covid-19 for geospatial technologies 431–433; data models 425–426; 5G and 6G mobile computing environments 430–431; GIS-T 425, 426; planners 424–425; smart geospatial technologies 426–427; TransCAD 424; 'triple bottom line' 424; world of Lyft and Uber 431
transportation network companies (TNCs) 431
'Trimble SiteVision' 381–382
Trip Advisor 134
Twitter 134, 494

underground mapping: acquisition and viewing data in field 363–365; cluster of cables and underground pipes 358, *359*; electronic markers for 358, *360*; geospatial data management with bentley systems *379*; GPR (*see* ground penetrating radar (GPR)); investigation route 363, *363*, *364*; Mobile Mapping system 366, *367*; network type and frequency 358, **360**; operating principles of GPR systems 360–363; pipelines visualization 365, *365*; principle of wave propagation in soil 360–361, *361*; of route of pipe with metallic tape 358, *359*;towards reliable map of underground utilities 379–382; types of underground utilities 358, *359*; utility networks and GIS 376–378, *376*, *377*
Underground R.R. 105
United Nations committee of experts on Global Geospatial Information Management (UN- GGIM) 61
United Nations Open GIS Initiative 251
Universal Transverse Mercator (UTM) coordinate system 177
unmanned aerial vehicles (UAVs) 2, 188–191, 202–203, 231, 520, 526
Unmanned Aircraft Systems (UAS) 178, *179*, 195
unmanned traffic management (UTM) systems 195–196
unsupervised learning (descriptive learning) 300–301
'urban grid information system' 55
URL (uniform resource locator) parameters 257
US Geological Survey's Earthquake Hazards map 522, *523*
US Office of Strategic Services (OSS) 40

vehicle miles travelled (VMT) 431
vehicle-mounted mobile mapping system 148, *148*
vehicle-to-everything (V2X) 429, 434
vehicle to infrastructure connectivity (V2I) 427
vehicle-to-vehicle connectivity (V2V) 427, 429
vehicular fog computing (VFC) 427, 428
'vGIS Utilities' application 379–380, *380*
virtual reality (VR) 1, 104, 236–237, *238*, 239, 242, 379, 403, 529, 535, 558

visualization: big data 293–294; 3D 373, *374*; electrical systems 444, *445*; spatial analysis and modelling 274, *275*; spatial anthropology 619–621; underground mapping 365, *365*
volunteered geographic information (VGI) 55, 69, 134–136, 467, 534, 579
Voronoi-based algorithm 303
VPN (virtual private network) 296
vulnerable road users (VRUs) 429

water management: case study 398–399; data 396–397; decision-making 399–400, *400*; future implications for society 400–401; models and data analytics 397–398, *398*; need for 394; technologies to support 394–396, *395*
wave piloting 415
weather information 68
Web mapping services (WMS) 521
Weibo 494
weighted overlay 280
Wi-Fi positioning systems (WFPS) 466

Wikimapia 134, 494
'Wild West' period of civilian drone experimentation 198
wireless sensor networks (WSNs) 428
Wittgenstein, Ludwig 140
work from home (WFH) 432
World Bank 61, 115
World Health Organization (WHO) 61, 450
World Reference Base of Soil Resources (WRB) 63
WorldView 351
World Wide Web Consortium (W3C) 62, 73, 296, 638
Wrico lettering guides 32, *34*

XML *(eXtensible Markup Language)* format 345

Yahoo Maps 162

zero meridian 7
zones and strata theory (Taylor) 87